Organic Chemistry

Volume I

Eighth Edition

John E. McMurry

Australia • Brazil • Japan • Korea • Mexico • Singapore • Spain • United Kingdom • United States

Organic Chemistry: Volume I, Eighth Edition

Organic Chemistry, 8th Edition
John E. McMurry

© 2012, 2008 Brooks/Cole, Cengage Learning. All rights reserved.

Executive Editors:
 Maureen Staudt
 Michael Stranz

Senior Project Development Manager:
 Linda deStefano

Marketing Specialist:
 Courtney Sheldon

Senior Production/Manufacturing Manager:
 Donna M. Brown

PreMedia Manager:
 Joel Brennecke

Sr. Rights Acquisition Account Manager:
 Todd Osborne

Cover Image: stock.xchng

ALL RIGHTS RESERVED. No part of this work covered by the copyright herein may be reproduced, transmitted, stored or used in any form or by any means graphic, electronic, or mechanical, including but not limited to photocopying, recording, scanning, digitizing, taping, Web distribution, information networks, or information storage and retrieval systems, except as permitted under Section 107 or 108 of the 1976 United States Copyright Act, without the prior written permission of the publisher.

> For product information and technology assistance, contact us at
> **Cengage Learning Customer & Sales Support, 1-800-354-9706**
> For permission to use material from this text or product,
> submit all requests online at cengage.com/permissions
> Further permissions questions can be emailed to
> **permissionrequest@cengage.com**

This book contains select works from existing Cengage Learning resources and was produced by Cengage Learning Custom Solutions for collegiate use. As such, those adopting and/or contributing to this work are responsible for editorial content accuracy, continuity and completeness.

Compilation © 2011 Cengage Learning

ISBN-13: 978-1-133-06741-2

ISBN-10: 1-133-06741-7

Cengage Learning
5191 Natorp Boulevard
Mason, Ohio 45040
USA

Cengage Learning is a leading provider of customized learning solutions with office locations around the globe, including Singapore, the United Kingdom, Australia, Mexico, Brazil, and Japan. Locate your local office at: **international.cengage.com/region.**

Cengage Learning products are represented in Canada by Nelson Education, Ltd.
For your lifelong learning solutions, visit **www.cengage.com /custom.**
Visit our corporate website at **www.cengage.com.**

Printed in the United States of America

Brief Contents

1. Structure and Bonding 1
2. Polar Covalent Bonds; Acids and Bases 34
3. Organic Compounds: Alkanes and Their Stereochemistry 74
4. Organic Compounds: Cycloalkanes and Their Stereochemistry 108
5. Stereochemistry at Tetrahedral Centers 142
6. An Overview of Organic Reactions 184
7. Alkenes: Structure and Reactivity 222
8. Alkenes: Reactions and Synthesis 262
9. Alkynes: An Introduction to Organic Synthesis 314
10. Organohalides 344
11. Reactions of Alkyl Halides: Nucleophilic Substitutions and Eliminations 372
12. Structure Determination: Mass Spectrometry and Infrared Spectroscopy 424

Appendix A: Nomenclature of Polyfunctional Organic Compounds A-1
Appendix B: Acidity Constants for Some Organic Compounds A-8
Appendix C: Glossary A-10
Appendix D: Answers to In-Text Problems A-28
Index I-1

1

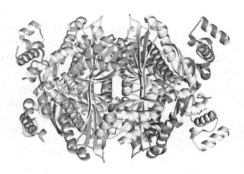

The enzyme HMG-CoA reductase, shown here as a so-called ribbon model, catalyzes a crucial step in the body's synthesis of cholesterol. Understanding how this enzyme functions has led to the development of drugs credited with saving millions of lives.

Structure and Bonding

1.1 Atomic Structure: The Nucleus
1.2 Atomic Structure: Orbitals
1.3 Atomic Structure: Electron Configurations
1.4 Development of Chemical Bonding Theory
1.5 Describing Chemical Bonds: Valence Bond Theory
1.6 sp^3 Hybrid Orbitals and the Structure of Methane
1.7 sp^3 Hybrid Orbitals and the Structure of Ethane
1.8 sp^2 Hybrid Orbitals and the Structure of Ethylene
1.9 sp Hybrid Orbitals and the Structure of Acetylene
1.10 Hybridization of Nitrogen, Oxygen, Phosphorus, and Sulfur
1.11 Describing Chemical Bonds: Molecular Orbital Theory
1.12 Drawing Chemical Structures

A Deeper Look—Organic Foods: Risk versus Benefit

OWL Sign in to OWL for Organic Chemistry at **www.cengage.com/owl** to view tutorials and simulations, develop problem-solving skills, and complete online homework assigned by your professor.

What is organic chemistry, and why should you study it? The answers to these questions are all around you. Every living organism is made of organic chemicals. The proteins that make up your hair, skin, and muscles; the DNA that controls your genetic heritage; the foods that nourish you; and the medicines that heal you are all organic chemicals. Anyone with a curiosity about life and living things, and anyone who wants to be a part of the remarkable advances now occurring in medicine and the biological sciences, must first understand organic chemistry. Look at the following drawings for instance, which show the chemical structures of some molecules whose names might be familiar to you. Although the drawings may appear unintelligible at this point, don't worry. Before long, they'll make perfectly good sense, and you'll soon be drawing similar structures for any substance you're interested in.

Rofecoxib (Vioxx)

Atorvastatin (Lipitor)

Oxycodone (OxyContin)

Cholesterol

Benzylpenicillin

The foundations of organic chemistry date from the mid-1700s, when chemistry was evolving from an alchemist's art into a modern science. Little was known about chemistry at that time, and the behavior of the "organic" substances isolated from plants and animals seemed different from that of the "inorganic" substances found in minerals. Organic compounds were generally low-melting solids and were usually more difficult to isolate, purify, and work with than high-melting inorganic compounds.

To many chemists, the simplest explanation for the difference in behavior between organic and inorganic compounds was that organic compounds contained a peculiar "vital force" as a result of their origin in living sources. Because of this vital force, chemists believed, organic compounds could not be prepared and manipulated in the laboratory as could inorganic compounds. As early as 1816, however, this vitalistic theory received a heavy blow when Michel Chevreul found that soap, prepared by the reaction of alkali with animal fat, could be separated into several pure organic compounds, which he termed *fatty acids*. For the first time, one organic substance (fat) was converted into others (fatty acids plus glycerin) without the intervention of an outside vital force.

$$\text{Animal fat} \xrightarrow[\text{H}_2\text{O}]{\text{NaOH}} \text{Soap} + \text{Glycerin}$$

$$\text{Soap} \xrightarrow{\text{H}_3\text{O}^+} \text{"Fatty acids"}$$

Little more than a decade later, the vitalistic theory suffered still further when Friedrich Wöhler discovered in 1828 that it was possible to convert the "inorganic" salt ammonium cyanate into the "organic" substance urea, which had previously been found in human urine.

$$\text{NH}_4^+ \ ^-\text{OCN} \xrightarrow{\text{Heat}} \text{H}_2\text{N}-\overset{\overset{\text{O}}{\|}}{\text{C}}-\text{NH}_2$$

Ammonium cyanate   **Urea**

By the mid-1800s, the weight of evidence was clearly against the vitalistic theory and it was clear that there was no fundamental difference between organic and inorganic compounds. The same fundamental principles explain the behaviors of all substances, regardless of origin or complexity. The only distinguishing characteristic of organic chemicals is that all contain the element carbon.

Organic chemistry, then, is the study of carbon compounds. But why is carbon special? Why, of the more than 50 million presently known chemical compounds, do most of them contain carbon? The answers to these questions come from carbon's electronic structure and its consequent position in the periodic table **(Figure 1.1)**. As a group 4A element, carbon can share four valence electrons and form four strong covalent bonds. Furthermore, carbon atoms can bond to one another, forming long chains and rings. Carbon, alone of all elements, is able to form an immense diversity of compounds, from the simple methane, with one carbon atom, to the staggeringly complex DNA, which can have more than *100 million* carbons.

Figure 1.1 The position of carbon in the periodic table. Other elements commonly found in organic compounds are shown in the colors typically used to represent them.

Group 1A																	8A
H	2A											3A	4A	5A	6A	7A	He
Li	Be											B	**C**	**N**	**O**	**F**	Ne
Na	Mg											Al	Si	**P**	**S**	**Cl**	Ar
K	Ca	Sc	Ti	V	Cr	Mn	Fe	Co	Ni	Cu	Zn	Ga	Ge	As	Se	**Br**	Kr
Rb	Sr	Y	Zr	Nb	Mo	Tc	Ru	Rh	Pd	Ag	Cd	In	Sn	Sb	Te	**I**	Xe
Cs	Ba	La	Hf	Ta	W	Re	Os	Ir	Pt	Au	Hg	Tl	Pb	Bi	Po	At	Rn
Fr	Ra	Ac															

Not all carbon compounds are derived from living organisms of course. Modern chemists have developed a remarkably sophisticated ability to design and synthesize new organic compounds in the laboratory—medicines, dyes, polymers, and a host of other substances. Organic chemistry touches the lives of everyone; its study can be a fascinating undertaking.

Why This Chapter?
We'll ease into the study of organic chemistry by first reviewing some ideas about atoms, bonds, and molecular geometry that you may recall from your general chemistry course. Much of the material in this chapter and the next is likely to be familiar to you, but it's nevertheless a good idea to make sure you understand it before going on.

1.1 Atomic Structure: The Nucleus

As you probably know from your general chemistry course, an atom consists of a dense, positively charged nucleus surrounded at a relatively large distance by negatively charged electrons **(Figure 1.2)**. The nucleus consists of subatomic particles called protons, which are positively charged, and neutrons, which are electrically neutral. Because an atom is neutral overall, the number of positive protons in the nucleus and the number of negative electrons surrounding the nucleus are the same.

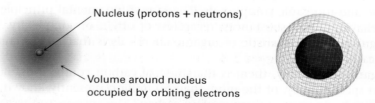

Figure 1.2 A schematic view of an atom. The dense, positively charged nucleus contains most of the atom's mass and is surrounded by negatively charged electrons. The three-dimensional view on the right shows calculated electron-density surfaces. Electron density increases steadily toward the nucleus and is 40 times greater at the blue solid surface than at the gray mesh surface.

Although extremely small—about 10^{-14} to 10^{-15} meter (m) in diameter—the nucleus nevertheless contains essentially all the mass of the atom. Electrons have negligible mass and circulate around the nucleus at a distance of approximately 10^{-10} m. Thus, the diameter of a typical atom is about 2×10^{-10} m, or

200 picometers (pm), where 1 pm = 10^{-12} m. To give you an idea of how small this is, a thin pencil line is about 3 million carbon atoms wide. Many organic chemists and biochemists, particularly in the United States, still use the unit *angstrom* (Å) to express atomic distances, where 1 Å = 100 pm = 10^{-10} m, but we'll stay with the SI unit picometer in this book.

A specific atom is described by its atomic number (Z), which gives the number of protons (or electrons) it contains, and its mass number (A), which gives the total number of protons plus neutrons in its nucleus. All the atoms of a given element have the same atomic number—1 for hydrogen, 6 for carbon, 15 for phosphorus, and so on—but they can have different mass numbers depending on how many neutrons they contain. Atoms with the same atomic number but different mass numbers are called **isotopes**.

The weighted average mass in atomic mass units (amu) of an element's naturally occurring isotopes is called the element's atomic mass (or atomic weight)—1.008 amu for hydrogen, 12.011 amu for carbon, 30.974 amu for phosphorus, and so on. Atomic masses of the elements are given in the periodic table in the front of this book.

1.2 Atomic Structure: Orbitals

How are the electrons distributed in an atom? You might recall from your general chemistry course that, according to the quantum mechanical model, the behavior of a specific electron in an atom can be described by a mathematical expression called a *wave equation*—the same type of expression used to describe the motion of waves in a fluid. The solution to a wave equation is called a *wave function*, or **orbital**, and is denoted by the Greek letter psi (ψ).

By plotting the square of the wave function, ψ^2, in three-dimensional space, an orbital describes the volume of space around a nucleus that an electron is most likely to occupy. You might therefore think of an orbital as looking like a photograph of the electron taken at a slow shutter speed. In such a photo, the orbital would appear as a blurry cloud, indicating the region of space where the electron has been. This electron cloud doesn't have a sharp boundary, but for practical purposes we can set the limits by saying that an orbital represents the space where an electron spends 90% to 95% of its time.

What do orbitals look like? There are four different kinds of orbitals, denoted *s*, *p*, *d*, and *f*, each with a different shape. Of the four, we'll be concerned primarily with *s* and *p* orbitals because these are the most common in organic and biological chemistry. An *s* orbital is spherical, with the nucleus at its center; a *p* orbital is dumbbell-shaped; and four of the five *d* orbitals are cloverleaf-shaped, as shown in **Figure 1.3**. The fifth *d* orbital is shaped like an elongated dumbbell with a doughnut around its middle.

An *s* orbital

A *p* orbital

A *d* orbital

Figure 1.3 Representations of *s*, *p*, and *d* orbitals. An *s* orbital is spherical, a *p* orbital is dumbbell-shaped, and four of the five *d* orbitals are cloverleaf-shaped. **Different lobes** of *p* and *d* orbitals are often drawn for convenience as teardrops, but their actual shape is more like that of a doorknob, as indicated.

The orbitals in an atom are organized into different **electron shells**, centered around the nucleus and having successively larger size and energy. Different shells contain different numbers and kinds of orbitals, and each orbital within a shell can be occupied by two electrons. The first shell contains only a single s orbital, denoted 1s, and thus holds only 2 electrons. The second shell contains one 2s orbital and three 2p orbitals and thus holds a total of 8 electrons. The third shell contains a 3s orbital, three 3p orbitals, and five 3d orbitals, for a total capacity of 18 electrons. These orbital groupings and their energy levels are shown in **Figure 1.4**.

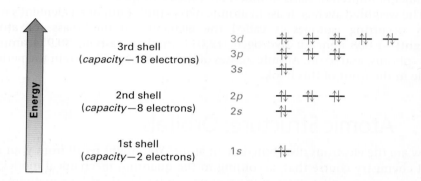

Figure 1.4 The energy levels of electrons in an atom. The first shell holds a maximum of 2 electrons in one 1s orbital; the second shell holds a maximum of 8 electrons in one 2s and three 2p orbitals; the third shell holds a maximum of 18 electrons in one 3s, three 3p, and five 3d orbitals; and so on. The two electrons in each orbital are represented by up and down arrows, ↑↓. Although not shown, the energy level of the 4s orbital falls between 3p and 3d.

The three different p orbitals within a given shell are oriented in space along mutually perpendicular directions, denoted p_x, p_y, and p_z. As shown in **Figure 1.5**, the two lobes of each p orbital are separated by a region of zero electron density called a **node**. Furthermore, the two orbital regions separated by the node have different algebraic signs, + and −, in the wave function, as represented by the different colors in Figure 1.5. We'll see in **Section 1.11** that these algebraic signs of different orbital lobes have important consequences with respect to chemical bonding and chemical reactivity.

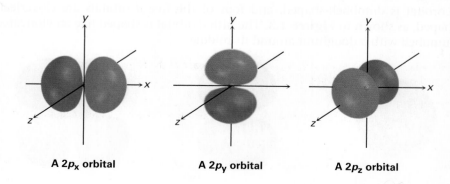

Figure 1.5 Shapes of the 2p orbitals. Each of the three mutually perpendicular, dumbbell-shaped orbitals has two lobes separated by a node. The two lobes have different algebraic signs in the corresponding wave function, as indicated by the different colors.

1.3 Atomic Structure: Electron Configurations

The lowest-energy arrangement, or **ground-state electron configuration**, of an atom is a listing of the orbitals occupied by its electrons. We can predict this arrangement by following three rules.

RULE 1
The lowest-energy orbitals fill up first, according to the order $1s \rightarrow 2s \rightarrow 2p \rightarrow 3s \rightarrow 3p \rightarrow 4s \rightarrow 3d$, a statement called the aufbau principle. Note that the $4s$ orbital lies between the $3p$ and $3d$ orbitals in energy.

RULE 2
Electrons act in some ways as if they were spinning around an axis, somewhat as the earth spins. This spin can have two orientations, denoted as up ($\uparrow$) and down ($\downarrow$). Only two electrons can occupy an orbital, and they must be of opposite spin, a statement called the Pauli exclusion principle.

RULE 3
If two or more empty orbitals of equal energy are available, one electron occupies each with spins parallel until all orbitals are half-full, a statement called Hund's rule.

Some examples of how these rules apply are shown in Table 1.1. Hydrogen, for instance, has only one electron, which must occupy the lowest-energy orbital. Thus, hydrogen has a $1s$ ground-state configuration. Carbon has six electrons and the ground-state configuration $1s^2\ 2s^2\ 2p_x^1\ 2p_y^1$, and so forth. Note that a superscript is used to represent the number of electrons in a particular orbital.

Table 1.1 Ground-State Electron Configurations of Some Elements

Element	Atomic number	Configuration	Element	Atomic number	Configuration
Hydrogen	1	$1s$ ⇅	Phosphorus	15	$3p$ ↑ ↑ ↑
Carbon	6	$2p$ ↑ ↑ —			$3s$ ⇅
		$2s$ ⇅			$2p$ ⇅ ⇅ ⇅
		$1s$ ⇅			$2s$ ⇅
					$1s$ ⇅

Problem 1.1
Give the ground-state electron configuration for each of the following elements:
(a) Oxygen **(b)** Nitrogen **(c)** Sulfur

Problem 1.2
How many electrons does each of the following elements have in its outermost electron shell?
(a) Magnesium **(b)** Cobalt **(c)** Selenium

1.4 Development of Chemical Bonding Theory

By the mid-1800s, the new science of chemistry was developing rapidly and chemists had begun to probe the forces holding compounds together. In 1858, August Kekulé and Archibald Couper independently proposed that, in all organic compounds, carbon is *tetravalent*—it always forms four bonds when it joins other elements to form stable compounds. Furthermore, said Kekulé, carbon atoms can bond to one another to form extended chains of linked atoms. In 1865, Kekulé provided another major advance when he suggested that carbon chains can double back on themselves to form *rings* of atoms.

Although Kekulé and Couper were correct in describing the tetravalent nature of carbon, chemistry was still viewed in a two-dimensional way until 1874. In that year, Jacobus van't Hoff and Joseph Le Bel added a third dimension to our ideas about organic compounds when they proposed that the four bonds of carbon are not oriented randomly but have specific spatial directions. Van't Hoff went even further and suggested that the four atoms to which carbon is bonded sit at the corners of a regular tetrahedron, with carbon in the center.

A representation of a tetrahedral carbon atom is shown in **Figure 1.6**. Note the conventions used to show three-dimensionality: solid lines represent bonds in the plane of the page, the heavy wedged line represents a bond coming out of the page toward the viewer, and the dashed line represents a bond receding back behind the page, away from the viewer. These representations will be used throughout the text.

Figure 1.6 A representation of a tetrahedral carbon atom. The solid lines represent bonds in the plane of the paper, the heavy wedged line represents a bond coming out of the plane of the page, and the dashed line represents a bond going back behind the plane of the page.

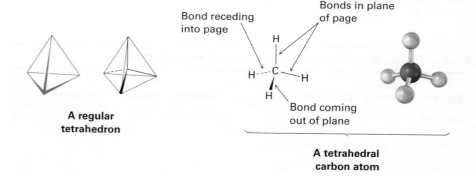

Why, though, do atoms bond together, and how can bonds be described electronically? The *why* question is relatively easy to answer: atoms bond together because the compound that results is more stable and lower in energy than the separate atoms. Energy—usually as heat—always flows out of the chemical system when a bond forms. Conversely, energy must be put into the chemical system to break a bond. Making bonds always releases energy, and breaking bonds always absorbs energy. The *how* question is more difficult. To answer it, we need to know more about the electronic properties of atoms.

We know through observation that eight electrons (an electron *octet*) in an atom's outermost shell, or **valence shell**, impart special stability to the noble-gas elements in group 8A of the periodic table: Ne (2 + 8); Ar (2 + 8 + 8); Kr (2 + 8 + 18 + 8). We also know that the chemistry of main-group elements is governed by their tendency to take on the electron configuration of the nearest

1.4 | Development of Chemical Bonding Theory

noble gas. The alkali metals in group 1A, for example, achieve a noble-gas configuration by losing the single s electron from their valence shell to form a cation, while the halogens in group 7A achieve a noble-gas configuration by gaining a p electron to fill their valence shell and form an anion. The resultant ions are held together in compounds like $Na^+ Cl^-$ by an electrostatic attraction that we call an *ionic bond*.

But how do elements closer to the middle of the periodic table form bonds? Look at methane, CH_4, the main constituent of natural gas, for example. The bonding in methane is not ionic because it would take too much energy for carbon ($1s^2\ 2s^2\ 2p^2$) either to gain or lose four electrons to achieve a noble-gas configuration. As a result, carbon bonds to other atoms, not by gaining or losing electrons, but by sharing them. Such a shared-electron bond, first proposed in 1916 by G. N. Lewis, is called a **covalent bond**. The neutral collection of atoms held together by covalent bonds is called a **molecule**.

A simple way of indicating the covalent bonds in molecules is to use what are called *Lewis structures*, or **electron-dot structures**, in which the valence-shell electrons of an atom are represented as dots. Thus, hydrogen has one dot representing its 1s electron, carbon has four dots ($2s^2\ 2p^2$), oxygen has six dots ($2s^2\ 2p^4$), and so on. A stable molecule results whenever a noble-gas configuration is achieved for all the atoms—eight dots (an octet) for main-group atoms or two dots for hydrogen. Simpler still is the use of *Kekulé structures*, or **line-bond structures**, in which a two-electron covalent bond is indicated as a line drawn between atoms.

The number of covalent bonds an atom forms depends on how many additional valence electrons it needs to reach a noble-gas configuration. Hydrogen has one valence electron (1s) and needs one more to reach the helium configuration ($1s^2$), so it forms one bond. Carbon has four valence electrons ($2s^2\ 2p^2$) and needs four more to reach the neon configuration ($2s^2\ 2p^6$), so it forms four bonds. Nitrogen has five valence electrons ($2s^2\ 2p^3$), needs three more, and forms three bonds; oxygen has six valence electrons ($2s^2\ 2p^4$), needs two more, and forms two bonds; and the halogens have seven valence electrons, need one more, and form one bond.

Valence electrons that are not used for bonding are called **lone-pair electrons**, or *nonbonding electrons*. The nitrogen atom in ammonia, NH_3, for instance, shares six valence electrons in three covalent bonds and has its remaining two valence electrons in a nonbonding lone pair. As a time-saving shorthand, nonbonding electrons are often omitted when drawing line-bond structures, but you still have to keep them in mind since they're often crucial in chemical reactions.

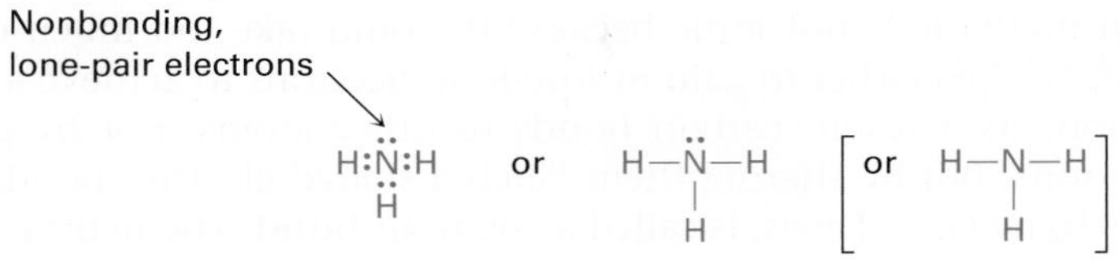

Nonbonding, lone-pair electrons

$$H:\ddot{N}:H \quad \text{or} \quad H-\ddot{N}-H \quad \text{or} \quad \left[H-N-H \right]$$
$$\quad\quad H \quad\quad\quad\quad\quad H \quad\quad\quad\quad\quad H$$

Ammonia

Worked Example 1.1 — Predicting the Number of Bonds Formed by an Atom

How many hydrogen atoms does phosphorus bond to in forming phosphine, $PH_?$?

Strategy
Identify the periodic group of phosphorus, and tell from that how many electrons (bonds) are needed to make an octet.

Solution
Phosphorus is in group 5A of the periodic table and has five valence electrons. It thus needs to share three more electrons to make an octet and therefore bonds to three hydrogen atoms, giving PH_3.

Worked Example 1.2 — Drawing Electron-Dot and Line-Bond Structures

Draw both electron-dot and line-bond structures for chloromethane, CH_3Cl.

Strategy
Remember that a bond—that is, a pair of shared electrons—is represented as a line between atoms.

Solution
Hydrogen has one valence electron, carbon has four valence electrons, and chlorine has seven valence electrons. Thus, chloromethane is represented as

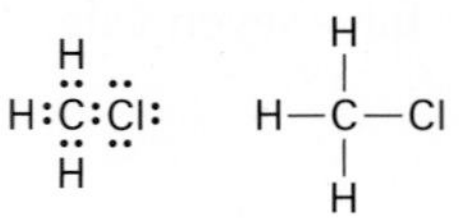

$$H:\overset{H}{\underset{H}{\ddot{C}}}:\ddot{Cl}: \quad\quad H-\overset{H}{\underset{H}{C}}-Cl \quad\quad \textbf{Chloromethane}$$

Problem 1.3
Draw a molecule of chloroform, $CHCl_3$, using solid, wedged, and dashed lines to show its tetrahedral geometry.

Problem 1.4
Convert the following representation of ethane, C_2H_6, into a conventional drawing that uses solid, wedged, and dashed lines to indicate tetrahedral geometry around each carbon (gray = C, ivory = H).

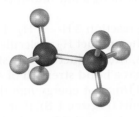

Ethane

Problem 1.5
What are likely formulas for the following substances?
(a) $CCl_?$ (b) $AlH_?$ (c) $CH_?Cl_2$ (d) $SiF_?$ (e) $CH_3NH_?$

Problem 1.6
Write line-bond structures for the following substances, showing all nonbonding electrons:
(a) $CHCl_3$, chloroform
(b) H_2S, hydrogen sulfide
(c) CH_3NH_2, methylamine
(d) CH_3Li, methyllithium

Problem 1.7
Why can't an organic molecule have the formula C_2H_7?

1.5 Describing Chemical Bonds: Valence Bond Theory

How does electron sharing lead to bonding between atoms? Two models have been developed to describe covalent bonding: *valence bond theory* and *molecular orbital theory*. Each model has its strengths and weaknesses, and chemists tend to use them interchangeably depending on the circumstances. Valence bond theory is the more easily visualized of the two, so most of the descriptions we'll use in this book derive from that approach.

According to **valence bond theory**, a covalent bond forms when two atoms approach each other closely and a singly occupied orbital on one atom overlaps a singly occupied orbital on the other atom. The electrons are now paired in the overlapping orbitals and are attracted to the nuclei of both atoms, thus bonding the atoms together. In the H_2 molecule, for instance, the H–H bond results from the overlap of two singly occupied hydrogen 1s orbitals.

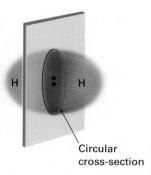

Figure 1.7 The cylindrical symmetry of the H−H σ bond in an H₂ molecule. The intersection of a plane cutting through the σ bond is a circle.

The overlapping orbitals in the H₂ molecule have the elongated egg shape we might get by pressing two spheres together. If a plane were to pass through the middle of the bond, the intersection of the plane and the overlapping orbitals would be a circle. In other words, the H−H bond is cylindrically symmetrical, as shown in **Figure 1.7**. Such bonds, which are formed by the head-on overlap of two atomic orbitals along a line drawn between the nuclei, are called **sigma (σ) bonds**.

During the bond-forming reaction 2 H· → H₂, 436 kJ/mol (104 kcal/mol) of energy is released. Because the product H₂ molecule has 436 kJ/mol less energy than the starting 2 H· atoms, the product is more stable than the reactant and we say that the H−H bond has a **bond strength** of 436 kJ/mol. In other words, we would have to put 436 kJ/mol of energy *into* the H−H bond to break the H₂ molecule apart into H atoms **(Figure 1.8)**. [For convenience, we'll generally give energies in both kilocalories (kcal) and the SI unit kilojoules (kJ): 1 kJ = 0.2390 kcal; 1 kcal = 4.184 kJ.]

Figure 1.8 Relative energy levels of two H atoms and the H₂ molecule. The H₂ molecule has 436 kJ/mol (104 kcal/mol) less energy than the two H atoms, so 436 kJ/mol of energy is *released* when the H−H bond *forms*. Conversely, 436 kJ/mol is *absorbed* when the H−H bond *breaks*.

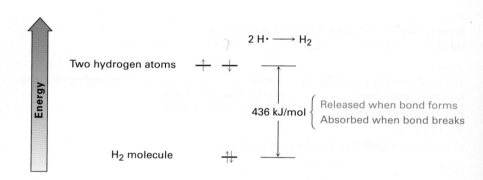

How close are the two nuclei in the H₂ molecule? If they are too close, they will repel each other because both are positively charged, yet if they're too far apart, they won't be able to share the bonding electrons. Thus, there is an optimum distance between nuclei that leads to maximum stability **(Figure 1.9)**. Called the **bond length**, this distance is 74 pm in the H₂ molecule. Every covalent bond has both a characteristic bond strength and bond length.

Figure 1.9 A plot of energy versus internuclear distance for two H atoms. The distance between nuclei at the minimum energy point is the bond length.

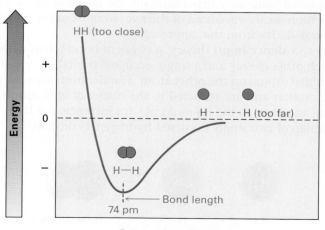

1.6 sp^3 Hybrid Orbitals and the Structure of Methane

The bonding in the hydrogen molecule is fairly straightforward, but the situation is more complicated in organic molecules with tetravalent carbon atoms. Take methane, CH_4, for instance. As we've seen, carbon has four valence electrons ($2s^2\ 2p^2$) and forms four bonds. Because carbon uses two kinds of orbitals for bonding, $2s$ and $2p$, we might expect methane to have two kinds of C–H bonds. In fact, though, all four C–H bonds in methane are identical and are spatially oriented toward the corners of a regular tetrahedron (Figure 1.6). How can we explain this?

An answer was provided in 1931 by Linus Pauling, who showed mathematically how an s orbital and three p orbitals on an atom can combine, or *hybridize*, to form four equivalent atomic orbitals with tetrahedral orientation. Shown in **Figure 1.10**, these tetrahedrally oriented orbitals are called **sp^3 hybrids**. Note that the superscript 3 in the name sp^3 tells how many of each type of atomic orbital combine to form the hybrid, not how many electrons occupy it.

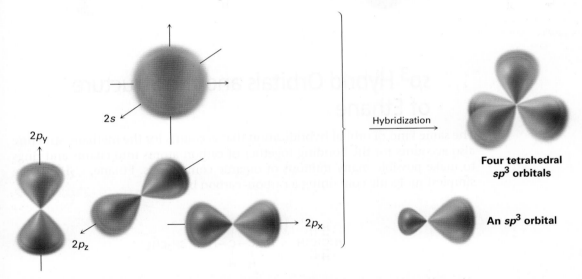

Figure 1.10 Four sp^3 hybrid orbitals, oriented to the corners of a regular tetrahedron, are formed by combination of an s orbital and three p orbitals (**red/blue**). The sp^3 hybrids have two lobes and are unsymmetrical about the nucleus, giving them a directionality and allowing them to form strong bonds when they overlap an orbital from another atom.

The concept of hybridization explains how carbon forms four equivalent tetrahedral bonds but not why it does so. The shape of the hybrid orbital suggests the answer. When an s orbital hybridizes with three p orbitals, the resultant sp^3 hybrid orbitals are unsymmetrical about the nucleus. One of the two lobes is larger than the other and can therefore overlap more effectively with an orbital from another atom to form a bond. As a result, sp^3 hybrid orbitals form stronger bonds than do unhybridized s or p orbitals.

Figure 1.11 The structure of methane, showing its 109.5° bond angles.

The asymmetry of sp^3 orbitals arises because, as noted previously, the two lobes of a p orbital have different algebraic signs, + and −, in the wave function. Thus, when a p orbital hybridizes with an s orbital, the positive p lobe adds to the s orbital but the negative p lobe subtracts from the s orbital. The resultant hybrid orbital is therefore unsymmetrical about the nucleus and is strongly oriented in one direction.

When each of the four identical sp^3 hybrid orbitals of a carbon atom overlaps with the 1s orbital of a hydrogen atom, four identical C−H bonds are formed and methane results. Each C−H bond in methane has a strength of 439 kJ/mol (105 kcal/mol) and a length of 109 pm. Because the four bonds have a specific geometry, we also can define a property called the **bond angle**. The angle formed by each H−C−H is 109.5°, the so-called tetrahedral angle. Methane thus has the structure shown in **Figure 1.11**.

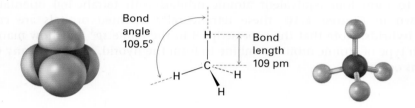

1.7 sp^3 Hybrid Orbitals and the Structure of Ethane

The same kind of orbital hybridization that accounts for the methane structure also accounts for the bonding together of carbon atoms into chains and rings to make possible many millions of organic compounds. Ethane, C_2H_6, is the simplest molecule containing a carbon–carbon bond.

Some representations of ethane

We can picture the ethane molecule by imagining that the two carbon atoms bond to each other by σ overlap of an sp^3 hybrid orbital from each **(Figure 1.12)**. The remaining three sp^3 hybrid orbitals on each carbon overlap with the 1s orbitals of three hydrogens to form the six C−H bonds. The C−H bonds in ethane are similar to those in methane, although a bit weaker—421 kJ/mol (101 kcal/mol) for ethane versus 439 kJ/mol for methane. The C−C bond is 154 pm long and has a strength of 377 kJ/mol (90 kcal/mol). All the bond angles of ethane are near, although not exactly at, the tetrahedral value of 109.5°.

1.8 | sp² Hybrid Orbitals and the Structure of Ethylene

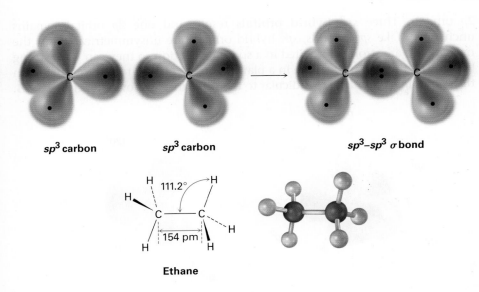

Figure 1.12 The structure of ethane. The carbon–carbon bond is formed by σ overlap of sp³ hybrid orbitals. For clarity, the smaller lobes of the sp³ hybrid orbitals are not shown.

Problem 1.8
Draw a line-bond structure for propane, CH₃CH₂CH₃. Predict the value of each bond angle, and indicate the overall shape of the molecule.

Problem 1.9
Convert the following molecular model of hexane, a component of gasoline, into a line-bond structure (gray = C, ivory = H).

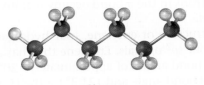

Hexane

1.8 sp² Hybrid Orbitals and the Structure of Ethylene

The bonds we've seen in methane and ethane are called *single bonds* because they result from the sharing of one electron pair between bonded atoms. It was recognized nearly 150 years ago, however, that carbon atoms can also form *double bonds* by sharing *two* electron pairs between atoms or *triple bonds* by sharing *three* electron pairs. Ethylene, for instance, has the structure H₂C=CH₂ and contains a carbon–carbon double bond, while acetylene has the structure HC≡CH and contains a carbon–carbon triple bond.

How are multiple bonds described by valence bond theory? When we discussed sp³ hybrid orbitals in **Section 1.6**, we said that the four valence-shell atomic orbitals of carbon combine to form four equivalent sp³ hybrids. Imagine instead that the 2s orbital combines with only *two* of the three available

$2p$ orbitals. Three **sp^2 hybrid orbitals** result, and one $2p$ orbital remains unchanged. Like sp^3 hybrids, sp^2 hybrid orbitals are unsymmetrical about the nucleus and are strongly oriented in a specific direction so they can form strong bonds. The three sp^2 orbitals lie in a plane at angles of 120° to one another, with the remaining p orbital perpendicular to the sp^2 plane, as shown in **Figure 1.13**.

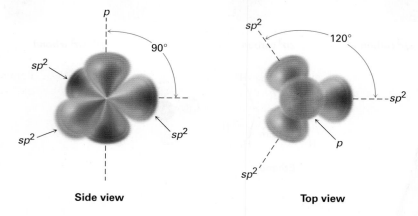

Figure 1.13 sp^2 Hybridization. The three equivalent sp^2 hybrid orbitals lie in a plane at angles of 120° to one another, and a single unhybridized p orbital (red/blue) is perpendicular to the sp^2 plane.

When two carbons with sp^2 hybridization approach each other, they form a strong σ bond by sp^2–sp^2 head-on overlap. At the same time, the unhybridized p orbitals interact by sideways overlap to form what is called a **pi (π) bond**. The combination of an sp^2–sp^2 σ bond and a $2p$–$2p$ π bond results in the sharing of four electrons and the formation of a carbon–carbon double bond (**Figure 1.14**). Note that the electrons in the σ bond occupy the region centered between nuclei, while the electrons in the π bond occupy regions above and below a line drawn between nuclei.

To complete the structure of ethylene, four hydrogen atoms form σ bonds with the remaining four sp^2 orbitals. Ethylene thus has a planar structure, with H–C–H and H–C–C bond angles of approximately 120°. (The actual values are 117.4° for the H–C–H bond angle and 121.3° for the H–C–C bond angle.) Each C–H bond has a length of 108.7 pm and a strength of 464 kJ/mol (111 kcal/mol).

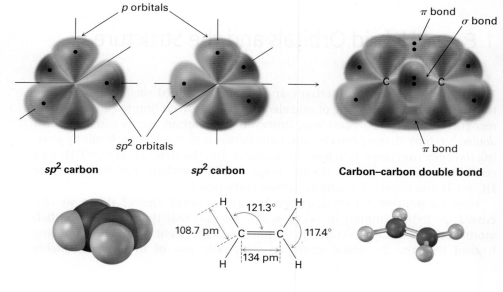

Figure 1.14 The structure of ethylene. One part of the double bond in ethylene results from σ (head-on) overlap of sp^2 orbitals, and the other part results from π (sideways) overlap of unhybridized p orbitals (red/blue). The π bond has regions of electron density above and below a line drawn between nuclei.

As you might expect, the carbon–carbon double bond in ethylene is both shorter and stronger than the single bond in ethane because it has four electrons bonding the nuclei together rather than two. Ethylene has a C=C bond length of 134 pm and a strength of 728 kJ/mol (174 kcal/mol) versus a C–C length of 154 pm and a strength of 377 kJ/mol for ethane. The carbon–carbon double bond is less than twice as strong as a single bond because the sideways overlap in the π part of the double bond is not as great as the head-on overlap in the σ part.

Worked Example 1.3

Drawing Electron-Dot and Line-Bond Structures

Commonly used in biology as a tissue preservative, formaldehyde, CH_2O, contains a carbon–*oxygen* double bond. Draw electron-dot and line-bond structures of formaldehyde, and indicate the hybridization of the carbon orbitals.

Strategy
We know that hydrogen forms one covalent bond, carbon forms four, and oxygen forms two. Trial and error, combined with intuition, is needed to fit the atoms together.

Solution
There is only one way that two hydrogens, one carbon, and one oxygen can combine:

:O: :O:
 ::   ‖
H:C:H   H—C—H

Electron-dot   Line-bond
structure   structure

Like the carbon atoms in ethylene, the carbon atom in formaldehyde is in a double bond and its orbitals are therefore sp^2-hybridized.

Problem 1.10
Draw a line-bond structure for propene, $CH_3CH=CH_2$. Indicate the hybridization of the orbitals on each carbon, and predict the value of each bond angle.

Problem 1.11
Draw a line-bond structure for 1,3-butadiene, $H_2C=CH-CH=CH_2$. Indicate the hybridization of the orbitals on each carbon, and predict the value of each bond angle.

Problem 1.12
Following is a molecular model of aspirin (acetylsalicylic acid). Identify the hybridization of the orbitals on each carbon atom in aspirin, and tell which atoms have lone pairs of electrons (gray = C, red = O, ivory = H).

Aspirin
(acetylsalicylic acid)

1.9 *sp* Hybrid Orbitals and the Structure of Acetylene

In addition to forming single and double bonds by sharing two and four electrons, respectively, carbon also can form a *triple* bond by sharing six electrons. To account for the triple bond in a molecule such as acetylene, H—C≡C—H, we need a third kind of hybrid orbital, an **sp hybrid**. Imagine that, instead of combining with two or three *p* orbitals, a carbon 2s orbital hybridizes with only a single *p* orbital. Two *sp* hybrid orbitals result, and two *p* orbitals remain unchanged. The two *sp* orbitals are oriented 180° apart on the *x*-axis, while the remaining two *p* orbitals are perpendicular on the *y*-axis and the *z*-axis, as shown in **Figure 1.15**.

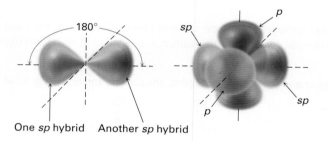

Figure 1.15 *sp* Hybridization. The two *sp* hybrid orbitals are oriented 180° away from each other, perpendicular to the two remaining *p* orbitals (red/blue).

When two *sp* carbon atoms approach each other, *sp* hybrid orbitals on each carbon overlap head-on to form a strong *sp*–*sp* σ bond. At the same time, the p_z orbitals from each carbon form a p_z–p_z π bond by sideways overlap, and the p_y orbitals overlap similarly to form a p_y–p_y π bond. The net effect is the sharing of six electrons and formation of a carbon–carbon triple bond. The two remaining *sp* hybrid orbitals each form a σ bond with hydrogen to complete the acetylene molecule (**Figure 1.16**).

Figure 1.16 The structure of acetylene. The two carbon atoms are joined by one *sp*–*sp* σ bond and two *p*–*p* π bonds.

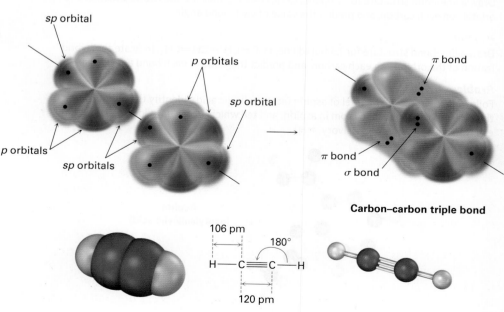

As suggested by *sp* hybridization, acetylene is a linear molecule with H−C−C bond angles of 180°. The C−H bonds have a length of 106 pm and a strength of 558 kJ/mol (133 kcal/mol). The C−C bond length in acetylene is 120 pm, and its strength is about 965 kJ/mol (231 kcal/mol), making it the shortest and strongest of any carbon–carbon bond. A comparison of sp, sp^2, and sp^3 hybridization is given in Table 1.2.

Table 1.2 Comparison of C−C and C−H Bonds in Methane, Ethane, Ethylene, and Acetylene

Molecule	Bond	Bond strength (kJ/mol)	Bond strength (kcal/mol)	Bond length (pm)
Methane, CH_4	(sp^3) C−H	439	105	109
Ethane, CH_3CH_3	(sp^3) C−C (sp^3)	377	90	154
	(sp^3) C−H	421	101	109
Ethylene, $H_2C=CH_2$	(sp^2) C=C (sp^2)	728	174	134
	(sp^2) C−H	464	111	109
Acetylene, HC≡CH	(sp) C≡C (sp)	965	231	120
	(sp) C−H	558	133	106

Problem 1.13
Draw a line-bond structure for propyne, $CH_3C \equiv CH$. Indicate the hybridization of the orbitals on each carbon, and predict a value for each bond angle.

1.10 Hybridization of Nitrogen, Oxygen, Phosphorus, and Sulfur

The valence-bond concept of orbital hybridization described in the previous four sections is not limited to carbon. Covalent bonds formed by other elements can also be described using hybrid orbitals. Look, for instance, at the nitrogen atom in methylamine (CH_3NH_2), an organic derivative of ammonia (NH_3) and the substance responsible for the odor of rotting fish.

The experimentally measured H−N−H bond angle in methylamine is 107.1°, and the C−N−H bond angle is 110.3°, both of which are close to the 109.5° tetrahedral angle found in methane. We therefore assume that nitrogen forms four sp^3-hybridized orbitals, just as carbon does. One of the four sp^3 orbitals is occupied by two nonbonding electrons, and the other three hybrid orbitals have one electron each. Overlap of these three half-filled nitrogen orbitals with half-filled orbitals from other atoms (C or H) gives methylamine. Note that the unshared lone pair of electrons in the fourth sp^3 hybrid orbital of nitrogen occupies as much space as an N−H bond does and is very important to the chemistry of methylamine and other nitrogen-containing organic molecules.

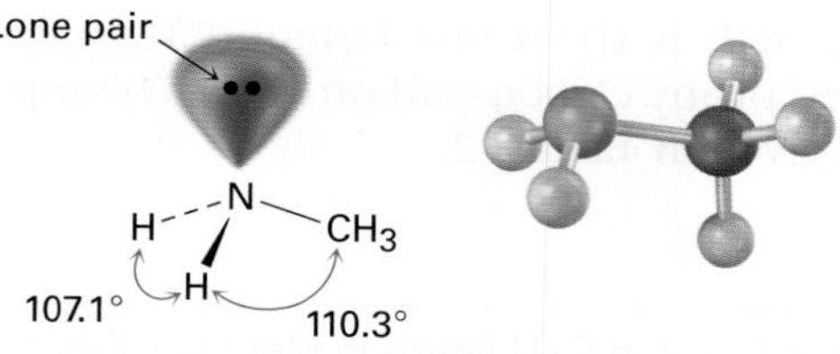

Methylamine

Like the carbon atom in methane and the nitrogen atom in methylamine, the oxygen atom in methanol (methyl alcohol) and many other organic molecules can be described as *sp*³-hybridized. The C–O–H bond angle in methanol is 108.5°, very close to the 109.5° tetrahedral angle. Two of the four *sp*³ hybrid orbitals on oxygen are occupied by nonbonding electron lone pairs, and two are used to form bonds.

Methanol (methyl alcohol)

Phosphorus and sulfur are the third-row analogs of nitrogen and oxygen, and the bonding in both can be described using hybrid orbitals. Because of their positions in the third row, however, both phosphorus and sulfur can expand their outer-shell octets and form more than the typical number of covalent bonds. Phosphorus, for instance, often forms five covalent bonds, and sulfur often forms four.

Phosphorus is most commonly encountered in biological molecules in *organophosphates*, compounds that contain a phosphorus atom bonded to four oxygens, with one of the oxygens also bonded to carbon. Methyl phosphate, $CH_3OPO_3^{2-}$, is the simplest example. The O–P–O bond angle in such compounds is typically in the range 110 to 112°, implying *sp*³ hybridization for the phosphorus orbitals.

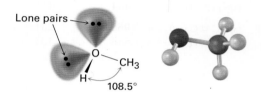

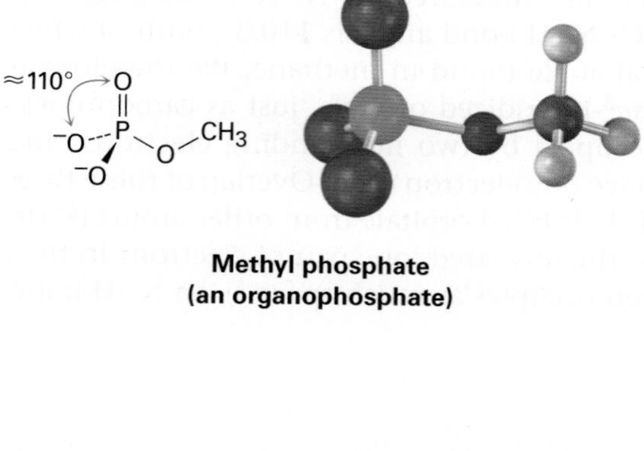

Methyl phosphate (an organophosphate)

Sulfur is most commonly encountered in biological molecules either in compounds called *thiols*, which have a sulfur atom bonded to one hydrogen and one carbon, or in *sulfides*, which have a sulfur atom bonded to two carbons. Produced by some bacteria, methanethiol (CH_3SH) is the simplest example of a thiol, and dimethyl sulfide [$(CH_3)_2S$] is the simplest example of a sulfide. Both can be described by approximate sp^3 hybridization around sulfur, although both have significant deviation from the 109.5° tetrahedral angle.

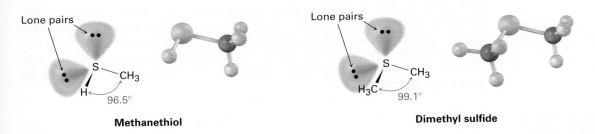

Methanethiol

Dimethyl sulfide

Problem 1.14
Identify all nonbonding lone pairs of electrons in the following molecules, and tell what geometry you expect for each of the indicated atoms.
(a) The oxygen atom in dimethyl ether, CH_3-O-CH_3
(b) The nitrogen atom in trimethylamine, $H_3C-N-CH_3$
   $\quad\quad\quad\quad\quad\quad\quad\quad\quad\quad\quad\quad\quad\quad\quad\ \ |$
   $\quad\quad\quad\quad\quad\quad\quad\quad\quad\quad\quad\quad\quad\quad\ CH_3$
(c) The phosphorus atom in phosphine, PH_3
(d) The sulfur atom in the amino acid methionine, $CH_3-S-CH_2CH_2\overset{\overset{\displaystyle O}{\|}}{\underset{\underset{\displaystyle NH_2}{|}}{C}}HCOH$

1.11 Describing Chemical Bonds: Molecular Orbital Theory

We said in **Section 1.5** that chemists use two models for describing covalent bonds: valence bond theory and molecular orbital theory. Having now seen the valence bond approach, which uses hybrid atomic orbitals to account for geometry and assumes the overlap of atomic orbitals to account for electron sharing, let's look briefly at the molecular orbital approach to bonding. We'll return to the topic in Chapters 14, 15, and 30 for a more in-depth discussion.

Molecular orbital (MO) theory describes covalent bond formation as arising from a mathematical combination of atomic orbitals (wave functions) on different atoms to form *molecular orbitals*, so called because they belong to the entire *molecule* rather than to an individual atom. Just as an *atomic* orbital, whether unhybridized or hybridized, describes a region of space around an atom where an electron is likely to be found, so a molecular orbital describes a region of space in a *molecule* where electrons are most likely to be found.

Like an atomic orbital, a molecular orbital has a specific size, shape, and energy. In the H_2 molecule, for example, two singly occupied $1s$ atomic orbitals combine to form two molecular orbitals. There are two ways for the orbital combination to occur—an additive way and a subtractive way. The additive combination leads to formation of a molecular orbital that is lower in energy and roughly egg-shaped, while the subtractive combination leads to formation of a molecular orbital that is higher in energy and has a node between nuclei **(Figure 1.17)**. Note that the additive combination is a *single,* egg-shaped, molecular orbital; it is not the same as the two overlapping $1s$ atomic orbitals of the valence bond description. Similarly, the subtractive combination is a single molecular orbital with the shape of an elongated dumbbell.

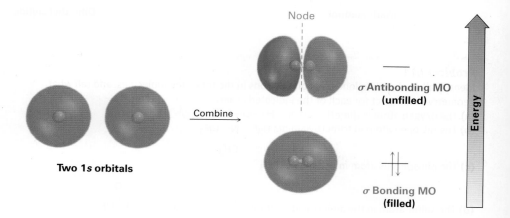

Figure 1.17 Molecular orbitals of H_2. Combination of two hydrogen $1s$ atomic orbitals leads to two H_2 molecular orbitals. The lower-energy, bonding MO is filled, and the higher-energy, antibonding MO is unfilled.

The additive combination is lower in energy than the two hydrogen $1s$ atomic orbitals and is called a **bonding MO** because electrons in this MO spend most of their time in the region between the two nuclei, thereby bonding the atoms together. The subtractive combination is higher in energy than the two hydrogen $1s$ orbitals and is called an **antibonding MO** because any electrons it contains *can't* occupy the central region between the nuclei, where there is a node, and can't contribute to bonding. The two nuclei therefore repel each other.

Just as bonding and antibonding σ molecular orbitals result from the head-on combination of two s atomic orbitals in H_2, so bonding and antibonding π molecular orbitals result from the sideways combination of two p atomic orbitals in ethylene. As shown in **Figure 1.18**, the lower-energy, π bonding MO has no node between nuclei and results from combination of p orbital lobes with the same algebraic sign. The higher-energy, π antibonding MO has a node between nuclei and results from combination of lobes with opposite algebraic signs. Only the bonding MO is occupied; the higher-energy, antibonding MO is vacant. We'll see in Chapters 14, 15, and 30 that molecular orbital theory is particularly useful for describing π bonds in compounds that have more than one double bond.

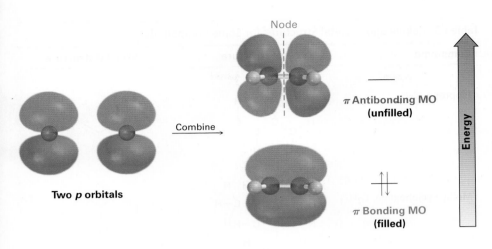

Figure 1.18 A molecular orbital description of the C–C π bond in ethylene. The lower-energy, π bonding MO results from an additive combination of p orbital lobes with the same algebraic sign and is filled. The higher-energy, π antibonding MO results from a subtractive combination of p orbital lobes with the opposite algebraic signs and is unfilled.

1.12 Drawing Chemical Structures

Let's cover just one more point before ending this introductory chapter. In the structures we've been drawing until now, a line between atoms has represented the two electrons in a covalent bond. Drawing every bond and every atom is tedious, however, so chemists have devised several shorthand ways for writing structures. In **condensed structures**, carbon–hydrogen and carbon–carbon single bonds aren't shown; instead, they're understood. If a carbon has three hydrogens bonded to it, we write CH_3; if a carbon has two hydrogens bonded to it, we write CH_2; and so on. The compound called 2-methylbutane, for example, is written as follows:

Notice that the horizontal bonds between carbons aren't shown in the condensed structures—the CH_3, CH_2, and CH units are simply placed next to each other—but the vertical carbon–carbon bond in the first of the condensed structures drawn above is shown for clarity. Notice also in the second of the condensed structures that the two CH_3 units attached to the CH carbon are grouped together as $(CH_3)_2$.

Even simpler than condensed structures are **skeletal structures**, such as those shown in Table 1.3. The rules for drawing skeletal structures are straightforward.

> **RULE 1**
> Carbon atoms aren't usually shown. Instead, a carbon atom is assumed to be at each intersection of two lines (bonds) and at the end of each line. Occasionally, a carbon atom might be indicated for emphasis or clarity.

Table 1.3 Kekulé and Skeletal Structures for Some Compounds

Compound	Kekulé structure	Skeletal structure
Isoprene, C_5H_8	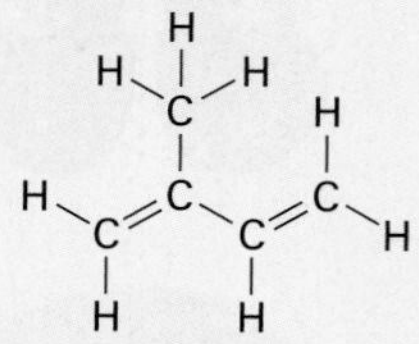	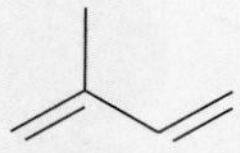
Methylcyclohexane, C_7H_{14}	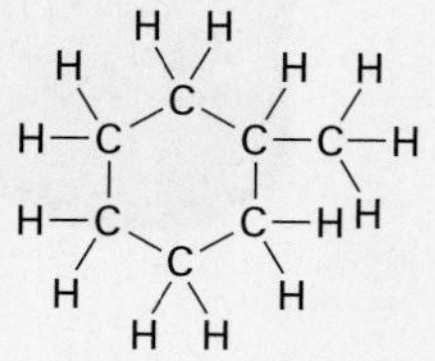	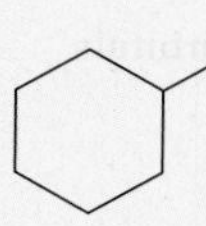
Phenol, C_6H_6O	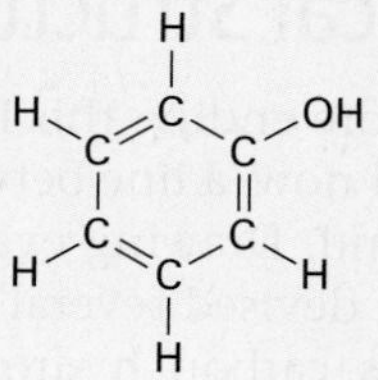	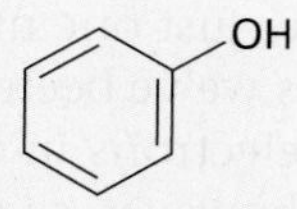

RULE 2

Hydrogen atoms bonded to carbon aren't shown. Because carbon always has a valence of 4, we mentally supply the correct number of hydrogen atoms for each carbon.

RULE 3

Atoms other than carbon and hydrogen *are* shown.

One further comment: although such groupings as –CH₃, –OH, and –NH₂ are usually written with the C, O, or N atom first and the H atom second, the order of writing is sometimes inverted to H₃C–, HO–, and H₂N– if needed to make the bonding connections in a molecule clearer. Larger units such as –CH₂CH₃ are not inverted, though; we don't write H₃CH₂C– because it would be confusing. There are, however, no well-defined rules that cover all cases; it's largely a matter of preference.

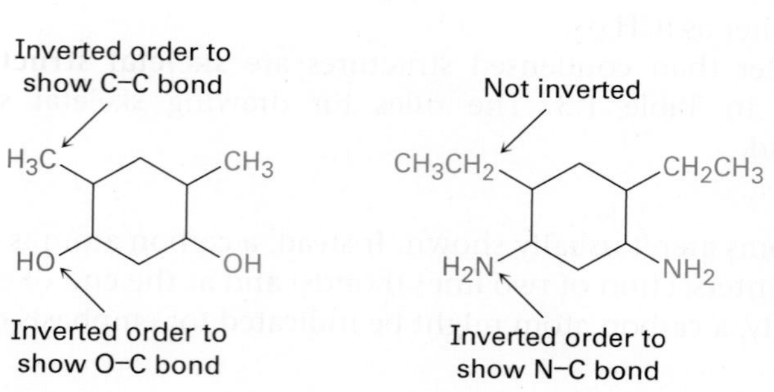

Interpreting a Line-Bond Structure

Worked Example 1.4

Carvone, a substance responsible for the odor of spearmint, has the following structure. Tell how many hydrogens are bonded to each carbon, and give the molecular formula of carvone.

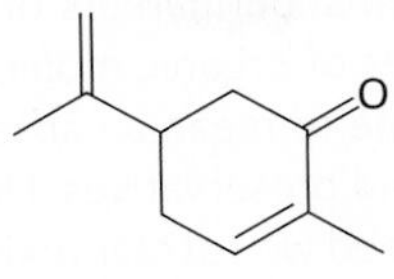

Carvone

Strategy
The end of a line represents a carbon atom with 3 hydrogens, CH_3; a two-way intersection is a carbon atom with 2 hydrogens, CH_2; a three-way intersection is a carbon atom with 1 hydrogen, CH; and a four-way intersection is a carbon atom with no attached hydrogens.

Solution

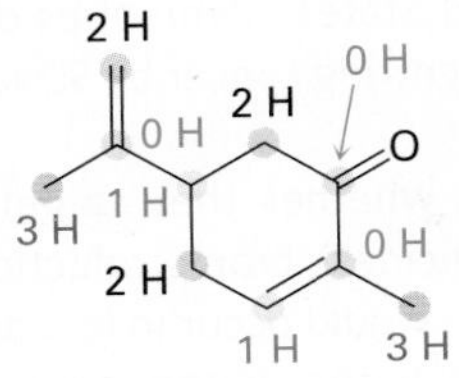

Carvone ($C_{10}H_{14}O$)

Problem 1.15
Tell how many hydrogens are bonded to each carbon in the following compounds, and give the molecular formula of each substance:

(a) Adrenaline

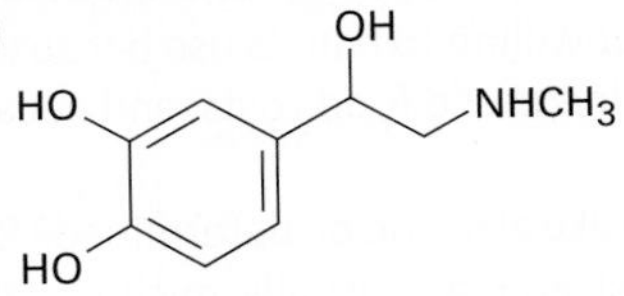

(b) 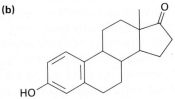 Estrone (a hormone)

Problem 1.16
Propose skeletal structures for compounds that satisfy the following molecular formulas. There is more than one possibility in each case.
(a) C_5H_{12} **(b)** C_2H_7N **(c)** C_3H_6O **(d)** C_4H_9Cl

Problem 1.17
The following molecular model is a representation of *para*-aminobenzoic acid (PABA), the active ingredient in many sunscreens. Indicate the positions of the multiple bonds, and draw a skeletal structure (gray = C, red = O, blue = N, ivory = H).

para-Aminobenzoic acid (PABA)

Organic Foods: Risk versus Benefit | A DEEPER LOOK

How dangerous is the pesticide being sprayed on this crop?

Contrary to what you may hear in supermarkets or on television, *all* foods are organic—that is, complex mixtures of organic molecules. Even so, when applied to food, the word *organic* has come to mean an absence of synthetic chemicals, typically pesticides, antibiotics, and preservatives. How concerned should we be about traces of pesticides in the food we eat? Or toxins in the water we drink? Or pollutants in the air we breathe?

Life is not risk-free—we all take many risks each day without even thinking about it. We decide to ride a bike rather than drive, even though there is a ten times greater likelihood per mile of dying in a bicycling accident than in a car. We decide to walk down stairs rather than take an elevator, even though 7000 people die from falls each year in the United States. Some of us decide to smoke cigarettes, even though it increases our chance of getting cancer by 50%. But what about risks from chemicals like pesticides?

One thing is certain: without pesticides, whether they target weeds (herbicides), insects (insecticides), or molds and fungi (fungicides), crop production would drop significantly, food prices would increase, and famines would occur in less developed parts of the world. Take the herbicide atrazine, for instance. In the United States alone, approximately 100 million pounds of atrazine are used each year to kill weeds in corn, sorghum, and sugarcane fields, greatly improving the yields of these crops. Nevertheless, the use of atrazine continues to be a concern because traces persist in the environment. Indeed, heavy atrazine exposure *can* pose health risks to humans and some animals, but the United States Environmental Protection Agency (EPA) is unwilling to ban its use because doing so would result in significantly lower crop yields and increased food costs, and because there is no suitable alternative herbicide available.

How can the potential hazards from a chemical like atrazine be determined? Risk evaluation of chemicals is carried out by exposing test animals, usually mice or rats, to the chemical and then monitoring the animals for signs of harm. To limit the expense and time needed, the amounts administered are typically hundreds or thousands of times greater than those a person might normally encounter. The results obtained in animal tests are then distilled into a single number called an LD_{50}, the amount of substance per kilogram body weight that is a lethal dose for 50% of the test animals. For atrazine, the LD_{50} value is between 1 and 4 g/kg depending on the animal species. Aspirin, for comparison, has an LD_{50} of 1.1 g/kg, and ethanol (ethyl alcohol) has an LD_{50} of 10.6 g/kg.

Table 1.4 lists values for some other familiar substances. The lower the value, the more toxic the substance. Note, though, that LD_{50} values tell only about the effects of heavy

Table 1.4 Some LD_{50} Values

Substance	LD_{50} (g/kg)	Substance	LD_{50} (g/kg)
Strychnine	0.005	Chloroform	1.2
Arsenic trioxide	0.015	Iron(II) sulfate	1.5
DDT	0.115	Ethyl alcohol	10.6
Aspirin	1.1	Sodium cyclamate	17

(continued)

exposure for a relatively short time. They say nothing about the risks of long-term exposure, such as whether the substance can cause cancer or interfere with development in the unborn.

So, should we still use atrazine? All decisions involve tradeoffs, and the answer is rarely obvious. Does the benefit of increased food production outweigh possible health risks of a pesticide? Do the beneficial effects of a new drug outweigh a potentially dangerous side effect in a small number of users? Different people will have different opinions, but an honest evaluation of facts is surely the best way to start. At present, atrazine is approved for continued use in the United States because the EPA believes that the benefits of increased food production outweigh possible health risks. At the same time, though, the use of atrazine is being phased out in Europe.

Summary

The purpose of this chapter has been to get you up to speed—to review some ideas about atoms, bonds, and molecular geometry. As we've seen, **organic chemistry** is the study of carbon compounds. Although a division into organic and inorganic chemistry occurred historically, there is no scientific reason for the division.

An atom consists of a positively charged nucleus surrounded by one or more negatively charged electrons. The electronic structure of an atom can be described by a quantum mechanical wave equation, in which electrons are considered to occupy **orbitals** around the nucleus. Different orbitals have different energy levels and different shapes. For example, s orbitals are spherical and p orbitals are dumbbell-shaped. The **ground-state electron configuration** of an atom can be found by assigning electrons to the proper orbitals, beginning with the lowest-energy ones.

A **covalent bond** is formed when an electron pair is shared between atoms. According to **valence bond theory**, electron sharing occurs by overlap of two atomic orbitals. According to **molecular orbital (MO) theory**, bonds result from the mathematical combination of atomic orbitals to give molecular orbitals, which belong to the entire molecule. Bonds that have a circular cross-section and are formed by head-on interaction are called **sigma (σ) bonds**; bonds formed by sideways interaction of p orbitals are called **pi (π) bonds**.

In the valence bond description, carbon uses hybrid orbitals to form bonds in organic molecules. When forming only single bonds with tetrahedral geometry, carbon uses four equivalent sp^3 **hybrid orbitals**. When forming a double bond with planar geometry, carbon uses three equivalent sp^2 **hybrid orbitals** and one unhybridized p orbital. When forming a triple bond with linear geometry, carbon uses two equivalent sp **hybrid orbitals** and two unhybridized p orbitals. Other atoms such as nitrogen, phosphorus, oxygen, and sulfur also use hybrid orbitals to form strong, oriented bonds.

Organic molecules are usually drawn using either condensed structures or skeletal structures. In **condensed structures**, carbon–carbon and carbon–hydrogen bonds aren't shown. In **skeletal structures**, only the bonds and not the atoms are shown. A carbon atom is assumed to be at the ends and at the junctions of lines (bonds), and the correct number of hydrogens is mentally supplied.

Key words

antibonding MO, 20
bond angle, 12
bond length, 10
bond strength, 10
bonding MO, 20
condensed structure, 21
covalent bond, 7
electron-dot structure, 7
electron shell, 4
ground-state electron configuration, 5
isotope, 3
line-bond structure, 7
lone-pair electrons, 8
molecular orbital (MO) theory, 19
molecule, 7
node, 4
orbital, 3
organic chemistry, 1
pi (π) bond, 14
sigma (σ) bond, 10
skeletal structure, 21
sp hybrid orbital, 16
sp^2 hybrid orbital, 14
sp^3 hybrid orbital, 11
valence bond theory, 9
valence shell, 6

Working Problems

There's no surer way to learn organic chemistry than by working problems. Although careful reading and rereading of this text are important, reading alone isn't enough. You must also be able to use the information you've read and be able to apply your knowledge in new situations. Working problems gives you practice at doing this.

Each chapter in this book provides many problems of different sorts. The in-chapter problems are placed for immediate reinforcement of ideas just learned, while end-of-chapter problems provide additional practice and are of several types. They begin with a short section called "Visualizing Chemistry," which helps you "see" the microscopic world of molecules and provides practice for working in three dimensions. After the visualizations are many "Additional Problems," which are organized by topic. Early problems are primarily of the drill type, providing an opportunity for you to practice your command of the fundamentals. Later problems tend to be more thought-provoking, and some are real challenges.

As you study organic chemistry, take the time to work the problems. Do the ones you can, and ask for help on the ones you can't. If you're stumped by a particular problem, check the accompanying *Study Guide and Solutions Manual* for an explanation that will help clarify the difficulty. Working problems takes effort, but the payoff in knowledge and understanding is immense.

Exercises

OWL Interactive versions of these problems are assignable in OWL for Organic Chemistry.

Visualizing Chemistry

(Problems 1.1–1.17 appear within the chapter.)

1.18 Convert each of the following molecular models into a skeletal structure, and give the formula of each. Only the connections between atoms are shown; multiple bonds are not indicated (gray = C, red = O, blue = N, ivory = H).

(a)

Coniine (the toxic substance in poison hemlock)

(b)

Alanine (an amino acid)

1.19 The following model is a representation of citric acid, the key substance in the so-called citric acid cycle by which food molecules are metabolized in the body. Only the connections between atoms are shown; multiple bonds are not indicated. Complete the structure by indicating the positions of multiple bonds and lone-pair electrons (gray = C, red = O, ivory = H).

1.20 The following model is a representation of acetaminophen, a pain reliever sold in drugstores under a variety of names, including Tylenol. Identify the hybridization of each carbon atom in acetaminophen, and tell which atoms have lone pairs of electrons (gray = C, red = O, blue = N, ivory = H).

1.21 The following model is a representation of aspartame, $C_{14}H_{18}N_2O_5$, known commercially under many names, including NutraSweet. Only the connections between atoms are shown; multiple bonds are not indicated. Complete the structure for aspartame, and indicate the positions of multiple bonds (gray = C, red = O, blue = N, ivory = H).

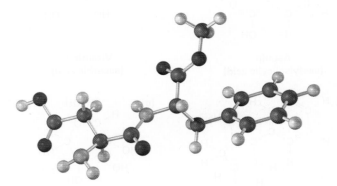

Additional Problems

Electron Configurations

1.22 How many valence electrons does each of the following dietary trace elements have?
(a) Zinc (b) Iodine (c) Silicon (d) Iron

1.23 Give the ground-state electron configuration for each of the following elements:
(a) Potassium (b) Arsenic (c) Aluminum (d) Germanium

Electron-Dot and Line-Bond Structures

1.24 What are likely formulas for the following molecules?
(a) $NH_?OH$ (b) $AlCl_?$ (c) $CF_2Cl_?$ (d) $CH_?O$

1.25 Why can't molecules with the following formulas exist?
(a) CH_5 (b) C_2H_6N (c) $C_3H_5Br_2$

1.26 Draw an electron-dot structure for acetonitrile, C_2H_3N, which contains a carbon–nitrogen triple bond. How many electrons does the nitrogen atom have in its outer shell? How many are bonding, and how many are nonbonding?

1.27 Draw a line-bond structure for vinyl chloride, C_2H_3Cl, the starting material from which PVC [poly(vinyl chloride)] plastic is made.

1.28 Fill in any nonbonding valence electrons that are missing from the following structures:

(a) $H_3C-S-S-CH_3$
Dimethyl disulfide

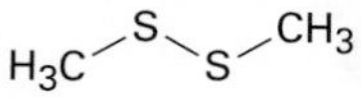

(b) Acetamide

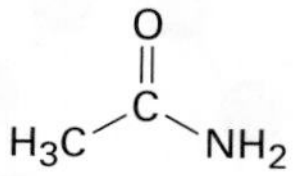

(c) Acetate ion

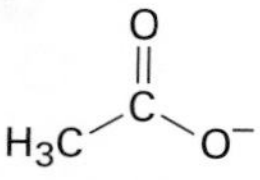

1.29 Convert the following line-bond structures into molecular formulas:

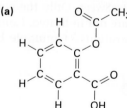

(a) Aspirin (acetylsalicylic acid) (b) Vitamin C (ascorbic acid)

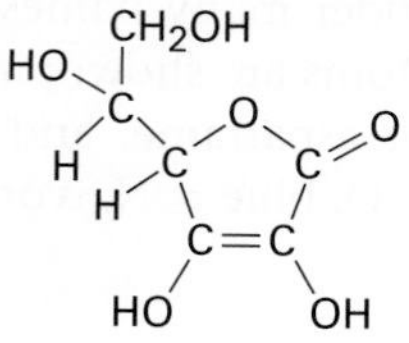

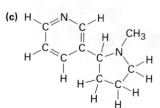

(c) Nicotine (d) Glucose

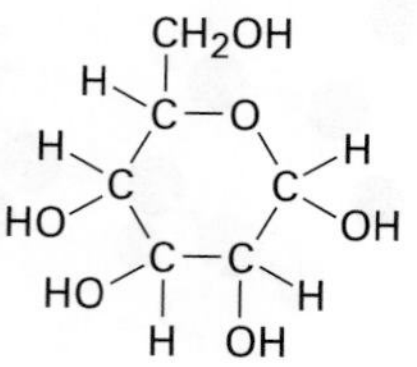

1.30 Convert the following molecular formulas into line-bond structures that are consistent with valence rules:
(a) C₃H₈
(b) CH₅N
(c) C₂H₆O (2 possibilities)
(d) C₃H₇Br (2 possibilities)
(e) C₂H₄O (3 possibilities)
(f) C₃H₉N (4 possibilities)

1.31 Draw a three-dimensional representation of the oxygen-bearing carbon atom in ethanol, CH₃CH₂OH, using the standard convention of solid, wedged, and dashed lines.

1.32 Oxaloacetic acid, an important intermediate in food metabolism, has the formula C₄H₄O₅ and contains three C=O bonds and two O–H bonds. Propose two possible structures.

1.33 Draw structures for the following molecules, showing lone pairs:
(a) Acrylonitrile, C₃H₃N, which contains a carbon–carbon double bond and a carbon–nitrogen triple bond
(b) Ethyl methyl ether, C₃H₈O, which contains an oxygen atom bonded to two carbons
(c) Butane, C₄H₁₀, which contains a chain of four carbon atoms
(d) Cyclohexene, C₆H₁₀, which contains a ring of six carbon atoms and one carbon–carbon double bond

1.34 Potassium methoxide, KOCH₃, contains both covalent and ionic bonds. Which do you think is which?

Hybridization

1.35 What is the hybridization of each carbon atom in acetonitrile (Problem 1.26)?

1.36 What kind of hybridization do you expect for each carbon atom in the following molecules?

(a) Propane, CH₃CH₂CH₃
(b) 2-Methylpropene,
$$CH_3C(CH_3)=CH_2$$

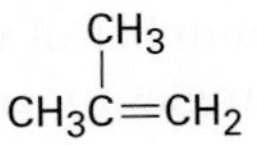

(c) 1-Butene-3-yne, H₂C=CH−C≡CH
(d) Acetic acid,
$$CH_3COOH$$

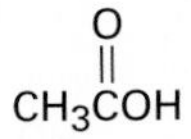

1.37 What is the shape of benzene, and what hybridization do you expect for each carbon?

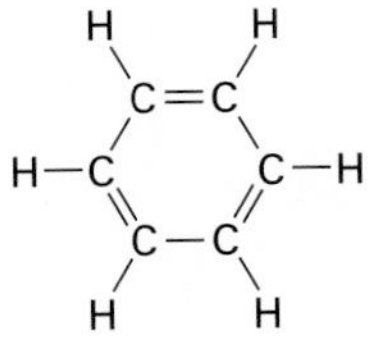

Benzene

1.38 What bond angles do you expect for each of the following, and what kind of hybridization do you expect for the central atom in each?

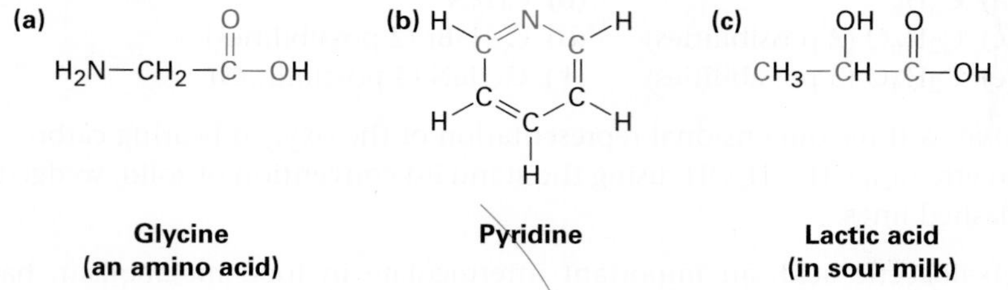

(a) H₂N—CH₂—C(=O)—OH

Glycine
(an amino acid)

(b) Pyridine

Pyridine

(c) CH₃—CH(OH)—C(=O)—OH

Lactic acid
(in sour milk)

1.39 Propose structures for molecules that meet the following descriptions:
(a) Contains two sp^2-hybridized carbons and two sp^3-hybridized carbons
(b) Contains only four carbons, all of which are sp^2-hybridized
(c) Contains two sp-hybridized carbons and two sp^2-hybridized carbons

1.40 What kind of hybridization do you expect for each carbon atom in the following molecules?

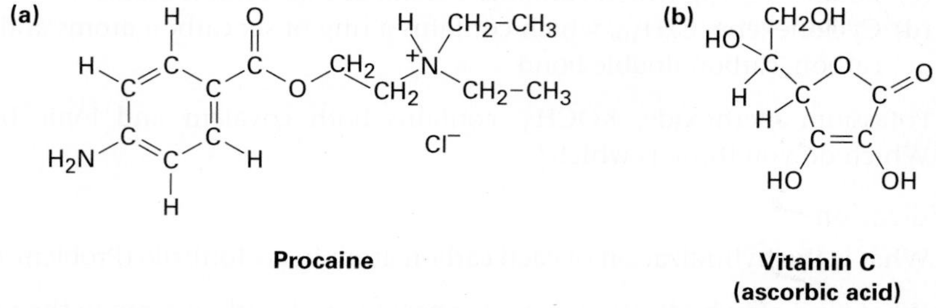

(a) **Procaine**

(b) **Vitamin C (ascorbic acid)**

1.41 Pyridoxal phosphate, a close relative of vitamin B₆, is involved in a large number of metabolic reactions. Tell the hybridization, and predict the bond angles for each nonterminal atom.

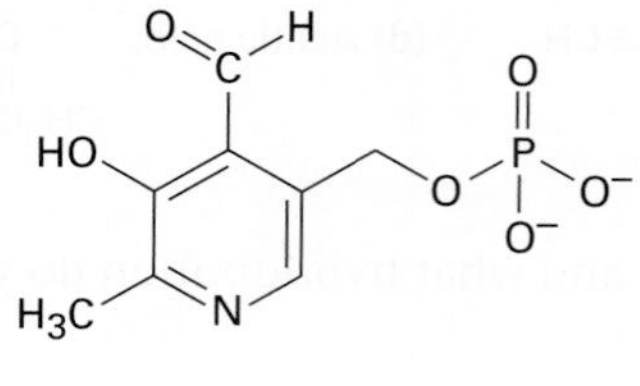

Pyridoxal phosphate

Skeletal Structures

1.42 Convert the following structures into skeletal drawings:

(a) Indole

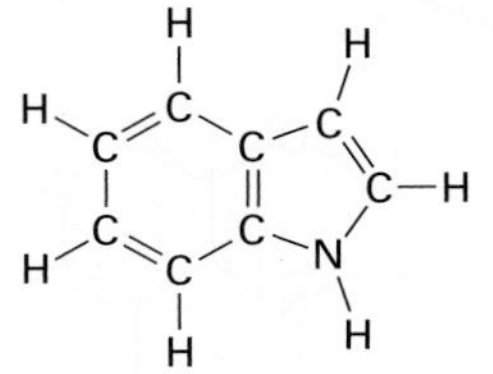

(b) 1,3-Pentadiene

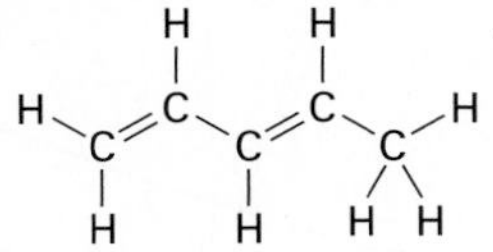

(c)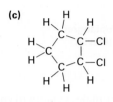
1,2-Dichlorocyclopentane

(d)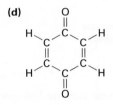
Benzoquinone

1.43 Tell the number of hydrogens bonded to each carbon atom in the following substances, and give the molecular formula of each:

(a)

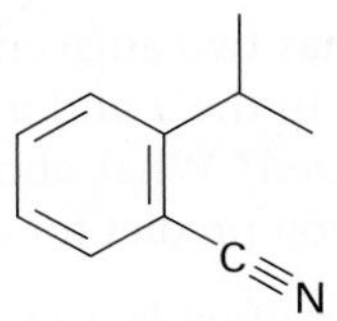

(b)

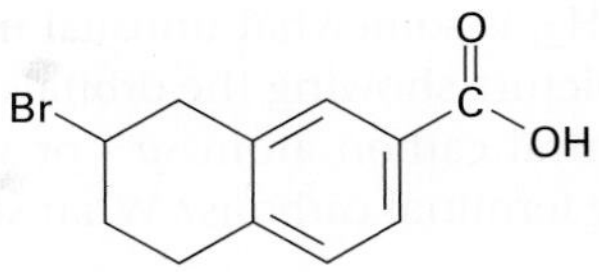

(c)

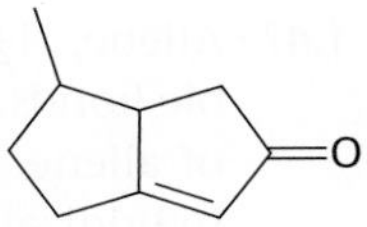

1.44 Quetiapine, marketed as Seroquel, is a heavily prescribed antipsychotic drug used in the treatment of schizophrenia and bipolar disorder. Convert the following representation into a skeletal structure, and give the molecular formula of quetiapine.

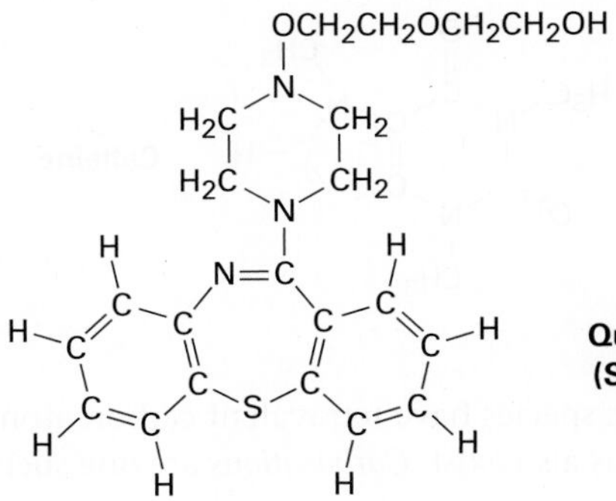

Quetiapine (Seroquel)

1.45 Tell the number of hydrogens bonded to each carbon atom in **(a)** the anti-influenza agent oseltamivir, marketed as Tamiflu, and **(b)** the platelet aggregation inhibitor clopidogrel, marketed as Plavix. Give the molecular formula of each.

(a) Oseltamivir (Tamiflu)

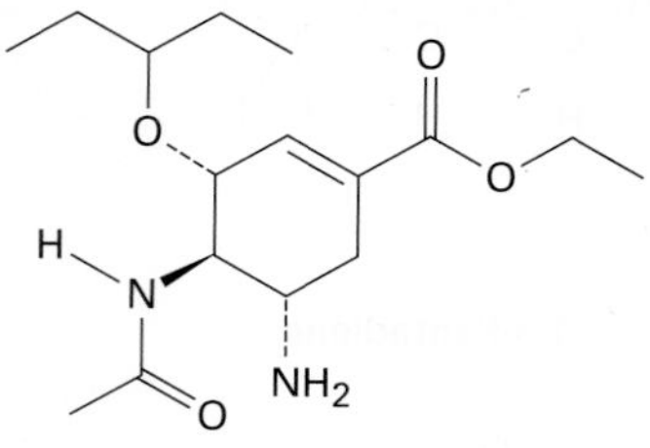

(b) Clopidogrel (Plavix)

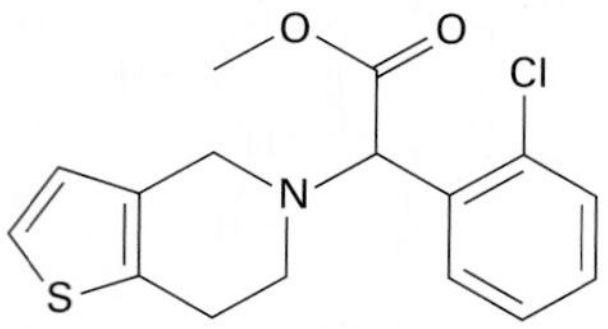

General Problems

1.46 Why do you suppose no one has ever been able to make cyclopentyne as a stable molecule?

Cyclopentyne

1.47 Allene, $H_2C=C=CH_2$, is somewhat unusual in that it has two adjacent double bonds. Draw a picture showing the orbitals involved in the σ and π bonds of allene. Is the central carbon atom sp^2- or sp-hybridized? What about the hybridization of the terminal carbons? What shape do you predict for allene?

1.48 Allene (see Problem 1.47) is related structurally to carbon dioxide, CO_2. Draw a picture showing the orbitals involved in the σ and π bonds of CO_2, and identify the likely hybridization of carbon.

1.49 Complete the electron-dot structure of caffeine, showing all lone-pair electrons, and identify the hybridization of the indicated atoms.

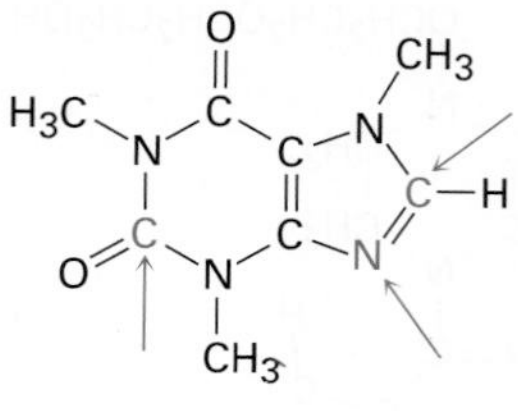

Caffeine

1.50 Most stable organic species have tetravalent carbon atoms, but species with trivalent carbon atoms also exist. *Carbocations* are one such class of compounds.

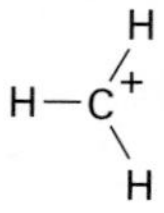

A carbocation

(a) How many valence electrons does the positively charged carbon atom have?
(b) What hybridization do you expect this carbon atom to have?
(c) What geometry is the carbocation likely to have?

1.51 A *carbanion* is a species that contains a negatively charged, trivalent carbon.

$$\text{H}-\overset{\overset{\text{H}}{|}}{\underset{\underset{\text{H}}{|}}{\text{C}}}{:}^{-} \quad \text{A carbanion}$$

(a) What is the electronic relationship between a carbanion and a trivalent nitrogen compound such as NH_3?
(b) How many valence electrons does the negatively charged carbon atom have?
(c) What hybridization do you expect this carbon atom to have?
(d) What geometry is the carbanion likely to have?

1.52 Divalent carbon species called *carbenes* are capable of fleeting existence. For example, methylene, $:CH_2$, is the simplest carbene. The two unshared electrons in methylene can be either paired in a single orbital or unpaired in different orbitals. Predict the type of hybridization you expect carbon to adopt in singlet (spin-paired) methylene and triplet (spin-unpaired) methylene. Draw a picture of each, and identify the valence orbitals on carbon.

1.53 There are two different substances with the formula C_4H_{10}. Draw both, and tell how they differ.

1.54 There are two different substances with the formula C_3H_6. Draw both, and tell how they differ.

1.55 There are two different substances with the formula C_2H_6O. Draw both, and tell how they differ.

1.56 There are three different substances that contain a carbon–carbon double bond and have the formula C_4H_8. Draw them, and tell how they differ.

1.57 Among the most common over-the-counter drugs you might find in a medicine cabinet are mild pain relievers such ibuprofen (Advil, Motrin), naproxen (Aleve), and acetaminophen (Tylenol).

Ibuprofen   **Naproxen**   **Acetaminophen**

(a) How many sp^3-hybridized carbons does each molecule have?
(b) How many sp^2-hybridized carbons does each molecule have?
(c) Can you spot any similarities in their structures?

The opium poppy is the source of morphine, one of the first "vegetable alkali," or *alkaloids*, to be isolated.
Image copyright Igor Plotnikov, 2010. Used under license from Shutterstock.com

2 Polar Covalent Bonds; Acids and Bases

2.1 Polar Covalent Bonds: Electronegativity
2.2 Polar Covalent Bonds: Dipole Moments
2.3 Formal Charges
2.4 Resonance
2.5 Rules for Resonance Forms
2.6 Drawing Resonance Forms
2.7 Acids and Bases: The Brønsted–Lowry Definition
2.8 Acid and Base Strength
2.9 Predicting Acid–Base Reactions from pK_a Values
2.10 Organic Acids and Organic Bases
2.11 Acids and Bases: The Lewis Definition
2.12 Noncovalent Interactions Between Molecules
  A Deeper Look—Alkaloids: From Cocaine to Dental Anesthetics

OWL Sign in to OWL for Organic Chemistry at www.cengage.com/owl to view tutorials and simulations, develop problem-solving skills, and complete online homework assigned by your professor.

We saw in the last chapter how covalent bonds between atoms are described, and we looked at the valence bond model, which uses hybrid orbitals to account for the observed shapes of organic molecules. Before going on to a systematic study of organic chemistry, however, we still need to review a few fundamental topics. In particular, we need to look more closely at how electrons are distributed in covalent bonds and at some of the consequences that arise when the electrons in a bond are not shared equally between atoms.

Why This Chapter? Understanding organic and biological chemistry means knowing not just what happens but also why and how it happens at the molecular level. In this chapter, we'll look at some of the ways that chemists describe and account for chemical reactivity, thereby providing a foundation to understand the specific reactions discussed in subsequent chapters. Topics such as bond polarity, the acid–base behavior of molecules, and hydrogen-bonding are a particularly important part of that foundation.

2.1 Polar Covalent Bonds: Electronegativity

Up to this point, we've treated chemical bonds as either ionic or covalent. The bond in sodium chloride, for instance, is ionic. Sodium transfers an electron to chlorine to give Na^+ and Cl^- ions, which are held together in the solid by electrostatic attractions between unlike charges. The C–C bond in ethane, however, is covalent. The two bonding electrons are shared equally by the two equivalent carbon atoms, resulting in a symmetrical electron distribution in the bond. Most bonds, however, are neither fully ionic nor fully covalent but are somewhere between the two extremes. Such bonds are called **polar covalent bonds**, meaning that the bonding electrons are attracted more strongly by one atom than the other so that the electron distribution between atoms is not symmetrical **(Figure 2.1)**.

2.1 | Polar Covalent Bonds: Electronegativity

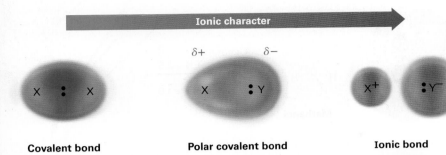

Figure 2.1 The continuum in bonding from covalent to ionic is a result of an unequal distribution of bonding electrons between atoms. The symbol δ (lowercase Greek delta) means *partial* charge, either partial positive (δ+) for the electron-poor atom or partial negative (δ−) for the electron-rich atom.

Bond polarity is due to differences in **electronegativity (EN)**, the intrinsic ability of an atom to attract the shared electrons in a covalent bond. As shown in **Figure 2.2**, electronegativities are based on an arbitrary scale, with fluorine the most electronegative (EN = 4.0) and cesium the least (EN = 0.7). Metals on the left side of the periodic table attract electrons weakly and have lower electronegativities, while oxygen, nitrogen, and halogens on the right side of the periodic table attract electrons strongly and have higher electronegativities. Carbon, the most important element in organic compounds, has an electronegativity value of 2.5.

H 2.1																	He
Li 1.0	Be 1.6											B 2.0	C 2.5	N 3.0	O 3.5	F 4.0	Ne
Na 0.9	Mg 1.2											Al 1.5	Si 1.8	P 2.1	S 2.5	Cl 3.0	Ar
K 0.8	Ca 1.0	Sc 1.3	Ti 1.5	V 1.6	Cr 1.6	Mn 1.5	Fe 1.8	Co 1.9	Ni 1.9	Cu 1.9	Zn 1.6	Ga 1.6	Ge 1.8	As 2.0	Se 2.4	Br 2.8	Kr
Rb 0.8	Sr 1.0	Y 1.2	Zr 1.4	Nb 1.6	Mo 1.8	Tc 1.9	Ru 2.2	Rh 2.2	Pd 2.2	Ag 1.9	Cd 1.7	In 1.7	Sn 1.8	Sb 1.9	Te 2.1	I 2.5	Xe
Cs 0.7	Ba 0.9	La 1.0	Hf 1.3	Ta 1.5	W 1.7	Re 1.9	Os 2.2	Ir 2.2	Pt 2.2	Au 2.4	Hg 1.9	Tl 1.8	Pb 1.9	Bi 1.9	Po 2.0	At 2.1	Rn

Figure 2.2 Electronegativity values and trends. Electronegativity generally increases from left to right across the periodic table and decreases from top to bottom. The values are on an arbitrary scale, with F = 4.0 and Cs = 0.7. Elements in red are the most electronegative, those in yellow are medium, and those in green are the least electronegative.

As a rough guide, bonds between atoms whose electronegativities differ by less than 0.5 are nonpolar covalent, bonds between atoms whose electronegativities differ by 0.5–2 are polar covalent, and bonds between atoms whose electronegativities differ by more than 2 are largely ionic. Carbon–hydrogen bonds, for example, are relatively nonpolar because carbon (EN = 2.5) and hydrogen (EN = 2.1) have similar electronegativities. Bonds between carbon and *more* electronegative elements such as oxygen (EN = 3.5) and nitrogen (EN = 3.0), by contrast, are polarized so that the bonding electrons are drawn away from carbon toward the electronegative atom. This leaves carbon with a partial positive charge, denoted by δ+, and the electronegative atom with a partial negative charge, δ− (δ is the lowercase Greek letter delta). An example, is the C–O bond in methanol, CH_3OH (**Figure 2.3a**). Bonds between carbon and *less* electronegative elements are polarized so that carbon bears a partial negative charge and the other atom bears a partial positive charge. An example is the C–Li bond in methyllithium, CH_3Li (**Figure 2.3b**).

Figure 2.3 (a) Methanol, CH₃OH, has a polar covalent C–O bond, and (b) methyllithium, CH₃Li, has a polar covalent C–Li bond. The computer-generated representations, called electrostatic potential maps, use color to show calculated charge distributions, ranging from **red (electron-rich; δ−)** to **blue (electron-poor; δ+)**.

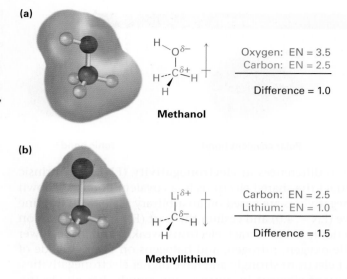

Note in the representations of methanol and methyllithium in Figure 2.3 that a crossed arrow $\mapsto$ is used to indicate the direction of bond polarity. By convention, *electrons are displaced in the direction of the arrow*. The tail of the arrow (which looks like a plus sign) is electron-poor (δ+), and the head of the arrow is electron-rich (δ−).

Note also in Figure 2.3 that calculated charge distributions in molecules can be displayed visually with what are called *electrostatic potential maps*, which use color to indicate electron-rich (red; δ−) and electron-poor (blue; δ+) regions. In methanol, oxygen carries a partial negative charge and is colored red, while the carbon and hydrogen atoms carry partial positive charges and are colored blue-green. In methyllithium, lithium carries a partial positive charge (blue), while carbon and the hydrogen atoms carry partial negative charges (red). Electrostatic potential maps are useful because they show at a glance the electron-rich and electron-poor atoms in molecules. We'll make frequent use of these maps throughout the text and will see many examples of how electronic structure correlates with chemical reactivity.

When speaking of an atom's ability to polarize a bond, we often use the term *inductive effect*. An **inductive effect** is simply the shifting of electrons in a σ bond in response to the electronegativity of nearby atoms. Metals, such as lithium and magnesium, inductively donate electrons, whereas reactive nonmetals, such as oxygen and nitrogen, inductively withdraw electrons. Inductive effects play a major role in understanding chemical reactivity, and we'll use them many times throughout this text to explain a variety of chemical observations.

Problem 2.1
Which element in each of the following pairs is more electronegative?
(a) Li or H **(b)** B or Br **(c)** Cl or I **(d)** C or H

Problem 2.2
Use the δ+/δ− convention to indicate the direction of expected polarity for each of the bonds indicated.
(a) H_3C-Cl **(b)** H_3C-NH_2 **(c)** H_2N-H
(d) H_3C-SH **(e)** $H_3C-MgBr$ **(f)** H_3C-F

Problem 2.3
Use the electronegativity values shown in Figure 2.2 to rank the following bonds from least polar to most polar: H₃C–Li, H₃C–K, H₃C–F, H₃C–MgBr, H₃C–OH

Problem 2.4
Look at the following electrostatic potential map of chloromethane, and tell the direction of polarization of the C–Cl bond:

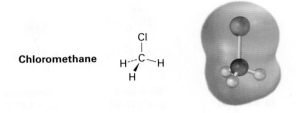

Chloromethane

2.2 Polar Covalent Bonds: Dipole Moments

Just as individual bonds are often polar, molecules as a whole are often polar also. Molecular polarity results from the vector summation of all individual bond polarities and lone-pair contributions in the molecule. As a practical matter, strongly polar substances are often soluble in polar solvents like water, whereas less polar substances are insoluble in water.

Net molecular polarity is measured by a quantity called the *dipole moment* and can be thought of in the following way: assume that there is a center of mass of all positive charges (nuclei) in a molecule and a center of mass of all negative charges (electrons). If these two centers don't coincide, then the molecule has a net polarity.

The **dipole moment**, μ (Greek mu), is defined as the magnitude of the charge Q at either end of the molecular dipole times the distance r between the charges, $\mu = Q \times r$. Dipole moments are expressed in *debyes* (D), where 1 D = 3.336×10^{-30} coulomb meter (C · m) in SI units. For example, the unit charge on an electron is 1.60×10^{-19} C. Thus, if one positive charge and one negative charge are separated by 100 pm (a bit less than the length of a typical covalent bond), the dipole moment is 1.60×10^{-29} C · m, or 4.80 D.

$$\mu = Q \times r$$
$$\mu = (1.60 \times 10^{-19} \text{ C})(100 \times 10^{-12} \text{ m})\left(\frac{1 \text{ D}}{3.336 \times 10^{-30} \text{ C} \cdot \text{m}}\right) = 4.80 \text{ D}$$

Dipole moments for some common substances are given in Table 2.1. Of the compounds shown in the table, sodium chloride has the largest dipole moment (9.00 D) because it is ionic. Even small molecules like water (μ = 1.85 D), methanol (CH₃OH; μ = 1.70 D), and ammonia (μ = 1.47 D), have substantial dipole moments, however, both because they contain strongly electronegative atoms (oxygen and nitrogen) and because all three molecules have lone-pair electrons. The lone-pair electrons on oxygen and nitrogen atom stick out into space away from the positively charged nuclei, giving rise to a considerable charge separation and making a large contribution to the dipole moment.

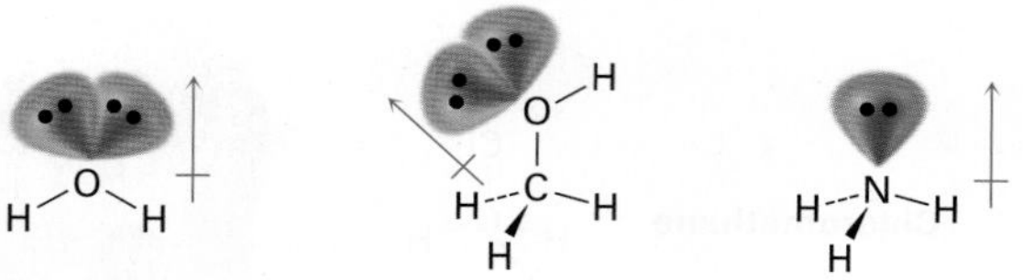

Water
($\mu = 1.85$ D)

Methanol
($\mu = 1.70$ D)

Ammonia
($\mu = 1.47$ D)

Table 2.1 Dipole Moments of Some Compounds

Compound	Dipole moment (D)	Compound	Dipole moment (D)
NaCl	9.00	NH_3	1.47
CH_2O	2.33	CH_3NH_2	1.31
CH_3Cl	1.87	CO_2	0
H_2O	1.85	CH_4	0
CH_3OH	1.70	CH_3CH_3	0
CH_3CO_2H	1.70	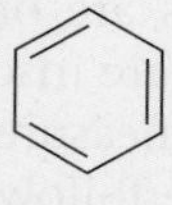 Benzene	0
CH_3SH	1.52		

In contrast with water, methanol, and ammonia, molecules such as carbon dioxide, methane, ethane, and benzene have zero dipole moments. Because of the symmetrical structures of these molecules, the individual bond polarities and lone-pair contributions exactly cancel.

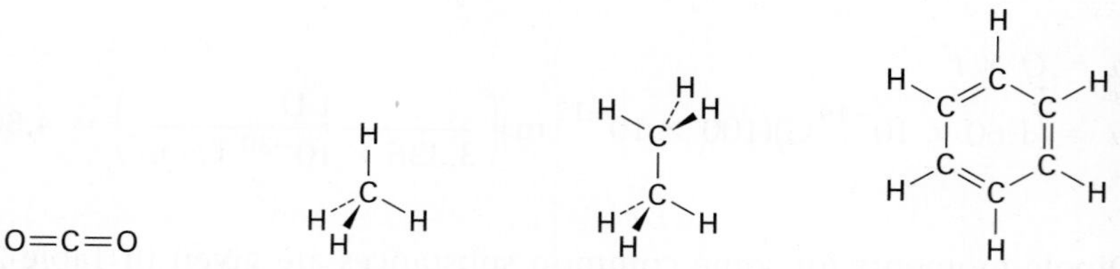

Carbon dioxide
($\mu = 0$)

Methane
($\mu = 0$)

Ethane
($\mu = 0$)

Benzene
($\mu = 0$)

Worked Example 2.1

Predicting the Direction of a Dipole Moment

Make a three-dimensional drawing of methylamine, CH$_3$NH$_2$, a substance responsible for the odor of rotting fish, and show the direction of its dipole moment ($\mu = 1.31$).

Strategy
Look for any lone-pair electrons, and identify any atom with an electronegativity substantially different from that of carbon. (Usually, this means O, N, F, Cl, or Br.) Electron density will be displaced in the general direction of the electronegative atoms and the lone pairs.

Solution
Methylamine contains an electronegative nitrogen atom with a lone-pair electrons. The dipole moment thus points generally from −CH$_3$ toward the lone pair.

Methylamine
($\mu = 1.31$)

Problem 2.5
Ethylene glycol, HOCH$_2$CH$_2$OH, has zero dipole moment even though carbon–oxygen bonds are strongly polar and oxygen has two lone-pairs of electrons. Explain.

Problem 2.6
Make three-dimensional drawings of the following molecules, and predict whether each has a dipole moment. If you expect a dipole moment, show its direction.
(a) H$_2$C=CH$_2$ (b) CHCl$_3$ (c) CH$_2$Cl$_2$ (d) H$_2$C=CCl$_2$

2.3 Formal Charges

Closely related to the ideas of bond polarity and dipole moment is the concept of assigning *formal charges* to specific atoms within a molecule, particularly atoms that have an apparently "abnormal" number of bonds. Look at dimethyl sulfoxide (CH$_3$SOCH$_3$), for instance, a solvent commonly used for preserving biological cell lines at low temperature. The sulfur atom in dimethyl sulfoxide has three bonds rather than the usual two and has a formal positive charge. The oxygen atom, by contrast, has one bond rather than the usual two and has a formal negative charge. Note that an electrostatic potential map of dimethyl sulfoxide shows the oxygen as negative (red) and the sulfur as relatively positive (blue), in accordance with the formal charges.

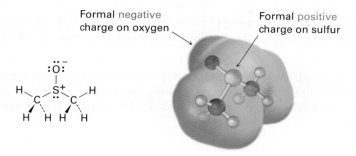

Dimethyl sulfoxide

Formal charges, as the name suggests, are a formalism and don't imply the presence of actual ionic charges in a molecule. Instead, they're a device for electron "bookkeeping" and can be thought of in the following way: a typical covalent bond is formed when each atom donates one electron. Although the bonding electrons are shared by both atoms, each atom can still be considered to "own" one electron for bookkeeping purposes. In methane, for instance, the carbon atom owns one electron in each of the four C–H bonds, for a total of four. Because a neutral, isolated carbon atom has four valence electrons, and because the carbon atom in methane still owns four, the methane carbon atom is neutral and has no formal charge.

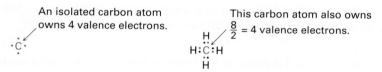

The same is true for the nitrogen atom in ammonia, which has three covalent N–H bonds and two nonbonding electrons (a lone pair). Atomic nitrogen has five valence electrons, and the ammonia nitrogen also has five—one in each of three shared N–H bonds plus two in the lone pair. Thus, the nitrogen atom in ammonia has no formal charge.

An isolated nitrogen atom owns 5 valence electrons.

This nitrogen atom also owns $\frac{6}{2} + 2 = 5$ valence electrons.

The situation is different in dimethyl sulfoxide. Atomic sulfur has six valence electrons, but the dimethyl sulfoxide sulfur owns only *five*—one in each of the two S–C single bonds, one in the S–O single bond, and two in a lone pair. Thus, the sulfur atom has formally lost an electron and therefore has a positive charge. A similar calculation for the oxygen atom shows that it has formally gained an electron and has a negative charge. Atomic oxygen has six valence electrons, but the oxygen in dimethyl sulfoxide has seven—one in the O–S bond and two in each of three lone pairs.

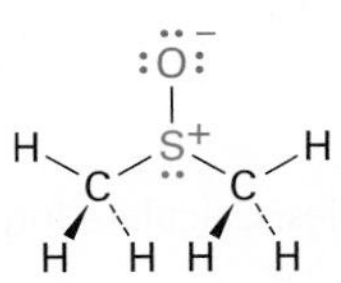

For sulfur:

Sulfur valence electrons = 6
Sulfur bonding electrons = 6
Sulfur nonbonding electrons = 2

Formal charge = 6 − 6/2 − 2 = +1

For oxygen:

Oxygen valence electrons = 6
Oxygen bonding electrons = 2
Oxygen nonbonding electrons = 6

Formal charge = 6 − 2/2 − 6 = −1

To express the calculations in a general way, the **formal charge** on an atom is equal to the number of valence electrons in a neutral, isolated atom minus the number of electrons owned by that bonded atom in a molecule. The number of electrons in the bonded atom, in turn, is equal to half the number of bonding electrons plus the nonbonding, lone-pair electrons.

$$\textbf{Formal charge} = \begin{pmatrix} \text{Number of} \\ \text{valence electrons} \\ \text{in free atom} \end{pmatrix} - \begin{pmatrix} \text{Number of} \\ \text{valence electrons} \\ \text{in bonded atom} \end{pmatrix}$$

$$= \begin{pmatrix} \text{Number of} \\ \text{valence electrons} \\ \text{in free atom} \end{pmatrix} - \begin{pmatrix} \dfrac{\text{Number of}}{\text{bonding electrons}} \\ 2 \end{pmatrix} - \begin{pmatrix} \text{Number of} \\ \text{nonbonding} \\ \text{electrons} \end{pmatrix}$$

A summary of commonly encountered formal charges and the bonding situations in which they occur is given in Table 2.2. Although only a bookkeeping device, formal charges often give clues about chemical reactivity, so it's helpful to be able to identify and calculate them correctly.

Table 2.2 A Summary of Common Formal Charges

Atom	C			N		O		S		P
Structure	—Ċ—	—C⁺—	—C̈—	—N⁺—	—N̈—	—O⁺—	—Ö:⁻	—S⁺—	—S̈:⁻	—P⁺—
Valence electrons	4	4	4	5	5	6	6	6	6	5
Number of bonds	3	3	3	4	2	3	1	3	1	4
Number of nonbonding electrons	1	0	2	0	4	2	6	2	6	0
Formal charge	0	+1	−1	+1	−1	+1	−1	+1	−1	+1

Problem 2.7
Calculate formal charges for the nonhydrogen atoms in the following molecules:
(a) Diazomethane, $H_2C=N=\ddot{N}:$
(b) Acetonitrile oxide, $H_3C-C\equiv N-\ddot{\underset{..}{O}}:$
(c) Methyl isocyanide, $H_3C-N\equiv C:$

Problem 2.8
Organic phosphate groups occur commonly in biological molecules. Calculate formal charges on the four O atoms in the methyl phosphate dianion.

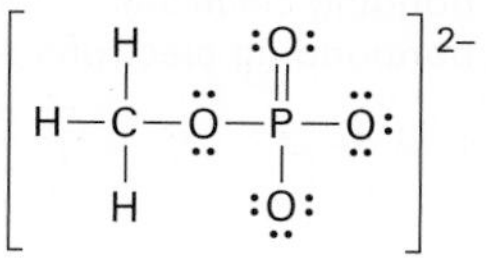

Methyl phosphate ion

2.4 Resonance

Most substances can be represented unambiguously by the Kekulé line-bond structures we've been using up to this point, but an interesting problem sometimes arises. Look at the acetate ion, for instance. When we draw a line-bond structure for acetate, we need to show a double bond to one oxygen and a single bond to the other. But which oxygen is which? Should we draw a double bond to the "top" oxygen and a single bond to the "bottom" oxygen, or vice versa?

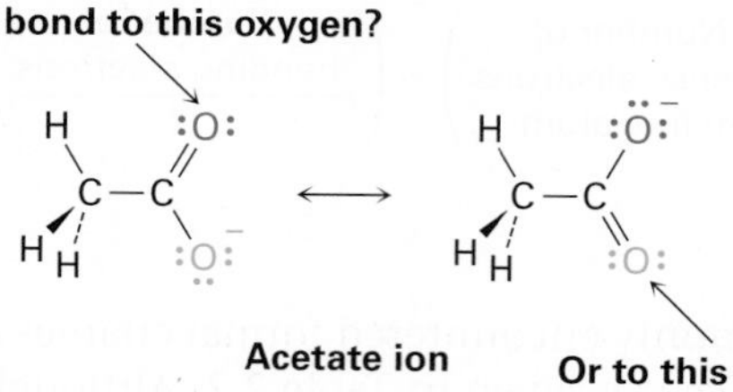

Double bond to this oxygen?

Acetate ion   **Or to this oxygen?**

Although the two oxygen atoms in the acetate ion appear different in line-bond structures, experiments show that they are equivalent. Both carbon–oxygen bonds, for example, are 127 pm in length, midway between the length of a typical C–O single bond (135 pm) and a typical C=O double bond (120 pm). In other words, *neither* of the two structures for acetate is correct by itself. The true structure is intermediate between the two, and an electrostatic potential map shows that both oxygen atoms share the negative charge and have equal electron densities (red).

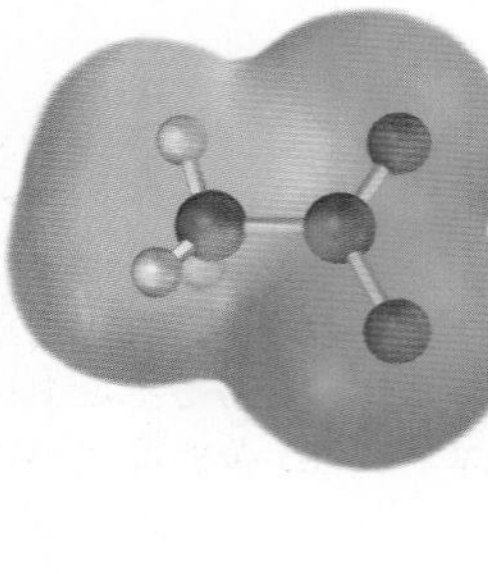

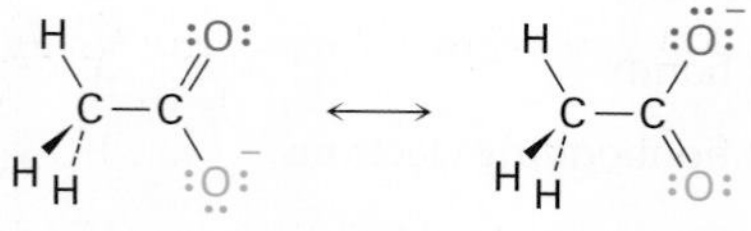

Acetate ion—two resonance forms

The two individual line-bond structures for acetate ion are called **resonance forms**, and their special resonance relationship is indicated by the double-headed arrow between them. *The only difference between resonance forms is the placement of the π and nonbonding valence electrons.* The atoms themselves occupy exactly the same place in both resonance forms, the connections between atoms are the same, and the three-dimensional shapes of the resonance forms are the same.

A good way to think about resonance forms is to realize that a substance like the acetate ion is the same as any other. Acetate doesn't jump back and forth between two resonance forms, spending part of the time looking like one and part of the time looking like the other. Rather, acetate has a single unchanging structure that we say is a **resonance hybrid** of the two individual forms and has characteristics of both. The only "problem" with acetate is that we can't draw it accurately using a familiar line-bond structure—line-bond structures just don't work well for resonance hybrids. The difficulty, however, is with the *representation* of acetate on paper, not with acetate itself.

Resonance is a very useful concept that we'll return to on numerous occasions throughout the rest of this book. We'll see in Chapter 15, for instance, that the six carbon–carbon bonds in aromatic compounds, such as benzene, are equivalent and that benzene is best represented as a hybrid of two resonance forms. Although each individual resonance form seems to imply that benzene has alternating single and double bonds, neither form is correct by itself. The true benzene structure is a hybrid of the two individual forms, and all six carbon–carbon bonds are equivalent. This symmetrical distribution of electrons around the molecule is evident in an electrostatic potential map.

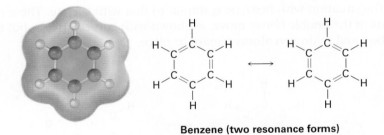

Benzene (two resonance forms)

2.5 Rules for Resonance Forms

When first dealing with resonance forms, it's useful to have a set of guidelines that describe how to draw and interpret them. The following rules should be helpful:

> **RULE 1**
> **Individual resonance forms are imaginary, not real.** The real structure is a composite, or resonance hybrid, of the different forms. Species such as the acetate ion and benzene are no different from any other. They have single, unchanging structures, and they do not switch back and forth between resonance forms. The only difference between these and other substances is in the way they must be represented in drawings on paper.

Key IDEAS

Test your knowledge of Key Ideas by answering end-of-chapter exercises marked with ▲.

RULE 2

Resonance forms differ only in the placement of their π or nonbonding electrons. Neither the position nor the hybridization of any atom changes from one resonance form to another. In the acetate ion, for instance, the carbon atom is sp^2-hybridized and the oxygen atoms remain in exactly the same place in both resonance forms. Only the positions of the π electrons in the C=O bond and the lone-pair electrons on oxygen differ from one form to another. This movement of electrons from one resonance structure to another can be indicated by using curved arrows. *A curved arrow always indicates the movement of electrons, not the movement of atoms.* An arrow shows that a pair of electrons moves *from* the atom or bond at the tail of the arrow *to* the atom or bond at the head of the arrow.

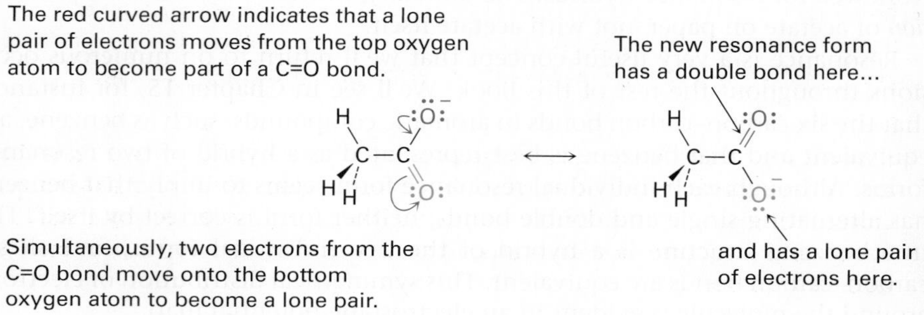

The red curved arrow indicates that a lone pair of electrons moves from the top oxygen atom to become part of a C=O bond.

The new resonance form has a double bond here...

Simultaneously, two electrons from the C=O bond move onto the bottom oxygen atom to become a lone pair.

and has a lone pair of electrons here.

The situation with benzene is similar to that with acetate. The π electrons in the double bonds move, as shown with curved arrows, but the carbon and hydrogen atoms remain in place.

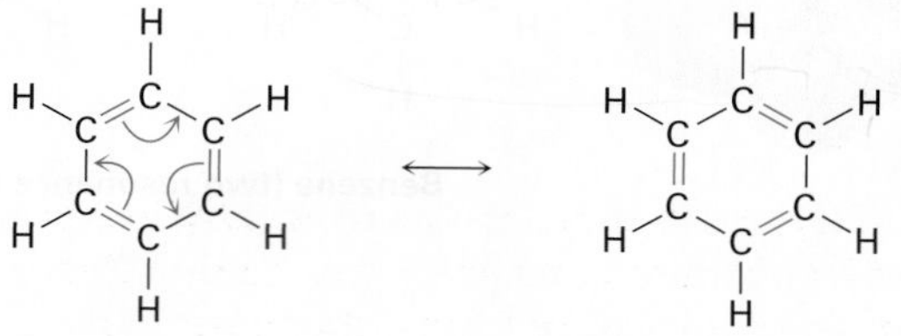

RULE 3

Different resonance forms of a substance don't have to be equivalent. As an example, we'll see in Chapter 22 that a compound such as acetone, which contains a C=O bond, can be converted into its anion by reaction with a strong base. The resultant anion has two resonance forms. One form contains a carbon–*oxygen* double bond and has a negative charge on *carbon;* the other contains a carbon–*carbon* double bond and has a negative charge on *oxygen.* Even though the two resonance forms aren't equivalent, both contribute to the overall resonance hybrid.

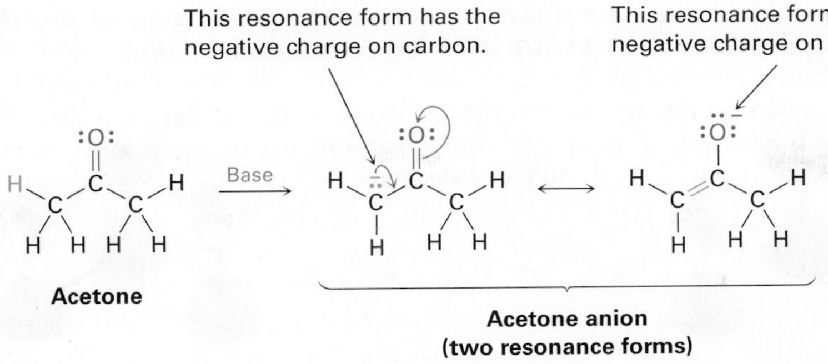

2.6 | Drawing Resonance Forms

This resonance form has the negative charge on carbon.

This resonance form has the negative charge on oxygen.

Acetone

Acetone anion (two resonance forms)

When two resonance forms are nonequivalent, the actual structure of the resonance hybrid resembles the more stable form more than it resembles the less stable form. Thus, we might expect the true structure of the acetone anion to be more like that of the form that places the negative charge on the electronegative oxygen atom rather than on carbon.

RULE 4
Resonance forms obey normal rules of valency. A resonance form is like any other structure: the octet rule still applies to second-row, main-group atoms. For example, one of the following structures for the acetate ion is not a valid resonance form because the carbon atom has five bonds and ten valence electrons:

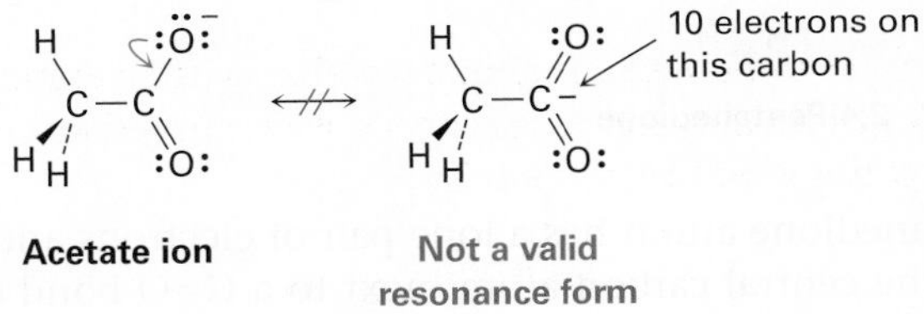

Acetate ion

Not a valid resonance form

10 electrons on this carbon

RULE 5
The resonance hybrid is more stable than any individual resonance form. In other words, resonance leads to stability. Generally speaking, the larger the number of resonance forms, the more stable a substance is because its electrons are spread out over a larger part of the molecule and are closer to more nuclei. We'll see in Chapter 15, for instance, that a benzene ring is more stable because of resonance than might otherwise be expected.

2.6 Drawing Resonance Forms

Look back at the resonance forms of the acetate ion and the acetone anion shown in the previous section. The pattern seen there is a common one that leads to a useful technique for drawing resonance forms. In general, *any three-atom grouping with a p orbital on each atom has two resonance forms:*

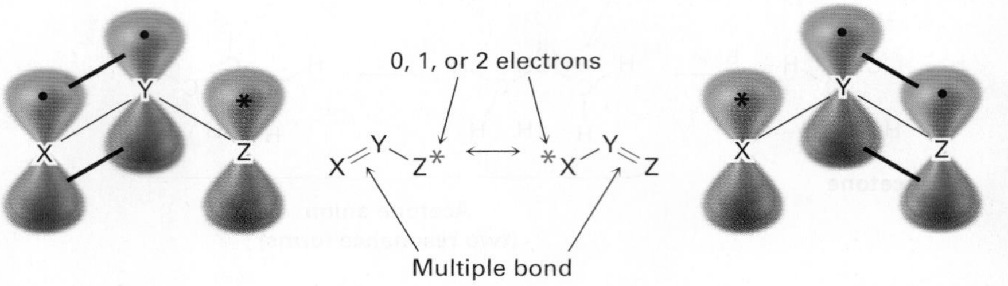

$$X\overset{Y}{=}Z^* \longleftrightarrow {}^*X\overset{Y}{-}Z$$

0, 1, or 2 electrons

Multiple bond

The atoms X, Y, and Z in the general structure might be C, N, O, P, S, or others, and the asterisk (*) might mean that the *p* orbital on atom Z is vacant, that it contains a single electron, or that it contains a lone pair of electrons. The two resonance forms differ simply by an exchange in position of the multiple bond and the asterisk from one end of the three-atom grouping to the other.

By learning to recognize such three-atom groupings within larger structures, resonance forms can be systematically generated. Look, for instance, at the anion produced when H⁺ is removed from 2,4-pentanedione by reaction with a base. How many resonance structures does the resultant anion have?

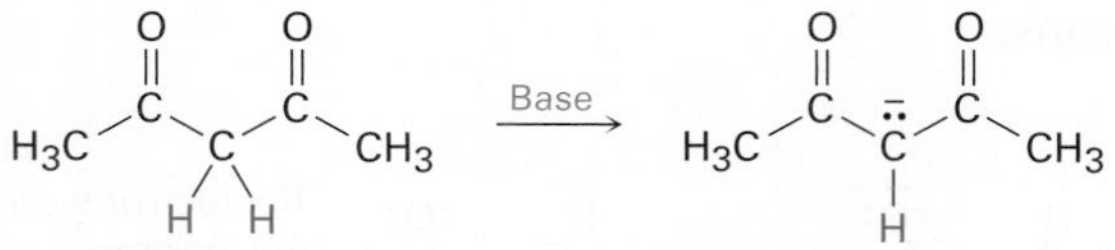

2,4-Pentanedione

The 2,4-pentanedione anion has a lone pair of electrons and a formal negative charge on the central carbon atom, next to a C=O bond on the left. The O=C–C:⁻ grouping is a typical one for which two resonance structures can be drawn.

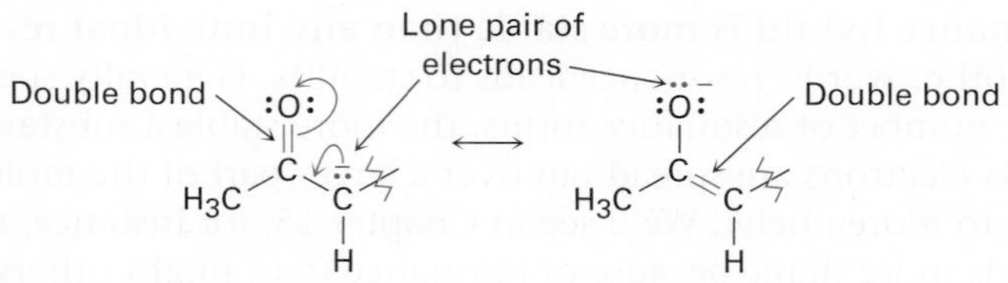

Just as there is a C=O bond to the left of the lone pair, there is a second C=O bond to the right. Thus, we can draw a total of three resonance structures for the 2,4-pentanedione anion.

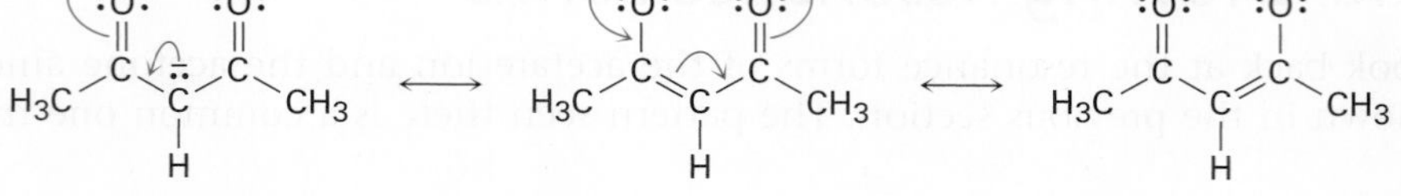

2.6 | Drawing Resonance Forms

Worked Example 2.2

Drawing Resonance Forms for an Anion

Draw three resonance structures for the carbonate ion, CO_3^{2-}.

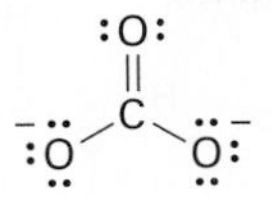

Carbonate ion

Strategy
Look for three-atom groupings that contain a multiple bond next to an atom with a *p* orbital. Then exchange the positions of the multiple bond and the electrons in the *p* orbital. In the carbonate ion, each of the singly bonded oxygen atoms with its lone pairs and negative charge is next to the C=O double bond, giving the grouping $O=C-O{:}^-$.

Solution
Exchanging the position of the double bond and an electron lone pair in each grouping generates three resonance structures.

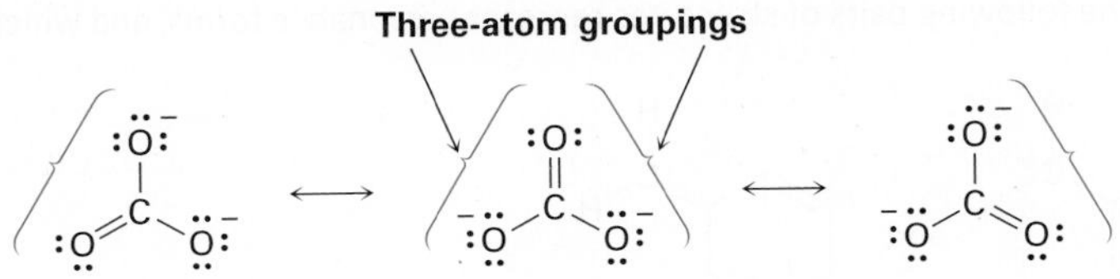

Three-atom groupings

Worked Example 2.3

Drawing Resonance Forms for a Radical

Draw three resonance forms for the pentadienyl radical, where a *radical* is a substance that contains a single, unpaired electron in one of its orbitals, denoted by a dot (·).

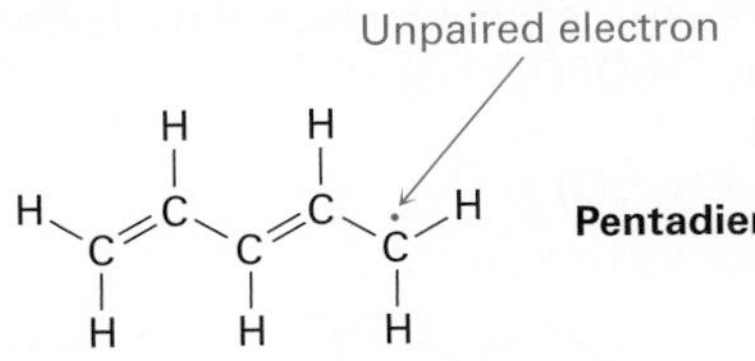

Pentadienyl radical

Unpaired electron

Strategy
Find the three-atom groupings that contain a multiple bond next to a *p* orbital.

Solution
The unpaired electron is on a carbon atom next to a C=C bond, giving a typical three-atom grouping that has two resonance forms.

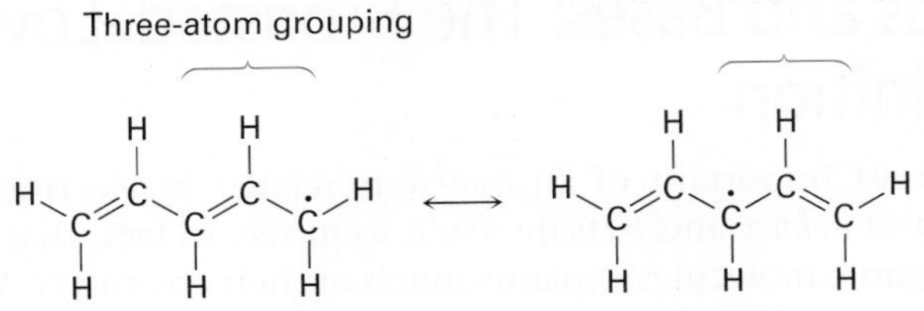

Three-atom grouping

In the second resonance form, the unpaired electron is next to another double bond, giving another three-atom grouping and leading to another resonance form.

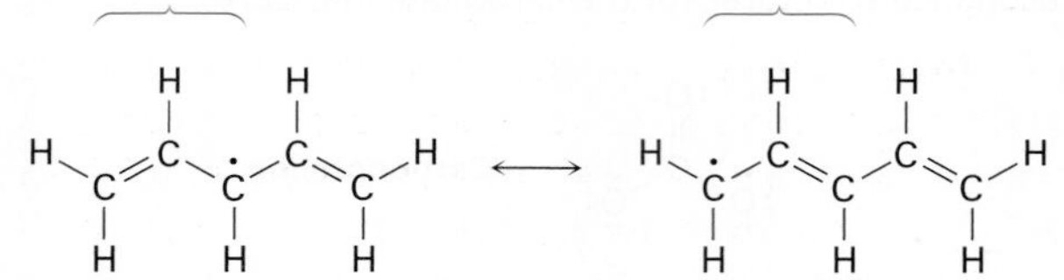

Thus, the three resonance forms for the pentadienyl radical are:

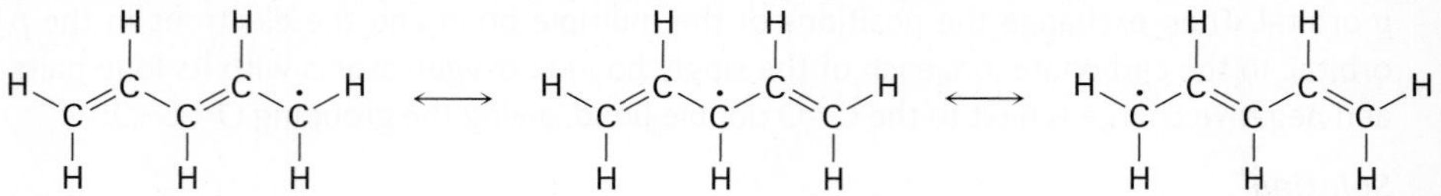

Problem 2.9
Which of the following pairs of structures represent resonance forms, and which do not? Explain.

(a)

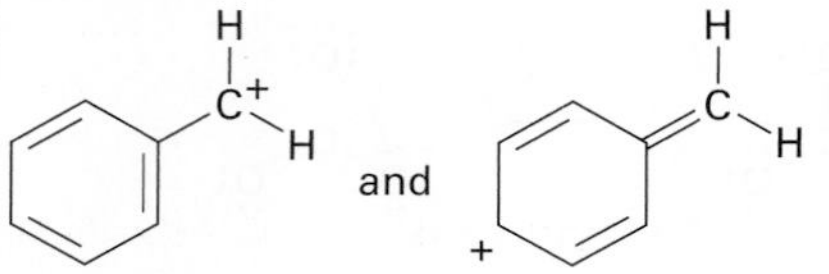

and

(b)

and

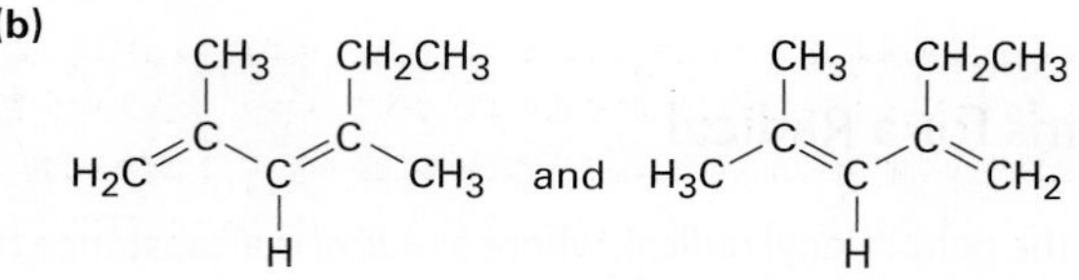

Problem 2.10
Draw the indicated number of resonance forms for each of the following species:
(a) The methyl phosphate anion, $CH_3OPO_3^{2-}$ (3)
(b) The nitrate anion, NO_3^- (3)
(c) The allyl cation, $H_2C=CH-CH_2^+$ (2)
(d) The benzoate anion (4)

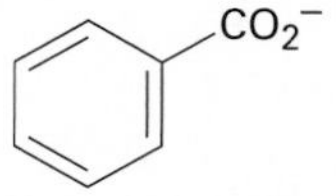

2.7 Acids and Bases: The Brønsted–Lowry Definition

Perhaps the most important of all concepts related to electronegativity and polarity is that of *acidity* and *basicity*. We'll soon see, in fact, that the acid–base behavior of organic molecules explains much of their chemistry. You may recall

2.7 | Acids and Bases: The Brønsted–Lowry Definition

from a course in general chemistry that two definitions of acidity are frequently used: the *Brønsted–Lowry definition* and the *Lewis definition*. We'll look at the Brønsted–Lowry definition in this and the following three sections and then discuss the Lewis definition in **Section 2.11**.

A **Brønsted–Lowry acid** is a substance that donates a hydrogen ion, H^+, and a **Brønsted–Lowry base** is a substance that accepts a hydrogen ion. (The name *proton* is often used as a synonym for H^+ because loss of the valence electron from a neutral hydrogen atom leaves only the hydrogen nucleus—a proton.) When gaseous hydrogen chloride dissolves in water, for example, a polar HCl molecule acts as an acid and donates a proton, while a water molecule acts as a base and accepts the proton, yielding chloride ion (Cl^-) and hydronium ion (H_3O^+). This and other acid–base reactions are reversible, so we'll write them with double, forward-and-backward arrows.

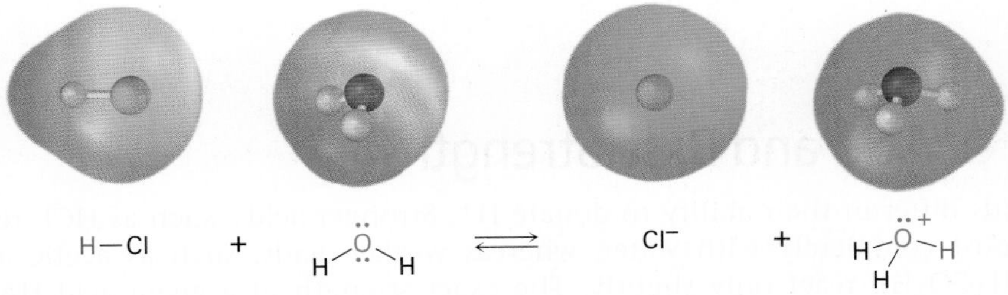

$$H\text{—}Cl \quad + \quad H_2O \quad \rightleftharpoons \quad Cl^- \quad + \quad H_3O^+$$

Acid Base Conjugate base Conjugate acid

Chloride ion, the product that results when the acid HCl loses a proton, is called the **conjugate base** of the acid, and hydronium ion, the product that results when the base H_2O gains a proton, is called the **conjugate acid** of the base. Other common mineral acids such as H_2SO_4 and HNO_3 behave similarly, as do organic acids such as acetic acid, CH_3CO_2H.

In a general sense,

$$H\text{—}A \quad + \quad :B \quad \rightleftharpoons \quad :A^- \quad + \quad H\text{—}B^+$$

Acid Base Conjugate base Conjugate acid

For example:

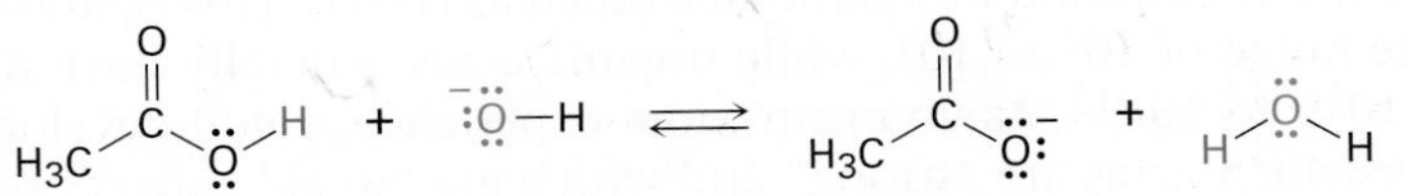

$$CH_3CO_2H \quad + \quad :\ddot{O}\text{—}H \quad \rightleftharpoons \quad CH_3CO_2^- \quad + \quad H_2O$$

Acid Base Conjugate base Conjugate acid

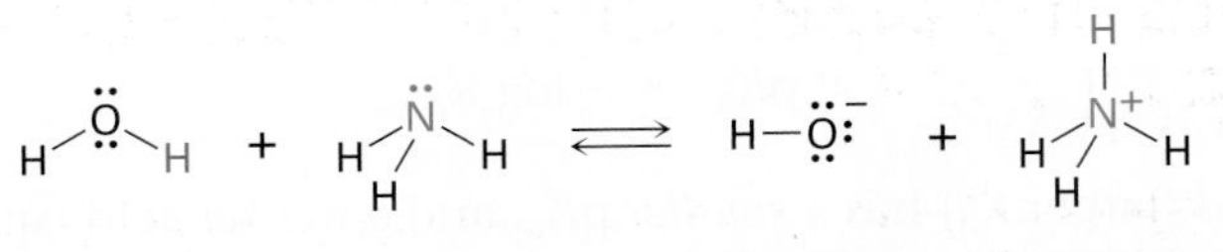

$$H_2O \quad + \quad NH_3 \quad \rightleftharpoons \quad HO^- \quad + \quad NH_4^+$$

Acid Base Conjugate base Conjugate acid

Notice that water can act either as an acid or as a base, depending on the circumstances. In its reaction with HCl, water is a base that accepts a proton to give the hydronium ion, H_3O^+. In its reaction with ammonia (NH_3), however, water is an acid that donates a proton to give ammonium ion (NH_4^+) and hydroxide ion, HO^-.

Problem 2.11
Nitric acid (HNO_3) reacts with ammonia (NH_3) to yield ammonium nitrate. Write the reaction, and identify the acid, the base, the conjugate acid product, and the conjugate base product.

2.8 Acid and Base Strength

Acids differ in their ability to donate H^+. Stronger acids, such as HCl, react almost completely with water, whereas weaker acids, such as acetic acid (CH_3CO_2H), react only slightly. The exact strength of a given acid HA in water solution is described using the **acidity constant (K_a)** for the acid-dissociation equilibrium. Remember from general chemistry that the concentration of solvent is ignored in the equilibrium expression and that brackets [ ] around a substance refer to the concentration of the enclosed species in moles per liter.

$$HA + H_2O \rightleftharpoons A^- + H_3O^+$$

$$K_a = \frac{[H_3O^+][A^-]}{[HA]}$$

Stronger acids have their equilibria toward the right and thus have larger acidity constants, whereas weaker acids have their equilibria toward the left and have smaller acidity constants. The range of K_a values for different acids is enormous, running from about 10^{15} for the strongest acids to about 10^{-60} for the weakest. The common inorganic acids such as H_2SO_4, HNO_3, and HCl have K_a's in the range of 10^2 to 10^9, while organic acids generally have K_a's in the range of 10^{-5} to 10^{-15}. As you gain more experience, you'll develop a rough feeling for which acids are "strong" and which are "weak" (always remembering that the terms are relative).

Acid strengths are normally expressed using pK_a values rather than K_a values, where the **pK_a** is the negative common logarithm of the K_a:

$$pK_a = -\log K_a$$

A *stronger* acid (larger K_a) has a *smaller* pK_a, and a *weaker* acid (smaller K_a) has a *larger* pK_a. Table 2.3 lists the pK_a's of some common acids in order of their strength, and a more comprehensive table is given in Appendix B.

Table 2.3 Relative Strengths of Some Common Acids and Their Conjugate Bases

	Acid	Name	pK_a	Conjugate base	Name	
Weaker acid ↓ Stronger acid	CH_3CH_2OH	Ethanol	16.00	$CH_3CH_2O^-$	Ethoxide ion	Stronger base ↑ Weaker base
	H_2O	Water	15.74	HO^-	Hydroxide ion	
	HCN	Hydrocyanic acid	9.31	CN^-	Cyanide ion	
	$H_2PO_4^-$	Dihydrogen phosphate ion	7.21	HPO_4^{2-}	Hydrogen phosphate ion	
	CH_3CO_2H	Acetic acid	4.76	$CH_3CO_2^-$	Acetate ion	
	H_3PO_4	Phosphoric acid	2.16	$H_2PO_4^-$	Dihydrogen phosphate ion	
	HNO_3	Nitric acid	−1.3	NO_3^-	Nitrate ion	
	HCl	Hydrochloric acid	−7.0	Cl^-	Chloride ion	

Notice that the pK_a value shown in Table 2.3 for water is 15.74, which results from the following calculation. Because water is both the acid and the solvent, the equilibrium expression is

$$H_2O + H_2O \rightleftharpoons OH^- + H_3O^+$$
(acid) (solvent)

$$K_a = \frac{[H_3O^+][A^-]}{[HA]} = \frac{[H_3O^+][OH^-]}{[H_2O]} = \frac{[1.0 \times 10^{-7}][1.0 \times 10^{-7}]}{[55.4]} = 1.8 \times 10^{-16}$$

$$pK_a = 15.74$$

The numerator in this expression is the so-called ion-product constant for water, $K_w = [H_3O^+][OH^-] = 1.00 \times 10^{-14}$, and the denominator is the molar concentration of pure water, $[H_2O] = 55.4$ M at 25 °C. The calculation is artificial in that the concentration of "solvent" water is ignored while the concentration of "acid" water is not, but it is nevertheless useful for making a comparison of water with other weak acids on a similar footing.

Notice also in Table 2.3 that there is an inverse relationship between the acid strength of an acid and the base strength of its conjugate base. A *strong* acid has a *weak* conjugate base, and a *weak* acid has a *strong* conjugate base. To understand this inverse relationship, think about what is happening to the acidic hydrogen in an acid–base reaction. A strong acid is one that loses H^+ easily, meaning that its conjugate base holds the H^+ weakly and is therefore a weak base. A weak acid is one that loses H^+ with difficulty, meaning that its conjugate base holds the proton tightly and is therefore a strong base. The fact that HCl is a strong acid, for example, means that Cl^- does not hold H^+ tightly and is thus a weak base. Water, on the other hand, is a weak acid, meaning that OH^- holds H^+ tightly and is a strong base.

Problem 2.12
The amino acid phenylalanine has $pK_a = 1.83$, and tryptophan has $pK_a = 2.83$. Which is the stronger acid?

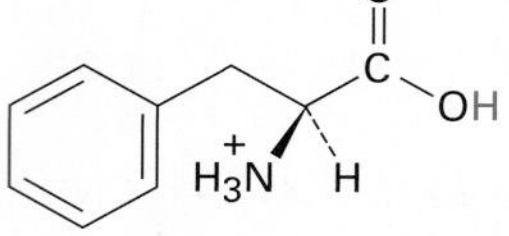

Phenylalanine
($pK_a = 1.83$)

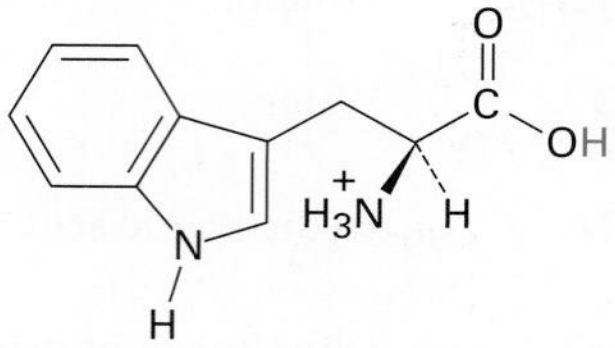

Tryptophan
($pK_a = 2.83$)

Problem 2.13
Amide ion, H_2N^-, is a much stronger base than hydroxide ion, HO^-. Which is the stronger acid, NH_3 or H_2O? Explain.

2.9 Predicting Acid–Base Reactions from pK_a Values

Compilations of pK_a values like those in Table 2.3 and Appendix B are useful for predicting whether a given acid–base reaction will take place because H^+ will always go *from* the stronger acid *to* the stronger base. That is, an acid will donate a proton to the conjugate base of a weaker acid, and the conjugate base of a weaker acid will remove the proton from a stronger acid. Since water ($pK_a = 15.74$) is a weaker acid than acetic acid ($pK_a = 4.76$), for example, hydroxide ion holds a proton more tightly than acetate ion does. Hydroxide ion will therefore react to a large extent with acetic acid, CH_3CO_2H, to yield acetate ion and H_2O.

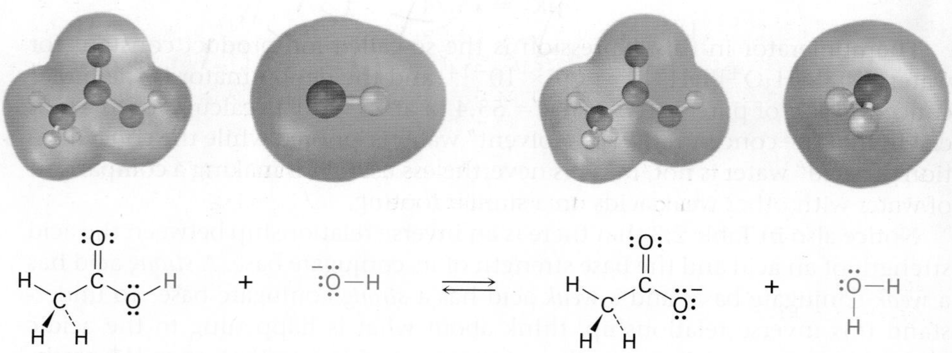

Acetic acid
($pK_a = 4.76$)

Hydroxide ion

Acetate ion

Water
($pK_a = 15.74$)

Another way to predict acid–base reactivity is to remember that the product conjugate acid in an acid–base reaction must be weaker and less reactive than the starting acid and the product conjugate base must be weaker and less reactive than the starting base. In the reaction of acetic acid with hydroxide ion, for example, the product conjugate acid (H₂O) is weaker than the starting acid (CH₃CO₂H), and the product conjugate base (CH₃CO₂⁻) is weaker than the starting base (OH⁻).

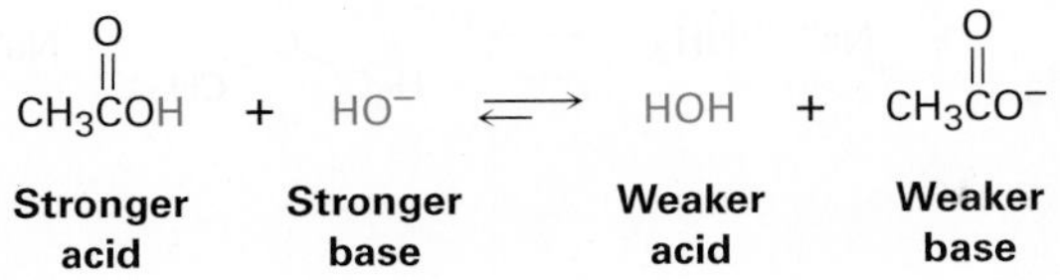

$$\text{CH}_3\text{COH} + \text{HO}^- \rightleftharpoons \text{HOH} + \text{CH}_3\text{CO}^-$$

Stronger acid   **Stronger base**   **Weaker acid**   **Weaker base**

Predicting Acid Strengths from pK_a Values

Worked Example 2.4

Water has pK_a = 15.74, and acetylene has pK_a = 25. Which is the stronger acid? Does hydroxide ion react to a significant extent with acetylene?

$$\text{H}-\text{C}\equiv\text{C}-\text{H} + \text{OH}^- \xrightarrow{?} \text{H}-\text{C}\equiv\text{C}:^- + \text{H}_2\text{O}$$

Acetylene

Strategy
In comparing two acids, the one with the lower pK_a is stronger. Thus, water is a stronger acid than acetylene and gives up H⁺ more easily.

Solution
Because water is a stronger acid and gives up H⁺ more easily than acetylene does, the HO⁻ ion must have less affinity for H⁺ than the HC≡C:⁻ ion has. In other words, the anion of acetylene is a stronger base than hydroxide ion, and the reaction will not proceed significantly as written.

Calculating K_a from pK_a

Worked Example 2.5

According to the data in Table 2.3, acetic acid has pK_a = 4.76. What is its K_a?

Strategy
Since pK_a is the negative logarithm of K_a, it's necessary to use a calculator with an ANTILOG or INV LOG function. Enter the value of the pK_a (4.76), change the sign (−4.76), and then find the antilog (1.74 × 10⁻⁵).

Solution
K_a = 1.74 × 10⁻⁵.

Problem 2.14
Will either of the following reactions take place to a significant extent as written, according to the data in Table 2.3?

(a) HCN + CH₃CO₂⁻ Na⁺ $\xrightarrow{?}$ Na⁺ ⁻CN + CH₃CO₂H

(b) CH₃CH₂OH + Na⁺ ⁻CN $\xrightarrow{?}$ CH₃CH₂O⁻ Na⁺ + HCN

Problem 2.15
Ammonia, NH_3, has $pK_a \approx 36$, and acetone has $pK_a \approx 19$. Will the following reaction take place to a significant extent?

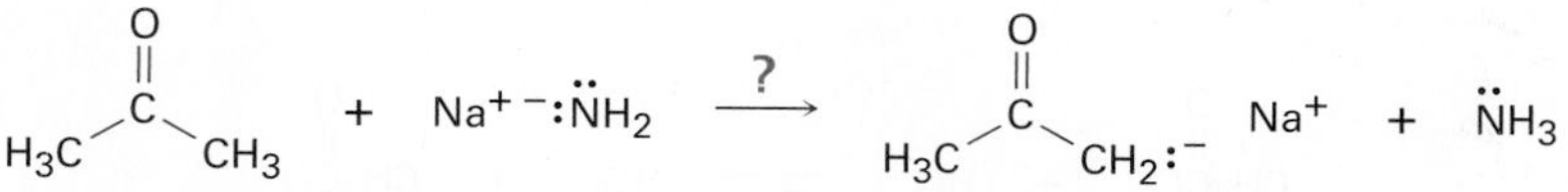

$$\underset{\text{Acetone}}{H_3C-\overset{\overset{O}{\|}}{C}-CH_3} + Na^+ \; {}^-{:}\ddot{N}H_2 \xrightarrow{?} H_3C-\overset{\overset{O}{\|}}{C}-CH_2{:}^- \; Na^+ + :\ddot{N}H_3$$

Problem 2.16
What is the K_a of HCN if its $pK_a = 9.31$?

2.10 Organic Acids and Organic Bases

Many of the reactions we'll be seeing in future chapters, including practically all biological reactions, involve organic acids and organic bases. Although it's too early to go into the details of these processes now, you might keep the following generalities in mind:

Organic Acids

Organic acids are characterized by the presence of a positively polarized hydrogen atom (blue in electrostatic potential maps) and are of two main kinds: those acids such as methanol and acetic acid that contain a hydrogen atom bonded to an electronegative oxygen atom (O–H) and those such as acetone (Section 2.5) that contain a hydrogen atom bonded to a carbon atom next to a C=O bond (O=C–C–H).

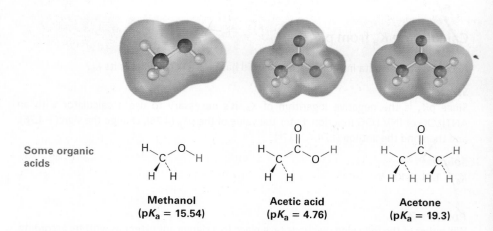

Some organic acids

Methanol ($pK_a = 15.54$)   Acetic acid ($pK_a = 4.76$)   Acetone ($pK_a = 19.3$)

Methanol contains an O–H bond and is a weak acid, while acetic acid also contains an O–H bond and is a somewhat stronger acid. In both cases, acidity is due to the fact that the conjugate base resulting from loss of H⁺ is stabilized by having its negative charge on a strongly electronegative oxygen atom. In

2.10 | Organic Acids and Organic Bases

addition, the conjugate base of acetic acid is stabilized by resonance (**Sections 2.4 and 2.5**)

Anion is stabilized by having negative charge on a highly electronegative atom.

Anion is stabilized both by having negative charge on a highly electronegative atom and by resonance.

The acidity of acetone and other compounds with C=O bonds is due to the fact that the conjugate base resulting from loss of H⁺ is stabilized by resonance. In addition, one of the resonance forms stabilizes the negative charge by placing it on an electronegative oxygen atom.

Anion is stabilized both by resonance and by having negative charge on a highly electronegative atom.

Electrostatic potential maps of the conjugate bases from methanol, acetic acid, and acetone are shown in **Figure 2.4**. As you might expect, all three show a substantial amount of negative charge (red) on oxygen.

Figure 2.4 Electrostatic potential maps of the conjugate bases of (a) methanol, (b) acetic acid, and (c) acetone. The electronegative oxygen atoms stabilize the negative charge in all three.

(a) CH_3O^-
(b) CH_3CO^- (with C=O)
(c) $CH_3CCH_2^-$ (with C=O)

Compounds called *carboxylic acids*, which contain the $-CO_2H$ grouping, occur abundantly in all living organisms and are involved in almost all metabolic pathways. Acetic acid, pyruvic acid, and citric acid are examples. You might note that at the typical pH of 7.3 found within cells, carboxylic acids are usually dissociated and exist as their carboxylate anions, $-CO_2^-$.

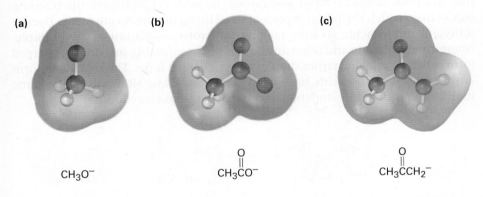

Acetic acid   Pyruvic acid   Citric acid

Organic Bases

Organic bases are characterized by the presence of an atom (reddish in electrostatic potential maps) with a lone pair of electrons that can bond to H$^+$. Nitrogen-containing compounds such as methylamine are the most common organic bases and are involved in almost all metabolic pathways, but oxygen-containing compounds can also act as bases when reacting with a sufficiently strong acid. Note that some oxygen-containing compounds can act both as acids and as bases depending on the circumstances, just as water can. Methanol and acetone, for instance, act as *acids* when they donate a proton but as *bases* when their oxygen atom accepts a proton.

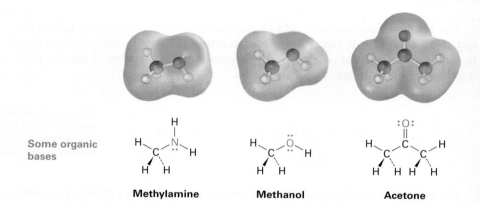

Some organic bases

Methylamine **Methanol** **Acetone**

We'll see in Chapter 26 that substances called *amino acids*, so-named because they are both amines (–NH$_2$) and carboxylic acids (–CO$_2$H), are the building blocks from which the proteins present in all living organisms are made. Twenty different amino acids go into making up proteins—alanine is an example. Interestingly, alanine and other amino acids exist primarily in a doubly charged form called a *zwitterion* rather than in the uncharged form. The zwitterion form arises because amino acids have both acidic and basic sites within the same molecule and therefore undergo an internal acid–base reaction.

Alanine
(uncharged form)

Alanine
(zwitterion form)

2.11 Acids and Bases: The Lewis Definition

The Lewis definition of acids and bases is broader and more encompassing than the Brønsted–Lowry definition because it's not limited to substances that donate or accept just protons. A **Lewis acid** is a substance that *accepts an electron pair*, and a **Lewis base** is a substance that *donates an electron pair*. The donated electron pair is shared between the acid and the base in a covalent bond.

2.11 | Acids and Bases: The Lewis Definition

[Diagram: Filled orbital (B:) + Vacant orbital (A) → B—A; labeled "Lewis base" and "Lewis acid"]

Lewis Acids and the Curved Arrow Formalism

The fact that a Lewis acid is able to accept an electron pair means that it must have either a vacant, low-energy orbital or a polar bond to hydrogen so that it can donate H^+ (which has an empty $1s$ orbital). Thus, the Lewis definition of acidity includes many species in addition to H^+. For example, various metal cations, such as Mg^{2+}, are Lewis acids because they accept a pair of electrons when they form a bond to a base. We'll also see in later chapters that certain metabolic reactions begin with an acid–base reaction between Mg^{2+} as a Lewis acid and an organic diphosphate or triphosphate ion as the Lewis base.

[Reaction scheme: Mg^{2+} (Lewis acid) + organodiphosphate ion (Lewis base) → Acid–base complex]

In the same way, compounds of group 3A elements, such as BF_3 and $AlCl_3$, are Lewis acids because they have unfilled valence orbitals and can accept electron pairs from Lewis bases, as shown in **Figure 2.5**. Similarly, many transition-metal compounds, such as $TiCl_4$, $FeCl_3$, $ZnCl_2$, and $SnCl_4$, are Lewis acids.

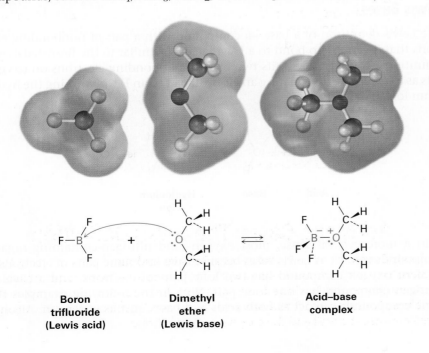

Figure 2.5 The reaction of boron trifluoride, a Lewis acid, with dimethyl ether, a Lewis base. The Lewis acid accepts a pair of electrons, and the Lewis base donates a pair of nonbonding electrons. Note how the movement of electrons *from* the Lewis base *to* the Lewis acid is indicated by a curved arrow. Note also how, in electrostatic potential maps, the boron becomes **more negative** after reaction because it has gained electrons and the oxygen atom becomes **more positive** because it has donated electrons.

Boron trifluoride (Lewis acid) + Dimethyl ether (Lewis base) ⇌ Acid–base complex

Look closely at the acid–base reaction in Figure 2.5, and note how it is shown. Dimethyl ether, the Lewis base, donates an electron pair to a vacant valence orbital of the boron atom in BF$_3$, a Lewis acid. The direction of electron-pair flow from the base to acid is shown using curved arrows, just as the direction of electron flow in going from one resonance structure to another was shown using curved arrows in **Section 2.5**. *A curved arrow always means that a pair of electrons moves* from *the atom at the tail of the arrow* to *the atom at the head of the arrow.* We'll use this curved-arrow notation throughout the remainder of this text to indicate electron flow during reactions.

Some further examples of Lewis acids follow:

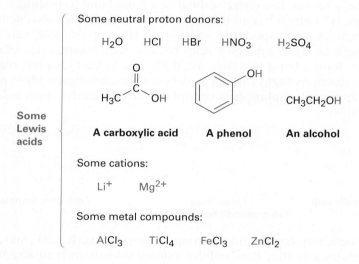

Lewis Bases

The Lewis definition of a base—a compound with a pair of nonbonding electrons that it can use to bond to a Lewis acid—is similar to the Brønsted–Lowry definition. Thus, H$_2$O, with its two pairs of nonbonding electrons on oxygen, acts as a Lewis base by donating an electron pair to an H$^+$ in forming the hydronium ion, H$_3$O$^+$.

$$\text{Cl}-\text{H} \;+\; :\!\ddot{\text{O}}\!:\!\text{H}_2 \;\rightleftharpoons\; \text{H}-\overset{+}{\ddot{\text{O}}}\!:\!\text{H}_2 \;+\; \text{Cl}^-$$

    Acid   Base   Hydronium ion

In a more general sense, most oxygen- and nitrogen-containing organic compounds can act as Lewis bases because they have lone pairs of electrons. A divalent oxygen compound has two lone pairs of electrons, and a trivalent nitrogen compound has one lone pair. Note in the following examples that some compounds can act as both acids and bases, just as water can. Alcohols and carboxylic acids, for instance, act as acids when they donate an H⁺ but as bases when their oxygen atom accepts an H⁺.

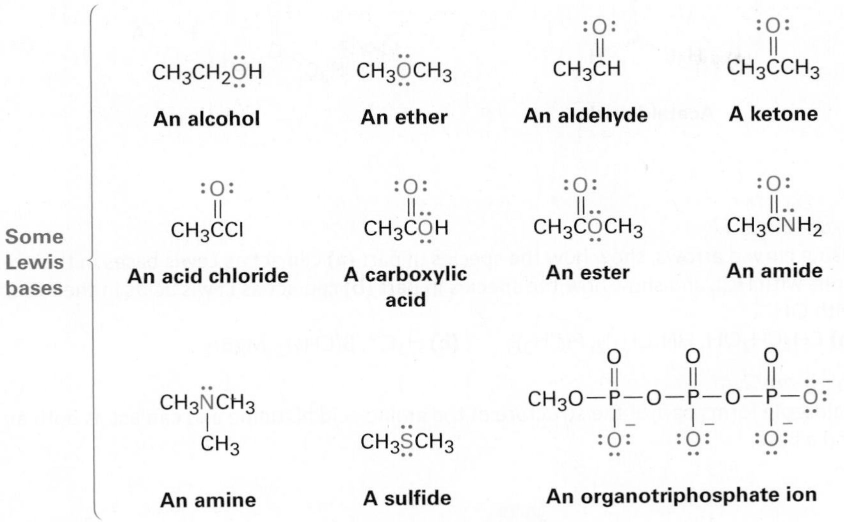

Some Lewis bases:

- CH₃CH₂ÖH — **An alcohol**
- CH₃ÖCH₃ — **An ether**
- CH₃CH(=O) — **An aldehyde**
- CH₃CCH₃(=O) — **A ketone**
- CH₃CCl(=O) — **An acid chloride**
- CH₃COH(=O) — **A carboxylic acid**
- CH₃COCH₃(=O) — **An ester**
- CH₃CNH₂(=O) — **An amide**
- CH₃N(CH₃)CH₃ — **An amine**
- CH₃SCH₃ — **A sulfide**
- CH₃O—P(=O)(O⁻)—O—P(=O)(O⁻)—O—P(=O)(O⁻)—Ö⁻ — **An organotriphosphate ion**

Notice in the list of Lewis bases just given that some compounds, such as carboxylic acids, esters, and amides, have more than one atom with a lone pair of electrons and can therefore react at more than one site. Acetic acid, for example, can be protonated either on the doubly bonded oxygen atom or on the singly bonded oxygen atom. Reaction normally occurs only once in such instances, and the more stable of the two possible protonation products is formed. For acetic acid, protonation by reaction with sulfuric acid occurs on the doubly bonded oxygen because that product is stabilized by two resonance forms.

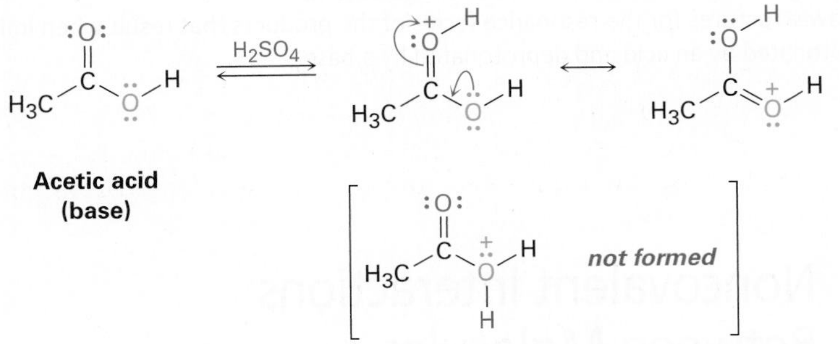

Acetic acid (base) + H₂SO₄ ⇌ [protonated on C=O, two resonance forms shown]

[protonation on singly-bonded O — **not formed**]

Worked Example 2.6

Using Curved Arrows to Show Electron Flow

Using curved arrows, show how acetaldehyde, CH₃CHO, can act as a Lewis base.

Strategy

A Lewis base donates an electron pair to a Lewis acid. We therefore need to locate the electron lone pairs on acetaldehyde and use a curved arrow to show the movement of a pair toward the H atom of the acid.

Solution

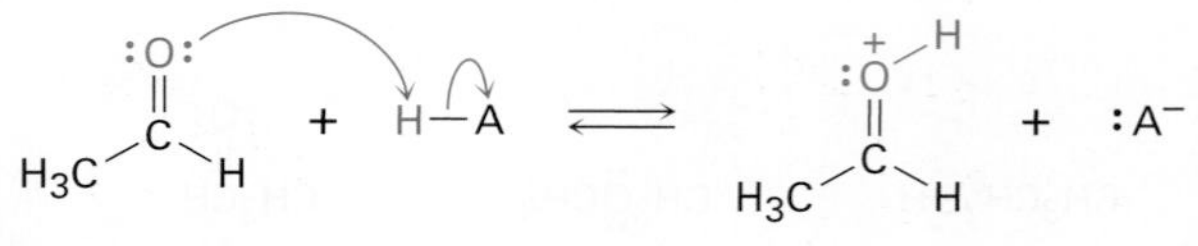

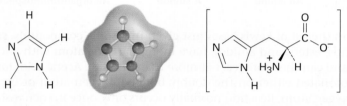

Acetaldehyde

Problem 2.17
Using curved arrows, show how the species in part **(a)** can act as Lewis bases in their reactions with HCl, and show how the species in part **(b)** can act as Lewis acids in their reaction with OH⁻.
(a) CH_3CH_2OH, $HN(CH_3)_2$, $P(CH_3)_3$ **(b)** H_3C^+, $B(CH_3)_3$, $MgBr_2$

Problem 2.18
Imidazole forms part of the structure of the amino acid histidine and can act as both an acid and a base.

Imidazole   **Histidine**

(a) Look at the electrostatic potential map of imidazole, and identify the most acidic hydrogen atom and the most basic nitrogen atom.
(b) Draw structures for the resonance forms of the products that result when imidazole is protonated by an acid and deprotonated by a base.

2.12 Noncovalent Interactions Between Molecules

When thinking about chemical reactivity, chemists usually focus their attention on bonds, the covalent interactions between atoms *within* molecules. Also important, however, particularly in large biomolecules like proteins and nucleic acids, are a variety of interactions *between* molecules that strongly affect molecular properties. Collectively called either *intermolecular forces, van der Waals forces,* or **noncovalent interactions**, they are of several different types: dipole–dipole forces, dispersion forces, and hydrogen bonds.

Dipole–dipole forces occur between polar molecules as a result of electrostatic interactions among dipoles. The forces can be either attractive or repulsive depending on the orientation of the molecules—attractive when unlike charges are together and repulsive when like charges are together. The attractive geometry is lower in energy and therefore predominates (**Figure 2.6**).

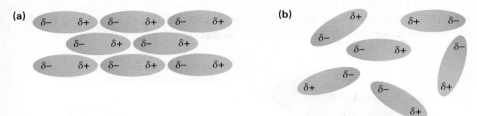

Figure 2.6 Dipole–dipole forces cause polar molecules **(a)** to attract one another when they orient with unlike charges together, but **(b)** to repel one another when they orient with like charges together.

Dispersion forces occur between all neighboring molecules and arise because the electron distribution within molecules is constantly changing. Although uniform on a time-averaged basis, the electron distribution even in nonpolar molecules is likely to be nonuniform at any given instant. One side of a molecule may, by chance, have a slight excess of electrons relative to the opposite side, giving the molecule a temporary dipole. This temporary dipole in one molecule causes a nearby molecule to adopt a temporarily opposite dipole, with the result that a tiny attraction is induced between the two (**Figure 2.7**). Temporary molecular dipoles have only a fleeting existence and are constantly changing, but their cumulative effect is often strong enough to hold molecules close together so that a substance is a liquid or solid rather than a gas.

Figure 2.7 Attractive dispersion forces in nonpolar molecules are caused by temporary dipoles, as shown in these models of pentane, C_5H_{12}.

Perhaps the most important noncovalent interaction in biological molecules is the **hydrogen bond**, an attractive interaction between a hydrogen bonded to an electronegative O or N atom and an unshared electron pair on another O or N atom. In essence, a hydrogen bond is a very strong dipole–dipole interaction involving polarized O–H or N–H bonds. Electrostatic potential maps of water and ammonia clearly show the positively polarized hydrogens (blue) and the negatively polarized oxygens and nitrogens (red).

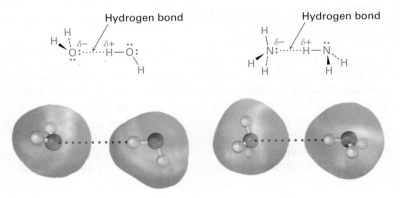

Hydrogen bonding has enormous consequences for living organisms. Hydrogen bonds cause water to be a liquid rather than a gas at ordinary temperatures, they hold enzymes in the shapes necessary for catalyzing biological reactions, and they cause strands of deoxyribonucleic acid (DNA) to pair up and coil into the double helix that stores genetic information.

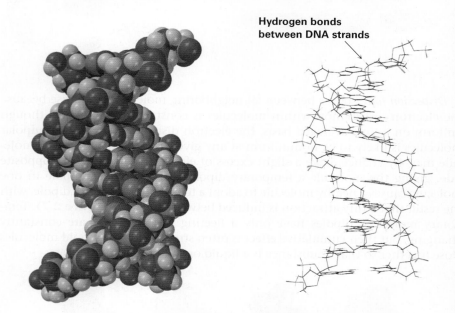

A deoxyribonucleic acid segment

One further point before leaving the subject of noncovalent interactions: biochemists frequently use the term *hydrophilic*, meaning "water-loving," to describe a substance that is strongly attracted to water and the term *hydrophobic*, meaning "water-fearing," to describe a substance that is not strongly attracted to water. Hydrophilic substances, such as table sugar, usually have a number of ionic charges or polar –OH groups in their structure so they can form hydrogen bonds, whereas hydrophobic substances, such as vegetable oil, do not have groups that form hydrogen bonds, so their attraction to water is limited to weak dispersion forces.

Problem 2.19
Of the two vitamins A and C, one is hydrophilic and water-soluble while the other is hydrophobic and fat-soluble. Which is which?

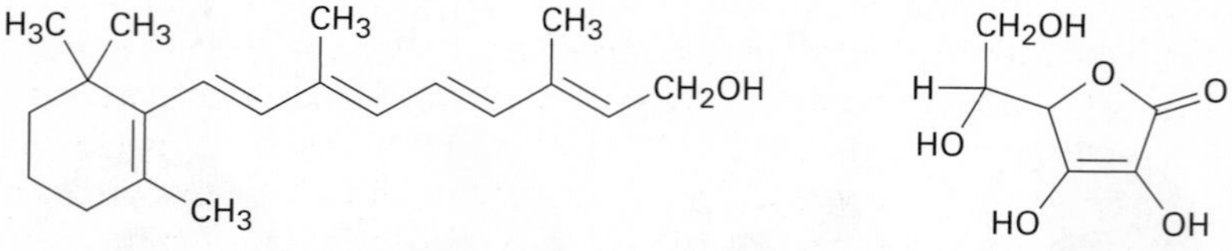

Vitamin A (retinol)

Vitamin C (ascorbic acid)

A DEEPER LOOK

Alkaloids: From Cocaine to Dental Anesthetics

Just as ammonia (NH₃) is a weak base, there are a large number of nitrogen-containing organic compounds called *amines* that are also weak bases. In the early days of organic chemistry, basic amines derived from natural sources were known as vegetable alkali, but they are now called *alkaloids*. More than 20,000 alkaloids are known. Their study provided much of the impetus for the growth of organic chemistry in the nineteenth century and remains today an active and fascinating area of research.

Alkaloids vary widely in structure, from the simple to the enormously complex. The odor of rotting fish, for example, is caused largely by methylamine, CH₃NH₂, a simple relative of ammonia in which one of the NH₃ hydrogens has been replaced by an organic CH₃ group. In fact, the use of lemon juice to mask fish odors is simply an acid–base reaction of the citric acid in lemons with methylamine base in the fish.

The coca bush *Erythroxylon coca*, native to upland rain forest areas of Colombia, Ecuador, Peru, Bolivia, and western Brazil, is the source of the alkaloid cocaine.

Many alkaloids have pronounced biological properties, and approximately 50% of the pharmaceutical agents used today are derived from naturally occurring amines. As just three examples, morphine, an analgesic agent, is obtained from the opium poppy *Papaver somniferum*. Ephedrine, a bronchodilator, decongestant, and appetite suppressant, is obtained from the Chinese plant *Ephedra sinica*. Cocaine, both an anesthetic and a stimulant, is obtained from the coca bush *Erythroxylon coca*, endemic to the upland rain forest areas of central South America. (And yes, there really was a small amount of cocaine in the original Coca-Cola recipe, although it was removed in 1906.)

Morphine　　　　**Ephedrine**　　　　**Cocaine**

Cocaine itself is no longer used as a medicine because it is too addictive, but its anesthetic properties provoked a search for related but nonaddictive compounds. This search ultimately resulted in the synthesis of the "caine" anesthetics that are commonly used today in dental and surgical anesthesia. Procaine, the first such compound, was synthesized in 1898 and marketed under the name Novocain. It was rapidly adopted and remains in use today as a topical anesthetic. Other related compounds with different activity profiles followed: Lidocaine, marketed as Xylocaine, was introduced in 1943, and mepivacaine (Carbocaine) in the early 1960s. More recently, bupivacaine (Marcaine) and prilocaine (Citanest) have gained popularity. Both are quick-acting, but the effects of

(continued)

bupivacaine last for 3 to 6 hours while those of prilocaine fade after 45 minutes. Note some structural similarity of all the caines to cocaine itself.

Procaine (Novocain)

Lidocaine (Xylocaine)

Mepivacaine (Carbocaine)

Bupivacaine (Marcaine)

Prilocaine (Citanest)

A recent report from the U.S. National Academy of Sciences estimates than less than 1% of all living species have been characterized. Thus, alkaloid chemistry remains today an active area of research, and innumerable substances with potentially useful properties remain to be discovered. Undoubtedly even the caine anesthetics will become obsolete at some point, perhaps supplanted by newly discovered alkaloids.

Key words

acidity constant (K_a), 50
Brønsted–Lowry acid, 49
Brønsted–Lowry base, 49
conjugate acid, 49
conjugate base, 49
dipole moment (μ), 37
electronegativity (EN), 35
formal charge, 41
hydrogen bond, 61
inductive effect, 36
Lewis acid, 56
Lewis base, 56
noncovalent interaction, 60
pK_a, 50
polar covalent bond, 34
resonance form, 43
resonance hybrid, 43

Summary

Understanding both organic and biological chemistry means knowing not just what happens but also why and how it happens at the molecular level. In this chapter, we've reviewed some of the ways that chemists describe and account for chemical reactivity, thereby providing a foundation for understanding the specific reactions that will be discussed in subsequent chapters.

Organic molecules often have **polar covalent bonds** as a result of unsymmetrical electron sharing caused by differences in the **electronegativity** of atoms. A carbon–oxygen bond is polar, for example, because oxygen attracts the shared electrons more strongly than carbon does. Carbon–hydrogen bonds are relatively nonpolar. Many molecules as a whole are also polar owing to the presence of individual polar bonds and electron lone pairs. The polarity of a molecule is measured by its **dipole moment**, μ.

Plus (+) and minus (−) signs are often used to indicate the presence of **formal charges** on atoms in molecules. Assigning formal charges to specific atoms is a bookkeeping technique that makes it possible to keep track of the valence electrons around an atom and offers some clues about chemical reactivity.

Some substances, such as acetate ion and benzene, can't be represented by a single line-bond structure and must be considered as a **resonance hybrid** of two or more structures, neither of which is correct by itself. The only difference between two **resonance forms** is in the location of their π and nonbonding electrons. The nuclei remain in the same places in both structures, and the hybridization of the atoms remains the same.

Acidity and basicity are closely related to the ideas of polarity and electronegativity. A **Brønsted–Lowry acid** is a compound that can donate a proton (hydrogen ion, H^+), and a **Brønsted–Lowry base** is a compound that can accept a proton. The strength of a Brønsted–Lowry acid or base is expressed by its **acidity constant**, K_a, or by the negative logarithm of the acidity constant, **pK_a**. The larger the pK_a, the weaker the acid. More useful is the Lewis definition of acids and bases. A **Lewis acid** is a compound that has a low-energy empty orbital that can accept an electron pair; Mg^{2+}, BF_3, $AlCl_3$, and H^+ are examples. A **Lewis base** is a compound that can donate an unshared electron pair; NH_3 and H_2O are examples. Most organic molecules that contain oxygen and nitrogen can act as Lewis bases toward sufficiently strong acids.

A variety of **noncovalent interactions** have a significant effect on the properties of large biomolecules. **Hydrogen bonding**—the attractive interaction between a positively polarized hydrogen atom bonded to an oxygen or nitrogen atom with an unshared electron pair on another O or N atom, is particularly important in giving proteins and nucleic acids their shapes.

Exercises

Visualizing Chemistry

(Problems 2.1–2.19 appear within the chapter.)

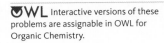 Interactive versions of these problems are assignable in OWL for Organic Chemistry.

▲ denotes problems linked to the Key Ideas in this chapter.

2.20 Fill in the multiple bonds in the following model of naphthalene, $C_{10}H_8$ (gray = C, ivory = H). How many resonance structures does naphthalene have? Draw them.

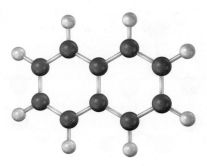

▲ Problems linked to Key Ideas in this chapter

2.21 The following model is a representation of ibuprofen, a common over-the-counter pain reliever. Indicate the positions of the multiple bonds, and draw a skeletal structure (gray = C, red = O, ivory = H).

2.22 *cis*-1,2-Dichloroethylene and *trans*-dichloroethylene are *isomers,* compounds with the same formula but different chemical structures. Look at the following electrostatic potential maps, and tell whether either compound has a dipole moment.

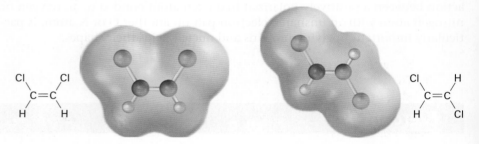

***cis*-1,2-Dichloroethylene** ***trans*-1,2-Dichloroethylene**

2.23 The following molecular models are representations of **(a)** adenine and **(b)** cytosine, constituents of DNA (deoxyribonucleic acid). Indicate the positions of multiple bonds and lone pairs for both, and draw skeletal structures (gray = C, red = O, blue = N, ivory = H).

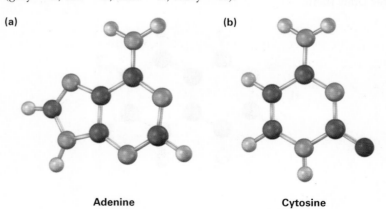

Adenine **Cytosine**

Additional Problems

Electronegativity and Dipole Moments

2.24 Identify the most electronegative element in each of the following molecules:
  (a) CH_2FCl
  (b) $FCH_2CH_2CH_2Br$
  (c) $HOCH_2CH_2NH_2$
  (d) CH_3OCH_2Li

2.25 Use the electronegativity table given in Figure 2.2 on page 35 to predict which bond in each of the following pairs is more polar, and indicate the direction of bond polarity for each compound.
  (a) H_3C-Cl or $Cl-Cl$
  (b) H_3C-H or $H-Cl$
  (c) $HO-CH_3$ or $(CH_3)_3Si-CH_3$
  (d) H_3C-Li or $Li-OH$

2.26 Which of the following molecules has a dipole moment? Indicate the expected direction of each.

(a) (b) (c) 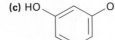 (d)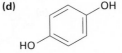

2.27 (a) The H–Cl bond length is 136 pm. What would the dipole moment of HCl be if the molecule were 100% ionic, $H^+ Cl^-$?
  (b) The actual dipole moment of HCl is 1.08 D. What is the percent ionic character of the H–Cl bond?

2.28 Phosgene, $Cl_2C=O$, has a smaller dipole moment than formaldehyde, $H_2C=O$, even though it contains electronegative chlorine atoms in place of hydrogen. Explain.

2.29 Fluoromethane (CH_3F, $\mu = 1.81$ D) has a smaller dipole moment than chloromethane (CH_3Cl, $\mu = 1.87$ D) even though fluorine is more electronegative than chlorine. Explain.

2.30 Methanethiol, CH_3SH, has a substantial dipole moment ($\mu = 1.52$) even though carbon and sulfur have identical electronegativities. Explain.

Formal Charges

2.31 Calculate the formal charges on the atoms shown in red.

(a) $(CH_3)_2\overset{..}{O}BF_3$
(b) $H_2\overset{..}{C}-N\equiv N:$
(c) $H_2C=N=\overset{..}{N}:$

(d) $:\overset{..}{\underset{..}{O}}=\overset{..}{O}-\overset{..}{\underset{..}{O}}:$
(e) $H_2\overset{..}{C}-\underset{\underset{CH_3}{|}}{\overset{\overset{CH_3}{|}}{P}}-CH_3$
(f)

2.32 Assign formal charges to the atoms in each of the following molecules:

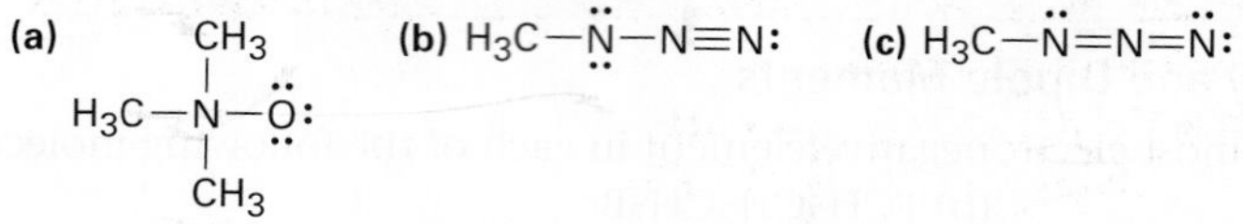

(a)
$$H_3C-\underset{\underset{CH_3}{|}}{\overset{\overset{CH_3}{|}}{N}}-\ddot{\underset{..}{O}}:$$

(b) $H_3C-\ddot{N}-N\equiv N:$

(c) $H_3C-\ddot{N}=N=\ddot{N}:$

Resonance

2.33 Which of the following pairs of structures represent resonance forms?

(a) [cyclobutabenzene] and [o-xylylene]

(b) [cyclohexenol with :Ö:⁻] and [cyclohexanone with :O: and ·⁻]

(c) [phenol with :Ö:⁻] and [cyclohexadienone with :O: and ·⁻]

(d) [phenol with :Ö:⁻] and [cyclohexadienone with :O:]

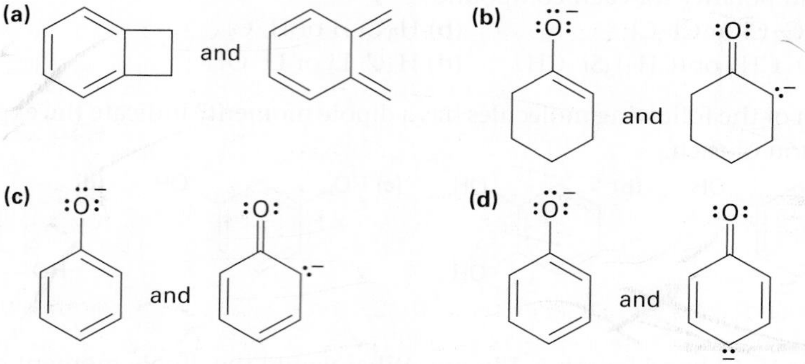

2.34 ▲ Draw as many resonance structures as you can for the following species:

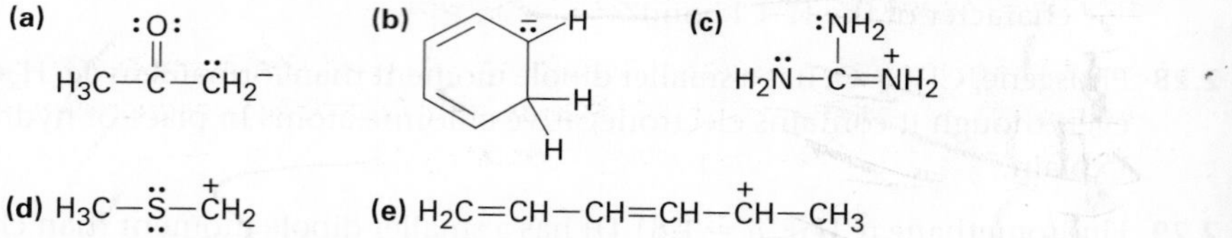

(a)
$$H_3C-\overset{\overset{:O:}{\|}}{C}-\ddot{C}H_2^-$$

(b) [cyclohexadienyl radical with H, H, and ·]

(c)
$$H_2\overset{:NH_2}{\underset{|}{N}}-C\overset{+}{=}NH_2$$

(d) $H_3C-\ddot{\underset{..}{S}}-\overset{+}{C}H_2$

(e) $H_2C=CH-CH=CH-\overset{+}{C}H-CH_3$

2.35 1,3-Cyclobutadiene is a rectangular molecule with two shorter double bonds and two longer single bonds. Why do the following structures *not* represent resonance forms?

[rectangle] ⇎ [rectangle]

Acids and Bases

2.36 Alcohols can act either as weak acids or as weak bases, just as water can. Show the reaction of methanol, CH_3OH, with a strong acid such as HCl and with a strong base such as $Na^+ \ ^-NH_2$.

2.37 ▲ The O–H hydrogen in acetic acid is more acidic than any of the C|H hydrogens. Explain this result using resonance structures.

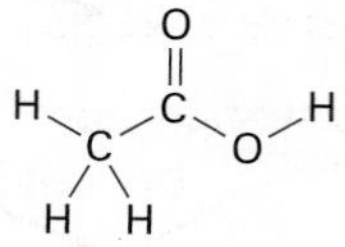

Acetic acid

▲ Problems linked to Key Ideas in this chapter

2.38 Draw electron-dot structures for the following molecules, indicating any unshared electron pairs. Which of the compounds are likely to act as Lewis acids and which as Lewis bases?
(a) $AlBr_3$ (b) $CH_3CH_2NH_2$ (c) BH_3
(d) HF (e) CH_3SCH_3 (f) $TiCl_4$

2.39 Write the products of the following acid–base reactions:
(a) $CH_3OH + H_2SO_4 \rightleftharpoons ?$
(b) $CH_3OH + NaNH_2 \rightleftharpoons ?$
(c) $CH_3NH_3^+ \; Cl^- + NaOH \rightleftharpoons ?$

2.40 Rank the following substances in order of increasing acidity:

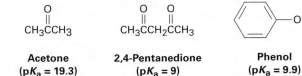

Acetone	2,4-Pentanedione	Phenol	Acetic acid
(pK_a = 19.3)	(pK_a = 9)	(pK_a = 9.9)	(pK_a = 4.76)

2.41 Which, if any, of the substances in Problem 2.40 is a strong enough acid to react almost completely with NaOH? (The pK_a of H_2O is 15.74.)

2.42 The ammonium ion (NH_4^+, pK_a = 9.25) has a lower pK_a than the methylammonium ion ($CH_3NH_3^+$, pK_a = 10.66). Which is the stronger base, ammonia (NH_3) or methylamine (CH_3NH_2)? Explain.

2.43 Is *tert*-butoxide anion a strong enough base to react significantly with water? In other words, can a solution of potassium *tert*-butoxide be prepared in water? The pK_a of *tert*-butyl alcohol is approximately 18.

$$K^+ \;\; ^-O-\underset{\underset{CH_3}{|}}{\overset{\overset{CH_3}{|}}{C}}-CH_3 \quad \text{Potassium } \textit{tert}\text{-butoxide}$$

2.44 Predict the structure of the product formed in the reaction of the organic base pyridine with the organic acid acetic acid, and use curved arrows to indicate the direction of electron flow.

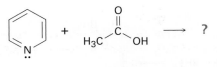

Pyridine   Acetic acid

2.45 Calculate K_a values from the following pK_a's:
(a) Acetone, pK_a = 19.3   (b) Formic acid, pK_a = 3.75

2.46 Calculate pK_a values from the following K_a's:
(a) Nitromethane, $K_a = 5.0 \times 10^{-11}$ (b) Acrylic acid, $K_a = 5.6 \times 10^{-5}$

▲ Problems linked to Key Ideas in this chapter

2.47 What is the pH of a 0.050 M solution of formic acid, pK_a = 3.75?

2.48 Sodium bicarbonate, $NaHCO_3$, is the sodium salt of carbonic acid (H_2CO_3), pK_a = 6.37. Which of the substances shown in Problem 2.40 will react significantly with sodium bicarbonate?

General Problems

2.49 Maleic acid has a dipole moment, but the closely related fumaric acid, a substance involved in the citric acid cycle by which food molecules are metabolized, does not. Explain.

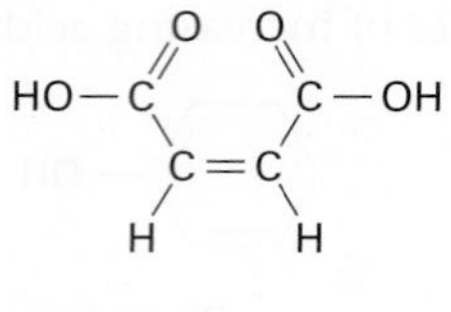

Maleic acid 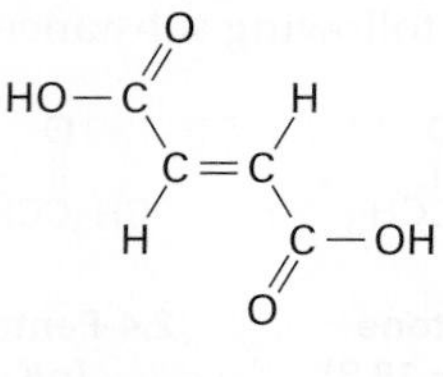 **Fumaric acid**

2.50 Assume that you have two unlabeled bottles, one of which contains phenol (pK_a = 9.9) and one of which contains acetic acid (pK_a = 4.76). In light of your answer to Problem 2.48, suggest a simple way to determine what is in each bottle.

2.51 Identify the acids and bases in the following reactions:

(a) $CH_3OH + H^+ \longrightarrow CH_3\overset{+}{O}H_2$

(b)

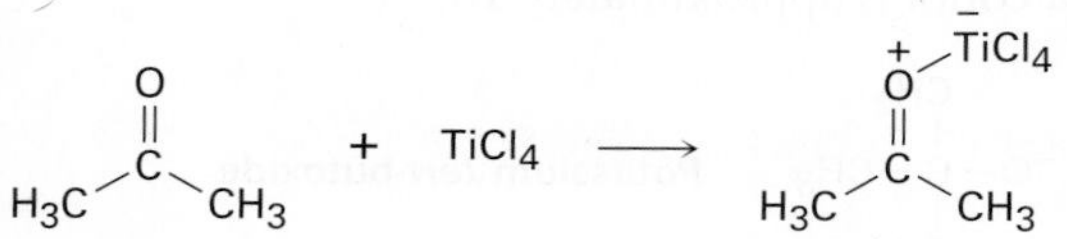

(c)

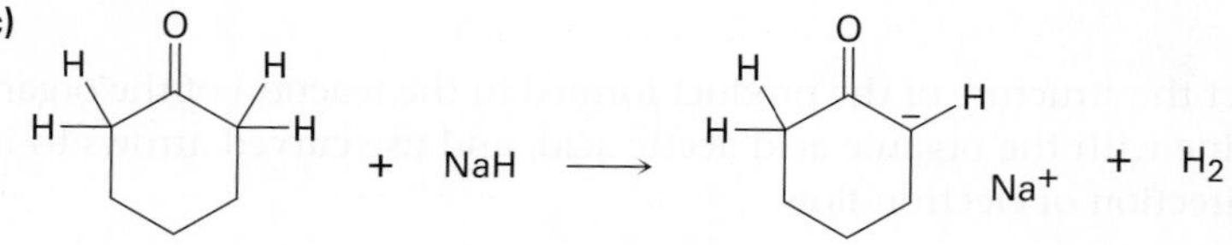

(d)

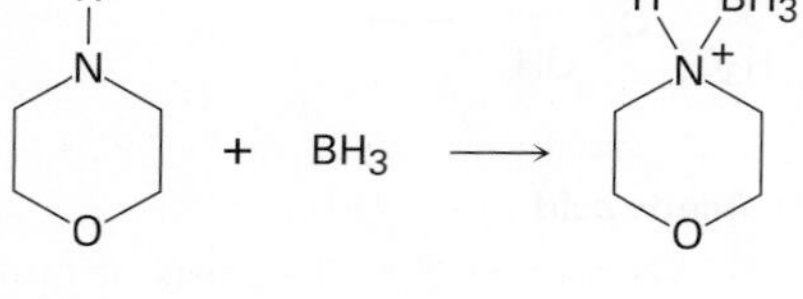

▲ Problems linked to Key Ideas in this chapter

2.52 ▲ Which of the following pairs represent resonance structures?

(a) $CH_3C \equiv \overset{+}{N} - \overset{..}{\underset{..}{O}}:^-$ and $CH_3\overset{+}{C} = \overset{..}{N} - \overset{..}{\underset{..}{O}}:^-$

(b)
$$CH_3\overset{\overset{:O:}{\|}}{C} - \overset{..}{\underset{..}{O}}:^- \quad \text{and} \quad :\bar{C}H_2\overset{\overset{:O:}{\|}}{C} - \overset{..}{\underset{..}{O}} - H$$

(c) [benzoyl-NH₃⁺ with :O:⁻ and benzoyl-NH₂ with :Ö–H⁺ resonance structures]

and

(d) $CH_2 = \overset{+}{\underset{\underset{..}{\underset{..}{O}:}}{N}}\overset{\overset{..}{\underset{..}{O}:^-}}{} \quad$ and $\quad :\bar{C}H_2 - \overset{+}{\underset{\underset{..}{\underset{..}{O}:^-}}{N}}\overset{\overset{..}{\underset{..}{O}}}{}$

2.53 ▲ Draw as many resonance structures as you can for the following species, adding appropriate formal charges to each:

(a) Nitromethane, $H_3C - \overset{+}{\underset{\underset{..}{\underset{..}{O}:^-}}{N}}\overset{\overset{:O:}{\|}}{}$

(b) Ozone, $:\overset{..}{\underset{..}{O}} = \overset{+}{\overset{..}{O}} - \overset{..}{\underset{..}{O}}:^-$

(c) Diazomethane, $H_2C = \overset{+}{N} = \overset{..}{N}:^-$

2.54 Carbocations, which contain a trivalent, positively charged carbon atom, react with water to give alcohols:

$$\underset{\text{A carbocation}}{H_3C\overset{\overset{H}{|}}{\underset{|}{\overset{+}{C}}}CH_3} \xrightarrow{H_2O} \underset{\text{An alcohol}}{H_3C\overset{\overset{H}{|}}{\underset{|}{\overset{OH}{C}}}CH_3} + H^+$$

How can you account for the fact that the following carbocation gives a mixture of two alcohols on reaction with water?

$$H_3C\overset{\overset{H}{|}}{\underset{|}{\overset{+}{C}}} - \overset{|}{\underset{H}{C}} = CH_2 \xrightarrow{H_2O} H_3C\overset{\overset{H}{|}}{\underset{|}{\overset{OH}{C}}} - \overset{|}{\underset{H}{C}} = CH_2 \quad + \quad H_3C\overset{\overset{H}{|}}{\underset{|}{C}} = \overset{|}{\underset{H}{C}} - CH_2OH$$

2.55 We'll see in the next chapter that organic molecules can be classified according to the *functional groups* they contain, where a functional group is a collection of atoms with a characteristic chemical reactivity. Use the electronegativity values given in Figure 2.2 on page 35 to predict the direction of polarization of the following functional groups.

(a) Ketone $\overset{\overset{O}{\|}}{C}$

(b) Alcohol $\overset{}{\underset{}{C}}-OH$

(c) Amide $\overset{\overset{O}{\|}}{C}-NH_2$

(d) Nitrile $-C \equiv N$

▲ Problems linked to Key Ideas in this chapter

2.56 The *azide* functional group (Problem 2.55), such as occurs in azidobenzene, contains three adjacent nitrogen atoms. One resonance structures for azidobenzene is shown. Draw three additional resonance structures, and assign appropriate formal charges to the atoms in all four.

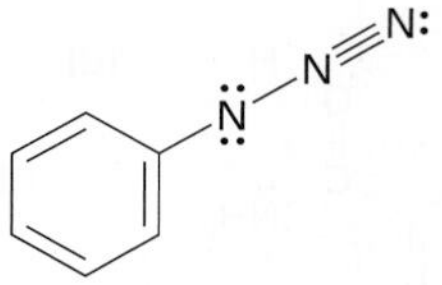

Azidobenzene

2.57 Phenol, C_6H_5OH, is a stronger acid than methanol, CH_3OH, even though both contain an O–H bond. Draw the structures of the anions resulting from loss of H^+ from phenol and methanol, and use resonance structures to explain the difference in acidity.

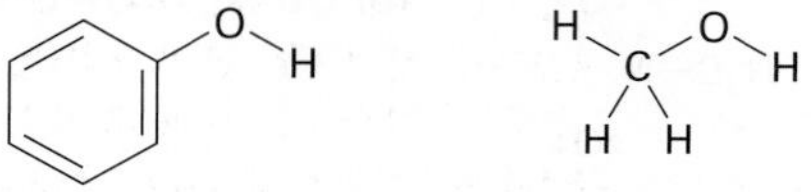

Phenol (pK_a = 9.89)   **Methanol (pK_a = 15.54)**

▲ Problems linked to Key Ideas in this chapter

2.58 Thiamin diphosphate (TPP), a derivative of vitamin B_1 required for glucose metabolism, is a weak acid that can be deprotonated by base. Assign formal charges to the appropriate atoms in both TPP and its deprotonation product.

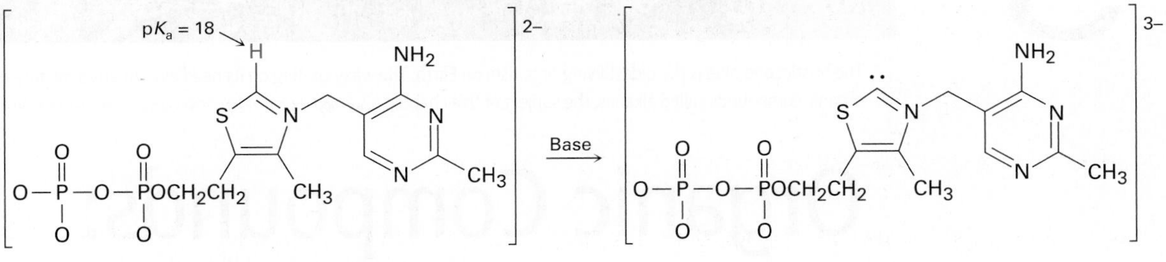

Thiamin diphosphate (TPP)

The bristlecone pine is the oldest living organism on Earth. The waxy coating on its needles contains a mixture of organic compounds called alkanes, the subject of this chapter. Image copyright Mike Norton, 2010. Used under license from Shutterstock.com

3 Organic Compounds: Alkanes and Their Stereochemistry

3.1 Functional Groups
3.2 Alkanes and Alkane Isomers
3.3 Alkyl Groups
3.4 Naming Alkanes
3.5 Properties of Alkanes
3.6 Conformations of Ethane
3.7 Conformations of Other Alkanes
  A Deeper Look—Gasoline

According to *Chemical Abstracts*, the publication that abstracts and indexes the chemical literature, there are more than 50 million known organic compounds. Each of these compounds has its own physical properties, such as melting point and boiling point, and each has its own chemical reactivity.

Chemists have learned through years of experience that organic compounds can be classified into families according to their structural features and that the members of a given family often have similar chemical behavior. Instead of 40 million compounds with random reactivity, there are a few dozen families of organic compounds whose chemistry is reasonably predictable. We'll study the chemistry of specific families throughout much of this book, beginning in this chapter with a look at the simplest family, the *alkanes*.

Why This Chapter? Alkanes are relatively unreactive and not often involved in chemical reactions, but they nevertheless provide a useful vehicle for introducing some important general ideas. In this chapter, we'll use alkanes to introduce the basic approach to naming organic compounds and to take an initial look at some of the three-dimensional aspects of molecules, a topic of particular importance in understanding biological organic chemistry.

3.1 Functional Groups

The structural features that make it possible to classify compounds into families are called *functional groups*. A **functional group** is a group of atoms within a molecule that has a characteristic chemical behavior. Chemically, a given functional group behaves in nearly the same way in every molecule it's a part of. For example, compare ethylene, a plant hormone that causes fruit to ripen, with menthene, a much more complicated molecule found in peppermint oil. Both substances contain a carbon–carbon double-bond functional group, and both therefore react with Br_2 in the same way to give a product in which a Br atom

OWL Sign in to OWL for Organic Chemistry at www.cengage.com/owl to view tutorials and simulations, develop problem-solving skills, and complete online homework assigned by your professor.

has added to each of the double-bond carbons **(Figure 3.1)**. This example is typical: *the chemistry of every organic molecule, regardless of size and complexity, is determined by the functional groups it contains.*

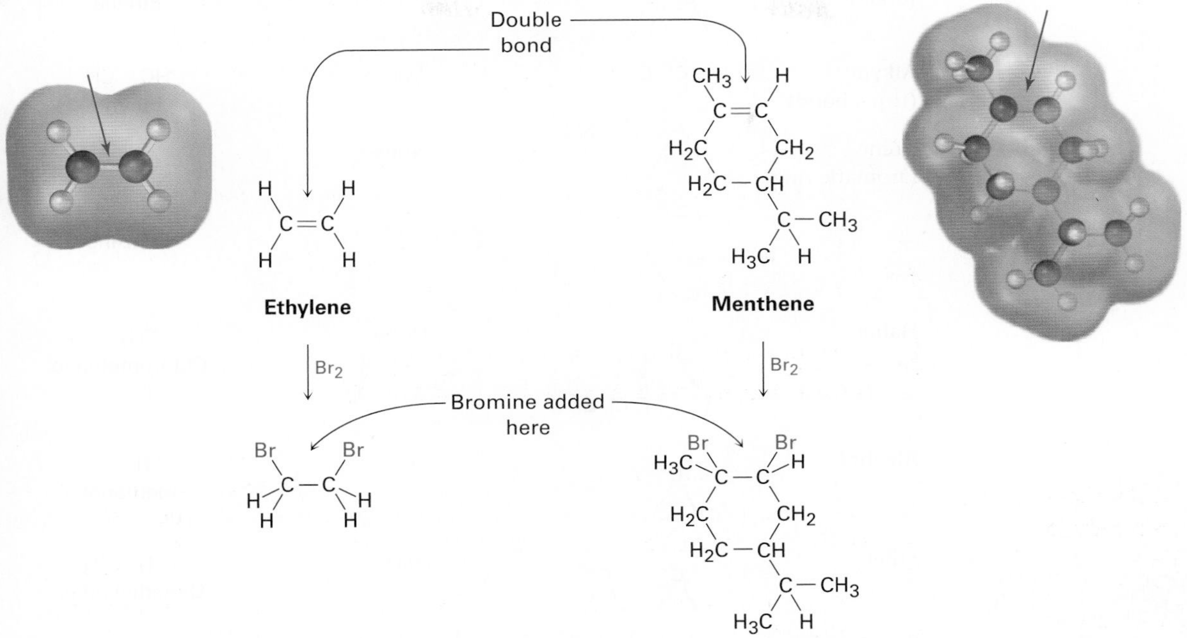

Figure 3.1 The reactions of ethylene and menthene with **bromine**. In both molecules, the carbon–carbon double-bond functional group has a similar polarity pattern, so both molecules react with Br_2 in the same way. The size and complexity of the molecules are not important.

Look at Table 3.1 on pages 76 and 77, which lists many of the common functional groups and gives simple examples of their occurrence. Some functional groups have only carbon–carbon double or triple bonds; others have halogen atoms; and still others contain oxygen, nitrogen, or sulfur. Much of the chemistry you'll be studying is the chemistry of these functional groups.

Functional Groups with Carbon–Carbon Multiple Bonds

Alkenes, alkynes, and arenes (aromatic compounds) all contain carbon–carbon multiple bonds. *Alkenes* have a double bond, *alkynes* have a triple bond, and *arenes* have alternating double and single bonds in a six-membered ring of carbon atoms. Because of their structural similarities, these compounds also have chemical similarities.

Table 3.1 Structures of Some Common Functional Groups

Name	Structure*	Name ending	Example
Alkene (double bond)	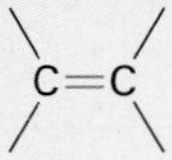 C=C	-ene	H₂C=CH₂ Ethene
Alkyne (triple bond)	—C≡C—	-yne	HC≡CH Ethyne
Arene (aromatic ring)	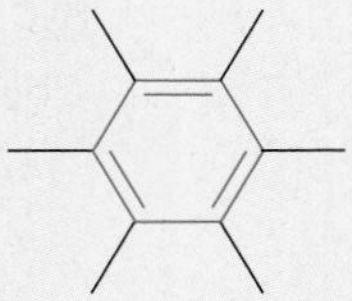 (benzene ring)	None	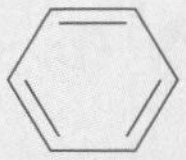 Benzene
Halide	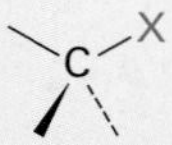 C—X (X = F, Cl, Br, I)	None	CH₃Cl Chloromethane
Alcohol	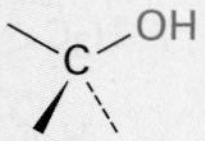 C—OH	-ol	CH₃OH Methanol
Ether	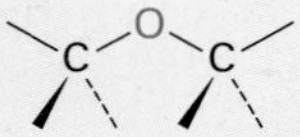 C—O—C	ether	CH₃OCH₃ Dimethyl ether
Monophosphate	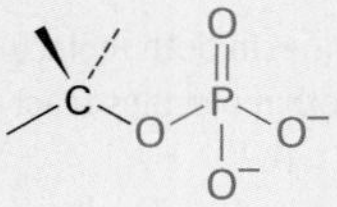 C—O—P(=O)(O⁻)(O⁻)	phosphate	CH₃OPO₃²⁻ Methyl phosphate
Diphosphate	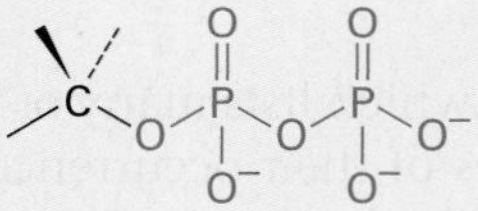 C—O—P(=O)(O⁻)—O—P(=O)(O⁻)(O⁻)	diphosphate	CH₃OP₂O₆³⁻ Methyl diphosphate
Amine	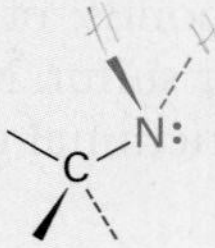 C—N(H)(H)	-amine	CH₃NH₂ Methylamine
Imine (Schiff base)	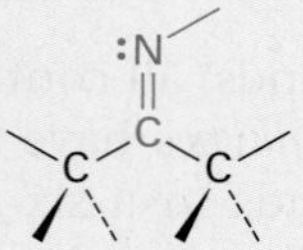 C=N—C	None	NH=CCH₃CH₃ Acetone imine
Nitrile	—C≡N	-nitrile	CH₃C≡N Ethanenitrile
Thiol	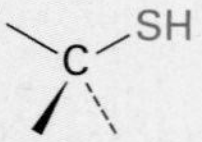 C—SH	-thiol	CH₃SH Methanethiol

*The bonds whose connections aren't specified are assumed to be attached to carbon or hydrogen atoms in the rest of the molecule.

Continued

Table 3.1 Structures of Some Common Functional Groups *(continued)*

Name	Structure*	Name ending	Example
Sulfide	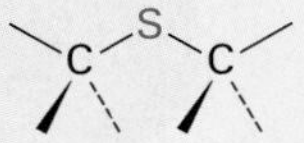 C–S–C	*sulfide*	CH_3SCH_3 Dimethyl sulfide
Disulfide	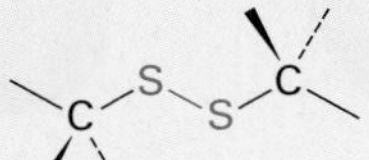 C–S–S–C	*disulfide*	CH_3SSCH_3 Dimethyl disulfide
Sulfoxide	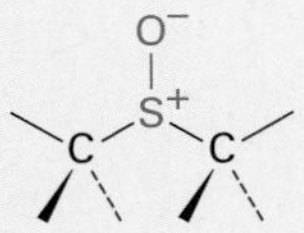 C–S⁺(O⁻)–C	*sulfoxide*	$CH_3\overset{O^-}{\underset{+}{S}}CH_3$ Dimethyl sulfoxide
Aldehyde	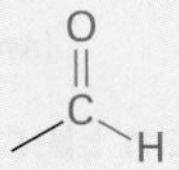 C(=O)H	*-al*	CH_3CHO Ethanal
Ketone	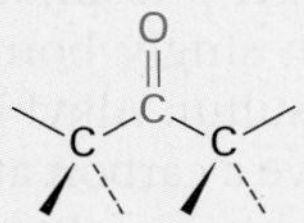 C–C(=O)–C	*-one*	CH_3COCH_3 Propanone
Carboxylic acid	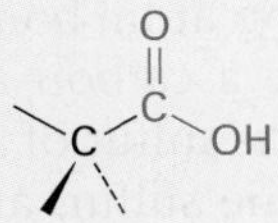 C–C(=O)–OH	*-oic acid*	CH_3COOH Ethanoic acid
Ester	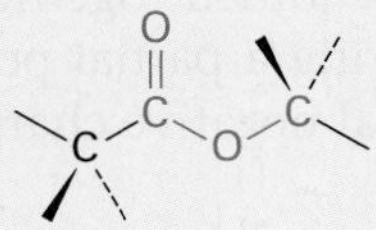 C–C(=O)–O–C	*-oate*	CH_3COOCH_3 Methyl ethanoate
Thioester	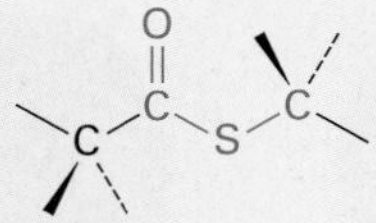 C–C(=O)–S–C	*-thioate*	CH_3COSCH_3 Methyl ethanethioate
Amide	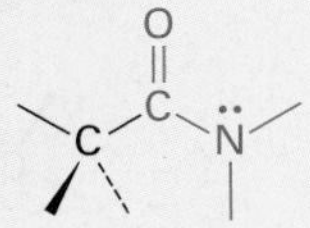 C–C(=O)–N	*-amide*	CH_3CONH_2 Ethanamide
Acid chloride	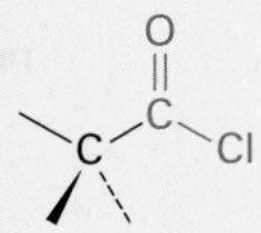 C–C(=O)–Cl	*-oyl chloride*	CH_3COCl Ethanoyl chloride
Carboxylic acid anhydride	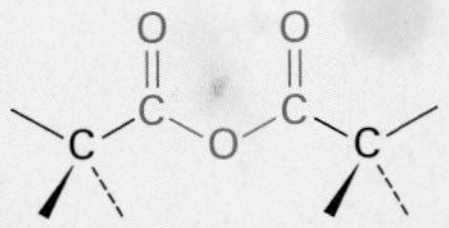 C–C(=O)–O–C(=O)–C	*-oic anhydride*	$CH_3COOCCH_3$ Ethanoic anhydride

*The bonds whose connections aren't specified are assumed to be attached to carbon or hydrogen atoms in the rest of the molecule.

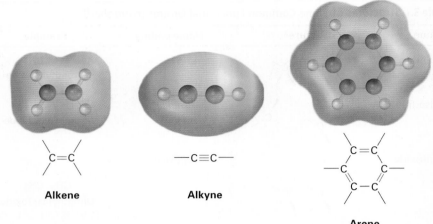

Alkene

Alkyne

Arene (aromatic ring)

Functional Groups with Carbon Singly Bonded to an Electronegative Atom
Alkyl halides (haloalkanes), alcohols, ethers, alkyl phosphates, amines, thiols, sulfides, and disulfides all have a carbon atom singly bonded to an electronegative atom—halogen, oxygen, nitrogen, or sulfur. Alkyl halides have a carbon atom bonded to halogen (−X), alcohols have a carbon atom bonded to the oxygen of a hydroxyl group (−OH), ethers have two carbon atoms bonded to the same oxygen, organophosphates have a carbon atom bonded to the oxygen of a phosphate group ($-OPO_3^{2-}$), amines have a carbon atom bonded to a nitrogen, thiols have a carbon atom bonded to the sulfur of an −SH group, sulfides have two carbon atoms bonded to the same sulfur, and disulfides have carbon atoms bonded to two sulfurs that are joined together. In all cases, the bonds are polar, with the carbon atom bearing a partial positive charge ($\delta+$) and the electronegative atom bearing a partial negative charge ($\delta-$).

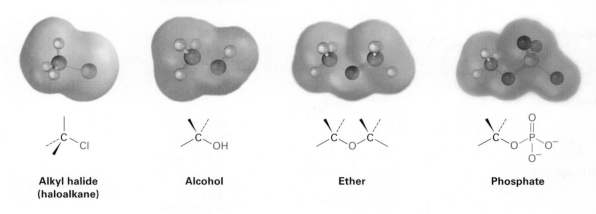

Alkyl halide (haloalkane)

Alcohol

Ether

Phosphate

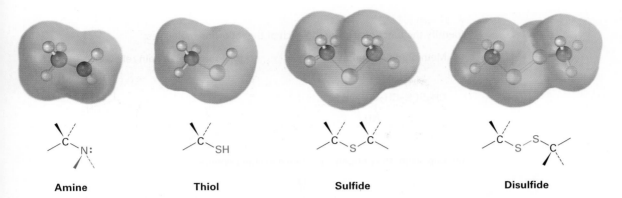

| Amine | Thiol | Sulfide | Disulfide |

Functional Groups with a Carbon–Oxygen Double Bond (Carbonyl Groups)

The *carbonyl group*, C=O (pronounced car-bo-**neel**) is common to many of the families listed in Table 3.1. Carbonyl groups are present in a large majority of organic compounds and in practically all biological molecules. These compounds behave similarly in many respects but differ depending on the identity of the atoms bonded to the carbonyl-group carbon. Aldehydes have at least one hydrogen bonded to the C=O, ketones have two carbons bonded to the C=O, carboxylic acids have an −OH group bonded to the C=O, esters have an ether-like oxygen bonded to the C=O, thioesters have a sulfide-like sulfur bonded to the C=O, amides have an amine-like nitrogen bonded to the C=O, acid chlorides have a chlorine bonded to the C=O, and so on. The carbonyl carbon atom bears a partial positive charge ($\delta+$), and the oxygen bears a partial negative charge ($\delta-$).

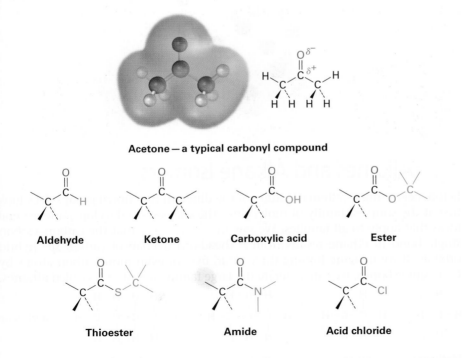

Acetone — a typical carbonyl compound

| Aldehyde | Ketone | Carboxylic acid | Ester |

| Thioester | Amide | Acid chloride |

Problem 3.1
Identify the functional groups in each of the following molecules:

(a) Methionine, an amino acid:

CH$_3$SCH$_2$CH$_2$CHCOH (with =O on C and NH$_2$ on CH)

(b) Ibuprofen, a pain reliever:

(structure with CO$_2$H, CH$_3$, and isobutyl group on benzene ring)

(c) Capsaicin, the pungent substance in chili peppers:

(structure with H$_3$C–O, HO, N–H, C=O, CH$_3$, CH$_3$ groups)

Problem 3.2
Propose structures for simple molecules that contain the following functional groups:
- **(a)** Alcohol **(b)** Aromatic ring **(c)** Carboxylic acid
- **(d)** Amine **(e)** Both ketone and amine **(f)** Two double bonds

Problem 3.3
Identify the functional groups in the following model of arecoline, a veterinary drug used to control worms in animals. Convert the drawing into a line-bond structure and a molecular formula (red = O, blue = N).

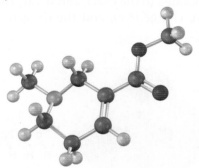

3.2 Alkanes and Alkane Isomers

Before beginning a systematic study of the different functional groups, let's look first at the simplest family of molecules—the *alkanes*—to develop some general ideas that apply to all families. We saw in **Section 1.7** that the carbon–carbon single bond in ethane results from σ (head-on) overlap of carbon sp^3 hybrid orbitals. If we imagine joining three, four, five, or even more carbon atoms by C–C single bonds, we can generate the large family of molecules called **alkanes**.

Methane **Ethane** **Propane** **Butane** . . . and so on

Alkanes are often described as *saturated hydrocarbons:* **hydrocarbons** because they contain only carbon and hydrogen; **saturated** because they have only C−C and C−H single bonds and thus contain the maximum possible number of hydrogens per carbon. They have the general formula C_nH_{2n+2}, where n is an integer. Alkanes are also occasionally called **aliphatic** compounds, a name derived from the Greek *aleiphas*, meaning "fat." We'll see in **Section 27.1** that many animal fats contain long carbon chains similar to alkanes.

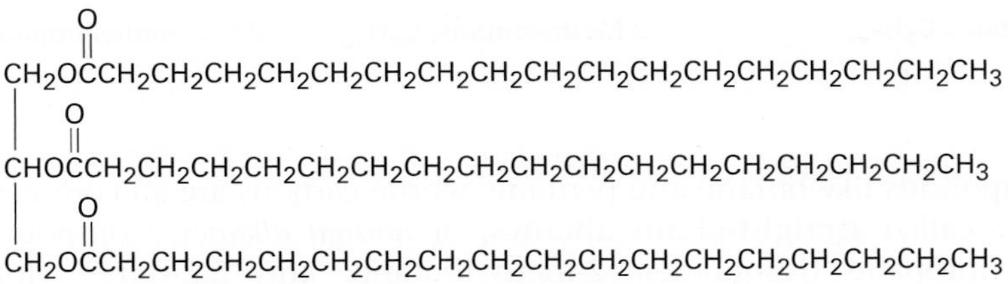

$$CH_2OCCH_2CH_2CH_2CH_2CH_2CH_2CH_2CH_2CH_2CH_2CH_2CH_2CH_2CH_2CH_3$$
$$\overset{O}{\underset{\|}{}}$$
$$CHOCCH_2CH_2CH_2CH_2CH_2CH_2CH_2CH_2CH_2CH_2CH_2CH_2CH_2CH_2CH_2CH_3$$
$$\overset{O}{\underset{\|}{}}$$
$$CH_2OCCH_2CH_2CH_2CH_2CH_2CH_2CH_2CH_2CH_2CH_2CH_2CH_2CH_2CH_2CH_2CH_3$$

A typical animal fat

Think about the ways that carbon and hydrogen might combine to make alkanes. With one carbon and four hydrogens, only one structure is possible: methane, CH_4. Similarly, there is only one combination of two carbons with six hydrogens (ethane, CH_3CH_3) and only one combination of three carbons with eight hydrogens (propane, $CH_3CH_2CH_3$). When larger numbers of carbons and hydrogens combine, however, more than one structure is possible. For example, there are *two* substances with the formula C_4H_{10}: the four carbons can all be in a row (butane), or they can branch (isobutane). Similarly, there are three C_5H_{12} molecules, and so on for larger alkanes.

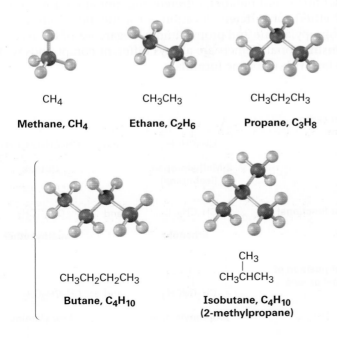

CH₄

Methane, CH₄

CH₃CH₃

Ethane, C₂H₆

CH₃CH₂CH₃

Propane, C₃H₈

CH₃CH₂CH₂CH₃

Butane, C₄H₁₀

$$\overset{CH_3}{\underset{|}{CH_3CHCH_3}}$$

**Isobutane, C₄H₁₀
(2-methylpropane)**

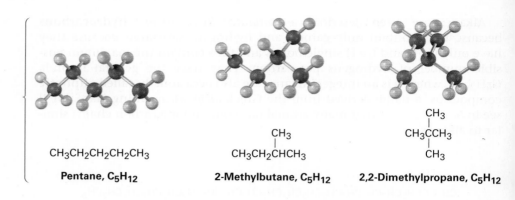

Pentane, C$_5$H$_{12}$: CH$_3$CH$_2$CH$_2$CH$_2$CH$_3$

2-Methylbutane, C$_5$H$_{12}$: CH$_3$CH$_2$CH(CH$_3$)CH$_3$

2,2-Dimethylpropane, C$_5$H$_{12}$: CH$_3$C(CH$_3$)$_2$CH$_3$

Compounds like butane and pentane, whose carbons are all connected in a row, are called **straight-chain alkanes**, or *normal alkanes*. Compounds like 2-methylpropane (isobutane), 2-methylbutane, and 2,2-dimethylpropane, whose carbon chains branch, are called **branched-chain alkanes**.

Compounds like the two C$_4$H$_{10}$ molecules and the three C$_5$H$_{12}$ molecules, which have the same formula but different structures, are called *isomers*, from the Greek *isos* + *meros*, meaning "made of the same parts." **Isomers** are compounds that have the same numbers and kinds of atoms but differ in the way the atoms are arranged. Compounds like butane and isobutane, whose atoms are connected differently, are called **constitutional isomers**. We'll see shortly that other kinds of isomers are also possible, even among compounds whose atoms are connected in the same order. As Table 3.2 shows, the number of possible alkane isomers increases dramatically as the number of carbon atoms increases.

Constitutional isomerism is not limited to alkanes—it occurs widely throughout organic chemistry. Constitutional isomers may have different carbon skeletons (as in isobutane and butane), different functional groups (as in ethanol and dimethyl ether), or different locations of a functional group along the chain (as in isopropylamine and propylamine). Regardless of the reason for the isomerism, constitutional isomers are always different compounds with different properties but with the same formula.

Table 3.2 Number of Alkane Isomers

Formula	Number of isomers
C$_6$H$_{14}$	5
C$_7$H$_{16}$	9
C$_8$H$_{18}$	18
C$_9$H$_{20}$	35
C$_{10}$H$_{22}$	75
C$_{15}$H$_{32}$	4,347
C$_{20}$H$_{42}$	366,319
C$_{30}$H$_{62}$	4,111,846,763

Different carbon skeletons
C$_4$H$_{10}$

CH$_3$CH(CH$_3$)CH$_3$ and CH$_3$CH$_2$CH$_2$CH$_3$
2-Methylpropane (isobutane)   Butane

Different functional groups
C$_2$H$_6$O

CH$_3$CH$_2$OH and CH$_3$OCH$_3$
Ethanol   Dimethyl ether

Different position of functional groups
C$_3$H$_9$N

CH$_3$CH(NH$_2$)CH$_3$ and CH$_3$CH$_2$CH$_2$NH$_2$
Isopropylamine   Propylamine

A given alkane can be drawn in many ways. For example, the straight-chain, four-carbon alkane called butane can be represented by any of the structures shown in **Figure 3.2**. These structures don't imply any particular three-dimensional geometry for butane; they indicate only the connections among atoms. In practice, as noted in **Section 1.12**, chemists rarely draw all the bonds in a molecule and usually refer to butane by the condensed structure, $CH_3CH_2CH_2CH_3$ or $CH_3(CH_2)_2CH_3$. Still more simply, butane can be represented as $n\text{-}C_4H_{10}$, where n denotes *normal* (straight-chain) butane.

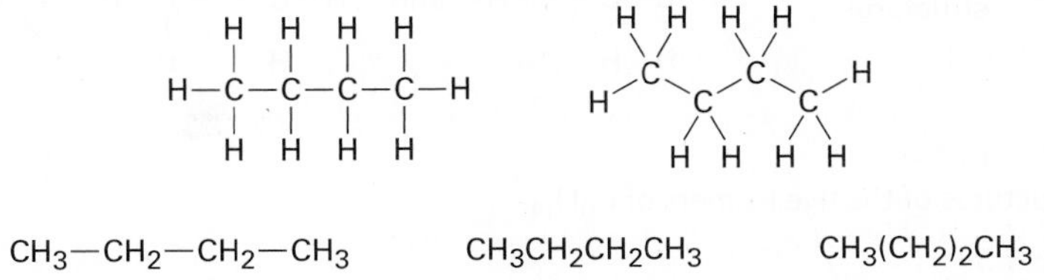

Figure 3.2 Some representations of butane, C_4H_{10}. The molecule is the same regardless of how it's drawn. These structures imply only that butane has a continuous chain of four carbon atoms; they do not imply any specific geometry.

Straight-chain alkanes are named according to the number of carbon atoms they contain, as shown in Table 3.3. With the exception of the first four compounds—methane, ethane, propane, and butane—whose names have historical roots, the alkanes are named based on Greek numbers. The suffix *-ane* is added to the end of each name to indicate that the molecule identified is an alkane. Thus, pent*ane* is the five-carbon alkane, hex*ane* is the six-carbon alkane, and so on. We'll soon see that these alkane names form the basis for naming all other organic compounds, so at least the first ten should be memorized.

Table 3.3 Names of Straight-Chain Alkanes

Number of carbons (n)	Name	Formula (C_nH_{2n+2})	Number of carbons (n)	Name	Formula (C_nH_{2n+2})
1	Methane	CH_4	9	Nonane	C_9H_{20}
2	Ethane	C_2H_6	10	Decane	$C_{10}H_{22}$
3	Propane	C_3H_8	11	Undecane	$C_{11}H_{24}$
4	Butane	C_4H_{10}	12	Dodecane	$C_{12}H_{26}$
5	Pentane	C_5H_{12}	13	Tridecane	$C_{13}H_{28}$
6	Hexane	C_6H_{14}	20	Icosane	$C_{20}H_{42}$
7	Heptane	C_7H_{16}	30	Triacontane	$C_{30}H_{62}$
8	Octane	C_8H_{18}			

Drawing the Structures of Isomers

Worked Example 3.1

Propose structures for two isomers with the formula C_2H_7N.

Strategy

We know that carbon forms four bonds, nitrogen forms three, and hydrogen forms one. Write down the carbon atoms first, and then use a combination of trial and error plus intuition to put the pieces together.

Solution

There are two isomeric structures. One has the connection C−C−N, and the other has the connection C−N−C.

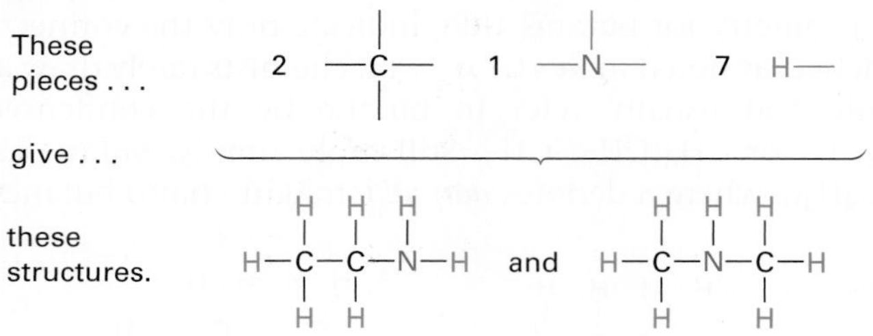

These pieces . . .   2 —C—   1 —N—   7 H—

give . . .

these structures.

H−C(H)(H)−C(H)(H)−N(H)−H   and   H−C(H)(H)−N(H)−C(H)(H)−H

Problem 3.4
Draw structures of the five isomers of C_6H_{14}.

Problem 3.5
Propose structures that meet the following descriptions:
(a) Two isomeric esters with the formula $C_5H_{10}O_2$
(b) Two isomeric nitriles with the formula C_4H_7N
(c) Two isomeric disulfides with the formula $C_4H_{10}S_2$

Problem 3.6
How many isomers are there with the following descriptions?
(a) Alcohols with the formula C_3H_8O
(b) Bromoalkanes with the formula C_4H_9Br
(c) Thioesters with the formula C_4H_8OS

3.3 Alkyl Groups

If you imagine removing a hydrogen atom from an alkane, the partial structure that remains is called an **alkyl group**. Alkyl groups are not stable compounds themselves, they are simply parts of larger compounds. Alkyl groups are named by replacing the *-ane* ending of the parent alkane with an *-yl* ending. For example, removal of a hydrogen from methane, CH_4, generates a *methyl* group, $-CH_3$, and removal of a hydrogen from ethane, CH_3CH_3, generates an *ethyl* group, $-CH_2CH_3$. Similarly, removal of a hydrogen atom from the end carbon of any straight-chain alkane gives the series of straight-chain alkyl groups shown in Table 3.4. Combining an alkyl group with any of the functional groups listed earlier makes it possible to generate and name many thousands of compounds. For example:

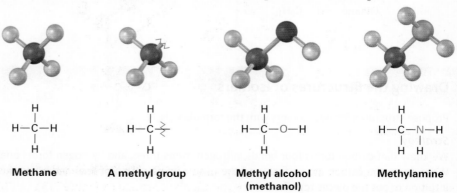

Methane   A methyl group   Methyl alcohol (methanol)   Methylamine

3.3 | Alkyl Groups

Table 3.4 Some Straight-Chain Alkyl Groups

Alkane	Name	Alkyl group	Name (abbreviation)
CH_4	Methane	$-CH_3$	Methyl (Me)
CH_3CH_3	Ethane	$-CH_2CH_3$	Ethyl (Et)
$CH_3CH_2CH_3$	Propane	$-CH_2CH_2CH_3$	Propyl (Pr)
$CH_3CH_2CH_2CH_3$	Butane	$-CH_2CH_2CH_2CH_3$	Butyl (Bu)
$CH_3CH_2CH_2CH_2CH_3$	Pentane	$-CH_2CH_2CH_2CH_2CH_3$	Pentyl, or amyl

Just as straight-chain alkyl groups are generated by removing a hydrogen from an end carbon, branched alkyl groups are generated by removing a hydrogen atom from an internal carbon. Two 3-carbon alkyl groups and four 4-carbon alkyl groups are possible **(Figure 3.3)**.

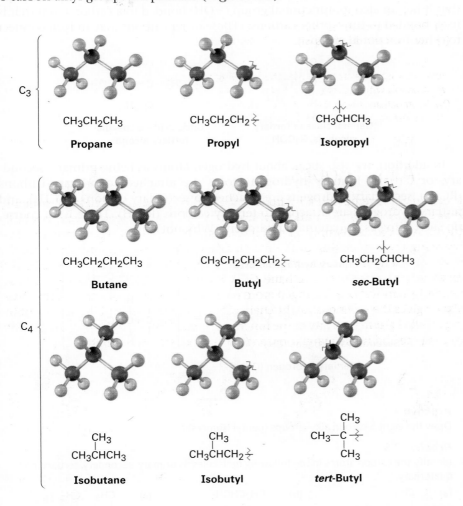

Figure 3.3 Alkyl groups generated from straight-chain alkanes.

One further comment about naming alkyl groups: the prefixes *sec-* (for secondary) and *tert-* (for tertiary) used for the C_4 alkyl groups in Figure 3.3 refer to the number of other carbon atoms attached to the branching carbon atom. There are four possibilities: primary (1°), secondary (2°), tertiary (3°), and quaternary (4°).

Primary carbon (1°) is bonded to one other carbon.	Secondary carbon (2°) is bonded to two other carbons.	Tertiary carbon (3°) is bonded to three other carbons.	Quaternary carbon (4°) is bonded to four other carbons.

The symbol **R** is used here and throughout organic chemistry to represent a *generalized* organic group. The R group can be methyl, ethyl, propyl, or any of a multitude of others. You might think of **R** as representing the **R**est of the molecule, which isn't specified.

The terms *primary, secondary, tertiary,* and *quaternary* are routinely used in organic chemistry, and their meanings need to become second nature. For example, if we were to say, "Citric acid is a tertiary alcohol," we would mean that it has an alcohol functional group (–OH) bonded to a carbon atom that is itself bonded to three other carbons. (These other carbons may in turn connect to other functional groups.)

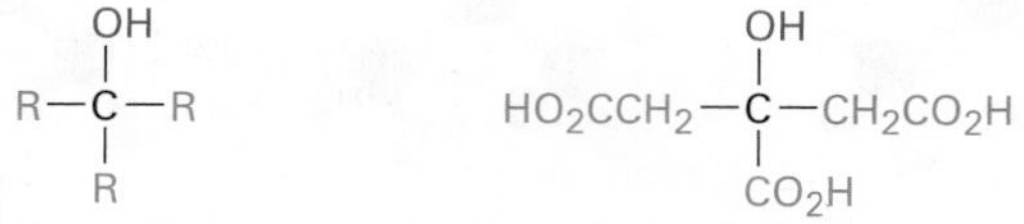

General class of tertiary alcohols, R₃COH

Citric acid—a specific tertiary alcohol

In addition, we also speak about hydrogen atoms as being primary, secondary, or tertiary. Primary hydrogen atoms are attached to primary carbons (RCH₃), secondary hydrogens are attached to secondary carbons (R₂CH₂), and tertiary hydrogens are attached to tertiary carbons (R₃CH). There is, of course, no such thing as a quaternary hydrogen. (Why not?)

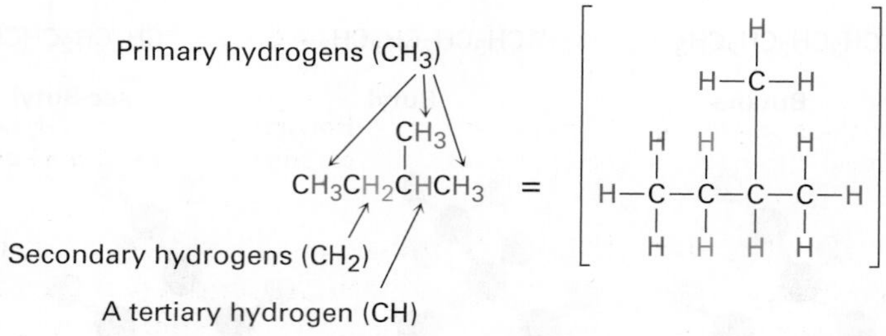

Primary hydrogens (CH₃)

Secondary hydrogens (CH₂)

A tertiary hydrogen (CH)

Problem 3.7
Draw the eight 5-carbon alkyl groups (pentyl isomers).

Problem 3.8
Identify the carbon atoms in the following molecules as primary, secondary, tertiary, or quaternary:

(a) 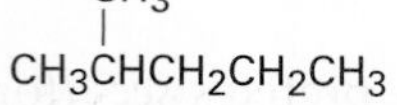   CH₃
      |
   CH₃CHCH₂CH₂CH₃

(b) 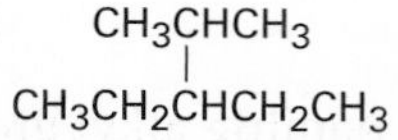   CH₃CHCH₃
      |
   CH₃CH₂CHCH₂CH₃

(c) 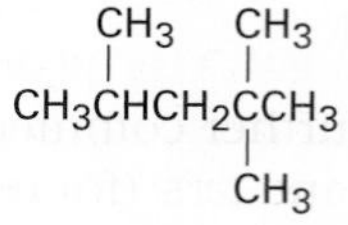   CH₃   CH₃
      | |
   CH₃CHCH₂CCH₃
            |
            CH₃

Problem 3.9
Identify the hydrogen atoms on the compounds shown in Problem 3.8 as primary, secondary, or tertiary.

Problem 3.10
Draw structures of alkanes that meet the following descriptions:
(a) An alkane with two tertiary carbons
(b) An alkane that contains an isopropyl group
(c) An alkane that has one quaternary and one secondary carbon

3.4 Naming Alkanes

In earlier times, when relatively few pure organic chemicals were known, new compounds were named at the whim of their discoverer. Thus, urea (CH_4N_2O) is a crystalline substance isolated from urine; morphine ($C_{17}H_{19}NO_3$) is an analgesic (painkiller) named after Morpheus, the Greek god of dreams; and acetic acid, the primary organic constituent of vinegar, is named from the Latin word for vinegar, *acetum*.

As the science of organic chemistry slowly grew in the 19th century, so too did the number of known compounds and the need for a systematic method of naming them. The system of nomenclature we'll use in this book is that devised by the International Union of Pure and Applied Chemistry (IUPAC, usually spoken as **eye**-you-pac).

A chemical name typically has four parts in the IUPAC system of nomenclature: prefix, parent, locant, and suffix. The prefix identifies the various **substituent** groups in the molecule, the parent selects a main part of the molecule and tells how many carbon atoms are in that part, the locants give the positions of the functional groups and substituents, and the suffix identifies the primary functional group.

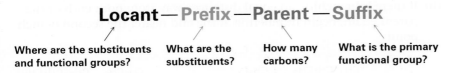

As we cover new functional groups in later chapters, the applicable IUPAC rules of nomenclature will be given. In addition, Appendix A at the back of this book gives an overall view of organic nomenclature and shows how compounds that contain more than one functional group are named. (If preferred, you can study that appendix now.) For the present, let's see how to name branched-chain alkanes and learn some general rules that are applicable to all compounds.

All but the most complex branched-chain alkanes can be named by following four steps. For a very few compounds, a fifth step is needed.

STEP 1
Find the parent hydrocarbon.
(a) Find the longest continuous chain of carbon atoms in the molecule, and use the name of that chain as the parent name. The longest chain

may not always be apparent from the manner of writing; you may have to "turn corners."

$$\begin{array}{c} \text{CH}_2\text{CH}_3 \\ | \\ \text{CH}_3\text{CH}_2\text{CH}_2\text{CH}-\text{CH}_3 \end{array}$$ Named as a substituted hexane

$$\begin{array}{c} \text{CH}_3 \\ | \\ \text{CH}_2 \\ | \\ \text{CH}_3-\text{CHCH}-\text{CH}_2\text{CH}_3 \\ | \\ \text{CH}_2\text{CH}_2\text{CH}_3 \end{array}$$ Named as a substituted heptane

(b) If two different chains of equal length are present, choose the one with the larger number of branch points as the parent.

$$\begin{array}{c} \text{CH}_3 \\ | \\ \text{CH}_3\text{CHCHCH}_2\text{CH}_2\text{CH}_3 \\ | \\ \text{CH}_2\text{CH}_3 \end{array}$$ NOT $$\begin{array}{c} \text{CH}_3 \\ | \\ \text{CH}_3\text{CH}-\text{CHCH}_2\text{CH}_2\text{CH}_3 \\ | \\ \text{CH}_2\text{CH}_3 \end{array}$$

Named as a hexane with *two* substituents

as a hexane with *one* substituent

STEP 2

Number the atoms in the longest chain.

(a) Beginning at the end nearer the first branch point, number each carbon atom in the parent chain.

$$\begin{array}{c} \overset{2\ \ 1}{\text{CH}_2\text{CH}_3} \\ | \\ \text{CH}_3-\underset{3\ \ \ |4}{\text{CHCH}}-\text{CH}_2\text{CH}_3 \\ \underset{5\ \ \ 6\ \ \ 7}{\text{CH}_2\text{CH}_2\text{CH}_3} \end{array}$$ NOT $$\begin{array}{c} \overset{6\ \ 7}{\text{CH}_2\text{CH}_3} \\ | \\ \text{CH}_3-\underset{5\ \ \ |4}{\text{CHCH}}-\text{CH}_2\text{CH}_3 \\ \underset{3\ \ \ 2\ \ \ 1}{\text{CH}_2\text{CH}_2\text{CH}_3} \end{array}$$

The first branch occurs at C3 in the proper system of numbering, not at C4.

(b) If there is branching an equal distance away from both ends of the parent chain, begin numbering at the end nearer the second branch point.

$$\begin{array}{c} \overset{8\ \ 9}{\text{CH}_2\text{CH}_3} \ \ \ \ \text{CH}_3 \ \ \text{CH}_2\text{CH}_3 \\ | \ \ \ \ \ \ \ \ \ \ | \ \ \ | \\ \text{CH}_3-\underset{7\ \ 6}{\text{CHCH}_2\text{CH}_2}\underset{5\ \ 4}{\text{CH}}-\underset{3\ \ 2\ \ 1}{\text{CHCH}_2\text{CH}_3} \end{array}$$ NOT $$\begin{array}{c} \overset{2\ \ 1}{\text{CH}_2\text{CH}_3} \ \ \ \ \text{CH}_3 \ \ \text{CH}_2\text{CH}_3 \\ | \ \ \ \ \ \ \ \ \ \ | \ \ \ | \\ \text{CH}_3-\underset{3\ \ 4}{\text{CHCH}_2\text{CH}_2}\underset{5\ \ 6}{\text{CH}}-\underset{7\ \ 8\ \ 9}{\text{CHCH}_2\text{CH}_3} \end{array}$$

STEP 3

Identify and number the substituents.

(a) Assign a number, or *locant*, to each substituent to locate its point of attachment to the parent chain.

$$\begin{array}{c} \overset{9\ \ 8}{\text{CH}_3\text{CH}_2} \ \ \ \ \text{H}_3\text{C} \ \ \text{CH}_2\text{CH}_3 \\ | \ \ \ \ \ \ \ \ \ \ | \ \ \ | \\ \text{CH}_3-\underset{7\ \ 6}{\text{CHCH}_2}\underset{5}{\text{CH}_2}\underset{4}{\text{CH}}\underset{3}{\text{CH}}\underset{2\ \ 1}{\text{CH}_2\text{CH}_3} \end{array}$$ Named as a nonane

Substituents: On C3, CH_2CH_3 (3-ethyl)
On C4, CH_3 (4-methyl)
On C7, CH_3 (7-methyl)

(b) If there are two substituents on the same carbon, give both the same number. There must be as many numbers in the name as there are substituents.

$$\underset{6}{CH_3}\underset{5}{CH_2}\underset{|3}{\overset{\overset{CH_3}{|4}}{C}}CH_2\underset{2}{\overset{\overset{CH_3}{|}}{C}}H\underset{1}{CH_3}$$
$$\underset{}{\overset{}{CH_2CH_3}}$$

Named as a hexane

Substituents: On C2, CH_3 (2-methyl)
On C4, CH_3 (4-methyl)
On C4, CH_2CH_3 (4-ethyl)

STEP 4
Write the name as a single word.
Use hyphens to separate the different prefixes, and use commas to separate numbers. If two or more different substituents are present, cite them in alphabetical order. If two or more identical substituents are present on the parent chain, use one of the multiplier prefixes *di-*, *tri-*, *tetra-*, and so forth, but don't use these prefixes for alphabetizing. Full names for some of the examples we have been using follow.

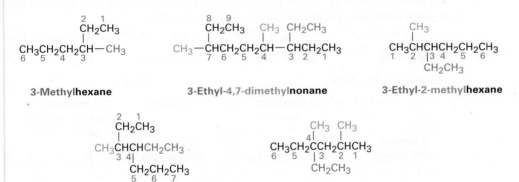

3-Methyl**hexane**   3-Ethyl-4,7-dimethyl**nonane**   3-Ethyl-2-methyl**hexane**

4-Ethyl-3-methyl**heptane**   4-Ethyl-2,4-dimethyl**hexane**

STEP 5
Name a complex substituent as though it were itself a compound.
In some particularly complex cases, a fifth step is necessary. It occasionally happens that a substituent on the main chain has sub-branching. In the following case, for instance, the substituent at C6 is a three-carbon chain with a methyl sub-branch. To name the compound fully, the complex substituent must first be named.

$$\underset{1}{CH_3}\underset{2|}{\overset{\overset{CH_3}{|}}{CH}}\underset{3}{CH}\underset{|}{\overset{}{CH_2}}\underset{4}{CH_2}\underset{5}{CH_2}\underset{6}{CH}-CH_2\underset{}{\overset{\overset{CH_3}{|}}{CH}}CH_3$$
$$\underset{}{CH_3}\quad\underset{7\ 8\ 9\ 10}{CH_2CH_2CH_2CH_3}$$

$\left[\ \underset{1}{\overset{}{\xi-CH_2}}\underset{2}{\overset{\overset{CH_3}{|}}{CH}}\underset{3}{CH_3}\ \right]$

Named as a 2,3,6-trisubstituted decane

A 2-methylpropyl group

Number the branched substituent beginning at its point of its attachment to the main chain, and identify it—in this case, a 2-methylpropyl group. The substituent is treated as a whole and is alphabetized according to the first letter of its complete name, including any numerical prefix. It is set off in parentheses when naming the entire molecule.

$$\underset{\text{2,3-Dimethyl-6-(2-methylpropyl)}\textbf{decane}}{\overset{\begin{array}{c}\text{CH}_3\\|\end{array}}{\underset{1\ 2\ 3\ 4\ 5\ 6}{\text{CH}_3\text{CHCHCH}_2\text{CH}_2\text{CH}}\underset{\underset{7\ 8\ 9\ 10}{\text{CH}_2\text{CH}_2\text{CH}_2\text{CH}_3}}{-}\overset{\text{CH}_3}{\underset{|}{\text{CH}_2\text{CHCH}_3}}}}$$

As a further example:

$$\underset{\text{5-(1,2-Dimethylpropyl)-2-methyl}\textbf{nonane}}{\underset{9\ 8\ 7\ 6\ 5}{\text{CH}_3\text{CH}_2\text{CH}_2\text{CH}_2\text{CH}}-\underset{\underset{\text{H}_3\text{C CH}_3}{|\ \ |}}{\text{CHCHCH}_3}\ \overset{\overset{\text{CH}_3}{|}}{\underset{4\ 3\ 2\ 1}{\text{CH}_2\text{CH}_2\text{CHCH}_3}}} \qquad \underset{\text{A 1,2-dimethylpropyl group}}{\left[\underset{\underset{\text{H}_3\text{C CH}_3}{|\ \ |}}{-\overset{1\ 2\ 3}{\text{CHCHCH}_3}}\right]}$$

For historical reasons, some of the simpler branched-chain alkyl groups also have nonsystematic, common names, as noted earlier.

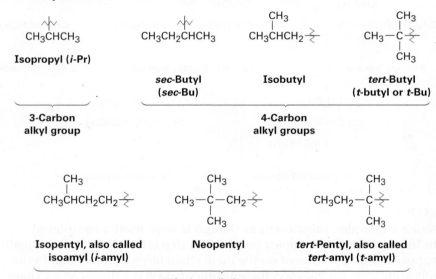

The common names of these simple alkyl groups are so well entrenched in the chemical literature that IUPAC rules make allowance for them. Thus, the following compound is properly named either 4-(1-methylethyl)heptane or 4-isopropylheptane. There's no choice but to memorize these common names; fortunately, there are only a few of them.

$$\underset{\textbf{4-(1-Methylethyl)}\textbf{heptane}\ \ \text{or}\ \ \textbf{4-Isopropyl}\textbf{heptane}}{\text{CH}_3\text{CH}_2\text{CH}_2\overset{\overset{\text{CH}_3\text{CHCH}_3}{|}}{\text{CHCH}_2\text{CH}_2\text{CH}_3}}$$

When writing an alkane name, the nonhyphenated prefix iso- is considered part of the alkyl-group name for alphabetizing purposes, but the hyphenated and italicized prefixes *sec-* and *tert-* are not. Thus, isopropyl and isobutyl are listed alphabetically under *i*, but *sec*-butyl and *tert*-butyl are listed under *b*.

Worked Example 3.2

Naming Alkanes

What is the IUPAC name of the following alkane?

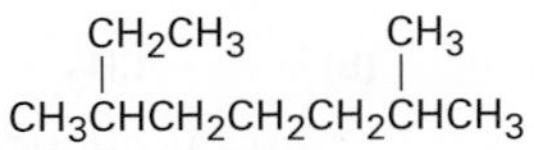

$$\begin{array}{c} CH_2CH_3 CH_3 \\ | | \\ CH_3CHCH_2CH_2CH_2CHCH_3 \end{array}$$

Strategy

Find the longest continuous carbon chain in the molecule, and use that as the parent name. This molecule has a chain of eight carbons—octane—with two methyl substituents. (You have to turn corners to see it.) Numbering from the end nearer the first methyl substituent indicates that the methyls are at C2 and C6

Solution

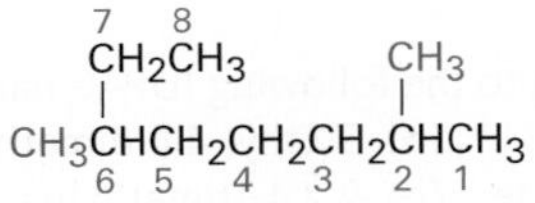

$$\begin{array}{c} \overset{78}{CH_2CH_3} CH_3 \\ | | \\ \underset{65}{CH_3CH}\underset{4}{CH_2}\underset{3}{CH_2}\underset{2}{CH_2}\underset{1}{CHCH_3} \end{array}$$

2,6-Dimethyloctane

Worked Example 3.3

Converting a Chemical Name into a Structure

Draw the structure of 3-isopropyl-2-methylhexane.

Strategy

This is the reverse of Worked Example 3.2 and uses a reverse strategy. Look at the parent name (hexane), and draw its carbon structure.

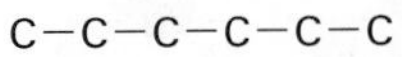

$$C-C-C-C-C-C \quad \textbf{Hexane}$$

Next, find the substituents (3-isopropyl and 2-methyl), and place them on the proper carbons.

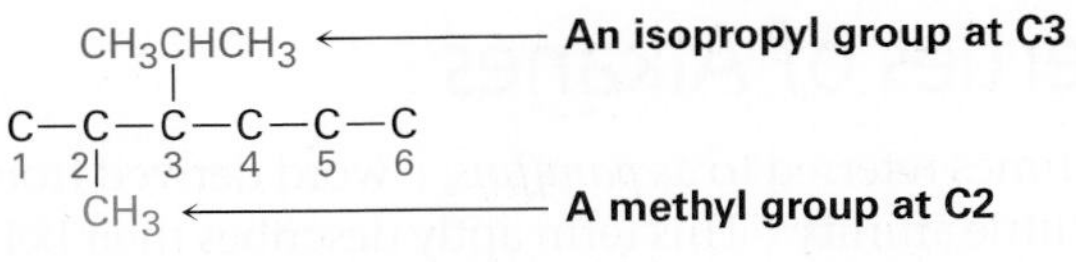

$$\begin{array}{c} CH_3CHCH_3 \longleftarrow \text{An isopropyl group at C3} \\ | \\ \underset{1}{C}-\underset{2|}{C}-\underset{3}{C}-\underset{4}{C}-\underset{5}{C}-\underset{6}{C} \\ CH_3 \longleftarrow \text{A methyl group at C2} \end{array}$$

Finally, add hydrogens to complete the structure.

Solution

$$CH_3CH(CH_3)$$
$$CH_3CHCHCH_2CH_2CH_3$$
$$CH_3$$

3-Isopropyl-2-methyl**hexane**

Problem 3.11
Give IUPAC names for the following compounds:

(a) The three isomers of C_5H_{12}

(b)
$$CH_3CH_2CHCHCH_3$$
with CH_3 and CH_3 substituents

(c) $(CH_3)_2CHCH_2CHCH_3$ with CH_3

(d) $(CH_3)_3CCH_2CH_2CH$ with CH_3 and CH_3

Problem 3.12
Draw structures corresponding to the following IUPAC names:
(a) 3,4-Dimethylnonane   **(b)** 3-Ethyl-4,4-dimethylheptane
(c) 2,2-Dimethyl-4-propyloctane   **(d)** 2,2,4-Trimethylpentane

Problem 3.13
Name the eight 5-carbon alkyl groups you drew in Problem 3.7.

Problem 3.14
Give the IUPAC name for the following hydrocarbon, and convert the drawing into a skeletal structure.

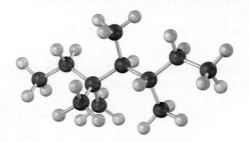

3.5 Properties of Alkanes

Alkanes are sometimes referred to as *paraffins,* a word derived from the Latin *parum affinis,* meaning "little affinity." This term aptly describes their behavior, for alkanes show little chemical affinity for other substances and are chemically inert to most laboratory reagents. They are also relatively inert biologically and are not often involved in the chemistry of living organisms. Alkanes do, however, react with oxygen, halogens, and a few other substances under appropriate conditions.

Reaction with oxygen occurs during combustion in an engine or furnace when the alkane is used as a fuel. Carbon dioxide and water are formed as products, and a large amount of heat is released. For example, methane (natural gas) reacts with oxygen according to the equation

$$CH_4 + 2\,O_2 \rightarrow CO_2 + 2\,H_2O + 890 \text{ kJ/mol (213 kcal/mol)}$$

The reaction of an alkane with Cl_2 occurs when a mixture of the two is irradiated with ultraviolet light (denoted $h\nu$, where ν is the Greek letter nu). Depending on the relative amounts of the two reactants and on the time allowed, a sequential substitution of the alkane hydrogen atoms by chlorine occurs, leading to a mixture of chlorinated products. Methane, for instance, reacts with Cl_2 to yield a mixture of CH_3Cl, CH_2Cl_2, $CHCl_3$, and CCl_4. We'll look at this reaction in more detail in **Section 6.3**.

$$CH_4 + Cl_2 \xrightarrow{h\nu} CH_3Cl + HCl$$
$$\xrightarrow{Cl_2} CH_2Cl_2 + HCl$$
$$\xrightarrow{Cl_2} CHCl_3 + HCl$$
$$\xrightarrow{Cl_2} CCl_4 + HCl$$

Alkanes show regular increases in both boiling point and melting point as molecular weight increases **(Figure 3.4)**, an effect due to the presence of weak dispersion forces between molecules **(Section 2.12)**. Only when sufficient energy is applied to overcome these forces does the solid melt or liquid boil. As you might expect, dispersion forces increase as molecule size increases, accounting for the higher melting and boiling points of larger alkanes.

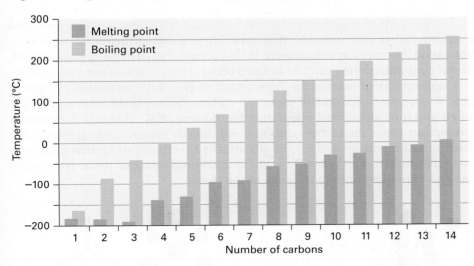

Figure 3.4 A plot of melting and boiling points versus number of carbon atoms for the C_1–C_{14} straight-chain alkanes. There is a regular increase with molecular size.

Another effect seen in alkanes is that increased branching lowers an alkane's boiling point. Thus, pentane has no branches and boils at 36.1 °C, isopentane (2-methylbutane) has one branch and boils at 27.85 °C, and neopentane (2,2-dimethylpropane) has two branches and boils at 9.5 °C. Similarly, octane boils at 125.7 °C, whereas isooctane (2,2,4-trimethylpentane) boils at 99.3 °C. Branched-chain alkanes are lower-boiling because they are more nearly spherical than straight-chain alkanes, have smaller surface areas, and consequently have smaller dispersion forces.

3.6 Conformations of Ethane

Up to now, we've viewed molecules primarily in a two-dimensional way and have given little thought to any consequences that might arise from the spatial arrangement of atoms in molecules. Now it's time to add a third dimension to our study. **Stereochemistry** is the branch of chemistry concerned with the three-dimensional aspects of molecules. We'll see on many occasions in future chapters that the exact three-dimensional structure of a molecule is often crucial to determining its properties and biological behavior.

We know from **Section 1.5** that σ bonds are cylindrically symmetrical. In other words, the intersection of a plane cutting through a carbon–carbon single-bond orbital looks like a circle. Because of this cylindrical symmetry, rotation is possible around carbon–carbon bonds in open-chain molecules. In ethane, for instance, rotation around the C–C bond occurs freely, constantly changing the spatial relationships between the hydrogens on one carbon and those on the other **(Figure 3.5)**.

Figure 3.5 Rotation occurs around the carbon–carbon single bond in ethane because of σ bond cylindrical symmetry.

The different arrangements of atoms that result from bond rotation are called **conformations**, and molecules that have different arrangements are called **conformational isomers**, or **conformers**. Unlike constitutional isomers, however, different conformers often can't be isolated because they interconvert too rapidly.

Conformational isomers are represented in two ways, as shown in **Figure 3.6**. A *sawhorse representation* views the carbon–carbon bond from an oblique angle and indicates spatial orientation by showing all C–H bonds. A **Newman projection** views the carbon–carbon bond directly end-on and represents the two carbon atoms by a circle. Bonds attached to the front carbon are represented by lines to the center of the circle, and bonds attached to the rear carbon are represented by lines to the edge of the circle.

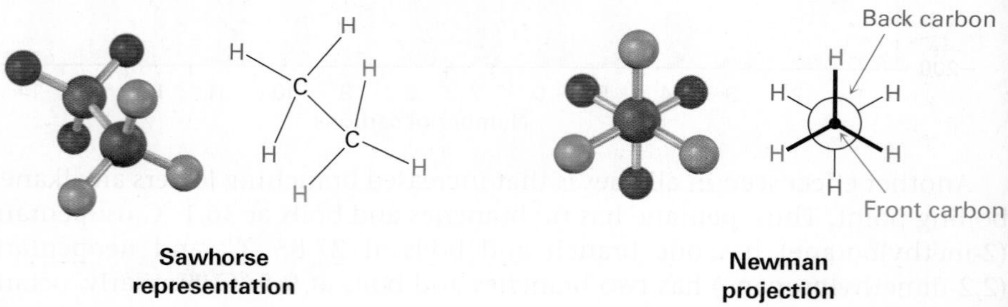

Figure 3.6 A sawhorse representation and a Newman projection of ethane. The sawhorse representation views the molecule from an oblique angle, while the Newman projection views the molecule end-on. Note that the molecular model of the Newman projection appears at first to have six atoms attached to a single carbon. Actually, the front carbon, with three attached **green atoms**, is directly in front of the rear carbon, with three attached **red atoms**.

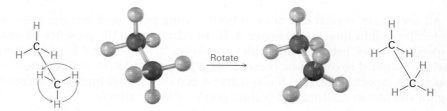

Sawhorse representation

Newman projection

3.6 | Conformations of Ethane

Despite what we've just said, we actually don't observe *perfectly* free rotation in ethane. Experiments show that there is a small (12 kJ/mol; 2.9 kcal/mol) barrier to rotation and that some conformations are more stable than others. The lowest-energy, most stable conformation is the one in which all six C−H bonds are as far away from one another as possible—**staggered** when viewed end-on in a Newman projection. The highest-energy, least stable conformation is the one in which the six C−H bonds are as close as possible—**eclipsed** in a Newman projection. At any given instant, about 99% of ethane molecules have an approximately staggered conformation and only about 1% are near the eclipsed conformation.

Ethane—staggered conformation

Ethane—eclipsed conformation

The extra 12 kJ/mol of energy present in the eclipsed conformation of ethane is called **torsional strain**. Its cause has been the subject of controversy, but the major factor is an interaction between C−H bonding orbitals on one carbon with antibonding orbitals on the adjacent carbon, which stabilizes the staggered conformation relative to the eclipsed one. Because the total strain of 12 kJ/mol arises from three equal hydrogen–hydrogen eclipsing interactions, we can assign a value of approximately 4.0 kJ/mol (1.0 kcal/mol) to each single interaction. The barrier to rotation that results can be represented on a graph of potential energy versus degree of rotation in which the angle between C−H bonds on front and back carbons as viewed end-on (the *dihedral angle*) goes full circle from 0 to 360°. Energy minima occur at staggered conformations, and energy maxima occur at eclipsed conformations, as shown in **Figure 3.7**.

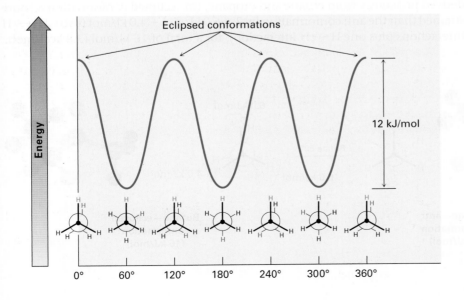

Figure 3.7 A graph of potential energy versus bond rotation in ethane. The staggered conformations are 12 kJ/mol lower in energy than the eclipsed conformations.

3.7 Conformations of Other Alkanes

Propane, the next higher member in the alkane series, also has a torsional barrier that results in hindered rotation around the carbon–carbon bonds. The barrier is slightly higher in propane than in ethane—a total of 14 kJ/mol (3.4 kcal/mol) versus 12 kJ/mol.

The eclipsed conformation of propane has three interactions—two ethane-type hydrogen–hydrogen interactions and one additional hydrogen–methyl interaction. Since each eclipsing H↔H interaction is the same as that in ethane and thus has an energy "cost" of 4.0 kJ/mol, we can assign a value of 14 − (2 × 4.0) = 6.0 kJ/mol (1.4 kcal/mol) to the eclipsing H↔CH_3 interaction (**Figure 3.8**).

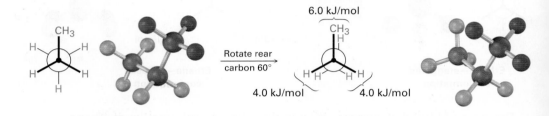

Figure 3.8 Newman projections of propane showing staggered and eclipsed conformations. The staggered conformer is lower in energy by 14 kJ/mol.

The conformational situation becomes more complex for larger alkanes because not all staggered conformations have the same energy and not all eclipsed conformations have the same energy. In butane, for instance, the lowest-energy arrangement, called the **anti conformation**, is the one in which the two methyl groups are as far apart as possible—180° away from each other. As rotation around the C2−C3 bond occurs, an eclipsed conformation is reached in which there are two CH_3↔H interactions and one H↔H interaction. Using the energy values derived previously from ethane and propane, this eclipsed conformation is more strained than the anti conformation by 2 × 6.0 kJ/mol + 4.0 kJ/mol (two CH_3↔H interactions plus one H↔H interaction), for a total of 16 kJ/mol (3.8 kcal/mol).

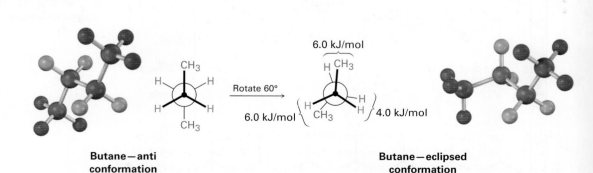

3.7 | Conformations of Other Alkanes

As bond rotation continues, an energy minimum is reached at the staggered conformation where the methyl groups are 60° apart. Called the **gauche conformation**, it lies 3.8 kJ/mol (0.9 kcal/mol) higher in energy than the anti conformation even though it has no eclipsing interactions. This energy difference occurs because the hydrogen atoms of the methyl groups are near one another in the gauche conformation, resulting in what is called *steric strain*. **Steric strain** is the repulsive interaction that occurs when atoms are forced closer together than their atomic radii allow. It's the result of trying to force two atoms to occupy the same space.

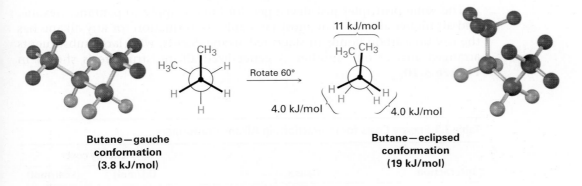

Butane—eclipsed conformation (16 kJ/mol) → Rotate 60° → **Butane—gauche conformation (3.8 kJ/mol)**

Steric strain 3.8 kJ/mol

As the dihedral angle between the methyl groups approaches 0°, an energy maximum is reached at a second eclipsed conformation. Because the methyl groups are forced even closer together than in the gauche conformation, both torsional strain and steric strain are present. A total strain energy of 19 kJ/mol (4.5 kcal/mol) has been estimated for this conformation, making it possible to calculate a value of 11 kJ/mol (2.6 kcal/mol) for the CH$_3$⟷CH$_3$ eclipsing interaction: total strain of 19 kJ/mol less the strain of two H⟷H eclipsing interactions (2 × 4.0 kcal/mol) equals 11 kJ/mol.

Butane—gauche conformation (3.8 kJ/mol) → Rotate 60° → **Butane—eclipsed conformation (19 kJ/mol)**

11 kJ/mol
4.0 kJ/mol 4.0 kJ/mol

After 0°, the rotation becomes a mirror image of what we've already seen: another gauche conformation is reached, another eclipsed conformation, and finally a return to the anti conformation. A plot of potential energy versus rotation about the C2–C3 bond is shown in **Figure 3.9**.

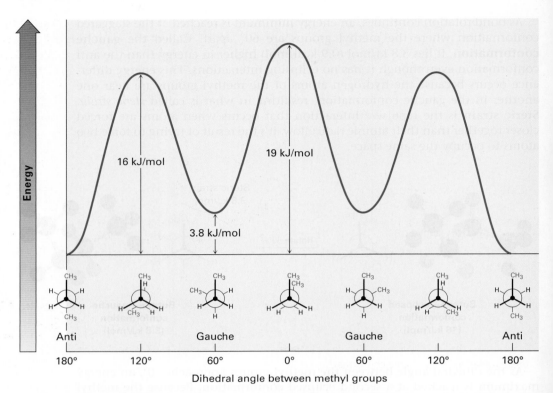

Figure 3.9 A plot of potential energy versus rotation for the C2−C3 bond in butane. The energy maximum occurs when the two methyl groups eclipse each other, and the energy minimum occurs when the two methyl groups are 180° apart (anti).

The notion of assigning definite energy values to specific interactions within a molecule is a very useful one that we'll return to in the next chapter. A summary of what we've seen thus far is given in Table 3.5.

The same principles just developed for butane apply to pentane, hexane, and all higher alkanes. The most favorable conformation for any alkane has the carbon–carbon bonds in staggered arrangements, with large substituents arranged anti to one another. A generalized alkane structure is shown in **Figure 3.10**.

Table 3.5 Energy Costs for Interactions in Alkane Conformers

Interaction	Cause	Energy cost (kJ/mol)	(kcal/mol)
H ⟷ H eclipsed	Torsional strain	4.0	1.0
H ⟷ CH$_3$ eclipsed	Mostly torsional strain	6.0	1.4
CH$_3$ ⟷ CH$_3$ eclipsed	Torsional and steric strain	11	2.6
CH$_3$ ⟷ CH$_3$ gauche	Steric strain	3.8	0.9

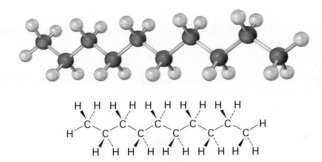

Figure 3.10 The most stable alkane conformation is the one in which all substituents are staggered and the carbon–carbon bonds are arranged anti, as shown in this model of decane.

One final point: saying that one particular conformer is "more stable" than another doesn't mean the molecule adopts and maintains only the more stable conformation. At room temperature, rotations around σ bonds occur so rapidly that all conformers are in equilibrium. At any given instant, however, a larger percentage of molecules will be found in a more stable conformation than in a less stable one.

Worked Example 3.4

Drawing Newman Projections

Sight along the C1—C2 bond of 1-chloropropane, and draw Newman projections of the most stable and least stable conformations.

Strategy
The most stable conformation of a substituted alkane is generally a staggered one in which large groups have an anti relationship. The least stable conformation is generally an eclipsed one in which large groups are as close as possible.

Solution

Most stable (staggered)   Least stable (eclipsed)

Problem 3.15
Make a graph of potential energy versus angle of bond rotation for propane, and assign values to the energy maxima.

Problem 3.16
Sight along the C2—C1 bond, 2-methylpropane (isobutane) and
(a) draw a Newman projection of the most stable conformation.
(b) draw a Newman projection of the least stable conformation.
(c) make a graph of energy versus angle of rotation around the C2—C1 bond.
(d) Since an H⟷H eclipsing interaction costs 4.0 kJ/mol and an H⟷CH₃ eclipsing interaction costs 6.0 kJ/mol, assign relative values to the maxima and minima in your graph.

Problem 3.17
Sight along the C2—C3 bond of 2,3-dimethylbutane, and draw a Newman projection of the most stable conformation.

Problem 3.18
Draw a Newman projection along the C2—C3 bond of the following conformation of 2,3-dimethylbutane, and calculate a total strain energy:

Gasoline | A DEEPER LOOK

Gasoline is a finite resource. It won't be around forever.

British Foreign Minister Ernest Bevin once said that "The Kingdom of Heaven runs on righteousness, but the Kingdom of Earth runs on alkanes." (Actually, he said "runs on oil" not "runs on alkanes," but they're essentially the same.) By far, the major sources of alkanes are the world's natural gas and petroleum deposits. Laid down eons ago, these deposits are thought to be derived primarily from the decomposition of tiny single-celled marine organisms called foraminifera. *Natural gas* consists chiefly of methane but also contains ethane, propane, and butane. *Petroleum* is a complex mixture of hydrocarbons that must be separated into fractions and then further refined before it can be used.

The petroleum era began in August 1859, when the world's first oil well was drilled by Edwin Drake near Titusville, Pennsylvania. The petroleum was distilled into fractions according to boiling point, but it was high-boiling kerosene, or lamp oil, rather than gasoline that was primarily sought. Literacy was becoming widespread at the time, and people wanted better light for reading than was available from candles. Gasoline was too volatile for use in lamps and was initially considered a waste by-product. The world has changed greatly since those early days, however, and it is now gasoline rather than lamp oil that is prized.

Petroleum refining begins by fractional distillation of crude oil into three principal cuts according to boiling point (bp): straight-run gasoline (bp 30–200 °C), kerosene (bp 175–300 °C), and heating oil, or diesel fuel (bp 275–400 °C). Further distillation under reduced pressure then yields lubricating oils and waxes and leaves a tarry residue of asphalt. The distillation of crude oil is only the first step in gasoline production, however. Straight-run gasoline turns out to be a poor fuel in automobiles because of engine knock, an uncontrolled combustion that can occur in a hot engine.

The *octane number* of a fuel is the measure by which its antiknock properties are judged. It was recognized long ago that straight-chain hydrocarbons are far more prone to induce

(continued)

engine knock than are highly branched compounds. Heptane, a particularly bad fuel, is assigned a base value of 0 octane number, and 2,2,4-trimethylpentane, commonly known as isooctane, has a rating of 100.

$$CH_3CH_2CH_2CH_2CH_2CH_2CH_3$$

Heptane
(octane number = 0)

$$CH_3\underset{\underset{CH_3}{|}}{\overset{\overset{CH_3}{|}}{C}}CH_2\overset{\overset{CH_3}{|}}{C}HCH_3$$

2,2,4-Trimethylpentane
(octane number = 100)

Because straight-run gasoline burns so poorly in engines, petroleum chemists have devised numerous methods for producing higher-quality fuels. One of these methods, *catalytic cracking*, involves taking the high-boiling kerosene cut (C_{11}–C_{14}) and "cracking" it into smaller branched molecules suitable for use in gasoline. Another process, called *reforming*, is used to convert C_6–C_8 alkanes to aromatic compounds such as benzene and toluene, which have substantially higher octane numbers than alkanes. The final product that goes in your tank has an approximate composition of 15% C_4–C_8 straight-chain alkanes, 25% to 40% C_4–C_{10} branched-chain alkanes, 10% cyclic alkanes, 10% straight-chain and cyclic alkenes, and 25% arenes (aromatics).

Summary

Even though alkanes are relatively unreactive and rarely involved in chemical reactions, they nevertheless provide a useful vehicle for introducing some important general ideas. In this chapter, we've used alkanes to introduce the basic approach to naming organic compounds and to take an initial look at some of the three-dimensional aspects of molecules.

A **functional group** is a group of atoms within a larger molecule that has a characteristic chemical reactivity. Because functional groups behave in approximately the same way in all molecules where they occur, the chemical reactions of an organic molecule are largely determined by its functional groups.

Alkanes are a class of **saturated hydrocarbons** with the general formula C_nH_{2n+2}. They contain no functional groups, are relatively inert, and can be either **straight-chain** (*normal*) or **branched**. Alkanes are named by a series of IUPAC rules of nomenclature. Compounds that have the same chemical formula but different structures are called **isomers**. More specifically, compounds such as butane and isobutane, which differ in their connections between atoms, are called **constitutional isomers**.

Carbon–carbon single bonds in alkanes are formed by σ overlap of carbon sp^3 hybrid orbitals. Rotation is possible around σ bonds because of their cylindrical symmetry, and alkanes therefore exist in a large number of rapidly interconverting **conformations**. **Newman projections** make it possible to visualize the spatial consequences of bond rotation by sighting directly along a carbon–carbon bond axis. Not all alkane conformations are equally stable. The **staggered** conformation of ethane is 12 kJ/mol (2.9 kcal/mol) more stable than the **eclipsed** conformation because of **torsional strain**. In general, any alkane is most stable when all its bonds are staggered.

Key words

aliphatic, 81
alkane, 80
alkyl group, 84
anti conformation, 96
branched-chain alkane, 82
conformation, 94
conformers, 94
constitutional isomers, 82
eclipsed conformation, 95
functional group, 74
gauche conformation, 97
hydrocarbon, 81
isomers, 82
Newman projection, 94
R group, 86
saturated, 81
staggered conformation, 95
stereochemistry, 94
steric strain, 97
straight-chain alkane, 82
substituent, 87
torsional strain, 95

Exercises

OWL Interactive versions of these problems are assignable in OWL for Organic Chemistry.

Visualizing Chemistry

(Problems 3.1–3.18 appear within the chapter.)

3.19 Identify the functional groups in the following substances, and convert each drawing into a molecular formula (red = O, blue = N).

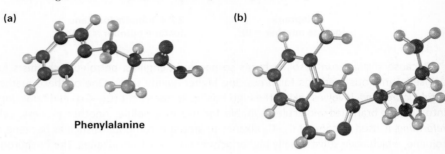

(a)

Phenylalanine

(b)

Lidocaine

3.20 Give IUPAC names for the following alkanes, and convert each drawing into a skeletal structure:

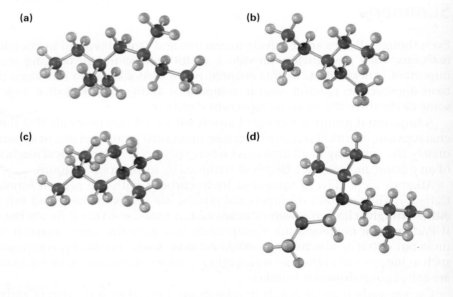

(a)

(b)

(c)

(d)

3.21 Draw a Newman projection along the C2–C3 bond of the following conformation of 2-butanol.

Additional Problems

Functional Groups

3.22 Locate and identify the functional groups in the following molecules.

(a) [structure: benzene ring with CH₂OH and NHCH₃ substituents]

(b) [structure: cyclohexenone]

(c) [structure: phenyl-N(H)-C(=O)-CH₃]

(d) CH₃CHCOH with =O and NH₂ substituent

(e) [bicyclic structure with ketone and isopropenyl group]

(f) [structure: (CH₃)₂CH-C≡C-C(=O)Cl]

3.23 Propose structures that meet the following descriptions:
  (a) A ketone with five carbons   (b) A four-carbon amide
  (c) A five-carbon ester   (d) An aromatic aldehyde
  (e) A keto ester   (f) An amino alcohol

3.24 Propose structures for the following:
  (a) A ketone, C_4H_8O (b) A nitrile, C_5H_9N
  (c) A dialdehyde, $C_4H_6O_2$ (d) A bromoalkene, $C_6H_{11}Br$
  (e) An alkane, C_6H_{14} (f) A *cyclic* saturated hydrocarbon, C_6H_{12}
  (g) A diene (dialkene), C_5H_8 (h) A keto alkene, C_5H_8O

3.25 Predict the hybridization of the carbon atom in each of the following functional groups:
  (a) Ketone   (b) Nitrile   (c) Carboxylic acid

3.26 Draw the structures of the following molecules:
  (a) *Biacetyl*, $C_4H_6O_2$, a substance with the aroma of butter; it contains no rings or carbon–carbon multiple bonds.
  (b) *Ethylenimine*, C_2H_5N, a substance used in the synthesis of melamine polymers; it contains no multiple bonds.
  (c) *Glycerol*, $C_3H_8O_3$, a substance isolated from fat and used in cosmetics; it has an $-OH$ group on each carbon.

Isomers

3.27 Draw structures that meet the following descriptions (there are many possibilities):
(a) Three isomers with the formula C_8H_{18}
(b) Two isomers with the formula $C_4H_8O_2$

3.28 Draw structures of the nine isomers of C_7H_{16}.

3.29 In each of the following sets, which structures represent the same compound and which represent different compounds?

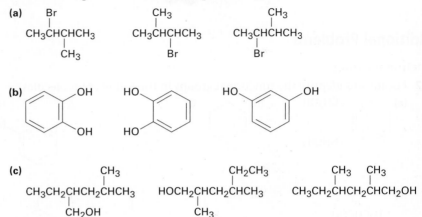

3.30 There are seven constitutional isomers with the formula $C_4H_{10}O$. Draw as many as you can.

3.31 Draw as many compounds as you can that fit the following descriptions:
(a) Alcohols with formula $C_4H_{10}O$
(b) Amines with formula $C_5H_{13}N$
(c) Ketones with formula $C_5H_{10}O$
(d) Aldehydes with formula $C_5H_{10}O$
(e) Esters with formula $C_4H_8O_2$
(f) Ethers with formula $C_4H_{10}O$

3.32 Draw compounds that contain the following:
(a) A primary alcohol
(b) A tertiary nitrile
(c) A secondary thiol
(d) Both primary and secondary alcohols
(e) An isopropyl group
(f) A quaternary carbon

Naming Compounds

3.33 Draw and name all monobromo derivatives of pentane, $C_5H_{11}Br$.

3.34 Draw and name all monochloro derivatives of 2,5-dimethylhexane, $C_8H_{17}Cl$.

3.35 Draw structures for the following:
(a) 2-Methylheptane
(b) 4-Ethyl-2,2-dimethylhexane
(c) 4-Ethyl-3,4-dimethyloctane
(d) 2,4,4-Trimethylheptane
(e) 3,3-Diethyl-2,5-dimethylnonane
(f) 4-Isopropyl-3-methylheptane

3.36 Draw a compound that:
(a) Has only primary and tertiary carbons
(b) Has no secondary or tertiary carbons
(c) Has four secondary carbons

3.37 Draw a compound that:
  (a) Has nine primary hydrogens
  (b) Has only primary hydrogens

3.38 Give IUPAC names for the following compounds:

(a)
$$\text{CH}_3\text{CHCH}_2\text{CH}_2\text{CH}_3$$
with CH₃ on the second carbon

(b)
$$\text{CH}_3\text{CH}_2\text{CCH}_3$$
with CH₃ groups on the third carbon

(c)
$$\text{CH}_3\text{CHCCH}_2\text{CH}_2\text{CH}_3$$
with H₃C, CH₃ on adjacent carbons and CH₃ below

(d)
$$\text{CH}_3\text{CH}_2\text{CHCH}_2\text{CH}_2\text{CHCH}_3$$
with CH₂CH₃ and CH₃ substituents

(e)
$$\text{CH}_3\text{CH}_2\text{CH}_2\text{CHCH}_2\text{CCH}_3$$
with CH₃, CH₂CH₃, and CH₃ substituents

(f)
$$\text{CH}_3\text{C}-\text{CCH}_2\text{CH}_2\text{CH}_3$$
with H₃C, CH₃ on the left carbon and H₃C, CH₃ on the right

3.39 Name the five isomers of C_6H_{14}.

3.40 Explain why each of the following names is incorrect:
  (a) 2,2-Dimethyl-6-ethylheptane
  (b) 4-Ethyl-5,5-dimethylpentane
  (c) 3-Ethyl-4,4-dimethylhexane
  (d) 5,5,6-Trimethyloctane
  (e) 2-Isopropyl-4-methylheptane

3.41 Propose structures and give IUPAC names for the following:
  (a) A diethyldimethylhexane
  (b) A (3-methylbutyl)-substituted alkane

Conformations

3.42 Consider 2-methylbutane (isopentane). Sighting along the C2–C3 bond:
  (a) Draw a Newman projection of the most stable conformation.
  (b) Draw a Newman projection of the least stable conformation.
  (c) If a $CH_3 \longleftrightarrow CH_3$ eclipsing interaction costs 11 kJ/mol (2.5 kcal/mol) and a $CH_3 \longleftrightarrow CH_3$ gauche interaction costs 3.8 kJ/mol (0.9 kcal/mol), make a quantitative plot of energy versus rotation about the C2–C3 bond.

3.43 What are the relative energies of the three possible staggered conformations around the C2–C3 bond in 2,3-dimethylbutane? (See Problem 3.42.)

3.44 Construct a qualitative potential-energy diagram for rotation about the C–C bond of 1,2-dibromoethane. Which conformation would you expect to be most stable? Label the anti and gauche conformations of 1,2-dibromoethane.

3.45 Which conformation of 1,2-dibromoethane (Problem 3.44) would you expect to have the largest dipole moment? The observed dipole moment of 1,2-dibromoethane is $\mu = 1.0$ D. What does this tell you about the actual conformation of the molecule?

3.46 Draw the most stable conformation of pentane, using wedges and dashes to represent bonds coming out of the paper and going behind the paper, respectively.

3.47 Draw the most stable conformation of 1,4-dichlorobutane, using wedges and dashes to represent bonds coming out of the paper and going behind the paper, respectively.

General Problems

3.48 For each of the following compounds, draw an isomer that has the same functional groups.

(a) CH₃CHCH₂CH₂Br with CH₃ substituent on first carbon

(b) 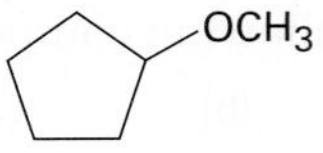 cyclopentyl-OCH₃

(c) $CH_3CH_2CH_2C\equiv N$

(d) 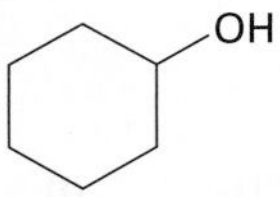 cyclohexyl-OH

(e) CH_3CH_2CHO

(f) 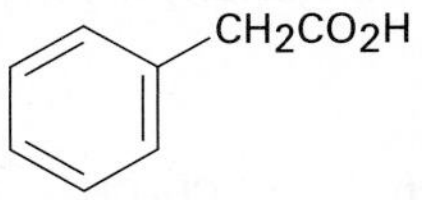 phenyl-CH_2CO_2H

3.49 Malic acid, $C_4H_6O_5$, has been isolated from apples. Because this compound reacts with 2 molar equivalents of base, it is a dicarboxylic acid.
(a) Draw at least five possible structures.
(b) If malic acid is a secondary alcohol, what is its structure?

3.50 Formaldehyde, $H_2C=O$, is known to all biologists because of its usefulness as a tissue preservative. When pure, formaldehyde *trimerizes* to give trioxane, $C_3H_6O_3$, which, surprisingly enough, has no carbonyl groups. Only one monobromo derivative ($C_3H_5BrO_3$) of trioxane is possible. Propose a structure for trioxane.

3.51 The barrier to rotation about the C—C bond in bromoethane is 15 kJ/mol (3.6 kcal/mol).
(a) What energy value can you assign to an H↔Br eclipsing interaction?
(b) Construct a quantitative diagram of potential energy versus bond rotation for bromoethane.

3.52 Increased substitution around a bond leads to increased strain. Take the four substituted butanes listed below, for example. For each compound, sight along the C2–C3 bond and draw Newman projections of the most stable and least stable conformations. Use the data in Table 3.5 to assign strain energy values to each conformation. Which of the eight conformations is most strained? Which is least strained?
(a) 2-Methylbutane   (b) 2,2-Dimethylbutane
(c) 2,3-Dimethylbutane   (d) 2,2,3-Trimethylbutane

3.53 The cholesterol-lowering agents called *statins*, such as simvastatin (Zocor) and pravastatin (Pravachol), are among the most widely prescribed drugs in the world, with annual sales estimated at approximately $15 billion. Identify the functional groups in both, and tell how the two substances differ.

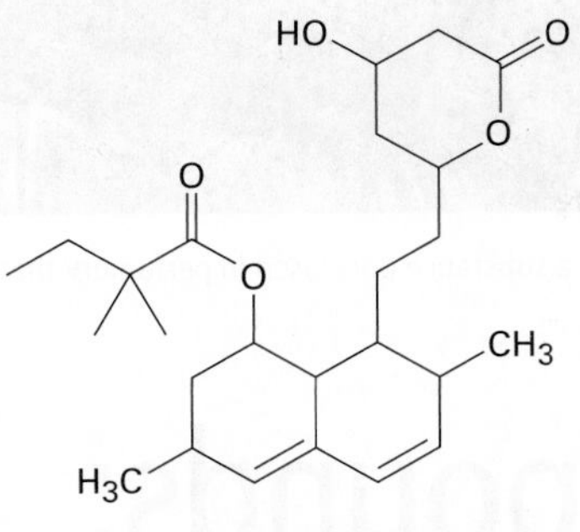

Simvastatin
(Zocor)

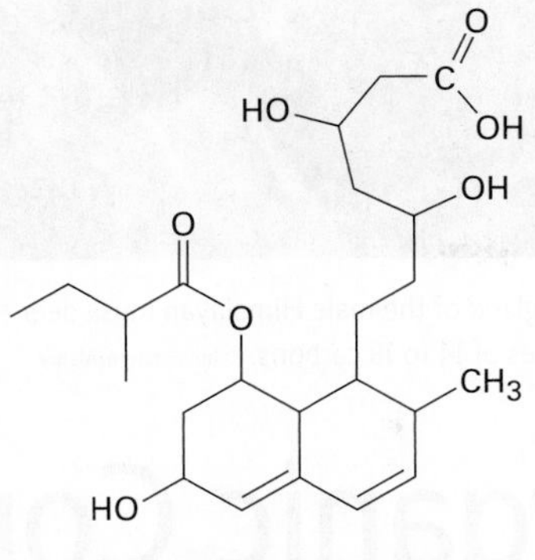

Pravastatin
(Pravachol)

3.54 We'll look in the next chapter at *cycloalkanes*—saturated cyclic hydrocarbons—and we'll see that the molecules generally adopt puckered, nonplanar conformations. Cyclohexane, for instance, has a puckered shape like a lounge chair rather than a flat shape. Why?

Nonplanar cyclohexane 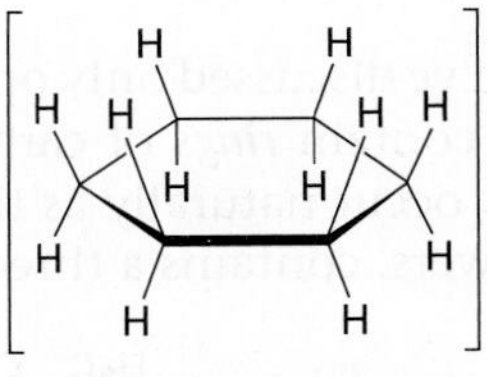   Planar cyclohexane

3.55 We'll see in the next chapter that there are two isomeric substances both named 1,2-dimethylcyclohexane. Explain.

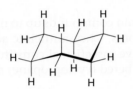

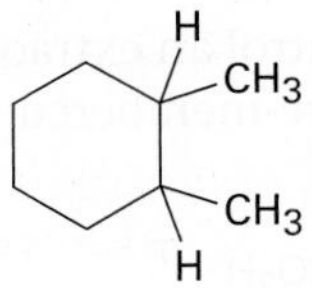

1,2-Dimethylcyclohexane

4

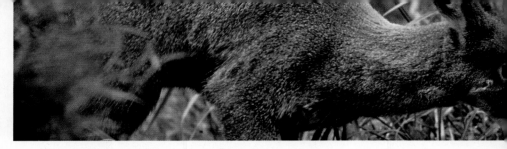

The musk gland of the male Himalayan musk deer secretes a substance once used in perfumery that contains cycloalkanes of 14 to 18 carbons. © Indiapicture/Alamy

Organic Compounds: Cycloalkanes and Their Stereochemistry

4.1 Naming Cycloalkanes
4.2 Cis–Trans Isomerism in Cycloalkanes
4.3 Stability of Cycloalkanes: Ring Strain
4.4 Conformations of Cycloalkanes
4.5 Conformations of Cyclohexane
4.6 Axial and Equatorial Bonds in Cyclohexane
4.7 Conformations of Monosubstituted Cyclohexanes
4.8 Conformations of Disubstituted Cyclohexanes
4.9 Conformations of Polycyclic Molecules
A Deeper Look— Molecular Mechanics

OWL Sign in to OWL for Organic Chemistry at **www.cengage.com/owl** to view tutorials and simulations, develop problem-solving skills, and complete online homework assigned by your professor.

Although we've discussed only open-chain compounds up to now, most organic compounds contain *rings* of carbon atoms. Chrysanthemic acid, for instance, whose esters occur naturally as the active insecticidal constituents of chrysanthemum flowers, contains a three-membered (cyclopropane) ring.

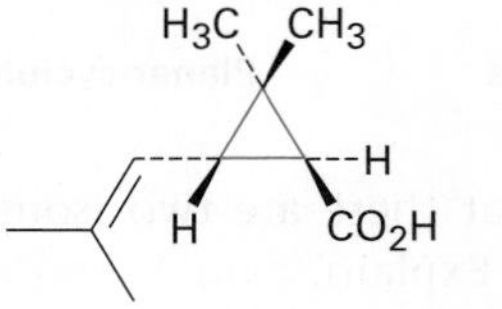

Chrysanthemic acid

Prostaglandins, potent hormones that control an extraordinary variety of physiological functions in humans, contain a five-membered (cyclopentane) ring.

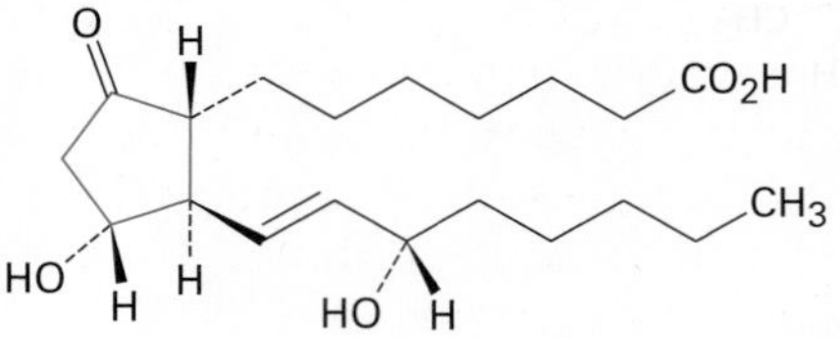

Prostaglandin E$_1$

Steroids, such as cortisone, contain four rings joined together—3 six-membered (cyclohexane) and 1 five-membered. We'll discuss steroids and their properties in more detail in **Sections 27.6 and 27.7**.

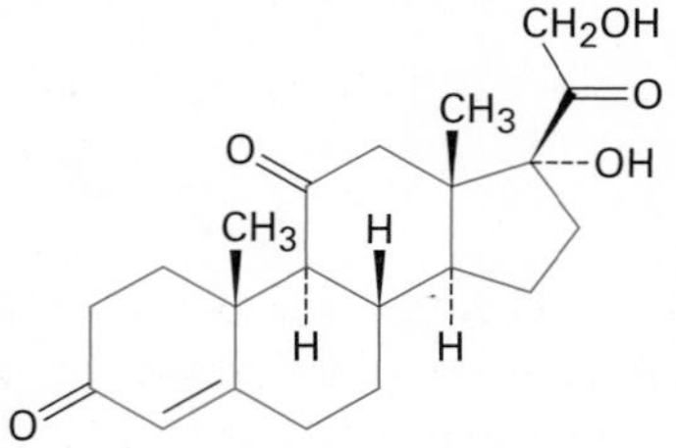

Cortisone

108

Why This Chapter?
We'll see numerous instances in future chapters where the chemistry of a given functional group is affected by being in a ring rather than an open chain. Because cyclic molecules are so commonly encountered in most pharmaceuticals and in all classes of biomolecules, including proteins, lipids, carbohydrates, and nucleic acids, it's important to understand the consequences of cyclic structures.

4.1 Naming Cycloalkanes

Saturated cyclic hydrocarbons are called **cycloalkanes**, or **alicyclic** compounds (**ali**phatic **cyclic**). Because cycloalkanes consist of rings of $-CH_2-$ units, they have the general formula $(CH_2)_n$, or C_nH_{2n}, and can be represented by polygons in skeletal drawings.

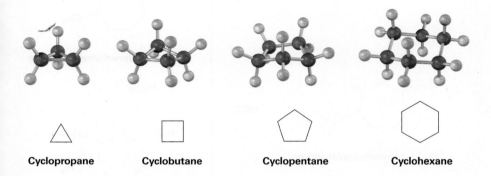

Cyclopropane   **Cyclobutane**   **Cyclopentane**   **Cyclohexane**

Substituted cycloalkanes are named by rules similar to those we saw in the previous chapter for open-chain alkanes **(Section 3.4)**. For most compounds, there are only two steps.

STEP 1
Find the parent.
Count the number of carbon atoms in the ring and the number in the largest substituent. If the number of carbon atoms in the ring is equal to or greater than the number in the substituent, the compound is named as an alkyl-substituted cycloalkane. If the number of carbon atoms in the largest substituent is greater than the number in the ring, the compound is named as a cycloalkyl-substituted alkane. For example:

Methylcyclopentane   **1-Cyclopropyl**butane

3 carbons   4 carbons

STEP 2
Number the substituents, and write the name.
For an alkyl- or halo-substituted cycloalkane, choose a point of attachment as carbon 1 and number the substituents on the ring so that the *second*

substituent has as low a number as possible. If ambiguity still exists, number so that the third or fourth substituent has as low a number as possible, until a point of difference is found.

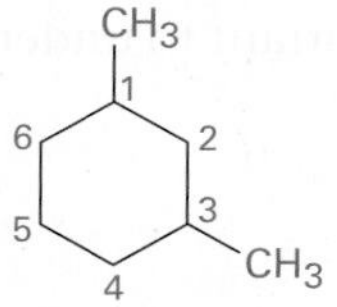

1,3-Dimethylcyclohexane ↑ Lower

NOT

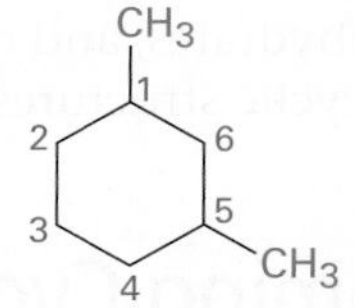

1,5-Dimethylcyclohexane ↑ Higher

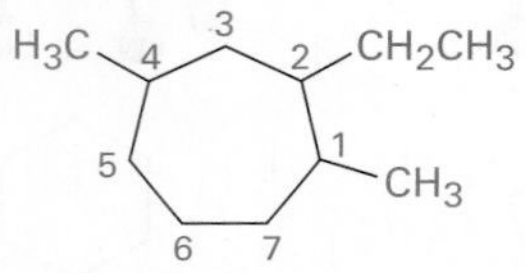

2-Ethyl-1,4-dimethylcycloheptane
↑ ↑
Lower Lower

NOT

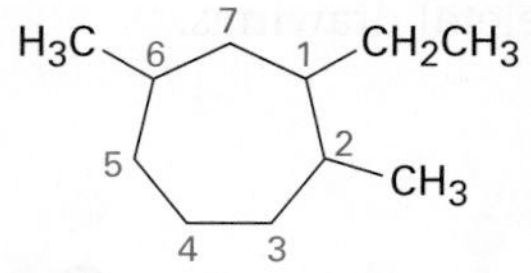

{ **1-Ethyl-2,6-dimethylcycloheptane** ↑ Higher

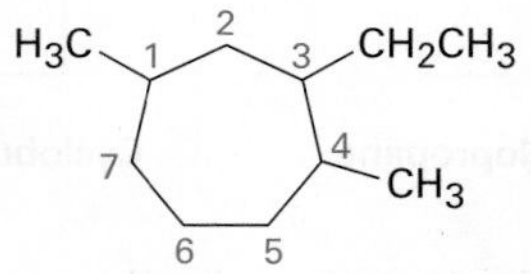

3-Ethyl-1,4-dimethylcycloheptane ↑ Higher }

(a) When two or more different alkyl groups that could potentially receive the same numbers are present, number them by alphabetical priority, ignoring numerical prefixes such as di- and tri-.

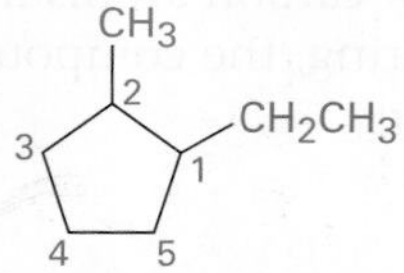

1-Ethyl-2-methylcyclopentane   NOT   **2-Ethyl-1-methylcyclopentane**

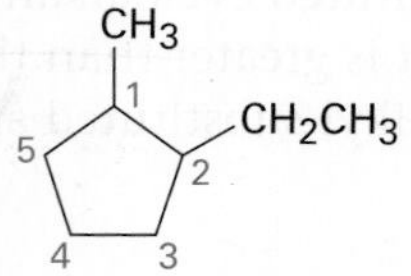

(b) If halogens are present, treat them just like alkyl groups.

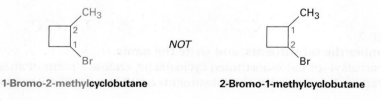

1-Bromo-2-methylcyclobutane   NOT   **2-Bromo-1-methylcyclobutane**

4.2 | Cis–Trans Isomerism in Cycloalkanes

Some additional examples follow:

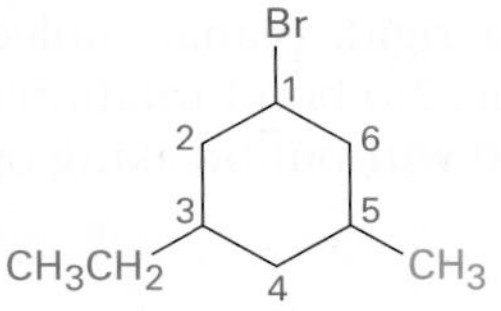

1-Bromo-3-ethyl-5-methyl-cyclohexane

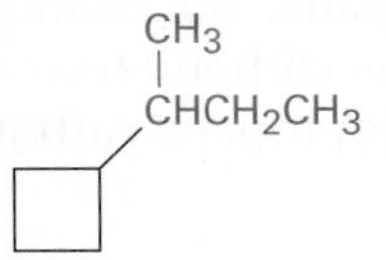

(1-Methylpropyl)cyclobutane or *sec*-butylcyclobutane

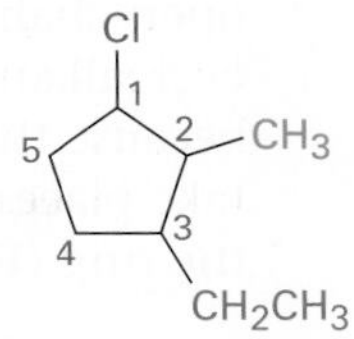

1-Chloro-3-ethyl-2-methyl-cyclopentane

Problem 4.1
Give IUPAC names for the following cycloalkanes:

(a) (b) (c)

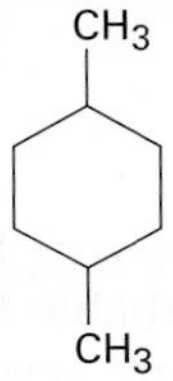

(d) 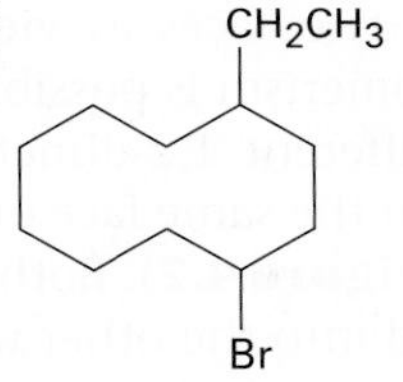 (e) 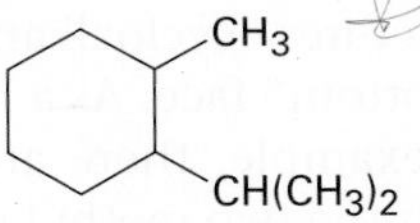 (f)

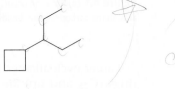

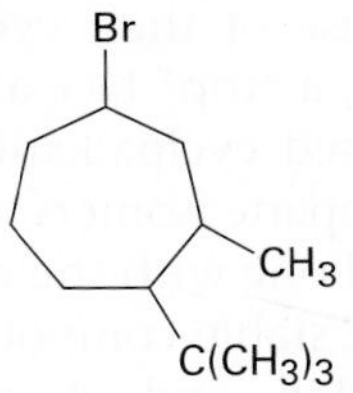

Problem 4.2
Draw structures corresponding to the following IUPAC names:
(a) 1,1-Dimethylcyclooctane (b) 3-Cyclobutylhexane
(c) 1,2-Dichlorocyclopentane (d) 1,3-Dibromo-5-methylcyclohexane

Problem 4.3
Name the following cycloalkane:

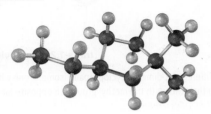

4.2 Cis–Trans Isomerism in Cycloalkanes

In many respects, the chemistry of cycloalkanes is like that of open-chain alkanes: both are nonpolar and fairly inert. There are, however, some important differences. One difference is that cycloalkanes are less flexible than open-chain alkanes. In contrast with the relatively free rotation around single bonds in open-chain alkanes **(Sections 3.6 and 3.7)**, there is much less freedom in cycloalkanes. Cyclopropane, for example, must be a rigid, planar molecule because three points (the carbon atoms) define a plane. No bond rotation can take place around a cyclopropane carbon–carbon bond without breaking open the ring **(Figure 4.1)**.

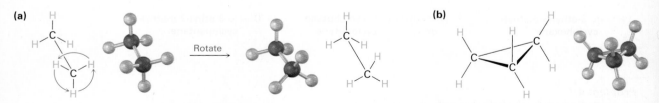

Figure 4.1 **(a)** Rotation occurs around the carbon–carbon bond in ethane, but **(b)** no rotation is possible around the carbon–carbon bonds in cyclopropane without breaking open the ring.

Larger cycloalkanes have increasing rotational freedom, and the very large rings (C_{25} and up) are so floppy that they are nearly indistinguishable from open-chain alkanes. The common ring sizes (C_3–C_7), however, are severely restricted in their molecular motions.

Because of their cyclic structures, cycloalkanes have two faces as viewed edge-on, a "top" face and a "bottom" face. As a result, isomerism is possible in substituted cycloalkanes. For example, there are two different 1,2-dimethyl-cyclopropane isomers, one with the two methyl groups on the same face of the ring and one with the methyl groups on opposite faces **(Figure 4.2)**. Both isomers are stable compounds, and neither can be converted into the other without breaking and reforming chemical bonds.

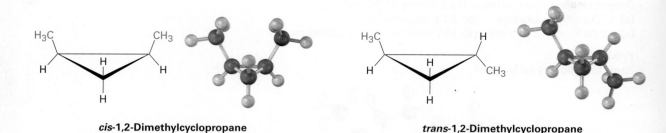

cis-1,2-Dimethylcyclopropane

trans-1,2-Dimethylcyclopropane

Figure 4.2 There are two different 1,2-dimethylcyclopropane isomers, one with the methyl groups on the same face of the ring (cis) and the other with the methyl groups on opposite faces of the ring (trans). The two isomers do not interconvert.

Unlike the constitutional isomers butane and isobutane, which have their atoms connected in a different order **(Section 3.2)**, the two 1,2-dimethyl-cyclopropanes have the same order of connections but differ in the spatial orientation of the atoms. Such compounds, which have their atoms connected in the same order but differ in three-dimensional orientation, are called stereochemical isomers, or **stereoisomers**. More generally, the term **stereochemistry** is used to refer to the three-dimensional aspects of chemical structure and reactivity.

Constitutional isomers (different connections between atoms)

$CH_3-CH(CH_3)-CH_3$ and $CH_3-CH_2-CH_2-CH_3$

Stereoisomers (same connections but different three-dimensional geometry)

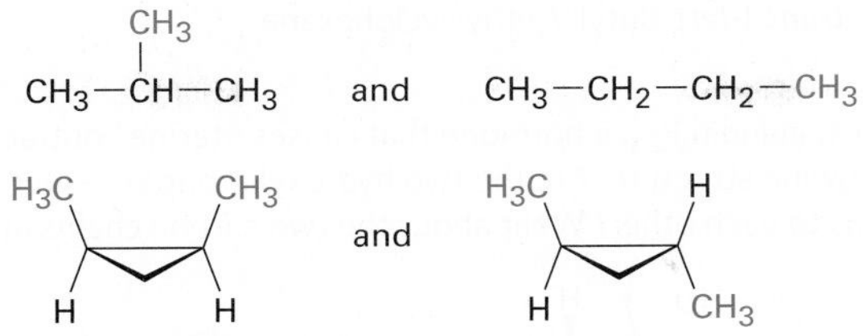

The 1,2-dimethylcyclopropanes are members of a subclass of stereoisomers called **cis–trans isomers**. The prefixes *cis-* (Latin "on the same side") and *trans-* (Latin "across") are used to distinguish between them. Cis–trans isomerism is a common occurrence in substituted cycloalkanes and in many cyclic biological molecules.

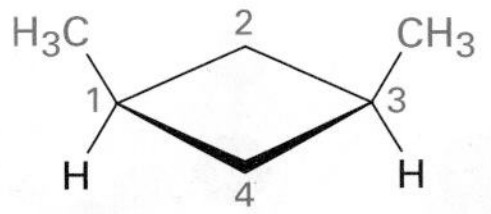

cis-1,3-Dimethylcyclobutane 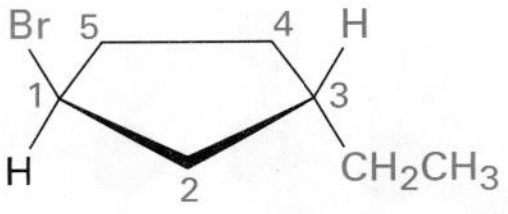   *trans*-1-Bromo-3-ethylcyclopentane

Worked Example 4.1

Naming Cycloalkanes

Name the following substances, including the *cis-* or *trans-* prefix:

(a) [structure] (b) [structure]

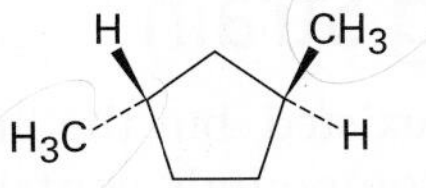

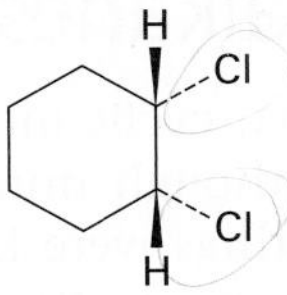

Strategy

In these views, the ring is roughly in the plane of the page, a wedged bond protrudes out of the page, and a dashed bond recedes into the page. Two substituents are cis if they are both out of or both into the page, and they are trans if one is out of and one is into the page.

Solution

(a) *trans*-1,3-Dimethylcyclopentane   (b) *cis*-1,2-Dichlorocyclohexane

---

Problem 4.4

Name the following substances, including the *cis-* or *trans-* prefix:

(a) [structure]   (b) [structure]

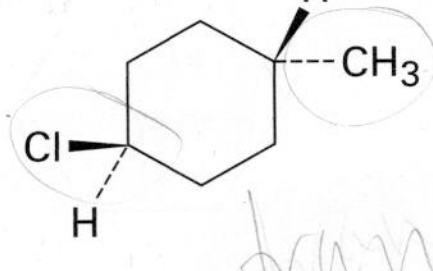

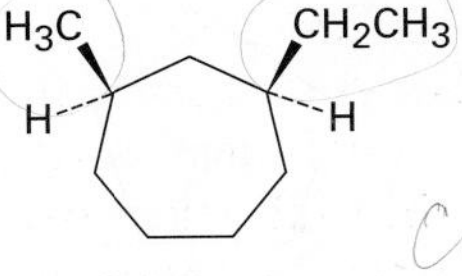

Problem 4.5
Draw the structures of the following molecules:
(a) *trans*-1-Bromo-3-methylcyclohexane (b) *cis*-1,2-Dimethylcyclobutane
(c) *trans*-1-*tert*-Butyl-2-ethylcyclohexane

Problem 4.6
Prostaglandin F$_{2\alpha}$, a hormone that causes uterine contraction during childbirth, has the following structure. Are the two hydroxyl groups (—OH) on the cyclopentane ring cis or trans to each other? What about the two carbon chains attached to the ring?

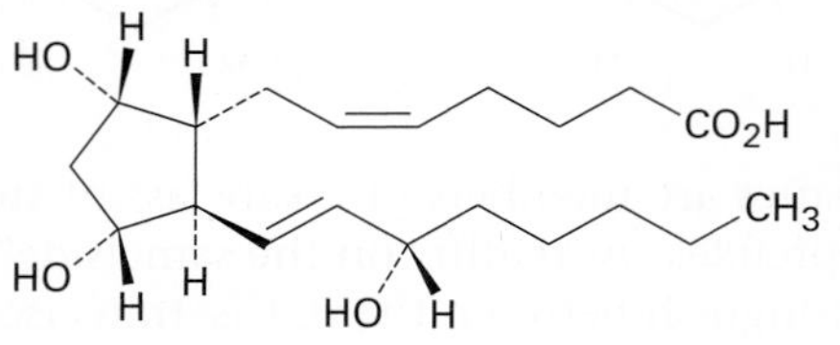

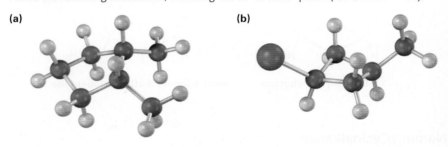

Prostaglandin F$_{2\alpha}$

Problem 4.7
Name the following substances, including the *cis*- or *trans*- prefix (red-brown = Br):

(a)   (b)

4.3 Stability of Cycloalkanes: Ring Strain

Chemists in the late 1800s knew that cyclic molecules existed, but the limitations on ring size were unclear. Although numerous compounds containing five-membered and six-membered rings were known, smaller and larger ring sizes had not been prepared, despite many efforts.

A theoretical interpretation of this observation was proposed in 1885 by Adolf von Baeyer, who suggested that small and large rings might be unstable due to **angle strain**—the strain induced in a molecule when bond angles are forced to deviate from the ideal 109° tetrahedral value. Baeyer based his suggestion on the simple geometric notion that a three-membered ring (cyclopropane) should be an equilateral triangle with bond angles of 60° rather than 109°, a four-membered ring (cyclobutane) should be a square with bond angles of 90°, a five-membered ring should be a regular pentagon with bond angles of 108°, and so on. Continuing this argument, large rings should be strained by having bond angles that are much greater than 109°.

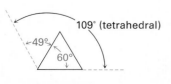

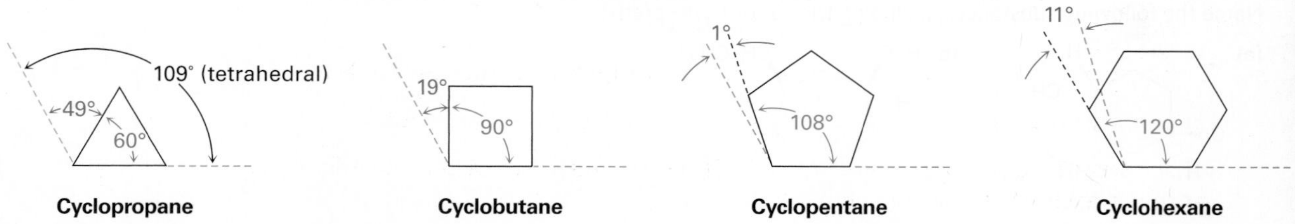

Cyclopropane

Cyclobutane

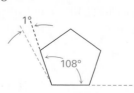

Cyclopentane

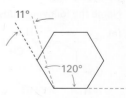

Cyclohexane

What are the facts? To measure the amount of strain in a compound, we have to measure the total energy of the compound and then subtract the energy of a strain-free reference compound. The difference between the two values should represent the amount of extra energy in the molecule due to strain. The simplest experimental way to do this for a cycloalkane is to measure its *heat of combustion,* the amount of heat released when the compound burns completely with oxygen. The more energy (strain) the compound contains, the more energy (heat) is released on combustion.

$$(CH_2)_n + 3n/2\ O_2 \longrightarrow n\ CO_2 + n\ H_2O + \text{Heat}$$

Because the heat of combustion of a cycloalkane depends on size, we need to look at heats of combustion per CH_2 unit. Subtracting a reference value derived from a strain-free acyclic alkane and then multiplying by the number of CH_2 units in the ring gives the overall strain energy. **Figure 4.3** shows the results.

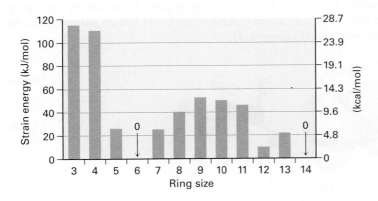

Figure 4.3 Cycloalkane strain energies, calculated by taking the difference between cycloalkane heat of combustion per CH_2 and acyclic alkane heat of combustion per CH_2, and multiplying by the number of CH_2 units in a ring. Small and medium rings are strained, but cyclohexane rings and very large rings are strain-free.

The data in Figure 4.3 show that Baeyer's theory is only partially correct. Cyclopropane and cyclobutane are indeed strained, just as predicted, but cyclopentane is more strained than predicted, and cyclohexane is strain-free. Cycloalkanes of intermediate size have only modest strain, and rings of 14 carbons or more are strain-free. Why is Baeyer's theory wrong?

Baeyer's theory is wrong for the simple reason that he assumed all cycloalkanes to be flat. In fact, as we'll see in the next section, most cycloalkanes are *not* flat; they adopt puckered three-dimensional conformations that allow bond angles to be nearly tetrahedral. As a result, angle strain occurs only in three- and four-membered rings, which have little flexibility. For most ring sizes, particularly the medium-ring (C_7–C_{11}) cycloalkanes, torsional strain caused by H↔H eclipsing interactions on adjacent carbons **(Section 3.6)** and steric strain caused by the repulsion between nonbonded atoms that approach too closely **(Section 3.7)** are the most important factors. Thus, three kinds of strain contribute to the overall energy of a cycloalkane.

- Angle strain—the strain due to expansion or compression of bond angles
- Torsional strain—the strain due to eclipsing of bonds on neighboring atoms
- Steric strain—the strain due to repulsive interactions when atoms approach each other too closely

Problem 4.8
Each H⟷H eclipsing interaction in ethane costs about 4.0 kJ/mol. How many such interactions are present in cyclopropane? What fraction of the overall 115 kJ/mol (27.5 kcal/mol) strain energy of cyclopropane is due to torsional strain?

Problem 4.9
cis-1,2-Dimethylcyclopropane has more strain than *trans*-1,2-dimethylcyclopropane. How can you account for this difference? Which of the two compounds is more stable?

4.4 Conformations of Cycloalkanes

Cyclopropane

Cyclopropane is the most strained of all rings, primarily because of the angle strain caused by its 60° C−C−C bond angles. In addition, cyclopropane has considerable torsional strain because the C−H bonds on neighboring carbon atoms are eclipsed **(Figure 4.4)**.

Figure 4.4 The structure of cyclopropane, showing the eclipsing of neighboring C−H bonds that gives rise to torsional strain. Part **(b)** is a Newman projection along a C−C bond.

How can the hybrid-orbital model of bonding account for the large distortion of bond angles from the normal 109° tetrahedral value to 60° in cyclopropane? The answer is that cyclopropane has *bent bonds*. In an unstrained alkane, maximum bonding is achieved when two atoms have their overlapping orbitals pointing directly toward each other. In cyclopropane, though, the orbitals can't point directly toward each other; rather, they overlap at a slight angle. The result is that cyclopropane bonds are weaker and more reactive than typical alkane bonds— 255 kJ/mol (61 kcal/mol) for a C−C bond in cyclopropane versus 370 kJ/mol (88 kcal/mol) for a C−C bond in open-chain propane.

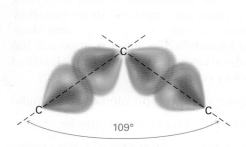
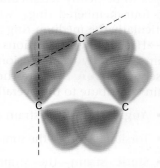

Typical alkane C–C bonds   **Typical bent cyclopropane C–C bonds**

Cyclobutane

Cyclobutane has less angle strain than cyclopropane but has more torsional strain because of its larger number of ring hydrogens. As a result, the total strain for the two compounds is nearly the same—110 kJ/mol (26.4 kcal/mol) for cyclobutane versus 115 kJ/mol (27.5 kcal/mol) for cyclopropane. Cyclobutane is not quite flat but is slightly bent so that one carbon atom lies about 25° above the plane of the other three **(Figure 4.5)**. The effect of this slight bend is to *increase* angle strain but to *decrease* torsional strain, until a minimum-energy balance between the two opposing effects is achieved.

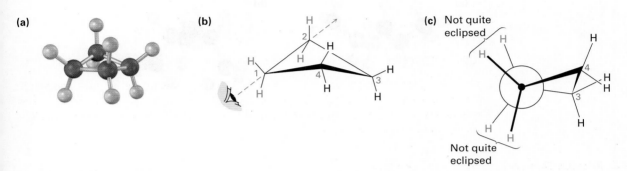

Figure 4.5 The conformation of cyclobutane. Part **(c)** is a Newman projection along a C—C bond, showing that neighboring C—H bonds are not quite eclipsed.

Cyclopentane

Cyclopentane was predicted by Baeyer to be nearly strain-free, but it actually has a total strain energy of 26 kJ/mol (6.2 kcal/mol). Although planar cyclopentane has practically no angle strain, it has a large amount of torsional strain. Cyclopentane therefore twists to adopt a puckered, nonplanar conformation that strikes a balance between increased angle strain and decreased torsional strain. Four of the cyclopentane carbon atoms are in approximately the same plane, with the fifth carbon atom bent out of the plane. Most of the hydrogens are nearly staggered with respect to their neighbors **(Figure 4.6)**.

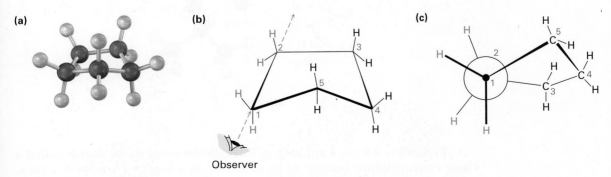

Figure 4.6 The conformation of cyclopentane. Carbons 1, 2, 3, and 4 are nearly planar, but carbon 5 is out of the plane. Part **(c)** is a Newman projection along the C1–C2 bond, showing that neighboring C—H bonds are nearly staggered.

Problem 4.10
How many H⟷H eclipsing interactions would be present if cyclopentane were planar? Assuming an energy cost of 4.0 kJ/mol for each eclipsing interaction, how much torsional strain would planar cyclopentane have? Since the measured total strain of cyclopentane is 26 kJ/mol, how much of the torsional strain is relieved by puckering?

Problem 4.11
Two conformations of *cis*-1,3-dimethylcyclobutane are shown. What is the difference between them, and which do you think is likely to be more stable?

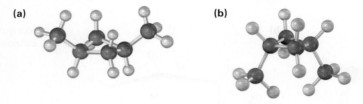

4.5 Conformations of Cyclohexane

Substituted cyclohexanes are the most common cycloalkanes and occur widely in nature. A large number of compounds, including steroids and many pharmaceutical agents, have cyclohexane rings. The flavoring agent menthol, for instance, has three substituents on a six-membered ring.

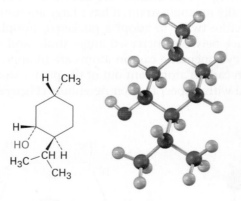

Menthol

Cyclohexane adopts a strain-free, three-dimensional shape that is called a **chair conformation** because of its similarity to a lounge chair, with a back, seat, and footrest **(Figure 4.7)**. Chair cyclohexane has neither angle strain nor torsional strain—all C−C−C bond angles are near the 109.5° tetrahedral value, and all neighboring C−H bonds are staggered.

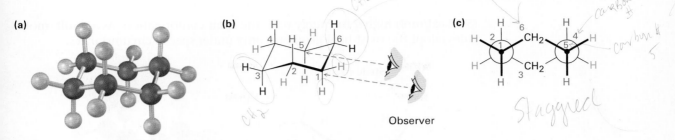

Figure 4.7 The strain-free chair conformation of cyclohexane. All C—C—C bond angles are 111.5°, close to the ideal 109.5° tetrahedral angle, and all neighboring C—H bonds are staggered.

The easiest way to visualize chair cyclohexane is to build a molecular model. (In fact, do it now if you have access to a model kit.) Two-dimensional drawings like that in Figure 4.7 are useful, but there's no substitute for holding, twisting, and turning a three-dimensional model in your own hands.

The chair conformation of cyclohexane can be drawn in three steps.

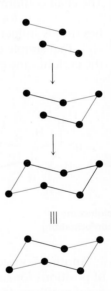

STEP 1
Draw two parallel lines, slanted downward and slightly offset from each other. This means that four of the cyclohexane carbons lie in a plane.

STEP 2
Place the topmost carbon atom above and to the right of the plane of the other four, and connect the bonds.

STEP 3
Place the bottommost carbon atom below and to the left of the plane of the middle four, and connect the bonds. Note that the bonds to the bottommost carbon atom are parallel to the bonds to the topmost carbon.

When viewing cyclohexane, it's helpful to remember that the lower bond is in front and the upper bond is in back. If this convention is not defined, an optical illusion can make it appear that the reverse is true. For clarity, all cyclohexane rings drawn in this book will have the front (lower) bond heavily shaded to indicate nearness to the viewer.

In addition to the chair conformation of cyclohexane, an alternative called the **twist-boat conformation** is also nearly free of angle strain. It does, however, have both steric strain and torsional strain and is about 23 kJ/mol

(5.5 kcal/mol) higher in energy than the chair conformation. As a result, molecules adopt the twist-boat geometry only under special circumstances.

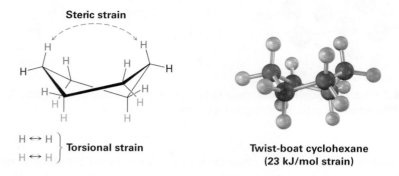

4.6 Axial and Equatorial Bonds in Cyclohexane

The chair conformation of cyclohexane leads to many consequences. We'll see in **Section 11.9**, for instance, that the chemical behavior of many substituted cyclohexanes is influenced by their conformation. In addition, we'll see in **Section 25.5** that simple carbohydrates, such as glucose, adopt a conformation based on the cyclohexane chair and that their chemistry is directly affected as a result.

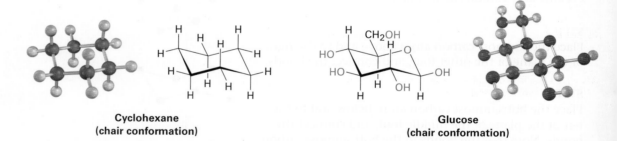

Another consequence of the chair conformation is that there are two kinds of positions for substituents on the cyclohexane ring: *axial* positions and *equatorial* positions **(Figure 4.8)**. The six **axial** positions are perpendicular to the ring, parallel to the ring axis, and the six **equatorial** positions are in the rough plane of the ring, around the ring equator.

Figure 4.8 Axial and **equatorial** positions in chair cyclohexane. The six axial hydrogens are parallel to the ring axis, and the six equatorial hydrogens are in a band around the ring equator.

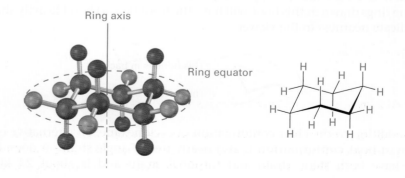

As shown in Figure 4.8, each carbon atom in chair cyclohexane has one axial and one equatorial hydrogen. Furthermore, each face of the ring has three axial and three equatorial hydrogens in an alternating arrangement. For example, if the top face of the ring has axial hydrogens on carbons 1, 3, and 5, then it has equatorial hydrogens on carbons 2, 4, and 6. Exactly the reverse is true for the bottom face: carbons 1, 3, and 5 have equatorial hydrogens, but carbons 2, 4, and 6 have axial hydrogens **(Figure 4.9)**.

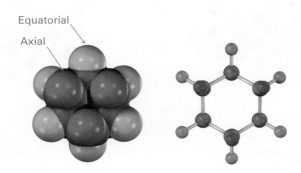

Figure 4.9 Alternating **axial** and **equatorial** positions in chair cyclohexane, as shown in a view looking directly down the ring axis. Each carbon atom has one axial and one equatorial position, and each face has alternating axial and equatorial positions.

Note that we haven't used the words *cis* and *trans* in this discussion of cyclohexane conformation. Two hydrogens on the same face of the ring are always cis, regardless of whether they're axial or equatorial and regardless of whether they're adjacent. Similarly, two hydrogens on opposite faces of the ring are always trans.

Axial and equatorial bonds can be drawn following the procedure in **Figure 4.10**. Look at a molecular model as you practice.

Axial bonds: The six axial bonds, one on each carbon, are parallel and alternate up–down.

Equatorial bonds: The six equatorial bonds, one on each carbon, come in three sets of two parallel lines. Each set is also parallel to two ring bonds. Equatorial bonds alternate between sides around the ring.

Completed cyclohexane

Figure 4.10 A procedure for drawing axial and equatorial bonds in chair cyclohexane.

Because chair cyclohexane has two kinds of positions—axial and equatorial—we might expect to find two isomeric forms of a monosubstituted cyclohexane. In fact, we don't. There is only *one* methylcyclohexane, *one* bromocyclohexane, *one* cyclohexanol (hydroxycyclohexane), and so on, because cyclohexane rings are *conformationally mobile* at room temperature. Different chair conformations readily interconvert, exchanging axial and equatorial positions. This interconversion, usually called a **ring-flip**, is shown in **Figure 4.11**.

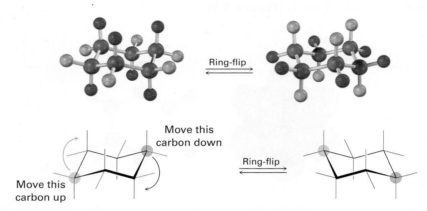

Figure 4.11 A ring-flip in chair cyclohexane interconverts axial and equatorial positions. What is **axial** in the starting structure becomes equatorial in the ring-flipped structure, and what is **equatorial** in the starting structure is axial after ring-flip.

As shown in Figure 4.11, a chair cyclohexane can be ring-flipped by keeping the middle four carbon atoms in place while folding the two end carbons in opposite directions. In so doing, an axial substituent in one chair form becomes an equatorial substituent in the ring-flipped chair form and vice versa. For example, axial bromocyclohexane becomes equatorial bromocyclohexane after ring-flip. Since the energy barrier to chair–chair interconversion is only about 45 kJ/mol (10.8 kcal/mol), the process is rapid at room temperature and we see what appears to be a single structure rather than distinct axial and equatorial isomers.

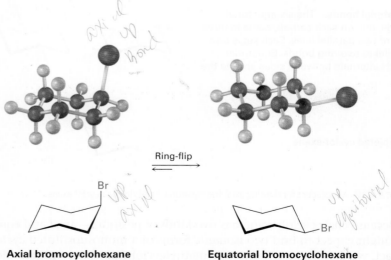

Axial bromocyclohexane   **Equatorial bromocyclohexane**

4.7 | Conformations of Monosubstituted Cyclohexanes

Worked Example 4.2

Drawing the Chair Conformation of a Substituted Cyclohexane

Draw 1,1-dimethylcyclohexane in a chair conformation, indicating which methyl group in your drawing is axial and which is equatorial.

Strategy

Draw a chair cyclohexane ring using the procedure in Figure 4.10, and then put two methyl groups on the same carbon. The methyl group in the rough plane of the ring is equatorial, and the one directly above or below the ring is axial.

Solution

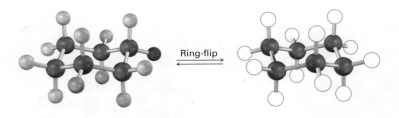

Problem 4.12
Draw two different chair conformations of cyclohexanol (hydroxycyclohexane), showing all hydrogen atoms. Identify each position as axial or equatorial.

Problem 4.13
Draw two different chair conformations of *trans*-1,4-dimethylcyclohexane, and label all positions as axial or equatorial.

Problem 4.14
Identify each of the colored positions—red, blue, and green—as axial or equatorial. Then carry out a ring-flip, and show the new positions occupied by each color.

4.7 Conformations of Monosubstituted Cyclohexanes

Even though cyclohexane rings flip rapidly between chair conformations at room temperature, the two conformations of a monosubstituted cyclohexane aren't equally stable. In methylcyclohexane, for instance, the equatorial conformation is more stable than the axial conformation by 7.6 kJ/mol (1.8 kcal/mol). The same is true of other monosubstituted cyclohexanes: a substituent is almost always more stable in an equatorial position than in an axial position.

You might recall from your general chemistry course that it's possible to calculate the percentages of two isomers at equilibrium using the equation

Key IDEAS

Test your knowledge of Key Ideas by answering end-of-chapter exercises marked with ▲.

$\Delta E = -RT \ln K$, where ΔE is the energy difference between isomers, R is the gas constant [8.315 J/(K·mol)], T is the Kelvin temperature, and K is the equilibrium constant between isomers. For example, an energy difference of 7.6 kJ/mol means that about 95% of methylcyclohexane molecules have the methyl group equatorial at any given instant and only 5% have the methyl group axial. **Figure 4.12** plots the relationship between energy and isomer percentages.

Figure 4.12 A plot of the percentages of two isomers at equilibrium versus the energy difference between them. The curves are calculated using the equation $\Delta E = -RT \ln K$.

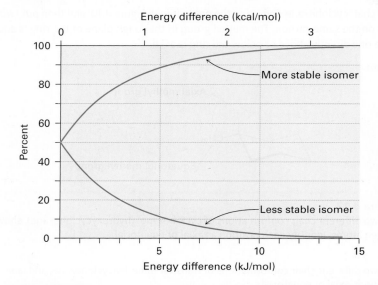

The energy difference between axial and equatorial conformations is due to steric strain caused by **1,3-diaxial interactions**. The axial methyl group on C1 is too close to the axial hydrogens three carbons away on C3 and C5, resulting in 7.6 kJ/mol of steric strain **(Figure 4.13)**.

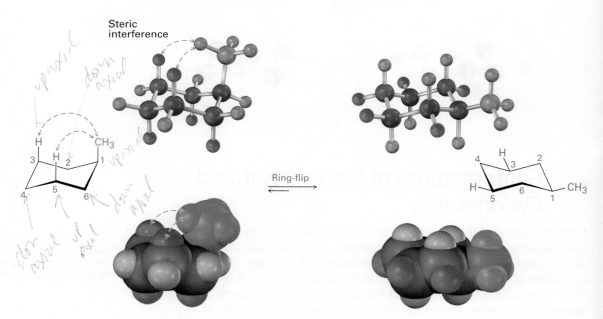

Figure 4.13 Interconversion of axial and equatorial methylcyclohexane, as represented in several formats. The equatorial conformation is more stable than the axial conformation by 7.6 kJ/mol.

4.7 | Conformations of Monosubstituted Cyclohexanes

The 1,3-diaxial steric strain in substituted methylcyclohexane is already familiar—we saw it previously as the steric strain between methyl groups in gauche butane. Recall from **Section 3.7** that gauche butane is less stable than anti butane by 3.8 kJ/mol (0.9 kcal/mol) because of steric interference between hydrogen atoms on the two methyl groups. Comparing a four-carbon fragment of axial methylcyclohexane with gauche butane shows that the steric interaction is the same in both cases **(Figure 4.14)**. Because axial methylcyclohexane has two such interactions, it has $2 \times 3.8 = 7.6$ kJ/mol of steric strain. Equatorial methylcyclohexane has no such interactions and is therefore more stable.

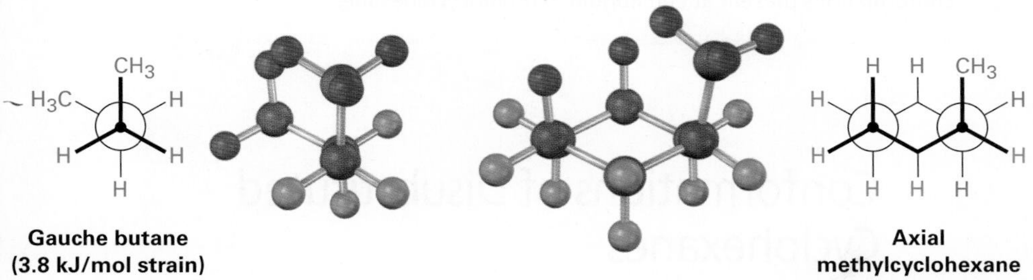

Gauche butane
(3.8 kJ/mol strain)

Axial methylcyclohexane
(7.6 kJ/mol strain)

Figure 4.14 The origin of 1,3-diaxial interactions in methylcyclohexane. The steric strain between an **axial methyl group** and **an axial hydrogen atom** three carbons away is identical to the steric strain in gauche butane. Note that the $-CH_3$ group in methylcyclohexane moves slightly away from a true axial position to minimize the strain.

The exact amount of 1,3-diaxial steric strain in a given substituted cyclohexane depends on the nature and size of the substituent, as indicated in Table 4.1. Not surprisingly, the amount of steric strain increases through the series $H_3C- < CH_3CH_2- < (CH_3)_2CH- << (CH_3)_3C-$, paralleling the increasing size of the alkyl groups. Note that the values in Table 4.1 refer to 1,3-diaxial interactions of the substituent with a single hydrogen atom. These values must be doubled to arrive at the amount of strain in a monosubstituted cyclohexane.

Table 4.1 Steric Strain in Monosubstituted Cyclohexanes

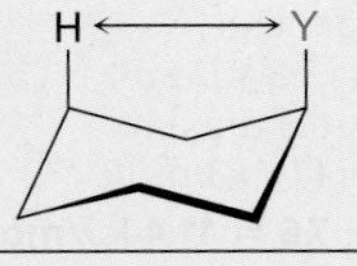

Y	1,3-Diaxial strain	
	(kJ/mol)	(kcal/mol)
F	0.5	0.12
Cl, Br	1.0	0.25
OH	2.1	0.5
CH_3	3.8	0.9
CH_2CH_3	4.0	0.95
$CH(CH_3)_2$	4.6	1.1
$C(CH_3)_3$	11.4	2.7
C_6H_5	6.3	1.5
CO_2H	2.9	0.7
CN	0.4	0.1

Problem 4.15
What is the energy difference between the axial and equatorial conformations of cyclohexanol (hydroxycyclohexane)?

Problem 4.16
Why do you suppose an axial cyano (–CN) substituent causes practically no 1,3-diaxial steric strain (0.4 kJ/mol)? Use molecular models to help with your answer.

Problem 4.17
Look at Figure 4.12 on page 124, and estimate the percentages of axial and equatorial conformations present at equilibrium in bromocyclohexane.

4.8 Conformations of Disubstituted Cyclohexanes

Monosubstituted cyclohexanes are always more stable with their substituent in an equatorial position, but the situation in disubstituted cyclohexanes is more complex because the steric effects of both substituents must be taken into account. All steric interactions in both possible chair conformations must be analyzed before deciding which conformation is favored.

Let's look at 1,2-dimethylcyclohexane as an example. There are two isomers, *cis*-1,2-dimethylcyclohexane and *trans*-1,2-dimethylcyclohexane, which must be considered separately. In the cis isomer, both methyl groups are on the same face of the ring and the compound can exist in either of the two chair conformations shown in **Figure 4.15**. (It may be easier for you to see whether a compound is cis- or trans-disubstituted by first drawing the ring as a flat representation and then converting to a chair conformation.)

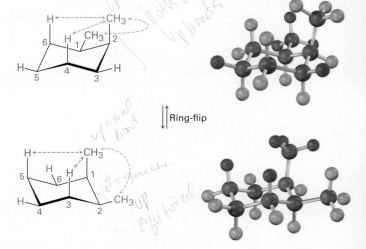

cis-1,2-Dimethylcyclohexane

One gauche interaction (3.8 kJ/mol)
Two CH$_3$ ↔ H diaxial interactions (7.6 kJ/mol)
Total strain: 3.8 + 7.6 = 11.4 kJ/mol

Ring-flip

One gauche interaction (3.8 kJ/mol)
Two CH$_3$ ↔ H diaxial interactions (7.6 kJ/mol)
Total strain: 3.8 + 7.6 = 11.4 kJ/mol

Figure 4.15 Conformations of *cis*-1,2-dimethylcyclohexane. The two chair conformations are equal in energy because each has one axial methyl group and one equatorial methyl group.

Both chair conformations of *cis*-1,2-dimethylcyclohexane have one axial methyl group and one equatorial methyl group. The top conformation in Figure 4.15 has an axial methyl group at C2, which has 1,3-diaxial interactions with hydrogens on C4 and C6. The ring-flipped conformation has an axial methyl group at C1, which has 1,3-diaxial interactions with hydrogens on C3 and C5. In addition, both conformations have gauche butane interactions between the two methyl groups. The two conformations are equal in energy, with a total steric strain of 3×3.8 kJ/mol $=$ 11.4 kJ/mol (2.7 kcal/mol).

In *trans*-1,2-dimethylcyclohexane, the two methyl groups are on opposite faces of the ring and the compound can exist in either of the two chair conformations shown in **Figure 4.16**. The situation here is quite different from that of the cis isomer. The top conformation in Figure 4.16 has both methyl groups equatorial and therefore has only a gauche butane interaction between them (3.8 kJ/mol) but no 1,3-diaxial interactions. The ring-flipped conformation, however, has both methyl groups axial. The axial methyl group at C1 interacts with axial hydrogens at C3 and C5, and the axial methyl group at C2 interacts with axial hydrogens at C4 and C6. These four 1,3-diaxial interactions produce a steric strain of 4×3.8 kJ/mol $=$ 15.2 kJ/mol and make the diaxial conformation $15.2 - 3.8 = 11.4$ kJ/mol less favorable than the diequatorial conformation. We therefore predict that *trans*-1,2-dimethylcyclohexane will exist almost exclusively in the diequatorial conformation.

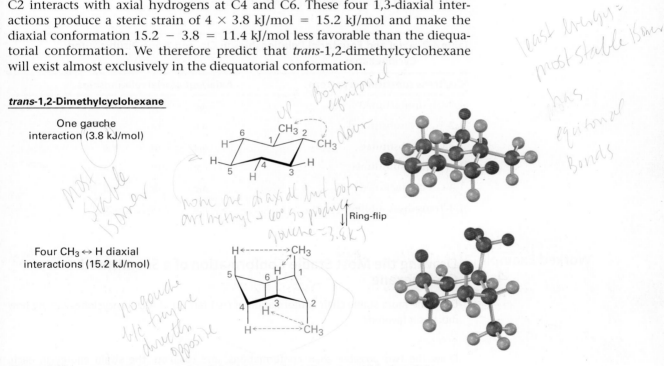

Figure 4.16 Conformations of *trans*-1,2-dimethylcyclohexane. The conformation with both methyl groups equatorial (top) is favored by 11.4 kJ/mol (2.7 kcal/mol) over the conformation with both methyl groups axial (bottom).

The same kind of **conformational analysis** just carried out for *cis*- and *trans*-1,2-dimethylcyclohexane can be done for any substituted cyclohexane, such as *cis*-1-*tert*-butyl-4-chlorocyclohexane (see Worked Example 4.3). As you might imagine, though, the situation becomes more complex as the number of substituents increases. For instance, compare glucose with mannose, a carbohydrate present in seaweed. Which do you think is more strained? In glucose, all substituents on the six-membered ring are equatorial, while in mannose, one of the −OH groups is axial, making mannose more strained.

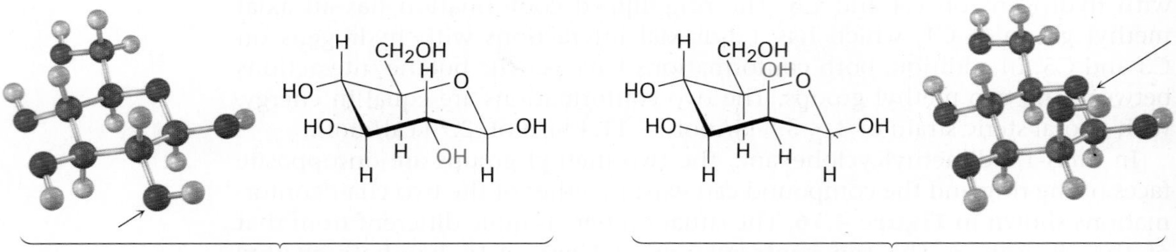

Glucose

Mannose

A summary of the various axial and equatorial relationships among substituent groups in the different possible cis and trans substitution patterns for disubstituted cyclohexanes is given in Table 4.2.

Table 4.2 Axial and Equatorial Relationships in Cis- and Trans-Disubstituted Cyclohexanes

Cis/trans substitution pattern	Axial/equatorial relationships		
1,2-Cis disubstituted	a,e	or	e,a
1,2-Trans disubstituted	a,a	or	e,e
1,3-Cis disubstituted	a,a	or	e,e
1,3-Trans disubstituted	a,e	or	e,a
1,4-Cis disubstituted	a,e	or	e,a
1,4-Trans disubstituted	a,a	or	e,e

Worked Example 4.3 Drawing the Most Stable Conformation of a Substituted Cyclohexane

Draw the more stable chair conformation of *cis*-1-*tert*-butyl-4-chlorocyclohexane. By how much is it favored?

Strategy
Draw the two possible chair conformations, and calculate the strain energy in each. Remember that equatorial substituents cause less strain than axial substituents.

Solution
First draw the two chair conformations of the molecule:

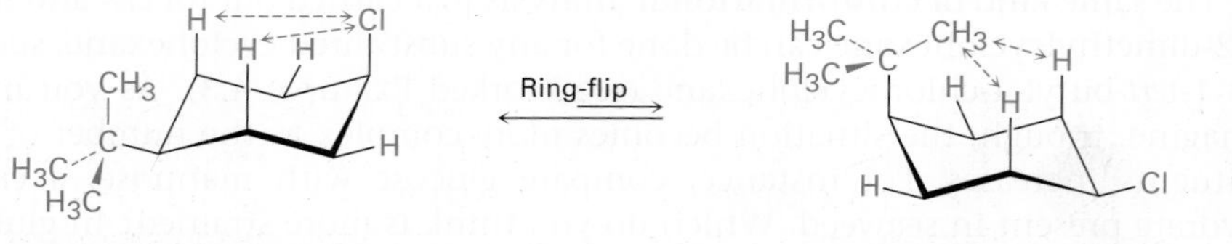

2 × 1.0 = 2.0 kJ/mol steric strain

2 × 11.4 = 22.8 kJ/mol steric strain

In the conformation on the left, the *tert*-butyl group is equatorial and the chlorine is axial. In the conformation on the right, the *tert*-butyl group is axial and the chlorine is equatorial. These conformations aren't of equal energy because an axial *tert*-butyl substituent and an axial chloro substituent produce different amounts of steric strain. Table 4.1 shows that the 1,3-diaxial interaction between a hydrogen and a *tert*-butyl group costs 11.4 kJ/mol (2.7 kcal/mol), whereas the interaction between a hydrogen and a chlorine costs only 1.0 kJ/mol (0.25 kcal/mol). An axial *tert*-butyl group therefore produces (2 × 11.4 kJ/mol) − (2 × 1.0 kJ/mol) = 20.8 kJ/mol (4.9 kcal/mol) more steric strain than does an axial chlorine, and the compound preferentially adopts the conformation with the chlorine axial and the *tert*-butyl equatorial.

Problem 4.18
Draw the more stable chair conformation of the following molecules, and estimate the amount of strain in each:
(a) *trans*-1-Chloro-3-methylcyclohexane (b) *cis*-1-Ethyl-2-methylcyclohexane
(c) *cis*-1-Bromo-4-ethylcyclohexane (d) *cis*-1-*tert*-Butyl-4-ethylcyclohexane

Problem 4.19
Identify each substituent in the following compound as axial or equatorial, and tell whether the conformation shown is the more stable or less stable chair form (green = Cl):

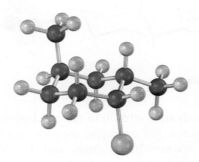

4.9 Conformations of Polycyclic Molecules

The final point we'll consider about cycloalkane stereochemistry is to see what happens when two or more cycloalkane rings are fused together along a common bond to construct a **polycyclic** molecule—for example, decalin.

Decalin—two fused cyclohexane rings

Decalin consists of two cyclohexane rings joined to share two carbon atoms (the *bridgehead* carbons, C1 and C6) and a common bond. Decalin can exist in either of two isomeric forms, depending on whether the rings are trans fused or cis fused. In *cis*-decalin, the hydrogen atoms at the bridgehead carbons are on the same face of the rings; in *trans*-decalin, the bridgehead hydrogens are on opposite faces. **Figure 4.17** shows how both compounds can be represented using chair cyclohexane conformations. Note that *cis*- and *trans*-decalin are not interconvertible by ring-flips or other rotations. They are cis–trans stereoisomers and have the same relationship to each other that *cis*- and *trans*-1,2-dimethylcyclohexane have.

Figure 4.17 Representations of *cis*- and *trans*-decalin. The **hydrogen atoms** at the bridgehead carbons are on the same face of the rings in the cis isomer but on opposite faces in the trans isomer.

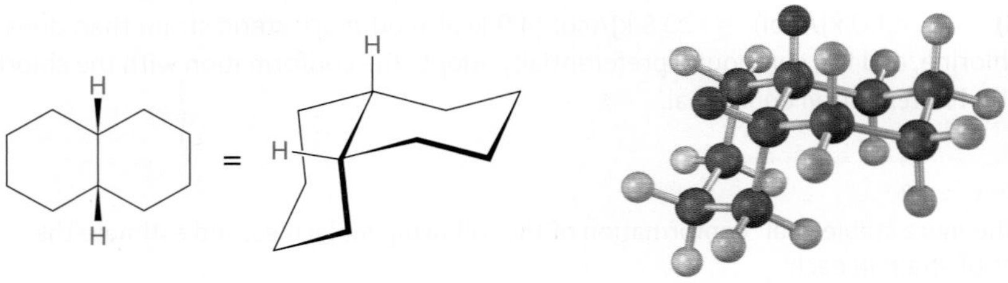

cis-Decalin

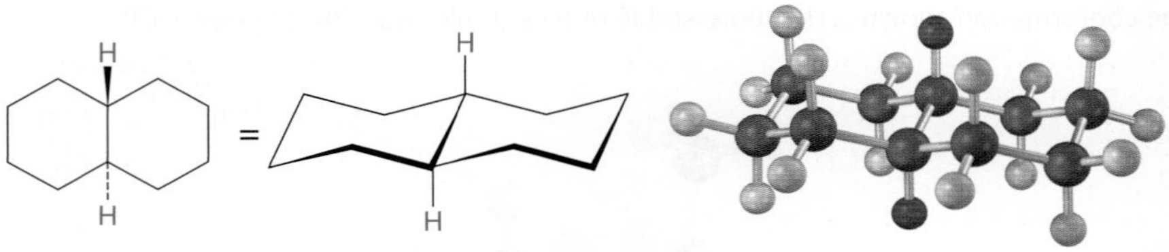

trans-Decalin

Polycyclic compounds are common in nature, and many valuable substances have fused-ring structures. For example, steroids, such as the male hormone testosterone, have 3 six-membered rings and 1 five-membered ring fused together. Although steroids look complicated compared with cyclohexane or decalin, the same principles that apply to the conformational analysis of simple cyclohexane rings apply equally well (and often better) to steroids.

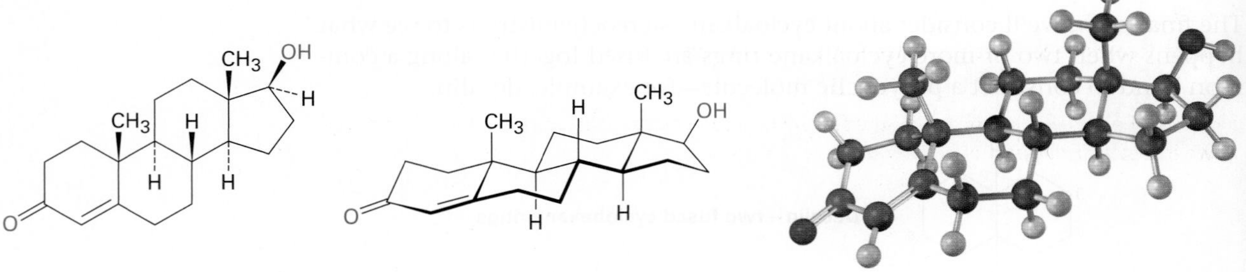

Testosterone (a steroid)

Another common ring system is the norbornane, or bicyclo[2.2.1]heptane, structure. Like decalin, norbornane is a *bicycloalkane,* so called because *two* rings would have to be broken open to generate an acyclic structure. Its systematic name, bicyclo[2.2.1]heptane, reflects the fact that the molecule has seven carbons, is bicyclic, and has three "bridges" of 2, 2, and 1 carbon atoms connecting the two bridgehead carbons.

**Norbornane
(bicyclo[2.2.1]heptane)**

Norbornane has a conformationally locked boat cyclohexane ring **(Section 4.5)** in which carbons 1 and 4 are joined by an additional CH_2 group. Note how, in drawing this structure, a break in the rear bond indicates that the vertical bond crosses in front of it. Making a molecular model is particularly helpful when trying to see the three-dimensionality of norbornane.

Substituted norbornanes, such as camphor, are found widely in nature, and many have been important historically in developing organic structural theories.

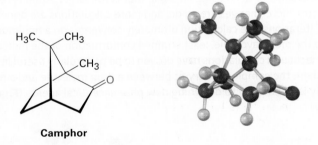

Camphor

Problem 4.20
Which isomer is more stable, *cis*-decalin or *trans*-decalin? Explain.

Problem 4.21
Look at the following structure of the female hormone estrone, and tell whether each of the two indicated ring-fusions is cis or trans.

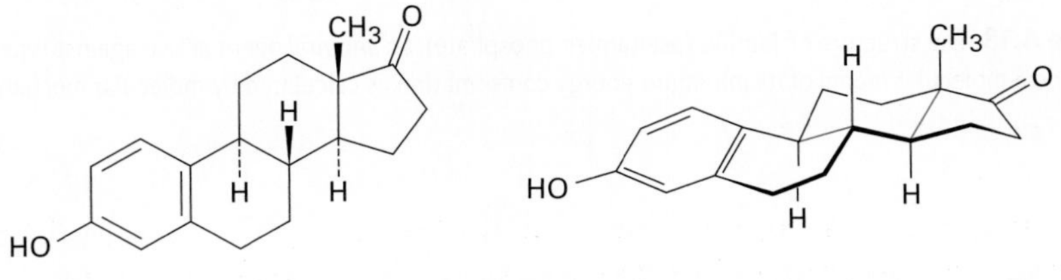

Estrone

Molecular Mechanics | A DEEPER LOOK

Computer programs make it possible to portray accurate representations of molecular geometry.

All the structural models in this book are computer-drawn. To make sure they accurately portray bond angles, bond lengths, torsional interactions, and steric interactions, the most stable geometry of each molecule has been calculated on a desktop computer using a commercially available *molecular mechanics* program based on work by N. L. Allinger of the University of Georgia.

The idea behind molecular mechanics is to begin with a rough geometry for a molecule and then calculate a total strain energy for that starting geometry, using mathematical equations that assign values to specific kinds of molecular interactions. Bond angles that are too large or too small cause angle strain; bond lengths that are too short or too long cause stretching or compressing strain; unfavorable eclipsing interactions around single bonds cause torsional strain; and nonbonded atoms that approach each other too closely cause steric, or *van der Waals*, strain.

$$E_{\text{total}} = E_{\text{bond stretching}} + E_{\text{angle strain}} + E_{\text{torsional strain}} + E_{\text{van der Waals}}$$

After calculating a total strain energy for the starting geometry, the program automatically changes the geometry slightly in an attempt to lower strain—perhaps by lengthening a bond that is too short or decreasing an angle that is too large. Strain is recalculated for the new geometry, more changes are made, and more calculations are done. After dozens or hundreds of iterations, the calculation ultimately converges on a minimum energy that corresponds to the most favorable, least strained conformation of the molecule.

Molecular mechanics calculations have proven to be particularly useful in pharmaceutical research, where the complementary fit between a drug molecule and a receptor molecule in the body is often a key to designing new pharmaceutical agents **(Figure 4.18)**.

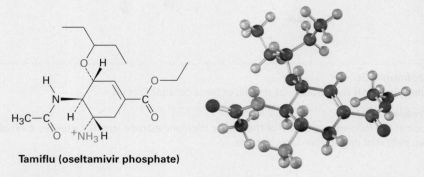

Tamiflu (oseltamivir phosphate)

Figure 4.18 The structure of Tamiflu (oseltamivir phosphate), an antiviral agent active against type A influenza, and a molecular model of its minimum-energy conformation as calculated by molecular mechanics.

Summary

Cyclic molecules are so commonly encountered throughout organic and biological chemistry that it's important to understand the consequences of their cyclic structures. Thus, we've taken a close look at cyclic structures in this chapter.

A **cycloalkane** is a saturated cyclic hydrocarbon with the general formula C_nH_{2n}. In contrast to open-chain alkanes, where nearly free rotation occurs around C−C bonds, rotation is greatly reduced in cycloalkanes. Disubstituted cycloalkanes can therefore exist as **cis–trans isomers**. The cis isomer has both substituents on the same face of the ring; the trans isomer has substituents on opposite faces. Cis–trans isomers are just one kind of **stereoisomers**—compounds that have the same connections between atoms but different three-dimensional arrangements.

Not all cycloalkanes are equally stable. Three kinds of strain contribute to the overall energy of a cycloalkane: (1) **angle strain** is the resistance of a bond angle to compression or expansion from the normal 109° tetrahedral value, (2) torsional strain is the energy cost of having neighboring C−H bonds eclipsed rather than staggered, and (3) steric strain is the repulsive interaction that arises when two groups attempt to occupy the same space.

Cyclopropane (115 kJ/mol strain) and cyclobutane (110.4 kJ/mol strain) have both angle strain and torsional strain. Cyclopentane is free of angle strain but has a substantial torsional strain due to its large number of eclipsing interactions. Both cyclobutane and cyclopentane pucker slightly away from planarity to relieve torsional strain.

Cyclohexane is strain-free because it adopts a puckered **chair conformation**, in which all bond angles are near 109° and all neighboring C−H bonds are staggered. Chair cyclohexane has two kinds of positions: **axial** and **equatorial**. Axial positions are oriented up and down, parallel to the ring axis, while equatorial positions lie in a belt around the equator of the ring. Each carbon atom has one axial and one equatorial position.

Chair cyclohexanes are conformationally mobile and can undergo a **ring-flip**, which interconverts axial and equatorial positions. Substituents on the ring are more stable in the equatorial position because axial substituents cause **1,3-diaxial interactions**. The amount of 1,3-diaxial steric strain caused by an axial substituent depends on its size.

Key words

alicyclic, 109
angle strain, 114
axial position, 120
chair conformation, 118
cis–trans isomers, 113
conformational analysis, 127
cycloalkane, 109
1,3-diaxial interaction, 124
equatorial position, 120
polycyclic compound, 129
ring-flip (cyclohexane), 122
stereochemistry, 112
stereoisomers, 112
twist-boat conformation, 119

Exercises

Visualizing Chemistry

(Problems 4.1–4.21 appear within the chapter.)

4.22 Name the following cycloalkanes:

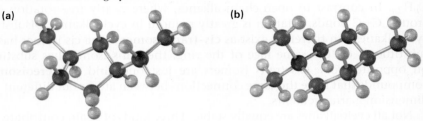

4.23 Name the following compound, identify each substituent as axial or equatorial, and tell whether the conformation shown is the more stable or less stable chair form (green = Cl):

4.24 ▲ A trisubstituted cyclohexane with three substituents—red, green, and blue—undergoes a ring-flip to its alternative chair conformation. Identify each substituent as axial or equatorial, and show the positions occupied by the three substituents in the ring-flipped form.

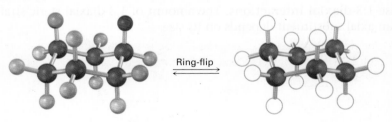

▲ Problems linked to Key Ideas in this chapter

4.25 The following cyclohexane derivative has three substituents—red, green, and blue. Identify each substituent as axial or equatorial, and identify each pair of relationships (red–blue, red–green, and blue–green) as cis or trans.

4.26 Glucose exists in two forms having a 36:64 ratio at equilibrium. Draw a skeletal structure of each, describe the difference between them, and tell which of the two you think is more stable (red = O).

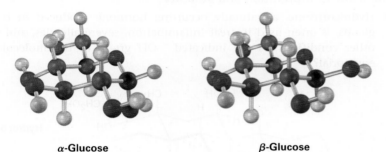

α-Glucose   β-Glucose

Additional Problems

Cycloalkane Isomers

4.27 Draw the five cycloalkanes with the formula C_5H_{10}.

4.28 Draw two constitutional isomers of *cis*-1,2-dibromocyclopentane.

4.29 Draw a stereoisomer of *trans*-1,3-dimethylcyclobutane.

4.30 Tell whether the following pairs of compounds are identical, constitutional isomers, stereoisomers, or unrelated.
  (a) *cis*-1,3-Dibromocyclohexane and *trans*-1,4-dibromocyclohexane
  (b) 2,3-Dimethylhexane and 2,3,3-trimethylpentane
  (c)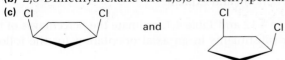

▲ Problems linked to Key Ideas in this chapter

4.31 Draw three isomers of *trans*-1,2-dichlorocyclobutane, and label them as either constitutional isomers or stereoisomers.

4.32 Identify each pair of relationships among the −OH groups in glucose (red–blue, red–green, red–black, blue–green, blue–black, green–black) as cis or trans.

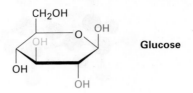

Glucose

4.33 ▲ Draw 1,3,5-trimethylcyclohexane using a hexagon to represent the ring. How many cis–trans stereoisomers are possible?

Cycloalkane Conformation and Stability

4.34 Hydrocortisone, a naturally occurring hormone produced in the adrenal glands, is often used to treat inflammation, severe allergies, and numerous other conditions. Is the indicated −OH group in the molecule axial or equatorial?

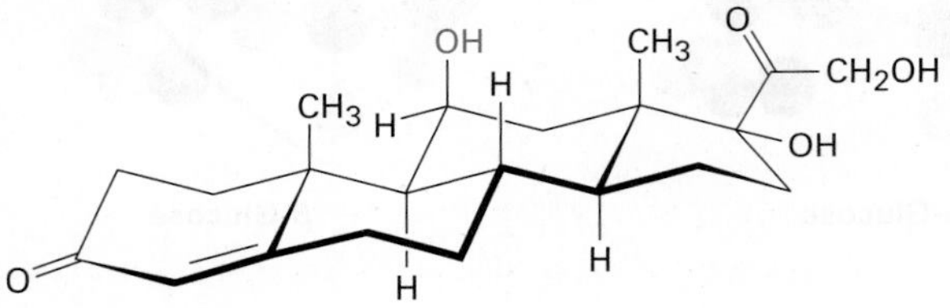

Hydrocortisone

4.35 A 1,2-cis disubstituted cyclohexane, such as *cis*-1,2-dichlorocyclohexane, must have one group axial and one group equatorial. Explain.

4.36 A 1,2-trans disubstituted cyclohexane must have either both groups axial or both groups equatorial. Explain.

4.37 Why is a 1,3-cis disubstituted cyclohexane more stable than its trans isomer?

4.38 Which is more stable, a 1,4-trans disubstituted cyclohexane or its cis isomer?

4.39 *cis*-1,2-Dimethylcyclobutane is less stable than its trans isomer, but *cis*-1,3-dimethylcyclobutane is more stable than its trans isomer. Draw the most stable conformations of both, and explain.

4.40 From the data in Figure 4.12 and Table 4.1, estimate the percentages of molecules that have their substituents in an axial orientation for the following compounds:
(a) Isopropylcyclohexane
(b) Fluorocyclohexane
(c) Cyclohexanecarbonitrile, $C_6H_{11}CN$

▲ Problems linked to Key Ideas in this chapter

4.41 ▲ Assume that you have a variety of cyclohexanes substituted in the positions indicated. Identify the substituents as either axial or equatorial. For example, a 1,2-cis relationship means that one substituent must be axial and one equatorial, whereas a 1,2-trans relationship means that both substituents are axial or both are equatorial.

(a) 1,3-Trans disubstituted
(b) 1,4-Cis disubstituted
(c) 1,3-Cis disubstituted
(d) 1,5-Trans disubstituted
(e) 1,5-Cis disubstituted
(f) 1,6-Trans disubstituted

Cyclohexane Conformational Analysis

4.42 Draw the two chair conformations of *cis*-1-chloro-2-methylcyclohexane. Which is more stable, and by how much?

4.43 Draw the two chair conformations of *trans*-1-chloro-2-methylcyclohexane. Which is more stable?

4.44 Galactose, a sugar related to glucose, contains a six-membered ring in which all the substituents except the —OH group indicated below in red are equatorial. Draw galactose in its more stable chair conformation.

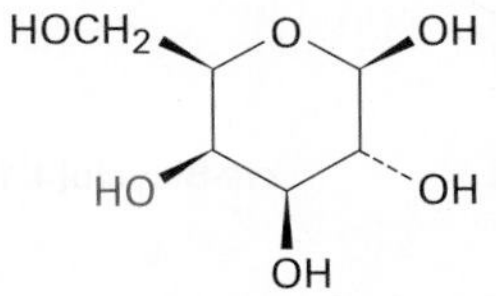

Galactose

4.45 Draw the two chair conformations of menthol, and tell which is more stable.

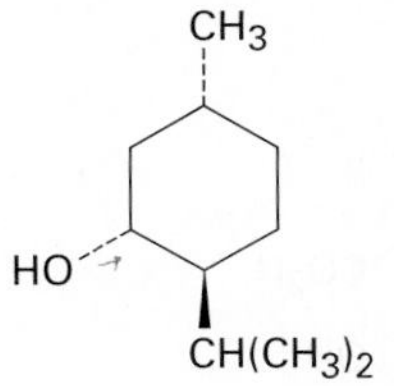

Menthol

4.46 There are four cis–trans isomers of menthol (Problem 4.45), including the one shown. Draw the other three.

4.47 ▲ The diaxial conformation of *cis*-1,3-dimethylcyclohexane is approximately 23 kJ/mol (5.4 kcal/mol) less stable than the diequatorial conformation. Draw the two possible chair conformations, and suggest a reason for the large energy difference.

4.48 Approximately how much steric strain does the 1,3-diaxial interaction between the two methyl groups introduce into the diaxial conformation of *cis*-1,3-dimethylcyclohexane? (See Problem 4.47.)

4.49 In light of your answer to Problem 4.48, draw the two chair conformations of 1,1,3-trimethylcyclohexane and estimate the amount of strain energy in each. Which conformation is favored?

▲ Problems linked to Key Ideas in this chapter

4.50 One of the two chair structures of *cis*-1-chloro-3-methylcyclohexane is more stable than the other by 15.5 kJ/mol (3.7 kcal/mol). Which is it? What is the energy cost of a 1,3-diaxial interaction between a chlorine and a methyl group?

General Problems

4.51 We saw in Problem 4.20 that *cis*-decalin is less stable than *trans*-decalin. Assume that the 1,3-diaxial interactions in *cis*-decalin are similar to those in axial methylcyclohexane [that is, one $CH_2 \longleftrightarrow H$ interaction costs 3.8 kJ/mol (0.9 kcal/mol)], and calculate the magnitude of the energy difference between *cis*- and *trans*-decalin.

4.52 Using molecular models as well as structural drawings, explain why *trans*-decalin is rigid and cannot ring-flip whereas *cis*-decalin can easily ring-flip.

4.53 *trans*-Decalin is more stable than its cis isomer, but *cis*-bicyclo[4.1.0]heptane is more stable than its trans isomer. Explain.

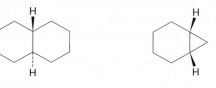

trans-Decalin   **cis-Bicyclo[4.1.0]heptane**

4.54 As mentioned in Problem 3.53, the statin drugs, such as simvastatin (Zocor), pravastatin (Pravachol), and atorvastatin (Lipitor) are the most widely prescribed drugs in the world.

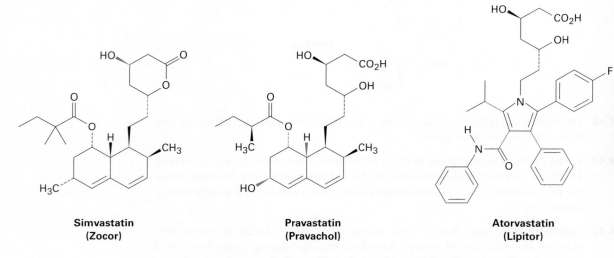

Simvastatin (Zocor)   **Pravastatin (Pravachol)**   **Atorvastatin (Lipitor)**

(a) Are the two indicated bonds on simvastatin cis or trans?
(b) What are the cis/trans relationships among the three indicated bonds on pravastatin?
(c) Why can't the three indicated bonds on atorvastatin be identified as cis or trans?

▲ Problems linked to Key Ideas in this chapter

4.55 ▲ *myo*-Inositol, one of the isomers of 1,2,3,4,5,6-hexahydroxycyclohexane, acts as a growth factor in both animals and microorganisms. Draw the most stable chair conformation of *myo*-inositol.

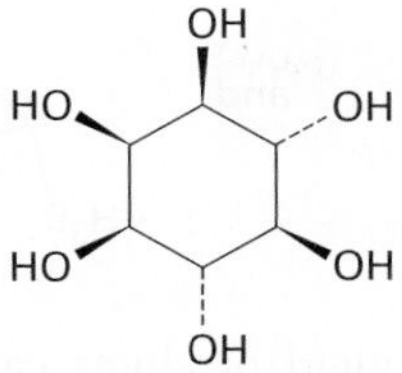

myo-Inositol

4.56 How many cis–trans stereoisomers of *myo*-inositol (Problem 4.55) are there? Draw the structure of the most stable isomer.

4.57 The German chemist J. Bredt proposed in 1935 that bicycloalkenes such as 1-norbornene, which have a double bond to the bridgehead carbon, are too strained to exist. Explain. (Making a molecular model will be helpful.)

1-Norbornene

4.58 Tell whether each of the following substituents on a steroid is axial or equatorial. (A substituent that is "up" is on the top face of the molecule as drawn, and a substituent that is "down" is on the bottom face.)
(a) Substituent up at C3
(b) Substituent down at C7
(c) Substituent down at C11

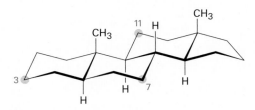

4.59 Amantadine is an antiviral agent that is active against influenza type A infection. Draw a three-dimensional representation of amantadine, showing the chair cyclohexane rings.

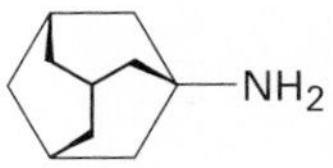

Amantadine

▲ Problems linked to Key Ideas in this chapter

4.60 Here's a difficult one. There are two different substances named *trans*-1,2-dimethylcyclopentane. What is the relationship between them? (We'll explore this kind of isomerism in the next chapter.)

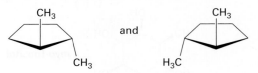

4.61 Ketones react with alcohols to yield products called *acetals*. Why does the all-cis isomer of 4-*tert*-butyl-1,3-cyclohexanediol react readily with acetone and an acid catalyst to form an acetal, but other stereoisomers do not react? In formulating your answer, draw the more stable chair conformations of all four stereoisomers and the product acetal from each.

4.62 Alcohols undergo an *oxidation* reaction to yield carbonyl compounds on treatment with CrO_3. For example, 2-*tert*-butylcyclohexanol gives 2-*tert*-butylcyclohexanone. If axial —OH groups are generally more reactive than their equatorial isomers, which do you think reacts faster, the cis isomer of 2-*tert*-butylcyclohexanol or the trans isomer? Explain.

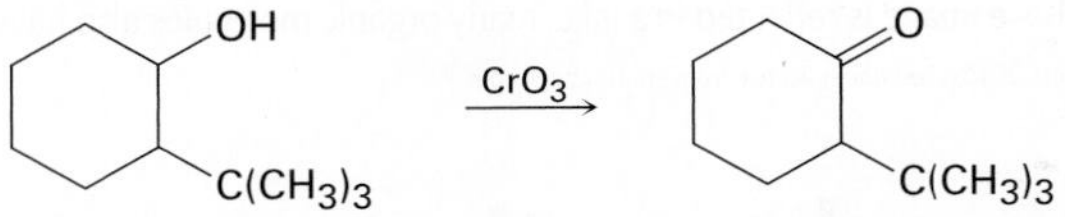

2-*tert*-Butylcyclohexanol $\xrightarrow{CrO_3}$ **2-*tert*-Butylcyclohexanone**

5 Stereochemistry at Tetrahedral Centers

Like the mountain whose image is reflected in a lake, many organic molecules also have mirror-image counterparts.
Image copyright Tischenko Irina, 2010. Used under license from Shutterstock.com

5.1 Enantiomers and the Tetrahedral Carbon
5.2 The Reason for Handedness in Molecules: Chirality
5.3 Optical Activity
5.4 Pasteur's Discovery of Enantiomers
5.5 Sequence Rules for Specifying Configuration
5.6 Diastereomers
5.7 Meso Compounds
5.8 Racemic Mixtures and the Resolution of Enantiomers
5.9 A Review of Isomerism
5.10 Chirality at Nitrogen, Phosphorus, and Sulfur
5.11 Prochirality
5.12 Chirality in Nature and Chiral Environments
A Deeper Look— Chiral Drugs

OWL Sign in to OWL for Organic Chemistry at www.cengage.com/owl to view tutorials and simulations, develop problem-solving skills, and complete online homework assigned by your professor.

Are you right-handed or left-handed? You may not spend much time thinking about it, but handedness plays a surprisingly large role in your daily activities. Many musical instruments, such as oboes and clarinets, have a handedness to them; the last available softball glove always fits the wrong hand; left-handed people write in a "funny" way. The reason for these difficulties is that our hands aren't identical; rather, they're *mirror images*. When you hold a *left* hand up to a mirror, the image you see looks like a *right* hand. Try it.

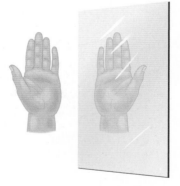

Left hand   Right hand

Handedness is also important in organic and biological chemistry, where it arises primarily as a consequence of the tetrahedral stereochemistry of sp^3-hybridized carbon atoms. Many drugs and almost all the molecules in our bodies—amino acids, carbohydrates, nucleic acids, and many more—are handed. Furthermore, molecular handedness makes possible the precise interactions between enzymes and their substrates that are involved in the hundreds of thousands of chemical reactions on which life is based.

Why This Chapter? Understanding the causes and consequences of molecular handedness is crucial to understanding organic and biological chemistry. The subject can be a bit complex at first, but the material covered in this chapter nevertheless forms the basis for much of the remainder of the book.

5.1 Enantiomers and the Tetrahedral Carbon

What causes molecular handedness? Look at generalized molecules of the type CH_3X, CH_2XY, and $CHXYZ$ shown in **Figure 5.1**. On the left are three molecules, and on the right are their images reflected in a mirror. The CH_3X and CH_2XY molecules are identical to their mirror images and thus are not handed. If you make a molecular model of each molecule and its mirror image, you find that you can superimpose one on the other so that all atoms coincide. The CHXYZ molecule, by contrast, is *not* identical to its mirror image. You can't superimpose a model of the molecule on a model of its mirror image for the same reason that you can't superimpose a left hand on a right hand: they simply aren't the same.

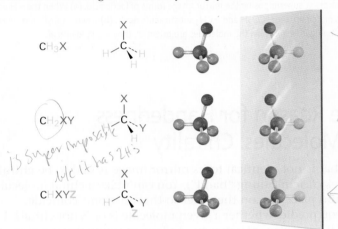

Figure 5.1 Tetrahedral carbon atoms and their mirror images. Molecules of the type CH_3X and CH_2XY are identical to their mirror images, but a molecule of the type CHXYZ is not. A CHXYZ molecule is related to its mirror image in the same way that a right hand is related to a left hand.

Molecules that are not identical to their mirror images are kinds of stereoisomers called **enantiomers** (Greek *enantio*, meaning "opposite"). Enantiomers are related to each other as a right hand is related to a left hand and result whenever a tetrahedral carbon is bonded to four different substituents (one need not be H). For example, lactic acid (2-hydroxypropanoic acid) exists as a pair of enantiomers because there are four different groups ($-H$, $-OH$, $-CH_3$, $-CO_2H$) bonded to the central carbon atom. The enantiomers are called (+)-lactic acid and (−)-lactic acid. Both are found in sour milk, but only the (+) enantiomer occurs in muscle tissue.

Lactic acid: a molecule of general formula CHXYZ

(+)-Lactic acid   (−)-Lactic acid

No matter how hard you try, you can't superimpose a molecule of (+)-lactic acid on a molecule of (−)-lactic acid. If any two groups match up, say —H and —CO₂H, the remaining two groups don't match **(Figure 5.2)**.

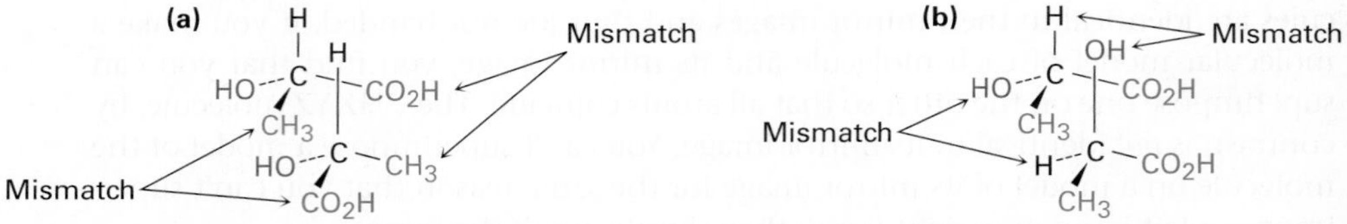

Figure 5.2 Attempts at superimposing the mirror-image forms of lactic acid. **(a)** When the —H and —OH substituents match up, the —CO₂H and —CH₃ substituents don't; **(b)** when —CO₂H and —CH₃ match up, —H and —OH don't. Regardless of how the molecules are oriented, they aren't identical.

5.2 The Reason for Handedness in Molecules: Chirality

Key IDEAS

Test your knowledge of Key Ideas by answering end-of-chapter exercises marked with ▲.

A molecule that is not identical to its mirror image is said to be **chiral** (**ky**-ral, from the Greek *cheir,* meaning "hand"). You can't take a chiral molecule and its enantiomer and place one on the other so that all atoms coincide.

How can you predict whether a given molecule is or is not chiral? *A molecule is not chiral if it has a plane of symmetry.* A plane of symmetry is a plane that cuts through the middle of a molecule (or any object) in such a way that one half of the molecule or object is a mirror image of the other half. A laboratory flask, for example, has a plane of symmetry. If you were to cut the flask in half, one half would be a mirror image of the other half. A hand, however, does not have a plane of symmetry. One "half" of a hand is not a mirror image of the other half **(Figure 5.3)**.

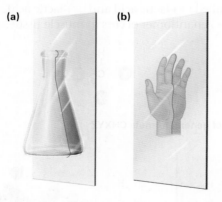

Figure 5.3 The meaning of *symmetry plane.* **(a)** An object like the flask has a symmetry plane cutting through it so that right and left halves are mirror images. **(b)** An object like a hand has no symmetry plane; the right "half" of a hand is not a mirror image of the left half.

A molecule that has a plane of symmetry in any conformation must be identical to its mirror image and hence must be nonchiral, or **achiral**. Thus, propanoic acid, CH₃CH₂CO₂H, has a plane of symmetry when lined up as shown in

Figure 5.4 and is achiral, while lactic acid, $CH_3CH(OH)CO_2H$, has no plane of symmetry in any conformation and is chiral.

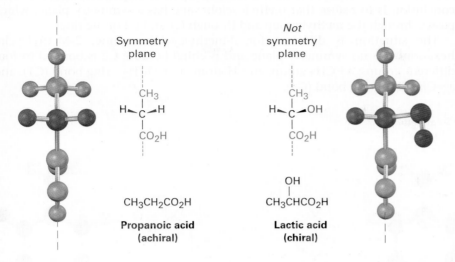

Figure 5.4 The achiral propanoic acid molecule versus the chiral lactic acid molecule. Propanoic acid has a plane of symmetry that makes one side of the molecule a mirror image of the other side. Lactic acid has no such symmetry plane.

The most common, although not the only, cause of chirality in an organic molecule is the presence of a tetrahedral carbon atom bonded to four different groups—for example, the central carbon atom in lactic acid. Such carbons are referred to as **chirality centers**, although other terms such as *stereocenter, asymmetric center,* and *stereogenic center* have also been used. Note that *chirality* is a property of the entire molecule, whereas a chirality *center* is the *cause* of chirality.

Detecting a chirality center in a complex molecule takes practice because it's not always immediately apparent that four different groups are bonded to a given carbon. The differences don't necessarily appear right next to the chirality center. For example, 5-bromodecane is a chiral molecule because four different groups are bonded to C5, the chirality center (marked with an asterisk). A butyl substituent is similar to a pentyl substituent, but it isn't identical. The difference isn't apparent until four carbon atoms away from the chirality center, but there's still a difference.

$$CH_3CH_2CH_2CH_2CH_2\overset{\overset{Br}{|}}{\underset{\underset{H}{|}}{C}}CH_2CH_2CH_2CH_3$$

5-Bromodecane (chiral)

Substituents on carbon 5

—H

—Br

—$CH_2CH_2CH_2CH_3$ (butyl)

—$CH_2CH_2CH_2CH_2CH_3$ (pentyl)

As other possible examples, look at methylcyclohexane and 2-methylcyclohexanone. Methylcyclohexane is achiral because no carbon atom in the molecule is bonded to four different groups. You can immediately eliminate all —CH_2— carbons and the —CH_3 carbon from consideration, but what about C1 on the ring? The C1 carbon atom is bonded to a —CH_3 group, to an —H atom, and to C2 and C6 of the ring. Carbons 2 and 6 are equivalent, however, as are carbons

3 and 5. Thus, the C6–C5–C4 "substituent" is equivalent to the C2–C3–C4 substituent, and methylcyclohexane is achiral. Another way of reaching the same conclusion is to realize that methylcyclohexane has a symmetry plane, which passes through the methyl group and through C1 and C4 of the ring.

The situation is different for 2-methylcyclohexanone. 2-Methylcyclohexanone has no symmetry plane and is chiral because C2 is bonded to four different groups: a –CH₃ group, an –H atom, a –COCH₂– ring bond (C1), and a –CH₂CH₂– ring bond (C3).

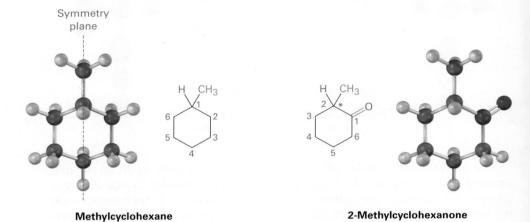

Methylcyclohexane
(achiral)

2-Methylcyclohexanone
(chiral)

Several more examples of chiral molecules are shown below. Check for yourself that the labeled carbons are chirality centers. You might note that carbons in –CH₂–, –CH₃, C=O, C=C, and C≡C groups can't be chirality centers. (Why not?)

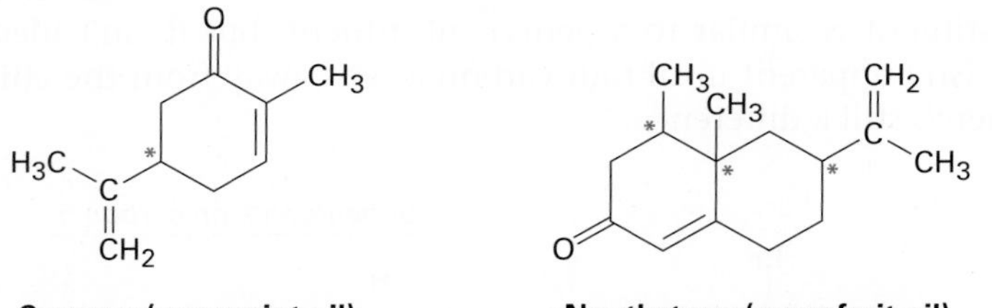

Carvone (spearmint oil)   **Nootkatone (grapefruit oil)**

Worked Example 5.1

Drawing the Three-Dimensional Structure of a Chiral Molecule

Draw the structure of a chiral alcohol.

Strategy
An alcohol is a compound that contains the –OH functional group. To make an alcohol chiral, we need to have four different groups bonded to a single carbon atom, say –H, –OH, –CH₃, and –CH₂CH₃.

Solution

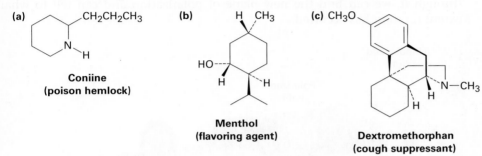

CH₃CH₂—C*(OH)(H)—CH₃   **2-Butanol (chiral)**

Problem 5.1
Which of the following objects are chiral?
(a) Soda can **(b)** Screwdriver **(c)** Screw **(d)** Shoe

Problem 5.2
Which of the following molecules are chiral? Identify the chirality center(s) in each.

(a) Coniine (poison hemlock) **(b)** Menthol (flavoring agent) **(c)** Dextromethorphan (cough suppressant)

Problem 5.3
Alanine, an amino acid found in proteins, is chiral. Draw the two enantiomers of alanine using the standard convention of solid, wedged, and dashed lines.

CH₃CH(NH₂)CO₂H   **Alanine**

Problem 5.4
Identify the chirality centers in the following molecules (green = Cl, yellow-green = F):

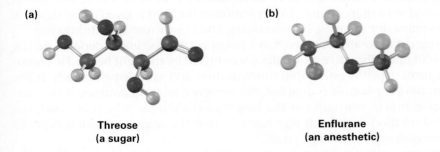

(a) Threose (a sugar) **(b)** Enflurane (an anesthetic)

5.3 Optical Activity

The study of chirality originated in the early 19th century during investigations by the French physicist Jean-Baptiste Biot into the nature of *plane-polarized light*. A beam of ordinary light consists of electromagnetic waves that oscillate in an infinite number of planes at right angles to the direction of light travel. When a beam of ordinary light passes through a device called a *polarizer,* however, only the light waves oscillating in a single plane pass through and the light is said to be plane-polarized. Light waves in all other planes are blocked out.

Biot made the remarkable observation that when a beam of plane-polarized light passes through a solution of certain organic molecules, such as sugar or camphor, the plane of polarization is *rotated* through an angle, α. Not all organic substances exhibit this property, but those that do are said to be **optically active**.

The angle of rotation can be measured with an instrument called a *polarimeter,* represented in **Figure 5.5**. A solution of optically active organic molecules is placed in a sample tube, plane-polarized light is passed through the tube, and rotation of the polarization plane occurs. The light then goes through a second polarizer called the *analyzer*. By rotating the analyzer until the light passes through *it*, we can find the new plane of polarization and can tell to what extent rotation has occurred.

Figure 5.5 Schematic representation of a polarimeter. Plane-polarized light passes through a solution of optically active molecules, which rotate the plane of polarization.

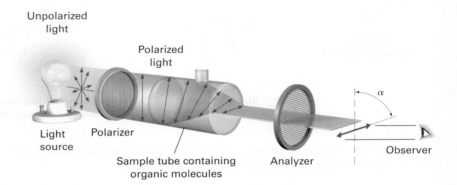

In addition to determining the extent of rotation, we can also find the direction. From the vantage point of the observer looking directly at the analyzer, some optically active molecules rotate polarized light to the left (counterclockwise) and are said to be **levorotatory**, whereas others rotate polarized light to the right (clockwise) and are said to be **dextrorotatory**. By convention, rotation to the left is given a minus sign (−) and rotation to the right is given a plus sign (+). (−)-Morphine, for example, is levorotatory, and (+)-sucrose is dextrorotatory.

The extent of rotation observed in a polarimetry experiment depends on the number of optically active molecules encountered by the light beam. This number, in turn, depends on sample concentration and sample pathlength. If the concentration of sample is doubled, the observed rotation doubles. If the concentration is kept constant but the length of the sample tube is doubled, the observed rotation doubles. It also happens that the angle of rotation depends on the wavelength of the light used.

To express optical rotations in a meaningful way so that comparisons can be made, we have to choose standard conditions. The **specific rotation, $[\alpha]_D$**, of a compound is defined as the observed rotation when light of 589.6 nanometer (nm; 1 nm = 10^{-9} m) wavelength is used with a sample pathlength l of 1 decimeter (dm; 1 dm = 10 cm) and a sample concentration c of 1 g/cm^3. (Light of 589.6 nm, the so-called sodium D line, is the yellow light emitted from common sodium street lamps.)

$$[\alpha]_D \times \frac{\text{Observed rotation (degrees)}}{\text{Pathlength, } l \text{ (dm)} \times \text{Concentration, } c \text{ (g/cm}^3)} = \frac{\alpha}{l \times c}$$

5.3 | Optical Activity

Table 5.1 Specific Rotation of Some Organic Molecules

Compound	$[\alpha]_D$	Compound	$[\alpha]_D$
Penicillin V	+233	Cholesterol	−31.5
Sucrose	+66.47	Morphine	−132
Camphor	+44.26	Cocaine	−16
Chloroform	0	Acetic acid	0

When optical rotation data are expressed in this standard way, the specific rotation, $[\alpha]_D$, is a physical constant characteristic of a given optically active compound. For example, (+)-lactic acid has $[\alpha]_D = +3.82$, and (−)-lactic acid has $[\alpha]_D = -3.82$. That is, the two enantiomers rotate plane-polarized light to exactly the same extent but in opposite directions. Note that the units of specific rotation are [(deg · cm²)/g] but that values are usually expressed without the units. Some additional examples are listed in Table 5.1.

Worked Example 5.2

Calculating an Optical Rotation

A 1.20 g sample of cocaine, $[\alpha]_D = -16$, was dissolved in 7.50 mL of chloroform and placed in a sample tube having a pathlength of 5.00 cm. What was the observed rotation?

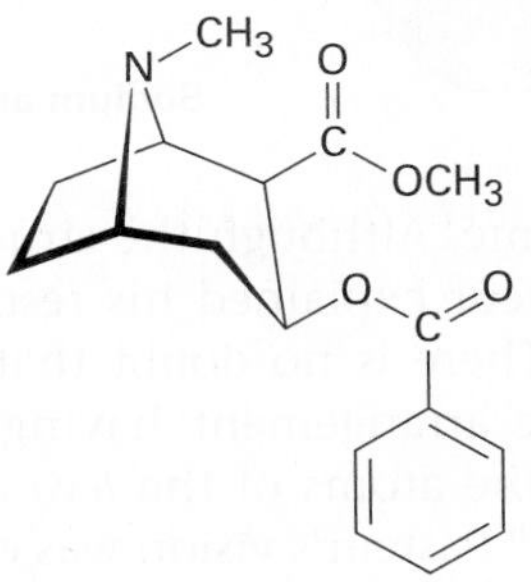

Strategy

Since $[\alpha]_D = \dfrac{\alpha}{l \times c}$

Then $\alpha = l \times c \times [\alpha]_D$

where $[\alpha]_D = -16$; $l = 5.00$ cm = 0.500 dm; $c = 1.20$ g/7.50 cm³ = 0.160 g/cm³

Solution

$\alpha = (-16)(0.500)(0.160) = -1.3°$.

Problem 5.5
Is cocaine (Worked Example 5.2) dextrorotatory or levorotatory?

Problem 5.6
A 1.50 g sample of coniine, the toxic extract of poison hemlock, was dissolved in 10.0 mL of ethanol and placed in a sample cell with a 5.00 cm pathlength. The observed rotation at the sodium D line was +1.21°. Calculate $[\alpha]_D$ for coniine.

5.4 Pasteur's Discovery of Enantiomers

Little was done after Biot's discovery of optical activity until 1848, when Louis Pasteur began work on a study of crystalline tartaric acid salts derived from wine. On crystallizing a concentrated solution of sodium ammonium tartrate below 28 °C, Pasteur made the surprising observation that two distinct kinds of crystals precipitated. Furthermore, the two kinds of crystals were nonsuperimposable mirror images and were related in the same way that a right hand is related to a left hand.

Working carefully with tweezers, Pasteur was able to separate the crystals into two piles, one of "right-handed" crystals and one of "left-handed" crystals, like those shown in **Figure 5.6**. Although the original sample, a 50:50 mixture of right and left, was optically inactive, solutions of the crystals from each of the sorted piles were optically active and their specific rotations were equal in amount but opposite in sign.

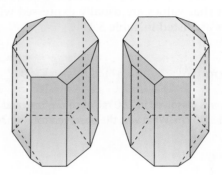

Sodium ammonium tartrate

Figure 5.6 Drawings of sodium ammonium tartrate crystals taken from Pasteur's original sketches. One of the crystals is dextrorotatory in solution, and the other is levorotatory.

Pasteur was far ahead of his time. Although the structural theory of Kekulé had not yet been proposed, Pasteur explained his results by speaking of the molecules themselves, saying, "There is no doubt that [in the *dextro* tartaric acid] there exists an asymmetric arrangement having a nonsuperimposable image. It is no less certain that the atoms of the *levo* acid have precisely the inverse asymmetric arrangement." Pasteur's vision was extraordinary, for it was not until 25 years later that his ideas regarding the asymmetric carbon atom were confirmed.

Today, we would describe Pasteur's work by saying that he had discovered enantiomers. Enantiomers, also called *optical isomers,* have identical physical properties, such as melting point and boiling point, but differ in the direction in which their solutions rotate plane-polarized light.

5.5 Sequence Rules for Specifying Configuration

Key IDEAS

Test your knowledge of Key Ideas by answering end-of-chapter exercises marked with ▲.

Structural drawings provide a visual representation of stereochemistry, but a written method for indicating the three-dimensional arrangement, or **configuration**, of substituents at a chirality center is also needed. The method used employs a set of *sequence rules* to rank the four groups attached to the chirality center and then looks at the handedness with which those groups are attached.

5.5 | Sequence Rules for Specifying Configuration

Called the **Cahn–Ingold–Prelog rules** after the chemists who proposed them, the sequence rules are as follows:

RULE 1
Look at the four atoms directly attached to the chirality center, and rank them according to atomic number. The atom with the highest atomic number has the highest ranking (first), and the atom with the lowest atomic number (usually hydrogen) has the lowest ranking (fourth). When different isotopes of the same element are compared, such as deuterium (^{2}H) and protium (^{1}H), the heavier isotope ranks higher than the lighter isotope. Thus, atoms commonly found in organic compounds have the following order.

Atomic number	35	17	16	15	8	7	6	(2)	(1)	
Higher ranking	Br >	Cl >	S >	P >	O >	N >	C >	^{2}H >	^{1}H	Lower ranking

RULE 2
If a decision can't be reached by ranking the first atoms in the substituent, look at the second, third, or fourth atoms away from the chirality center until the first difference is found. A $-CH_2CH_3$ substituent and a $-CH_3$ substituent are equivalent by rule 1 because both have carbon as the first atom. By rule 2, however, ethyl ranks higher than methyl because ethyl has a *carbon* as its highest second atom, while methyl has only *hydrogen* as its second atom. Look at the following pairs of examples to see how the rule works:

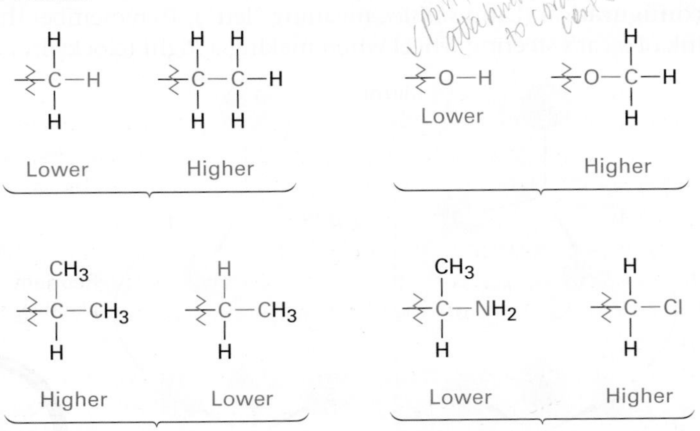

RULE 3
Multiple-bonded atoms are equivalent to the same number of single-bonded atoms. For example, an aldehyde substituent ($-CH=O$), which has a carbon atom *doubly* bonded to *one* oxygen, is equivalent to a substituent having a carbon atom *singly* bonded to *two* oxygens:

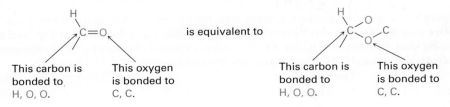

As further examples, the following pairs are equivalent:

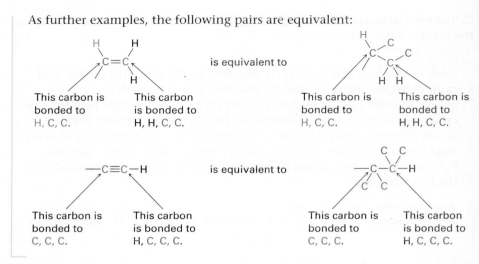

Having ranked the four groups attached to a chiral carbon, we describe the stereochemical configuration around the carbon by orienting the molecule so that the group with the lowest ranking (4) points directly back, away from us. We then look at the three remaining substituents, which now appear to radiate toward us like the spokes on a steering wheel **(Figure 5.7)**. If a curved arrow drawn from the highest to second-highest to third-highest ranked substituent (1 → 2 → 3) is clockwise, we say that the chirality center has the ***R* configuration** (Latin *rectus,* meaning "right"). If an arrow from 1 → 2 → 3 is counterclockwise, the chirality center has the ***S* configuration** (Latin *sinister,* meaning "left"). To remember these assignments, think of a car's steering wheel when making a *R*ight (clockwise) turn.

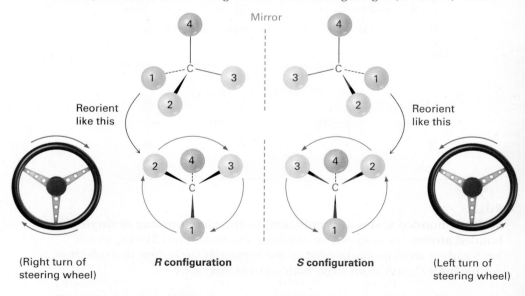

Figure 5.7 Assigning configuration to a chirality center. When the molecule is oriented so that the lowest-ranked group (**4**) is toward the rear, the remaining three groups radiate toward the viewer like the spokes of a steering wheel. If the direction of travel 1 → 2 → 3 is clockwise (right turn), the center has the *R* configuration. If the direction of travel 1 → 2 → 3 is counterclockwise (left turn), the center is *S*.

Look at (−)-lactic acid in **Figure 5.8** for an example of how to assign configuration. Sequence rule 1 says that −OH is ranked 1 and −H is ranked 4, but it doesn't allow us to distinguish between −CH$_3$ and −CO$_2$H because both groups have carbon as their first atom. Sequence rule 2, however, says that −CO$_2$H ranks higher than −CH$_3$ because O (the highest second atom in −CO$_2$H) outranks H (the highest second atom in −CH$_3$). Now, turn the molecule so that the fourth-ranked group (−H) is oriented toward the rear, away from the observer. Since a curved arrow from 1 (−OH) to 2 (−CO$_2$H) to 3 (−CH$_3$) is clockwise (right turn of the steering wheel), (−)-lactic acid has the *R* configuration. Applying the same procedure to (+)-lactic acid leads to the opposite assignment.

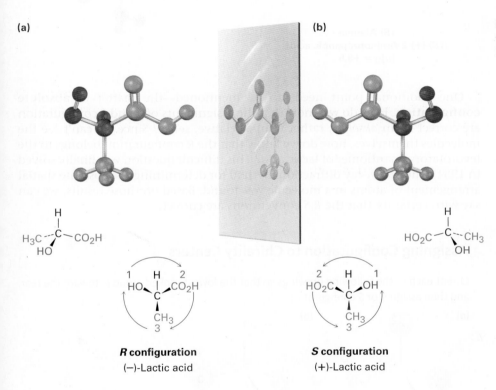

Figure 5.8 Assigning configuration to **(a)** (*R*)-(−)-lactic acid and **(b)** (*S*)-(+)-lactic acid.

Further examples are provided by naturally occurring (−)-glyceraldehyde and (+)-alanine, which both have the S configuration as shown in **Figure 5.9**. Note that the sign of optical rotation, (+) or (−), is not related to the *R*,*S* designation. (*S*)-Glyceraldehyde happens to be levorotatory (−), and (*S*)-alanine happens to be dextrorotatory (+). There is no simple correlation between *R*,*S* configuration and direction or magnitude of optical rotation.

Figure 5.9 Assigning configuration to (a) (−)-glyceraldehyde. (b) (+)-alanine. Both happen to have the S configuration, although one is levorotatory and the other is dextrorotatory.

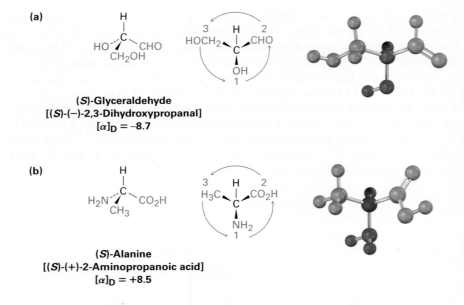

(a)

(S)-Glyceraldehyde
[(S)-(−)-2,3-Dihydroxypropanal]
[α]$_D$ = −8.7

(b)

(S)-Alanine
[(S)-(+)-2-Aminopropanoic acid]
[α]$_D$ = +8.5

One additional point needs to be mentioned—the matter of **absolute configuration**. How do we know that the assignments of R and S configuration are correct in an *absolute*, rather than a relative, sense? Since we can't see the molecules themselves, how do we know that the R configuration belongs to the levorotatory enantiomer of lactic acid? This difficult question was finally solved in 1951, when an X-ray diffraction method for determining the absolute spatial arrangement of atoms in a molecule was found. Based on those results, we can say with certainty that the R,S conventions are correct.

Worked Example 5.3 — Assigning Configuration to Chirality Centers

Orient each of the following drawings so that the lowest-ranked group is toward the rear, and then assign R or S configuration:

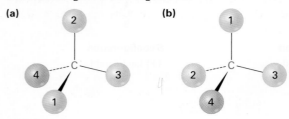

Strategy

It takes practice to be able to visualize and orient a chirality center in three dimensions. You might start by indicating where the observer must be located—180° opposite the lowest-ranked group. Then imagine yourself in the position of the observer, and redraw what you would see.

Solution

In **(a)**, you would be located in front of the page toward the top right of the molecule, and you would see group 2 to your left, group 3 to your right, and group 1 below you. This corresponds to an R configuration.

(a)

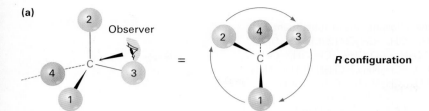

R configuration

In **(b)**, you would be located behind the page toward the top left of the molecule from your point of view, and you would see group 3 to your left, group 1 to your right, and group 2 below you. This also corresponds to an *R* configuration.

(b)

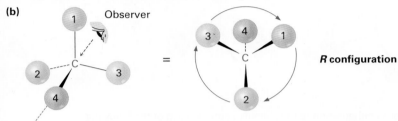

R configuration

Worked Example 5.4
Drawing the Three-Dimensional Structure of a Specific Enantiomer

Draw a tetrahedral representation of (*R*)-2-chlorobutane.

Strategy

Begin by ranking the four substituents bonded to the chirality center: (1) —Cl, (2) —CH$_2$CH$_3$, (3) —CH$_3$, (4) —H. To draw a tetrahedral representation of the molecule, orient the lowest-ranked group (—H) away from you and imagine that the other three groups are coming out of the page toward you. Then place the remaining three substituents such that the direction of travel 1 → 2 → 3 is clockwise (right turn), and tilt the molecule toward you to bring the rear hydrogen into view. Using molecular models is a great help in working problems of this sort.

Solution

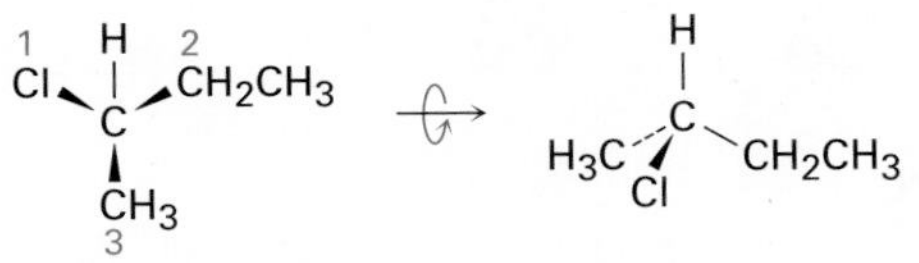

(*R*)-2-Chlorobutane

Problem 5.7

Which member in each of the following sets ranks higher?
(a) —H or —Br
(b) —Cl or —Br
(c) —CH$_3$ or —CH$_2$CH$_3$
(d) —NH$_2$ or —OH
(e) —CH$_2$OH or —CH$_3$
(f) —CH$_2$OH or —CH=O

Problem 5.8
Rank the following sets of substituents:
(a) –H, –OH, –CH$_2$CH$_3$, –CH$_2$CH$_2$OH
(b) –CO$_2$H, –CO$_2$CH$_3$, –CH$_2$OH, –OH
(c) –CN, –CH$_2$NH$_2$, –CH$_2$NHCH$_3$, –NH$_2$
(d) –SH, –CH$_2$SCH$_3$, –CH$_3$, –SSCH$_3$

Problem 5.9
Orient each of the following drawings so that the lowest-ranked group is toward the rear, and then assign R or S configuration:

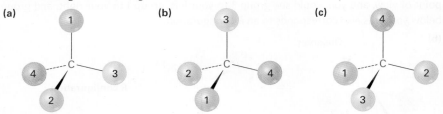

Problem 5.10
Assign R or S configuration to the chirality center in each of the following molecules:

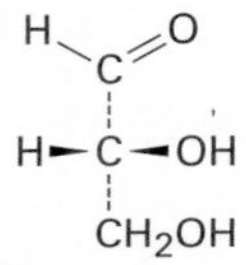

Problem 5.11
Draw a tetrahedral representation of (S)-2-pentanol (2-hydroxypentane).

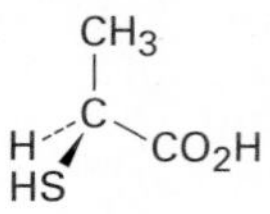

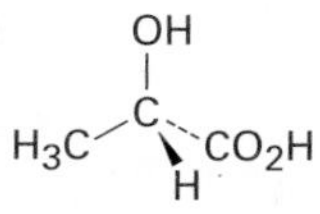

Problem 5.12
Assign R or S configuration to the chirality center in the following molecular model of the amino acid methionine (blue = N, yellow = S):

5.6 Diastereomers

Molecules like lactic acid, alanine, and glyceraldehyde are relatively simple because each has only one chirality center and only two stereoisomers. The situation becomes more complex, however, with molecules that have more than one chirality center. As a general rule, a molecule with n chirality centers can have up to 2^n stereoisomers (although it may have fewer, as we'll see below). Take the amino acid threonine (2-amino-3-hydroxybutanoic acid), for example. Since threonine has two chirality centers (C2 and C3), there are four possible stereoisomers, as shown in **Figure 5.10**. Check for yourself that the R,S configurations are correct.

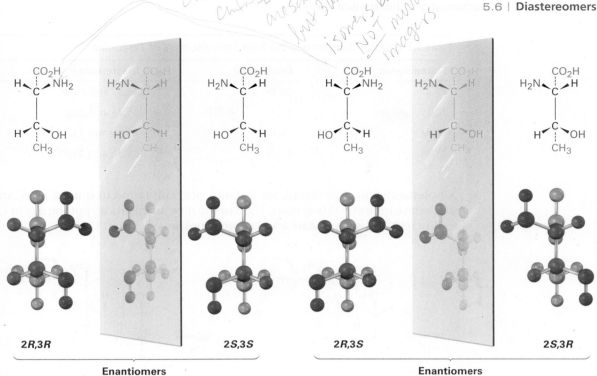

Figure 5.10 The four stereoisomers of 2-amino-3-hydroxybutanoic acid.

The four stereoisomers of 2-amino-3-hydroxybutanoic acid can be grouped into two pairs of enantiomers. The 2R,3R stereoisomer is the mirror image of 2S,3S, and the 2R,3S stereoisomer is the mirror image of 2S,3R. But what is the relationship between any two molecules that are not mirror images? What, for instance, is the relationship between the 2R,3R isomer and the 2R,3S isomer? They are stereoisomers, yet they aren't enantiomers. To describe such a relationship, we need a new term—*diastereomer*.

Diastereomers are stereoisomers that are not mirror images. Since we used the right-hand/left-hand analogy to describe the relationship between two enantiomers, we might extend the analogy by saying that the relationship between diastereomers is like that of hands from different people. Your hand and your friend's hand look *similar*, but they aren't identical and they aren't mirror images. The same is true of diastereomers: they're similar, but they aren't identical and they aren't mirror images.

Note carefully the difference between enantiomers and diastereomers: enantiomers have opposite configurations at *all* chirality centers, whereas diastereomers have opposite configurations at *some* (one or more) chirality centers but the same configuration at others. A full description of the four stereoisomers of threonine is given in Table 5.2. Of the four, only the 2S,3R isomer, $[\alpha]_D = -28.3$, occurs naturally in plants and animals and is an essential human nutrient. This result is typical: most biological molecules are chiral, and usually only one stereoisomer is found in nature.

In the special case where two diastereomers differ at only one chirality center but are the same at all others, we say that the compounds are **epimers**.

Table 5.2 Relationships among the Four Stereoisomers of Threonine

Stereoisomer	Enantiomer	Diastereomer
2R,3R	2S,3S	2R,3S and 2S,3R
2S,3S	2R,3R	2R,3S and 2S,3R
2R,3S	2S,3R	2R,3R and 2S,3S
2S,3R	2R,3S	2R,3R and 2S,3S

Cholestanol and coprostanol, for instance, are both found in human feces, and both have nine chirality centers. Eight of the nine are identical, but the one at C5 is different. Thus, cholestanol and coprostanol are *epimeric* at C5.

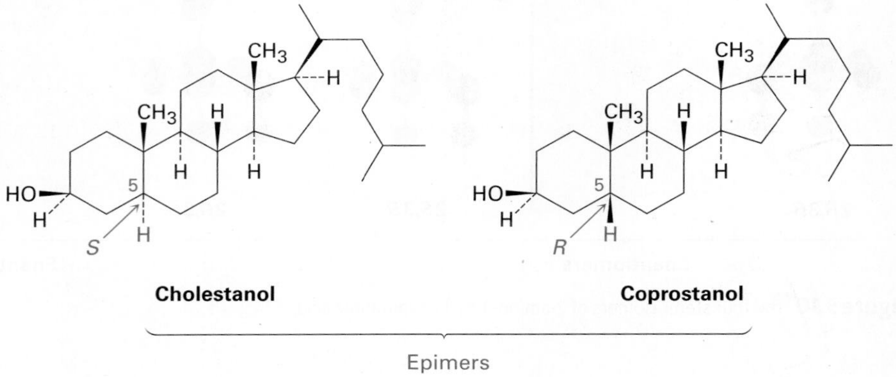

Cholestanol

Coprostanol

Epimers

Problem 5.13
One of the following molecules **(a)–(d)** is D-erythrose 4-phosphate, an intermediate in the Calvin photosynthetic cycle by which plants incorporate CO_2 into carbohydrates. If D-erythrose 4-phosphate has *R* stereochemistry at both chirality centers, which of the structures is it? Which of the remaining three structures is the enantiomer of D-erythrose 4-phosphate, and which are diastereomers?

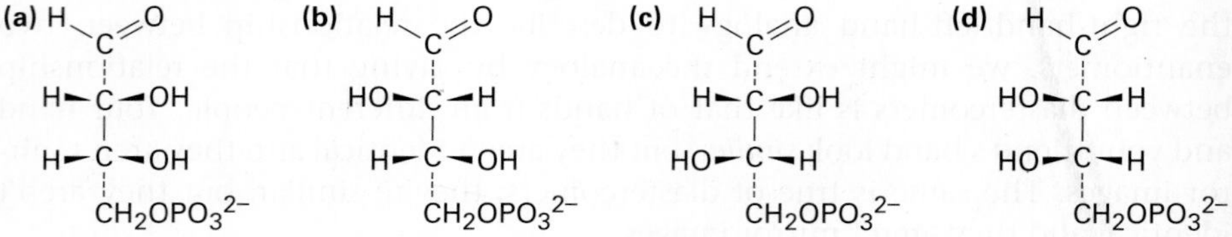

(a)
H\C=O
H—C—OH
H—C—OH
$CH_2OPO_3^{2-}$

(b)
H\C=O
HO—C—H
H—C—OH
$CH_2OPO_3^{2-}$

(c)
H\C=O
H—C—OH
HO—C—H
$CH_2OPO_3^{2-}$

(d)
H\C=O
HO—C—H
HO—C—H
$CH_2OPO_3^{2-}$

Problem 5.14
How many chirality centers does morphine have? How many stereoisomers of morphine are possible in principle?

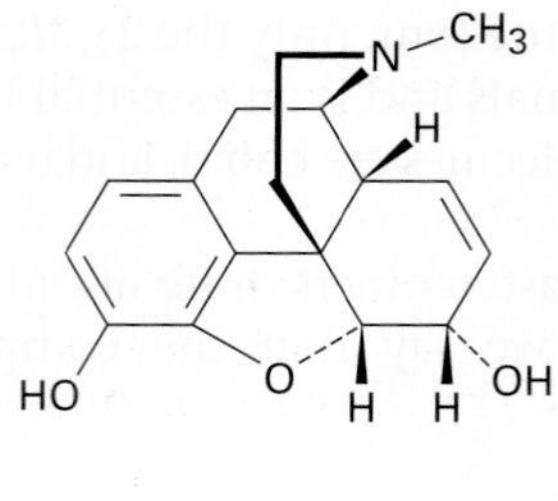

Morphine

Problem 5.15

Assign *R*,*S* configuration to each chirality center in the following molecular model of the amino acid isoleucine (blue = N):

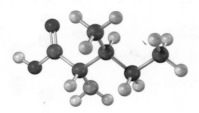

5.7 Meso Compounds

Let's look at another example of a compound with more than one chirality center: the tartaric acid used by Pasteur. The four stereoisomers can be drawn as follows:

2*R*,3*R* **2*S*,3*S*** **2*R*,3*S*** **2*S*,3*R***

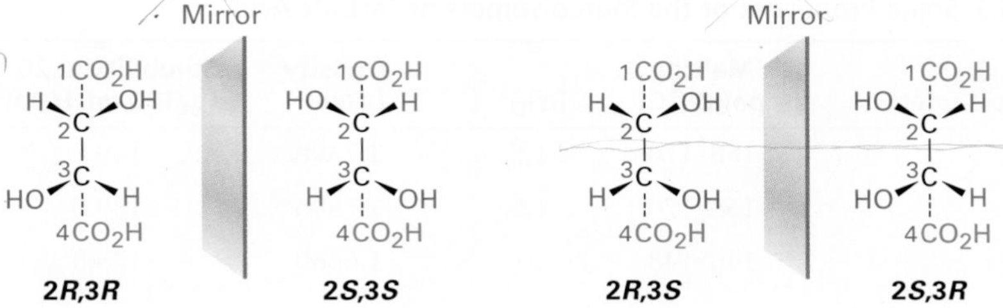

The 2*R*,3*R* and 2*S*,3*S* structures are nonsuperimposable mirror images and therefore represent a pair of enantiomers. A close look at the 2*R*,3*S* and 2*S*,3*R* structures, however, shows that they *are* superimposable, and thus identical, as can be seen by rotating one structure 180°.

2*R*,3*S* **2*S*,3*R***

Identical

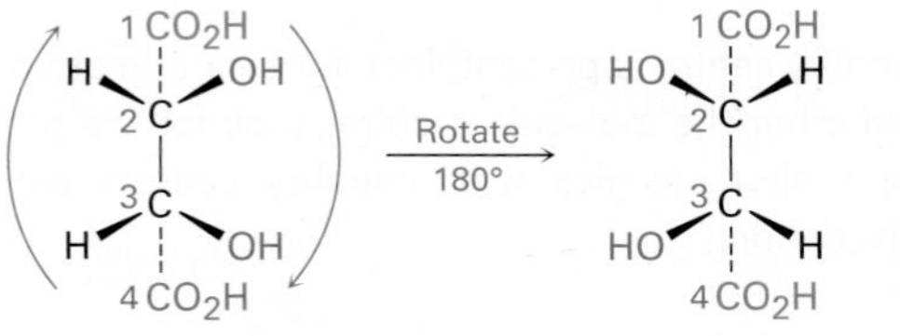

The 2*R*,3*S* and 2*S*,3*R* structures are identical because the molecule has a plane of symmetry and is therefore achiral. The symmetry plane cuts through the C2–C3 bond, making one half of the molecule a mirror image of the other half **(Figure 5.11)**. Because of the plane of symmetry, the molecule is achiral, despite the fact that it has two chirality centers. Compounds that are achiral, yet contain chirality centers, are called **meso compounds (me-zo)**. Thus,

Figure 5.11 A symmetry plane through the C2–C3 bond of *meso*-tartaric acid makes the molecule achiral.

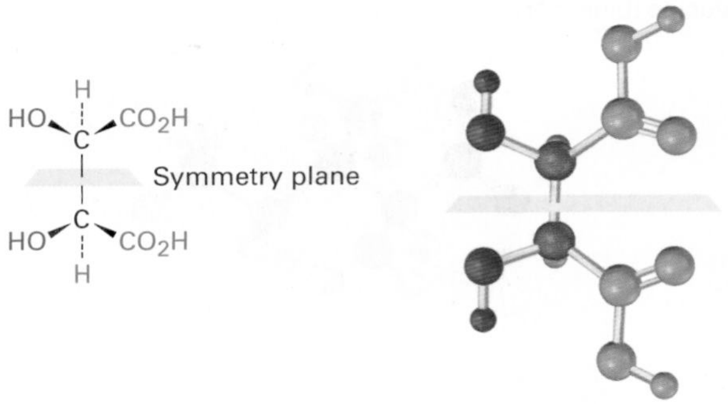

tartaric acid exists in three stereoisomeric forms: two enantiomers and one meso form.

Some physical properties of the three stereoisomers are listed in Table 5.3. The (+)- and (−)-tartaric acids have identical melting points, solubilities, and densities, but they differ in the sign of their rotation of plane-polarized light. The meso isomer, by contrast, is diastereomeric with the (+) and (−) forms. It has no mirror-image relationship to (+)- and (−)-tartaric acids, is a different compound altogether, and has different physical properties.

Table 5.3 Some Properties of the Stereoisomers of Tartaric Acid

Stereoisomer	Melting point (°C)	$[\alpha]_D$	Density (g/cm^3)	Solubility at 20 °C (g/100 mL H$_2$O)
(+)	168–170	+12	1.7598	139.0
(−)	168–170	−12	1.7598	139.0
Meso	146–148	0	1.6660	125.0

Worked Example 5.5 Distinguishing Chiral Compounds from Meso Compounds

Does *cis*-1,2-dimethylcyclobutane have any chirality centers? Is it chiral?

Strategy
To see whether a chirality center is present, look for a carbon atom bonded to four different groups. To see whether the molecule is chiral, look for the presence or absence of a symmetry plane. Not all molecules with chirality centers are chiral overall—meso compounds are an exception.

Solution
A look at the structure of *cis*-1,2-dimethylcyclobutane shows that both methyl-bearing ring carbons (C1 and C2) are chirality centers. Overall, though, the compound is achiral because there is a symmetry plane bisecting the ring between C1 and C2. Thus, the molecule is a meso compound.

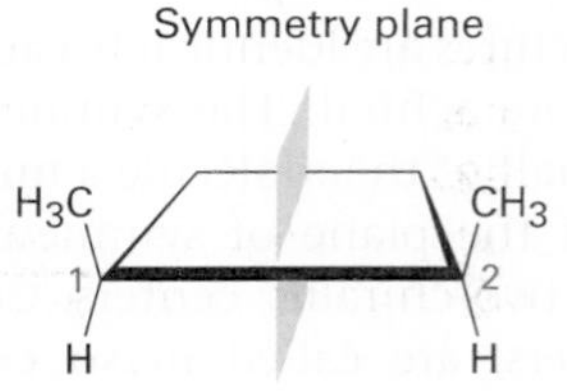

Problem 5.16
Which of the following structures represent meso compounds?

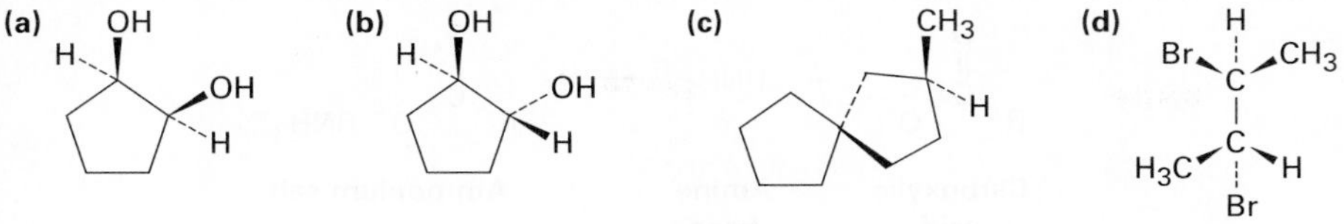

(a) [cyclopentane with OH groups] (b) [cyclopentane with OH groups] (c) [spiro compound with CH₃ and H] (d) [Br, CH₃, H₃C, H, Br structure]

Problem 5.17
Which of the following have a meso form? (Recall that the *-ol* suffix refers to an alcohol, ROH.)

(a) 2,3-Butanediol (b) 2,3-Pentanediol (c) 2,4-Pentanediol

Problem 5.18
Does the following structure represent a meso compound? If so, indicate the symmetry plane.

5.8 Racemic Mixtures and the Resolution of Enantiomers

To end this discussion of stereoisomerism, let's return for a last look at Pasteur's pioneering work, described in **Section 5.4**. Pasteur took an optically inactive tartaric acid salt and found that he could crystallize from it two optically active forms having what we would now call the 2*R*,3*R* and 2*S*,3*S* configurations. But what was the optically inactive form he started with? It couldn't have been *meso*-tartaric acid, because *meso*-tartaric acid is a different chemical compound and can't interconvert with the two chiral enantiomers without breaking and re-forming chemical bonds.

The answer is that Pasteur started with a 50 : 50 *mixture* of the two chiral tartaric acid enantiomers. Such a mixture is called a **racemate** (**raa**-suh-mate), or *racemic mixture*, and is denoted by either the symbol (±) or the prefix *d,l* to indicate an equal mixture of dextrorotatory and levorotatory forms. Racemates show no optical rotation because the (+) rotation from one enantiomer exactly cancels the (−) rotation from the other. Through luck, Pasteur was able to separate, or **resolve**, racemic tartaric acid into its (+) and (−) enantiomers. Unfortunately, the fractional crystallization technique he used doesn't work for most racemates, so other methods are needed.

The most common method of resolution uses an acid–base reaction between the racemate of a chiral carboxylic acid (RCO_2H) and an amine base (RNH_2) to yield an ammonium salt:

$$R-CO-OH + RNH_2 \longrightarrow R-CO-O^- \ RNH_3^+$$

Carboxylic acid   **Amine base**   **Ammonium salt**

To understand how this method of resolution works, let's see what happens when a racemic mixture of chiral acids, such as (+)- and (−)-lactic acids, reacts with an achiral amine base, such as methylamine, CH_3NH_2. Stereochemically, the situation is analogous to what happens when left and right hands (chiral) pick up a ball (achiral). Both left and right hands pick up the ball equally well, and the products—ball in right hand versus ball in left hand—are mirror images. In the same way, both (+)- and (−)-lactic acid react with methylamine equally well, and the product is a racemic mixture of the two enantiomers methylammonium (+)-lactate and methylammonium (−)-lactate (**Figure 5.12**).

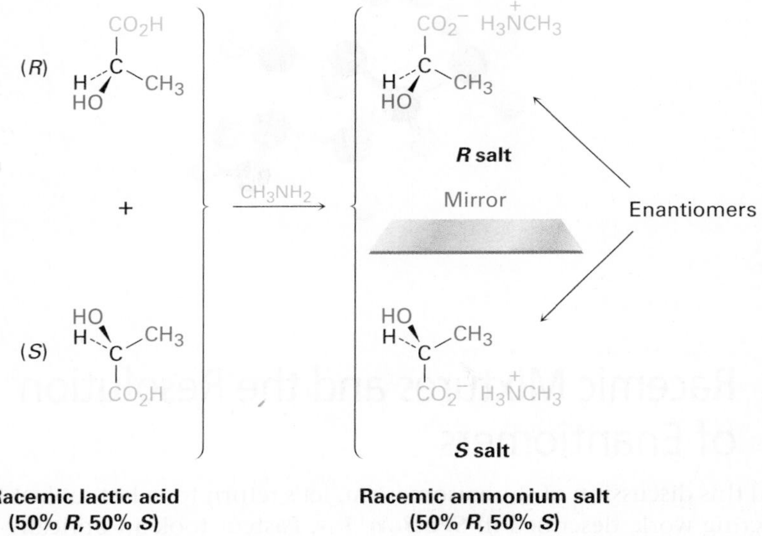

Figure 5.12 Reaction of racemic lactic acid with achiral methylamine leads to a racemic mixture of ammonium salts.

Racemic lactic acid (50% R, 50% S) → **Racemic ammonium salt (50% R, 50% S)**

Now let's see what happens when the racemic mixture of (+)- and (−)-lactic acids reacts with a single enantiomer of a chiral amine base, such as (R)-1-phenylethylamine. Stereochemically, the situation is analogous to what happens when left and right hands (chiral) put on a right-handed glove (*also chiral*). Left and right hands don't put on the right-handed glove in the same way, so the products—right hand in right glove versus left hand in right glove—are not mirror images; they're similar but different.

In the same way, (+)- and (−)-lactic acids react with (R)-1-phenylethylamine to give two different products (**Figure 5.13**). (R)-Lactic acid reacts with (R)-1-phenylethylamine to give the R,R salt, and (S)-lactic acid reacts with the R amine to give the S,R salt. *The two salts are diastereomers.* They have different chemical and physical properties, and it may therefore be possible to separate them by crystallization or some other means. Once separated, acidification of the two diastereomeric salts with a strong acid then allows us to isolate the two pure enantiomers of lactic acid and to recover the chiral amine for reuse.

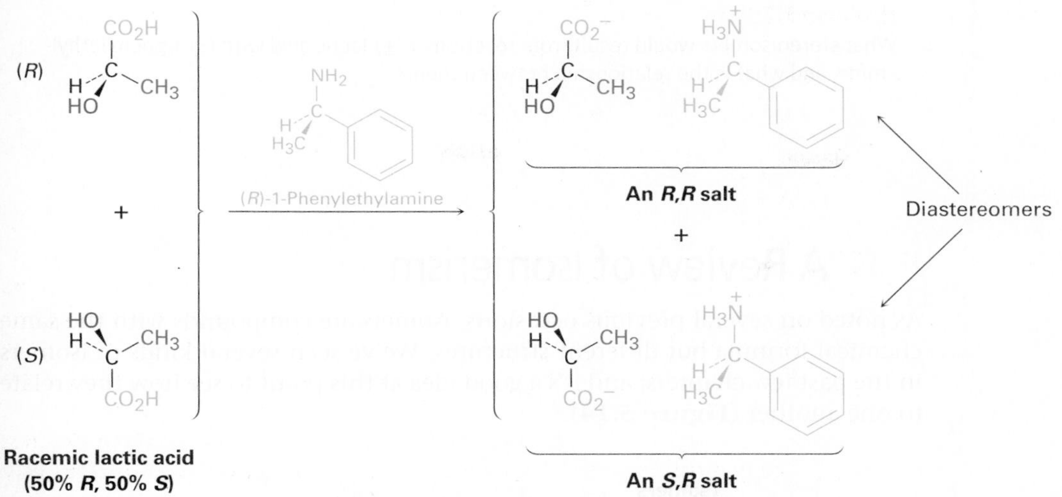

Figure 5.13 Reaction of racemic lactic acid with (R)-1-phenylethylamine yields a mixture of diastereomeric ammonium salts, which have different properties and can be separated.

Predicting the Chirality of a Reaction Product

Worked Example 5.6

We'll see in **Section 21.3** that carboxylic acids (RCO_2H) react with alcohols ($R'OH$) to form esters (RCO_2R'). Suppose that (±)-lactic acid reacts with CH_3OH to form the ester, methyl lactate. What stereochemistry would you expect the product(s) to have? What is the relationship of the products?

$$CH_3CHCOH + CH_3OH \xrightarrow{\text{Acid catalyst}} CH_3CHCOCH_3 + H_2O$$

Lactic acid **Methanol** **Methyl lactate**

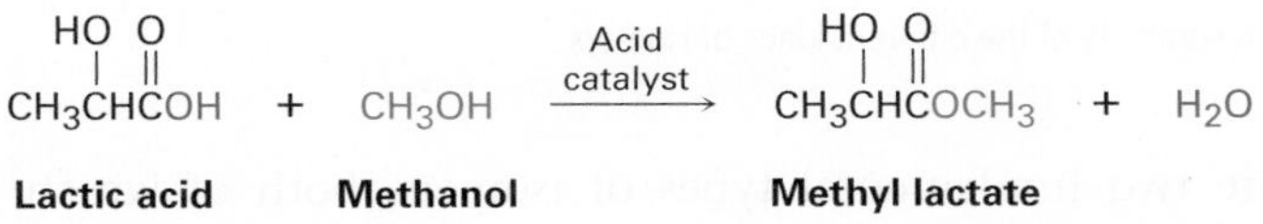

Solution
Reaction of a racemic acid with an achiral alcohol such as methanol yields a racemic mixture of mirror-image (enantiomeric) products.

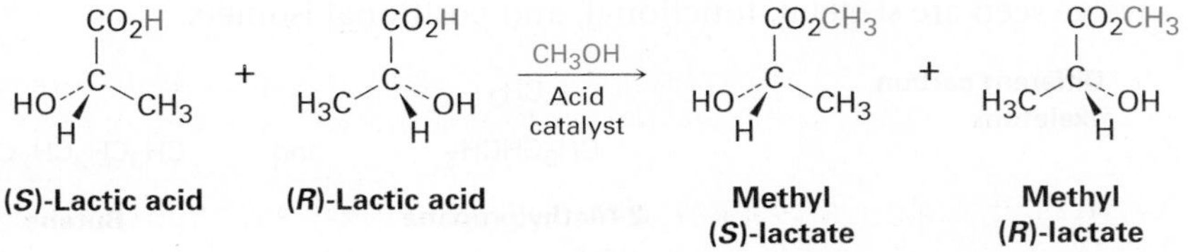

(S)-Lactic acid + (R)-Lactic acid $\xrightarrow[\text{catalyst}]{CH_3OH \text{ Acid}}$ Methyl (S)-lactate + Methyl (R)-lactate

Problem 5.19
Suppose that acetic acid (CH_3CO_2H) reacts with (S)-2-butanol to form an ester (see Worked Example 5.6). What stereochemistry would you expect the product(s) to have? What is the relationship of the products?

$$CH_3COH + CH_3CHCH_2CH_3 \xrightarrow{\text{Acid catalyst}} CH_3COCHCH_2CH_3 + H_2O$$

Acetic acid **2-Butanol** ***sec*-Butyl acetate**

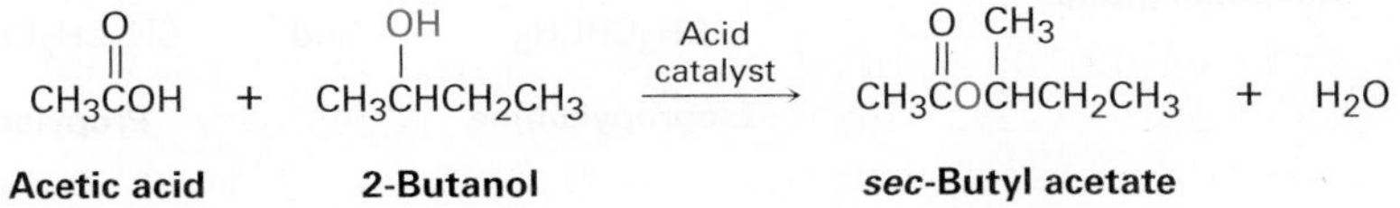

Problem 5.20
What stereoisomers would result from reaction of (±)-lactic acid with (S)-1-phenylethylamine, and what is the relationship between them?

5.9 A Review of Isomerism

As noted on several previous occasions, isomers are compounds with the same chemical formula but different structures. We've seen several kinds of isomers in the past few chapters, and it's a good idea at this point to see how they relate to one another **(Figure 5.14)**.

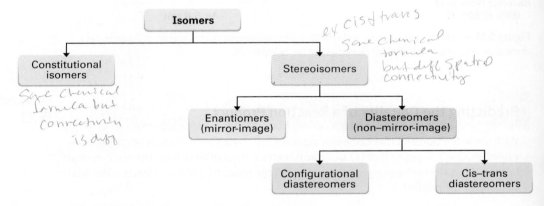

Figure 5.14 A summary of the different kinds of isomers.

There are two fundamental types of isomers, both of which we've now encountered: constitutional isomers and stereoisomers.

Constitutional isomers (Section 3.2) are compounds whose atoms are connected differently. Among the kinds of constitutional isomers we've seen are skeletal, functional, and positional isomers.

Different carbon skeletons	CH_3CHCH_3 with CH_3 branch	and	$CH_3CH_2CH_2CH_3$
	2-Methylpropane		**Butane**
Different functional groups	CH_3CH_2OH	and	CH_3OCH_3
	Ethyl alcohol		**Dimethyl ether**
Different position of functional groups	CH_3CHCH_3 with NH_2	and	$CH_3CH_2CH_2NH_2$
	Isopropylamine		**Propylamine**

Stereoisomers (Section 4.2) are compounds whose atoms are connected in the same order but with a different spatial arrangement. Among the kinds of stereoisomers we've seen are enantiomers, diastereomers, and cis–trans isomers of cycloalkanes. Actually, cis–trans isomers are just a subclass of diastereomers because they are non–mirror-image stereoisomers:

Enantiomers (nonsuperimposable mirror-image stereoisomers)

(*R*)-Lactic acid

(*S*)-Lactic acid

Diastereomers (nonsuperimposable non–mirror-image stereoisomers)

Configurational diastereomers

(2*R*,3*R*-2)-Amino-3-hydroxybutanoic acid

(2*R*,3*S*-2)-Amino-3-hydroxybutanoic acid

Cis–trans diastereomers (substituents on same side or opposite side of double bond or ring)

trans-1,3-Dimethylcyclopentane

and

cis-1,3-Dimethylcyclopentane

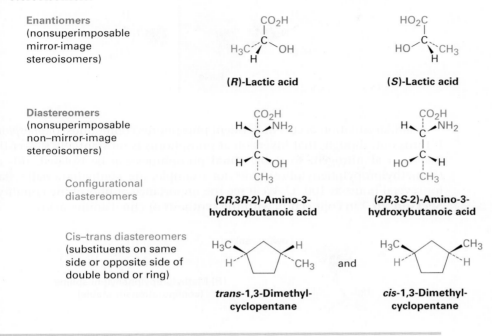

Problem 5.21

What kinds of isomers are the following pairs?

(a) (*S*)-5-Chloro-2-hexene and chlorocyclohexane

(b) (2*R*,3*R*)-Dibromopentane and (2*S*,3*R*)-dibromopentane

5.10 Chirality at Nitrogen, Phosphorus, and Sulfur

Although the most common cause of chirality is the presence of four different substituents bonded to a tetrahedral atom, that atom doesn't necessarily have to be carbon. Nitrogen, phosphorus, and sulfur are all commonly encountered in organic molecules, and all can be chirality centers. We know, for instance, that trivalent nitrogen is tetrahedral, with its lone pair of electrons acting as the fourth "substituent" **(Section 1.10)**. Is trivalent nitrogen chiral? Does a compound such as ethylmethylamine exist as a pair of enantiomers?

The answer is both yes and no. Yes in principle, but no in practice. Most trivalent nitrogen compounds undergo a rapid umbrella-like inversion that interconverts enantiomers, so we can't isolate individual enantiomers except in special cases.

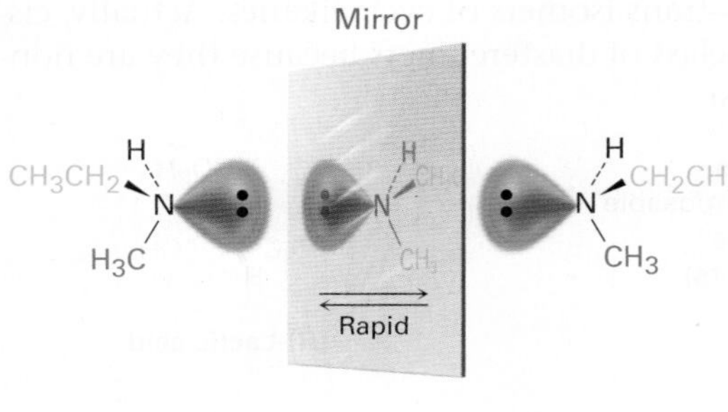

A similar situation occurs in trivalent phosphorus compounds, or *phosphines*. It turns out, though, that inversion at phosphorus is substantially slower than inversion at nitrogen, so stable chiral phosphines *can* be isolated. (*R*)- and (*S*)-methylpropylphenylphosphine, for example, are configurationally stable for several hours at 100 °C. We'll see the importance of phosphine chirality in **Section 26.7** in connection with the synthesis of chiral amino acids.

(*R*)-Methylpropylphenylphosphine (configurationally stable)

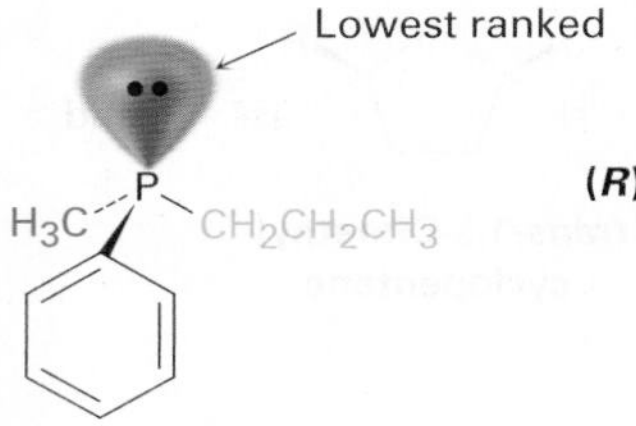

Divalent sulfur compounds are achiral, but trivalent sulfur compounds called *sulfonium salts* (R_3S^+) can be chiral. Like phosphines, sulfonium salts undergo relatively slow inversion, so chiral sulfonium salts are configurationally stable and can be isolated. Perhaps the best known example is the coenzyme *S*-adenosylmethionine, the so-called biological methyl donor, which is involved in many metabolic pathways as a source of CH_3 groups. (The "*S*" in the name *S*-adenosylmethionine stands for *sulfur* and means that the adenosyl group is attached to the sulfur atom of the amino acid methionine.) The molecule has *S* stereochemistry at sulfur and is configurationally stable for several days at room temperature. Its *R* enantiomer is also known but is not biologically active.

(*S*)-*S*-Adenosylmethionine

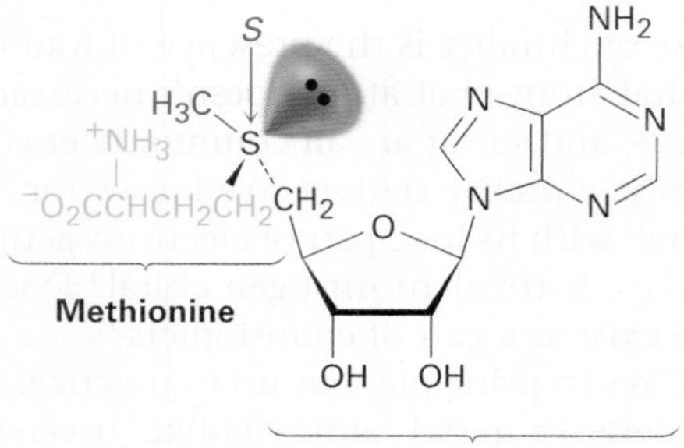

5.11 Prochirality

Closely related to the concept of chirality, and particularly important in biological chemistry, is the notion of *prochirality*. A molecule is said to be **prochiral** if it can be converted from achiral to chiral in a single chemical step. For instance, an unsymmetrical ketone like 2-butanone is prochiral because it can be converted to the chiral alcohol 2-butanol by addition of hydrogen, as we'll see in **Section 17.4**.

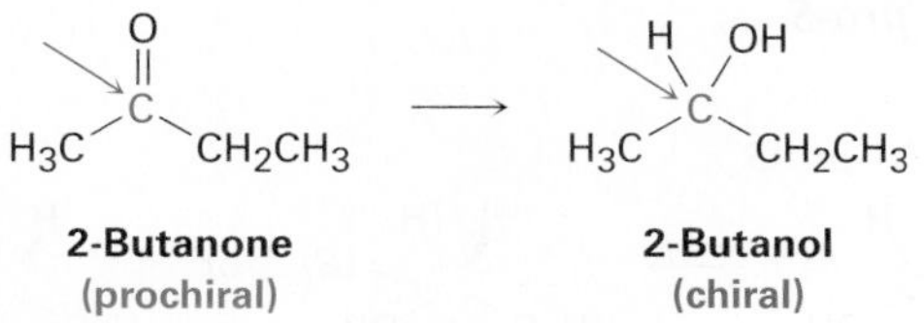

2-Butanone (prochiral) → **2-Butanol** (chiral)

Which enantiomer of 2-butanol is produced depends on which face of the planar carbonyl group undergoes reaction. To distinguish between the possibilities, we use the stereochemical descriptors *Re* and *Si*. Rank the three groups attached to the trigonal, sp^2-hybridized carbon, and imagine curved arrows from the highest to second-highest to third-highest ranked substituents. The face on which the arrows curve clockwise is designated **Re** (similar to R), and the face on which the arrows curve counterclockwise is designated **Si** (similar to S). In this particular example, addition of hydrogen from the *Re* faces gives (*S*)-butan-2-ol, and addition from the *Si* face gives (*R*)-butan-2-ol.

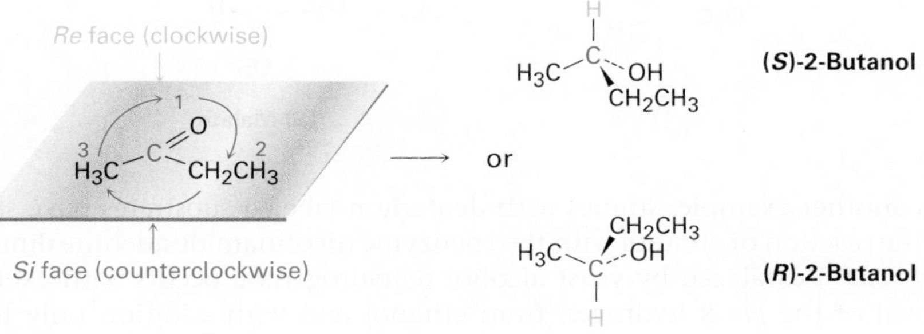

Re face (clockwise) → (**S**)-2-Butanol

or

Si face (counterclockwise) → (**R**)-2-Butanol

In addition to compounds with planar, sp^2-hybridized atoms, compounds with tetrahedral, sp^3-hybridized atoms can also be prochiral. An sp^3-hybridized atom is said to be a **prochirality center** if, by changing one of its attached groups, it becomes a chirality center. The $-CH_2OH$ carbon atom of ethanol, for instance, is a prochirality center because changing one of its attached $-H$ atoms converts it into a chirality center.

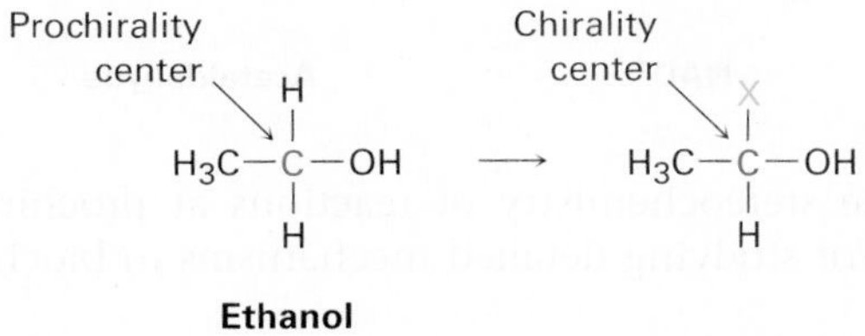

Prochirality center → Chirality center

Ethanol

To distinguish between the two identical atoms (or groups of atoms) on a prochirality center, we imagine a change that will raise the ranking of one atom over the other without affecting its rank with respect to other attached groups. On the –CH$_2$OH carbon of ethanol, for instance, we might imagine replacing one of the ^{1}H atoms (protium) by ^{2}H (deuterium). The newly introduced ^{2}H atom ranks higher than the remaining ^{1}H atom, but it remains lower than other groups attached to the carbon. Of the two identical atoms in the original compound, that atom whose replacement leads to an *R* chirality center is said to be ***pro-R*** and that atom whose replacement leads to an *S* chirality center is ***pro-S***.

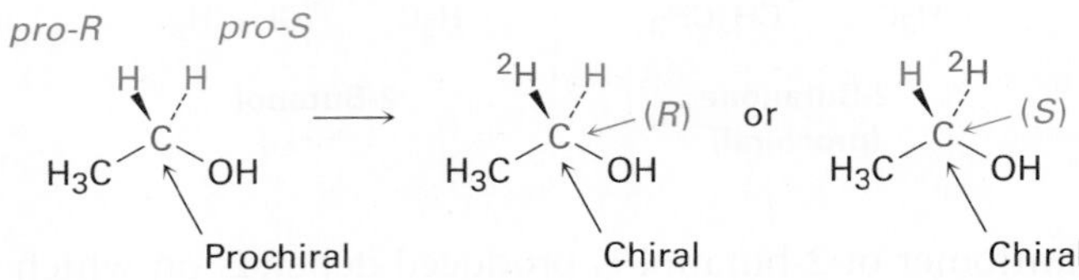

A large number of biological reactions involve prochiral compounds. One of the steps in the citric acid cycle by which food is metabolized, for instance, is the addition of H$_2$O to fumarate to give malate. Addition of –OH occurs on the *Si* face of a fumarate carbon and gives (*S*)-malate as product.

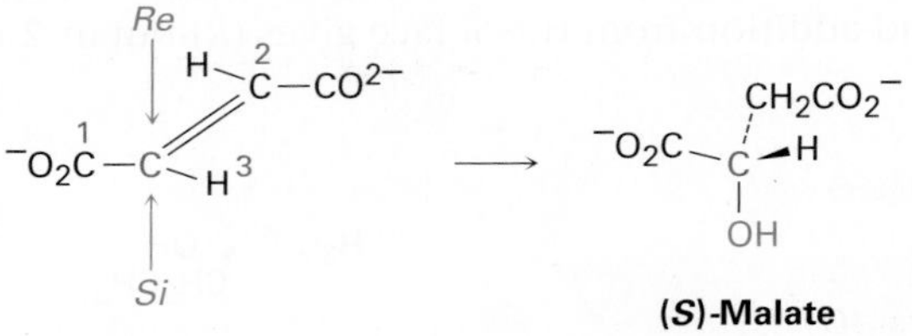

As another example, studies with deuterium-labeled substrates have shown that the reaction of ethanol with the coenzyme nicotinamide adenine dinucleotide (NAD$^+$) catalyzed by yeast alcohol dehydrogenase occurs with exclusive removal of the *pro-R* hydrogen from ethanol and with addition only to the *Re* face of NAD$^+$.

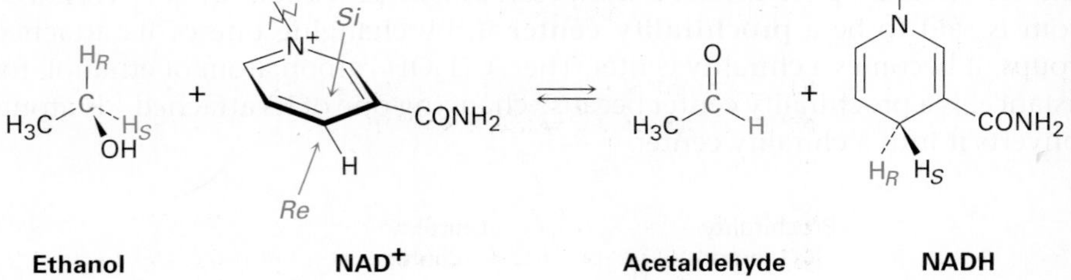

Determining the stereochemistry of reactions at prochirality centers is a powerful method for studying detailed mechanisms in biochemical reactions.

5.11 | Prochirality

As just one example, the conversion of citrate to (*cis*)-aconitate in the citric acid cycle has been shown to occur with loss of a *pro-R* hydrogen, implying that the OH and H groups leave from opposite sides of the molecule.

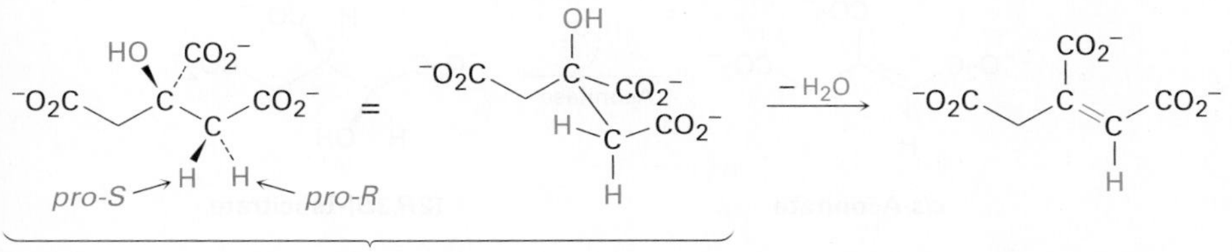

[Structures of Citrate and cis-Aconitate with pro-S and pro-R hydrogens labeled, showing loss of H₂O]

Citrate → **cis-Aconitate**

Note that when drawing compounds like threonine, cholestanol, and coprostanol, which have more than one chiral center, the wedges and dashes in a structure are used only to imply *relative* stereochemistry within the molecule rather than absolute stereochemistry, unless stated otherwise.

Problem 5.22
Identify the indicated hydrogens in the following molecules as *pro-R* or *pro-S*:

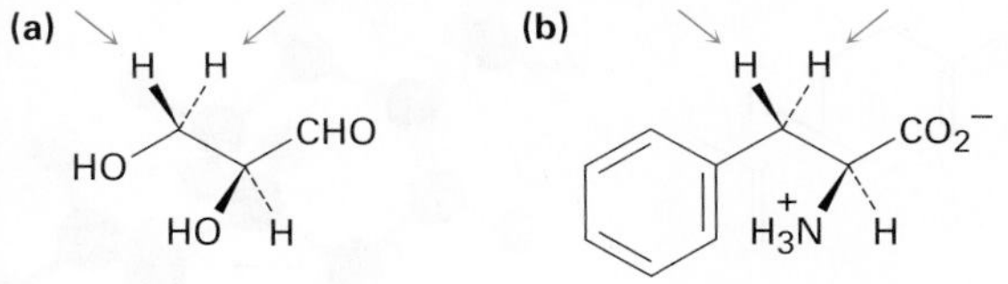

(a) (*S*)-Glyceraldehyde
(b) Phenylalanine

Problem 5.23
Identify the indicated faces of carbon atoms in the following molecules as *Re* or *Si*:

(a) Hydroxyacetone
(b) Crotyl alcohol

Problem 5.24
The lactic acid that builds up in tired muscles is formed from pyruvate. If the reaction occurs with addition of hydrogen to the *Re* face of pyruvate, what is the stereochemistry of the product?

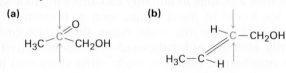

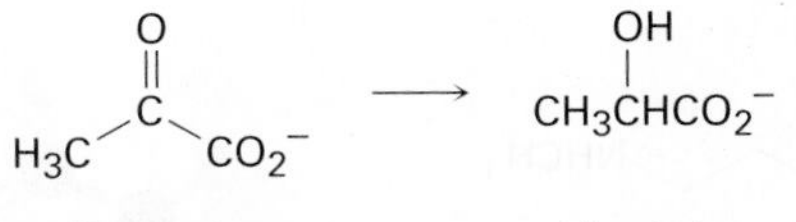

Pyruvate → Lactate

Problem 5.25
The aconitase-catalyzed addition of water to *cis*-aconitate in the citric acid cycle occurs with the following stereochemistry. Does the addition of the OH group occur on the *Re* or

the *Si* face of the substrate? What about the addition of the H? Do the H and OH groups adds from the same side of the double bond or from opposite sides?

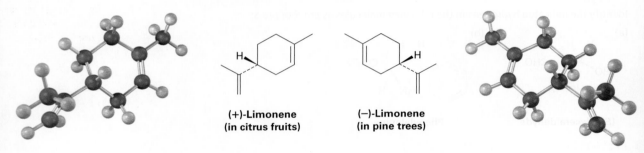

5.12 Chirality in Nature and Chiral Environments

Although the different enantiomers of a chiral molecule have the same physical properties, they usually have different biological properties. For example, the (+) enantiomer of limonene has the odor of oranges and lemons, but the (−) enantiomer has the odor of pine trees.

More dramatic examples of how a change in chirality can affect the biological properties of a molecule are found in many drugs, such as fluoxetine, a heavily prescribed medication sold under the trade name Prozac. Racemic fluoxetine is an extraordinarily effective antidepressant but has no activity against migraine. The pure *S* enantiomer, however, works remarkably well in preventing migraine. Other examples of how chirality affects biological properties are given in *A Deeper Look* at the end of this chapter.

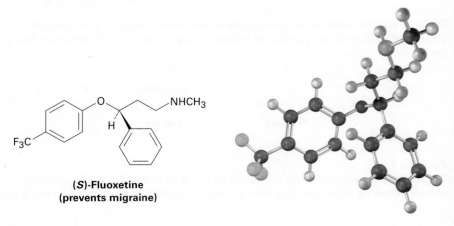

Why do different enantiomers have different biological properties? To have a biological effect, a substance typically must fit into an appropriate receptor that has an exactly complementary shape. But because biological receptors are chiral, only one enantiomer of a chiral substrate can fit in, just as only a right hand can fit into right-handed glove. The mirror-image enantiomer will be a misfit, like a left hand in a right-handed glove. A representation of the interaction between a chiral molecule and a chiral biological receptor is shown in **Figure 5.15**: one enantiomer fits the receptor perfectly, but the other does not.

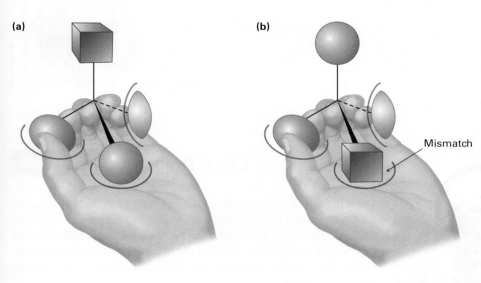

Figure 5.15 Imagine that a left hand interacts with a chiral object, much as a biological receptor interacts with a chiral molecule. **(a)** One enantiomer fits into the hand perfectly: green thumb, red palm, and gray pinkie finger, with the blue substituent exposed. **(b)** The other enantiomer, however, can't fit into the hand. When the green thumb and gray pinkie finger interact appropriately, the palm holds a blue substituent rather than a red one, with the red substituent exposed.

The hand-in-glove fit of a chiral substrate into a chiral receptor is relatively straightforward, but it's less obvious how a prochiral substrate can undergo a selective reaction. Take the reaction of ethanol with NAD^+ catalyzed by yeast alcohol dehydrogenase. As we saw at the end of **Section 5.11**, the reaction occurs with exclusive removal of the *pro-R* hydrogen from ethanol and with addition only to the *Re* face of the NAD^+ carbon.

We can understand this result by imagining that the chiral enzyme receptor again has three binding sites, as was previously the case in Figure 5.15. When green and gray substituents of a prochiral substrate are held appropriately, however, only one of the two red substituents—say, the *pro-S* one—is also held while the other, *pro-R*, substituent is exposed for reaction.

We describe the situation by saying that the receptor provides a **chiral environment** for the substrate. In the absence of a chiral environment, the two red substituents are chemically identical, but in the presence of the chiral environment, they are chemically distinctive **(Figure 5.16a)**. The situation is similar to what happens when you pick up a coffee mug. By itself, the mug has a plane of symmetry and is achiral. When you pick up the mug, however, your hand provides a chiral environment so that one side becomes much more accessible and easier to drink from than the other **(Figure 5.16b)**.

Figure 5.16 (a) When a prochiral molecule is held in a chiral environment, the **two seemingly identical substituents** are distinguishable. (b) Similarly, when an achiral coffee mug is held in the chiral environment of your hand, it's much easier to drink from one side than the other because the two sides of the mug are now distinguishable.

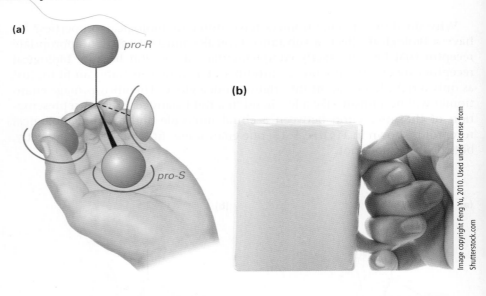

Chiral Drugs | A DEEPER LOOK

The hundreds of different pharmaceutical agents approved for use by the U.S. Food and Drug Administration come from many sources. Many drugs are isolated directly from plants or bacteria, and others are made by chemical modification of naturally occurring compounds. An estimated 33%, however, are made entirely in the laboratory and have no relatives in nature.

Those drugs that come from natural sources, either directly or after chemical modification, are usually chiral and are generally found only as a single enantiomer rather than as a racemate. Penicillin V, for example, an antibiotic isolated from the *Penicillium* mold, has the 2S,5R,6R configuration. Its enantiomer, which does not occur naturally but can be made in the laboratory, has no antibiotic activity.

The *S* enantiomer of ibuprofen soothes the aches and pains of athletic injuries much more effectively than the *R* enantiomer.

Penicillin V (2*S*,5*R*,6*R* configuration)

In contrast to drugs from natural sources, those drugs that are made entirely in the laboratory either are achiral or, if chiral, are often produced and sold as racemates. Ibuprofen, for example, has one chirality center and is sold commercially under such trade names as Advil, Nuprin, and Motrin as a 50 : 50 mixture of *R* and *S*. It turns out, however, that only

(continued)

the *S* enantiomer is active as an analgesic and anti-inflammatory agent. The *R* enantiomer of ibuprofen is inactive, although it is slowly converted in the body to the active *S* form.

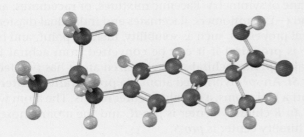

(*S*)-Ibuprofen
(an active analgesic agent)

Not only is it chemically wasteful to synthesize and administer an enantiomer that does not serve the intended purpose, many instances are now known where the presence of the "wrong" enantiomer in a racemic mixture either affects the body's ability to utilize the "right" enantiomer or has unintended pharmacological effects of its own. The presence of (*R*)-ibuprofen in the racemic mixture, for instance, slows the rate at which the *S* enantiomer takes effect in the body, from 12 minutes to 38 minutes.

To get around this problem, pharmaceutical companies attempt to devise methods of *enantioselective synthesis*, which allow them to prepare only a single enantiomer rather than a racemic mixture. Viable methods have been developed for the preparation of (*S*)-ibuprofen, which is now being marketed in Europe. We'll look further into enantioselective synthesis in the Chapter 19 *A Deeper Look*.

Summary

In this chapter, we've looked at some of the causes and consequences of molecular handedness—a topic of particular importance in understanding biological chemistry. The subject can be a bit complex but is so important that it's worthwhile spending the time needed to become familiar with it.

An object or molecule that is not superimposable on its mirror image is said to be **chiral**, meaning "handed." A chiral molecule is one that does not have a plane of symmetry cutting through it so that one half is a mirror image of the other half. The most common cause of chirality in organic molecules is the presence of a tetrahedral, sp^3-hybridized carbon atom bonded to four different groups—a so-called **chirality center**. Chiral compounds can exist as a pair of nonsuperimposable mirror-image stereoisomers called **enantiomers**. Enantiomers are identical in all physical properties except for their **optical activity**, or direction in which they rotate plane-polarized light.

The stereochemical **configuration** of a chirality center can be specified as either ***R*** (*rectus*) or ***S*** (*sinister*) by using the **Cahn–Ingold–Prelog rules**. First

Key words

absolute configuration, 154
achiral, 144
Cahn–Ingold–Prelog rules, 151
chiral, 144
chiral environment, 171
chirality center, 145
configuration, 150
dextrorotatory, 148
diastereomers, 157
enantiomers, 143
epimers, 157
levorotatory, 148

Key words—cont'd

meso compound, 159
optically active, 148
pro-R configuration, 168
pro-S configuration, 168
prochiral, 167
prochirality center, 167
R configuration, 152
racemate, 161
Re face, 167
resolution, 161
S configuration, 152
Si face, 167
specific rotation, [α]$_D$, 148

rank the four substituents on the chiral carbon atom, and then orient the molecule so that the lowest-ranked group points directly back. If a curved arrow drawn in the direction of decreasing rank (1 → 2 → 3) for the remaining three groups is clockwise, the chirality center has the *R* configuration. If the direction is counterclockwise, the chirality center has the *S* configuration.

Some molecules have more than one chirality center. Enantiomers have opposite configuration at all chirality centers, whereas **diastereomers** have the same configuration in at least one center but opposite configurations at the others. **Epimers** are diastereomers that differ in configuration at only one chirality center. A compound with *n* chirality centers can have a maximum of 2n stereoisomers.

Meso compounds contain chirality centers but are achiral overall because they have a plane of symmetry. Racemic mixtures, or **racemates**, are 50:50 mixtures of (+) and (−) enantiomers. Racemates and individual diastereomers differ in their physical properties, such as solubility, melting point, and boiling point.

A molecule is **prochiral** if it can be converted from achiral to chiral in a single chemical step. A prochiral sp^2-hybridized atom has two faces, described as either ***Re*** or ***Si***. An sp^3-hybridized atom is a **prochirality center** if, by changing one of its attached atoms, a chirality center results. The atom whose replacement leads to an *R* chirality center is ***pro-R***, and the atom whose replacement leads to an *S* chirality center is ***pro-S***.

Exercises

OWL Interactive versions of these problems are assignable in OWL for Organic Chemistry.

▲ denotes problems linked to the Key Ideas in this chapter.

Visualizing Chemistry

(Problems 5.1–5.25 appear within the chapter.)

5.26 Which of the following structures are identical? (Green = Cl.)

(a)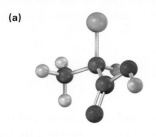

(b)

(c)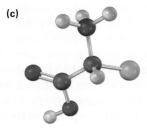

(d)

▲ Problems linked to Key Ideas in this chapter

5.27 ▲ Assign *R* or *S* configuration to the chirality centers in the following molecules (blue = N):

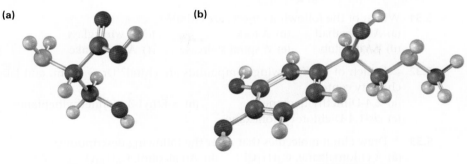

(a) Serine   (b) Adrenaline

5.28 Which, if any, of the following structures represent meso compounds? (Blue = N, green = Cl.)

(a)  (b)  (c)

5.29 ▲ Assign *R* or *S* configuration to each chirality center in pseudoephedrine, an over-the-counter decongestant found in cold remedies (blue = N).

5.30 Orient each of the following drawings so that the lowest-ranked group is toward the rear, and then assign *R* or *S* configuration:

(a) 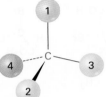 (b) 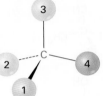 (c)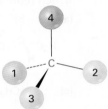

▲ Problems linked to Key Ideas in this chapter

Additional Problems

Chirality and Optical Activity

5.31 Which of the following objects are chiral?
   (a) A basketball   (b) A fork   (c) A wine glass
   (d) A golf club   (e) A spiral staircase   (f) A snowflake

5.32 ▲ Which of the following compounds are chiral? Draw them, and label the chirality centers.
   (a) 2,4-Dimethylheptane   (b) 5-Ethyl-3,3-dimethylheptane
   (c) *cis*-1,4-Dichlorocyclohexane

5.33 ▲ Draw chiral molecules that meet the following descriptions:
   (a) A chloroalkane, $C_5H_{11}Cl$ (b) An alcohol, $C_6H_{14}O$
   (c) An alkene, C_6H_{12} (d) An alkane, C_8H_{18}

5.34 ▲ Eight alcohols have the formula $C_5H_{12}O$. Draw them. Which are chiral?

5.35 Draw compounds that fit the following descriptions:
   (a) A chiral alcohol with four carbons
   (b) A chiral carboxylic acid with the formula $C_5H_{10}O_2$
   (c) A compound with two chirality centers
   (d) A chiral aldehyde with the formula C_3H_5BrO

5.36 Erythronolide B is the biological precursor of erythromycin, a broad-spectrum antibiotic. How many chirality centers does erythronolide B have? Identify them.

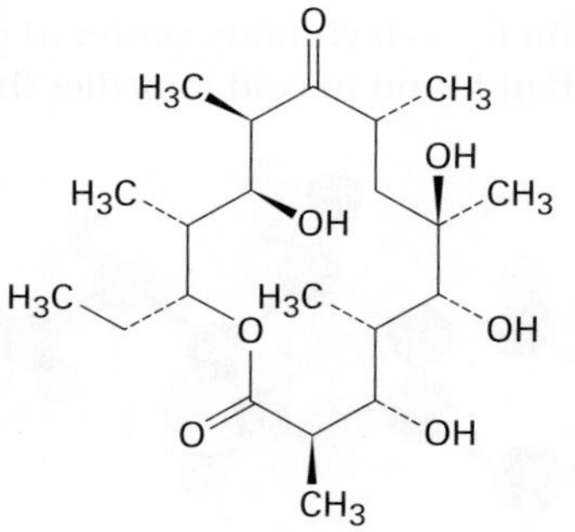

Erythronolide B

Assigning Configuration to Chirality Centers

5.37 Which of the following pairs of structures represent the same enantiomer, and which represent different enantiomers?

▲ Problems linked to Key Ideas in this chapter

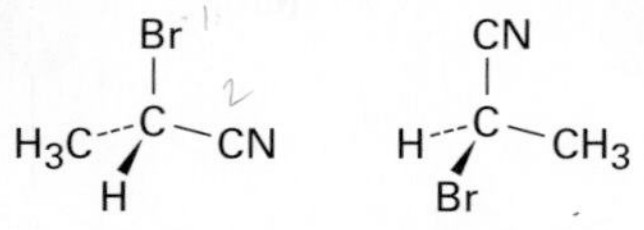

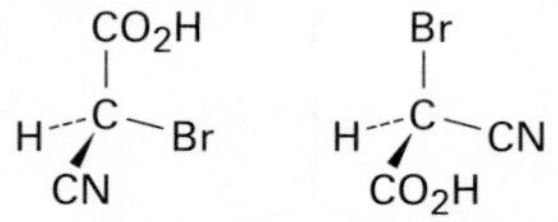

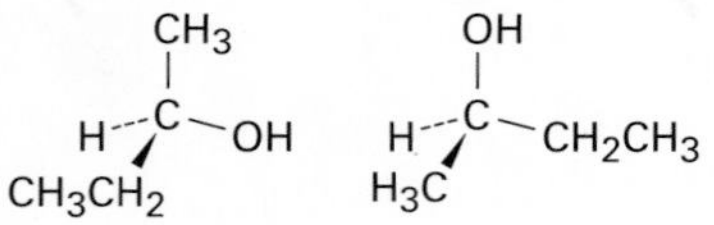

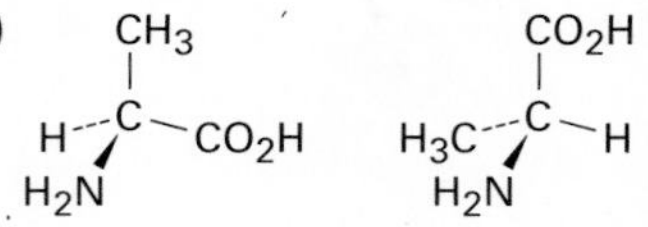

5.38 What is the relationship between the specific rotations of (2R,3R)-dichloropentane and (2S,3S)-dichloropentane? Between (2R,3S)-dichloropentane and (2R,3R)-dichloropentane?

5.39 What is the stereochemical configuration of the enantiomer of (2S,4R)-2,4-octanediol? (A diol is a compound with two —OH groups.)

5.40 What are the stereochemical configurations of the two diastereomers of (2S,4R)-2,4-octanediol? (A diol is a compound with two —OH groups.)

5.41 Orient each of the following drawings so that the lowest-ranked group is toward the rear, and then assign R or S configuration:

(a) 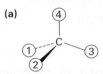 (b) 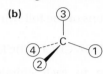 (c)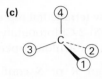

5.42 Assign Cahn–Ingold–Prelog rankings to the following sets of substituents:

(a) —CH=CH$_2$, —CH(CH$_3$)$_2$, —C(CH$_3$)$_3$, —CH$_2$CH$_3$

(b) —C≡CH, —CH=CH$_2$, —C(CH$_3$)$_3$,

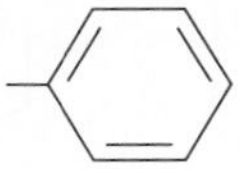

(c) —CO$_2$CH$_3$, —COCH$_3$, —CH$_2$OCH$_3$, —CH$_2$CH$_3$

(d) —C≡N, —CH$_2$Br, —CH$_2$CH$_2$Br, —Br

5.43 Assign R or S configurations to the chirality centers in the following molecules:

(a) (b) 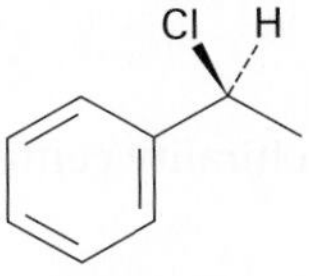 (c) H OCH$_3$
                                                    HOCH$_2$ CO$_2$H

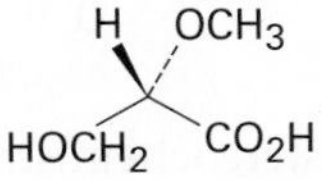

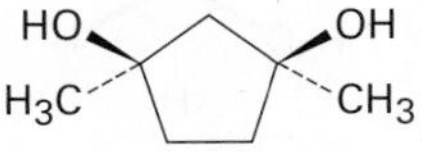

5.44 Assign R or S configuration to each chirality center in the following molecules:

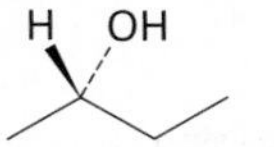

(a) OH, H, Cl, H (on cyclohexane)

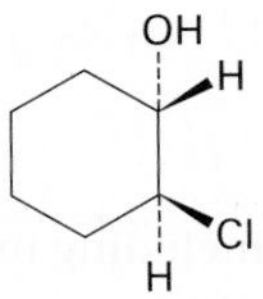

(b) H, CH$_3$, CH$_3$CH$_2$, H (on cyclohexane)

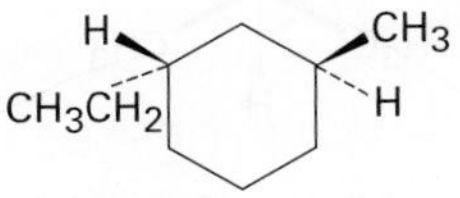

(c) HO, OH, H$_3$C, CH$_3$ (on cyclopentane)

5.45 Assign *R* or *S* configuration to each chirality center in the following biological molecules:

(a) Biotin

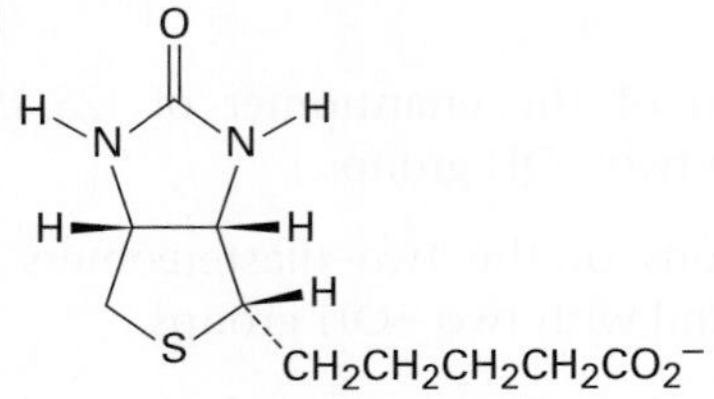

(b) Prostaglandin E$_1$

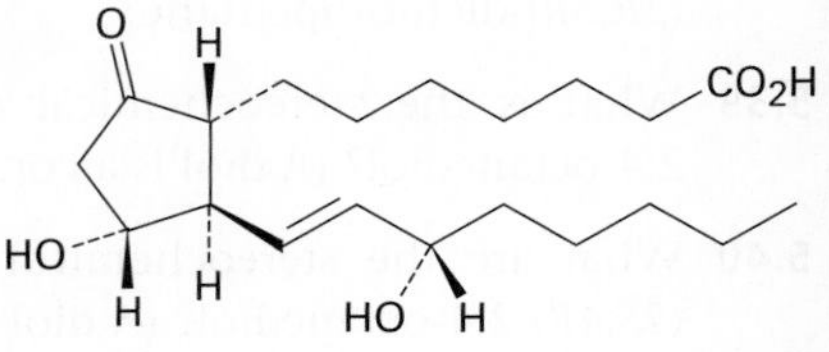

5.46 Draw tetrahedral representations of the following molecules:
(a) (*S*)-2-Chlorobutane
(b) (*R*)-3-Chloro-1-pentene [H$_2$C=CHCH(Cl)CH$_2$CH$_3$]

5.47 Assign *R* or *S* configuration to each chirality center in the following molecules:

(a)

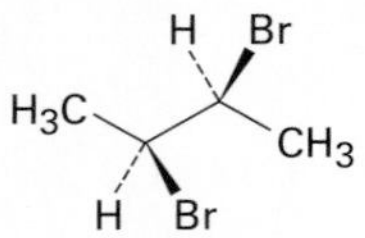

(b)

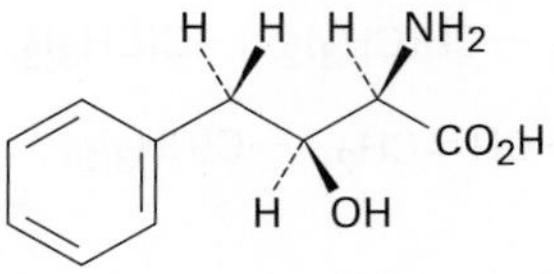

5.48 Assign *R* or *S* configurations to the chirality centers in ascorbic acid (vitamin C).

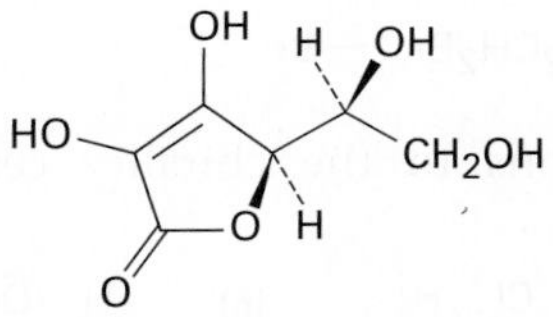

Ascorbic acid

5.49 Assign *R* or *S* stereochemistry to the chirality centers in the following Newman projections:

(a)

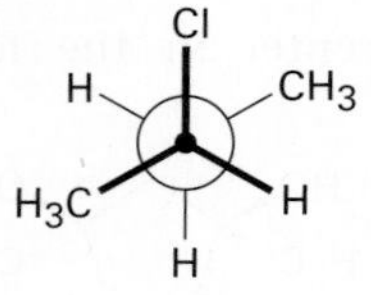

(b)

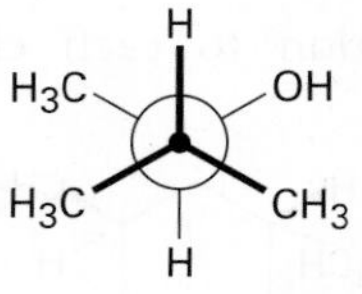

5.50 Xylose is a common sugar found in many types of wood, including maple and cherry. Because it is much less prone to cause tooth decay than sucrose, xylose has been used in candy and chewing gum. Assign *R* or *S* configurations to the chirality centers in xylose.

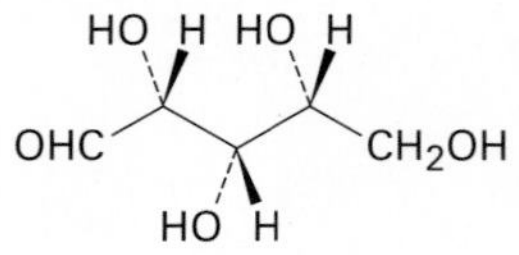

(+)-Xylose

▲ Problems linked to Key Ideas in this chapter

Meso Compounds

5.51 Draw examples of the following:
(a) A meso compound with the formula C_8H_{18}
(b) A meso compound with the formula C_9H_{20}
(c) A compound with two chirality centers, one R and the other S

5.52 Draw the meso form of each of the following molecules, and indicate the plane of symmetry in each:

(a) $CH_3\overset{OH}{\underset{|}{C}}HCH_2CH_2\overset{OH}{\underset{|}{C}}HCH_3$

(b) cyclohexane with CH₃ groups at 1,3 positions

(c) cyclopentane with H₃C, H₃C, and OH substituents

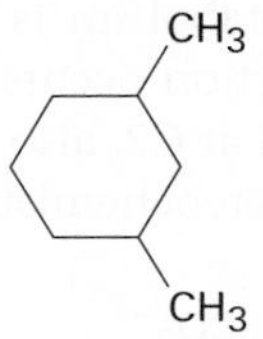

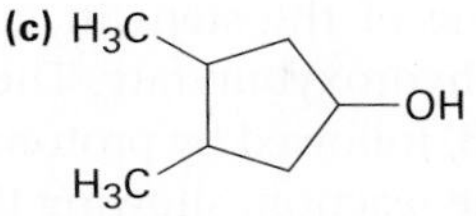

5.53 Draw the structure of a meso compound that has five carbons and three chirality centers.

5.54 Ribose, an essential part of ribonucleic acid (RNA), has the following structure:

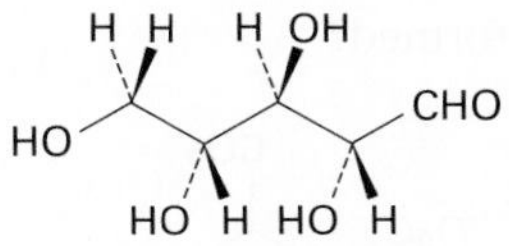

Ribose structure: HO-CH₂-CH(OH)-CH(OH)-CH(OH)-CHO

(a) How many chirality centers does ribose have? Identify them.
(b) How many stereoisomers of ribose are there?
(c) Draw the structure of the enantiomer of ribose.
(d) Draw the structure of a diastereomer of ribose.

5.55 On reaction with hydrogen gas with a platinum catalyst, ribose (Problem 5.54) is converted into ribitol. Is ribitol optically active or inactive? Explain.

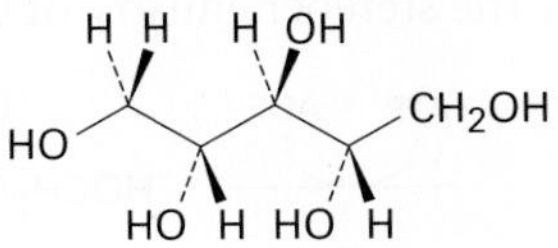

Ribitol structure

Prochirality

5.56 Identify the indicated hydrogens in the following molecules as *pro-R* or *pro-S*:

(a) Malic acid — $HO_2C-CH_2-CH(OH)-CO_2H$ (indicated H H on CH₂)

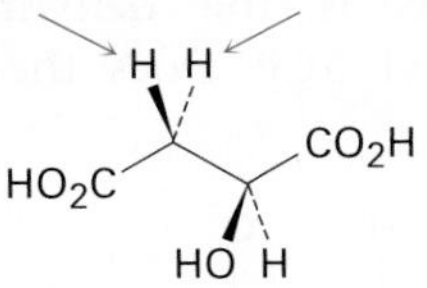

(b) Methionine — $CH_3S-CH_2-CH_2-CH(NH_3^+)-CO_2^-$ (indicated H H on CH₂)

(c) Cysteine — $HS-CH_2-CH(NH_3^+)-CO_2^-$ (indicated H H on CH₂)

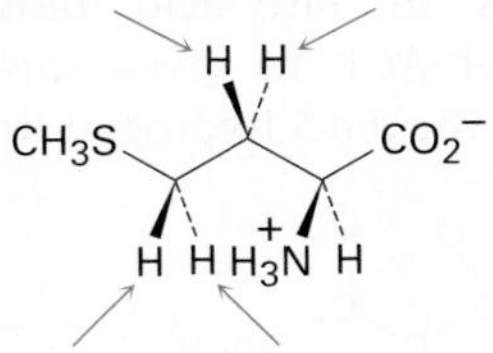

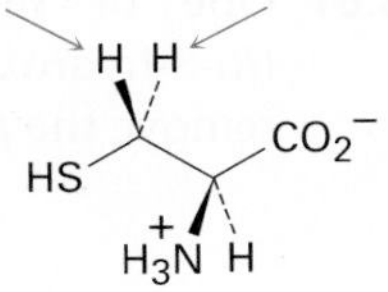

▲ Problems linked to Key Ideas in this chapter

5.57 Identify the indicated faces in the following molecules as *Re* or *Si*:

(a) Pyruvate: $H_3C-C(=O)-CO_2^-$

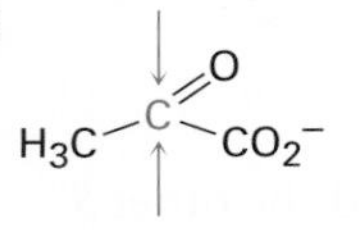

(b) Crotonate: $^-O_2C-CH=CH-CH_3$

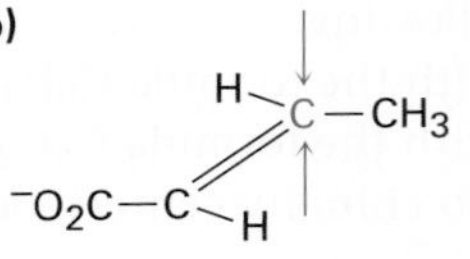

5.58 One of the steps in fat metabolism is the hydration of crotonate to yield 3-hydroxybutyrate. The reaction occurs by addition of –OH to the *Si* face at C3, followed by protonation at C2, also from the *Si* face. Draw the product of the reaction, showing the stereochemistry of each step.

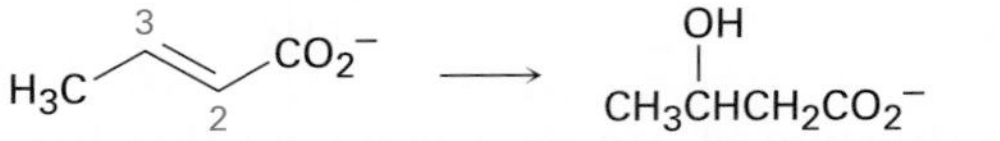

$H_3C-CH=CH-CO_2^-$ (Crotonate, C3 and C2 labeled) $\longrightarrow$ $CH_3CH(OH)CH_2CO_2^-$ (3-Hydroxybutyrate)

5.59 The dehydration of citrate to yield *cis*-aconitate, a step in the citric acid cycle, involves the *pro-R* "arm" of citrate rather than the *pro-S* arm. Which of the following two products is formed?

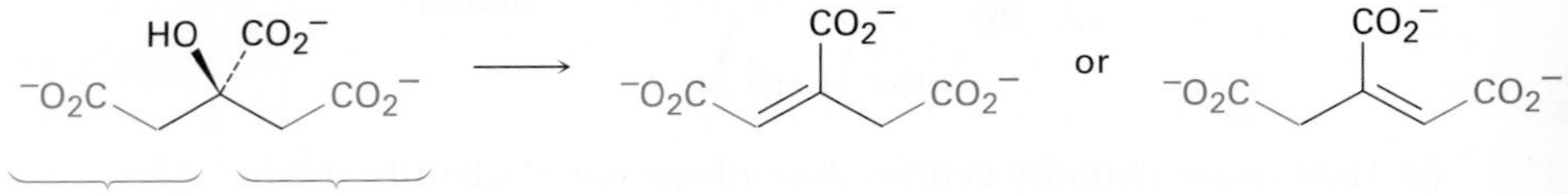

Citrate: $^-O_2C-CH_2-C(OH)(CO_2^-)-CH_2-CO_2^-$ $\longrightarrow$ *cis*-Aconitate (two possible isomers shown)

5.60 The first step in the metabolism of glycerol, formed by digestion of fats, is phosphorylation of the *pro-R* –CH$_2$OH group by reaction with adenosine triphosphate (ATP) to give the corresponding glycerol phosphate plus adenosine diphosphate (ADP). Show the stereochemistry of the product.

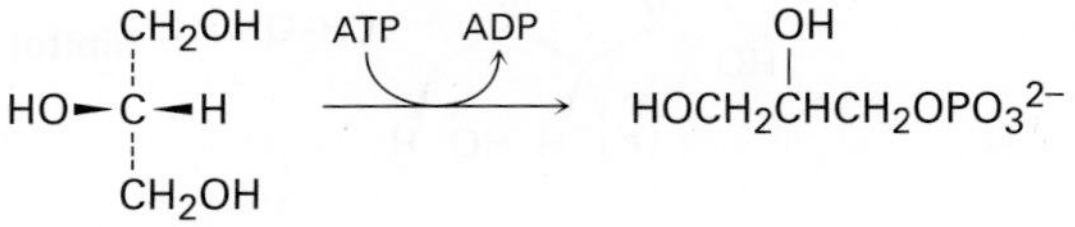

Glycerol: $CH_2OH-C(H)(OH)-CH_2OH$ $\xrightarrow{\text{ATP} \rightarrow \text{ADP}}$ Glycerol phosphate: $HOCH_2CH(OH)CH_2OPO_3^{2-}$

5.61 One of the steps in fatty-acid biosynthesis is the dehydration of (*R*)-3-hydroxybutyryl ACP to give *trans*-crotonyl ACP. Does the reaction remove the *pro-R* or the *pro-S* hydrogen from C2?

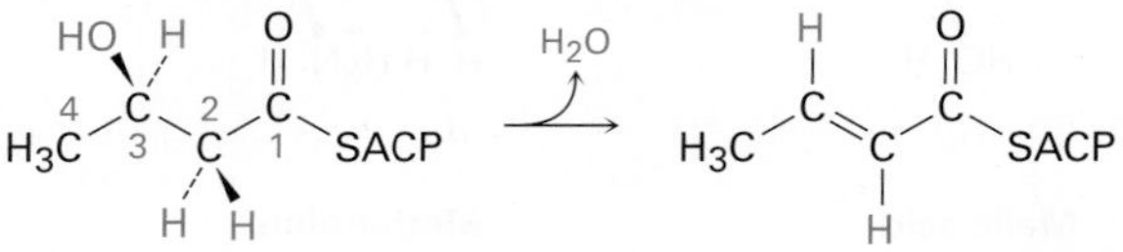

(*R*)-3-Hydroxybutyryl ACP $\xrightarrow{H_2O}$ *trans*-Crotonyl ACP

▲ Problems linked to Key Ideas in this chapter

General Problems

5.62 Draw all possible stereoisomers of 1,2-cyclobutanedicarboxylic acid, and indicate the interrelationships. Which, if any, are optically active? Do the same for 1,3-cyclobutanedicarboxylic acid.

5.63 Draw tetrahedral representations of the two enantiomers of the amino acid cysteine, $HSCH_2CH(NH_2)CO_2H$, and identify each as *R* or *S*.

5.64 The naturally occurring form of the amino acid cysteine (Problem 5.63) has the *S* configuration at its chirality center. On treatment with a mild oxidizing agent, two cysteines join to give cystine, a disulfide. Assuming that the chirality center is not affected by the reaction, is cystine optically active? Explain.

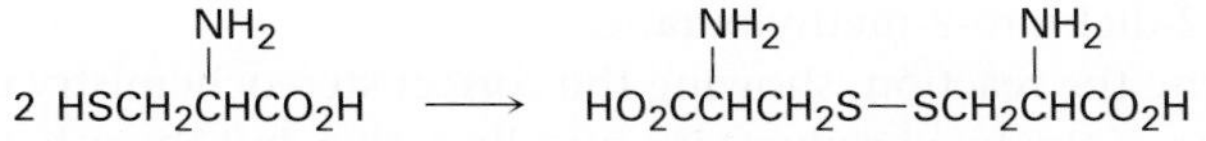

$$2 \; HSCH_2\overset{NH_2}{\underset{|}{C}}HCO_2H \longrightarrow HO_2C\overset{NH_2}{\underset{|}{C}}HCH_2S-SCH_2\overset{NH_2}{\underset{|}{C}}HCO_2H$$

Cysteine   **Cystine**

5.65 Draw tetrahedral representations of the following molecules:
(a) The 2*S*,3*R* enantiomer of 2,3-dibromopentane
(b) The meso form of 3,5-heptanediol

5.66 Assign *R*,*S* configurations to the chiral centers in cephalexin, trade-named Keflex, the most widely prescribed antibiotic in the United States.

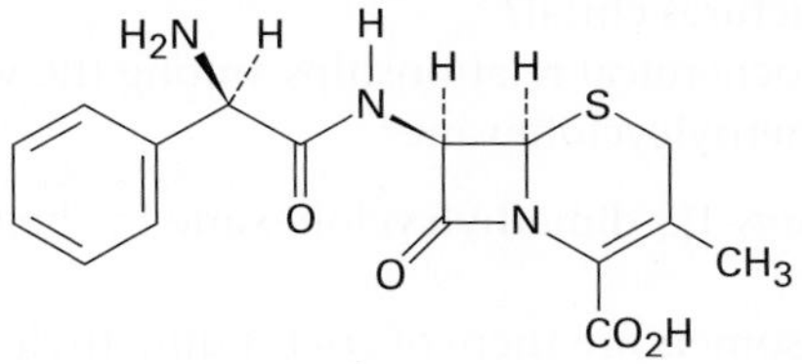

Cephalexin

5.67 Chloramphenicol, a powerful antibiotic isolated in 1949 from the *Streptomyces venezuelae* bacterium, is active against a broad spectrum of bacterial infections and is particularly valuable against typhoid fever. Assign *R*,*S* configurations to the chirality centers in chloramphenicol.

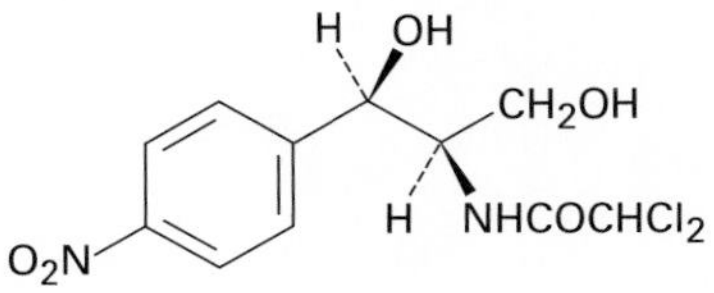

Chloramphenicol

5.68 *Allenes* are compounds with adjacent carbon–carbon double bonds. Many allenes are chiral, even though they don't contain chirality centers. Mycomycin, for example, a naturally occurring antibiotic isolated from the bacterium *Nocardia acidophilus*, is chiral and has $[\alpha]_D = -130$. Explain why mycomycin is chiral.

$$HC\equiv C-C\equiv C-CH=C=CH-CH=CH-CH=CH-CH_2CO_2H$$

Mycomycin

▲ Problems linked to Key Ideas in this chapter

5.69 Long before chiral allenes were known (Problem 5.68), the resolution of 4-methylcyclohexylideneacetic acid into two enantiomers had been carried out. Why is it chiral? What geometric similarity does it have to allenes?

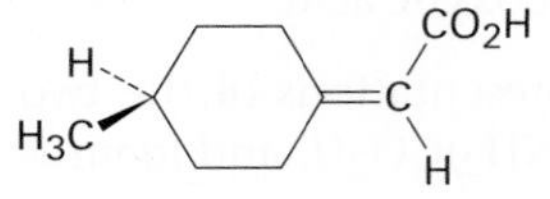

$$\text{4-Methylcyclohexylideneacetic acid}$$

5.70 (S)-1-Chloro-2-methylbutane undergoes light-induced reaction with Cl_2 to yield a mixture of products, among which are 1,4-dichloro-2-methylbutane and 1,2-dichloro-2-methylbutane.
  (a) Write the reaction, showing the correct stereochemistry of the reactant.
  (b) One of the two products is optically active, but the other is optically inactive. Which is which?

5.71 How many stereoisomers of 2,4-dibromo-3-chloropentane are there? Draw them, and indicate which are optically active.

5.72 Draw both *cis*- and *trans*-1,4-dimethylcyclohexane in their more stable chair conformations.
  (a) How many stereoisomers are there of *cis*-1,4-dimethylcyclohexane, and how many of *trans*-1,4-dimethylcyclohexane?
  (b) Are any of the structures chiral?
  (c) What are the stereochemical relationships among the various stereoisomers of 1,4-dimethylcyclohexane?

5.73 Draw both *cis*- and *trans*-1,3-dimethylcyclohexane in their more stable chair conformations.
  (a) How many stereoisomers are there of *cis*-1,3-dimethylcyclohexane, and how many of *trans*-1,3-dimethylcyclohexane?
  (b) Are any of the structures chiral?
  (c) What are the stereochemical relationships among the various stereoisomers of 1,3-dimethylcyclohexane?

5.74 *cis*-1,2-Dimethylcyclohexane is optically inactive even though it has two chirality centers. Explain.

5.75 We'll see in Chapter 11 that alkyl halides react with hydrosulfide ion (HS⁻) to give a product whose stereochemistry is *inverted* from that of the reactant.

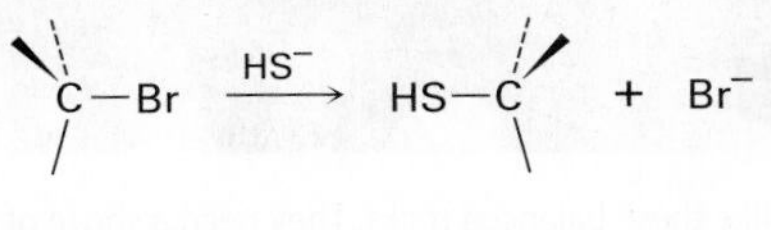

$$\text{C-Br} \xrightarrow{\text{HS}^-} \text{HS-C} + \text{Br}^-$$

An alkyl bromide

Draw the reaction of (S)-2-bromobutane with HS⁻ ion to yield 2-butanethiol, CH$_3$CH$_2$CH(SH)CH$_3$. Is the stereochemistry of the product R or S?

5.76 Ketones react with sodium acetylide (the sodium salt of acetylene, Na$^+$ $^-$:C≡CH) to give alcohols. For example, the reaction of sodium acetylide with 2-butanone yields 3-methyl-1-pentyn-3-ol:

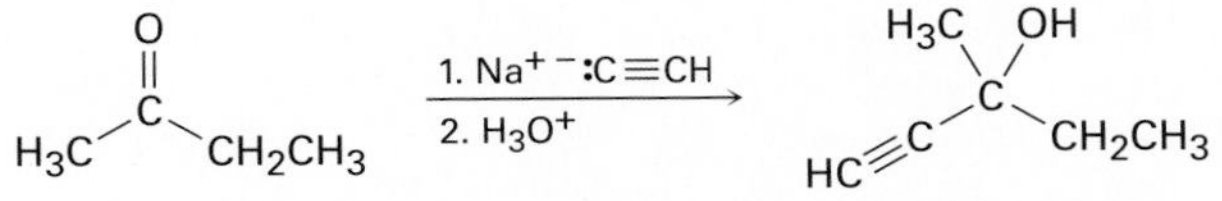

2-Butanone → (1. Na$^+$ $^-$:C≡CH, 2. H$_3$O$^+$) → 3-Methyl-1-pentyn-3-ol

(a) Is the product chiral?
(b) Assuming that the reaction takes place with equal likelihood from both *Re* and *Si* faces of the carbonyl group, is the product optically active? Explain.

5.77 Imagine that a reaction similar to that in Problem 5.76 is carried out between sodium acetylide and (R)-2-phenylpropanal to yield 4-phenyl-1-pentyn-3-ol:

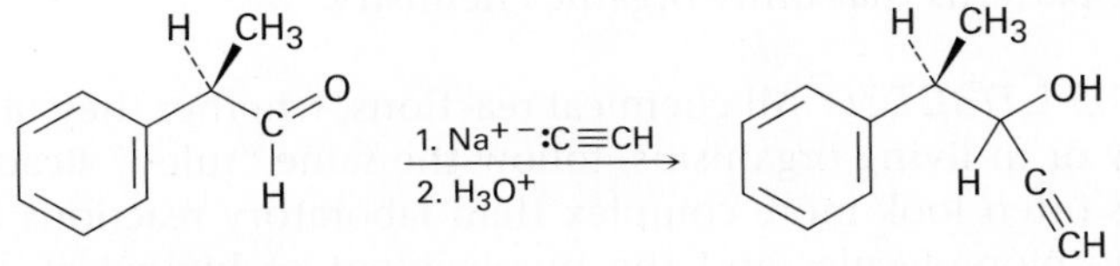

(R)-2-Phenylpropanal → (1. Na$^+$ $^-$:C≡CH, 2. H$_3$O$^+$) → 4-Phenyl-1-pentyn-3-ol

(a) Is the product chiral?
(b) Draw both major and minor reaction products, assuming that the reaction takes place preferentially from the *Re* face of the carbonyl group. Is the product mixture optically active? Explain.

▲ Problems linked to Key Ideas in this chapter

6

Many chemical reactions are like these balanced rocks. They need a shove of energy to get them started moving.
© Mira/Alamy

An Overview of Organic Reactions

6.1 Kinds of Organic Reactions
6.2 How Organic Reactions Occur: Mechanisms
6.3 Radical Reactions
6.4 Polar Reactions
6.5 An Example of a Polar Reaction: Addition of HBr to Ethylene
6.6 Using Curved Arrows in Polar Reaction Mechanisms
6.7 Describing a Reaction: Equilibria, Rates, and Energy Changes
6.8 Describing a Reaction: Bond Dissociation Energies
6.9 Describing a Reaction: Energy Diagrams and Transition States
6.10 Describing a Reaction: Intermediates
6.11 A Comparison Between Biological Reactions and Laboratory Reactions
A Deeper Look—Where Do Drugs Come From?

OWL Sign in to OWL for Organic Chemistry at **www.cengage.com/owl** to view tutorials and simulations, develop problem-solving skills, and complete online homework assigned by your professor.

When first approached, organic chemistry might seem overwhelming. It's not so much that any one part is difficult to understand, it's that there are so many parts: tens of millions of compounds, dozens of functional groups, and an apparently endless number of reactions. With study, though, it becomes evident that there are only a few fundamental ideas that underlie all organic reactions. Far from being a collection of isolated facts, organic chemistry is a beautifully logical subject that is unified by a few broad themes. When these themes are understood, learning organic chemistry becomes much easier and memorization is minimized. The aim of this book is to describe the themes and clarify the patterns that unify organic chemistry.

Why This Chapter? All chemical reactions, whether they take place in the laboratory or in living organisms, follow the same "rules." Reactions in living organisms often look more complex than laboratory reactions because of the size of the biomolecules and the involvement of biological catalysts called *enzymes*, but the principles governing all reactions are the same.

To understand both organic and biological chemistry, it's necessary to know not just *what* occurs but also *why* and *how* chemical reactions take place. In this chapter, we'll start with an overview of the fundamental kinds of organic reactions, we'll see why reactions occur, and we'll see how reactions can be described. Once this background is out of the way, we'll then be ready to begin studying the details of organic chemistry.

6.1 Kinds of Organic Reactions

Organic chemical reactions can be organized broadly in two ways—by *what kinds* of reactions occur and by *how* those reactions occur. Let's look first at the kinds of reactions that take place. There are four general types of organic reactions: *additions, eliminations, substitutions,* and *rearrangements.*

* **Addition reactions** occur when two reactants add together to form a single product with no atoms "left over." An example that we'll be

studying soon is the reaction of an alkene, such as ethylene, with HBr to yield an alkyl bromide.

These two reactants... Ethylene (an alkene) + H—Br → Bromoethane (an alkyl halide) ...add to give this product.

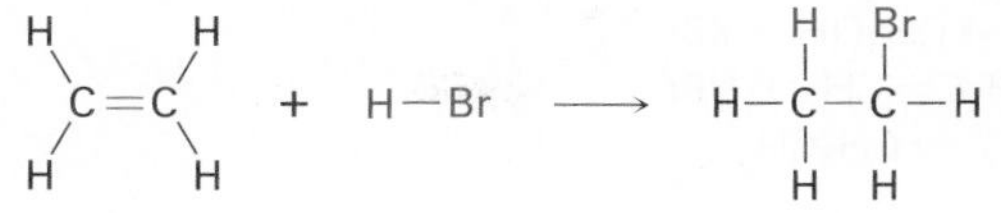

* **Elimination reactions** are, in a sense, the opposite of addition reactions. They occur when a single reactant splits into two products, often with formation of a small molecule such as water or HBr. An example is the acid-catalyzed reaction of an alcohol to yield water and an alkene.

This one reactant... Ethanol (an alcohol) ⇌ (Acid catalyst) Ethylene (an alkene) + H_2O ...gives these two products.

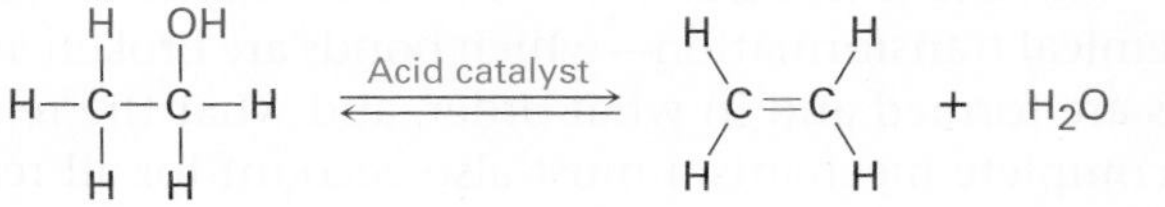

* **Substitution reactions** occur when two reactants exchange parts to give two new products. An example is the reaction of an ester such as methyl acetate with water to yield a carboxylic acid plus an alcohol. Similar reactions occur in many biological pathways, including the metabolism of dietary fats.

These two reactants... Methyl acetate (an ester) + H—O—H → (Acid catalyst) Acetic acid (a carboxylic acid) + Methanol (an alcohol) ...give these two products.

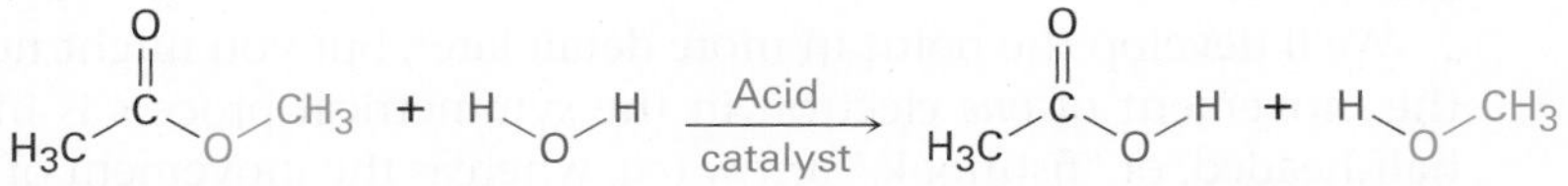

* **Rearrangement reactions** occur when a single reactant undergoes a reorganization of bonds and atoms to yield an isomeric product. An example is the conversion of dihydroxyacetone phosphate into its constitutional isomer glyceraldehyde 3-phosphate, a step in the glycolysis pathway by which carbohydrates are metabolized.

This reactant... $^{2-}O_3PO$—Dihydroxyacetone phosphate → $^{2-}O_3PO$—Glyceraldehyde 3-phosphate ...gives this isomeric product.

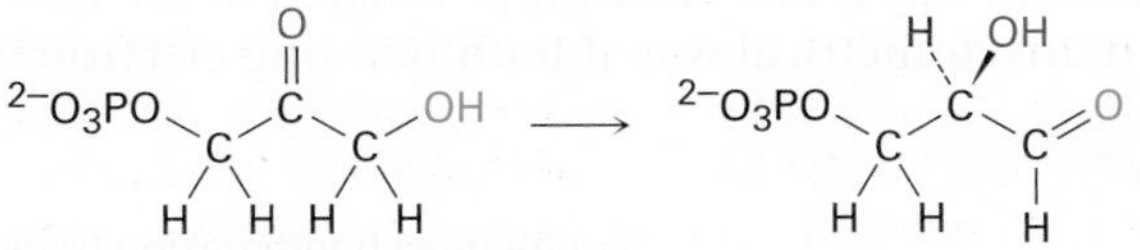

Problem 6.1
Classify each of the following reactions as an addition, elimination, substitution, or rearrangement:
(a) $CH_3Br + KOH \rightarrow CH_3OH + KBr$
(b) $CH_3CH_2Br \rightarrow H_2C=CH_2 + HBr$
(c) $H_2C=CH_2 + H_2 \rightarrow CH_3CH_3$

6.2 How Organic Reactions Occur: Mechanisms

Having looked at the kinds of reactions that take place, let's now see how reactions occur. An overall description of how a reaction occurs is called a **reaction mechanism**. A mechanism describes in detail exactly what takes place at each stage of a chemical transformation—which bonds are broken and in what order, which bonds are formed and in what order, and what the relative rates of the steps are. A complete mechanism must also account for all reactants used and all products formed.

All chemical reactions involve bond-breaking and bond-making. When two molecules come together, react, and yield products, specific bonds in the reactant molecules are broken and specific bonds in the product molecules are formed. Fundamentally, there are two ways in which a covalent two-electron bond can break. A bond can break in an electronically *symmetrical* way so that one electron remains with each product fragment, or a bond can break in an electronically *unsymmetrical* way so that both bonding electrons remain with one product fragment, leaving the other with a vacant orbital. The symmetrical cleavage is said to be *homolytic,* and the unsymmetrical cleavage is said to be *heterolytic.*

We'll develop the point in more detail later, but you might note for now that the movement of *one* electron in the symmetrical process is indicated using a half-headed, or "fishhook," arrow (⌒), whereas the movement of *two* electrons in the unsymmetrical process is indicated using a full-headed curved arrow (⌒).

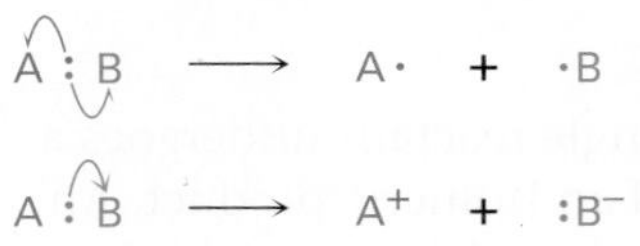

A:B ⟶ A· + ·B   Symmetrical bond-breaking (radical): one bonding electron stays with each product.

A:B ⟶ A⁺ + :B⁻   Unsymmetrical bond-breaking (polar): two bonding electrons stay with one product.

Just as there are two ways in which a bond can break, there are two ways in which a covalent two-electron bond can form. A bond can form in an electronically symmetrical way if one electron is donated to the new bond by each reactant or in an unsymmetrical way if both bonding electrons are donated by one reactant.

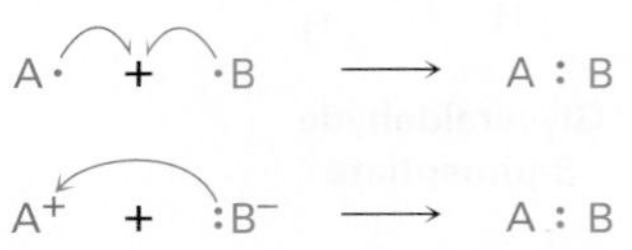

A· + ·B ⟶ A:B   Symmetrical bond-making (radical): one bonding electron is donated by each reactant.

A⁺ + :B⁻ ⟶ A:B   Unsymmetrical bond-making (polar): two bonding electrons are donated by one reactant.

Processes that involve symmetrical bond-breaking and bond-making are called **radical reactions**. A **radical**, often called a *"free radical,"* is a neutral chemical species that contains an odd number of electrons and thus has a single, unpaired electron in one of its orbitals. Processes that involve unsymmetrical bond-breaking and bond-making are called **polar reactions**. Polar reactions involve species that have an even number of electrons and thus have only electron pairs in their orbitals. Polar processes are by far the more common reaction type in both organic and biological chemistry, and a large part of this book is devoted to their description.

In addition to polar and radical reactions, there is a third, less commonly encountered process called a *pericyclic reaction*. Rather than explain pericyclic reactions now, though, we'll look at them more carefully in Chapter 30.

6.3 Radical Reactions

Radical reactions are not as common as polar reactions but are nevertheless important in some industrial processes and biological pathways. Let's see briefly how they occur.

A radical is highly reactive because it contains an atom with an odd number of electrons (usually seven) in its valence shell, rather than a stable, noble-gas octet. A radical can achieve a valence-shell octet in several ways. For example, the radical might abstract an atom and one bonding electron from another reactant, leaving behind a new radical. The net result is a radical substitution reaction.

$$\text{Rad} \cdot \;+\; A{:}B \;\longrightarrow\; \text{Rad}{:}A \;+\; \cdot B$$

Reactant radical   **Substitution product**   **Product radical**

Alternatively, a reactant radical might add to a double bond, taking one electron from the double bond and yielding a new radical. The net result is a radical addition reaction.

$$\text{Rad} \cdot \;+\; \text{C}{=}\text{C} \;\longrightarrow\; \overset{\text{Rad}}{-\text{C}-\text{C}\cdot}$$

Reactant radical   **Alkene**   **Addition product radical**

An example of an industrially useful radical reaction is the chlorination of methane to yield chloromethane. This substitution reaction is the first step in the preparation of the solvents dichloromethane (CH_2Cl_2) and chloroform ($CHCl_3$).

$$CH_4 \;+\; Cl-Cl \;\xrightarrow{\text{Light}}\; CH_3Cl \;+\; H-Cl$$

Methane   **Chlorine**   **Chloromethane**

Like many radical reactions in the laboratory, methane chlorination requires three kinds of steps: *initiation, propagation,* and *termination.*

Initiation   Irradiation with ultraviolet light begins the reaction by breaking the relatively weak Cl–Cl bond of a small number of Cl_2 molecules to give a few reactive chlorine radicals.

$$:\ddot{Cl}:\ddot{Cl}: \xrightarrow{Light} 2\ :\ddot{Cl}\cdot$$

Propagation   Once produced, a reactive chlorine radical collides with a methane molecule in a propagation step, abstracting a hydrogen atom to give HCl and a methyl radical ($\cdot CH_3$). This methyl radical reacts further with Cl_2 in a second propagation step to give the product chloromethane plus a new chlorine radical, which cycles back and repeats the first propagation step. Thus, once the sequence has been initiated, it becomes a self-sustaining cycle of repeating steps (a) and (b), making the overall process a *chain reaction.*

(a) $:\ddot{Cl}\cdot\ +\ H:CH_3\ \longrightarrow\ H:\ddot{Cl}:\ +\ \cdot CH_3$

(b) $:\ddot{Cl}:\ddot{Cl}:\ +\ \cdot CH_3\ \longrightarrow\ :\ddot{Cl}\cdot\ +\ :\ddot{Cl}:CH_3$

Termination   Occasionally, two radicals might collide and combine to form a stable product. When that happens, the reaction cycle is broken and the chain is ended. Such termination steps occur infrequently, however, because the concentration of radicals in the reaction at any given moment is very small. Thus, the likelihood that two radicals will collide is also small.

$:\ddot{Cl}\cdot\ +\ \cdot\ddot{Cl}:\ \longrightarrow\ :\ddot{Cl}:\ddot{Cl}:$

$:\ddot{Cl}\cdot\ +\ \cdot CH_3\ \longrightarrow\ :\ddot{Cl}:CH_3$ **Possible termination steps**

$H_3C\cdot\ +\ \cdot CH_3\ \longrightarrow\ H_3C:CH_3$

As a biological example of a radical reaction, look at the synthesis of *prostaglandins,* a large class of molecules found in virtually all body tissues and fluids. A number of pharmaceuticals are based on or derived from prostaglandins, including medicines that induce labor during childbirth, reduce intraocular pressure in glaucoma, control bronchial asthma, and help treat congenital heart defects.

Prostaglandin biosynthesis is initiated by abstraction of a hydrogen atom from arachidonic acid by an iron–oxygen radical, thereby generating a new, carbon radical in a substitution reaction. Don't be intimidated by the size of the molecules; focus on the changes occurring in each step. (To help you do that, the unchanged part of the molecule is "ghosted," with only the reactive part clearly visible.)

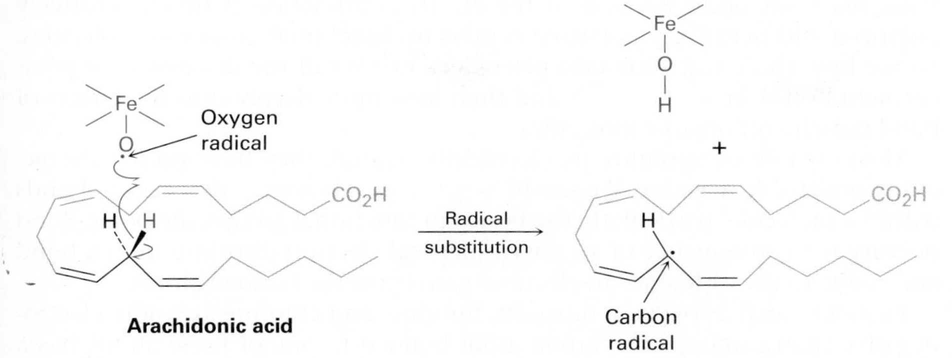

Following the initial abstraction of a hydrogen atom, the carbon radical then reacts with O_2 to give an oxygen radical, which reacts with a C=C bond within the same molecule in an addition reaction. Several further transformations ultimately yield prostaglandin H_2.

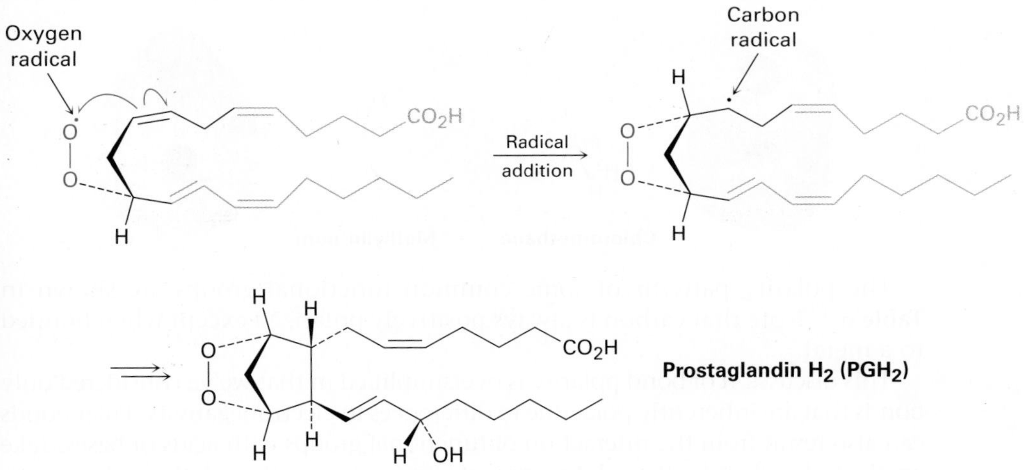

Problem 6.2
Radical chlorination of alkanes is not generally useful because mixtures of products often result when more than one kind of C–H bond is present in the substrate. Draw and name all monochloro substitution products $C_6H_{13}Cl$ you might obtain by reaction of 2-methylpentane with Cl_2.

Problem 6.3
Using a curved fishhook arrow, propose a mechanism for formation of the cyclopentane ring of prostaglandin H_2.

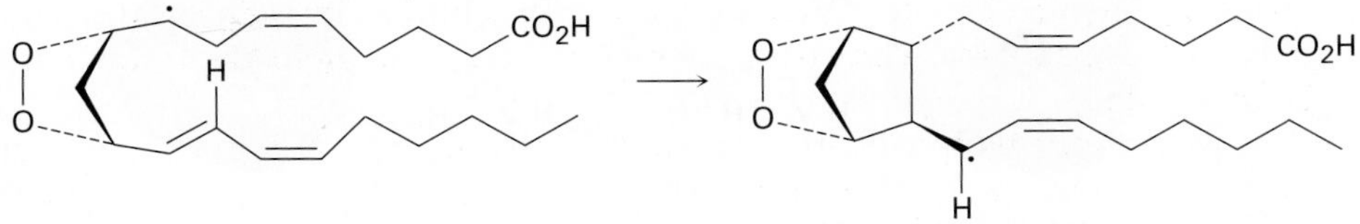

6.4 Polar Reactions

Polar reactions occur because of the electrical attraction between positively polarized and negatively polarized centers on functional groups in molecules. To see how these reactions take place, let's first recall the discussion of polar covalent bonds in **Section 2.1** and then look more deeply into the effects of bond polarity on organic molecules.

Most organic compounds are electrically neutral; they have no net charge, either positive or negative. We saw in **Section 2.1**, however, that certain bonds within a molecule, particularly the bonds in functional groups, are polar. Bond polarity is a consequence of an unsymmetrical electron distribution in a bond and is due to the difference in electronegativity of the bonded atoms.

Elements such as oxygen, nitrogen, fluorine, and chlorine are more electronegative than carbon, so a carbon atom bonded to one of these atoms has a partial positive charge ($\delta+$). Conversely, metals are less electronegative than carbon, so a carbon atom bonded to a metal has a partial negative charge ($\delta-$). Electrostatic potential maps of chloromethane and methyllithium illustrate these charge distributions, showing that the carbon atom in chloromethane is electron-poor (blue) while the carbon in methyllithium is electron-rich (red).

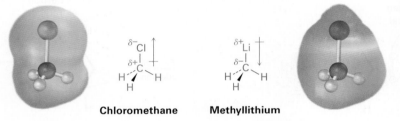

Chloromethane   Methyllithium

The polarity patterns of some common functional groups are shown in Table 6.1. Note that carbon is always positively polarized except when bonded to a metal.

This discussion of bond polarity is oversimplified in that we've considered only bonds that are inherently polar due to differences in electronegativity. Polar bonds can also result from the interaction of functional groups with acids or bases. Take an alcohol such as methanol, for example. In neutral methanol, the carbon atom is somewhat electron-poor because the electronegative oxygen attracts the electrons in the C–O bond. On protonation of the methanol oxygen by an acid, however, a full positive charge on oxygen attracts the electrons in the C–O bond much more strongly and makes the carbon much more electron-poor. We'll see numerous examples throughout this book of reactions that are catalyzed by acids because of the resultant increase in bond polarity on protonation.

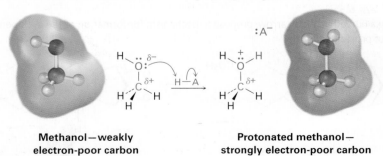

Methanol—weakly electron-poor carbon   Protonated methanol—strongly electron-poor carbon

Table 6.1 Polarity Patterns in Some Common Functional Groups

Compound type	Functional group structure	Compound type	Functional group structure
Alcohol	—C—OH ($\delta+$, $\delta-$)	Carbonyl	C=O ($\delta+$, $\delta-$)
Alkene	C=C (Symmetrical, nonpolar)	Carboxylic acid	—C(=O)($\delta-$)—OH ($\delta-$), C is $\delta+$
Alkyl halide	—C—X ($\delta+$, $\delta-$)	Carboxylic acid chloride	—C(=O)($\delta-$)—Cl ($\delta-$), C is $\delta+$
Amine	—C—NH$_2$ ($\delta+$, $\delta-$)	Thioester	—C(=O)($\delta-$)—S—C ($\delta-$), C is $\delta+$
Ether	—C—O—C— ($\delta+$, $\delta-$, $\delta+$)	Aldehyde	—C(=O)($\delta-$)—H, C is $\delta+$
Thiol	—C—SH ($\delta+$, $\delta-$)	Ester	—C(=O)($\delta-$)—O—C ($\delta-$), C is $\delta+$
Nitrile	—C≡N ($\delta+$, $\delta-$)	Ketone	—C(=O)($\delta-$)—C, C is $\delta+$
Grignard reagent	—C—MgBr ($\delta-$, $\delta+$)		
Alkyllithium	—C—Li ($\delta-$, $\delta+$)		

Yet a further consideration is the *polarizability* (as opposed to polarity) of atoms in a molecule. As the electric field around a given atom changes because of changing interactions with solvent or other polar molecules nearby, the electron distribution around that atom also changes. The measure of this response to an external electrical influence is called the polarizability of the atom. Larger atoms with more loosely held electrons are more polarizable, and smaller atoms with fewer, tightly held electrons are less polarizable. Thus, sulfur is more polarizable than oxygen, and iodine is more polarizable than chlorine. The effect of this higher polarizability for sulfur and iodine is that carbon–sulfur and carbon–iodine bonds, although nonpolar according to electronegativity values (Figure 2.2 on page 35), nevertheless usually react as if they were polar.

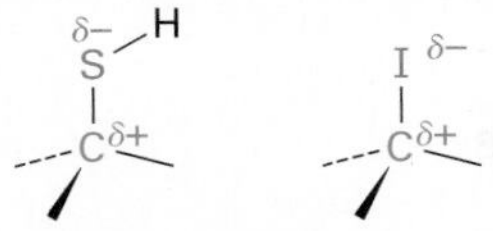

What does functional-group polarity mean with respect to chemical reactivity? Because unlike charges attract, the fundamental characteristic of all polar organic reactions is that electron-rich sites react with electron-poor sites. Bonds are made when an electron-rich atom donates a pair of electrons to an electron-poor atom, and bonds are broken when one atom leaves with both electrons from the former bond.

As we saw in **Section 2.11**, chemists indicate the movement of an electron pair during a polar reaction by using a curved, full-headed arrow. A curved arrow shows where electrons move when reactant bonds are broken and product bonds are formed. It means that an electron pair moves *from* the atom (or bond) at the tail of the arrow *to* the atom at the head of the arrow during the reaction.

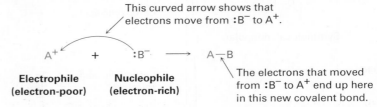

In referring to the electron-rich and electron-poor species involved in polar reactions, chemists use the words *nucleophile* and *electrophile*. A **nucleophile** is a substance that is "nucleus-loving." (Remember that a nucleus is positively charged.) A nucleophile has a negatively polarized, electron-rich atom and can form a bond by donating a pair of electrons to a positively polarized, electron-poor atom. Nucleophiles can be either neutral or negatively charged; ammonia, water, hydroxide ion, and chloride ion are examples. An **electrophile**, by contrast, is "electron-loving." An electrophile has a positively polarized, electron-poor atom and can form a bond by accepting a pair of electrons from a nucleophile. Electrophiles can be either neutral or positively charged. Acids (H^+ donors), alkyl halides, and carbonyl compounds are examples (**Figure 6.1**).

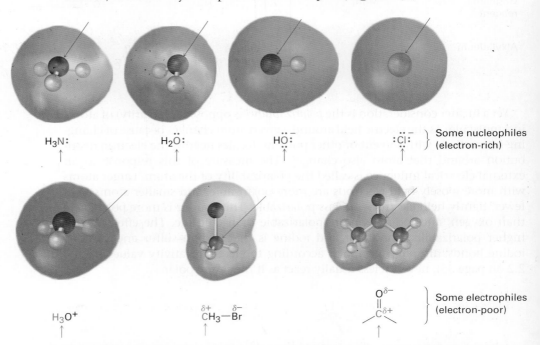

Figure 6.1 Some nucleophiles and electrophiles. Electrostatic potential maps identify the nucleophilic (**negative**) and electrophilic (**positive**) atoms.

Note that neutral compounds can often react either as nucleophiles or as electrophiles, depending on the circumstances. After all, if a compound is neutral yet has an electron-*rich* nucleophilic site, it must also have a corresponding electron-*poor* electrophilic site. Water, for instance, acts as an electrophile when it donates H^+ but acts as a nucleophile when it donates a nonbonding pair of electrons. Similarly, a carbonyl compound acts as an electrophile when it reacts at its positively polarized carbon atom, yet acts as a nucleophile when it reacts at its negatively polarized oxygen atom.

If the definitions of nucleophiles and electrophiles sound similar to those given in **Section 2.11** for Lewis acids and Lewis bases, that's because there is indeed a correlation. Lewis bases are electron donors and behave as nucleophiles, whereas Lewis acids are electron acceptors and behave as electrophiles. Thus, much of organic chemistry is explainable in terms of acid–base reactions. The main difference is that the words *acid* and *base* are used broadly in all fields of chemistry, while the words *nucleophile* and *electrophile* are used primarily in organic chemistry when bonds to carbon are involved.

Worked Example 6.1

Identifying Electrophiles and Nucleophiles

Which of the following species is likely to behave as a nucleophile and which as an electrophile?
(a) NO_2^+ **(b)** CN^- **(c)** CH_3NH_2 **(d)** $(CH_3)_3S^+$

Strategy
A nucleophile has an electron-rich site, either because it is negatively charged or because it has a functional group containing an atom that has a lone pair of electrons. An electrophile has an electron-poor site, either because it is positively charged or because it has a functional group containing an atom that is positively polarized.

Solution
(a) NO_2^+ (nitronium ion) is likely to be an electrophile because it is positively charged.
(b) :C≡N⁻ (cyanide ion) is likely to be a nucleophile because it is negatively charged.
(c) CH_3NH_2 (methylamine) might be either a nucleophile or an electrophile depending on the circumstances. The lone pair of electrons on the nitrogen atom makes methylamine a potential nucleophile, while positively polarized N–H hydrogens make methylamine a potential acid (electrophile).
(d) $(CH_3)_3S^+$ (trimethylsulfonium ion) is likely to be an electrophile because it is positively charged.

Problem 6.4
Which of the following species are likely to be nucleophiles and which electrophiles? Which might be both?

(a) CH_3Cl **(b)** CH_3S^- **(c)** imidazole ring with $N-CH_3$ **(d)** $CH_3CH=O$

Problem 6.5
An electrostatic potential map of boron trifluoride is shown. Is BF$_3$ likely to be a nucleophile or an electrophile? Draw a Lewis structure for BF$_3$, and explain your answer.

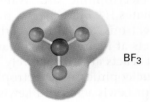

6.5 An Example of a Polar Reaction: Addition of HBr to Ethylene

Let's look at a typical polar process—the addition reaction of an alkene, such as ethylene, with hydrogen bromide. When ethylene is treated with HBr at room temperature, bromoethane is produced. Overall, the reaction can be formulated as

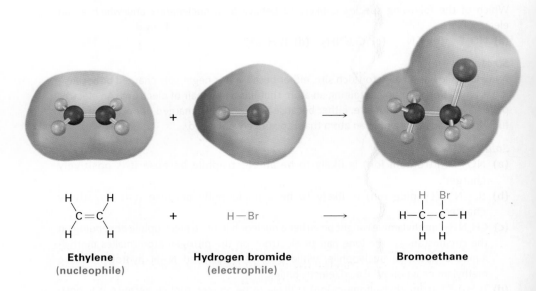

The reaction is an example of a polar reaction type known as an *electrophilic addition reaction* and can be understood using the general ideas discussed in the previous section. Let's begin by looking at the two reactants.

What do we know about ethylene? We know from **Section 1.8** that a carbon–carbon double bond results from orbital overlap of two sp^2-hybridized carbon atoms. The σ part of the double bond results from sp^2–sp^2 overlap, and the π part results from p–p overlap.

6.5 | An Example of a Polar Reaction: Addition of HBr to Ethylene

What kind of chemical reactivity might we expect of a C=C bond? We know that alkanes, such as ethane, are relatively inert because all valence electrons are tied up in strong, nonpolar, C–C and C–H bonds. Furthermore, the bonding electrons in alkanes are relatively inaccessible to approaching reactants because they are sheltered in σ bonds between nuclei. The electronic situation in alkenes is quite different, however. For one thing, double bonds have a greater electron density than single bonds—four electrons in a double bond versus only two in a single bond. In addition, the electrons in the π bond are accessible to approaching reactants because they are located above and below the plane of the double bond rather than being sheltered between the nuclei **(Figure 6.2)**. As a result, the double bond is nucleophilic and the chemistry of alkenes is dominated by reactions with electrophiles.

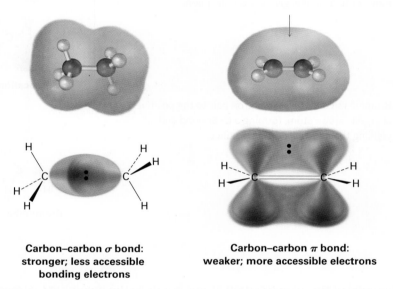

Carbon–carbon σ bond:
stronger; less accessible
bonding electrons

Carbon–carbon π bond:
weaker; more accessible electrons

Figure 6.2 A comparison of carbon–carbon single and double bonds. A double bond is both more accessible to approaching reactants than a single bond and more electron-rich (more nucleophilic). An electrostatic potential map of ethylene indicates that the double bond is the region of **highest negative charge**.

What about the second reactant, HBr? As a strong acid, HBr is a powerful proton (H$^+$) donor and electrophile. Thus, the reaction between HBr and ethylene is a typical electrophile–nucleophile combination, characteristic of all polar reactions.

We'll see more details about alkene electrophilic addition reactions shortly, but for the present we can imagine the reaction as taking place by the pathway shown in **Figure 6.3**. The reaction begins when the alkene nucleophile donates a pair of electrons from its C=C bond to HBr to form a new C–H bond plus Br$^-$, as indicated by the path of the curved arrows in the first step of Figure 6.3. One curved arrow begins at the middle of the double bond (the source of the electron pair) and points to the hydrogen atom in HBr (the atom to which a bond will form). This arrow indicates that a new C–H bond forms using electrons from the former C=C bond. Simultaneously, a second curved arrow begins in the middle of the H–Br bond and points to the Br, indicating that the H–Br bond breaks and the electrons remain with the Br atom, giving Br$^-$.

Figure 6.3 MECHANISM

The electrophilic addition reaction of ethylene and HBr. The reaction takes place in two steps, both of which involve electrophile–nucleophile interactions.

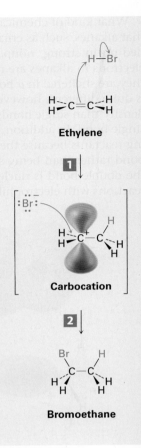

1 A hydrogen atom on the electrophile HBr is attacked by π electrons from the nucleophilic double bond, forming a new C–H bond. This leaves the other carbon atom with a + charge and a vacant *p* orbital. Simultaneously, two electrons from the H–Br bond move onto bromine, giving bromide anion.

2 Bromide ion donates an electron pair to the positively charged carbon atom, forming a C–Br bond and yielding the neutral addition product.

When one of the alkene carbon atoms bonds to the incoming hydrogen, the other carbon atom, having lost its share of the double-bond electrons, now has only six valence electrons and is left with a positive charge. This positively charged species—a carbon-cation, or **carbocation**—is itself an electrophile that can accept an electron pair from nucleophilic Br⁻ anion in a second step, forming a C–Br bond and yielding the observed addition product. Once again, a curved arrow in Figure 6.3 shows the electron-pair movement from Br⁻ to the positively charged carbon.

The electrophilic addition of HBr to ethylene is only one example of a polar process; there are many others that we'll study in detail in later chapters. But regardless of the details of individual reactions, all polar reactions take place between an electron-poor site and an electron-rich site and involve the donation of an electron pair from a nucleophile to an electrophile.

Problem 6.6
What product would you expect from reaction of cyclohexene with HBr? With HCl?

Problem 6.7
Reaction of HBr with 2-methylpropene yields 2-bromo-2-methylpropane. What is the structure of the carbocation formed during the reaction? Show the mechanism of the reaction.

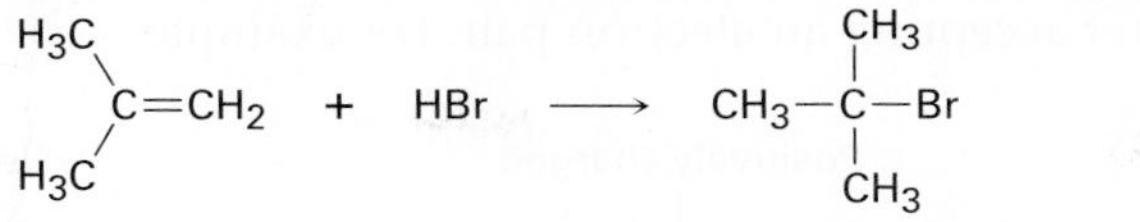

$$\underset{\text{2-Methylpropene}}{(H_3C)_2C=CH_2} + HBr \longrightarrow \underset{\text{2-Bromo-2-methylpropane}}{CH_3-C(CH_3)_2-Br}$$

6.6 Using Curved Arrows in Polar Reaction Mechanisms

It takes practice to use curved arrows properly in reaction mechanisms, but there are a few rules and a few common patterns you should look for that will help you become more proficient:

Key IDEAS
Test your knowledge of Key Ideas by answering end-of-chapter exercises marked with ▲.

RULE 1
Electrons move *from* a nucleophilic source (Nu: or Nu:⁻) *to* an electrophilic sink (E or E⁺). The nucleophilic source must have an electron pair available, usually either as a lone pair or in a multiple bond. For example:

Electrons usually flow *from* one of these nucleophiles.

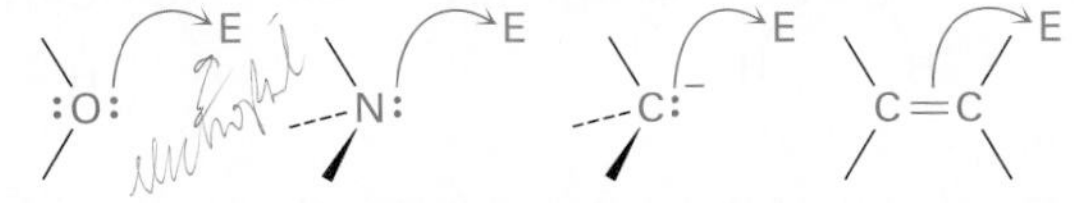

The electrophilic sink must be able to accept an electron pair, usually because it has either a positively charged atom or a positively polarized atom in a functional group. For example:

Electrons usually flow *to* one of these electrophiles.

RULE 2
The nucleophile can be either negatively charged or neutral. If the nucleophile is negatively charged, the atom that donates an electron pair becomes neutral. For example:

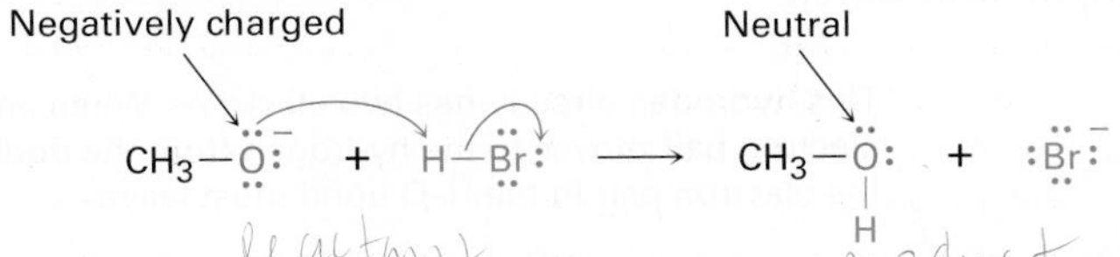

Negatively charged → Neutral

$$CH_3-\ddot{O}:^- + H-\ddot{B}\ddot{r}: \longrightarrow CH_3-\underset{H}{\ddot{O}}: + :\ddot{B}\ddot{r}:^-$$

If the nucleophile is neutral, the atom that donates the electron pair acquires a positive charge. For example:

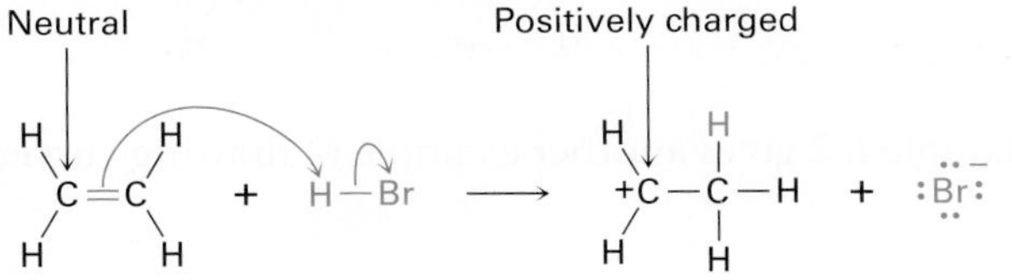

Neutral → Positively charged

$$H_2C=CH_2 + H-Br \longrightarrow {^+}CH_2-CH_3 + :\ddot{B}\ddot{r}:^-$$

RULE 3

The electrophile can be either positively charged or neutral. If the electrophile is positively charged, the atom bearing that charge becomes neutral after accepting an electron pair. For example:

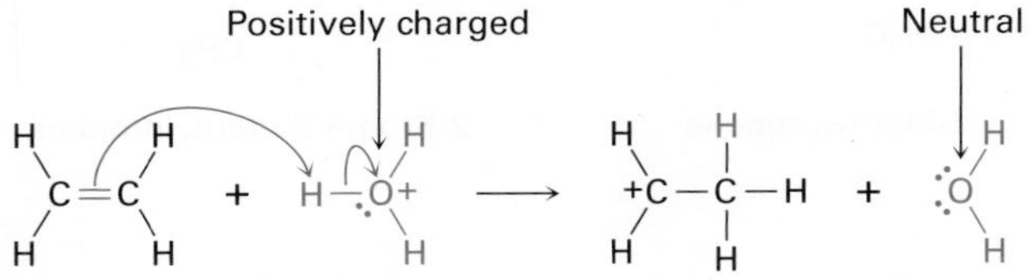

[Reaction: ethylene + H_3O^+ → protonated ethyl cation + H_2O; the O is labeled "Positively charged" in the reactant and "Neutral" in the product.]

If the electrophile is neutral, the atom that ultimately accepts the electron pair acquires a negative charge. For this to happen, however, the negative charge must be stabilized by being on an electronegative atom such as oxygen, nitrogen, or a halogen. Carbon and hydrogen do not typically stabilize a negative charge. For example:

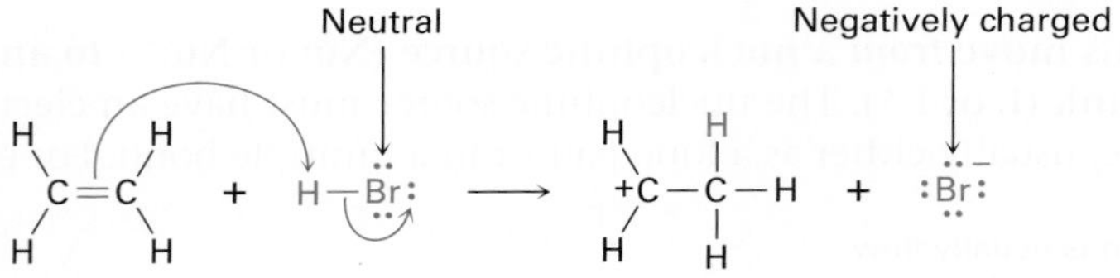

[Reaction: ethylene + H–Br → protonated ethyl cation + Br^-; HBr labeled "Neutral", Br^- labeled "Negatively charged".]

The result of Rules 2 and 3 together is that charge is conserved during the reaction. A negative charge in one of the reactants gives a negative charge in one of the products, and a positive charge in one of the reactants gives a positive charge in one of the products.

RULE 4

The octet rule must be followed. That is, no second-row atom can be left with ten electrons (or four for hydrogen). If an electron pair moves *to* an atom that already has an octet (or two for hydrogen), another electron pair must simultaneously move *from* that atom to maintain the octet. When two electrons move from the C=C bond of ethylene to the hydrogen atom of H_3O^+, for instance, two electrons must leave that hydrogen. This means that the H–O bond must break and the electrons must stay with the oxygen, giving neutral water.

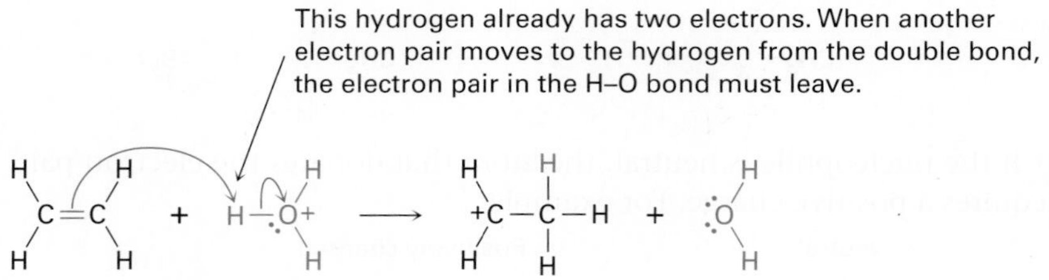

[Reaction diagram with annotation: "This hydrogen already has two electrons. When another electron pair moves to the hydrogen from the double bond, the electron pair in the H–O bond must leave."]

Worked Example 6.2 gives another example of drawing curved arrows.

Using Curved Arrows in Reaction Mechanisms

Worked Example 6.2

Add curved arrows to the following polar reaction to show the flow of electrons:

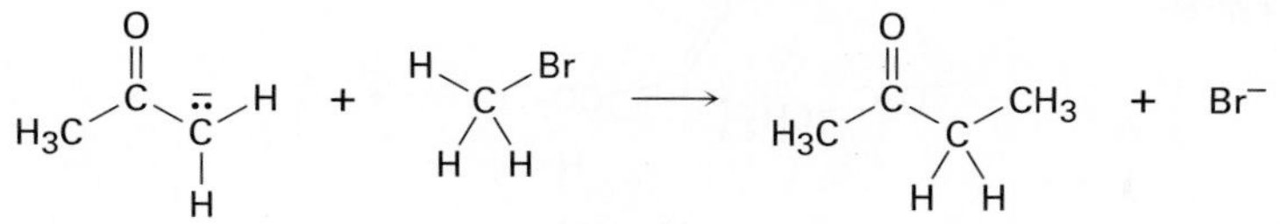

$$H_3C-\overset{O}{\overset{\|}{C}}-\overset{\ominus}{\underset{H}{C}}H_2 \; + \; CH_3Br \; \longrightarrow \; H_3C-\overset{O}{\overset{\|}{C}}-\underset{H\;H}{C}-CH_3 \; + \; Br^-$$

Strategy
Look at the reaction, and identify the bonding changes that have occurred. In this case, a C–Br bond has broken and a C–C bond has formed. The formation of the C–C bond involves donation of an electron pair from the nucleophilic carbon atom of the reactant on the left to the electrophilic carbon atom of CH_3Br, so we draw a curved arrow originating from the lone pair on the negatively charged C atom and pointing to the C atom of CH_3Br. At the same time that the C–C bond forms, the C–Br bond must break so that the octet rule is not violated. We therefore draw a second curved arrow from the C–Br bond to Br. The bromine is now a stable Br^- ion.

Solution

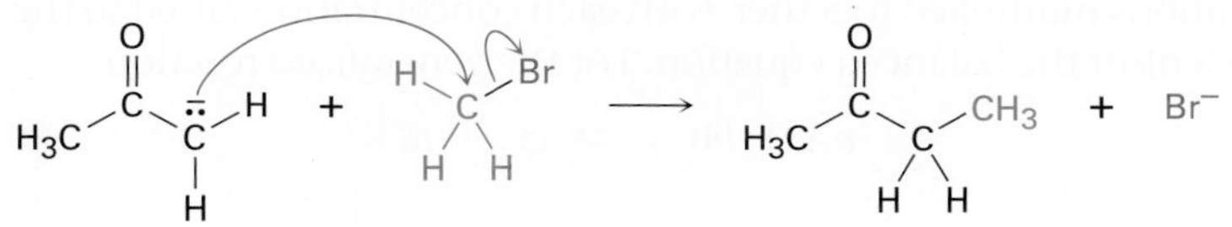

$$H_3C-\overset{O}{\overset{\|}{C}}-\overset{\ominus}{\underset{H}{C}}H_2 \; + \; CH_3Br \; \longrightarrow \; H_3C-\overset{O}{\overset{\|}{C}}-\underset{H\;H}{C}-CH_3 \; + \; Br^-$$

Problem 6.8
Add curved arrows to the following polar reactions to indicate the flow of electrons in each:

(a)
$$:\!\ddot{C}l-\ddot{C}l\!: \; + \; H-\underset{H}{\overset{H}{N}}-H \; \longrightarrow \; H-\underset{H}{\overset{:\ddot{C}l:}{\overset{|+}{N}}}-H \; + \; :\ddot{C}l:^-$$

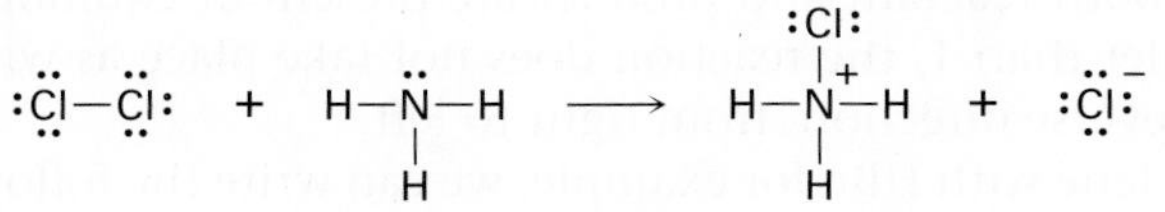

(b)
$$CH_3-\ddot{O}:^- \; + \; H-\underset{H}{\overset{H}{C}}-\ddot{B}r: \; \longrightarrow \; CH_3-\ddot{O}-CH_3 \; + \; :\ddot{B}r:^-$$

(c)
$$H_3C-\underset{Cl}{\overset{:\ddot{O}:^-}{C}}-OCH_3 \; \longrightarrow \; H_3C-\overset{:O:}{\overset{\|}{C}}-OCH_3 \; + \; :\ddot{C}l:^-$$

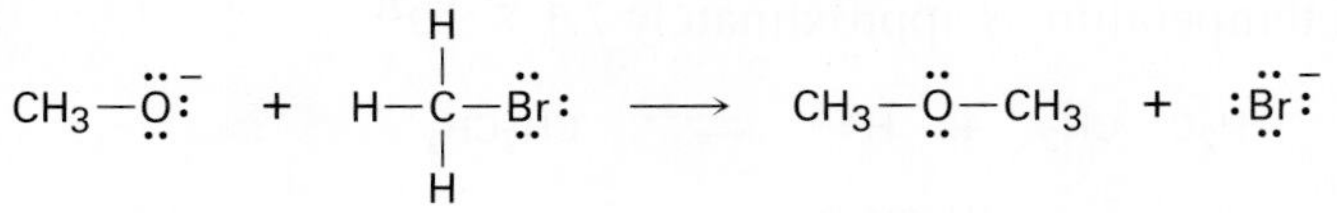

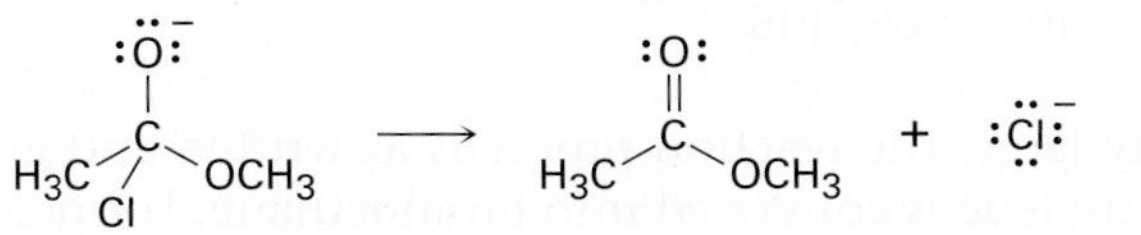

Problem 6.9
Predict the products of the following polar reaction, a step in the citric acid cycle for food metabolism, by interpreting the flow of electrons indicated by the curved arrows:

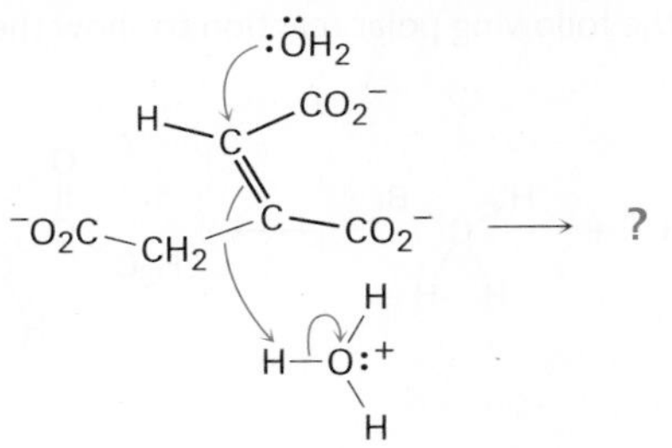

6.7 Describing a Reaction: Equilibria, Rates, and Energy Changes

Every chemical reaction can go in either forward or reverse direction. Reactants can go forward to products, and products can revert to reactants. As you may remember from your general chemistry course, the position of the resulting chemical equilibrium is expressed by an equation in which K_{eq}, the equilibrium constant, is equal to the product concentrations multiplied together, divided by the reactant concentrations multiplied together, with each concentration raised to the power of its coefficient in the balanced equation. For the generalized reaction

$$aA + bB \rightleftharpoons cC + dD$$

we have

$$K_{eq} = \frac{[C]^c [D]^d}{[A]^a [B]^b}$$

The value of the equilibrium constant tells which side of the reaction arrow is energetically favored. If K_{eq} is much larger than 1, then the product concentration term $[C]^c [D]^d$ is much larger than the reactant concentration term $[A]^a [B]^b$, and the reaction proceeds as written from left to right. If K_{eq} is near 1, appreciable amounts of both reactant and product are present at equilibrium. And if K_{eq} is much smaller than 1, the reaction does not take place as written but instead goes in the reverse direction, from right to left.

In the reaction of ethylene with HBr, for example, we can write the following equilibrium expression and determine experimentally that the equilibrium constant at room temperature is approximately 7.1×10^7:

$$H_2C=CH_2 + HBr \rightleftharpoons CH_3CH_2Br$$

$$K_{eq} = \frac{[CH_3CH_2Br]}{[H_2C=CH_2][HBr]} = 7.1 \times 10^7$$

Because K_{eq} is relatively large, the reaction proceeds as written and greater than 99.999 99% of the ethylene is converted into bromoethane. For practical purposes, an equilibrium constant greater than about 10^3 means that the amount of reactant left over will be barely detectable (less than 0.1%).

What determines the magnitude of the equilibrium constant? For a reaction to have a favorable equilibrium constant and proceed as written, the energy of the products must be lower than the energy of the reactants. In other words, energy must be released. The situation is analogous to that of a rock poised precariously in a high-energy position near the top of a hill. When it rolls downhill, the rock releases energy until it reaches a more stable, low-energy position at the bottom.

The energy change that occurs during a chemical reaction is called the **Gibbs free-energy change (ΔG)**, which is equal to the free energy of the products minus the free energy of the reactants: $\Delta G = G_{products} - G_{reactants}$. For a favorable reaction, ΔG has a negative value, meaning that energy is lost by the chemical system and released to the surroundings, usually as heat. Such reactions are said to be **exergonic**. For an unfavorable reaction, ΔG has a positive value, meaning that energy is absorbed by the chemical system *from* the surroundings. Such reactions are said to be **endergonic**.

You might also recall from general chemistry that the *standard* free-energy change for a reaction is denoted $\Delta G°$, where the superscript ° means that the reaction is carried out under standard conditions, with pure substances in their most stable form at 1 atm pressure and a specified temperature, usually 298 K. For biological reactions, the standard free-energy change is symbolized $\Delta G°'$ and refers to a reaction carried out at pH = 7.0 with solute concentrations of 1.0 M.

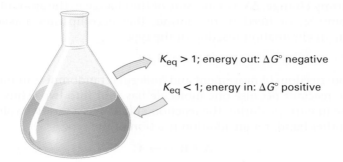

$K_{eq} > 1$; energy out: $\Delta G°$ negative

$K_{eq} < 1$; energy in: $\Delta G°$ positive

Because the equilibrium constant, K_{eq}, and the standard free-energy change, $\Delta G°$, both measure whether a reaction is favorable, they are mathematically related by the equation

$$\Delta G° = -RT \ln K_{eq} \quad \text{or} \quad K_{eq} = e^{-\Delta G°/RT}$$

where $R = 8.314 \text{ J/(K} \cdot \text{mol)} = 1.987 \text{ cal/(K} \cdot \text{mol)}$
T = Kelvin temperature
$e = 2.718$
$\ln K_{eq}$ = natural logarithm of K_{eq}

For example, the reaction of ethylene with HBr has $K_{eq} = 7.1 \times 10^7$, so $\Delta G° = -44.8$ kJ/mol (-10.7 kcal/mol) at 298 K:

$$K_{eq} = 7.1 \times 10^7 \quad \text{and} \quad \ln K_{eq} = 18.08$$
$$\Delta G° = -RT \ln K_{eq} = -[8.314 \text{ J/(K} \cdot \text{mol)}] (298 \text{ K}) (18.08)$$
$$= -44{,}800 \text{ J/mol} = -44.8 \text{ kJ/mol}$$

The free-energy change ΔG is made up of two terms, an *enthalpy* term, ΔH, and a temperature-dependent *entropy* term, $T\Delta S$. Of the two terms, the enthalpy term is often larger and more dominant.

$$\Delta G° = \Delta H° - T\Delta S°$$

For the reaction of ethylene with HBr at room temperature (298 K), the approximate values are

$$H_2C=CH_2 + HBr \rightleftharpoons CH_3CH_2Br$$

$\Delta G° = -44.8$ kJ/mol
$\Delta H° = -84.1$ kJ/mol
$\Delta S° = -0.132$ kJ/(K · mol)
$K_{eq} = 7.1 \times 10^7$

The **enthalpy change**, **ΔH**, also called the **heat of reaction**, is a measure of the change in total bonding energy during a reaction. If ΔH is negative, as in the reaction of HBr with ethylene, the products have less energy than the reactants. Thus, the products are more stable and have stronger bonds than the reactants, heat is released, and the reaction is said to be **exothermic**. If ΔH is positive, the products are less stable and have weaker bonds than the reactants, heat is absorbed, and the reaction is said to be **endothermic**. For example, if a reaction breaks reactant bonds with a total strength of 380 kJ/mol and forms product bonds with a total strength of 400 kJ/mol, then ΔH for the reaction is -20 kJ/mol and the reaction is exothermic.

The **entropy change**, **ΔS**, is a measure of the change in the amount of molecular randomness, or freedom of motion, that accompanies a reaction. For example, in an elimination reaction of the type

$$A \longrightarrow B + C$$

there is more freedom of movement and molecular randomness in the products than in the reactant because one molecule has split into two. Thus, there is a net increase in entropy during the reaction and ΔS has a positive value.

On the other hand, for an addition reaction of the type

$$A + B \longrightarrow C$$

the opposite is true. Because such reactions restrict the freedom of movement of two molecules by joining them together, the product has less randomness than the reactants and ΔS has a negative value. The reaction of ethylene and HBr to yield bromoethane, which has $\Delta S° = -0.132$ kJ/(K · mol), is an example. Table 6.2 describes the thermodynamic terms more fully.

Knowing the value of K_{eq} for a reaction is useful, but it's important to realize the limitations. An equilibrium constant tells only the *position* of the equilibrium, or how much product is theoretically possible. It doesn't tell the *rate* of reaction, or how fast the equilibrium is established. Some reactions are extremely slow even though they have favorable equilibrium constants. Gasoline is stable at room temperature, for instance, because the rate of its reaction with oxygen is slow at 298 K. Only at higher temperatures, such as contact with a lighted match, does gasoline react rapidly with oxygen and undergoes complete conversion to the equilibrium products water and carbon dioxide. Rates (*how fast* a reaction occurs) and equilibria (*how much* a reaction occurs) are entirely different.

Rate $\longrightarrow$ **Is the reaction fast or slow?**

Equilibrium $\longrightarrow$ **In what direction does the reaction proceed?**

Table 6.2 Explanation of Thermodynamic Quantities: $\Delta G° = \Delta H° - T\Delta S°$

Term	Name	Explanation
$\Delta G°$	Gibbs free-energy change	The energy difference between reactants and products. When $\Delta G°$ is negative, the reaction is **exergonic**, has a favorable equilibrium constant, and can occur spontaneously. When $\Delta G°$ is positive, the reaction is **endergonic**, has a unfavorable equilibrium constant, and cannot occur spontaneously.
$\Delta H°$	Enthalpy change	The heat of reaction, or difference in strength between the bonds broken in a reaction and the bonds formed. When $\Delta H°$ is negative, the reaction releases heat and is **exothermic**. When $\Delta H°$ is positive, the reaction absorbs heat and is **endothermic**.
$\Delta S°$	Entropy change	The change in molecular randomness during a reaction. When $\Delta S°$ is negative, randomness decreases. When $\Delta S°$ is positive, randomness increases.

Problem 6.10
Which reaction is more energetically favored, one with $\Delta G° = -44$ kJ/mol or one with $\Delta G° = +44$ kJ/mol?

Problem 6.11
Which reaction is likely to be more exergonic, one with $K_{eq} = 1000$ or one with $K_{eq} = 0.001$?

6.8 Describing a Reaction: Bond Dissociation Energies

We've just seen that heat is released (negative ΔH) when a bond is formed because the products are more stable and have stronger bonds than the reactants. Conversely, heat is absorbed (positive ΔH) when a bond is broken because the products are less stable and have weaker bonds than the reactants. The amount of energy needed to break a given bond to produce two radical fragments when the molecule is in the gas phase at 25 °C is a quantity called *bond strength*, or **bond dissociation energy (D)**.

$$A:B \xrightarrow{\text{Bond dissociation energy}} A\cdot + \cdot B$$

Each specific bond has its own characteristic strength, and extensive tables of data are available. For example, a C–H bond in methane has a bond dissociation energy $D = 439.3$ kJ/mol (105.0 kcal/mol), meaning that 439.3 kJ/mol must be added to break a C–H bond of methane to give the two radical fragments ·CH$_3$ and ·H. Conversely, 439.3 kJ/mol of energy is released when a methyl radical and a hydrogen atom combine to form methane. Table 6.3 lists some other bond strengths.

Think again about the connection between bond strengths and chemical reactivity. In an exothermic reaction, more heat is released than is absorbed. But because making bonds in the products releases heat and breaking bonds in the reactants absorbs heat, the bonds in the products must be stronger than the

Table 6.3 Some Bond Dissociation Energies, D

Bond	D (kJ/mol)	Bond	D (kJ/mol)	Bond	D (kJ/mol)
H—H	436	$(CH_3)_3C$—I	227	$(CH_3)_2CH$—CH_3	369
H—F	570	H_2C=CH—H	464	$(CH_3)_3C$—CH_3	363
H—Cl	431	H_2C=CH—Cl	396	H_2C=CH—CH_3	426
H—Br	366	H_2C=CHCH$_2$—H	369	H_2C=CHCH$_2$—CH_3	318
H—I	298	H_2C=CHCH$_2$—Cl	298	H_2C=CH_2	728
Cl—Cl	242	Ph—H	472	Ph—CH_3	427
Br—Br	194	Ph—Cl	400	Ph—CH_2CH_3	325
I—I	152	Ph—CH_2—H	375	$CH_3C(=O)$—H	374
CH_3—H	439	Ph—CH_2—Cl	300	HO—H	497
CH_3—Cl	350	Ph—Br	336	HO—OH	211
CH_3—Br	294	Ph—OH	464	CH_3O—H	440
CH_3—I	239	HC≡C—H	558	CH_3S—H	366
CH_3—OH	385	CH_3—CH_3	377	C_2H_5O—H	441
CH_3—NH_2	386	C_2H_5—CH_3	370	$CH_3C(=O)$—CH_3	352
C_2H_5—H	421			CH_3CH_2O—CH_3	355
C_2H_5—Cl	352			NH_2—H	450
C_2H_5—Br	293			H—CN	528
C_2H_5—I	233				
C_2H_5—OH	391				
$(CH_3)_2CH$—H	410				
$(CH_3)_2CH$—Cl	354				
$(CH_3)_2CH$—Br	299				
$(CH_3)_3C$—H	400				
$(CH_3)_3C$—Cl	352				
$(CH_3)_3C$—Br	293				

bonds in the reactants. In other words, exothermic reactions are favored by products with strong bonds and by reactants with weak, easily broken bonds.

Sometimes, particularly in biochemistry, reactive substances that undergo highly exothermic reactions, such as ATP (adenosine triphosphate), are referred to as "energy-rich" or "high-energy" compounds. Such a label doesn't mean that ATP is special or different from other compounds, it only means that ATP has relatively weak bonds that require a relatively small amount of heat to break, thus leading to a larger release of heat when a strong new bond forms in a reaction. When a typical organic phosphate such as glycerol 3-phosphate reacts with water, for instance, only 9 kJ/mol of heat is released ($\Delta H = -9$ kJ/mol), but when ATP reacts with water, 30 kJ/mol of heat is released ($\Delta H = -30$ kJ/mol). The difference between the two reactions is due to the fact that the bond broken in ATP is substantially weaker than the bond broken in glycerol 3-phosphate. We'll see the metabolic importance of this reaction in later chapters.

$\Delta H°' = -9$ kJ/mol

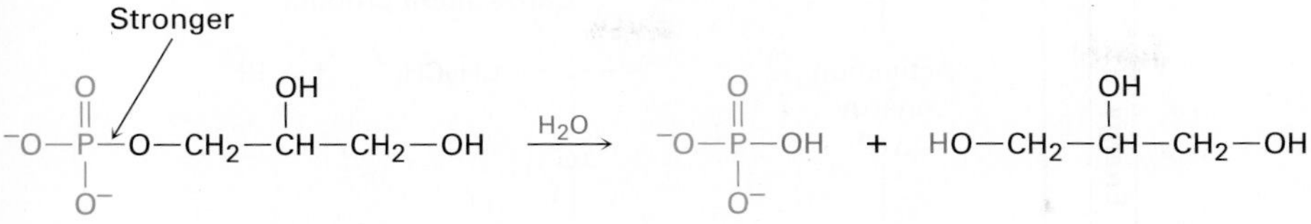

Stronger

⁻O—P(=O)(O⁻)—O—CH₂—CH(OH)—CH₂—OH →(H₂O) ⁻O—P(=O)(O⁻)—OH + HO—CH₂—CH(OH)—CH₂—OH

Glycerol 3-phosphate   **Glycerol**

$\Delta H°' = -30$ kJ/mol

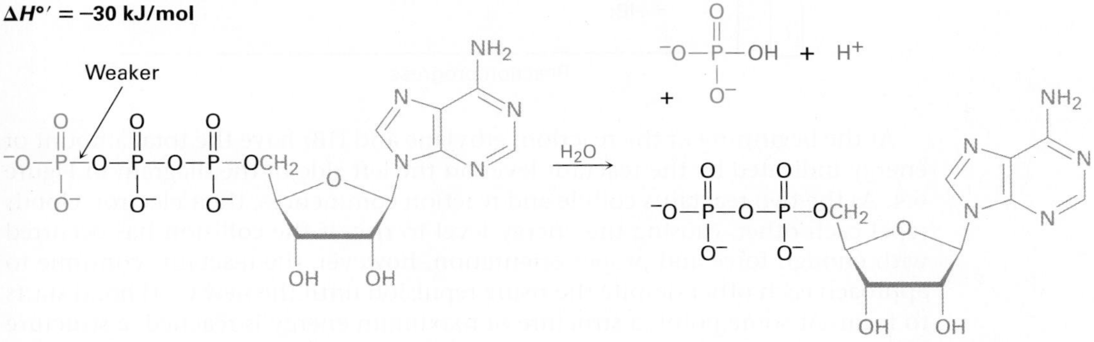

Weaker

Adenosine triphosphate (ATP) →(H₂O) Adenosine diphosphate (ADP) + ⁻O—P(=O)(OH)—O⁻ + H⁺

Adenosine triphosphate (ATP)   **Adenosine diphosphate (ADP)**

6.9 Describing a Reaction: Energy Diagrams and Transition States

For a reaction to take place, reactant molecules must collide and reorganization of atoms and bonds must occur. Let's again look at the addition reaction of HBr and ethylene.

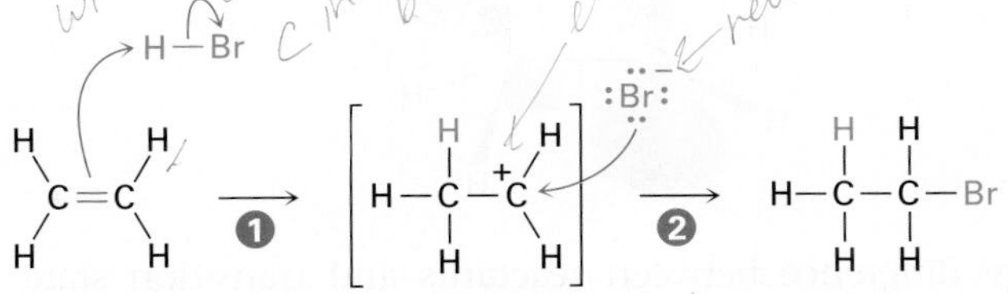

H₂C=CH₂ + H—Br →① [H₃C—CH₂]⁺ (Carbocation) + :Br:⁻ →② H₃C—CH₂—Br

As the reaction proceeds, ethylene and HBr must approach each other, the ethylene π bond and the H–Br bond must break, a new C–H bond must form in step ①, and a new C–Br bond must form in step ②.

To depict graphically the energy changes that occur during a reaction, chemists use energy diagrams, such as that shown in **Figure 6.4**. The vertical axis of the diagram represents the total energy of all reactants, and the horizontal axis, called the *reaction coordinate*, represents the progress of the reaction from beginning to end. Let's see how the addition of HBr to ethylene can be described in an energy diagram.

Figure 6.4 An energy diagram for the first step in the reaction of ethylene with HBr. The energy difference between reactants and transition state, $\Delta G^{\ddagger}$, defines the reaction rate. The energy difference between reactants and carbocation product, $\Delta G°$, defines the position of the equilibrium.

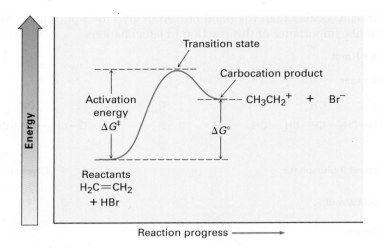

At the beginning of the reaction, ethylene and HBr have the total amount of energy indicated by the reactant level on the left side of the diagram in Figure 6.4. As the two reactants collide and reaction commences, their electron clouds repel each other, causing the energy level to rise. If the collision has occurred with enough force and proper orientation, however, the reactants continue to approach each other despite the rising repulsion until the new C–H bond starts to form. At some point, a structure of maximum energy is reached, a structure called the *transition state*.

The **transition state** represents the highest-energy structure involved in this step of the reaction. It is unstable and can't be isolated, but we can nevertheless imagine it to be an activated complex of the two reactants in which both the C=C π bond and H–Br bond are partially broken and the new C–H bond is partially formed **(Figure 6.5)**.

Figure 6.5 A hypothetical transition-state structure for the first step of the reaction of ethylene with HBr. The C=C π bond and H–Br bond are just beginning to break, and the C–H bond is just beginning to form.

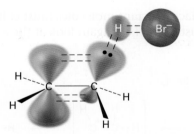

The energy difference between reactants and transition state is called the **activation energy, $\Delta G^{\ddagger}$**, and determines how rapidly the reaction occurs at a given temperature. (The double-dagger superscript, $\ddagger$, always refers to the transition state.) A large activation energy results in a slow reaction because few collisions occur with enough energy for the reactants to reach the transition state. A small activation energy results in a rapid reaction because almost all collisions occur with enough energy for the reactants to reach the transition state.

As an analogy, you might think of reactants that need enough energy to climb the activation barrier to the transition state as similar to hikers who need enough energy to climb to the top of a mountain pass. If the pass is a high one, the hikers need a lot of energy and surmount the barrier with difficulty. If the pass is low, however, the hikers need less energy and reach the top easily.

As a rough generalization, many organic reactions have activation energies in the range 40 to 150 kJ/mol (10–35 kcal/mol). The reaction of ethylene with HBr, for example, has an activation energy of approximately 140 kJ/mol (34 kcal/mol). Reactions with activation energies less than 80 kJ/mol take place at or below room temperature, while reactions with higher activation energies normally require a higher temperature to give the reactants enough energy to climb the activation barrier.

Once the transition state is reached, the reaction can either continue on to give the carbocation product or revert back to reactants. When reversion to reactants occurs, the transition-state structure comes apart and an amount of free energy corresponding to $-\Delta G^{\ddagger}$ is released. When the reaction continues on to give the carbocation, the new C–H bond forms fully and an amount of energy corresponding to the difference between transition state and carbocation product is released. The net energy change for the step, $\Delta G°$, is represented in the diagram as the difference in level between reactant and product. Since the carbocation is higher in energy than the starting alkene, the step is endergonic, has a positive value of $\Delta G°$, and absorbs energy.

Not all energy diagrams are like that shown for the reaction of ethylene and HBr. Each reaction has its own energy profile. Some reactions are fast (small $\Delta G^{\ddagger}$) and some are slow (large $\Delta G^{\ddagger}$); some have a negative $\Delta G°$, and some have a positive $\Delta G°$. **Figure 6.6** illustrates some different possibilities.

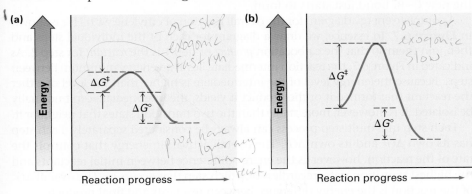

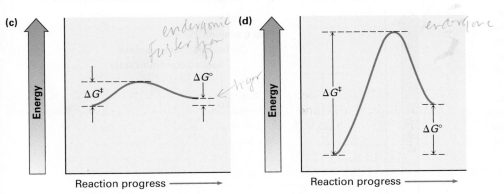

Figure 6.6 Some hypothetical energy diagrams: **(a)** a fast exergonic reaction (small $\Delta G^{\ddagger}$, negative $\Delta G°$); **(b)** a slow exergonic reaction (large $\Delta G^{\ddagger}$, negative $\Delta G°$); **(c)** a fast endergonic reaction (small $\Delta G^{\ddagger}$, small positive $\Delta G°$); **(d)** a slow endergonic reaction (large $\Delta G^{\ddagger}$, positive $\Delta G°$).

Problem 6.12
Which reaction is faster, one with $\Delta G^{\ddagger} = +45$ kJ/mol or one with $\Delta G^{\ddagger} = +70$ kJ/mol?

6.10 Describing a Reaction: Intermediates

How can we describe the carbocation formed in the first step of the reaction of ethylene with HBr? The carbocation is clearly different from the reactants, yet it isn't a transition state and it isn't a final product.

$$H_2C=CH_2 + H-Br \longrightarrow \left[H-\overset{H}{\underset{H}{C}}-\overset{+}{\underset{H}{C}}\overset{H}{\underset{}{}} \quad :\!\ddot{B}\ddot{r}\!:^- \right] \longrightarrow H-\overset{H}{\underset{H}{C}}-\overset{H}{\underset{H}{C}}-Br$$

Reaction intermediate

We call the carbocation, which exists only transiently during the course of the multistep reaction, a **reaction intermediate**. As soon as the intermediate is formed in the first step by reaction of ethylene with H^+, it reacts further with Br^- in a second step to give the final product, bromoethane. This second step has its own activation energy ($\Delta G^‡$), its own transition state, and its own energy change ($\Delta G°$). We can picture the second transition state as an activated complex between the electrophilic carbocation intermediate and the nucleophilic bromide anion, in which Br^- donates a pair of electrons to the positively charged carbon atom as the new C–Br bond just starts to form.

A complete energy diagram for the overall reaction of ethylene with HBr is shown in **Figure 6.7**. In essence, we draw a diagram for each of the individual steps and then join them so that the carbocation *product* of step 1 is the *reactant* for step 2. As indicated in Figure 6.7, the reaction intermediate lies at an energy minimum between steps. Because the energy level of the intermediate is higher than the level of either the reactant that formed it or the product it yields, the intermediate can't normally be isolated. It is, however, more stable than the two transition states that neighbor it.

Each step in a multistep process can always be considered separately. Each step has its own $\Delta G^‡$ and its own $\Delta G°$. The overall activation energy that controls the rate of the reaction, however, is the energy difference between initial reactants and the highest transition state, regardless of which step that occurs in. The overall $\Delta G°$ of the reaction is the energy difference between reactants and final products.

Figure 6.7 An energy diagram for the reaction of ethylene with HBr. Two separate steps are involved, each with its own activation energy ($\Delta G^‡$) and free-energy change ($\Delta G°$). The overall $\Delta G^‡$ for the complete reaction is the energy difference between reactants and the highest transition state (which corresponds to $\Delta G_1^‡$ in this case), and the overall $\Delta G°$ for the reaction is the energy difference between reactants and final products.

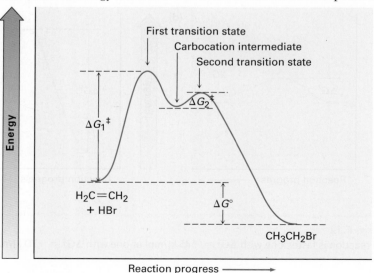

The biological reactions that take place in living organisms have the same energy requirements as reactions that take place in the laboratory and can be described in similar ways. They are, however, constrained by the fact that they must have low enough activation energies to occur at moderate temperatures, and they must release energy in relatively small amounts to avoid overheating the organism. These constraints are generally met through the use of large, structurally complex, enzyme catalysts that change the mechanism of a reaction to an alternative pathway that proceeds through a series of small steps rather than one or two large steps. Thus, a typical energy diagram for a biological reaction might look like that in **Figure 6.8**.

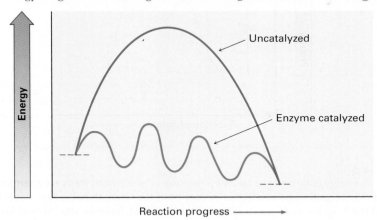

Figure 6.8 An energy diagram for a typical, **enzyme-catalyzed biological reaction** versus an **uncatalyzed laboratory reaction**. The biological reaction involves many steps, each of which has a relatively small activation energy and small energy change. The end result is the same, however.

Drawing an Energy Diagram for a Reaction

Worked Example 6.3

Sketch an energy diagram for a one-step reaction that is fast and highly exergonic.

Strategy
A fast reaction has a small $\Delta G^{\ddagger}$, and a highly exergonic reaction has a large negative $\Delta G°$.

Solution

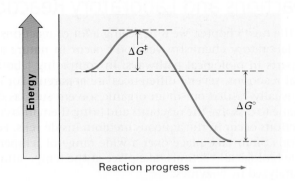

Drawing an Energy Diagram for a Reaction

Worked Example 6.4

Sketch an energy diagram for a two-step exergonic reaction whose second step has a higher-energy transition state than its first step. Show $\Delta G^{\ddagger}$ and $\Delta G°$ for the overall reaction.

Strategy

A two-step reaction has two transition states and an intermediate between them. The $\Delta G^{\ddagger}$ for the overall reaction is the energy change between reactants and the highest-energy transition state—the second one in this case. An exergonic reaction has a negative overall $\Delta G°$.

Solution

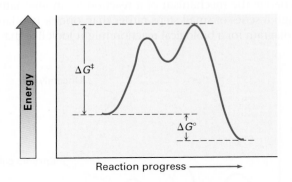

Problem 6.13
Sketch an energy diagram for a two-step reaction in which both steps are exergonic and in which the second step has a higher-energy transition state than the first. Label the parts of the diagram corresponding to reactant, product, intermediate, overall $\Delta G^{\ddagger}$, and overall $\Delta G°$.

6.11 A Comparison Between Biological Reactions and Laboratory Reactions

Beginning in the next chapter, we'll be seeing a lot of reactions, some that are important in laboratory chemistry yet don't occur in nature and others that have counterparts in biological pathways. In comparing laboratory reactions with biological reactions, several differences are apparent. For one, laboratory reactions are usually carried out in an organic solvent such as diethyl ether or dichloromethane to dissolve the reactants and bring them into contact, whereas biological reactions occur in the aqueous medium inside cells. For another, laboratory reactions often take place over a wide range of temperatures without catalysts, while biological reactions take place at the temperature of the organism and are catalyzed by enzymes.

We'll look at enzymes in more detail in **Section 26.10**, but you may already be aware that an enzyme is a large, globular, protein molecule that contains in its structure a protected pocket called its *active site*. The active site is lined by acidic or basic groups as needed for catalysis and has precisely the right shape to bind and hold a substrate molecule in the orientation necessary for reaction. **Figure 6.9** shows a molecular model of hexokinase, along with an X-ray crystal structure of the glucose substrate and adenosine diphosphate (ADP) bound in the active site. Hexokinase is an enzyme that catalyzes the initial step of glucose metabolism—the transfer of a phosphate group from ATP to glucose, giving glucose 6-phosphate and ADP. The structures of ATP and ADP were shown at the end of **Section 6.8**.

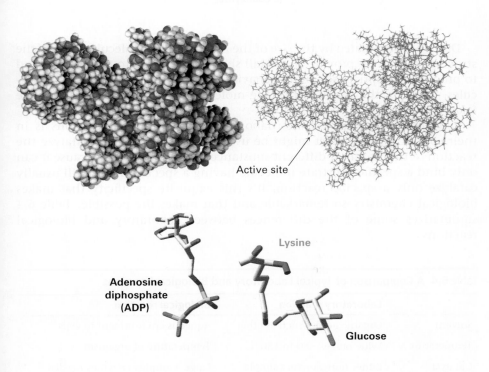

Note how the hexokinase-catalyzed phosphorylation reaction of glucose is written. It's common when writing biological equations to show only the structures of the primary reactant and product, while abbreviating the structures of various biological "reagents" and by-products such as ATP and ADP. A curved arrow intersecting the straight reaction arrow indicates that ATP is also a reactant and ADP also a product.

Figure 6.9 Models of hexokinase in space-filling and wire-frame formats, showing the cleft that contains the active site where substrate binding and reaction catalysis occur. At the bottom is an X-ray crystal structure of the enzyme active site, showing the positions of both glucose and ADP as well as a lysine amino acid that acts as a base to deprotonate glucose.

Yet another difference between laboratory and biological reactions is that laboratory reactions are often done using relatively small, simple reagents such as Br_2, HCl, $NaBH_4$, CrO_3, and so forth, while biological reactions usually involve relatively complex "reagents" called *coenzymes*. In the hexokinase-catalyzed phosphorylation of glucose just shown, ATP is the coenzyme. As another example, compare the H_2 molecule, a laboratory reagent that adds to a carbon–carbon double bond to yield an alkane, with the reduced nicotinamide adenine dinucleotide (NADH) molecule, a coenzyme that effects an analogous addition of hydrogen to a double bond in many biological pathways. Of all the atoms in the entire coenzyme, only the one hydrogen atom shown in red is transferred to the double-bond substrate.

**Reduced nicotinamide adenine dinucleotide, NADH
(a coenzyme)**

Don't be intimidated by the size of the ATP or NADH molecule; most of the structure is there to provide an overall shape for binding to the enzyme and to provide appropriate solubility behavior. When looking at biological molecules, focus on the small part of the molecule where the chemical change takes place.

One final difference between laboratory and biological reactions is in their specificity. A catalyst might be used in the laboratory to catalyze the reaction of thousands of different substances, but an enzyme, because it can only bind a specific substrate molecule having a specific shape, will usually catalyze only a specific reaction. It's this exquisite specificity that makes biological chemistry so remarkable and that makes life possible. Table 6.4 summarizes some of the differences between laboratory and biological reactions.

Table 6.4 A Comparison of Typical Laboratory and Biological Reactions

	Laboratory reaction	Biological reaction
Solvent	Organic liquid, such as ether	Aqueous environment in cells
Temperature	Wide range; −80 to 150 °C	Temperature of organism
Catalyst	Either none, or very simple	Large, complex enzymes needed
Reagent size	Usually small and simple	Relatively complex coenzymes
Specificity	Little specificity for substrate	Very high specificity for substrate

A DEEPER LOOK — Where Do Drugs Come From?

It has been estimated that major pharmaceutical companies in the United States spend some $33 billion per year on drug research and development, while government agencies and private foundations spend another $28 billion. What does this money buy? For the period 1981 to 2008, the money resulted in a total of 989 new molecular entities (NMEs)—new biologically active chemical substances approved for sale as drugs by the U.S. Food and Drug Administration (FDA). That's an average of only 35 new drugs each year, spread over all diseases and conditions, and the number is steadily falling. In 2008, only 20 NMEs were approved.

Where do the new drugs come from? According to a study carried out at the U.S. National Cancer Institute, only about 33% of new drugs are entirely synthetic and completely unrelated to any naturally occurring substance. The remaining 67% take their lead, to a greater or lesser extent, from nature. Vaccines and genetically engineered proteins of biological origin account for 15% of NMEs, but most new drugs come from *natural products,* a catchall term generally taken to mean small molecules found in bacteria, plants, and other living organisms. Unmodified natural products isolated directly from the producing organism account for 24% of NMEs, while natural products that have been chemically modified in the laboratory account for the remaining 28%.

Introduced in June, 2006, Gardasil is the first vaccine ever approved for the prevention of cancer. Where do new drugs like this come from?

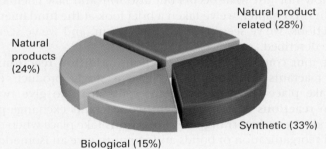

Origin of New Drugs 1981–2002

- Natural product related (28%)
- Synthetic (33%)
- Biological (15%)
- Natural products (24%)

Many years of work go into screening many thousands of substances to identify a single compound that might ultimately gain approval as an NME. But after that single compound has been identified, the work has just begun because it takes an average of 9 to 10 years for a drug to make it through the approval process. First, the safety of the drug in animals must be demonstrated and an economical method of manufacture must be devised. With these preliminaries out of the way, an Investigational New Drug (IND) application is submitted to the FDA for permission to begin testing in humans.

Human testing takes 5 to 7 years and is divided into three phases. Phase I clinical trials are carried out on a small group of healthy volunteers to establish safety and look for side effects. Several months to a year are needed, and only about 70% of drugs pass at this point. Phase II clinical trials next test the drug for 1 to 2 years in several hundred patients with the target disease or condition, looking both for safety and for efficacy, and only about 33% of the original group pass. Finally, phase III trials are undertaken on a large sample of patients to document definitively the drug's safety, dosage, and efficacy. If the

(continued)

drug is one of the 25% of the original group that make it to the end of phase III, all the data are then gathered into a New Drug Application (NDA) and sent to the FDA for review and approval, which can take another 2 years. Ten years have elapsed and at least $500 million has been spent, with only a 20% success rate for the drugs that began testing. Finally, though, the drug will begin to appear in medicine cabinets. The following timeline shows the process.

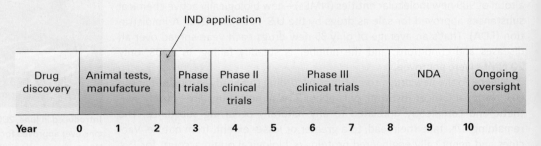

Summary

Key words

activation energy ($\Delta G^{\ddagger}$), 206
addition reaction, 184
bond dissociation energy (*D*), 203
carbocation, 196
electrophile, 192
elimination reaction, 185
endergonic, 201
endothermic, 202
enthalpy change (ΔH), 202
entropy change (ΔS), 202
exergonic, 201
exothermic, 202
Gibbs free-energy change (ΔG), 201
heat of reaction, 202
nucleophile, 192
polar reaction, 187
radical, 187
radical reaction, 187
reaction intermediate, 208
reaction mechanism, 186
rearrangement reaction, 185
substitution reaction, 185
transition state, 206

All chemical reactions, whether in the laboratory or in living organisms, follow the same "rules." To understand both organic and biological chemistry, it's necessary to know not just *what* occurs but also *why* and *how* chemical reactions take place. In this chapter, we've taken a brief look at the fundamental kinds of organic reactions, we've seen why reactions occur, and we've seen how reactions can be described.

There are four common kinds of reactions: **addition reactions** take place when two reactants add together to give a single product; **elimination reactions** take place when one reactant splits apart to give two products; **substitution reactions** take place when two reactants exchange parts to give two new products; and **rearrangement reactions** take place when one reactant undergoes a reorganization of bonds and atoms to give an isomeric product.

A full description of how a reaction occurs is called its **mechanism**. There are two general kinds of mechanisms by which most reactions take place: **radical** mechanisms and **polar** mechanisms. Polar reactions, the more common type, occur because of an attractive interaction between a **nucleophilic** (electron-rich) site in one molecule and an **electrophilic** (electron-poor) site in another molecule. A bond is formed in a polar reaction when the nucleophile donates an electron pair to the electrophile. This movement of electrons is indicated by a curved arrow showing the direction of electron travel from the nucleophile to the electrophile. Radical reactions involve species that have an odd number of electrons. A bond is formed when each reactant donates one electron.

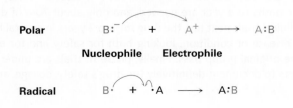

The energy changes that take place during reactions can be described by considering both rates (how fast the reactions occur) and equilibria (how much the reactions occur). The position of a chemical equilibrium is determined by the value of the **free-energy change (ΔG)** for the reaction, where $\Delta G = \Delta H - T\Delta S$. The **enthalpy** term (ΔH) corresponds to the net change in strength of chemical bonds broken and formed during reaction; the **entropy** term (ΔS) corresponds to the change in the amount of molecular randomness during the reaction. Reactions that have negative values of ΔG release energy, are said to be **exergonic**, and have favorable equilibria. Reactions that have positive values of ΔG absorb energy, are said to be **endergonic**, and have unfavorable equilibria.

A reaction can be described pictorially using an energy diagram that follows the reaction course from reactant through transition state to product. The **transition state** is an activated complex occurring at the highest-energy point of a reaction. The amount of energy needed by reactants to reach this high point is the **activation energy, $\Delta G^{\ddagger}$**. The higher the activation energy, the slower the reaction.

Many reactions take place in more than one step and involve the formation of a **reaction intermediate**. An intermediate is a species that lies at an energy minimum between steps on the reaction curve and is formed briefly during the course of a reaction.

Exercises

Visualizing Chemistry

(Problems 6.1–6.13 appear within the chapter.)

▲ OWL Interactive versions of these problems are assignable in OWL for Organic Chemistry.

▲ denotes problems linked to the Key Ideas in this chapter.

6.14 The following alkyl halide can be prepared by addition of HBr to two different alkenes. Draw the structures of both (reddish-brown = Br).

6.15 The following structure represents the carbocation intermediate formed in the addition reaction of HBr to two different alkenes. Draw the structures of both.

▲ Problems linked to Key Ideas in this chapter

6.16 Electrostatic potential maps of **(a)** formaldehyde (CH_2O) and **(b)** methanethiol (CH_3SH) are shown. Is the formaldehyde carbon atom likely to be electrophilic or nucleophilic? What about the methanethiol sulfur atom? Explain.

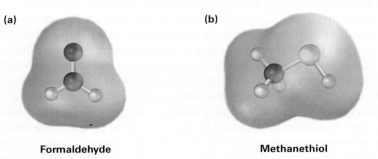

(a) Formaldehyde **(b)** Methanethiol

6.17 Look at the following energy diagram:

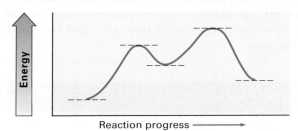

(a) Is $\Delta G°$ for the reaction positive or negative? Label it on the diagram.
(b) How many steps are involved in the reaction?
(c) How many transition states are there? Label them on the diagram.

6.18 Look at the following energy diagram for an enzyme-catalyzed reaction:

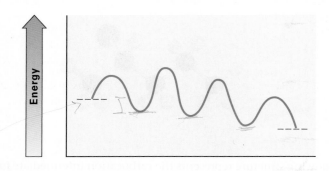

(a) How many steps are involved?
(b) Which step is most exergonic?
(c) Which step is slowest?

▲ Problems linked to Key Ideas in this chapter

Additional Problems

Polar Reactions

6.19 Identify the functional groups in the following molecules, and show the polarity of each:

(a) $CH_3CH_2C\equiv N$

(b) 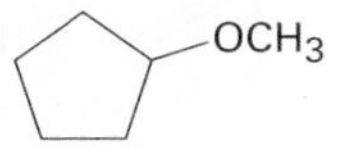 cyclopentyl-OCH_3

(c) 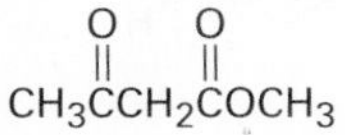 $CH_3\overset{O}{\overset{\|}{C}}CH_2\overset{O}{\overset{\|}{C}}OCH_3$

(d) 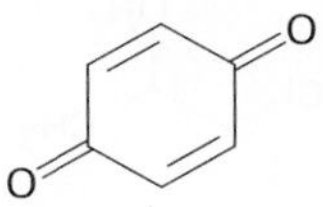 1,4-benzoquinone

(e) 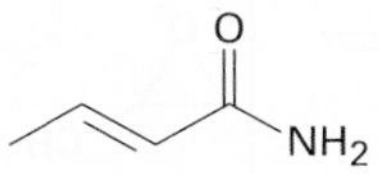 $CH_3CH=CHC(=O)NH_2$ (acrylamide derivative)

(f) 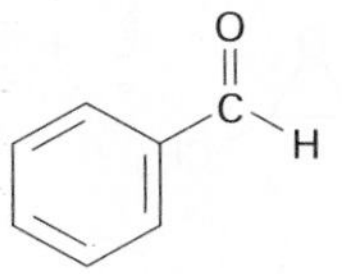 benzaldehyde

6.20 Identify the following reactions as additions, eliminations, substitutions, or rearrangements:

(a) $CH_3CH_2Br + NaCN \longrightarrow CH_3CH_2CN \;(+ NaBr)$

(b) 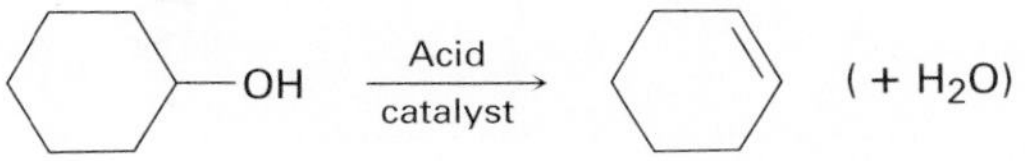 cyclohexanol $\xrightarrow{\text{Acid catalyst}}$ cyclohexene $(+ H_2O)$

(c) 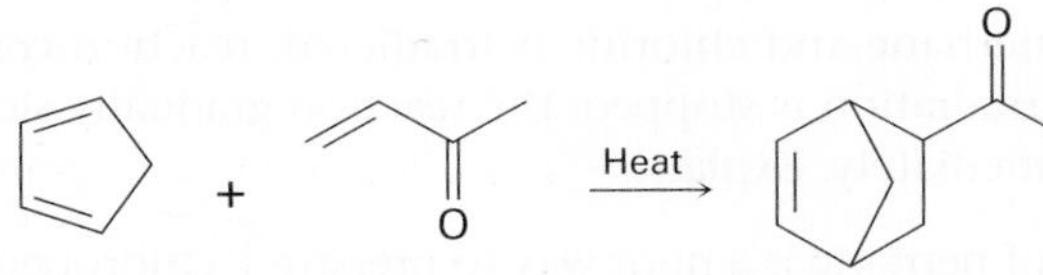 cyclopentadiene + methyl vinyl ketone $\xrightarrow{\text{Heat}}$ norbornene acetyl product

(d) 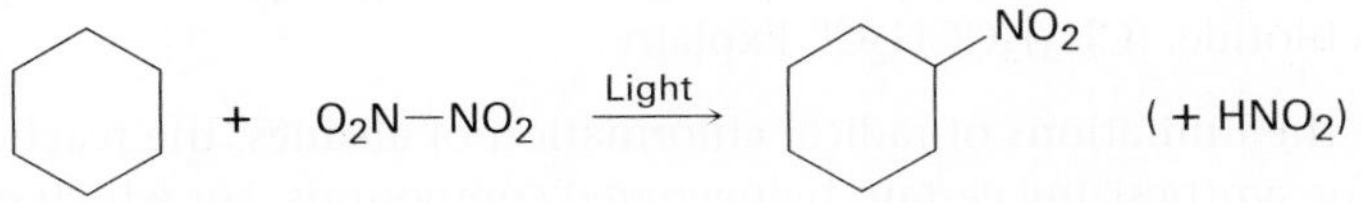 cyclohexane + O_2N-NO_2 $\xrightarrow{\text{Light}}$ nitrocyclohexane $(+ HNO_2)$

6.21 Identify the likely electrophilic and nucleophilic sites in each of the following molecules:

(a) 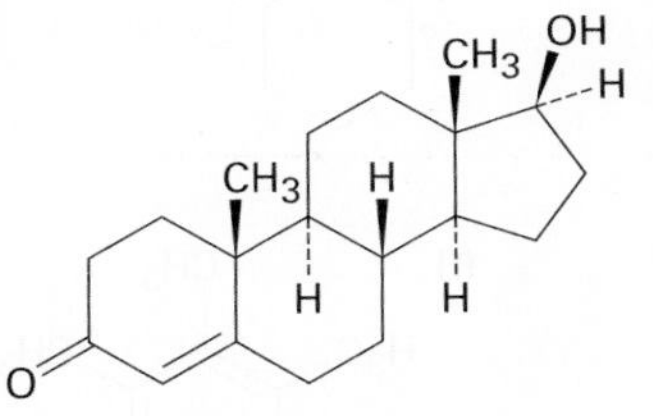 Testosterone

(b) 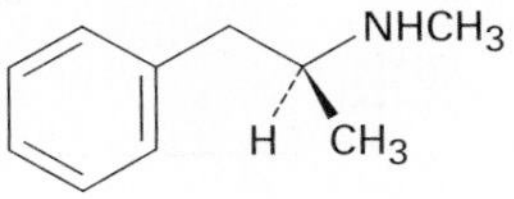 Amphetamine

▲ Problems linked to Key Ideas in this chapter

6.22 ▲ Add curved arrows to the following polar reactions to indicate the flow of electrons in each:

(a) [reaction of benzene with D–Cl through arenium intermediate yielding deuterated benzene + H–Cl]

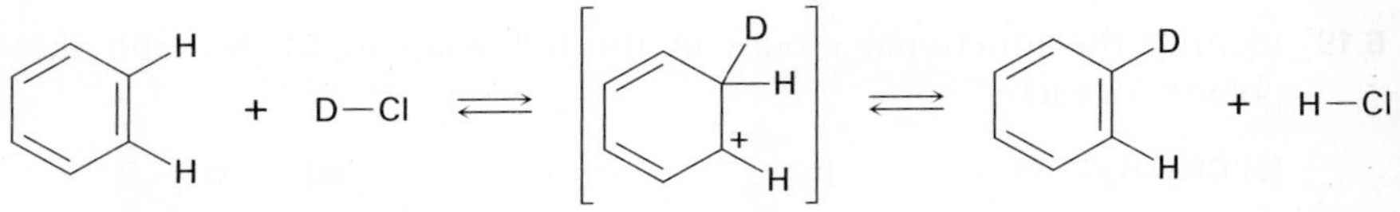

(b) [reaction of methyloxirane with H–Cl through protonated epoxide yielding 1-chloro-2-propanol]

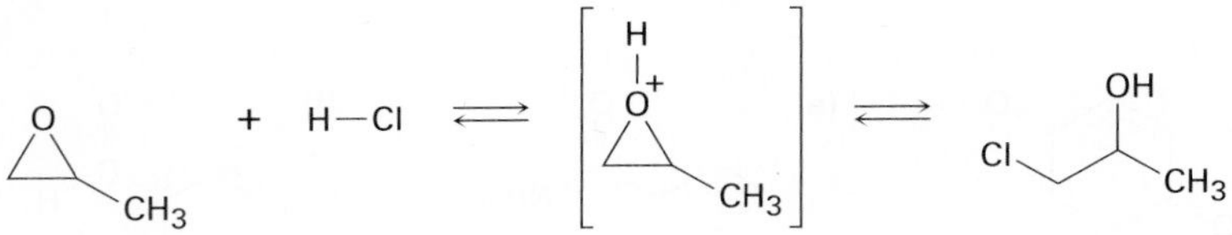

6.23 ▲ Follow the flow of electrons indicated by the curved arrows in each of the following polar reactions, and predict the products that result:

(a) [(CH₃)₂C(OCH₃)O⁻ with :Ö–H, arrows shown] ⇌ ?

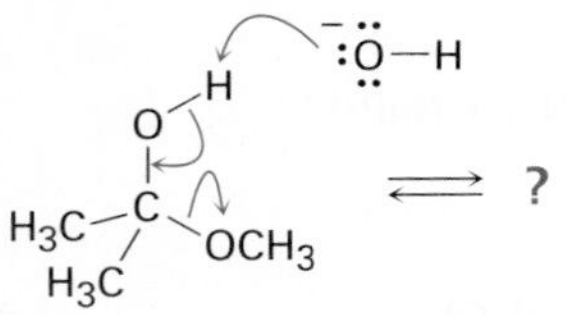

(b) [H–Ö: attacking CH₃–C(=O)–H with arrows] ⇌ ?

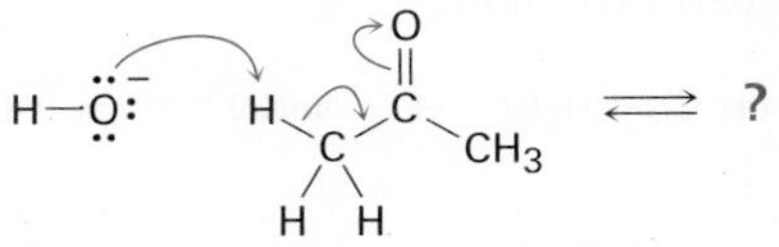

Radical Reactions

6.24 When a mixture of methane and chlorine is irradiated, reaction commences immediately. When irradiation is stopped, the reaction gradually slows down but does not stop immediately. Explain.

6.25 Radical chlorination of pentane is a poor way to prepare 1-chloropentane, but radical chlorination of neopentane, $(CH_3)_4C$, is a good way to prepare neopentyl chloride, $(CH_3)_3CCH_2Cl$. Explain.

6.26 Despite the limitations of radical chlorination of alkanes, the reaction is still useful for synthesizing certain halogenated compounds. For which of the following compounds does radical chlorination give a single monochloro product?

(a) CH_3CH_3

(b) $CH_3CH_2CH_3$

(c) [cyclohexane]

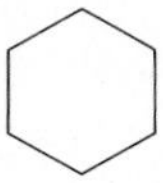

(d) $CH_3CCH_2CH_3$ with two CH_3 substituents on central C (2,2-dimethylbutane)

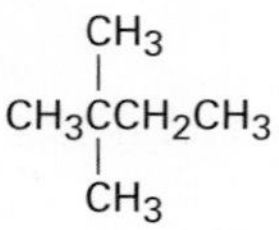

(e) $CH_3C\equiv CCH_3$

(f) [hexamethylbenzene]

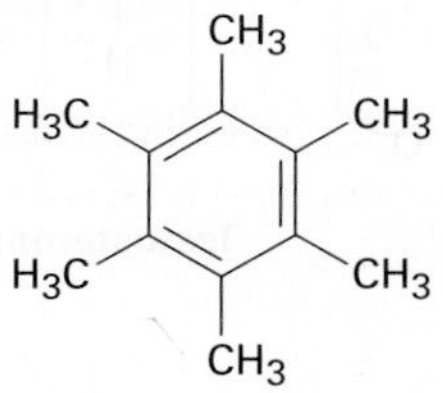

▲ Problems linked to Key Ideas in this chapter

Energy Diagrams and Reaction Mechanisms

6.27 What is the difference between a transition state and an intermediate?

6.28 Draw an energy diagram for a one-step reaction with $K_{eq} < 1$. Label the parts of the diagram corresponding to reactants, products, transition state, $\Delta G°$, and $\Delta G^{\ddagger}$. Is $\Delta G°$ positive or negative?

6.29 Draw an energy diagram for a two-step reaction with $K_{eq} > 1$. Label the overall $\Delta G°$, transition states, and intermediate. Is $\Delta G°$ positive or negative?

6.30 Draw an energy diagram for a two-step exergonic reaction whose second step is faster than its first step.

6.31 Draw an energy diagram for a reaction with $K_{eq} = 1$. What is the value of $\Delta G°$ in this reaction?

6.32 The addition of water to ethylene to yield ethanol has the following thermodynamic parameters:

$$H_2C=CH_2 + H_2O \rightleftharpoons CH_3CH_2OH \quad \begin{cases} \Delta H° = -44 \text{ kJ/mol} \\ \Delta S° = -0.12 \text{ kJ/(K} \cdot \text{mol)} \\ K_{eq} = 24 \end{cases}$$

(a) Is the reaction exothermic or endothermic?
(b) Is the reaction favorable (spontaneous) or unfavorable (nonspontaneous) at room temperature (298 K)?

6.33 When isopropylidenecyclohexane is treated with strong acid at room temperature, isomerization occurs by the mechanism shown below to yield 1-isopropylcyclohexene:

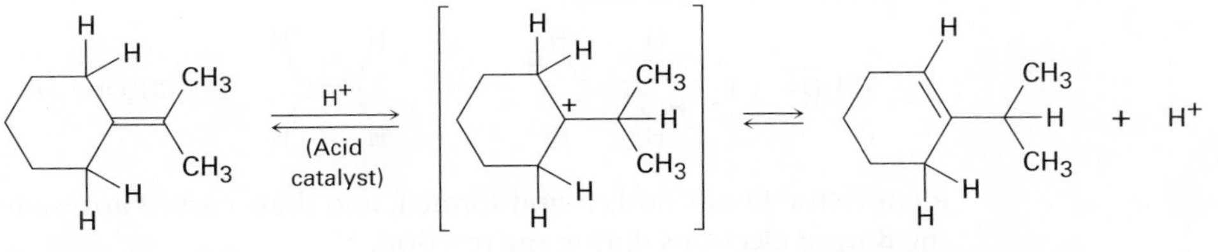

Isopropylidenecyclohexane ⇌ (H⁺, Acid catalyst) ⇌ **1-Isopropylcyclohexene** + H⁺

At equilibrium, the product mixture contains about 30% isopropylidenecyclohexane and about 70% 1-isopropylcyclohexene.
(a) What is an approximate value of K_{eq} for the reaction?
(b) Since the reaction occurs slowly at room temperature, what is its approximate $\Delta G^{\ddagger}$?
(c) Draw an energy diagram for the reaction.

6.34 ▲ Add curved arrows to the mechanism shown in Problem 6.33 to indicate the electron movement in each step.

General Problems

6.35 2-Chloro-2-methylpropane reacts with water in three steps to yield 2-methyl-2-propanol. The first step is slower than the second, which in turn is much

▲ Problems linked to Key Ideas in this chapter

slower than the third. The reaction takes place slowly at room temperature, and the equilibrium constant is near 1.

$$H_3C-\underset{\underset{CH_3}{|}}{\overset{\overset{CH_3}{|}}{C}}-Cl \rightleftharpoons \left[H_3C-\underset{\underset{CH_3}{|}}{\overset{\overset{CH_3}{|}}{C^+}} \underset{\rightleftharpoons}{\overset{H_2O}{\rightleftharpoons}} H_3C-\underset{\underset{CH_3}{|}}{\overset{\overset{CH_3}{|}}{C}}-\overset{+}{\underset{H}{O}}{}^{\diagdown H} \right] \underset{\rightleftharpoons}{\overset{H_2O}{\rightleftharpoons}} H_3C-\underset{\underset{CH_3}{|}}{\overset{\overset{CH_3}{|}}{C}}-O-H + H_3O^+ + Cl^-$$

2-Chloro-2-methylpropane

2-Methyl-2-propanol

(a) Give approximate values for $\Delta G^\ddagger$ and $\Delta G°$ that are consistent with the above information.

(b) Draw an energy diagram for the reaction, labeling all points of interest and making sure that the relative energy levels on the diagram are consistent with the information given.

6.36 ▲ Add curved arrows to the mechanism shown in Problem 6.35 to indicate the electron movement in each step.

6.37 The reaction of hydroxide ion with chloromethane to yield methanol and chloride ion is an example of a general reaction type called a *nucleophilic substitution reaction*:

$$HO^- + CH_3Cl \rightleftharpoons CH_3OH + Cl^-$$

The value of $\Delta H°$ for the reaction is -75 kJ/mol, and the value of $\Delta S°$ is $+54$ J/(K·mol). What is the value of $\Delta G°$ (in kJ/mol) at 298 K? Is the reaction exothermic or endothermic? Is it exergonic or endergonic?

6.38 Methoxide ion (CH_3O^-) reacts with bromoethane in a single step according to the following equation:

$$CH_3\ddot{\underset{..}{O}}{:}^- + \underset{\underset{H}{|}}{\overset{\overset{H}{|}}{C}}\underset{Br}{\overset{H}{-}}\overset{H}{\underset{H}{C}} \longrightarrow \underset{\underset{H}{}}{\overset{\overset{H}{}}{C}}=\underset{\underset{H}{}}{\overset{\overset{H}{}}{C}} + CH_3OH + :\ddot{\underset{..}{Br}}{:}^-$$

Identify the bonds broken and formed, and draw curved arrows to represent the flow of electrons during the reaction.

6.39 ▲ Ammonia reacts with acetyl chloride (CH_3COCl) to give acetamide (CH_3CONH_2). Identify the bonds broken and formed in each step of the reaction, and draw curved arrows to represent the flow of electrons in each step.

$$\underset{\textbf{Acetyl chloride}}{H_3C\overset{\overset{:O:}{\|}}{\underset{}{C}}{\diagdown}Cl} \xrightarrow{:NH_3} H_3C\overset{\overset{:\ddot{O}:^-}{|}}{\underset{Cl}{C}}{\diagdown}NH_3^+ \longrightarrow H_3C\overset{\overset{:O:}{\|}}{\underset{}{C}}{\diagdown}NH_3^+ \longrightarrow$$

$$\xrightarrow{:NH_3} \underset{\textbf{Acetamide}}{H_3C\overset{\overset{:O:}{\|}}{\underset{}{C}}{\diagdown}\ddot{N}H_2} + NH_4^+ Cl^-$$

▲ Problems linked to Key Ideas in this chapter

6.40 The naturally occurring molecule α-terpineol is biosynthesized by a route that includes the following step:

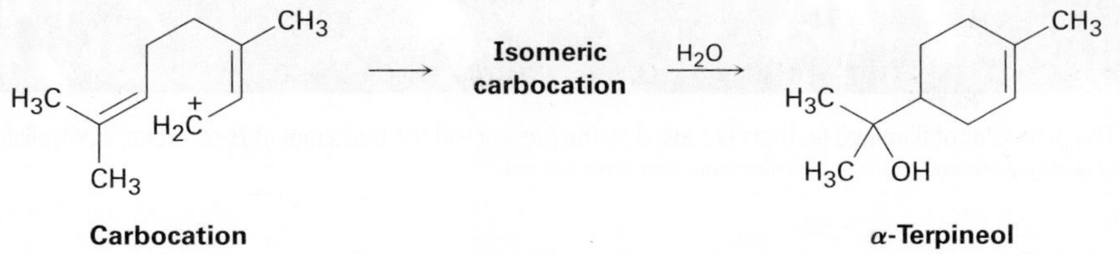

Carbocation → Isomeric carbocation →[H₂O] α-Terpineol

(a) Propose a likely structure for the isomeric carbocation intermediate.
(b) Show the mechanism of each step in the biosynthetic pathway, using curved arrows to indicate electron flow.

6.41 Predict the product(s) of each of the following biological reactions by interpreting the flow of electrons as indicated by the curved arrows:

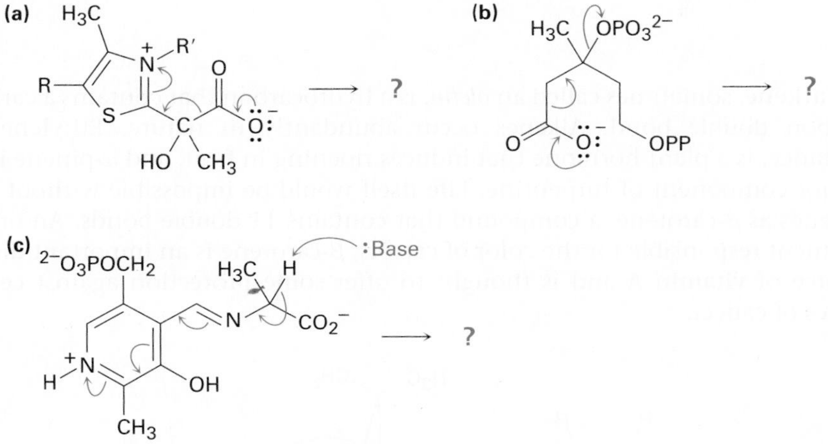

(a), (b), (c)

6.42 Reaction of 2-methylpropene with HBr might, in principle, lead to a mixture of two alkyl bromide addition products. Name them, and draw their structures.

6.43 Draw the structures of the two carbocation intermediates that might form during the reaction of 2-methylpropene with HBr (Problem 6.42). We'll see in the next chapter that the stability of carbocations depends on the number of alkyl substituents attached to the positively charged carbon—the more alkyl substituents there are, the more stable the cation. Which of the two carbocation intermediates you drew is more stable?

▲ Problems linked to Key Ideas in this chapter

7

The pink color of flamingo feathers is caused by the presence in the bird's diet of β-carotene, a polyalkene.
Image copyright George Burba, 2010. Used under license from Shutterstock.com

Alkenes: Structure and Reactivity

7.1 Industrial Preparation and Use of Alkenes
7.2 Calculating Degree of Unsaturation
7.3 Naming Alkenes
7.4 Cis–Trans Isomerism in Alkenes
7.5 Alkene Stereochemistry and the E,Z Designation
7.6 Stability of Alkenes
7.7 Electrophilic Addition Reactions of Alkenes
7.8 Orientation of Electrophilic Additions: Markovnikov's Rule
7.9 Carbocation Structure and Stability
7.10 The Hammond Postulate
7.11 Evidence for the Mechanism of Electrophilic Additions: Carbocation Rearrangements

A Deeper Look— Bioprospecting: Hunting for Natural Products

OWL Sign in to OWL for Organic Chemistry at www.cengage.com/owl to view tutorials and simulations, develop problem-solving skills, and complete online homework assigned by your professor.

An **alkene**, sometimes called an *olefin,* is a hydrocarbon that contains a carbon–carbon double bond. Alkenes occur abundantly in nature. Ethylene, for instance, is a plant hormone that induces ripening in fruit, and α-pinene is the major component of turpentine. Life itself would be impossible without such alkenes as β-carotene, a compound that contains 11 double bonds. An orange pigment responsible for the color of carrots, β-carotene is an important dietary source of vitamin A and is thought to offer some protection against certain types of cancer.

Why This Chapter? Carbon–carbon double bonds are present in most organic and biological molecules, so a good understanding of their behavior is needed. In this chapter, we'll look at some consequences of alkene stereoisomerism and then focus on the broadest and most general class of alkene reactions, the electrophilic addition reaction.

7.1 Industrial Preparation and Use of Alkenes

Ethylene and propylene, the simplest alkenes, are the two most important organic chemicals produced industrially. Approximately 127 million metric tons of ethylene and 54 million metric tons of propylene are produced worldwide each year for use in the synthesis of polyethylene, polypropylene, ethylene glycol, acetic acid, acetaldehyde, and a host of other substances (**Figure 7.1**).

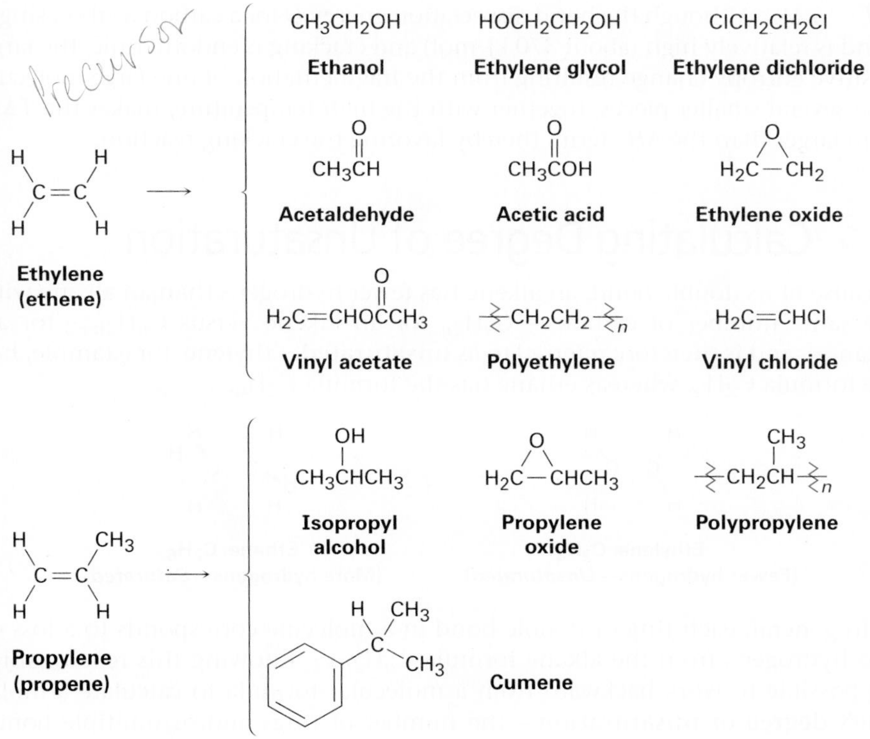

Figure 7.1 Compounds derived industrially from ethylene and propylene.

Ethylene (ethene) [precursor] →
- CH_3CH_2OH — Ethanol
- $HOCH_2CH_2OH$ — Ethylene glycol
- $ClCH_2CH_2Cl$ — Ethylene dichloride
- CH_3CHO — Acetaldehyde
- CH_3COOH — Acetic acid
- $H_2C\text{—}CH_2$ (epoxide) — Ethylene oxide
- $H_2C=CHOCCH_3$ — Vinyl acetate
- $\{CH_2CH_2\}_n$ — Polyethylene
- $H_2C=CHCl$ — Vinyl chloride

Propylene (propene) →
- $CH_3CHOHCH_3$ — Isopropyl alcohol
- $H_2C\text{—}CHCH_3$ (epoxide) — Propylene oxide
- $\{CH_2CH(CH_3)\}_n$ — Polypropylene
- Cumene (isopropylbenzene)

Ethylene, propylene, and butene are synthesized industrially by steam cracking of light (C_2–C_8) alkanes.

$$CH_3(CH_2)_nCH_3 \quad [n = 0\text{–}6] \xrightarrow{850\text{–}900\ °C,\ \text{steam}} H_2 + H_2C=CH_2 + CH_3CH=CH_2 + CH_3CH_2CH=CH_2$$

Steam cracking takes place without a catalyst at temperatures up to 900 °C. The process is complex, although it undoubtedly involves radical reactions. The high-temperature reaction conditions cause spontaneous homolytic breaking of C–C and C–H bonds, with resultant formation of smaller fragments. We might imagine, for instance, that a molecule of butane splits into two ethyl radicals, each of which then loses a hydrogen atom to generate two molecules of ethylene.

$$\text{CH}_3\text{CH}_2\text{CH}_2\text{CH}_3 \xrightarrow{900\ °C} \left[2\ \text{H}_3\text{C}-\overset{\cdot}{\text{C}}\text{H}_2 \right] \longrightarrow 2\ \text{H}_2\text{C}=\text{CH}_2 + \text{H}_2$$

Steam cracking is an example of a reaction whose energetics are dominated by entropy ($\Delta S°$) rather than by enthalpy ($\Delta H°$) in the free-energy equation $\Delta G° = \Delta H° - T\Delta S°$. Although the bond dissociation energy D for a carbon–carbon single bond is relatively high (about 370 kJ/mol) and cracking is endothermic, the large positive entropy change resulting from the fragmentation of one large molecule into several smaller pieces, together with the high temperature, makes the $T\Delta S°$ term larger than the $\Delta H°$ term, thereby favoring the cracking reaction.

7.2 Calculating Degree of Unsaturation

Because of its double bond, an alkene has fewer hydrogens than an alkane with the same number of carbons—C_nH_{2n} for an alkene versus C_nH_{2n+2} for an alkane—and is therefore referred to as **unsaturated**. Ethylene, for example, has the formula C_2H_4, whereas ethane has the formula C_2H_6.

Ethylene: C_2H_4
(Fewer hydrogens—*Unsaturated*)

Ethane: C_2H_6
(More hydrogens—*Saturated*)

In general, each ring or double bond in a molecule corresponds to a loss of two hydrogens from the alkane formula C_nH_{2n+2}. Knowing this relationship, it's possible to work backward from a molecular formula to calculate a molecule's **degree of unsaturation**—the number of rings and/or multiple bonds present in the molecule.

Let's assume that we want to find the structure of an unknown hydrocarbon. A molecular weight determination on the unknown yields a value of 82 amu, which corresponds to a molecular formula of C_6H_{10}. Since the saturated C_6 alkane (hexane) has the formula C_6H_{14}, the unknown compound has two fewer pairs of hydrogens ($H_{14} - H_{10} = H_4 = 2\ H_2$) so its degree of unsaturation is 2. The unknown therefore contains two double bonds, one ring and one double bond, two rings, or one triple bond. There's still a long way to go to establish structure, but the simple calculation has told us a lot about the molecule.

4-Methyl-1,3-pentadiene
(two double bonds)

Cyclohexene
(one ring, one double bond)

Bicyclo[3.1.0]hexane
(two rings)

4-Methyl-2-pentyne
(one triple bond)

C_6H_{10}

7.2 | Calculating Degree of Unsaturation

Similar calculations can be carried out for compounds containing elements other than just carbon and hydrogen.

- **Organohalogen compounds (C, H, X, where X = F, Cl, Br, or I)** A halogen substituent acts as a replacement for hydrogen in an organic molecule, so we can add the number of halogens and hydrogens to arrive at an equivalent hydrocarbon formula from which the degree of unsaturation can be found. For example, the formula $C_4H_6Br_2$ is equivalent to the hydrocarbon formula C_4H_8 and thus corresponds to one degree of unsaturation.

$$\text{BrCH}_2\text{CH=CHCH}_2\text{Br} \quad \underset{\text{Replace 2 Br by 2 H}}{=} \quad \text{HCH}_2\text{CH=CHCH}_2\text{H}$$

$$\underbrace{C_4H_6Br_2}_{\text{Add}} = \text{"}C_4H_8\text{"} \quad \text{One unsaturation: one double bond}$$

- **Organooxygen compounds (C, H, O)** Oxygen forms two bonds, so it doesn't affect the formula of an equivalent hydrocarbon and can be ignored when calculating the degree of unsaturation. You can convince yourself of this by seeing what happens when an oxygen atom is inserted into an alkane bond: C–C becomes C–O–C or C–H becomes C–O–H, and there is no change in the number of hydrogen atoms. For example, the formula C_5H_8O is equivalent to the hydrocarbon formula C_5H_8 and thus corresponds to two degrees of unsaturation.

$$\text{H}_2\text{C=CHCH=CHCH}_2\text{OH} \quad \underset{\text{O removed from here}}{=} \quad \text{H}_2\text{C=CHCH=CHCH}_2\text{–H}$$

$$C_5H_8O = \text{"}C_5H_8\text{"} \quad \text{Two unsaturations: two double bonds}$$

- **Organonitrogen compounds (C, H, N)** Nitrogen forms three bonds, so an organonitrogen compound has one more hydrogen than a related hydrocarbon. We therefore subtract the number of nitrogens from the number of hydrogens to arrive at the equivalent hydrocarbon formula. Again, you can convince yourself of this by seeing what happens when a nitrogen atom is inserted into an alkane bond: C–C becomes C–NH–C or C–H becomes C–NH$_2$, meaning that one additional hydrogen atom has been added. We must therefore subtract this extra hydrogen atom to arrive at the equivalent hydrocarbon formula. For example, the formula C_5H_9N is equivalent to C_5H_8 and thus has two degrees of unsaturation.

$$C_5H_9N = \text{"}C_5H_8\text{"} \quad \text{Two unsaturations: one ring and one double bond}$$

To summarize:

- **Add** the number of **halogens** to the number of hydrogens.
- **Ignore** the number of **oxygens**.
- **Subtract** the number of **nitrogens** from the number of hydrogens.

Problem 7.1
Calculate the degree of unsaturation in each of the following formulas, and then draw as many structures as you can for each:
(a) C_4H_8 (b) C_4H_6 (c) C_3H_4

Problem 7.2
Calculate the degree of unsaturation in each of the following formulas:
(a) C_6H_5N (b) $C_6H_5NO_2$ (c) $C_8H_9Cl_3$
(d) $C_9H_{16}Br_2$ (e) $C_{10}H_{12}N_2O_3$ (f) $C_{20}H_{32}ClN$

Problem 7.3
Diazepam, marketed as an antianxiety medication under the name Valium, has three rings, eight double bonds, and the formula $C_{16}H_?ClN_2O$. How many hydrogens does diazepam have? (Calculate the answer; don't count hydrogens in the structure.)

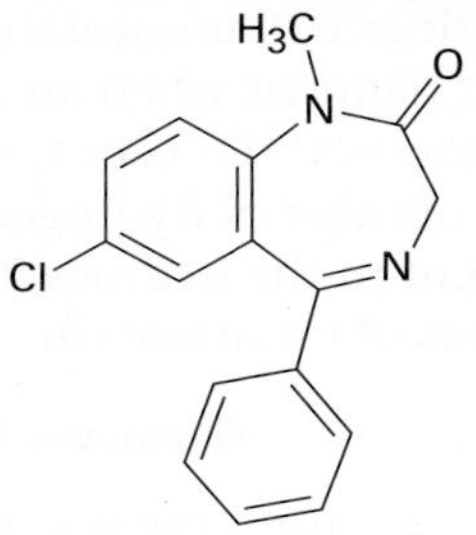

Diazepam

7.3 Naming Alkenes

Alkenes are named using a series of rules similar to those for alkanes (**Section 3.4**), with the suffix *-ene* used instead of *-ane* to identify the functional group. There are three steps.

STEP 1
Name the parent hydrocarbon. Find the longest carbon chain containing the double bond, and name the compound accordingly, using the suffix *-ene*:

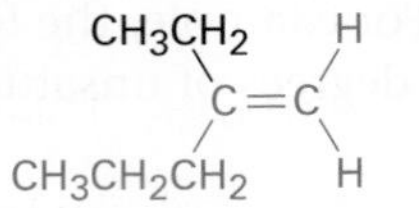

Named as a *pentene*   NOT   as a hexene, since the double bond is not contained in the six-carbon chain

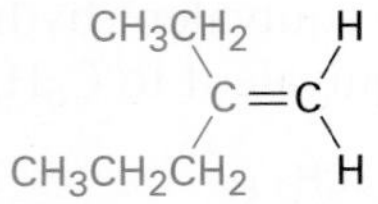

STEP 2
Number the carbon atoms in the chain. Begin at the end nearer the double bond or, if the double bond is equidistant from the two ends, begin at

the end nearer the first branch point. This rule ensures that the double-bond carbons receive the lowest possible numbers.

$$\underset{654321}{CH_3CH_2CH_2CH=CHCH_3} \qquad \underset{123456}{\underset{|}{\overset{CH_3}{CH_3CHCH=CHCH_2CH_3}}}$$

STEP 3
Write the full name. Number the substituents according to their positions in the chain, and list them alphabetically. Indicate the position of the double bond by giving the number of the first alkene carbon and placing that number directly before the parent name. If more than one double bond is present, indicate the position of each and use one of the suffixes -*diene*, -*triene*, and so on.

$$\underset{654321}{CH_3CH_2CH_2CH=CHCH_3} \qquad \underset{123456}{\underset{|}{\overset{CH_3}{CH_3CHCH=CHCH_2CH_3}}}$$
2-**Hex**ene 2-Methyl-3-**hex**ene

2-Ethyl-1-**pent**ene   2-Methyl-1,3-**buta**diene

We should also note that IUPAC changed their naming recommendations in 1993 to place the locant indicating the position of the double bond immediately before the -*ene* suffix rather than before the parent name: but-2-ene rather than 2-butene, for instance. This change has not been widely accepted by the chemical community in the United States, however, so we'll stay with the older but more commonly used names. Be aware, though, that you may occasionally encounter the newer system.

Older naming system: 2,5-Dimethyl-3-**hept**ene   3-Propyl-1,4-**hexa**diene

(Newer naming system: 2,5-Dimethyl**hept**-3-ene   3-Propyl**hexa**-1,4-diene)

Cycloalkenes are named similarly, but because there is no chain end to begin from, we number the cycloalkene so that the double bond is between C1 and C2 and the first substituent has as low a number as possible. It's not necessary to indicate the position of the double bond in the name because it's always between C1 and C2. As with open-chain alkenes, newer but not yet widely accepted naming rules place the locant immediately before the suffix in a diene.

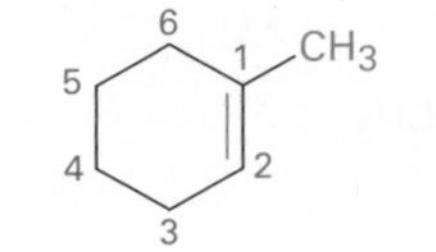

1-Methylcyclohexene

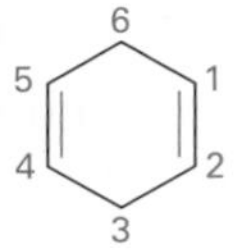

1,4-Cyclohexadiene
(New: Cyclohexa-1,4-diene)

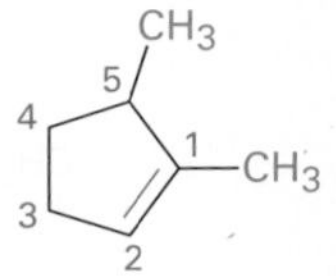

1,5-Dimethylcyclopentene

For historical reasons, there are a few alkenes whose names are firmly entrenched in common usage but don't conform to the rules. For example, the alkene derived from ethane should be called *ethene*, but the name *ethylene* has been used so long that it is accepted by IUPAC. Table 7.1 lists several other common names that are often used and are recognized by IUPAC. Note also that a =CH$_2$ substituent is called a **methylene group**, a H$_2$C=CH– substituent is called a **vinyl group**, and a H$_2$C=CHCH$_2$– substituent is called an **allyl group**.

H$_2$C=   H$_2$C=CH–   H$_2$C=CH–CH$_2$–

A methylene group   A vinyl group   An allyl group

Table 7.1 Common Names of Some Alkenes

Compound	Systematic name	Common name
H$_2$C=CH$_2$	Ethene	Ethylene
CH$_3$CH=CH$_2$	Propene	Propylene
(CH$_3$)$_2$C=CH$_2$	2-Methylpropene	Isobutylene
H$_2$C=C(CH$_3$)–CH=CH$_2$	2-Methyl-1,3-butadiene	Isoprene

Problem 7.4
Give IUPAC names for the following compounds:

(a) 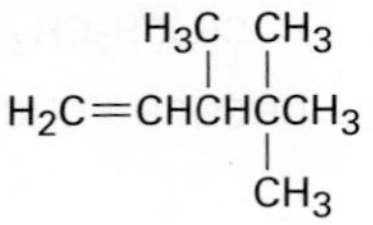 H$_2$C=CHCH(CH$_3$)C(CH$_3$)$_2$CH$_3$

(b) CH$_3$CH$_2$CH=C(CH$_3$)CH$_2$CH$_3$

(c) CH$_3$CH=CHCH(CH$_3$)CH=CHCH(CH$_3$)CH$_3$

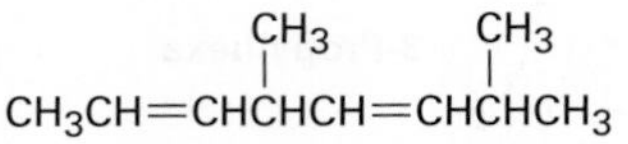

(d) 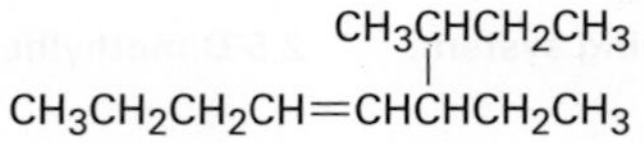 CH$_3$CH$_2$CH$_2$CH=CHCH(CH(CH$_3$)CH$_2$CH$_3$)CH$_2$CH$_3$

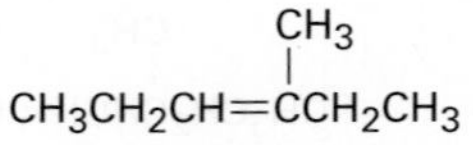

Problem 7.5
Draw structures corresponding to the following IUPAC names:
(a) 2-Methyl-1,5-hexadiene
(b) 3-Ethyl-2,2-dimethyl-3-heptene
(c) 2,3,3-Trimethyl-1,4,6-octatriene
(d) 3,4-Diisopropyl-2,5-dimethyl-3-hexene

Problem 7.6
Name the following cycloalkenes:
(a)   (b)   (c)

Problem 7.7
Change the following old names to new, post-1993 names, and draw the structure of each compound:
(a) 2,5,5-Trimethyl-2-hexene   (b) 2,3-Dimethyl-1,3-cyclohexadiene

7.4 Cis–Trans Isomerism in Alkenes

We saw in Chapter 1 that the carbon–carbon double bond can be described in two ways. In valence bond language (Section 1.8), the carbons are sp^2-hybridized and have three equivalent hybrid orbitals that lie in a plane at angles of 120° to one another. The carbons form a σ bond by head-on overlap of sp^2 orbitals and a π bond by sideways overlap of unhybridized p orbitals oriented perpendicular to the sp^2 plane, as shown in Figure 1.14 on page 14.

In molecular orbital language (Section 1.11), interaction between the p orbitals leads to one bonding and one antibonding π molecular orbital. The π bonding MO has no node between nuclei and results from a combination of p orbital lobes with the same algebraic sign. The π antibonding MO has a node between nuclei and results from a combination of lobes with different algebraic signs, as shown in Figure 1.18, page 21.

Although essentially free rotation is possible around single bonds (Section 3.6), the same is not true of double bonds. For rotation to occur around a double bond, the π bond must break and re-form (Figure 7.2). Thus, the barrier to double-bond rotation must be at least as great as the strength of the π bond itself, an estimated 350 kJ/mol (84 kcal/mol). Recall that the barrier to bond rotation in ethane is only 12 kJ/mol.

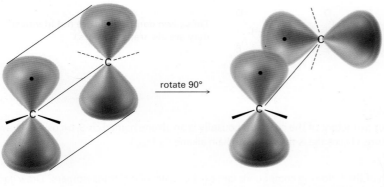

Figure 7.2 The π bond must break for rotation to take place around a carbon–carbon double bond.

The lack of rotation around carbon–carbon double bonds is of more than just theoretical interest; it also has chemical consequences. Imagine the situation for a disubstituted alkene such as 2-butene. (*Disubstituted* means that two substituents other than hydrogen are bonded to the double-bond carbons.) The two methyl groups in 2-butene can be either on the same side of the double bond or on opposite sides, a situation similar to that in disubstituted cycloalkanes (Section 4.2).

Since bond rotation can't occur, the two 2-butenes can't spontaneously interconvert; they are different, isolable compounds. As with disubstituted cycloalkanes, we call such compounds *cis–trans stereoisomers*. The compound with substituents on the same side of the double bond is called *cis*-2-butene, and the isomer with substituents on opposite sides is *trans*-2-butene (**Figure 7.3**).

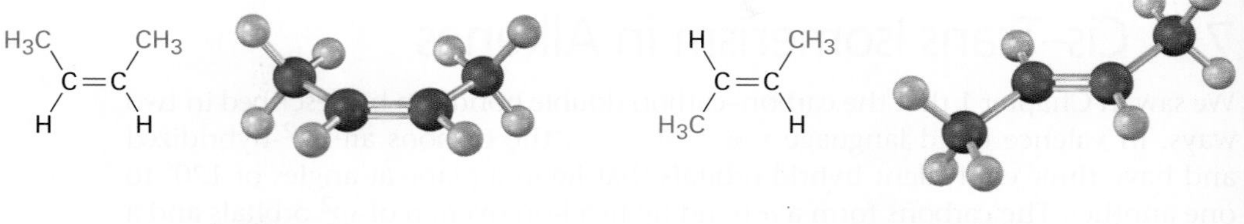

Figure 7.3 Cis and trans isomers of 2-butene. The cis isomer has the two methyl groups on the same side of the double bond, and the trans isomer has the methyl groups on opposite sides.

Cis–trans isomerism is not limited to disubstituted alkenes. It can occur whenever both double-bond carbons are attached to two different groups. If one of the double-bond carbons is attached to two identical groups, however, then cis–trans isomerism is not possible (**Figure 7.4**).

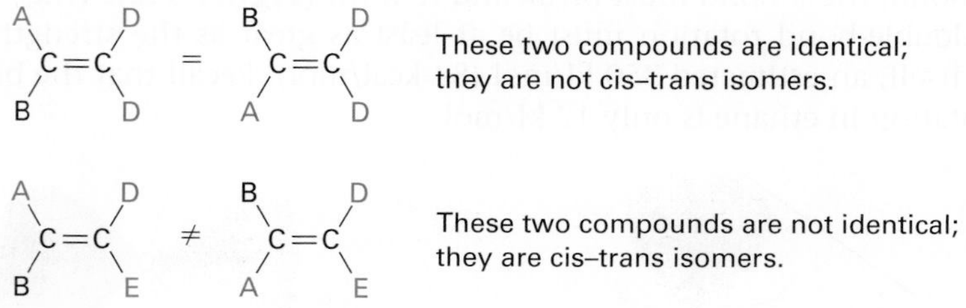

Figure 7.4 The requirement for cis–trans isomerism in alkenes. Compounds that have one of their carbons bonded to two identical groups can't exist as cis–trans isomers. Only when both carbons are bonded to two different groups is cis–trans isomerism possible.

Problem 7.8
The sex attractant of the common housefly is an alkene named *cis*-9-tricosene. Draw its structure. (Tricosane is the straight-chain alkane $C_{23}H_{48}$.)

Problem 7.9
Which of the following compounds can exist as pairs of cis–trans isomers? Draw each cis–trans pair, and indicate the geometry of each isomer.
(a) $CH_3CH=CH_2$
(b) $(CH_3)_2C=CHCH_3$
(c) $CH_3CH_2CH=CHCH_3$
(d) $(CH_3)_2C=C(CH_3)CH_2CH_3$
(e) $ClCH=CHCl$
(f) $BrCH=CHCl$

Problem 7.10
Name the following alkenes, including the cis or trans designation:

(a)   (b)

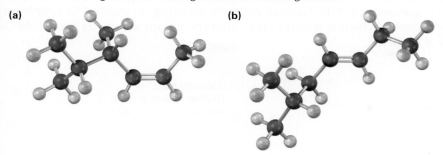

7.5 Alkene Stereochemistry and the E,Z Designation

The cis–trans naming system used in the previous section works only with disubstituted alkenes—compounds that have two substituents other than hydrogen on the double bond. With trisubstituted and tetrasubstituted double bonds, a more general method is needed for describing double-bond geometry. (*Trisubstituted* means three substituents other than hydrogen on the double bond; *tetrasubstituted* means four substituents other than hydrogen.)

The method used for describing alkene stereochemistry is called the **E,Z system** and employs the same Cahn–Ingold–Prelog sequence rules given in **Section 5.5** for specifying the configuration of a chirality center. Let's briefly review the sequence rules and then see how they're used to specify double-bond geometry. For a more thorough review, you should reread **Section 5.5**.

> **RULE 1**
> Considering each of the double-bond carbons separately, look at the two substituents attached and rank them according to the atomic number of the first atom in each. An atom with higher atomic number ranks higher than an atom with lower atomic number.
>
> **RULE 2**
> If a decision can't be reached by ranking the first atoms in the two substituents, look at the second, third, or fourth atoms away from the double-bond until the first difference is found.
>
> **RULE 3**
> Multiple-bonded atoms are equivalent to the same number of single-bonded atoms.

Once the two groups attached to each doubly bonded carbon atom have been ranked as either higher or lower, look at the entire molecule. If the higher-ranked groups on each carbon are on the same side of the double

Key IDEAS

Test your knowledge of Key Ideas by answering end-of-chapter exercises marked with ▲.

bond, the alkene is said to have **Z geometry**, for the German *zusammen*, meaning "together." If the higher-ranked groups are on opposite sides, the alkene has **E geometry**, for the German *entgegen*, meaning "opposite." (A simple way to remember which is which to note that the groups are on "ze zame zide" in the Z isomer.)

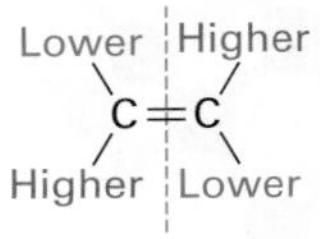

Lower Higher
   C=C
Higher Lower

E double bond
(Higher-ranked groups are on opposite sides.)

Higher Higher
   C=C
Lower Lower

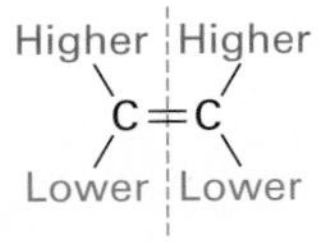

Z double bond
(Higher-ranked groups are on the same side.)

For further practice, work through each of the following examples to convince yourself that the assignments are correct:

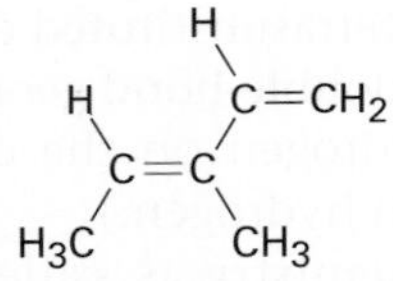

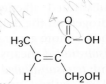

(E)-3-Methyl-1,3-pentadiene

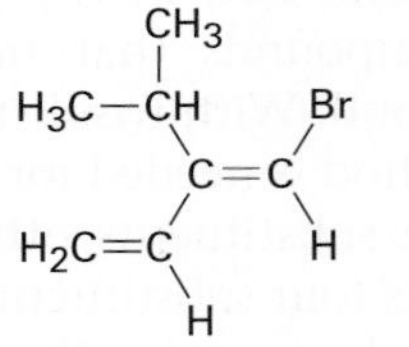

(E)-1-Bromo-2-isopropyl-1,3-butadiene

(Z)-2-Hydroxymethyl-2-butenoic acid

Worked Example 7.1 — Assigning *E* and *Z* Configurations to Substituted Alkenes

Assign *E* or *Z* configuration to the double bond in the following compound:

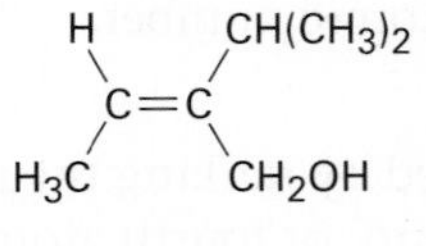

Strategy
Look at the two substituents connected to each double-bond carbon, and determine their ranking using the Cahn–Ingold–Prelog rules. Then see whether the two higher-ranked groups are on the same or opposite sides of the double bond.

Solution
The left-hand carbon has —H and —CH₃ substituents, of which —CH₃ ranks higher by sequence rule 1. The right-hand carbon has —CH(CH₃)₂ and —CH₂OH substituents, which are equivalent by rule 1. By rule 2, however, —CH₂OH ranks higher than —CH(CH₃)₂ because the substituent —CH₂OH has an *oxygen* as its highest second atom,

7.5 | Alkene Stereochemistry and the E,Z Designation

but —CH(CH₃)₂ has a *carbon* as its highest second atom. The two higher-ranked groups are on the same side of the double bond, so we assign Z configuration.

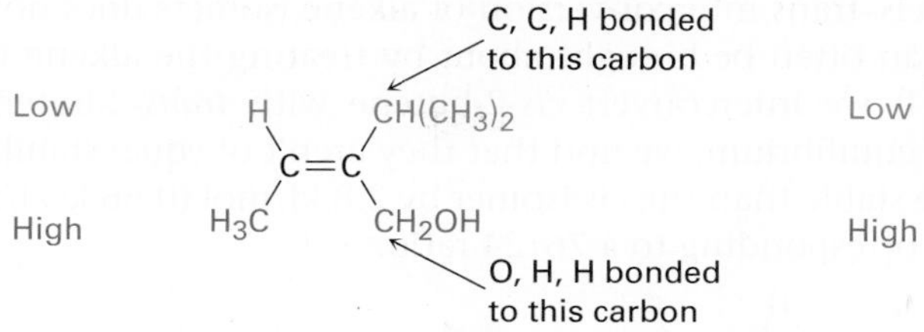

$$\underset{\text{High}}{\overset{\text{Low}}{}}\quad \underset{H_3C}{\overset{H}{}}C=C\underset{CH_2OH}{\overset{CH(CH_3)_2}{}}\quad \underset{\text{High}}{\overset{\text{Low}}{}}$$

C, C, H bonded to this carbon

O, H, H bonded to this carbon

Z configuration

Problem 7.11
Which member in each of the following sets ranks higher?
(a) —H or —CH₃
(b) —Cl or —CH₂Cl
(c) —CH₂CH₂Br or —CH=CH₂
(d) —NHCH₃ or —OCH₃
(e) —CH₂OH or —CH=O
(f) —CH₂OCH₃ or —CH=O

Problem 7.12
Rank the substituents in each of the following sets according to the sequence rules:
(a) —CH₃, —OH, —H, —Cl
(b) —CH₃, —CH₂CH₃, —CH=CH₂, —CH₂OH
(c) —CO₂H, —CH₂OH, —C≡N, —CH₂NH₂
(d) —CH₂CH₃, —C≡CH, —C≡N, —CH₂OCH₃

Problem 7.13
Assign E or Z configuration to the following alkenes:

(a) H₃C, CH₂OH, CH₃CH₂, Cl on C=C
(b) Cl, CH₂CH₃, CH₃O, CH₂CH₂CH₃ on C=C
(c) CH₃ (cyclopentane), CO₂H, CH₂OH on C=C 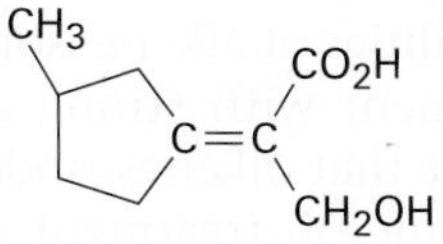
(d) H, CN, H₃C, CH₂NH₂ on C=C

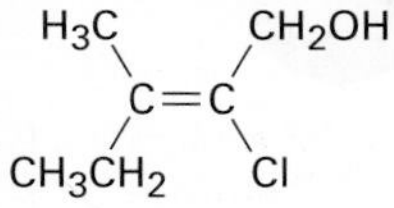

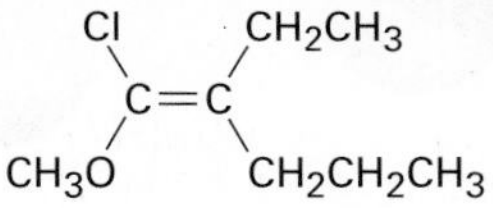

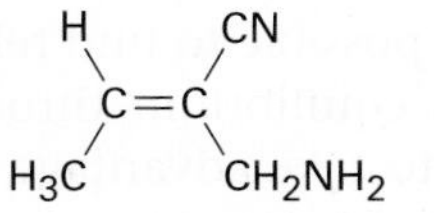

Problem 7.14
Assign stereochemistry (E or Z) to the double bond in the following compound, and convert the drawing into a skeletal structure (red = O):

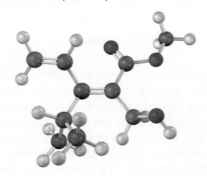

7.6 Stability of Alkenes

Although the cis–trans interconversion of alkene isomers does not occur spontaneously, it can often be brought about by treating the alkene with a strong acid catalyst. If we interconvert *cis*-2-butene with *trans*-2-butene and allow them to reach equilibrium, we find that they aren't of equal stability. The trans isomer is more stable than the cis isomer by 2.8 kJ/mol (0.66 kcal/mol) at room temperature, corresponding to a 76:24 ratio.

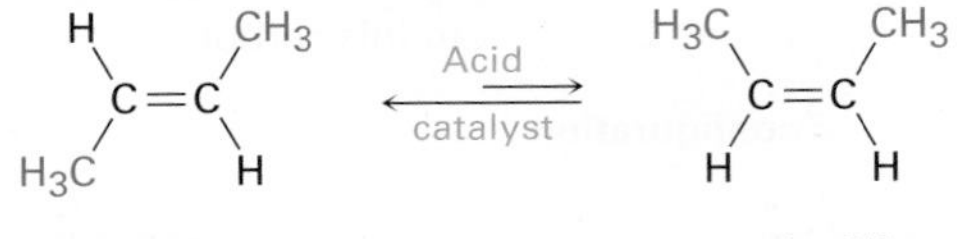

Cis alkenes are less stable than their trans isomers because of steric strain between the two larger substituents on the same side of the double bond. This is the same kind of steric interference that we saw previously in the axial conformation of methylcyclohexane **(Section 4.7)**.

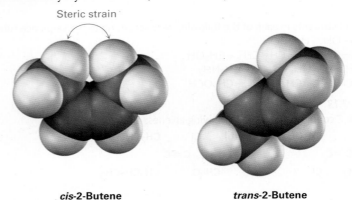

cis-2-Butene trans-2-Butene

Although it's sometimes possible to find relative stabilities of alkene isomers by establishing a cis–trans equilibrium through treatment with strong acid, a more general method is to take advantage of the fact that alkenes undergo a *hydrogenation* reaction to give the corresponding alkane on treatment with H_2 gas in the presence of a catalyst such as palladium or platinum.

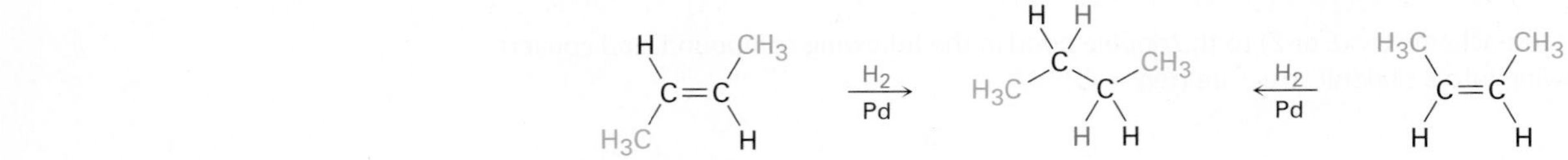

trans-2-Butene   Butane   *cis*-2-Butene

Energy diagrams for the hydrogenation reactions of *cis*- and *trans*-2-butene are shown in **Figure 7.5**. Because *cis*-2-butene is less stable than *trans*-2-butene by 2.8 kJ/mol, the energy diagram shows the cis alkene at a higher energy level. After reaction, however, both curves are at the same energy level (butane). It therefore follows that $\Delta G°$ for reaction of the cis isomer must be larger than $\Delta G°$ for reaction of the trans isomer by 2.8 kJ/mol. In other words, more energy is released in the hydrogenation of the cis isomer than the trans isomer because the cis isomer has more energy to begin with.

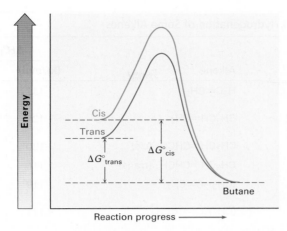

Figure 7.5 Energy diagrams for hydrogenation of *cis-* and *trans-*2-butene. The cis isomer is higher in energy than the trans isomer by about 2.8 kJ/mol and therefore releases more energy in the reaction.

If we were to measure the so-called heats of hydrogenation ($\Delta H°_{hydrog}$) for two double-bond isomers and find their difference, we could determine the relative stabilities of cis and trans isomers without having to measure an equilibrium position. *cis*-2-Butene, for instance, has $\Delta H°_{hydrog} = -120$ kJ/mol (-28.6 kcal/mol), while *trans*-2-butene has $\Delta H°_{hydrog} = -116$ kJ/mol (-27.6 kcal/mol)—a difference of 4 kJ/mol.

Cis isomer
$\Delta H°_{hydrog} = -120$ kJ/mol

Trans isomer
$\Delta H°_{hydrog} = -116$ kJ/mol

The 4 kJ/mol energy difference between the 2-butene isomers calculated from heats of hydrogenation agrees reasonably well with the 2.8 kcal/mol energy difference calculated from equilibrium data, but the numbers aren't exactly the same for two reasons. First, there is probably some experimental error, since heats of hydrogenation are difficult to measure accurately. Second, heats of reaction and equilibrium constants don't measure exactly the same thing. Heats of reaction measure enthalpy changes, $\Delta H°$, whereas equilibrium constants measure free-energy changes, $\Delta G°$, so we might expect a slight difference between the two.

Table 7.2 lists some representative data for the hydrogenation of different alkenes and shows that alkenes become more stable with increasing substitution. That is, alkenes follow the stability order:

Tetrasubstituted > Trisubstituted > Disubstituted > Monosubstituted

The stability order of substituted alkenes is due to a combination of two factors. One is a stabilizing interaction between the C=C π bond and adjacent C–H σ bonds on substituents. In valence-bond language, the interaction is called **hyperconjugation**. In a molecular orbital description, there is a bonding MO that extends over the four-atom C=C–C–H grouping, as shown in **Figure 7.6**. The more substituents that are present on the double bond, the more hyperconjugation there is and the more stable the alkene.

Table 7.2 Heats of Hydrogenation of Some Alkenes

Substitution	Alkene	ΔH°hydrog (kJ/mol)	(kcal/mol)
Ethylene	$H_2C=CH_2$	−137	−32.8
Monosubstituted	$CH_3CH=CH_2$	−126	−30.1
Disubstituted	$CH_3CH=CHCH_3$ (cis)	−120	−28.6
	$CH_3CH=CHCH_3$ (trans)	−116	−27.6
	$(CH_3)_2C=CH_2$	−119	−28.4
Trisubstituted	$(CH_3)_2C=CHCH_3$	−113	−26.9
Tetrasubstituted	$(CH_3)_2C=C(CH_3)_2$	−111	−26.6

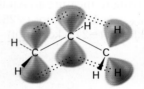

Figure 7.6 Hyperconjugation is a stabilizing interaction between the C=C π bond and adjacent C–H σ bonds on substituents. The more substituents there are, the greater the stabilization of the alkene.

A second factor that contributes to alkene stability involves bond strengths. A bond between an sp^2 carbon and an sp^3 carbon is somewhat stronger than a bond between two sp^3 carbons. Thus, in comparing 1-butene and 2-butene, the monosubstituted isomer has one sp^3–sp^3 bond and one sp^3–sp^2 bond, while the disubstituted isomer has two sp^3–sp^2 bonds. More highly substituted alkenes always have a higher ratio of sp^3–sp^2 bonds to sp^3–sp^3 bonds than less highly substituted alkenes and are therefore more stable.

$$sp^3\text{-}sp^2 \quad sp^2\text{-}sp^3$$
$$\downarrow \quad \downarrow$$
$$CH_3-CH=CH-CH_3$$
2-Butene (more stable)

$$sp^3\text{-}sp^3 \quad sp^3\text{-}sp^2$$
$$\downarrow \quad \downarrow$$
$$CH_3-CH_2-CH=CH_2$$
1-Butene (less stable)

Problem 7.15
Name the following alkenes, and tell which compound in each pair is more stable:

(a) $H_2C=CHCH_2CH_3$ or $H_2C=C(CH_3)CH_3$ (with CH$_3$ substituent)

(b)
$$\underset{H_3C}{\overset{H}{}}C=C\underset{CH_2CH_2CH_3}{\overset{H}{}} \quad \text{or} \quad \underset{H_3C}{\overset{H}{}}C=C\underset{H}{\overset{CH_2CH_2CH_3}{}}$$

(c) 1-methylcyclohexene or 4-methylcyclohexene

7.7 Electrophilic Addition Reactions of Alkenes

Before beginning a detailed discussion of alkene reactions, let's review briefly some conclusions from the previous chapter. We said in **Section 6.5** that alkenes behave as nucleophiles (Lewis bases) in polar reactions, donating a pair of electrons from their electron-rich C=C bond to an electrophile (Lewis acid). For example, reaction of 2-methylpropene with HBr yields 2-bromo-2-methylpropane. A careful study of this and similar reactions by Christopher Ingold and others in the 1930s led to the generally accepted mechanism shown in **Figure 7.7** for **electrophilic addition reactions**.

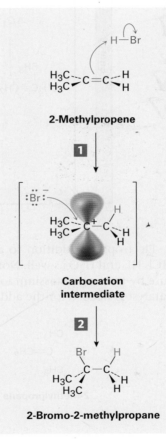

Figure 7.7 | MECHANISM

Mechanism of the electrophilic addition of HBr to 2-methylpropene. The reaction occurs in two steps, protonation and bromide addition, and involves a carbocation intermediate.

1 A hydrogen atom on the electrophile HBr is attacked by π electrons from the nucleophilic double bond, forming a new C–H bond. This leaves the other carbon atom with a + charge and a vacant p orbital. Simultaneously, two electrons from the H–Br bond move onto bromine, giving bromide anion.

2 Bromide ion donates an electron pair to the positively charged carbon atom, forming a C–Br bond and yielding the neutral addition product.

The reaction begins with an attack on the hydrogen of the electrophile HBr by the electrons of the nucleophilic π bond. Two electrons from the π bond form a new σ bond between the entering hydrogen and an alkene carbon, as shown by the curved arrow at the top of Figure 7.7. The carbocation intermediate that results is itself an electrophile, which can accept an electron pair from nucleophilic Br⁻ ion to form a C–Br bond and yield a neutral addition product.

An energy diagram for the overall electrophilic addition reaction (**Figure 7.8**) has two peaks (transition states) separated by a valley (carbocation intermediate). The energy level of the intermediate is higher than that of the starting alkene, but the reaction as a whole is exergonic (negative $\Delta G°$). The first step, protonation of the alkene to yield the intermediate cation, is relatively slow but, once formed, the cation intermediate rapidly reacts further to yield the final alkyl bromide product. The relative rates of the two steps are indicated in Figure 7.8 by the fact that $\Delta G_1^\ddagger$ is larger than $\Delta G_2^\ddagger$.

Figure 7.8 Energy diagram for the two-step electrophilic addition of HBr to 2-methylpropene. The first step is slower than the second step.

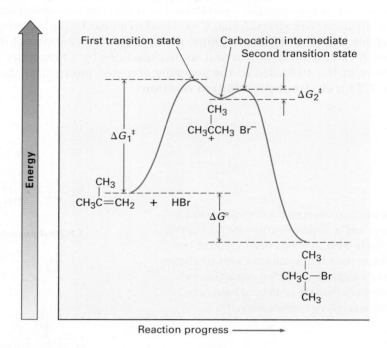

Electrophilic addition to alkenes is successful not only with HBr but with HCl, HI, and H₂O as well. Note that HI is usually generated in the reaction mixture by treating potassium iodide with phosphoric acid and that a strong acid catalyst is needed for the addition of water.

Writing Organic Reactions

This is a good time to mention that organic reaction equations are sometimes written in different ways to emphasize different points. In describing a laboratory process, for instance, the reaction of 2-methylpropene with HCl might be written in the format A + B → C to emphasize that both reactants are equally important for the purposes of the discussion. The solvent and notes about other reaction conditions such as temperature are written either above or below the reaction arrow.

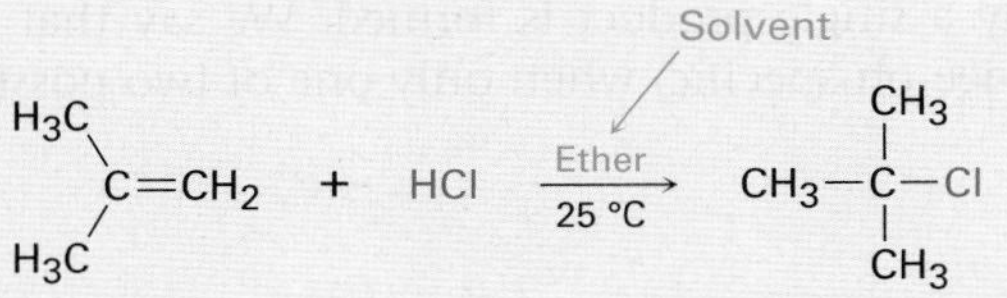

$$(CH_3)_2C{=}CH_2 + HCl \xrightarrow[25\ °C]{Ether} CH_3{-}C(CH_3)_2{-}Cl$$

2-Methylpropene   **2-Chloro-2-methylpropane**

Alternatively, we might write the same reaction in a format to emphasize that 2-methylpropene is the reactant whose chemistry is of greater interest. The second reactant, HCl, is placed above the reaction arrow together with notes about solvent and reaction conditions.

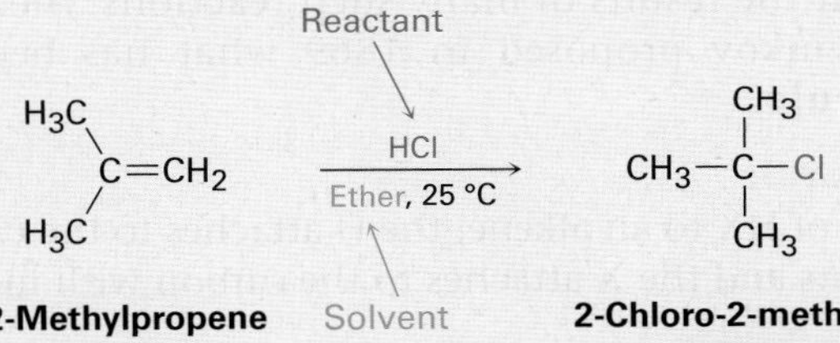

$$(CH_3)_2C{=}CH_2 \xrightarrow[Ether,\ 25\ °C]{HCl} CH_3{-}C(CH_3)_2{-}Cl$$

2-Methylpropene   **2-Chloro-2-methylpropane**

In describing a biological process, the reaction is usually written to show only the structures of the primary reactant and product, while abbreviating the structures of various biological "reagents" and by-products by using a curved arrow that intersects the straight reaction arrow. As discussed in **Section 6.11**, the reaction of glucose with ATP to give glucose 6-phosphate plus ADP would be written as

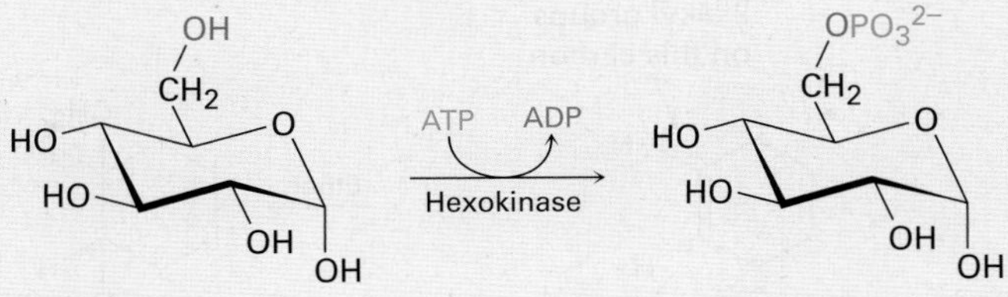

Glucose $\xrightarrow[\text{Hexokinase}]{\text{ATP} \quad \text{ADP}}$ Glucose 6-phosphate

7.8 Orientation of Electrophilic Additions: Markovnikov's Rule

Key IDEAS

Test your knowledge of Key Ideas by answering end-of-chapter exercises marked with ▲.

Look carefully at the electrophilic addition reactions shown in the previous section. In each case, an unsymmetrically substituted alkene gives a single addition product rather than the mixture that might be expected. For example, 2-methylpropene *might* react with HCl to give both 2-chloro-2-methylpropane and 1-chloro-2-methylpropane, but it doesn't. It gives only 2-chloro-2-methylpropane as the sole product. Similarly, it's invariably the case in biological alkene addition reactions that only a single product is formed. We say that such reactions are **regiospecific** (**ree**-jee-oh-specific) when only one of two possible orientations of addition occurs.

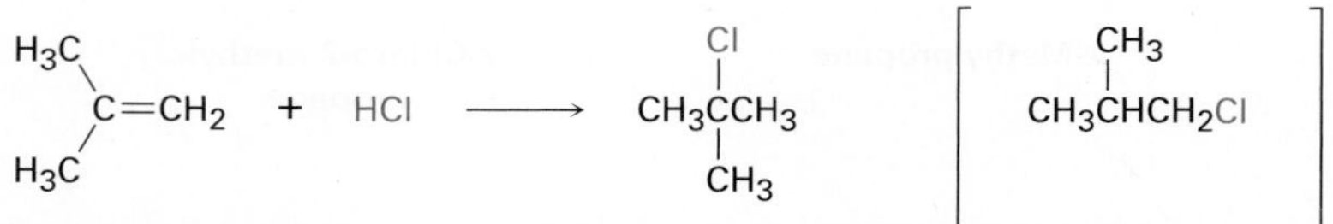

2-Methylpropene + HCl ⟶ 2-Chloro-2-methylpropane (Sole product) [1-Chloro-2-methylpropane (Not formed)]

After looking at the results of many such reactions, the Russian chemist Vladimir Markovnikov proposed in 1869 what has become known as **Markovnikov's rule**.

> Markovnikov's rule
> In the addition of HX to an alkene, the H attaches to the carbon with fewer alkyl substituents and the X attaches to the carbon with more alkyl substituents.

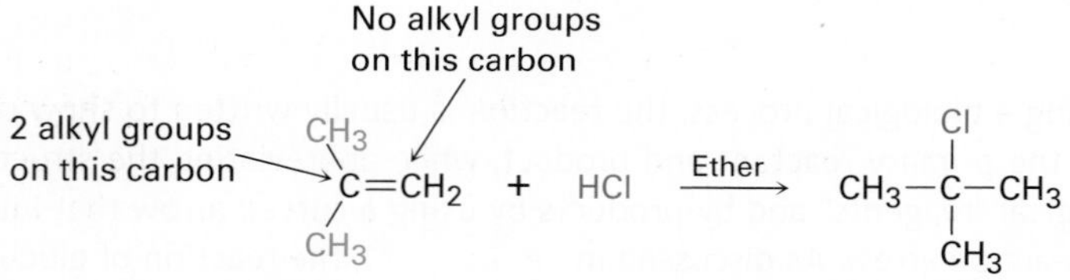

2-Methylpropene (2 alkyl groups on this carbon; No alkyl groups on this carbon) + HCl →(Ether) 2-Chloro-2-methylpropane

1-Methylcyclohexene (2 alkyl groups on this carbon; 1 alkyl group on this carbon) + HBr →(Ether) 1-Bromo-1-methylcyclohexane

7.8 | Orientation of Electrophilic Additions: Markovnikov's Rule

When both double-bond carbon atoms have the same degree of substitution, a mixture of addition products results.

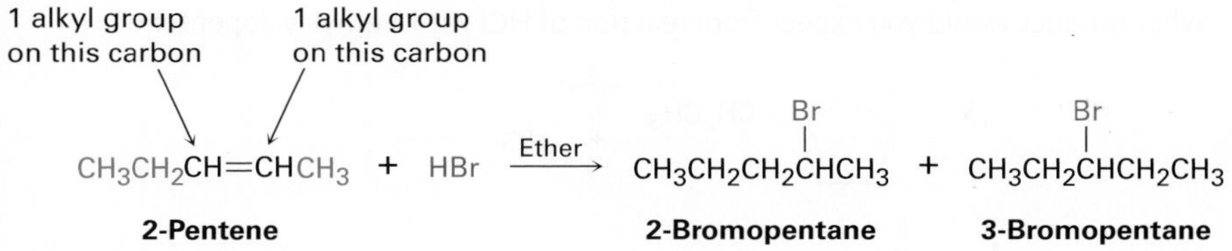

1 alkyl group on this carbon
1 alkyl group on this carbon

$$CH_3CH_2CH=CHCH_3 + HBr \xrightarrow{Ether} CH_3CH_2CH_2CHBrCH_3 + CH_3CH_2CHBrCH_2CH_3$$

2-Pentene → **2-Bromopentane** + **3-Bromopentane**

Because carbocations are involved as intermediates in these electrophilic addition reactions, Markovnikov's rule can be restated in the following way:

> **Markovnikov's rule (restated)**
> In the addition of HX to an alkene, the more highly substituted carbocation is formed as the intermediate rather than the less highly substituted one.

For example, addition of H$^+$ to 2-methylpropene yields the intermediate *tertiary* carbocation rather than the alternative primary carbocation, and addition to 1-methylcyclohexene yields a tertiary cation rather than a secondary one. Why should this be?

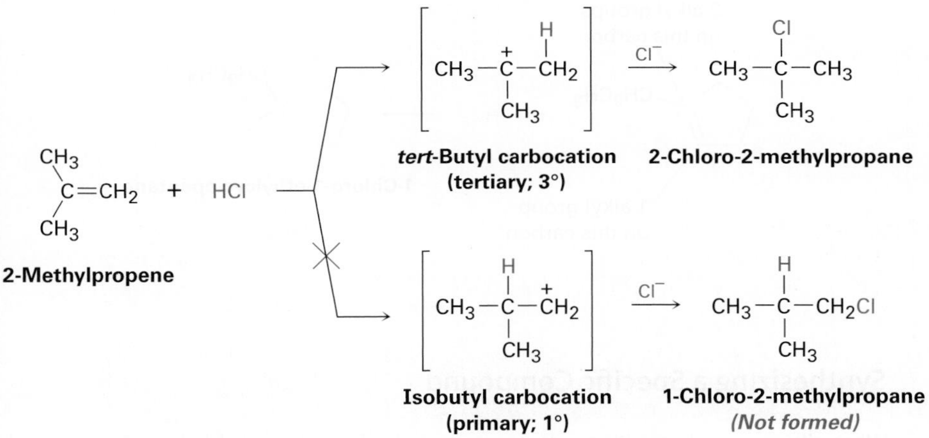

2-Methylpropene + HCl →

- [(CH$_3$)$_3$C$^+$—H shown as CH$_3$—C$^+$(CH$_3$)—CH$_2$H] **tert-Butyl carbocation (tertiary; 3°)** $\xrightarrow{Cl^-}$ (CH$_3$)$_3$C—Cl **2-Chloro-2-methylpropane**

- [CH$_3$—CH(CH$_3$)—CH$_2^+$] **Isobutyl carbocation (primary; 1°)** $\xrightarrow{Cl^-}$ CH$_3$—CH(CH$_3$)—CH$_2$Cl **1-Chloro-2-methylpropane** *(Not formed)*

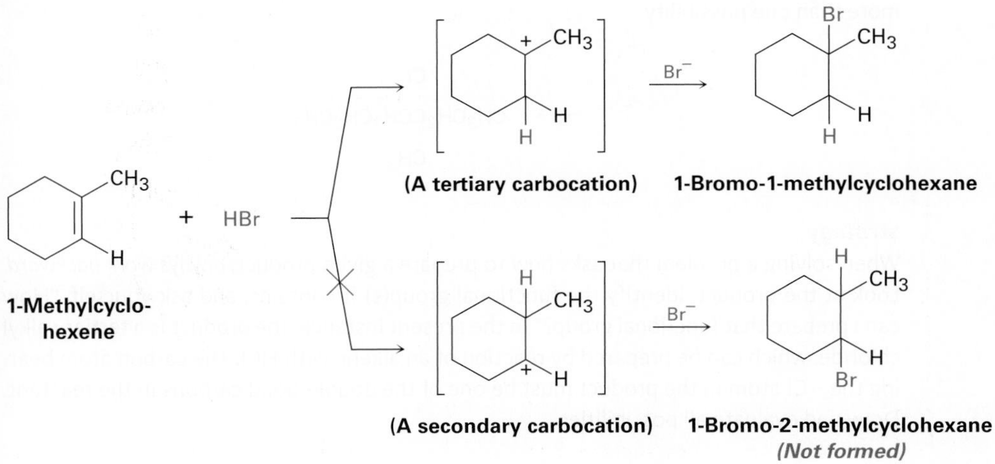

1-Methylcyclohexene + HBr →

- (A tertiary carbocation) $\xrightarrow{Br^-}$ **1-Bromo-1-methylcyclohexane**

- (A secondary carbocation) $\xrightarrow{Br^-}$ **1-Bromo-2-methylcyclohexane** *(Not formed)*

Worked Example 7.2 Predicting the Product of an Electrophilic Addition Reaction

What product would you expect from reaction of HCl with 1-ethylcyclopentene?

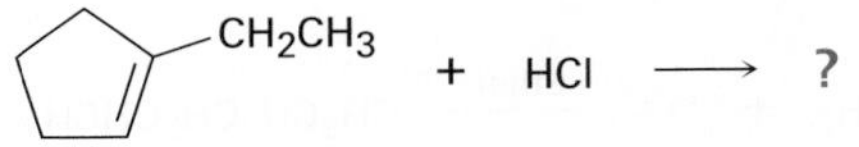

Strategy
When solving a problem that asks you to predict a reaction product, begin by looking at the functional group(s) in the reactants and deciding what kind of reaction is likely to occur. In the present instance, the reactant is an alkene that will probably undergo an electrophilic addition reaction with HCl. Next, recall what you know about electrophilic addition reactions and use your knowledge to predict the product. You know that electrophilic addition reactions follow Markovnikov's rule, so H$^+$ will add to the double-bond carbon that has one alkyl group (C2 on the ring) and the Cl will add to the double-bond carbon that has two alkyl groups (C1 on the ring).

Solution
The expected product is 1-chloro-1-ethylcyclopentane.

1-Chloro-1-ethylcyclopentane

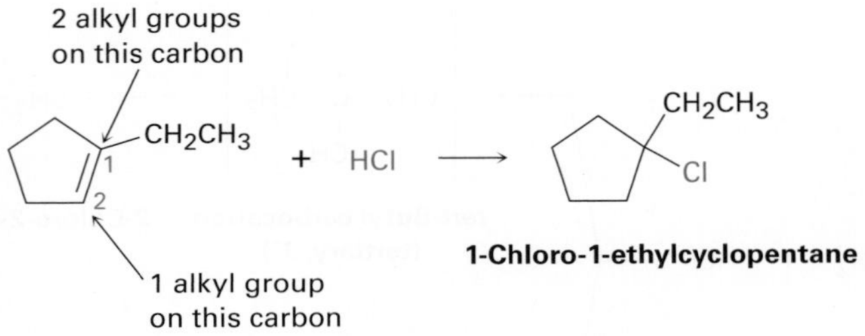

Worked Example 7.3 Synthesizing a Specific Compound

What alkene would you start with to prepare the following alkyl halide? There may be more than one possibility.

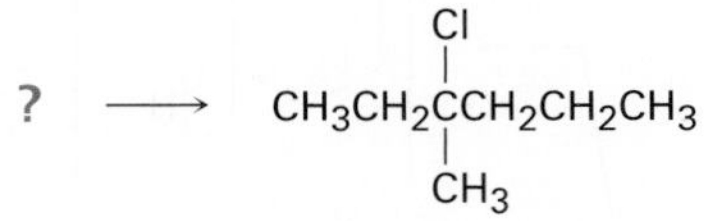

$$? \longrightarrow CH_3CH_2\underset{\underset{CH_3}{|}}{\overset{\overset{Cl}{|}}{C}}CH_2CH_2CH_3$$

Strategy
When solving a problem that asks how to prepare a given product, *always work backward*. Look at the product, identify the functional group(s) it contains, and ask yourself, "How can I prepare that functional group?" In the present instance, the product is a tertiary alkyl chloride, which can be prepared by reaction of an alkene with HCl. The carbon atom bearing the —Cl atom in the product must be one of the double-bond carbons in the reactant. Draw and evaluate all possibilities.

Solution

There are three possibilities, any one of which could give the desired product according to Markovnikov's rule.

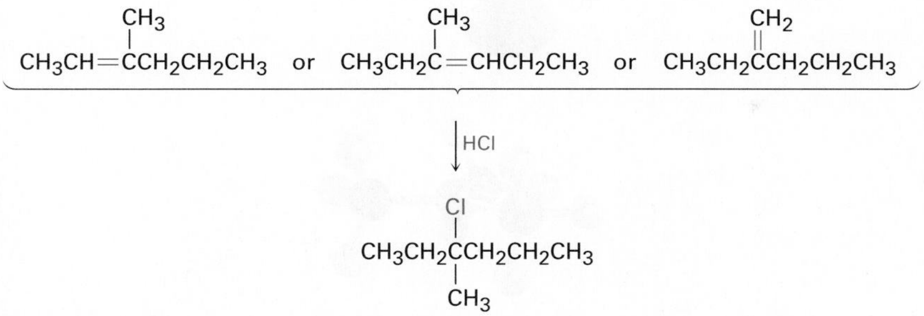

$$\text{CH}_3\text{CH}=\overset{\overset{\text{CH}_3}{|}}{\text{C}}\text{CH}_2\text{CH}_3 \quad \text{or} \quad \text{CH}_3\text{CH}_2\overset{\overset{\text{CH}_3}{|}}{\text{C}}=\text{CHCH}_2\text{CH}_3 \quad \text{or} \quad \text{CH}_3\text{CH}_2\overset{\overset{\text{CH}_2}{||}}{\text{C}}\text{CH}_2\text{CH}_2\text{CH}_3$$

$$\downarrow \text{HCl}$$

$$\text{CH}_3\text{CH}_2\overset{\overset{\text{Cl}}{|}}{\underset{\underset{\text{CH}_3}{|}}{\text{C}}}\text{CH}_2\text{CH}_2\text{CH}_3$$

Problem 7.16
Predict the products of the following reactions:

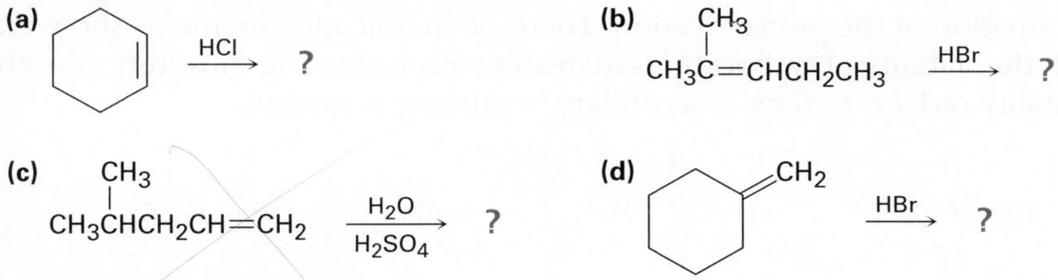

(a) cyclohexene $\xrightarrow{\text{HCl}}$?

(b) $\text{CH}_3\overset{\overset{\text{CH}_3}{|}}{\text{C}}=\text{CHCH}_2\text{CH}_3 \xrightarrow{\text{HBr}}$?

(c) $\text{CH}_3\overset{\overset{\text{CH}_3}{|}}{\text{CH}}\text{CH}_2\text{CH}=\text{CH}_2 \xrightarrow[\text{H}_2\text{SO}_4]{\text{H}_2\text{O}}$?

(Addition of H₂O occurs.)

(d) methylenecyclohexane $\xrightarrow{\text{HBr}}$?

Problem 7.17
What alkenes would you start with to prepare the following products?

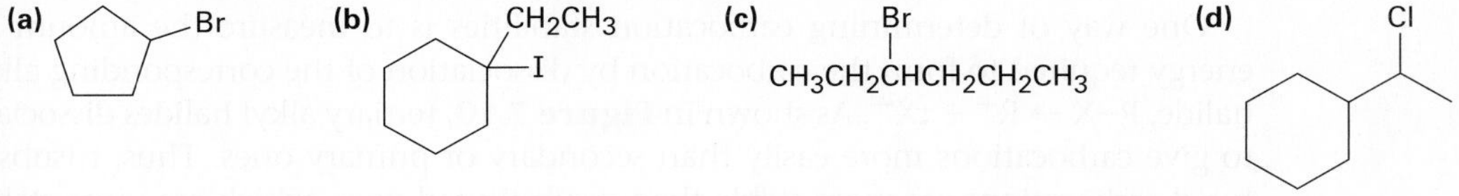

(a) cyclopentyl–Br (b) 1-ethyl-1-iodocyclohexane (CH₂CH₃, I) (c) $\text{CH}_3\text{CH}_2\overset{\overset{\text{Br}}{|}}{\text{CH}}\text{CH}_2\text{CH}_2\text{CH}_3$ (d) 1-chloroethylcyclohexane (Cl)

7.9 Carbocation Structure and Stability

To understand why Markovnikov's rule works, we need to learn more about the structure and stability of carbocations and about the general nature of reactions and transition states. The first point to explore involves structure.

A great deal of experimental evidence has shown that carbocations are planar. The trivalent carbon is sp^2-hybridized, and the three substituents are oriented toward the corners of an equilateral triangle, as indicated in **Figure 7.9**. Because there are only six valence electrons on carbon and all six are used in the three σ bonds, the *p* orbital extending above and below the plane is unoccupied.

Figure 7.9 The structure of a carbocation. The trivalent carbon is sp^2-hybridized and has a vacant p orbital perpendicular to the plane of the carbon and three attached groups.

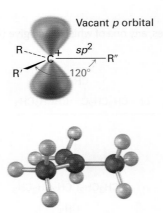

The second point to explore involves carbocation stability. 2-Methylpropene might react with H^+ to form a carbocation having three alkyl substituents (a tertiary ion, 3°), or it might react to form a carbocation having one alkyl substituent (a primary ion, 1°). Since the tertiary alkyl chloride, 2-chloro-2-methylpropane, is the only product observed, formation of the tertiary cation is evidently favored over formation of the primary cation. Thermodynamic measurements show that, indeed, the stability of carbocations increases with increasing substitution so that the stability order is tertiary > secondary > primary > methyl.

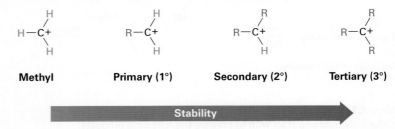

One way of determining carbocation stabilities is to measure the amount of energy required to form the carbocation by dissociation of the corresponding alkyl halide, $R-X \rightarrow R^+ + :X^-$. As shown in **Figure 7.10**, tertiary alkyl halides dissociate to give carbocations more easily than secondary or primary ones. Thus, trisubstituted carbocations are more stable than disubstituted ones, which are more stable than monosubstituted ones. The data in Figure 7.10 are taken from measurements made in the gas phase, but a similar stability order is found for carbocations in solution. The dissociation enthalpies are much lower in solution because polar solvents can stabilize the ions, but the order of carbocation stability remains the same.

Figure 7.10 A plot of dissociation enthalpy versus substitution pattern for the gas-phase dissociation of alkyl chlorides to yield carbocations. More highly substituted alkyl halides dissociate more easily than less highly substituted ones.

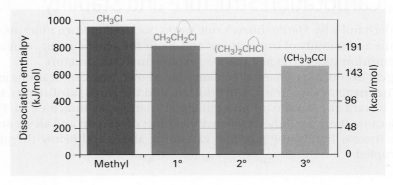

Why are more highly substituted carbocations more stable than less highly substituted ones? There are at least two reasons. Part of the answer has to do with inductive effects, and part has to do with hyperconjugation. Inductive effects, discussed in **Section 2.1** in connection with polar covalent bonds, result from the shifting of electrons in a σ bond in response to the electronegativity of nearby atoms. In the present instance, electrons from a relatively larger and more polarizable alkyl group can shift toward a neighboring positive charge more easily than the electron from a hydrogen. Thus, the more alkyl groups there are attached to the positively charged carbon, the more electron density shifts toward the charge and the more inductive stabilization of the cation occurs **(Figure 7.11)**.

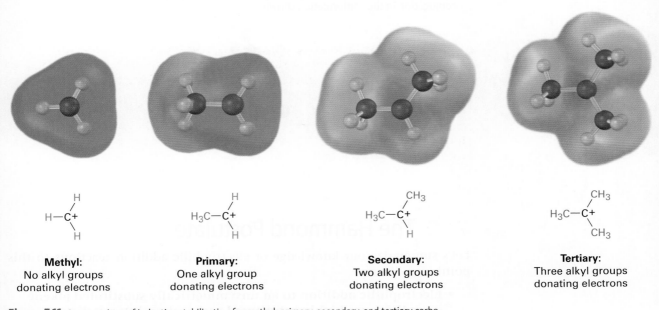

Figure 7.11 A comparison of inductive stabilization for methyl, primary, secondary, and tertiary carbocations. The more alkyl groups that are bonded to the positively charged carbon, the more electron density shifts toward the charge, making the charged carbon less electron poor (blue in electrostatic potential maps).

Hyperconjugation, discussed in **Section 7.6** in connection with the stabilities of substituted alkenes, is the stabilizing interaction between a p orbital and properly oriented C–H σ bonds on neighboring carbons that are roughly parallel to the p orbital. The more alkyl groups there are on the carbocation, the more possibilities there are for hyperconjugation and the more stable the carbocation. **Figure 7.12** shows the molecular orbital for the ethyl carbocation, $CH_3CH_2^+$, and indicates the difference between the C–H bond perpendicular to the cation p orbital and the two C–H bonds more nearly parallel to the cation p orbital. Only the roughly parallel C–H bonds are oriented properly to take part in hyperconjugation.

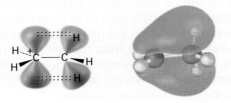

Figure 7.12 Stabilization of the ethyl carbocation, $CH_3CH_2^+$, through hyperconjugation. Interaction of neighboring C–H σ bonds with the vacant p orbital stabilizes the cation and lowers its energy. The molecular orbital shows that only the two C–H bonds more nearly parallel to the cation p orbital are oriented properly. The C–H bond perpendicular to the cation p orbital cannot take part.

Problem 7.18
Show the structures of the carbocation intermediates you would expect in the following reactions:

(a) 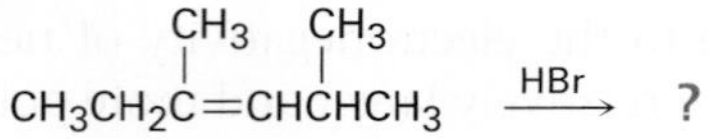
$$CH_3CH_2\underset{\underset{CH_3}{|}}{C}=\underset{\underset{CH_3}{|}}{C}HCHCH_3 \xrightarrow{HBr} ?$$

(b)

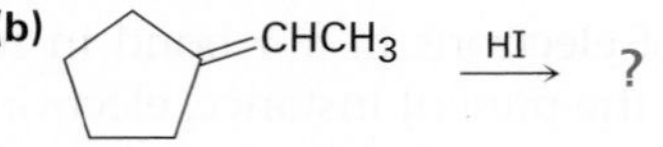

Problem 7.19
Draw a skeletal structure of the following carbocation. Identify it as primary, secondary, or tertiary, and identify the hydrogen atoms that have the proper orientation for hyperconjugation in the conformation shown.

7.10 The Hammond Postulate

Let's summarize our knowledge of electrophilic addition reactions to this point:

- **Electrophilic addition to an unsymmetrically substituted alkene gives the more highly substituted carbocation intermediate.** A more highly substituted carbocation forms faster than a less highly substituted one and, once formed, rapidly goes on to give the final product.

- **A more highly substituted carbocation is more stable than a less highly substituted one.** That is, the stability order of carbocations is tertiary > secondary > primary > methyl.

What we have not yet seen is how these two points are related. Why does the *stability* of the carbocation intermediate affect the *rate* at which it's formed and thereby determine the structure of the final product? After all, carbocation stability is determined by the free-energy change $\Delta G°$, but reaction rate is determined by the activation energy $\Delta G^{\ddagger}$. The two quantities aren't directly related.

Although there is no simple quantitative relationship between the stability of a carbocation intermediate and the rate of its formation, there *is* an intuitive relationship. It's generally true when comparing two similar reactions that the more stable intermediate forms faster than the less stable one. The situation is shown graphically in **Figure 7.13**, where the energy profile in part (a) represents the typical situation rather than the profile in part (b). That is, the curves for two similar reactions don't cross one another.

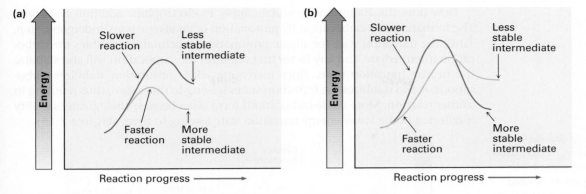

Figure 7.13 Energy diagrams for two similar competing reactions. In **(a)**, the faster reaction yields the more stable intermediate. In **(b)**, the slower reaction yields the more stable intermediate. The curves shown in **(a)** represent the typical situation.

Called the **Hammond postulate**, the explanation of the relationship between reaction rate and intermediate stability goes like this: Transition states represent energy maxima. They are high-energy activated complexes that occur transiently during the course of a reaction and immediately go on to a more stable species. Although we can't actually observe transition states because they have no finite lifetime, the Hammond postulate says that we can get an idea of a particular transition state's structure by looking at the structure of the nearest stable species. Imagine the two cases shown in **Figure 7.14**, for example. The reaction profile in part (a) shows the energy curve for an endergonic reaction step, and the profile in part (b) shows the curve for an exergonic step.

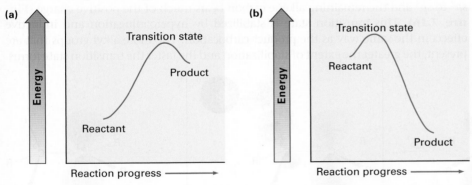

Figure 7.14 Energy diagrams for endergonic and exergonic steps. **(a)** In an endergonic step, the energy levels of transition state and *product* are closer. **(b)** In an exergonic step, the energy levels of transition state and *reactant* are closer.

In an endergonic reaction (Figure 7.14a), the energy level of the transition state is closer to that of the product than to that of the reactant. Since the transition state is closer energetically to the product, we make the natural assumption that it's also closer structurally. In other words, *the transition state for an endergonic reaction step structurally resembles the product of that step*. Conversely, the transition state for an exergonic reaction (Figure 7.14b) is closer energetically, and thus structurally, to the reactant than to the product. We therefore say that *the transition state for an exergonic reaction step structurally resembles the reactant for that step*.

> **Hammond postulate**
> The structure of a transition state resembles the structure of the nearest stable species. Transition states for endergonic steps structurally resemble products, and transition states for exergonic steps structurally resemble reactants.

How does the Hammond postulate apply to electrophilic addition reactions? The formation of a carbocation by protonation of an alkene is an endergonic step. Thus, the transition state for alkene protonation structurally resembles the carbocation intermediate, and any factor that stabilizes the carbocation will also stabilize the nearby transition state. Since increasing alkyl substitution stabilizes carbocations, it also stabilizes the transition states leading to those ions, thus resulting in a faster reaction. More stable carbocations form faster because their greater stability is reflected in the lower-energy transition state leading to them **(Figure 7.15)**.

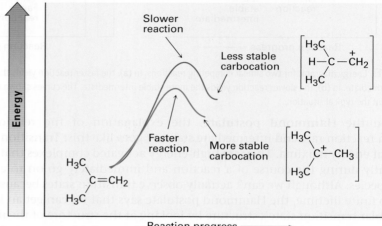

Figure 7.15 Energy diagrams for carbocation formation. The more stable tertiary carbocation is formed faster (green curve) because its increased stability lowers the energy of the transition state leading to it.

We can imagine the transition state for alkene protonation to be a structure in which one of the alkene carbon atoms has almost completely rehybridized from sp^2 to sp^3 and the remaining alkene carbon bears much of the positive charge **(Figure 7.16)**. This transition state is stabilized by hyperconjugation and inductive effects in the same way as the product carbocation. The more alkyl groups that are present, the greater the extent of stabilization and the faster the transition state forms.

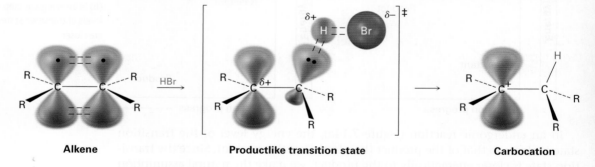

Figure 7.16 The hypothetical structure of a transition state for alkene protonation. The transition state is closer in both energy and structure to the carbocation than to the alkene. Thus, an increase in carbocation stability (lower $\Delta G°$) also causes an increase in transition-state stability (lower $\Delta G^{\ddagger}$), thereby increasing the rate of its formation.

Problem 7.20
What about the second step in the electrophilic addition of HCl to an alkene—the reaction of chloride ion with the carbocation intermediate? Is this step exergonic or endergonic? Does the transition state for this second step resemble the reactant (carbocation) or product (alkyl chloride)? Make a rough drawing of what the transition-state structure might look like.

7.11 Evidence for the Mechanism of Electrophilic Additions: Carbocation Rearrangements

How do we know that the carbocation mechanism for electrophilic addition reactions of alkenes is correct? The answer is that we *don't* know it's correct; at least we don't know with complete certainty. Although an incorrect reaction mechanism can be disproved by demonstrating that it doesn't account for observed data, a correct reaction mechanism can never be entirely proven. The best we can do is to show that a proposed mechanism is consistent with all known facts. If enough facts are accounted for, the mechanism is probably correct.

One of the best pieces of evidence supporting the carbocation mechanism for the electrophilic addition reaction was discovered during the 1930s by F. C. Whitmore of the Pennsylvania State University, who found that structural rearrangements often occur during the reaction of HX with an alkene. For example, reaction of HCl with 3-methyl-1-butene yields a substantial amount of 2-chloro-2-methylbutane in addition to the "expected" product, 2-chloro-3-methylbutane.

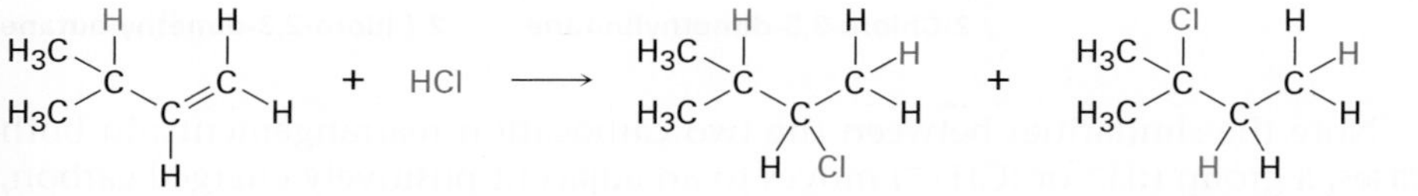

3-Methyl-1-butene + HCl ⟶ **2-Chloro-3-methylbutane** (approx. 50%) + **2-Chloro-2-methylbutane** (approx. 50%)

If the reaction takes place in a single step, it would be difficult to account for rearrangement, but if the reaction takes place in several steps, rearrangement is more easily explained. Whitmore suggested that it is a carbocation intermediate that undergoes rearrangement. The secondary carbocation intermediate formed by protonation of 3-methyl-1-butene rearranges to a more stable tertiary carbocation by a **hydride shift**—the shift of a hydrogen atom and its electron pair (a hydride ion, :H⁻) between neighboring carbons.

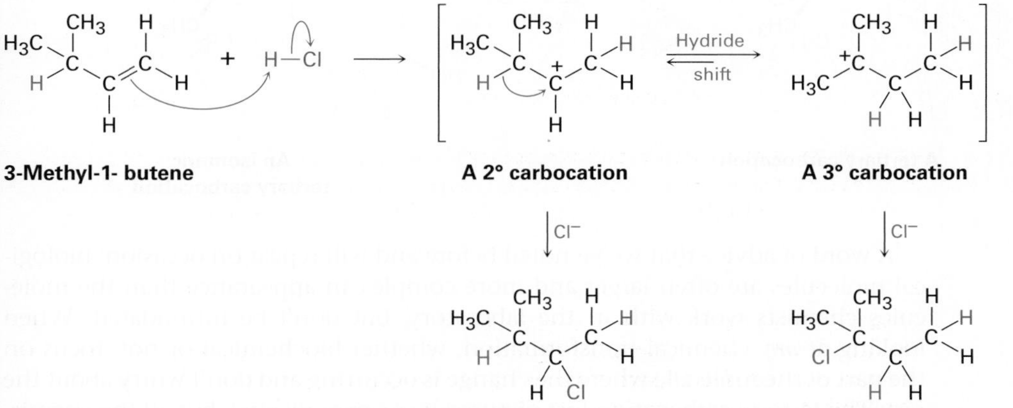

3-Methyl-1-butene + H–Cl ⟶ [**A 2° carbocation** ⇌ (Hydride shift) **A 3° carbocation**]

↓ Cl⁻   ↓ Cl⁻

2-Chloro-3-methylbutane   **2-Chloro-2-methylbutane**

Carbocation rearrangements can also occur by the shift of an alkyl group with its electron pair. For example, reaction of 3,3-dimethyl-1-butene with HCl leads to an equal mixture of unrearranged 2-chloro-3,3-dimethylbutane and rearranged 2-chloro-2,3-dimethylbutane. In this instance, a secondary carbocation rearranges to a more stable tertiary carbocation by the shift of a methyl group.

3,3-Dimethyl-1-butene — A 2° carbocation — A 3° carbocation

2-Chloro-3,3-dimethylbutane — 2-Chloro-2,3-dimethylbutane

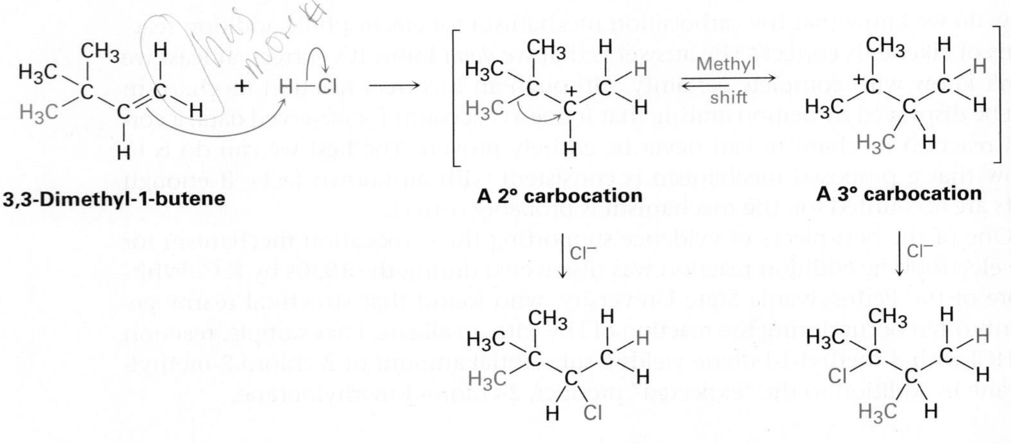

Note the similarities between the two carbocation rearrangements: in both cases, a group (:H^- or :CH_3^-) moves to an adjacent positively charged carbon, taking its bonding electron pair with it. Also in both cases, a less stable carbocation rearranges to a more stable ion. Rearrangements of this kind are a common feature of carbocation chemistry and are particularly important in the biological pathways by which steroids and related substances are synthesized. An example is the following hydride shift that occurs during the biosynthesis of cholesterol.

A tertiary carbocation — An isomeric tertiary carbocation

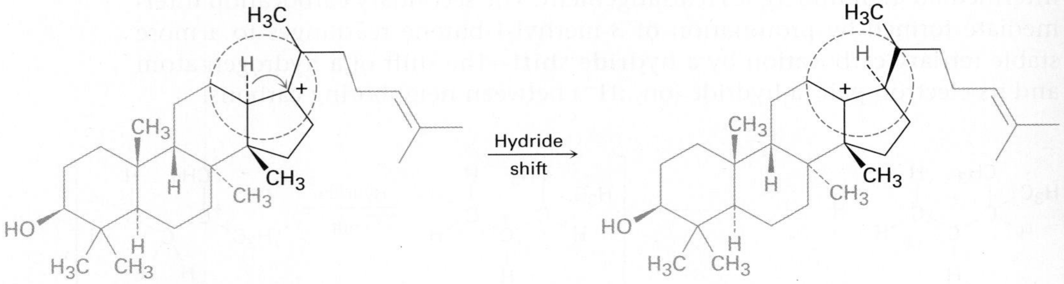

A word of advice that we've noted before and will repeat on occasion: biological molecules are often larger and more complex in appearance than the molecules chemists work with in the laboratory, but don't be intimidated. When looking at *any* chemical transformation, whether biochemical or not, focus on the part of the molecule where the change is occurring and don't worry about the rest. The tertiary carbocation just pictured looks complicated, but all the chemistry is taking place in the small part of the molecule inside the red circle.

Problem 7.21

On treatment with HBr, vinylcyclohexane undergoes addition and rearrangement to yield 1-bromo-1-ethylcyclohexane. Using curved arrows, propose a mechanism to account for this result.

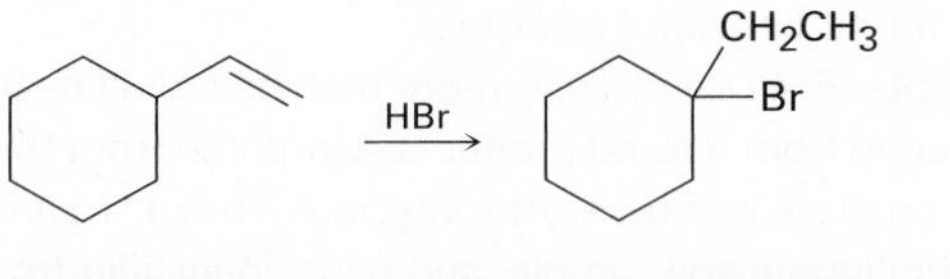

Vinylcyclohexane → (HBr) → 1-Bromo-1-ethylcyclohexane

A DEEPER LOOK
Bioprospecting: Hunting for Natural Products

Most people know the names of the common classes of biomolecules—proteins, carbohydrates, lipids, and nucleic acids—but there are far more kinds of compounds in living organisms than just those four. All living organisms also contain a vast diversity of substances usually grouped under the heading *natural products*. The term natural product really refers to *any* naturally occurring substance but is generally taken to mean a so-called secondary metabolite—a small molecule that is not essential to the growth and development of the producing organism and is not classified by structure.

It has been estimated that well over 300,000 secondary metabolites exist, and it's thought that their primary function is to increase the likelihood of an organism's survival by repelling or attracting other organisms. Alkaloids, such as morphine; antibiotics, such as erythromycin and the penicillins; and immunosuppressive agents, such as rapamycin (sirolimus) prescribed for liver transplant recipients, are examples.

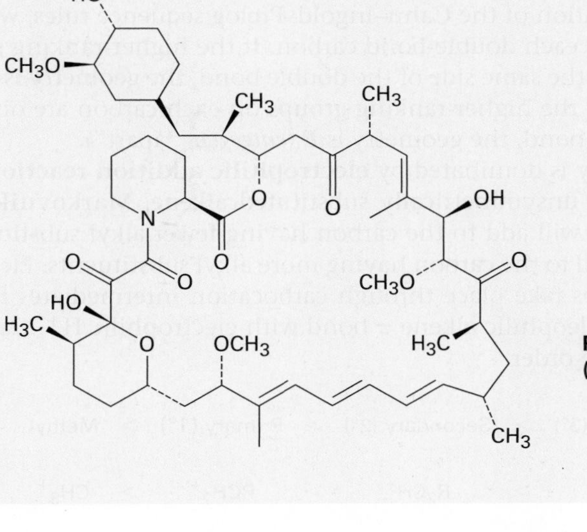

Rapamycin, an immunosuppressant natural product used during organ transplants, was originally isolated from a soil sample found on Easter Island, or Rapa Nui, an island 2200 miles off the coast of Chile known for its giant Moai statues.

Rapamycin (Sirolimus)

(continued)

(continued)

Where do these natural products come from, and how are they found? Although most chemists and biologists spend most of their time in the laboratory, a few spend their days scuba diving on South Pacific islands or trekking through the rainforests of South America and Southeast Asia at work as bioprospectors. Their job is to hunt for new and unusual natural products that might be useful as drugs.

As noted in the Chapter 6 *A Deeper Look*, more than half of all new drug candidates come either directly or indirectly from natural products. Morphine from the opium poppy, prostaglandin E_1 from sheep prostate glands, erythromycin A from a *Streptomyces erythreus* bacterium cultured from a Philippine soil sample, and benzylpenicillin from the mold *Penicillium notatum* are examples. The immunosuppressive agent rapamycin, whose structure is shown on the previous page, was first isolated from a *Streptomyces hygroscopicus* bacterium found in a soil sample from Easter Island (Rapa Nui), located 2200 miles off the coast of Chile.

With less than 1% of living organisms yet investigated, bioprospectors have a lot of work to do. But there is a race going on. Rainforests throughout the world are being destroyed at an alarming rate, causing many species of both plants and animals to become extinct before they can even be examined. Fortunately, the governments in many countries seem aware of the problem, but there is as yet no international treaty on biodiversity that could help preserve vanishing species.

Summary

Key words

alkene ($R_2C=CR_2$), 222
allyl group, 228
degree of unsaturation, 224
E geometry, 232
E,Z system, 231
electrophilic addition reaction, 237
Hammond postulate, 247
hydride shift, 249
hyperconjugation, 235
Markovnikov's rule, 240
methylene group, 228
regiospecific, 240
unsaturated, 224
vinyl group, 228
Z geometry, 232

Carbon–carbon double bonds are present in most organic and biological molecules, so a good understanding of their behavior is needed. In this chapter, we've looked at some consequences of alkene stereoisomerism and at the details of the broadest and most general class of alkene reactions—the electrophilic addition reaction.

An **alkene** is a hydrocarbon that contains a carbon–carbon double bond. Because they contain fewer hydrogens than alkanes with the same number of carbons, alkenes are said to be **unsaturated**.

Because rotation around the double bond can't occur, substituted alkenes can exist as cis–trans stereoisomers. The geometry of a double bond can be specified by application of the Cahn–Ingold–Prelog sequence rules, which rank the substituents on each double-bond carbon. If the higher-ranking groups on each carbon are on the same side of the double bond, the geometry is *Z* (*zusammen*, "together"); if the higher-ranking groups on each carbon are on opposite sides of the double bond, the geometry is *E* (*entgegen*, "apart").

Alkene chemistry is dominated by **electrophilic addition reactions**. When HX reacts with an unsymmetrically substituted alkene, **Markovnikov's rule** predicts that the H will add to the carbon having fewer alkyl substituents and the X group will add to the carbon having more alkyl substituents. Electrophilic additions to alkenes take place through carbocation intermediates formed by reaction of the nucleophilic alkene π bond with electrophilic H^+. Carbocation stability follows the order

Tertiary (3°) > Secondary (2°) > Primary (1°) > Methyl

R_3C^+ > R_2CH^+ > RCH_2^+ > CH_3^+

Markovnikov's rule can be restated by saying that, in the addition of HX to an alkene, the more stable carbocation intermediate is formed. This result is explained by the **Hammond postulate**, which says that the transition state of an exergonic reaction step structurally resembles the reactant, whereas the transition state of an endergonic reaction step structurally resembles the product. Since an alkene protonation step is endergonic, the stability of the more highly substituted carbocation is reflected in the stability of the transition state leading to its formation.

Evidence in support of a carbocation mechanism for electrophilic additions comes from the observation that structural rearrangements often take place during reaction. Rearrangements occur by shift of either a hydride ion, $:H^-$ (a **hydride shift**), or an alkyl anion, $:R^-$, from a carbon atom to the neighboring positively charged carbon. The result is isomerization of a less stable carbocation to a more stable one.

Exercises

Visualizing Chemistry

(Problems 7.1–7.21 appear within the chapter.)

7.22 Name the following alkenes, and convert each drawing into a skeletal structure:

(a)   (b)

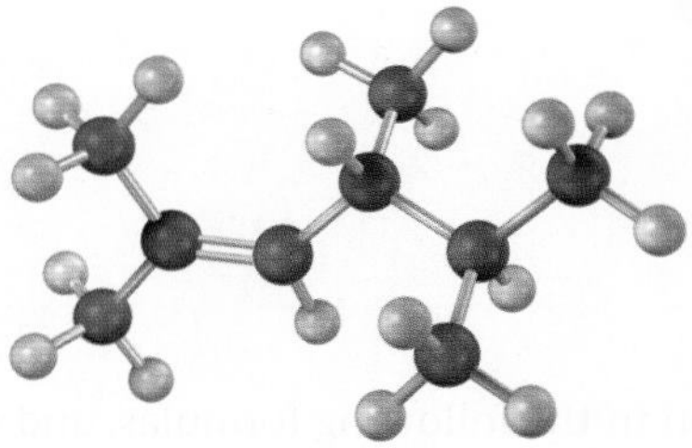

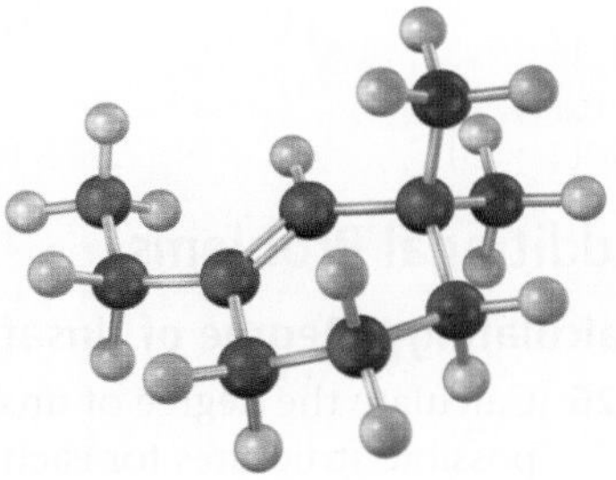

7.23 Assign *E* or *Z* stereochemistry to the double bonds in each of the following alkenes, and convert each drawing into a skeletal structure (red = O, green = Cl):

(a)   (b)

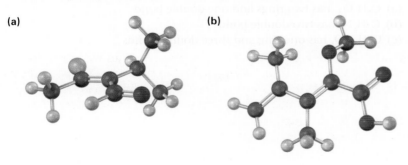

OWL Interactive versions of these problems are assignable in OWL for Organic Chemistry.

▲ denotes problems linked to the Key Ideas in this chapter.

▲ Problems linked to Key Ideas in this chapter

7.24 The following carbocation is an intermediate in the electrophilic addition reaction of HCl with two different alkenes. Identify both, and tell which C–H bonds in the carbocation are aligned for hyperconjugation with the vacant *p* orbital on the positively charged carbon.

7.25 The following alkyl bromide can be made by HBr addition to three different alkenes. Show their structures.

Additional Problems

Calculating a Degree of Unsaturation

7.26 Calculate the degree of unsaturation in the following formulas, and draw five possible structures for each:
  (a) $C_{10}H_{16}$ (b) C_8H_8O (c) $C_7H_{10}Cl_2$
  (d) $C_{10}H_{16}O_2$ (e) $C_5H_9NO_2$ (f) $C_8H_{10}ClNO$

7.27 How many hydrogens does each of the following compounds have?
  (a) $C_8H_?O_2$, has two rings and one double bond
  (b) $C_7H_?N$, has two double bonds
  (c) $C_9H_?NO$, has one ring and three double bonds

▲ Problems linked to Key Ideas in this chapter

Exercises

7.28 Loratadine, marketed as an antiallergy medication under the name Claritin, has four rings, eight double bonds, and the formula $C_{22}H_?ClN_2O_2$. How many hydrogens does loratadine have? (Calculate your answer; don't count hydrogens in the structure.)

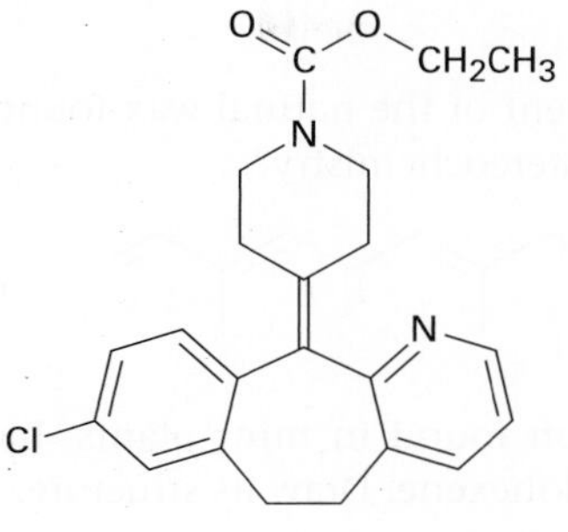

Loratadine

Naming Alkenes

7.29 Name the following alkenes:

(a), (b), 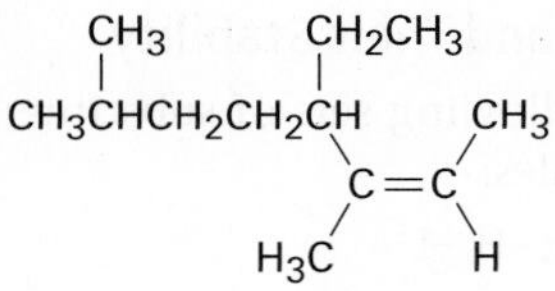 (c)

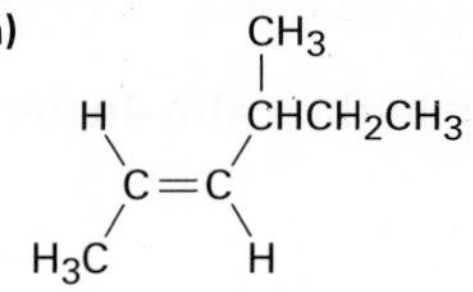

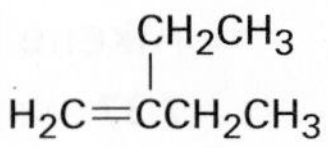

(d), 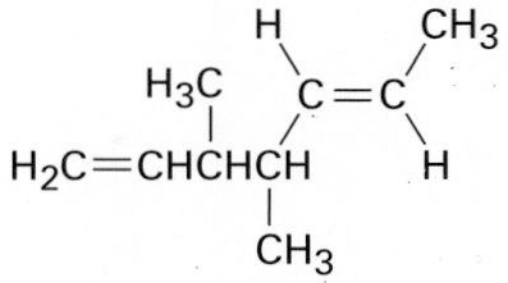 (e), 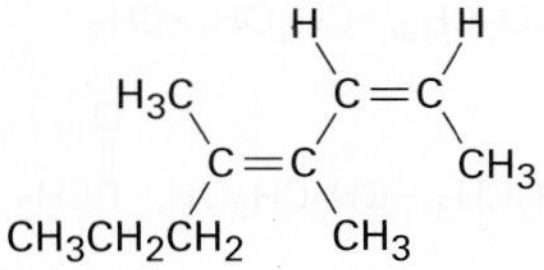 (f) $H_2C=C=CHCH_3$

7.30 Draw structures corresponding to the following systematic names:
(a) (4E)-2,4-Dimethyl-1,4-hexadiene
(b) cis-3,3-Dimethyl-4-propyl-1,5-octadiene
(c) 4-Methyl-1,2-pentadiene
(d) (3E,5Z)-2,6-Dimethyl-1,3,5,7-octatetraene
(e) 3-Butyl-2-heptene
(f) trans-2,2,5,5-Tetramethyl-3-hexene

7.31 Name the following cycloalkenes:

(a) CH₃ (b) 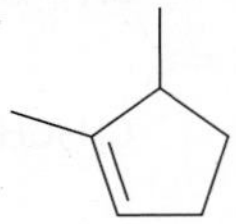 (c)

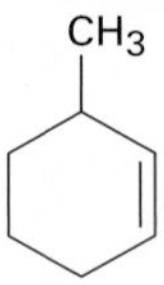

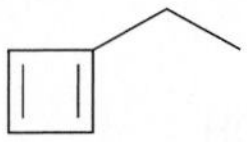

(d) 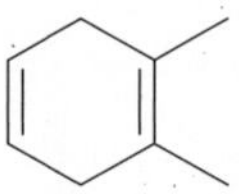 (e) 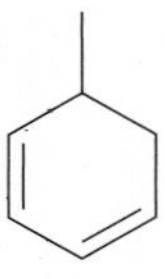 (f)

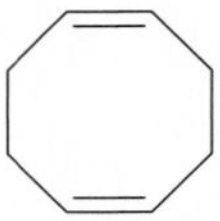

▲ Problems linked to Key Ideas in this chapter

7.32 Ocimene is a triene found in the essential oils of many plants. What is its IUPAC name, including stereochemistry?

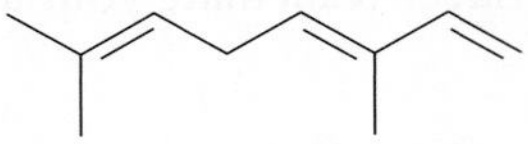

Ocimene

7.33 α-Farnesene is a constituent of the natural wax found on apples. What is its IUPAC name, including stereochemistry?

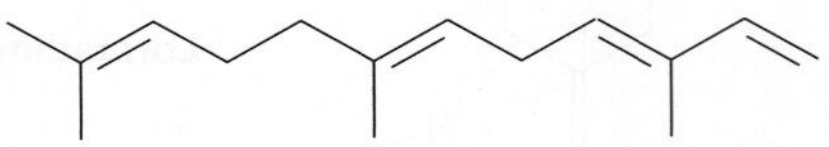

α-Farnesene

7.34 Menthene, a hydrocarbon found in mint plants, has the systematic name 1-isopropyl-4-methylcyclohexene. Draw its structure.

7.35 Draw and name the six alkene isomers, C_5H_{10}, including E,Z isomers.

7.36 Draw and name the 17 alkene isomers, C_6H_{12}, including E,Z isomers.

Alkene Isomers and Their Stability

7.37 Rank the following sets of substituents according to the Cahn–Ingold–Prelog sequence rules:

(a) –CH₃, –Br, –H, –I

(b) –OH, –OCH₃, –H, –CO₂H

(c) –CO₂H, –CO₂CH₃, –CH₂OH, –CH₃

(d) –CH₃, –CH₂CH₃, –CH₂CH₂OH, –C(=O)CH₃

(e) –CH=CH₂, –CN, –CH₂NH₂, –CH₂Br

(f) –CH=CH₂, –CH₂CH₃, –CH₂OCH₃, –CH₂OH

7.38 ▲ Assign E or Z configuration to each of the following compounds:

(a) 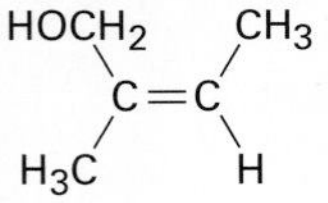 HOCH₂, CH₃ / H₃C, H on C=C

(b) 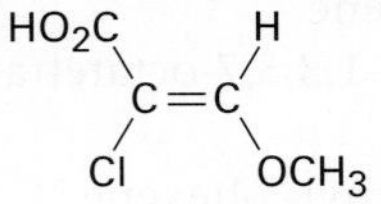 HO₂C, H / Cl, OCH₃ on C=C

(c) NC, CH₃ / CH₃CH₂, CH₂OH on C=C

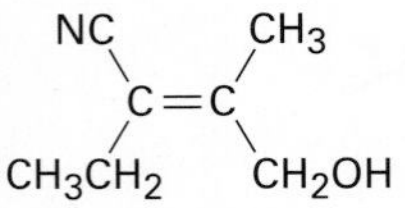

(d) 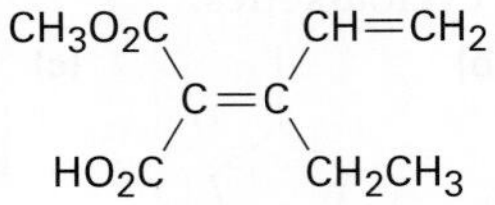 CH₃O₂C, CH=CH₂ / HO₂C, CH₂CH₃ on C=C

▲ Problems linked to Key Ideas in this chapter

7.39 ▲ Which of the following *E,Z* designations are correct, and which are incorrect?

(a) 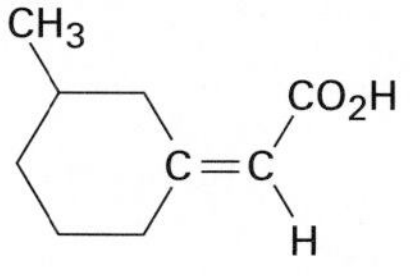
CH₃–cyclohexenyl C=C with CO₂H and H
Z

(b) 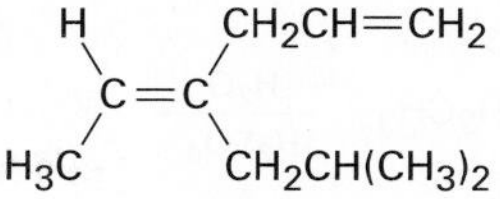
H, H₃C / C=C / CH₂CH=CH₂, CH₂CH(CH₃)₂
E

(c) 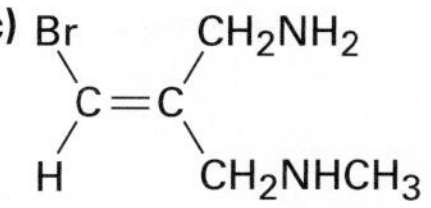
Br, H / C=C / CH₂NH₂, CH₂NHCH₃
Z

(d) 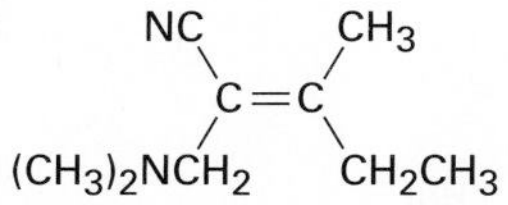
NC, (CH₃)₂NCH₂ / C=C / CH₃, CH₂CH₃
E

(e) 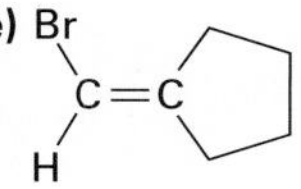
Br, H / C=C / cyclopentyl
Z

(f) 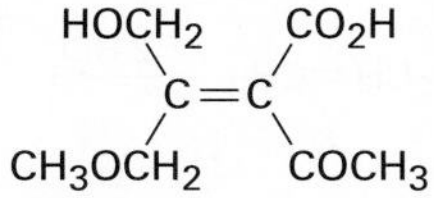
HOCH₂, CH₃OCH₂ / C=C / CO₂H, COCH₃
E

7.40 *trans*-2-Butene is more stable than *cis*-2-butene by only 4 kJ/mol, but *trans*-2,2,5,5-tetramethyl-3-hexene is more stable than its cis isomer by 39 kJ/mol. Explain.

7.41 Cyclodecene can exist in both cis and trans forms, but cyclohexene cannot. Explain. (Making molecular models is helpful.)

7.42 Normally, a trans alkene is *more* stable than its cis isomer. *trans*-Cyclooctene, however, is *less* stable than *cis*-cyclooctene by 38.5 kJ/mol. Explain.

7.43 *trans*-Cyclooctene is less stable than *cis*-cyclooctene by 38.5 kJ/mol, but *trans*-cyclononene is less stable than *cis*-cyclononene by only 12.2 kJ/mol. Explain.

7.44 Tamoxifen, a drug used in the treatment of breast cancer, and clomiphene, a drug used as a fertility treatment, have similar structures but very different effects. Assign *E* or *Z* configuration to the double bonds in both compounds.

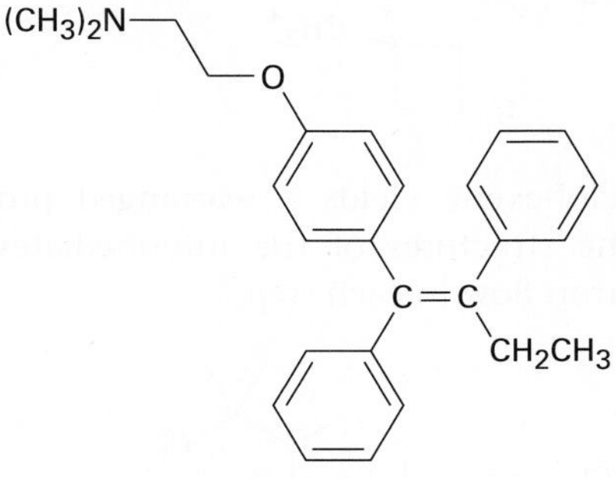

Tamoxifen (anticancer)

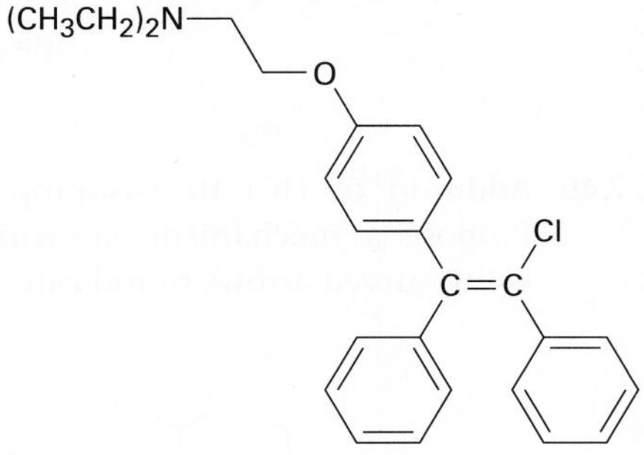

Clomiphene (fertility treatment)

▲ Problems linked to Key Ideas in this chapter

Carbocations and Electrophilic Addition Reactions

7.45 Predict the major product in each of the following reactions:

(a) $CH_3CH_2CH=C(CH_3)CH_2CH_3 \xrightarrow[H_2SO_4]{H_2O}$?

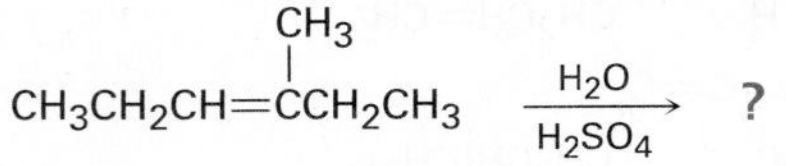

(Addition of H_2O occurs.)

(b) 1-ethylcyclopentene $\xrightarrow{HBr}$?

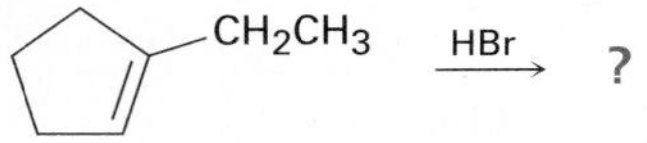

(c) 3-methylcyclohexene (with methyl group) $\xrightarrow{HBr}$?

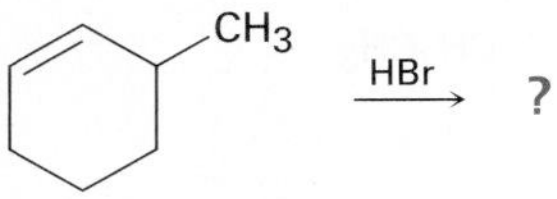

(d) $H_2C=CHCH_2CH_2CH_2CH=CH_2 \xrightarrow{2\ HCl}$?

7.46 ▲ Predict the major product from addition of HBr to each of the following alkenes:

(a) methylenecyclohexane

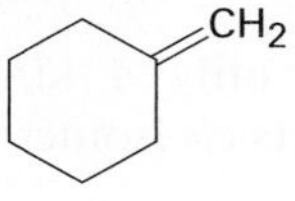

(b) octahydronaphthalene with double bond at ring junction

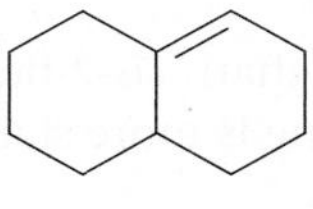

(c) $CH_3CH=CHCH(CH_3)CH_3$

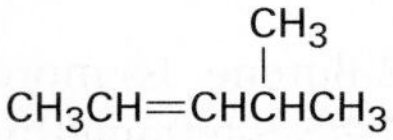

7.47 ▲ Alkenes can be converted into alcohols by acid-catalyzed addition of water. Assuming that Markovnikov's rule is valid, predict the major alcohol product from each of the following alkenes.

(a) $CH_3CH_2C(CH_3)=CHCH_3$

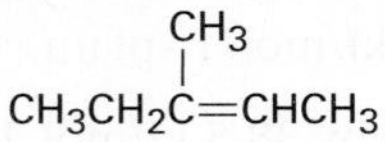

(b) methylenecyclohexane

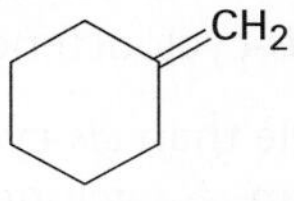

(c) $CH_3CH(CH_3)CH_2CH=CH_2$

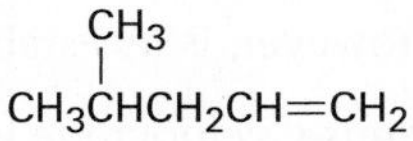

7.48 Each of the following carbocations can rearrange to a more stable ion. Propose structures for the likely rearrangement products.

(a) $CH_3CH_2CH_2CH_2^+$ (b) $CH_3CH(CH_3)\overset{+}{C}HCH_3$ (c) cyclobutyl-CH_2^+

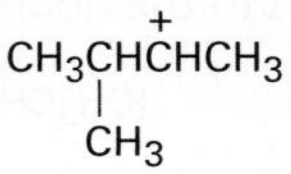

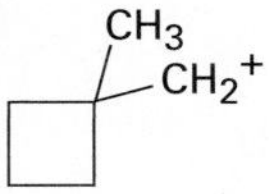

7.49 Addition of HCl to 1-isopropylcyclohexene yields a rearranged product. Propose a mechanism, showing the structures of the intermediates and using curved arrows to indicate electron flow in each step.

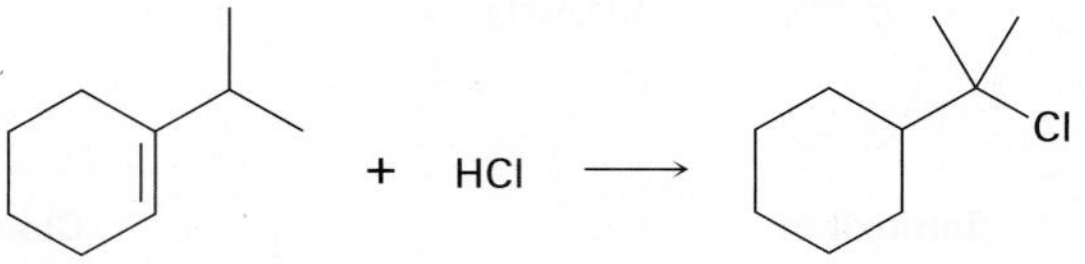

1-isopropylcyclohexene + HCl ⟶ 1-chloro-1,1-dimethyl-substituted cyclohexane

General Problems

7.50 Allene (1,2-propadiene), $H_2C=C=CH_2$, has two adjacent double bonds. What kind of hybridization must the central carbon have? Sketch the bonding π orbitals in allene. What shape do you predict for allene?

7.51 The heat of hydrogenation for allene (Problem 7.50) to yield propane is -295 kJ/mol, and the heat of hydrogenation for a typical monosubstituted alkene such as propene is -126 kJ/mol. Is allene more stable or less stable than you might expect for a diene? Explain.

7.52 Retin A, or retinoic acid, is a medication commonly used to reduce wrinkles and treat severe acne. How many different isomers arising from double-bond isomerizations are possible?

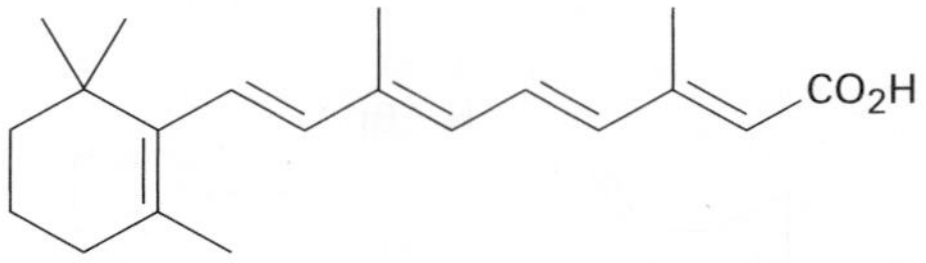

Retin A (retinoic acid)

7.53 Fucoserratene and ectocarpene are sex pheromones produced by marine brown algae. What are their systematic names? (Ectocarpene is a bit difficult; make your best guess, and then check your answer in the *Study Guide and Solutions Manual*.)

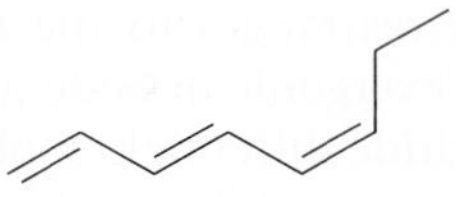

Fucoserratene 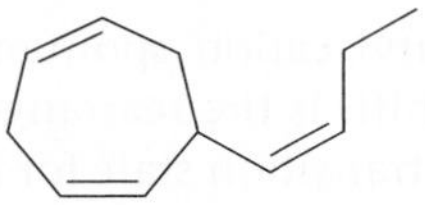   **Ectocarpene**

7.54 ▲ *tert*-Butyl esters [$RCO_2C(CH_3)_3$] are converted into carboxylic acids (RCO_2H) by reaction with trifluoroacetic acid, a reaction useful in protein synthesis (Section 26.7). Assign E,Z designation to the double bonds of both reactant and product in the following scheme, and explain why there is an apparent change of double-bond stereochemistry:

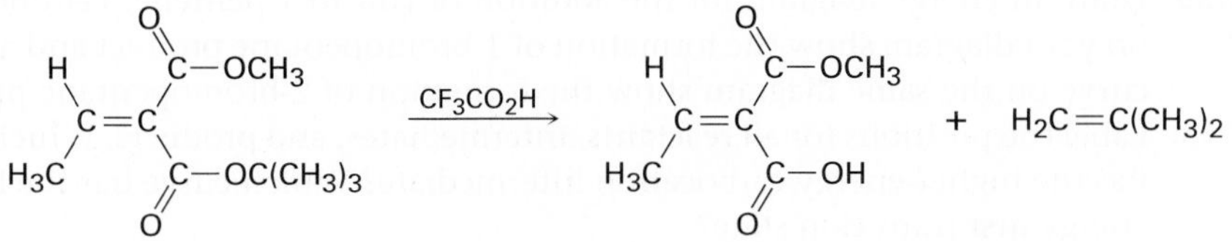

▲ Problems linked to Key Ideas in this chapter

7.55 Addition of HCl to 1-isopropenyl-1-methylcyclopentane yields 1-chloro-1,2,2-trimethylcyclohexane. Propose a mechanism, showing the structures of the intermediates and using curved arrows to indicate electron flow in each step.

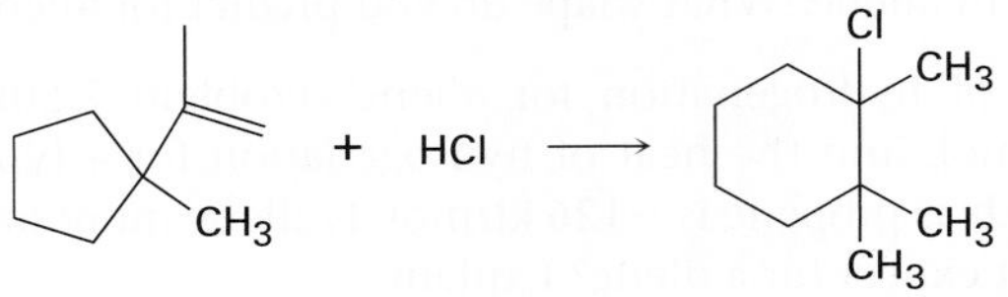

7.56 Vinylcyclopropane reacts with HBr to yield a rearranged alkyl bromide. Follow the flow of electrons as represented by the curved arrows, show the structure of the carbocation intermediate in brackets, and show the structure of the final product.

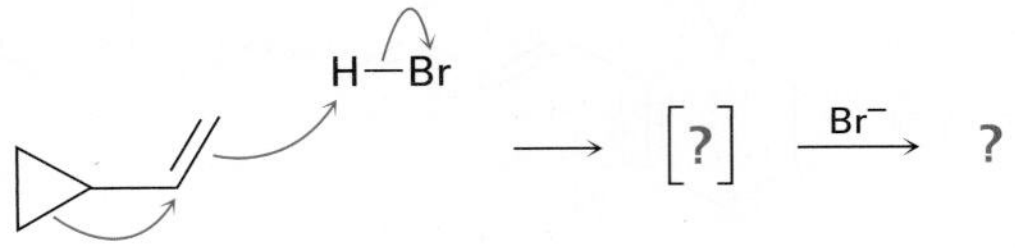

Vinylcyclopropane

7.57 Calculate the degree of unsaturation in each of the following formulas:
  (a) Cholesterol, $C_{27}H_{46}O$
  (b) DDT, $C_{14}H_9Cl_5$
  (c) Prostaglandin E_1, $C_{20}H_{34}O_5$
  (d) Caffeine, $C_8H_{10}N_4O_2$
  (e) Cortisone, $C_{21}H_{28}O_5$
  (f) Atropine, $C_{17}H_{23}NO_3$

7.58 The isobutyl cation spontaneously rearranges to the *tert*-butyl cation by a hydride shift. Is the rearrangement exergonic or endergonic? Draw what you think the transition state for the hydride shift might look like according to the Hammond postulate.

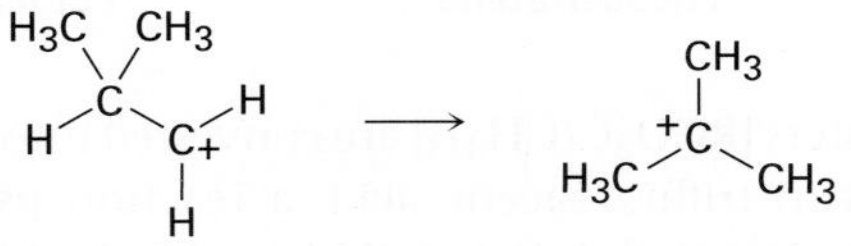

Isobutyl cation   ***tert*-Butyl cation**

7.59 Draw an energy diagram for the addition of HBr to 1-pentene. Let one curve on your diagram show the formation of 1-bromopentane product and another curve on the same diagram show the formation of 2-bromopentane product. Label the positions for all reactants, intermediates, and products. Which curve has the higher-energy carbocation intermediate? Which curve has the higher-energy first transition state?

7.60 Make sketches of the transition-state structures involved in the reaction of HBr with 1-pentene (Problem 7.59). Tell whether each structure resembles reactant or product.

7.61 Limonene, a fragrant hydrocarbon found in lemons and oranges, is biosynthesized from geranyl diphosphate by the following pathway. Add curved arrows to show the mechanism of each step. Which step involves an alkene

▲ Problems linked to Key Ideas in this chapter

electrophilic addition? (The ion $OP_2O_6^{4-}$ is the diphosphate ion, and "Base" is an unspecified base in the enzyme that catalyzes the reaction.)

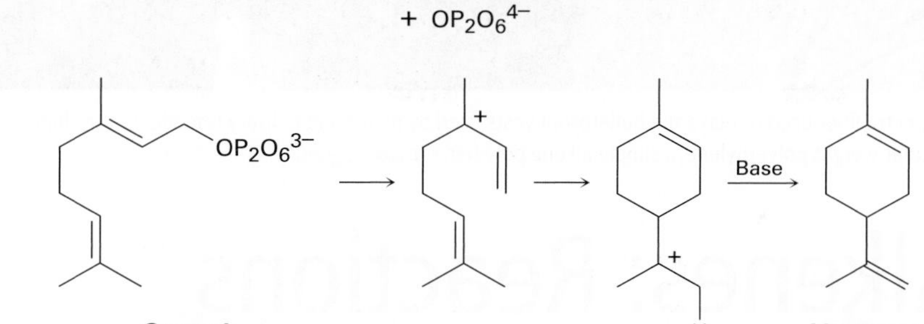

7.62 *epi*-Aristolochene, a hydrocarbon found in both pepper and tobacco, is biosynthesized by the following pathway. Add curved arrows to show the mechanism of each step. Which steps involve alkene electrophilic addition(s), and which involve carbocation rearrangement(s)? (The abbreviation H−A stands for an unspecified acid, and "Base" is an unspecified base in the enzyme.)

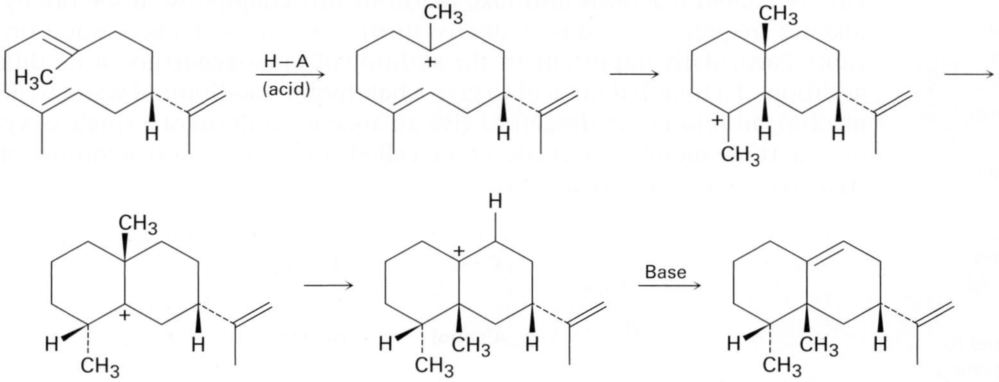

7.63 Aromatic compounds such as benzene react with alkyl chlorides in the presence of $AlCl_3$ catalyst to yield alkylbenzenes. The reaction occurs through a carbocation intermediate, formed by reaction of the alkyl chloride with $AlCl_3$ (R−Cl + $AlCl_3$ → R^+ + $AlCl_4^-$). How can you explain the observation that reaction of benzene with 1-chloropropane yields isopropylbenzene as the major product?

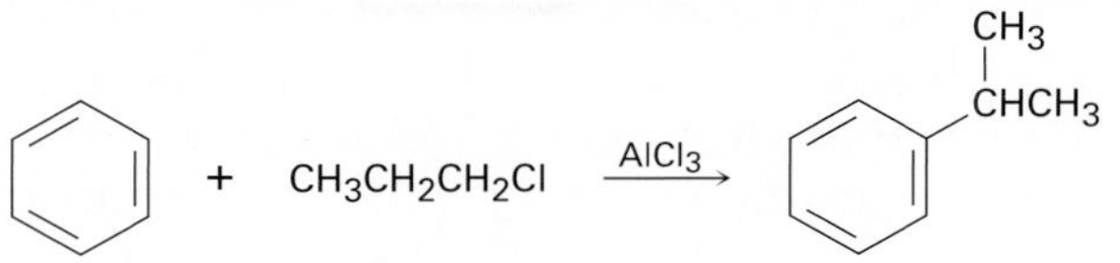

7.64 Reaction of 2,3-dimethyl-1-butene with HBr leads to an alkyl bromide, $C_6H_{13}Br$. On treatment of this alkyl bromide with KOH in methanol, elimination of HBr occurs and a hydrocarbon that is isomeric with the starting alkene is formed. What is the structure of this hydrocarbon, and how do you think it is formed from the alkyl bromide?

▲ Problems linked to Key Ideas in this chapter

8

The Spectra fiber used to make the bulletproof vests used by police and military is made of ultra-high-molecular-weight polyethylene, a simple alkene polymer. Ed Darack/Getty Images

Alkenes: Reactions and Synthesis

8.1 Preparing Alkenes: A Preview of Elimination Reactions
8.2 Halogenation of Alkenes: Addition of X_2
8.3 Halohydrins from Alkenes: Addition of HOX
8.4 Hydration of Alkenes: Addition of H_2O by Oxymercuration
8.5 Hydration of Alkenes: Addition of H_2O by Hydroboration
8.6 Reduction of Alkenes: Hydrogenation
8.7 Oxidation of Alkenes: Epoxidation and Hydroxylation
8.8 Oxidation of Alkenes: Cleavage to Carbonyl Compounds
8.9 Addition of Carbenes to Alkenes: Cyclopropane Synthesis
8.10 Radical Additions to Alkenes: Chain-Growth Polymers
8.11 Biological Additions of Radicals to Alkenes
8.12 Reaction Stereochemistry: Addition of H_2O to an Achiral Alkene
8.13 Reaction Stereochemistry: Addition of H_2O to a Chiral Alkene
A Deeper Look—Terpenes: Naturally Occurring Alkenes

OWL Sign in to OWL for Organic Chemistry at **www.cengage.com/owl** to view tutorials and simulations, develop problem-solving skills, and complete online homework assigned by your professor.

Alkene addition reactions occur widely, both in the laboratory and in living organisms. Although we've studied only the addition of HX thus far, many closely related reactions also take place. In this chapter, we'll see briefly how alkenes are prepared and we'll discuss further examples of alkene addition reactions. Particularly important are the addition of a halogen to give a 1,2-dihalide, addition of a hypohalous acid to give a halohydrin, addition of water to give an alcohol, addition of hydrogen to give an alkane, addition of a single oxygen to give a three-membered cyclic ether called an *epoxide,* and addition of two hydroxyl groups to give a 1,2-diol.

Why This Chapter? Much of the background needed to understand organic reactions has now been covered, and it's time to begin a systematic description of the major functional groups. In this chapter on alkenes and in future chapters on other functional groups, we'll discuss a variety of reactions but try to focus on the general principles and patterns of reactivity that tie organic chemistry together. There are no shortcuts: you have to know the reactions to understand organic and biological chemistry.

8.1 Preparing Alkenes: A Preview of Elimination Reactions

Before getting to the main subject of this chapter—the reactions of alkenes—let's take a brief look at how alkenes are prepared. The subject is a bit complex, though, so we'll return in Chapter 11 for a more detailed study. For the present, it's enough to realize that alkenes are readily available from simple precursors—usually alcohols in biological systems and either alcohols or alkyl halides in the laboratory.

Just as the chemistry of alkenes is dominated by addition reactions, the preparation of alkenes is dominated by elimination reactions. Additions and eliminations are, in many respects, two sides of the same coin. That is, an addition reaction might involve the addition of HBr or H$_2$O to an alkene to form an alkyl halide or alcohol, whereas an elimination reaction might involve the loss of HBr or H$_2$O from an alkyl halide or alcohol to form an alkene.

$$\text{C=C} + \text{X-Y} \underset{\text{Elimination}}{\overset{\text{Addition}}{\rightleftharpoons}} \text{X-C-C-Y}$$

The two most common elimination reactions are *dehydrohalogenation*—the loss of HX from an alkyl halide—and *dehydration*—the loss of water from an alcohol. Dehydrohalogenation usually occurs by reaction of an alkyl halide with strong base such as potassium hydroxide. For example, bromocyclohexane yields cyclohexene when treated with KOH in ethanol solution.

Bromocyclohexane → Cyclohexene (81%) + KBr + H$_2$O
(KOH, CH$_3$CH$_2$OH)

Dehydration is often carried out in the laboratory by treatment of an alcohol with a strong acid. For example, when 1-methylcyclohexanol is warmed with aqueous sulfuric acid in tetrahydrofuran (THF) solvent, loss of water occurs and 1-methylcyclohexene is formed.

1-Methylcyclohexanol → 1-Methylcyclohexene (91%) + H$_2$O
(H$_2$SO$_4$, H$_2$O, THF, 50 °C)
Solvent

[Tetrahydrofuran (THF)—a common solvent]

In biological pathways, dehydrations rarely occur with isolated alcohols. Instead, they normally take place on substrates in which the −OH is positioned two carbons away from a carbonyl group. In the biosynthesis of fats, for instance, β-hydroxybutyryl ACP is converted by dehydration to *trans*-crotonyl ACP, where ACP is an abbreviation for *acyl carrier protein*. We'll see the reason for this requirement in **Section 11.10**.

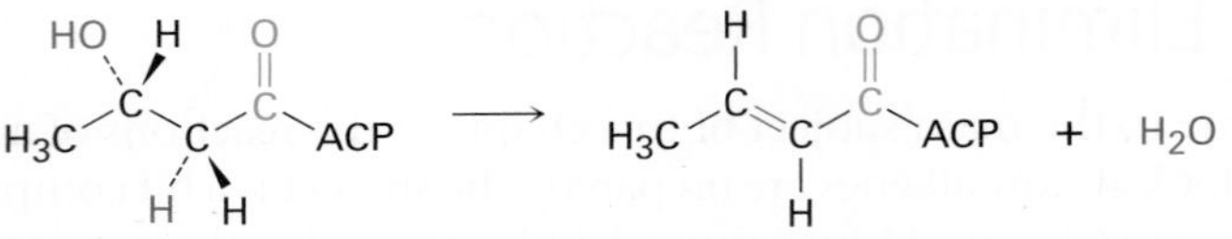

β-Hydroxybutyryl ACP → *trans*-Crotonyl ACP + H$_2$O

Problem 8.1
One problem with elimination reactions is that mixtures of products are often formed. For example, treatment of 2-bromo-2-methylbutane with KOH in ethanol yields a mixture of two alkene products. What are their likely structures?

Problem 8.2
How many alkene products, including E,Z isomers, might be obtained by dehydration of 3-methyl-3-hexanol with aqueous sulfuric acid?

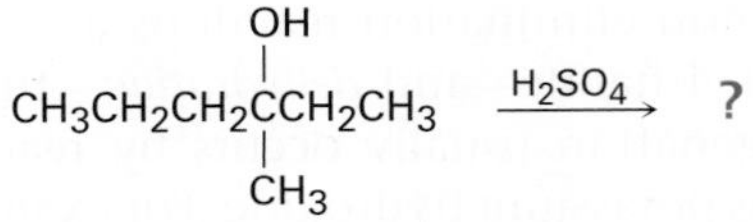

CH$_3$CH$_2$CH$_2$C(OH)(CH$_3$)CH$_2$CH$_3$ $\xrightarrow{H_2SO_4}$?

3-Methyl-3-hexanol

8.2 Halogenation of Alkenes: Addition of X$_2$

Bromine and chlorine add rapidly to alkenes to yield 1,2-dihalides, a process called *halogenation*. For example, more than 18 million tons 1,2-dichloroethane (ethylene dichloride) is synthesized worldwide each year, much of it by addition of Cl$_2$ to ethylene. The product is used both as a solvent and as starting material for the manufacture of poly(vinyl chloride), PVC. Fluorine is too reactive and difficult to control for most laboratory applications, and iodine does not react with most alkenes.

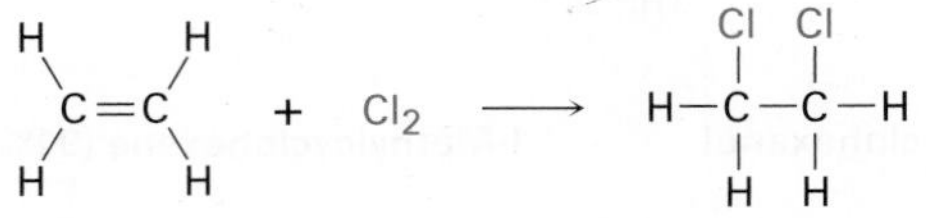

H$_2$C=CH$_2$ + Cl$_2$ ⟶ ClCH$_2$−CH$_2$Cl

Ethylene → **1,2-Dichloroethane (ethylene dichloride)**

8.2 | Halogenation of Alkenes: Addition of X_2

Based on what we've seen thus far, a possible mechanism for the reaction of bromine with alkenes might involve electrophilic addition of Br^+ to the alkene, giving a carbocation intermediate that could undergo further reaction with Br^- to yield the dibromo addition product.

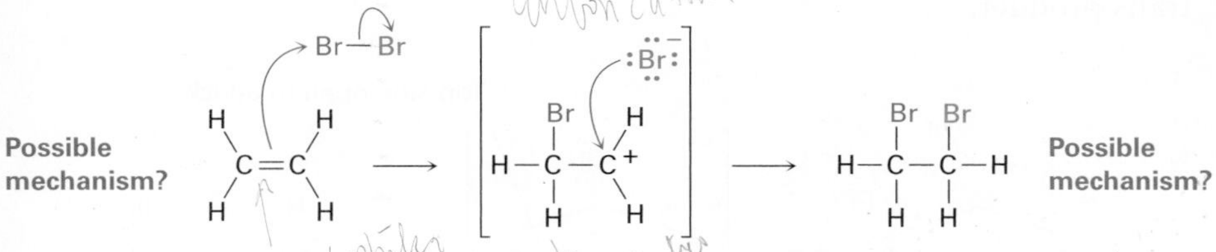

Possible mechanism? *(carbocation intermediate; nucleophile must go to electrophile)* **Possible mechanism?**

Although this mechanism seems plausible, it's not fully consistent with known facts. In particular, it doesn't explain the *stereochemistry* of the addition reaction. That is, the mechanism doesn't tell which product stereoisomer is formed.

When the halogenation reaction is carried out on a cycloalkene, such as cyclopentene, only the *trans* stereoisomer of the dihalide addition product is formed rather than the mixture of cis and trans isomers that might have been expected if a planar carbocation intermediate were involved. We say that the reaction occurs with **anti stereochemistry**, meaning that the two bromine atoms come from opposite faces of the double bond—one from the top face and one from the bottom face.

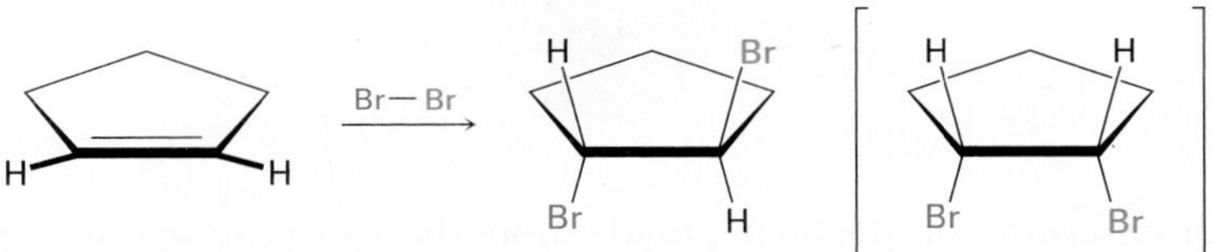

Cyclopentene → *trans*-1,2-Dibromocyclopentane (sole product) [*cis*-1,2-Dibromocyclopentane (Not formed)]

An explanation for the observed stereochemistry of addition was suggested in 1937 by George Kimball and Irving Roberts, who proposed that the reaction intermediate is not a carbocation but is instead a **bromonium ion**, R_2Br^+, formed by electrophilic addition of Br^+ to the alkene. (Similarly, a *chloronium ion* contains a positively charged, divalent chlorine, R_2Cl^+.) The bromonium ion is formed in a single step by interaction of the alkene with Br_2 and simultaneous loss of Br^-.

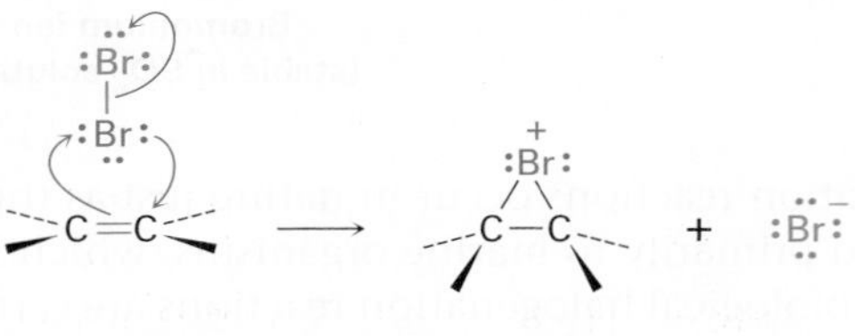

An alkene → **A bromonium ion** + $:Br:^-$

How does the formation of a bromonium ion account for the observed anti stereochemistry of addition to cyclopentene? If a bromonium ion is formed as an intermediate, we can imagine that the large bromine atom might "shield" one side of the molecule. Reaction with Br⁻ ion in the second step could then occur only from the opposite, unshielded side to give trans product.

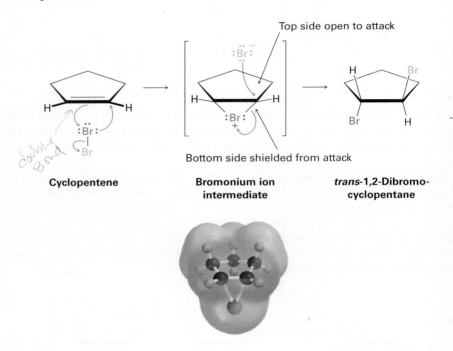

Cyclopentene **Bromonium ion intermediate** ***trans*-1,2-Dibromo-cyclopentane**

The bromonium ion postulate, made more than 75 years ago to explain the stereochemistry of halogen addition to alkenes, is a remarkable example of deductive logic in chemistry. Arguing from experimental results, chemists were able to make a hypothesis about the intimate mechanistic details of alkene electrophilic reactions. Subsequently, strong evidence supporting the mechanism came from the work of George Olah, who prepared and studied *stable* solutions of cyclic bromonium ions in liquid SO_2. There's no question that bromonium ions exist.

Bromonium ion (stable in SO_2 solution)

Alkene halogenation reactions occur in nature just as they do in the laboratory but are limited primarily to marine organisms, which live in a halide-rich environment. The biological halogenation reactions are carried out by enzymes called *haloperoxidases*, which use H_2O_2 to oxidize Br^- or Cl^- ions to a biological equivalent of Br⁺ or Cl⁺. Electrophilic addition to the double bond of a substrate molecule then yields a bromonium or chloronium ion intermediate just as in the laboratory, and reaction with another halide ion completes the process. Halomon, for example, an anticancer pentahalide isolated from red alga, is thought to arise by a route that involves twofold addition of BrCl through the corresponding bromonium ions.

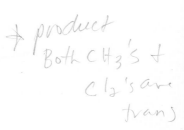

Halomon

Problem 8.3
What product would you expect to obtain from addition of Cl_2 to 1,2-dimethylcyclohexene? Show the stereochemistry of the product.

Problem 8.4
Addition of HCl to 1,2-dimethylcyclohexene yields a mixture of two products. Show the stereochemistry of each, and explain why a mixture is formed.

8.3 Halohydrins from Alkenes: Addition of HOX

Another example of an electrophilic addition is the reaction of alkenes with the hypohalous acids HO–Cl or HO–Br to yield 1,2-halo alcohols, called **halohydrins**. Halohydrin formation doesn't take place by direct reaction of an alkene with HOBr or HOCl, however. Rather, the addition is done indirectly by reaction of the alkene with either Br_2 or Cl_2 in the presence of water.

We saw in the previous section that when Br_2 reacts with an alkene, the cyclic bromonium ion intermediate reacts with the only nucleophile present, Br^- ion. If the reaction is carried out in the presence of an additional nucleophile, however, the intermediate bromonium ion can be intercepted by the added nucleophile and diverted to a different product. In the presence of a high concentration of water, for instance, water competes with Br^- ion as nucleophile and reacts with the bromonium ion intermediate to yield a bromohydrin. The net effect is addition of HO–Br to the alkene by the pathway shown in **Figure 8.1**.

Figure 8.1 | MECHANISM

Bromohydrin formation by reaction of an alkene with **Br₂** in the presence of **water**. Water acts as a nucleophile in step 2 to react with the intermediate bromonium ion.

1. Reaction of the alkene with Br₂ yields a bromonium ion intermediate, as previously discussed.

2. Water acts as a nucleophile, using a lone pair of electrons to open the bromonium ion ring and form a bond to carbon. Since oxygen donates its electrons in this step, it now has the positive charge.

3. Loss of a proton (H⁺) from oxygen then gives H₃O⁺ and the neutral bromohydrin addition product.

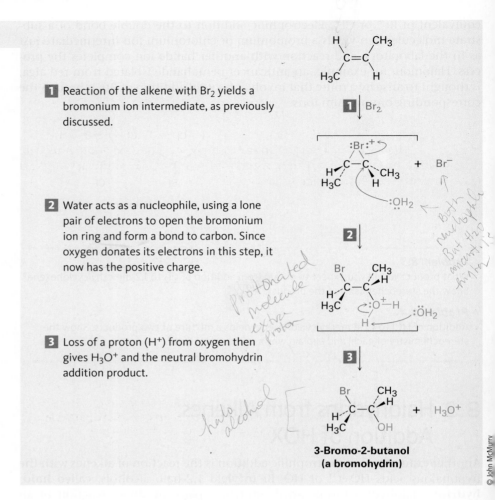

3-Bromo-2-butanol
(a bromohydrin)

In practice, few alkenes are soluble in water, and bromohydrin formation is often carried out in a solvent such as aqueous dimethyl sulfoxide, CH₃SOCH₃ (DMSO), using a reagent called *N*-bromosuccinimide (NBS) as a source of Br₂. NBS is a stable, easily handled compound that slowly decomposes in water to yield Br₂ at a controlled rate. Bromine itself can also be used in the addition reaction, but it is more dangerous and more difficult to handle than NBS.

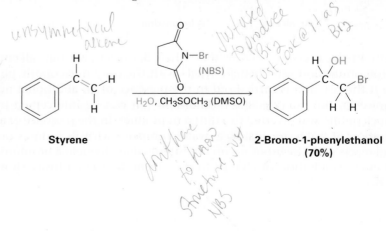

Styrene

2-Bromo-1-phenylethanol
(70%)

Note that the aromatic ring in the above example does not react with Br_2 under the conditions used, even though it appears to contain three carbon–carbon double bonds. As we'll see in Chapter 15, aromatic rings are a good deal more stable and less reactive than might be expected.

There are a number of biological examples of halohydrin formation, particularly in marine organisms. As with halogenation (Section 8.2), halohydrin formation is carried out by haloperoxidases, which function by oxidizing Br^- or Cl^- ions to the corresponding HOBr or HOCl bonded to a metal atom in the enzyme. Electrophilic addition to the double bond of a substrate molecule then yields a bromonium or chloronium ion intermediate, and reaction with water gives the halohydrin. For example:

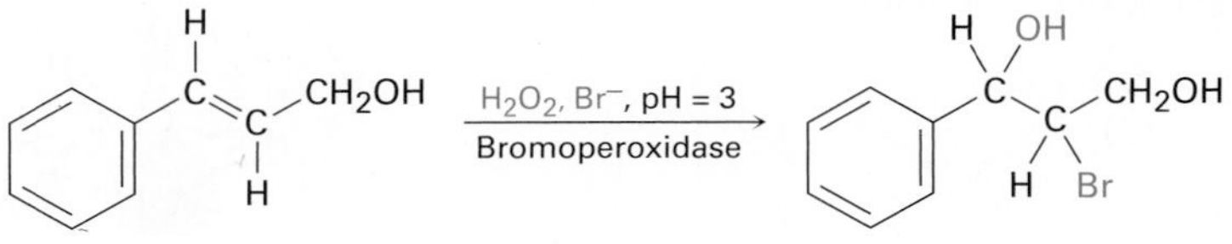

Problem 8.5
What product would you expect from the reaction of cyclopentene with NBS and water? Show the stereochemistry.

Problem 8.6
When an unsymmetrical alkene such as propene is treated with *N*-bromosuccinimide in aqueous dimethyl sulfoxide, the major product has the bromine atom bonded to the less highly substituted carbon atom. Is this Markovnikov or non-Markovnikov orientation? Explain.

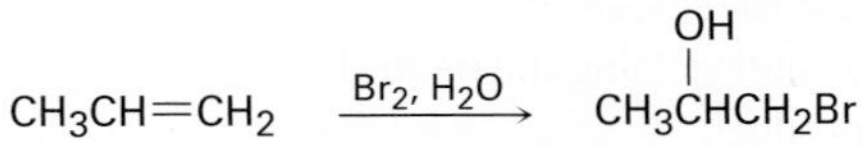

$$CH_3CH=CH_2 \xrightarrow{Br_2, H_2O} CH_3CHCH_2Br \text{ (with OH on middle C)}$$

8.4 Hydration of Alkenes: Addition of H_2O by Oxymercuration

Water adds to alkenes to yield alcohols, a process called *hydration*. The reaction takes place on treatment of the alkene with water and a strong acid catalyst, such as H_2SO_4, by a mechanism similar to that of HX addition. Thus, as shown in **Figure 8.2**, protonation of an alkene double bond yields a carbocation intermediate, which reacts with water to yield a protonated alcohol product, ROH_2^+. Loss of H^+ from this protonated alcohol gives the neutral alcohol and regenerates the acid catalyst.

Figure 8.2 | MECHANISM

Mechanism of the acid-catalyzed hydration of an alkene to yield an alcohol. Protonation of the alkene gives a carbocation intermediate, which reacts with water. The initial product is then deprotonated.

1 A hydrogen atom on the electrophile H_3O^+ is attacked by π electrons from the nucleophilic double bond, forming a new C–H bond. This leaves the other carbon atom with a + charge and a vacant p orbital. Simultaneously, two electrons from the H–O bond move onto oxygen, giving neutral water.

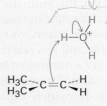

2-Methylpropene

2 The nucleophile H_2O donates an electron pair to the positively charged carbon atom, forming a C–O bond and leaving a positive charge on oxygen in the protonated alcohol addition product.

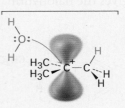

Carbocation

3 Water acts as a base to remove H^+, regenerating H_3O^+ and yielding the neutral alcohol addition product.

Protonated alcohol

2-Methyl-2-propanol

© John McMurry

Acid-catalyzed alkene hydration is particularly suited to large-scale industrial procedures, and approximately 300,000 tons of ethanol is manufactured each year in the United States by hydration of ethylene. The reaction is of little value in the typical laboratory, however, because it requires high temperatures—250 °C in the case of ethylene—and strongly acidic conditions.

$$CH_2=CH_2 + H_2O \xrightarrow[250\ °C]{H_3PO_4\ \text{catalyst}} CH_3CH_2OH$$

Ethylene → Ethanol

Acid-catalyzed hydration of isolated double bonds, although known, is also uncommon in biological pathways. More frequently, biological hydrations require that the double bond be adjacent to a carbonyl group for reaction to proceed. Fumarate, for instance, is hydrated to give malate as one step in the citric acid cycle of food metabolism. Note that the requirement for an adjacent carbonyl group in the addition of water is the same as that we saw in **Section 8.1** for the elimination of water. We'll see the reason for the requirement in **Section 19.13**, but might note for now that the reaction is not an electrophilic addition but instead occurs through a mechanism that involves formation of an anion intermediate followed by protonation by an acid HA.

Fumarate → Anion intermediate → Malate
(H_2O, pH = 7.4, Fumarase ; HA)

When it comes to circumventing problems like those with acid-catalyzed alkene hydrations, laboratory chemists have a great advantage over the cellular "chemists" in living organisms. Laboratory chemists are not constrained to carry out their reactions in water solution; they can choose from any of a large number of solvents. Laboratory reactions don't need to be carried out at a fixed temperature; they can take place over a wide range of temperatures. And laboratory reagents aren't limited to containing carbon, oxygen, nitrogen, and a few other elements; they can contain any element in the periodic table.

In the laboratory, alkenes are often hydrated by the **oxymercuration–demercuration** procedure. Oxymercuration involves electrophilic addition of Hg^{2+} to the alkene on reaction with mercury(II) acetate [$(CH_3CO_2)_2Hg$, often abbreviated $Hg(OAc)_2$] in aqueous tetrahydrofuran (THF) solvent. When the intermediate organomercury compound is then treated with sodium borohydride, $NaBH_4$, demercuration occurs to produce an alcohol. For example:

1-Methylcyclopentene → 1-Methylcyclopentanol (92%)
(1. $Hg(OAc)_2$, H_2O/THF ; 2. $NaBH_4$)

Alkene oxymercuration is closely analogous to halohydrin formation. The reaction is initiated by electrophilic addition of Hg^{2+} (mercuric) ion to the alkene to give an intermediate *mercurinium ion,* whose structure resembles that of a bromonium ion **(Figure 8.3)**. Nucleophilic addition of water as in halohydrin formation, followed by loss of a proton, then yields a stable organomercury product. The final step, demercuration of the organomercury compound by reaction with sodium borohydride, is complex and involves radicals. Note that the regiochemistry of the reaction corresponds to Markovnikov addition of water; that is, the —OH group attaches to the more highly substituted carbon atom, and the —H attaches to the less highly substituted carbon. The hydrogen that replaces mercury in the demercuration step can attach from either side of the molecule depending on the exact circumstances.

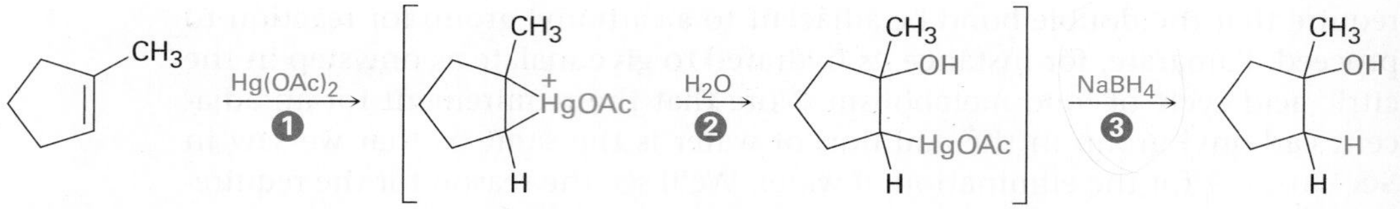

Figure 8.3 Mechanism of the oxymercuration of an alkene to yield an alcohol. (**1**) Electrophilic addition of Hg^{2+} gives a mercurinium ion, which (**2**) reacts with water as in halohydrin formation. Loss of a proton gives an organomercury product, and (**3**) reaction with NaBH$_4$ removes the mercury. The product of the reaction is the more highly substituted alcohol, corresponding to Markovnikov regiochemistry.

Problem 8.7
What products would you expect from oxymercuration–demercuration of the following alkenes?
(a) $CH_3CH_2CH_2CH=CH_2$
(b) $CH_3C(CH_3)=CHCH_2CH_3$

Problem 8.8
From what alkenes might the following alcohols have been prepared?
(a) $CH_3C(OH)(CH_3)CH_2CH_2CH_2CH_3$
(b) cyclohexyl-CH(OH)-CH$_3$

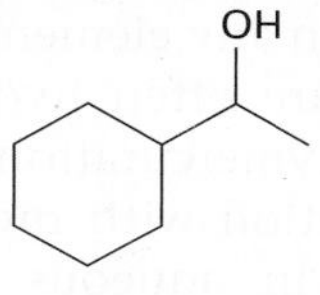

8.5 Hydration of Alkenes: Addition of H$_2$O by Hydroboration

In addition to the oxymercuration–demercuration method, which yields the Markovnikov product, a complementary method that yields the non-Markovnikov product is also useful. Discovered in 1959 by H.C. Brown and called **hydroboration**, the reaction involves addition of a B–H bond of borane, BH$_3$, to an alkene to yield an organoborane intermediate, RBH$_2$. Oxidation of the organoborane by reaction with basic hydrogen peroxide, H$_2$O$_2$, then gives an alcohol. For example:

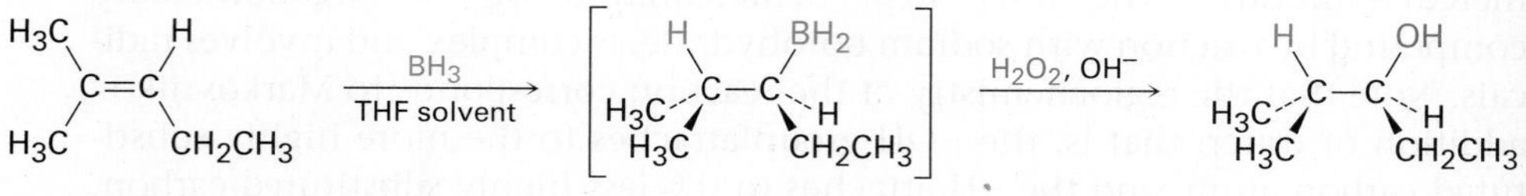

2-Methyl-2-pentene → (BH$_3$, THF solvent) → Organoborane intermediate → (H$_2$O$_2$, OH$^-$) → 2-Methyl-3-pentanol

Borane is very reactive as a Lewis acid because the boron atom has only six electrons in its valence shell. In tetrahydrofuran solution, BH_3 accepts an electron pair from a solvent molecule in a Lewis acid–base reaction to complete its octet and form a stable BH_3–THF complex.

Borane **THF** **BH_3–THF complex**

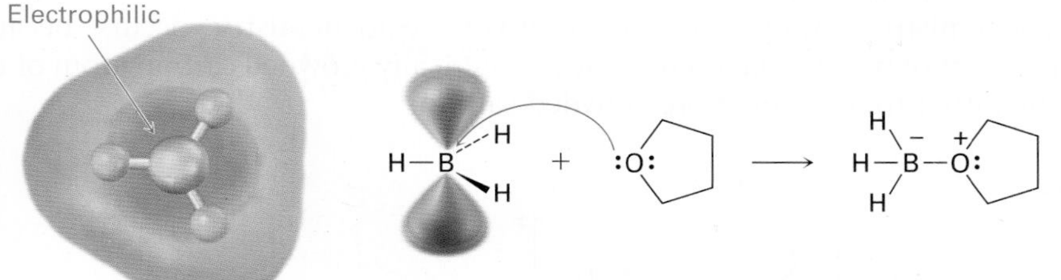

When an alkene reacts with BH_3 in THF solution, rapid addition to the double bond occurs three times and a trialkylborane, R_3B, is formed. For example, 1 molar equivalent of BH_3 adds to 3 molar equivalents of cyclohexene to yield tricyclohexylborane. When tricyclohexylborane is then treated with aqueous hydrogen H_2O_2 in basic solution, an oxidation takes place. The three C–B bonds are broken, –OH groups bond to the three carbons, and 3 equivalents of cyclohexanol are produced. The net effect of the two-step hydroboration–oxidation sequence is hydration of the alkene double bond.

Cyclohexene **Cyclohexanol (87%)**

Tricyclohexylborane

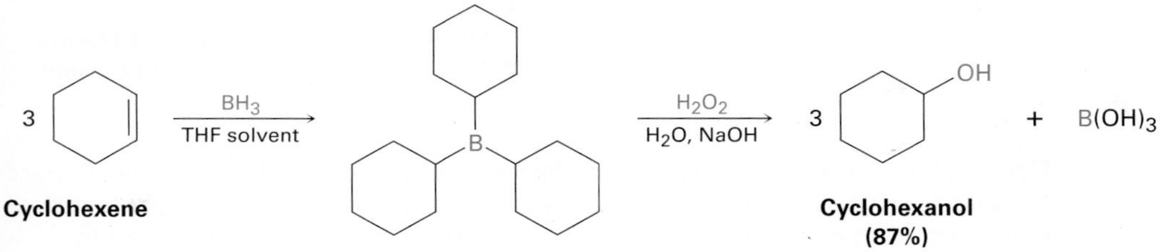

One of the features that makes the hydroboration reaction so useful is the regiochemistry that results when an unsymmetrical alkene is hydroborated. For example, hydroboration–oxidation of 1-methylcyclopentene yields *trans*-2-methylcyclopentanol. Boron and hydrogen add to the alkene from the same face of the double bond—that is, with **syn stereochemistry**, the opposite of anti—with boron attaching to the less highly substituted carbon. During the oxidation step, the boron is replaced by an –OH with the same stereochemistry, resulting in an overall syn non-Markovnikov addition of water. This stereochemical result is particularly useful because it is complementary to the Markovnikov regiochemistry observed for oxymercuration–demercuration.

1-Methylcyclopentene **Organoborane intermediate** ***trans*-2-Methylcyclopentanol (85% yield)**

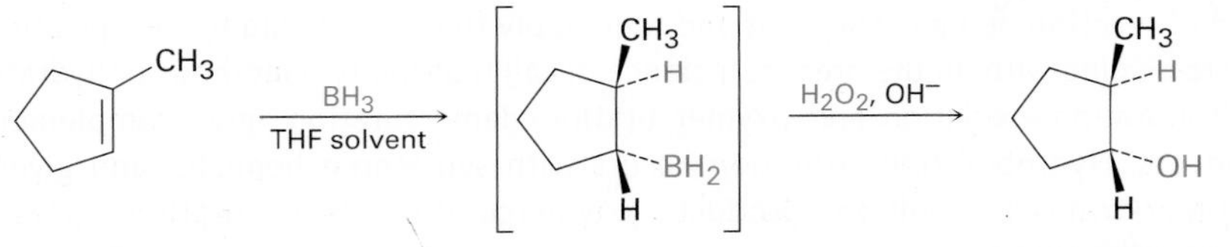

Why does alkene hydroboration take place with syn, non-Markovnikov regiochemistry to yield the less highly substituted alcohol? Hydroboration differs from many other alkene addition reactions in that it occurs in a single step without a carbocation intermediate (**Figure 8.4**). Because both C–H and C–B bonds form at the same time and from the same face of the alkene, syn stereochemistry results. Non-Markovnikov regiochemistry occurs because attachment of boron is favored at the less sterically crowded carbon atom of the alkene rather than at the more crowded carbon.

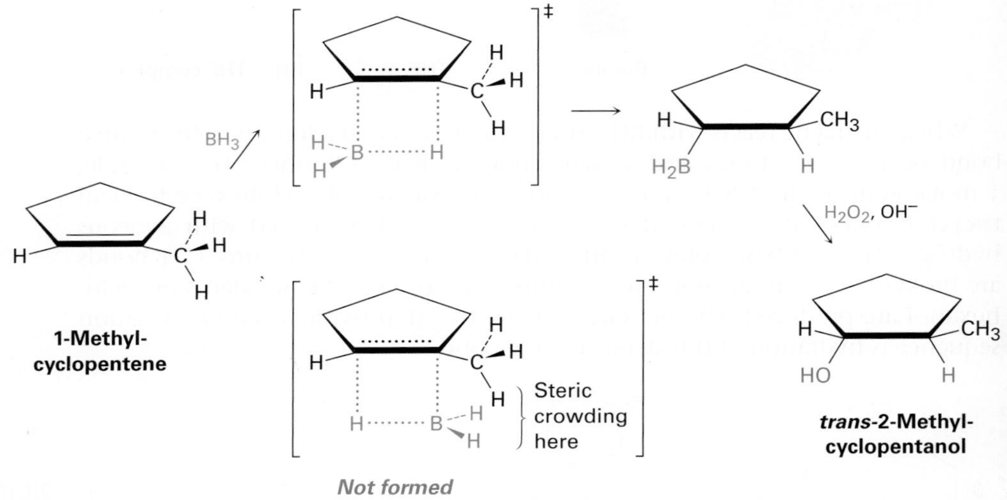

Figure 8.4 Mechanism of alkene hydroboration. The reaction occurs in a single step in which both C–H and C–B bonds form at the same time and on the same face of the double bond. The lower energy, more rapidly formed transition state is the one with less steric crowding, leading to non-Markovnikov regiochemistry.

Worked Example 8.1 Predicting the Products Formed in a Reaction

What products would you obtain from reaction of 2,4-dimethyl-2-pentene with:
(a) BH_3, followed by H_2O_2, OH^- (b) $Hg(OAc)_2$, followed by $NaBH_4$

Strategy
When predicting the product of a reaction, you have to recall what you know about the kind of reaction being carried out and then apply that knowledge to the specific case you're dealing with. In the present instance, recall that the two methods of hydration—hydroboration–oxidation and oxymercuration–demercuration—give complementary products. Hydroboration–oxidation occurs with syn stereochemistry and gives the non-Markovnikov addition product; oxymercuration–demercuration gives the Markovnikov product.

Solution

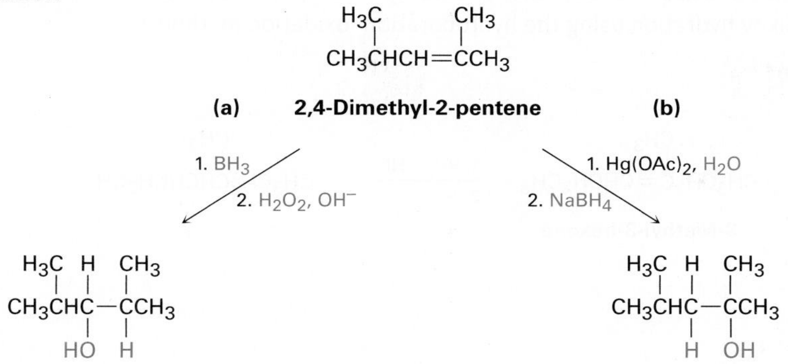

(a)
$$CH_3CHCH=CCH_3$$
with substituents H_3C and CH_3

2,4-Dimethyl-2-pentene

Route (a): 1. BH_3; 2. H_2O_2, OH^- →

$$CH_3CHC-CCH_3$$ with H_3C, H, CH_3 on top and HO, H on bottom

2,4-Dimethyl-3-pentanol

Route (b): 1. $Hg(OAc)_2$, H_2O; 2. $NaBH_4$ →

$$CH_3CHC-CCH_3$$ with H_3C, H, CH_3 on top and H, OH on bottom

2,4-Dimethyl-2-pentanol

Worked Example 8.2

Synthesizing an Alcohol

How might you prepare the following alcohol?

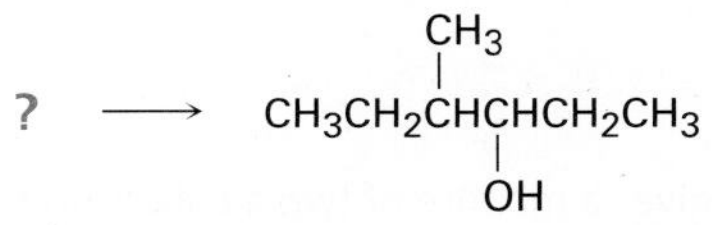

$$? \longrightarrow CH_3CH_2\underset{OH}{\overset{CH_3}{C}}HCHCH_2CH_3$$

Strategy

Problems that require the synthesis of a specific target molecule should always be worked backward. Look at the target, identify its functional group(s), and ask yourself "What are the methods for preparing that functional group?" In the present instance, the target molecule is a secondary alcohol (R_2CHOH), and we've seen that alcohols can be prepared from alkenes by either hydroboration–oxidation or oxymercuration. The —OH bearing carbon in the product must have been a double-bond carbon in the alkene reactant, so there are two possibilities: 4-methyl-2-hexene and 3-methyl-3-hexene.

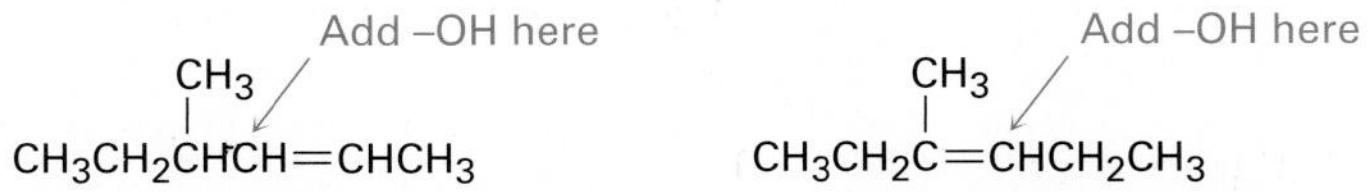

Add —OH here
$$CH_3CH_2\overset{CH_3}{C}HCH=CHCH_3$$
4-Methyl-2-hexene

Add —OH here
$$CH_3CH_2\overset{CH_3}{C}=CHCH_2CH_3$$
3-Methyl-3-hexene

4-Methyl-2-hexene has a disubstituted double bond, RCH=CHR′, and will probably give a mixture of two alcohols with either hydration method since Markovnikov's rule does not apply to symmetrically substituted alkenes. 3-Methyl-3-hexene, however, has a trisubstituted double bond, and should give only the desired product on non-Markovnikov hydration using the hydroboration–oxidation method.

Solution

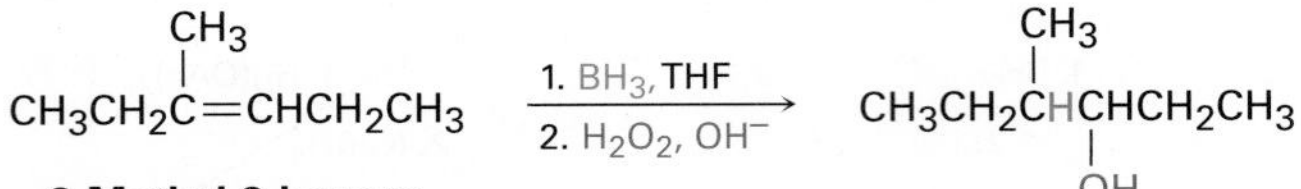

$$\underset{\textbf{3-Methyl-3-hexene}}{CH_3CH_2\underset{\underset{CH_3}{|}}{C}=CHCH_2CH_3} \xrightarrow[\text{2. } H_2O_2, OH^-]{\text{1. } BH_3, THF} CH_3CH_2\underset{\underset{OH}{|}}{\overset{\overset{CH_3}{|}}{C}H}CHCH_2CH_3$$

Problem 8.9

Show the structures of the products you would obtain by hydroboration–oxidation of the following alkenes:

(a) 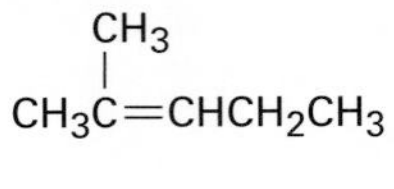 $CH_3\underset{\underset{}{|}}{\overset{\overset{CH_3}{|}}{C}}=CHCH_2CH_3$

(b)

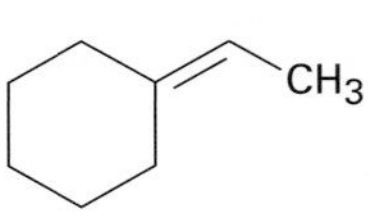

Problem 8.10

What alkenes might be used to prepare the following alcohols by hydroboration–oxidation?

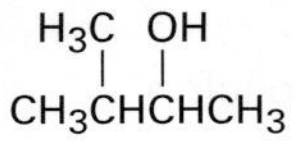

(a) 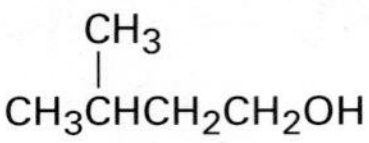 $CH_3\underset{\underset{}{|}}{\overset{\overset{CH_3}{|}}{C}H}CH_2CH_2OH$

(b) $CH_3\overset{\overset{H_3C}{|}}{C}H\overset{\overset{OH}{|}}{C}HCH_3$

(c) cyclohexyl—CH_2OH

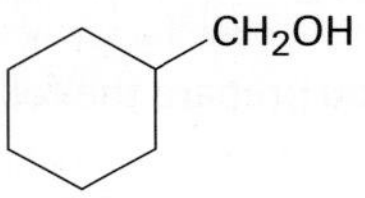

Problem 8.11

The following cycloalkene gives a mixture of two alcohols on hydroboration followed by oxidation. Draw the structures of both, and explain the result.

8.6 Reduction of Alkenes: Hydrogenation

Alkenes react with H_2 in the presence of a metal catalyst such as palladium or platinum to yield the corresponding saturated alkane addition products. We describe the result by saying that the double bond has been **hydrogenated**, or *reduced*. Note that the word *reduction* is used somewhat differently

8.6 | Reduction of Alkenes: Hydrogenation

in organic chemistry from what you might have learned previously. In general chemistry, a reduction is defined as the gain of one or more electrons by an atom. In organic chemistry, however, a **reduction** is a reaction that results in a gain of electron density by carbon, caused either by bond formation between carbon and a less electronegative atom—usually hydrogen—or by bond-breaking between carbon and a more electronegative atom—usually oxygen, nitrogen, or a halogen. We'll explore the topic in more detail in **Section 10.8**.

Reduction   Increases electron density on carbon by:
– forming this: C–H
– or breaking one of these: C–O   C–N   C–X

A reduction:

$$\text{C=C} + H_2 \xrightarrow{\text{Catalyst}} \text{H}_3\text{C-CH}_3$$

An alkene → An alkane

Platinum and palladium are the most common laboratory catalysts for alkene hydrogenations. Palladium is normally used as a very fine powder "supported" on an inert material such as charcoal (Pd/C) to maximize surface area. Platinum is normally used as PtO_2, a reagent known as *Adams' catalyst* after its discoverer, Roger Adams.

Catalytic hydrogenation, unlike most other organic reactions, is a *heterogeneous* process rather than a homogeneous one. That is, the hydrogenation reaction does not occur in a homogeneous solution but instead takes place on the surface of solid catalyst particles. Hydrogenation usually occurs with syn stereochemistry: both hydrogens add to the double bond from the same face.

1,2-Dimethylcyclohexene → (H$_2$, PtO$_2$, CH$_3$CO$_2$H solvent) → *cis*-1,2-Dimethylcyclohexane (82%)

As shown in **Figure 8.5**, hydrogenation begins with adsorption of H_2 onto the catalyst surface. Complexation between catalyst and alkene then occurs as a vacant orbital on the metal interacts with the filled alkene π orbital. In the final steps, hydrogen is inserted into the double bond and the saturated product diffuses away from the catalyst. The stereochemistry of hydrogenation is syn because both hydrogens add to the double bond from the same catalyst surface.

Figure 8.5 | MECHANISM

Mechanism of alkene hydrogenation. The reaction takes place with syn stereochemistry on the surface of insoluble catalyst particles.

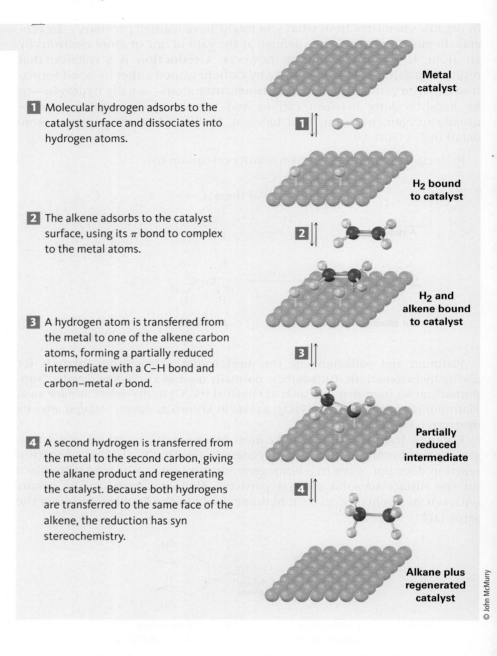

1 Molecular hydrogen adsorbs to the catalyst surface and dissociates into hydrogen atoms.

2 The alkene adsorbs to the catalyst surface, using its π bond to complex to the metal atoms.

3 A hydrogen atom is transferred from the metal to one of the alkene carbon atoms, forming a partially reduced intermediate with a C–H bond and carbon–metal σ bond.

4 A second hydrogen is transferred from the metal to the second carbon, giving the alkane product and regenerating the catalyst. Because both hydrogens are transferred to the same face of the alkene, the reduction has syn stereochemistry.

© John McMurry

An interesting feature of catalytic hydrogenation is that the reaction is extremely sensitive to the steric environment around the double bond. As a result, the catalyst usually approaches only the more accessible face of an alkene, giving rise to a single product. In α-pinene, for example, one of the methyl groups attached to the four-membered ring hangs over the top face of the double bond and blocks approach of the hydrogenation catalyst from that side. Reduction therefore occurs exclusively from the bottom face to yield the product shown.

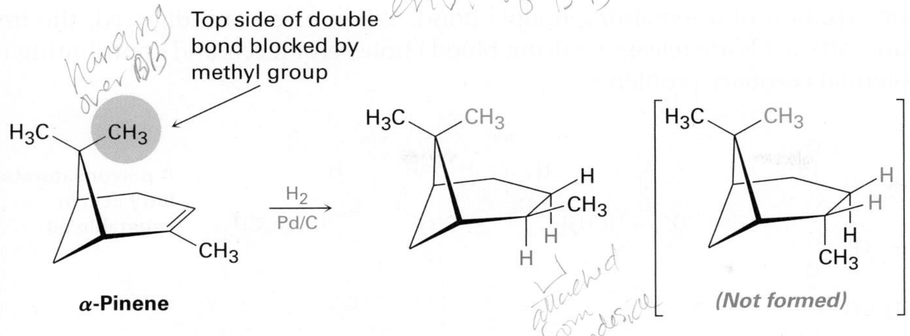

8.6 | Reduction of Alkenes: Hydrogenation

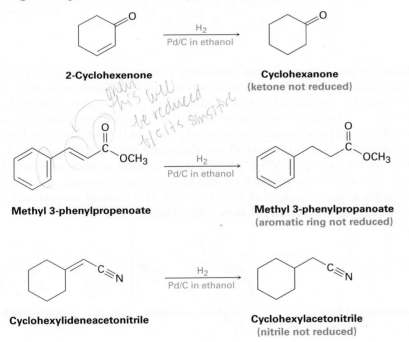

Alkenes are much more reactive than most other unsaturated functional groups toward catalytic hydrogenation, and the reaction is therefore quite selective. Other functional groups, such as aldehydes, ketones, esters, and nitriles, often survive alkene hydrogenation conditions unchanged, although reaction with these groups does occur under more vigorous conditions. Note particularly in the hydrogenation of methyl 3-phenylpropenoate shown below that the aromatic ring is not reduced by hydrogen and palladium even though it contains apparent double bonds.

In addition to its usefulness in the laboratory, catalytic hydrogenation is also important in the food industry, where unsaturated vegetable oils are reduced on a large scale to produce the saturated fats used in margarine and cooking products **(Figure 8.6)**. As we'll see in **Section 27.1**, vegetable oils are triesters of glycerol, HOCH$_2$CH(OH)CH$_2$OH, with three long-chain carboxylic acids called *fatty acids*. The fatty acids are generally polyunsaturated, and their double bonds have cis stereochemistry. Complete hydrogenation yields the corresponding saturated fatty acids, but incomplete hydrogenation often results in partial cis–trans

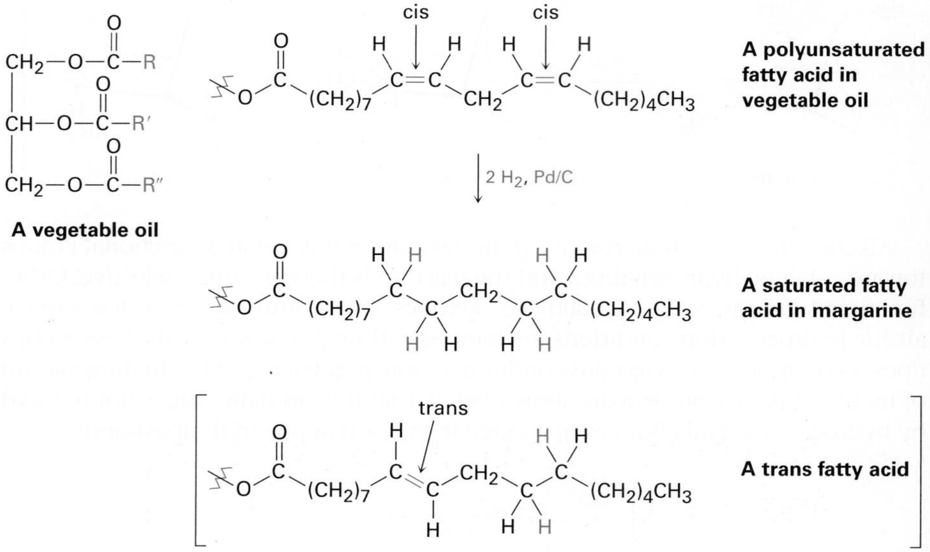

Figure 8.6 Catalytic hydrogenation of polyunsaturated fats leads to saturated products, along with a small amount of isomerized trans fats.

isomerization of a remaining double bond. When eaten and digested, the free trans fatty acids are released, raising blood cholesterol levels and contributing to potential coronary problems.

Double-bond reductions are extremely common in biological pathways, although the mechanism of the process is of course different from that of laboratory catalytic hydrogenation over palladium. As with biological hydrations (**Section 8.4**), biological reductions usually occur in two steps and require that the double bond be adjacent to a carbonyl group. In the first step, the biological reducing agent NADPH (reduced nicotinamide adenine dinucleotide phosphate), adds a hydride ion (H:$^-$) to the double bond to give an anion. In the second, the anion is protonated by acid HA, leading to overall addition of H_2. An example is the reduction of *trans*-crotonyl ACP to yield butyryl ACP, a step involved in the biosynthesis of fatty acids (**Figure 8.7**).

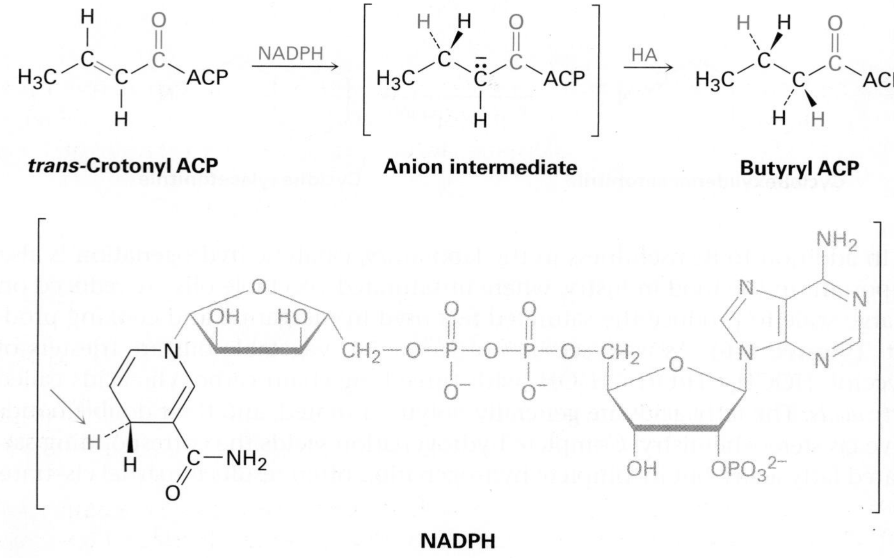

Figure 8.7 Reduction of the carbon–carbon double bond in *trans*-crotonyl ACP, a step in the biosynthesis of fatty acids. **One hydrogen** is delivered from NADPH as a hydride ion, H:$^-$; the **other hydrogen** is delivered by protonation of the anion intermediate with an acid, HA.

Problem 8.12
What product would you obtain from catalytic hydrogenation of the following alkenes?

(a)
$$CH_3C(CH_3)=CHCH_2CH_3$$

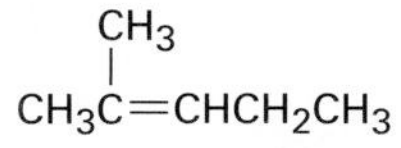

(b) 1-methyl-1-... cyclopentene with CH$_3$, CH$_3$ substituents

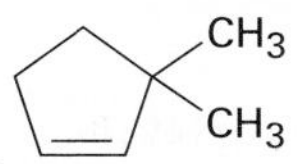

(c)
3-(2-methylpropan-2-yl)cyclohexene = the same drawn in another way

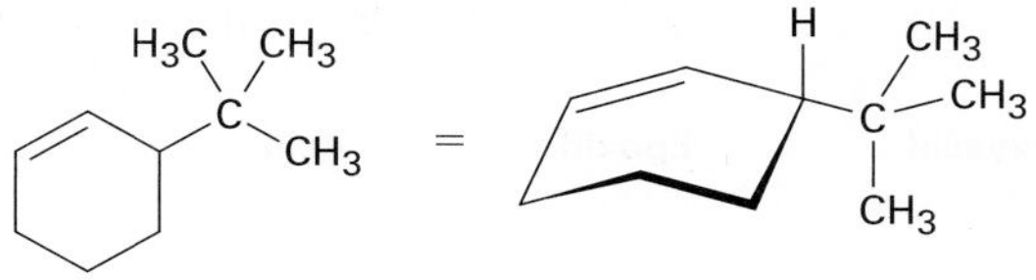

8.7 Oxidation of Alkenes: Epoxidation and Hydroxylation

Like the word *reduction* used in the previous section for the addition of hydrogen to a double bond, the word *oxidation* has a slightly different meaning in organic chemistry from what you might have previously learned. In general chemistry, an oxidation is defined as the loss of one or more electrons by an atom. In organic chemistry, however, an **oxidation** is a reaction that results in a loss of electron density by carbon, caused either by bond formation between carbon and a more electronegative atom—usually oxygen, nitrogen, or a halogen—or by bond-breaking between carbon and a less electronegative atom—usually hydrogen. Note that an *oxidation* often adds oxygen, while a *reduction* often adds hydrogen.

Oxidation Decreases electron density on carbon by:
– forming one of these: C–O   C–N   C–X
– or breaking this: C–H

In the laboratory, alkenes are oxidized to give *epoxides* on treatment with a peroxyacid, RCO₃H, such as *meta*-chloroperoxybenzoic acid. An **epoxide**, also called an *oxirane*, is a cyclic ether with an oxygen atom in a three-membered ring. For example:

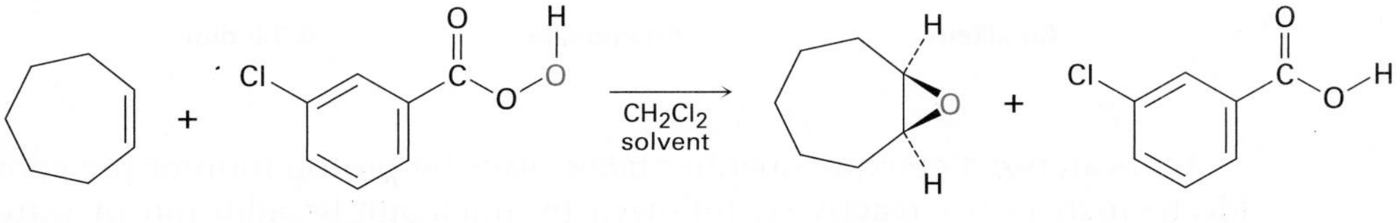

Cycloheptene + *meta*-Chloroperoxybenzoic acid →(CH₂Cl₂ solvent)→ 1,2-Epoxycycloheptane + *meta*-Chlorobenzoic acid

Peroxyacids transfer an oxygen atom to the alkene with syn stereochemistry—both C–O bonds form on the same face of the double bond—through a one-step mechanism without intermediates. The oxygen atom farthest from the carbonyl group is the one transferred.

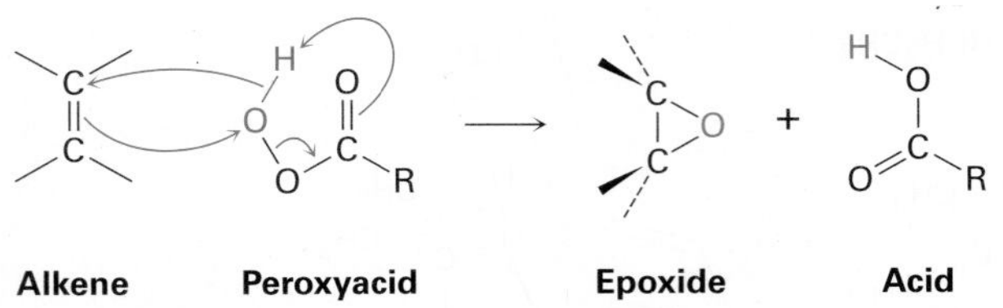

Alkene   **Peroxyacid**   **Epoxide**   **Acid**

Another method for the synthesis of epoxides is through the use of halohydrins, prepared by electrophilic addition of HO−X to alkenes (Section 8.3). When a halohydrin is treated with base, HX is eliminated and an epoxide is produced.

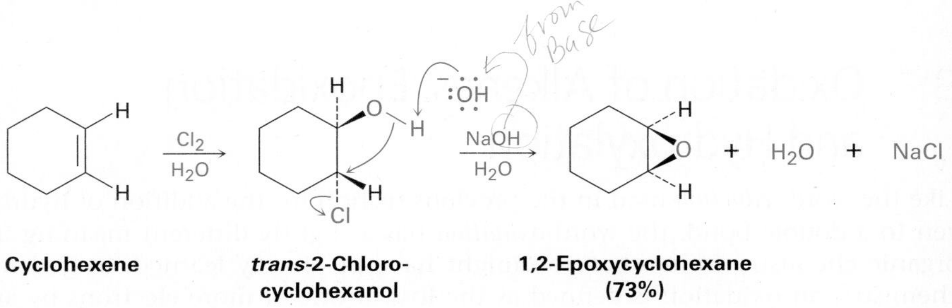

Cyclohexene   ***trans*-2-Chloro-cyclohexanol**   **1,2-Epoxycyclohexane (73%)**

Epoxides undergo an acid-catalyzed ring-opening reaction with water (a *hydrolysis*) to give the corresponding 1,2-dialcohol, or *diol*, also called a **glycol**. Thus, the net result of the two-step alkene epoxidation/hydrolysis is **hydroxylation**—the addition of an −OH group to each of the two double-bond carbons. In fact, approximately 18 million metric tons of ethylene glycol, $HOCH_2CH_2OH$, most of it used for automobile antifreeze, is produced worldwide each year by epoxidation of ethylene followed by hydrolysis.

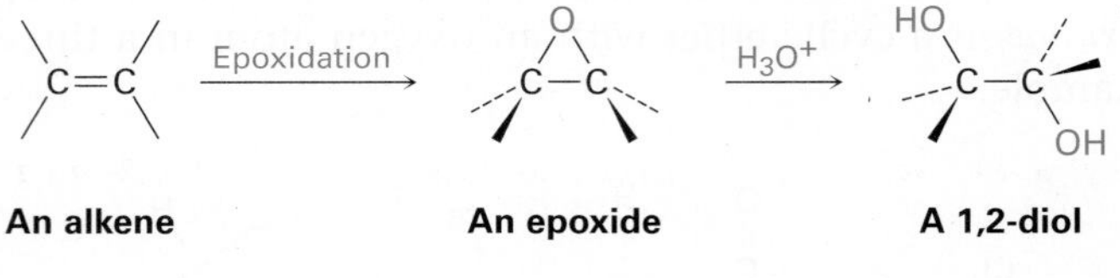

An alkene   **An epoxide**   **A 1,2-diol**

Acid-catalyzed epoxide opening takes place by protonation of the epoxide to increase its reactivity, followed by nucleophilic addition of water. This nucleophilic addition is analogous to the final step of alkene bromination, in which a cyclic bromonium ion is opened by a nucleophile (Section 8.2). That is, a *trans*-1,2-diol results when an epoxycycloalkane is

8.7 | Oxidation of Alkenes: Epoxidation and Hydroxylation

opened by aqueous acid, just as a *trans*-1,2-dibromide results when a cycloalkene is brominated. We'll look at epoxide chemistry in more detail in **Section 18.6**.

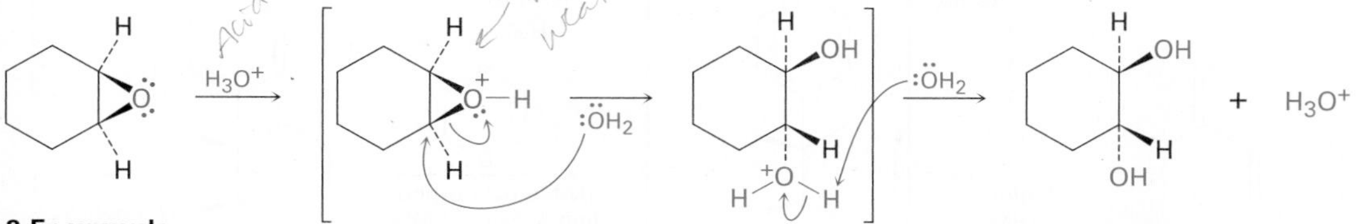

1,2-Epoxycyclohexane → (H$_3$O$^+$) → [protonated epoxide intermediate, *very weak bond*] → → **trans-1,2-Cyclohexanediol (86%)** + H$_3$O$^+$

Recall the following:

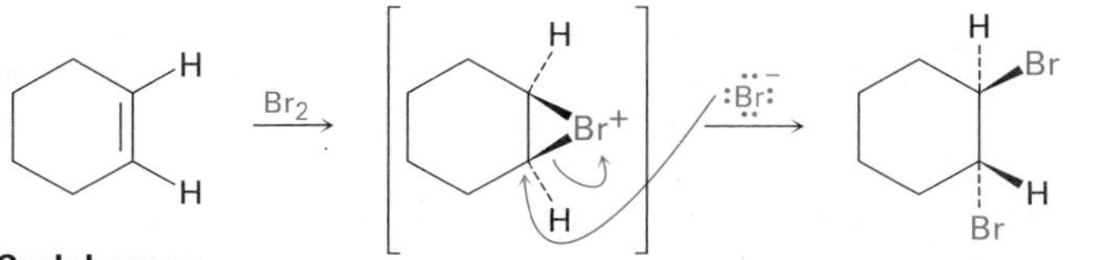

Cyclohexene → (Br$_2$) → [bromonium ion intermediate] → **trans-1,2-Dibromocyclohexane**

Hydroxylation can be carried out directly without going through an intermediate epoxide by treating an alkene with osmium tetroxide, OsO$_4$. The reaction occurs with syn stereochemistry and does not involve a carbocation intermediate. Instead, it takes place through an intermediate cyclic *osmate*, which is formed in a single step by addition of OsO$_4$ to the alkene. This cyclic osmate is then cleaved using aqueous sodium bisulfite, NaHSO$_3$.

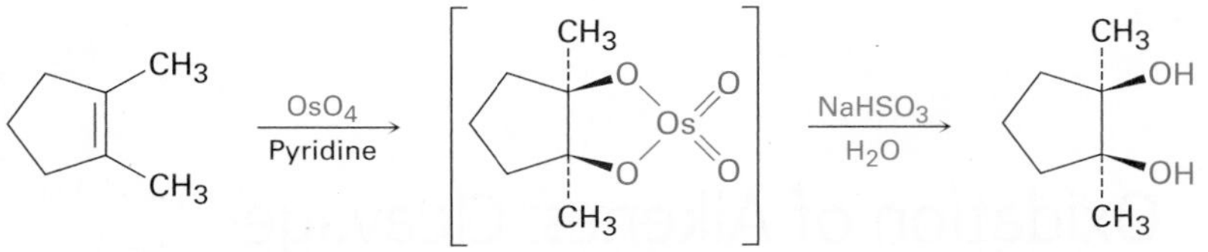

1,2-Dimethylcyclopentene → (OsO$_4$, Pyridine) → **A cyclic osmate intermediate** → (NaHSO$_3$, H$_2$O) → ***cis*-1,2-Dimethyl-1,2-cyclopentanediol (87%)**

Because OsO$_4$ is both very expensive and *very* toxic, the reaction is usually carried out using only a small, catalytic amount of OsO$_4$ in the presence of a stoichiometric amount of a safe and inexpensive co-oxidant such as *N*-methylmorpholine *N*-oxide, abbreviated NMO. The initially formed osmate intermediate reacts rapidly with NMO to yield the product diol plus

N-methylmorpholine and reoxidized OsO₄, which reacts with more alkene in a catalytic cycle.

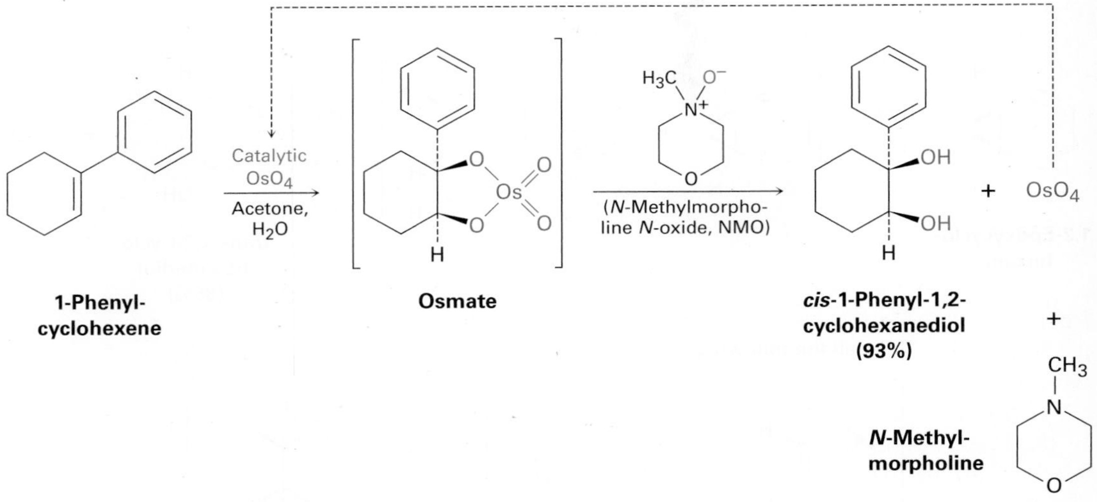

1-Phenyl-cyclohexene → (Catalytic OsO₄, Acetone, H₂O) → **Osmate** → (N-Methylmorpholine N-oxide, NMO) → ***cis*-1-Phenyl-1,2-cyclohexanediol (93%)** + OsO₄ + **N-Methylmorpholine**

Problem 8.13
What product would you expect from reaction of *cis*-2-butene with *meta*-chloroperoxybenzoic acid? Show the stereochemistry.

Problem 8.14
How would you prepare each of the following compounds starting with an alkene?
(a) cyclohexane with H, OH (up) and OH, CH₃ (down)
(b) 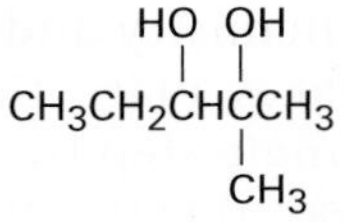 HO OH
   CH₃CH₂CHCCH₃
           |
           CH₃
(c) HO OH
    HOCH₂CHCHCH₂OH

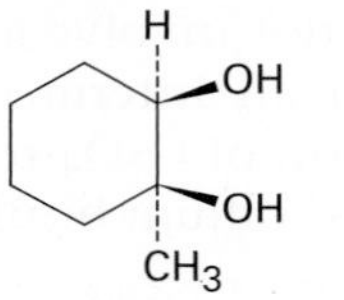

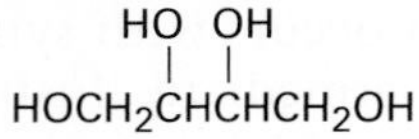

8.8 Oxidation of Alkenes: Cleavage to Carbonyl Compounds

In all the alkene addition reactions we've seen thus far, the carbon–carbon double bond has been converted into a single bond but the carbon skeleton has been unchanged. There are, however, powerful oxidizing reagents that will cleave C=C bonds and produce two carbonyl-containing fragments.

Ozone (O₃) is perhaps the most useful double-bond cleavage reagent. Prepared by passing a stream of oxygen through a high-voltage electrical discharge, ozone adds rapidly to a C=C bond at low temperature to give a cyclic intermediate called a *molozonide*. Once formed, the molozonide spontaneously rearranges to form an **ozonide**. Although we won't study the mechanism of

8.8 | Oxidation of Alkenes: Cleavage to Carbonyl Compounds

this rearrangement in detail, it involves the molozonide coming apart into two fragments that then recombine in a different way.

$$3 O_2 \xrightarrow{\text{Electric discharge}} 2 O_3$$

An alkene →[O_3, CH_2Cl_2, −78 °C]→ A molozonide → An ozonide →[Zn, CH_3CO_2H/H_2O]→ C=O + O=C

Low-molecular-weight ozonides are explosive and are therefore not isolated. Instead, the ozonide is immediately treated with a reducing agent such as zinc metal in acetic acid to convert it to carbonyl compounds. The net result of the ozonolysis/reduction sequence is that the C=C bond is cleaved and an oxygen atom becomes doubly bonded to each of the original alkene carbons. If an alkene with a tetrasubstituted double bond is ozonized, two ketone fragments result; if an alkene with a trisubstituted double bond is ozonized, one ketone and one aldehyde result; and so on.

Isopropylidenecyclohexane (tetrasubstituted) →[1. O_3; 2. Zn, H_3O^+]→ Cyclohexanone + Acetone (CH_3COCH_3)

84%; two ketones

$CH_3(CH_2)_7CH=CH(CH_2)_7COCH_3$
Methyl 9-octadecenoate (disubstituted) →[1. O_3; 2. Zn, H_3O^+]→ $CH_3(CH_2)_7CHO$ (Nonanal) + $HC(CH_2)_7COCH_3$ (Methyl 9-oxononanoate)

78%; two aldehydes

Several oxidizing reagents other than ozone also cause double-bond cleavage, although the reaction is not often used. For example, potassium permanganate ($KMnO_4$) in neutral or acidic solution cleaves alkenes to give carbonyl-containing products. If hydrogens are present on the double bond, carboxylic acids are produced; if two hydrogens are present on one carbon, CO_2 is formed.

$CH_3CHCH_2CH_2CH_2CHCH=CH_2$ (with CH_3 substituents)
3,7-Dimethyl-1-octene →[$KMnO_4$, H_3O^+]→ $CH_3CHCH_2CH_2CH_2CHCOOH$
2,6-Dimethylheptanoic acid (45%) + CO_2

In addition to direct cleavage with ozone or $KMnO_4$, an alkene can also be cleaved in a two-step process by initial hydroxylation to a 1,2-diol, as discussed in the previous section, followed by treatment of the diol with periodic acid, HIO_4. If the two −OH groups are in an open chain, two carbonyl compounds result. If the two −OH groups are on a ring, a single, open-chain dicarbonyl compound is formed. As indicated in the following examples, the cleavage reaction takes place through a cyclic periodate intermediate.

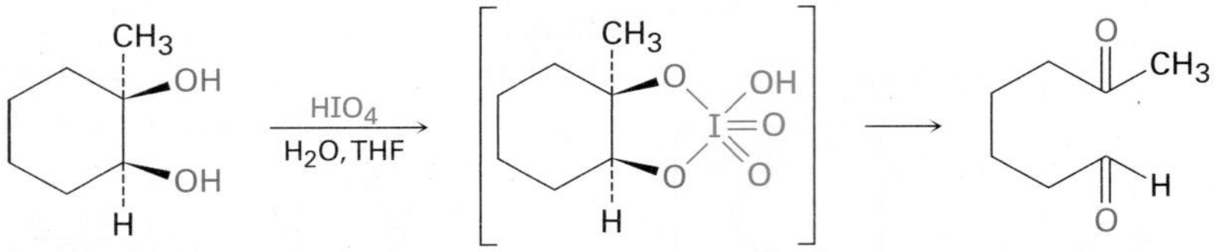

A 1,2-diol → [Cyclic periodate intermediate] → 6-Oxoheptanal (86%)

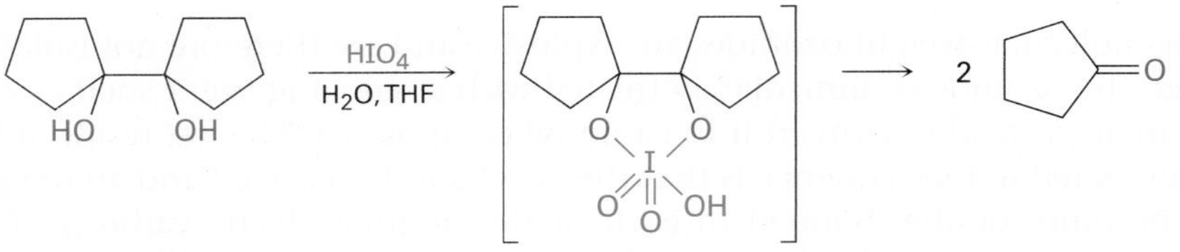

A 1,2-diol → [Cyclic periodate intermediate] → 2 Cyclopentanone (81%)

Worked Example 8.3

Predicting the Reactant in an Ozonolysis Reaction

What alkene would yield a mixture of cyclopentanone and propanal on treatment with ozone followed by reduction with zinc?

? →(1. O$_3$; 2. Zn, acetic acid)→ cyclopentanone=O + CH$_3$CH$_2$CHO

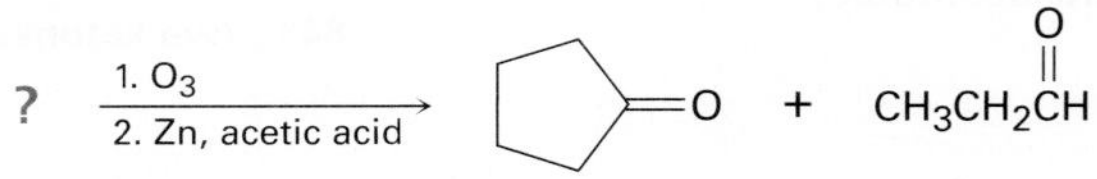

Strategy
Reaction of an alkene with ozone, followed by reduction with zinc, cleaves the C=C bond and gives two carbonyl-containing fragments. That is, the C=C bond becomes two C=O bonds. Working backward from the carbonyl-containing products, the alkene precursor can be found by removing the oxygen from each product and joining the two carbon atoms to form a double bond.

Solution

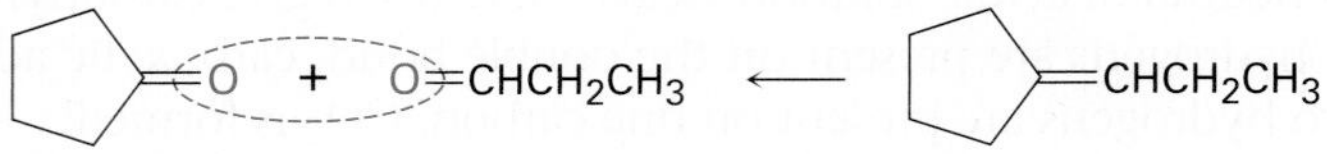

cyclopentanone=O + O=CHCH$_2$CH$_3$ ← cyclopentylidene=CHCH$_2$CH$_3$

Problem 8.15
What products would you expect from reaction of 1-methylcyclohexene with the following reagents?
(a) Aqueous acidic KMnO$_4$ **(b)** O$_3$, followed by Zn, CH$_3$CO$_2$H

Problem 8.16
Propose structures for alkenes that yield the following products on reaction with ozone followed by treatment with Zn:
(a) (CH$_3$)$_2$C=O + H$_2$C=O   **(b)** 2 equiv CH$_3$CH$_2$CH=O

8.9 Addition of Carbenes to Alkenes: Cyclopropane Synthesis

Yet another kind of alkene addition is the reaction with a *carbene* to yield a cyclopropane. A **carbene, $R_2C:$**, is a neutral molecule containing a divalent carbon with only six electrons in its valence shell. It is therefore highly reactive and is generated only as a reaction intermediate, rather than as an isolable molecule. Because they're electron-deficient, carbenes behave as electrophiles and react with nucleophilic C=C bonds. The reaction occurs in a single step without intermediates.

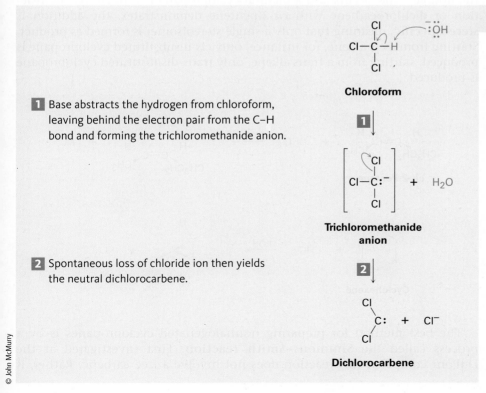

One of the simplest methods for generating a substituted carbene is by treatment of chloroform, $CHCl_3$, with a strong base such as KOH. As shown in **Figure 8.8**, loss of a proton from $CHCl_3$ gives the trichloromethanide anion, $^-:CCl_3$, which spontaneously expels a Cl^- ion to yield dichlorocarbene, $:CCl_2$.

Figure 8.8 | MECHANISM

Mechanism of the formation of dichlorocarbene by reaction of chloroform with strong base. Deprotonation of $CHCl_3$ gives the trichloromethanide anion, $^-:CCl_3$, which spontaneously expels a Cl^- ion.

1 Base abstracts the hydrogen from chloroform, leaving behind the electron pair from the C–H bond and forming the trichloromethanide anion.

2 Spontaneous loss of chloride ion then yields the neutral dichlorocarbene.

The dichlorocarbene carbon atom is sp^2-hybridized, with a vacant p orbital extending above and below the plane of the three atoms and with an unshared pair of electrons occupying the third sp^2 lobe. Note that this electronic description of dichlorocarbene is similar to that of a carbocation (Section 7.9) with respect to both the sp^2 hybridization of carbon and the vacant p orbital. Electrostatic potential maps further show this similarity (Figure 8.9).

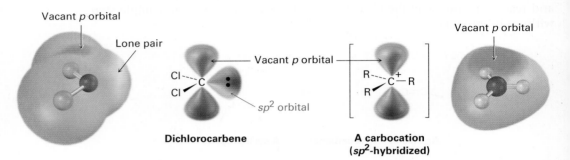

Figure 8.9 The structure of dichlorocarbene. Electrostatic potential maps show how the **positive region** coincides with the empty p orbital in both dichlorocarbene and a carbocation (CH_3^+). The **negative region** in the dichlorocarbene map coincides with the lone-pair electrons.

If dichlorocarbene is generated in the presence of an alkene, addition to the double bond occurs and a dichlorocyclopropane is formed. As the reaction of dichlorocarbene with *cis*-2-pentene demonstrates, the addition is **stereospecific**, meaning that only a single stereoisomer is formed as product. Starting from a cis alkene, for instance, only cis-disubstituted cyclopropane is produced; starting from a trans alkene, only trans-disubstituted cyclopropane is produced.

The best method for preparing nonhalogenated cyclopropanes is by a process called the **Simmons–Smith reaction**. First investigated at the DuPont company, this reaction does not involve a free carbene. Rather, it utilizes a *carbenoid*—a metal-complexed reagent with carbene-like reactivity. When diiodomethane is treated with a specially prepared zinc–copper mix, (iodomethyl)zinc iodide, ICH_2ZnI, is formed. In the presence of an alkene, (iodomethyl)zinc iodide transfers a CH_2 group to the double bond and yields the cyclopropane. For example, cyclohexene reacts cleanly and in good yield to give the corresponding cyclopropane. Although we won't discuss the mechanistic details, carbene addition to an alkene is one of a general class of reactions called *cycloadditions,* which we'll study more carefully in Chapter 30.

$$CH_2I_2 + Zn(Cu) \longrightarrow ICH_2-ZnI \quad [":CH_2"]$$

Diiodomethane (Iodomethyl)zinc iodide (a carbenoid)

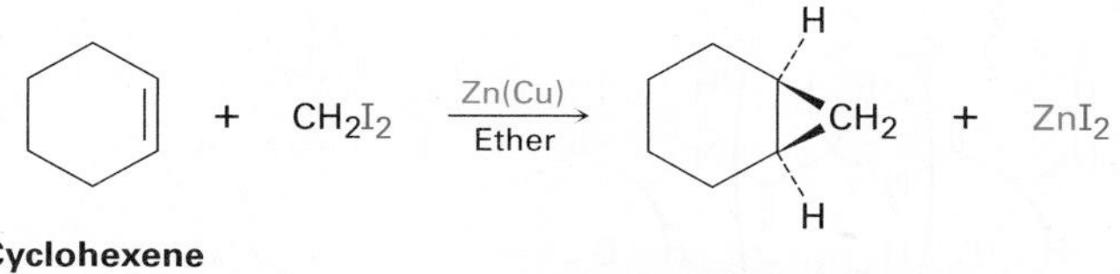

— divalent w/ lone pair
8 e−

Cyclohexene + CH_2I_2 $\xrightarrow{Zn(Cu)}{Ether}$ Bicyclo[4.1.0]heptane (92%) + ZnI_2

Problem 8.17
What products would you expect from the following reactions?

(a) methylenecyclohexane + $CHCl_3$ $\xrightarrow{KOH}$?

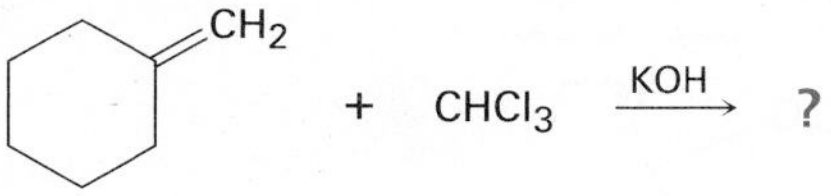

(b) $CH_3CHCH_2CH=CHCH_3$ (with CH_3 branch) + CH_2I_2 $\xrightarrow{Zn(Cu)}$?

8.10 Radical Additions to Alkenes: Chain-Growth Polymers

In our brief introduction to radical reactions in **Section 6.3**, we said that radicals can add to C=C bonds, taking one electron from the double bond and leaving one behind to yield a new radical. Let's now look at the process in more detail, focusing on the industrial synthesis of alkene polymers. A **polymer** is simply a large—sometimes *very* large—molecule built up by repetitive bonding together of many smaller molecules, called **monomers**.

Nature makes wide use of biological polymers. Cellulose, for instance, is a polymer built of repeating glucose monomer units; proteins are polymers built of repeating amino acid monomers; and nucleic acids are polymers built of repeating nucleotide monomers.

Cellulose—a glucose polymer

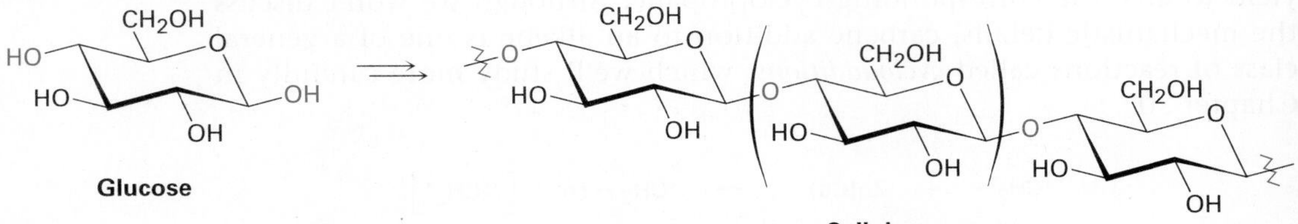

Glucose ⟹ Cellulose

Protein—an amino acid polymer

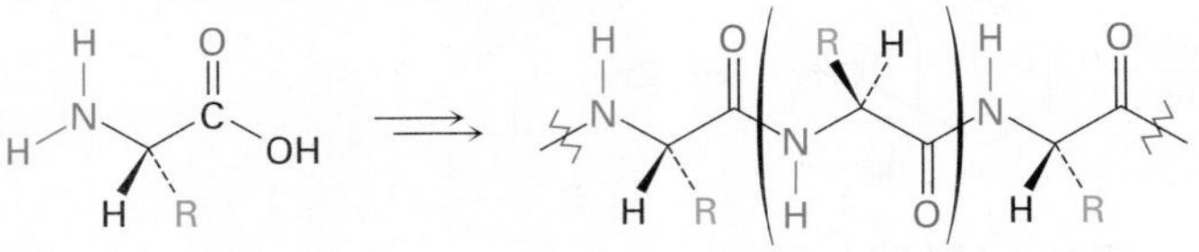

An amino acid ⟹ A protein

Nucleic acid—a nucleotide polymer

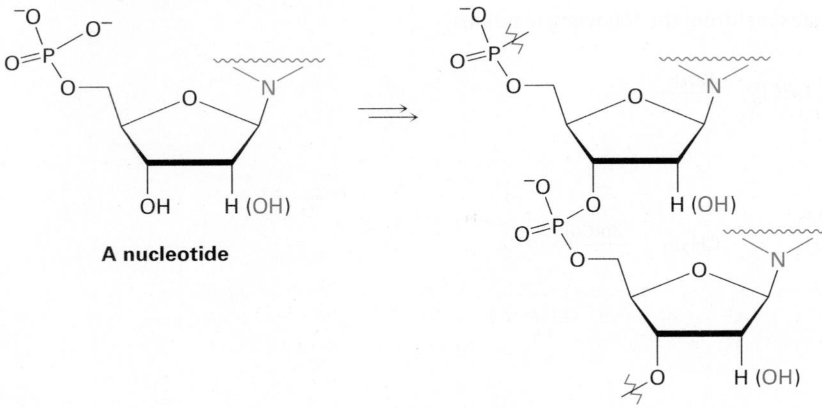

A nucleotide ⟹ A nucleic acid

Synthetic polymers, such as polyethylene, are chemically much simpler than biopolymers, but there is still a great diversity to their structures and properties, depending on the identity of the monomers and on the reaction conditions used for polymerization. The simplest synthetic polymers are those that result when an alkene is treated with a small amount of a suitable catalyst. Ethylene, for example, yields polyethylene, an enormous alkane that may have a molecular weight up to *6 million* amu and may contain as many as 200,000 monomer units incorporated into a gigantic hydrocarbon chain. Worldwide production of polyethylene is approximately 80 million metric tons per year.

Polyethylene—a synthetic alkene polymer

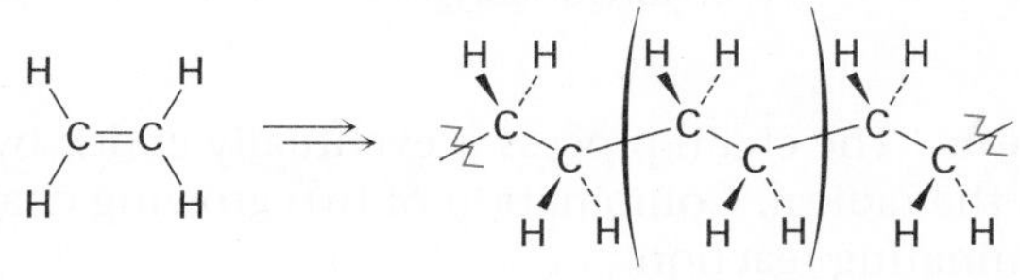

Ethylene → Polyethylene

Polyethylene and other simple alkene polymers are called **chain-growth polymers** because they are formed in a chain reaction process in which an initiator adds to a carbon–carbon double bond to yield a reactive intermediate. The intermediate then reacts with a second molecule of monomer to yield a new intermediate, which reacts with a third monomer unit, and so on.

Historically, ethylene polymerization was carried out at high pressure (1000–3000 atm) and high temperature (100–250 °C) in the presence of a radical catalyst such as benzoyl peroxide, although other catalysts and reaction conditions are now used. The key step is the addition of a radical to the ethylene double bond, a reaction similar in many respects to what takes place in the addition of an electrophile. In writing the mechanism, recall that a curved half-arrow, or "fishhook" ⁀, is used to show the movement of a single electron, as opposed to the full curved arrow used to show the movement of an electron pair in a polar reaction.

- **Initiation** The polymerization reaction is initiated when a few radicals are generated on heating a small amount of benzoyl peroxide catalyst to break the weak O–O bond. The initially formed benzoyloxy radical loses CO_2 and gives a phenyl radical (Ph·), which adds to the C=C bond of ethylene to start the polymerization process. One electron from the ethylene double bond pairs up with the odd electron on the phenyl radical to form a new C–C bond, and the other electron remains on carbon.

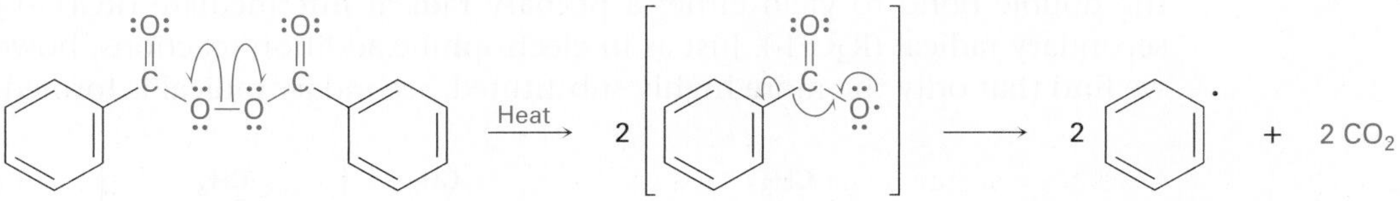

Benzoyl peroxide $\xrightarrow{\text{Heat}}$ 2 [Benzoyloxy radical] ⟶ 2 Phenyl radical (Ph·) + 2 CO_2

Ph· + $H_2C=CH_2$ ⟶ Ph—CH_2CH_2·

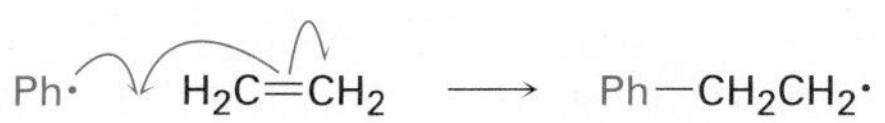

- **Propagation** Polymerization occurs when the carbon radical formed in the initiation step adds to another ethylene molecule to yield another radical. Repetition of the process for hundreds or thousands of times builds the polymer chain.

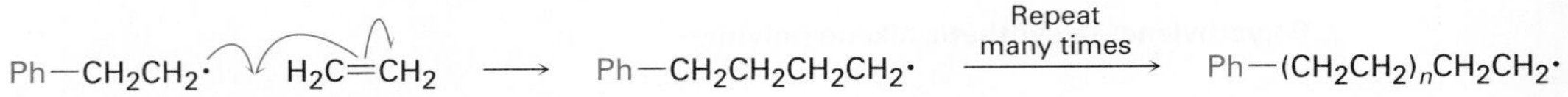

$$Ph-CH_2CH_2\cdot \quad H_2C=CH_2 \quad \longrightarrow \quad Ph-CH_2CH_2CH_2CH_2\cdot \quad \xrightarrow{\text{Repeat many times}} \quad Ph-(CH_2CH_2)_nCH_2CH_2\cdot$$

- **Termination** The chain process is eventually ended by a reaction that consumes the radical. Combination of two growing chains is one possible chain-terminating reaction.

$$2\ R-CH_2CH_2\cdot \quad \longrightarrow \quad R-CH_2CH_2CH_2CH_2-R$$

Ethylene is not unique in its ability to form a polymer. Many substituted ethylenes, called *vinyl monomers,* also undergo polymerization to yield polymers with substituent groups regularly spaced on alternating carbon atoms along the chain. Propylene, for example, yields polypropylene, and styrene yields polystyrene.

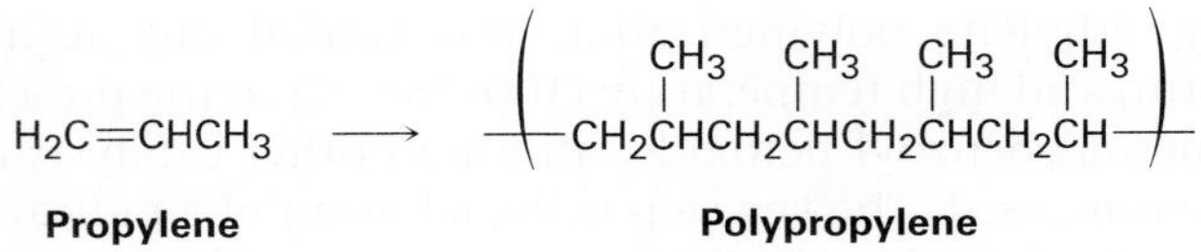

$$H_2C=CHCH_3 \longrightarrow \left(\begin{array}{c}\overset{CH_3}{|}\quad\overset{CH_3}{|}\quad\overset{CH_3}{|}\quad\overset{CH_3}{|}\\-CH_2CHCH_2CHCH_2CHCH_2CH-\end{array}\right)$$

Propylene → **Polypropylene**

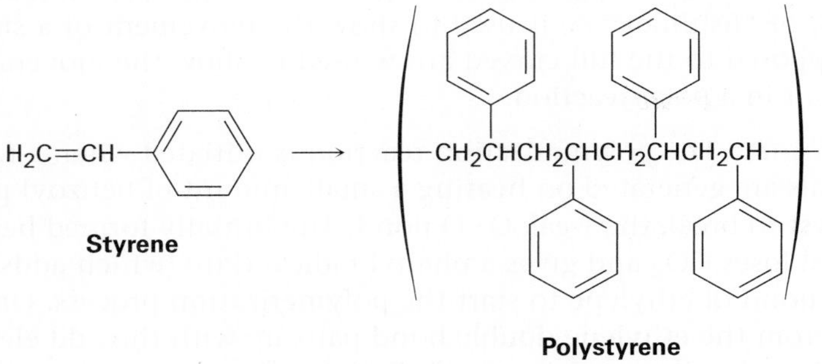

$$H_2C=CH-Ph \longrightarrow \left(-CH_2CHCH_2CHCH_2CHCH_2CH-\right)$$

Styrene → **Polystyrene**

When an unsymmetrically substituted vinyl monomer such as propylene or styrene is polymerized, the radical addition steps can take place at either end of the double bond to yield either a primary radical intermediate ($RCH_2\cdot$) or a secondary radical ($R_2CH\cdot$). Just as in electrophilic addition reactions, however, we find that only the more highly substituted, secondary radical is formed.

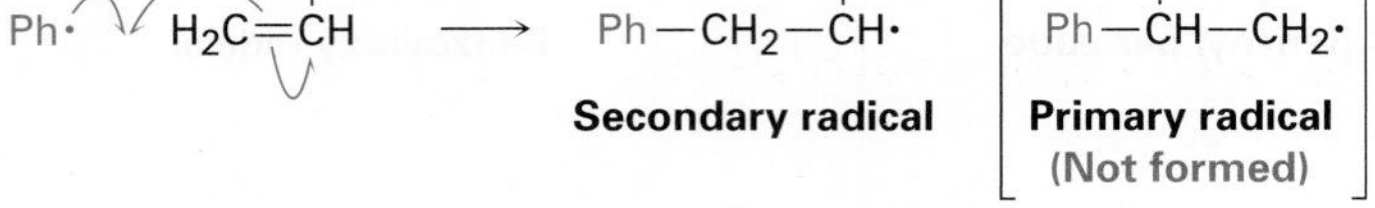

$$Ph\cdot \quad H_2C=\overset{CH_3}{\underset{|}{CH}} \quad \longrightarrow \quad Ph-CH_2-\overset{CH_3}{\underset{|}{CH}}\cdot \quad \left[Ph-\overset{CH_3}{\underset{|}{CH}}-CH_2\cdot\right]$$

Secondary radical **Primary radical (Not formed)**

Table 8.1 shows some commercially important alkene polymers, their uses, and the vinyl monomers from which they are made.

8.10 | Radical Additions to Alkenes: Chain-Growth Polymers

Table 8.1 Some Alkene Polymers and Their Uses

Monomer	Formula	Trade or common name of polymer	Uses
Ethylene	$H_2C=CH_2$	Polyethylene	Packaging, bottles
Propene (propylene)	$H_2C=CHCH_3$	Polypropylene	Moldings, rope, carpets
Chloroethylene (vinyl chloride)	$H_2C=CHCl$	Poly(vinyl chloride) Tedlar	Insulation, films, pipes
Styrene	$H_2C=CHC_6H_5$	Polystyrene	Foam, moldings
Tetrafluoroethylene	$F_2C=CF_2$	Teflon	Gaskets, nonstick coatings
Acrylonitrile	$H_2C=CHCN$	Orlon, Acrilan	Fibers
Methyl methacrylate	$H_2C=C(CH_3)CO_2CH_3$	Plexiglas, Lucite	Paint, sheets, moldings
Vinyl acetate	$H_2C=CHOCOCH_3$	Poly(vinyl acetate)	Paint, adhesives, foams

Worked Example 8.4
Predicting the Structure of a Polymer

Show the structure of poly(vinyl chloride), a polymer made from $H_2C=CHCl$, by drawing several repeating units.

Strategy
Mentally break the carbon–carbon double bond in the monomer unit, and form single bonds by connecting numerous units together.

Solution
The general structure of poly(vinyl chloride) is

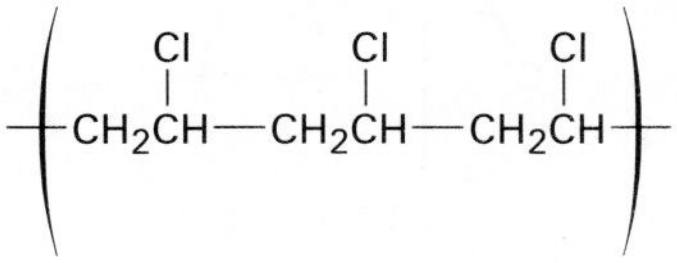

$$\pm(CH_2CH(Cl)-CH_2CH(Cl)-CH_2CH(Cl))\pm$$

Problem 8.18
Show the monomer units you would use to prepare the following polymers:

(a) $\pm(CH_2-CH(OCH_3)-CH_2-CH(OCH_3)-CH_2-CH(OCH_3))\pm$

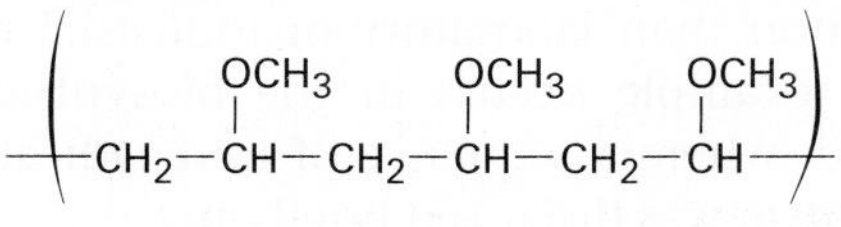

(b) $\pm(CH(Cl)-CH(Cl)-CH(Cl)-CH(Cl)-CH(Cl)-CH(Cl))\pm$

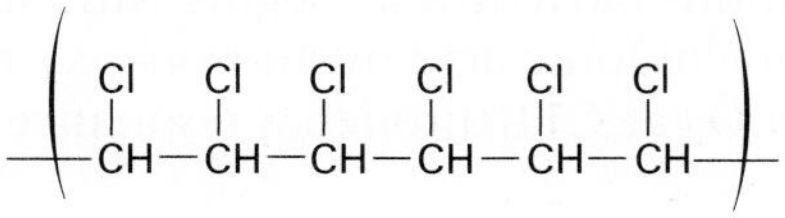

Problem 8.19
One of the chain-termination steps that sometimes occurs to interrupt polymerization is the following reaction between two radicals. Propose a mechanism for the reaction, using fishhook arrows to indicate electron flow.

$$2 \ \ {\sim}CH_2\dot{C}H_2 \longrightarrow {\sim}CH_2CH_3 + {\sim}CH=CH_2$$

8.11 Biological Additions of Radicals to Alkenes

The same high reactivity of radicals that makes possible the alkene polymerization we saw in the previous section also makes it difficult to carry out controlled radical reactions on complex molecules. As a result, there are severe limitations on the usefulness of radical addition reactions in the laboratory. In contrast to an electrophilic addition, where reaction occurs once and the reactive cation intermediate is rapidly quenched by a nucleophile, the reactive intermediate in a radical reaction is not usually quenched. Instead, it reacts again and again in a largely uncontrollable way.

Electrophilic addition
(Intermediate is quenched, so reaction stops.)

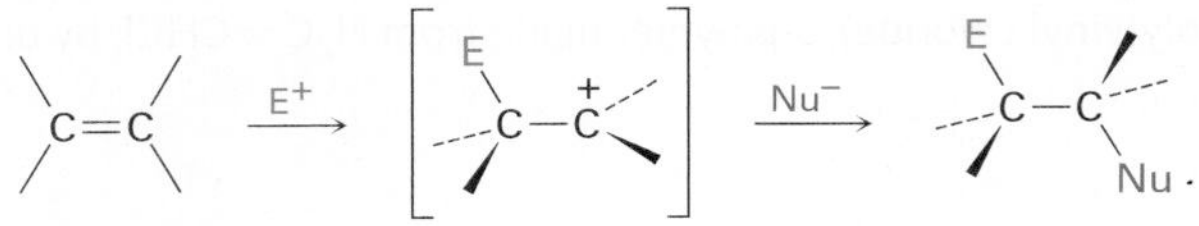

Radical addition
(Intermediate is not quenched, so reaction does not stop.)

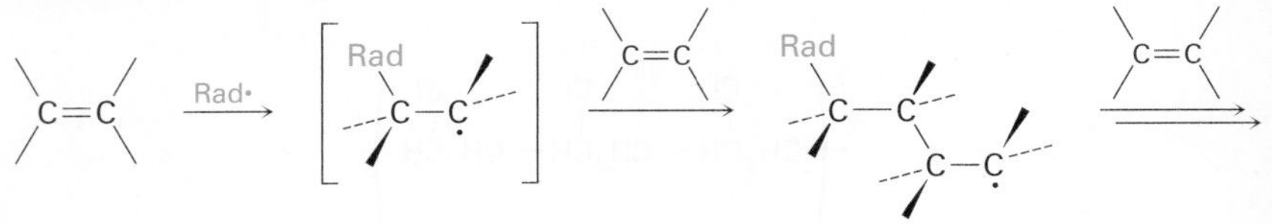

In biological reactions, the situation is different from that in the laboratory. Only one substrate molecule at a time is present in the active site of the enzyme where reaction takes place, and that molecule is held in a precise position, with other necessary reacting groups nearby. As a result, biological radical reactions are more controlled and more common than laboratory or industrial radical reactions. A particularly impressive example occurs in the biosynthesis of prostaglandins from arachidonic acid, where a sequence of four radical additions take place. The reaction mechanism was discussed briefly in **Section 6.3**.

As shown in **Figure 8.10**, prostaglandin biosynthesis begins with abstraction of a hydrogen atom from C13 of arachidonic acid by an iron–oxy radical to give a carbon radical that reacts with O_2 at C11 through a resonance form.

The oxygen radical that results adds to the C8–C9 double bond to give a carbon radical at C8, which adds to the C12–C13 double bond and gives a carbon radical at C13. A resonance form of this carbon radical adds at C15 to a second O_2 molecule, completing the prostaglandin skeleton. Reduction of the O–O bond then gives prostaglandin H_2, called PGH_2. The pathway looks complicated, but the entire process is catalyzed with exquisite control by a single enzyme.

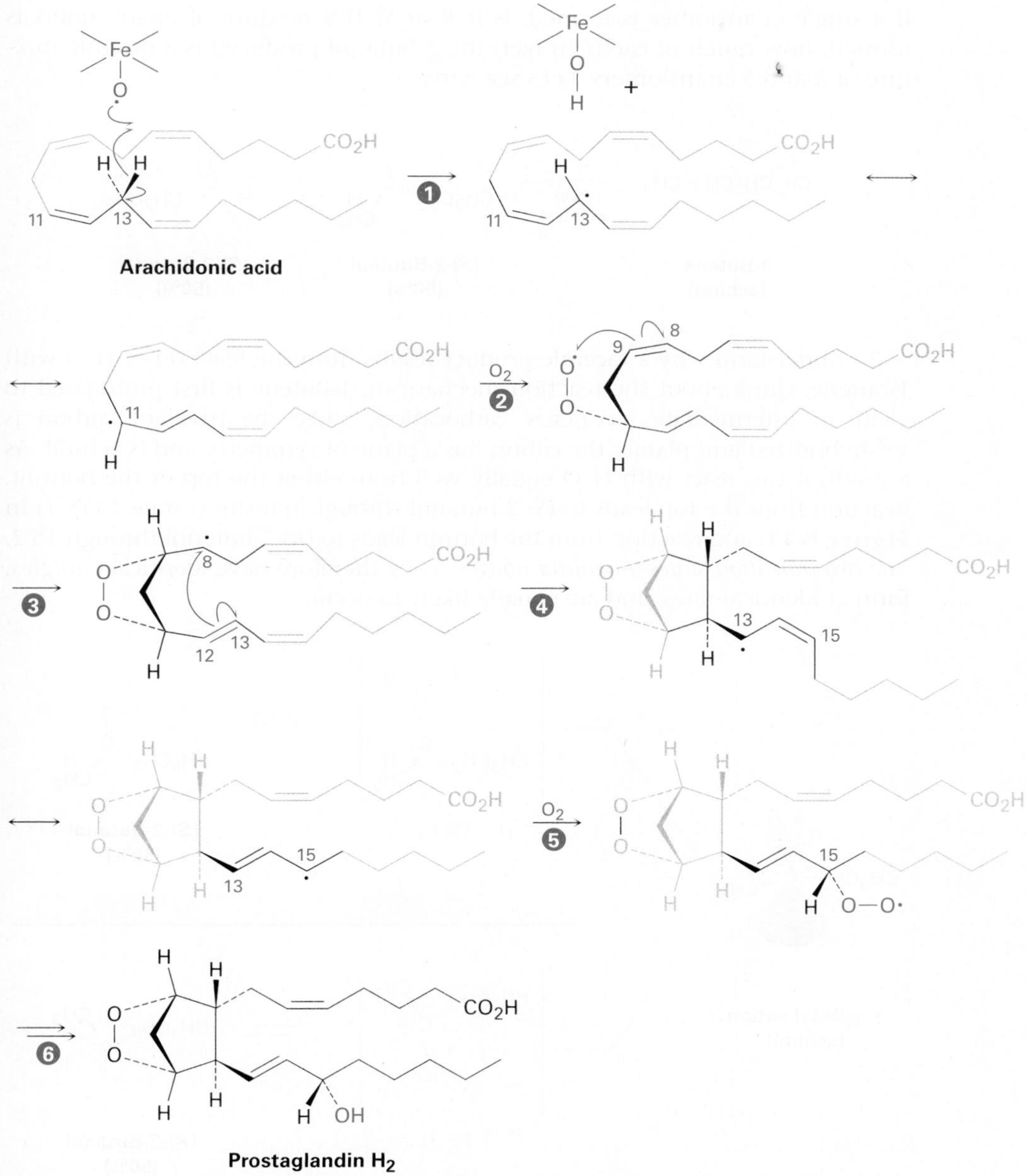

Figure 8.10 Pathway for the biosynthesis of prostaglandins from arachidonic acid. Steps ❷ and ❺ are radical addition reactions to O_2; steps ❸ and ❹ are radical additions to carbon–carbon double bonds.

8.12 Reaction Stereochemistry: Addition of H₂O to an Achiral Alkene

Most of the biochemical reactions that take place in the body, as well as many organic reactions in the laboratory, yield products with chirality centers. For example, acid-catalyzed addition of H_2O to 1-butene in the laboratory yields 2-butanol, a chiral alcohol. What is the stereochemistry of this chiral product? If a single enantiomer is formed, is it *R* or *S*? If a mixture of enantiomers is formed, how much of each? In fact, the 2-butanol produced is a racemic mixture of *R* and *S* enantiomers. Let's see why.

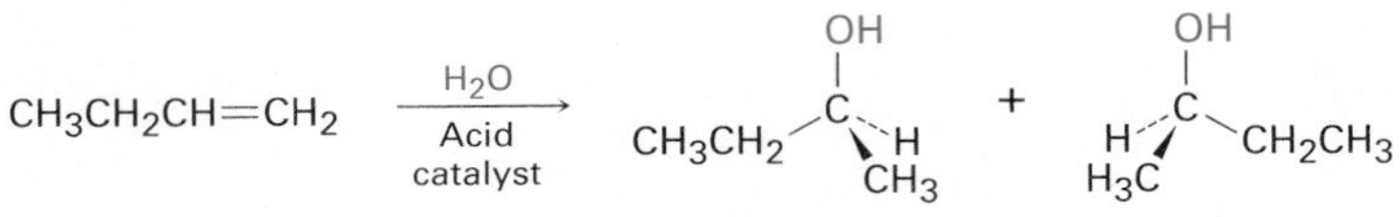

$$CH_3CH_2CH=CH_2 \xrightarrow[\text{Acid catalyst}]{H_2O} \text{(S)-2-Butanol (50\%)} + \text{(R)-2-Butanol (50\%)}$$

1-Butene (achiral)

To understand why a racemic product results from the reaction of H_2O with 1-butene, think about the reaction mechanism. 1-Butene is first protonated to yield an intermediate secondary carbocation. Since the trivalent carbon is sp^2-hybridized and planar, the cation has a plane of symmetry and is achiral. As a result, it can react with H_2O equally well from either the top or the bottom. Reaction from the top leads to (*S*)-2-butanol through transition state 1 (TS 1) in **Figure 8.11**, and reaction from the bottom leads to (*R*)-2-butanol through TS 2. *The two transition states are mirror images.* They therefore have identical energies, form at identical rates, and are equally likely to occur.

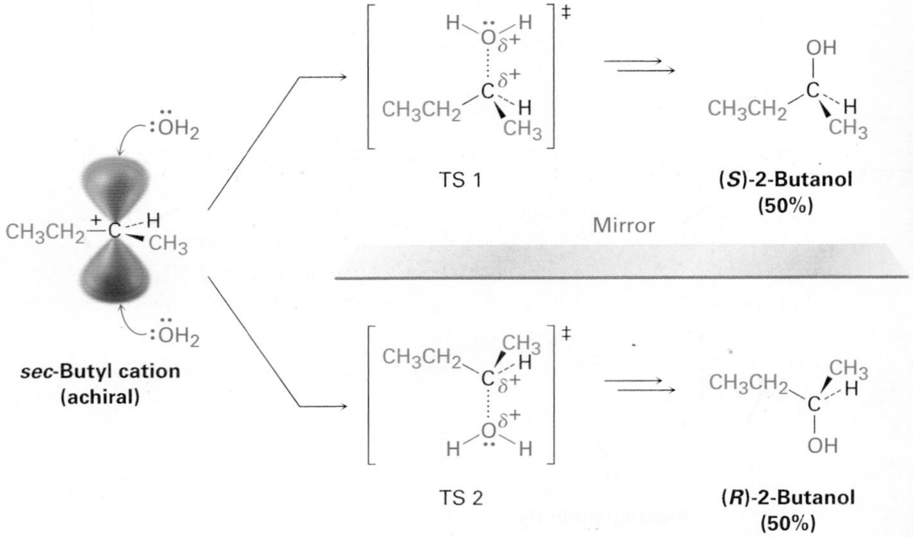

Figure 8.11 Reaction of H_2O with the carbocation resulting from protonation of 1-butene. Reaction from the top leads to *S* product and is the mirror image of reaction from the bottom, which leads to *R* product. Because they are energetically identical, they are equally likely and lead to a racemic mixture of products. The dotted C⋯O bond in the transition state indicates partial bond formation.

As a general rule, the formation of a new chirality center by reaction of achiral reactants always leads to a racemic mixture of enantiomeric products. Put another way, optical activity can't appear from nowhere; an optically active product can only result by starting with an optically active reactant or chiral environment (**Section 5.12**).

In contrast to laboratory reactions, enzyme-catalyzed biological reactions often give a single enantiomer of a chiral product, even when the substrate is achiral. One step in the citric acid cycle of food metabolism, for instance, is the aconitase-catalyzed addition of water to (Z)-aconitate (usually called *cis*-aconitate) to give isocitrate.

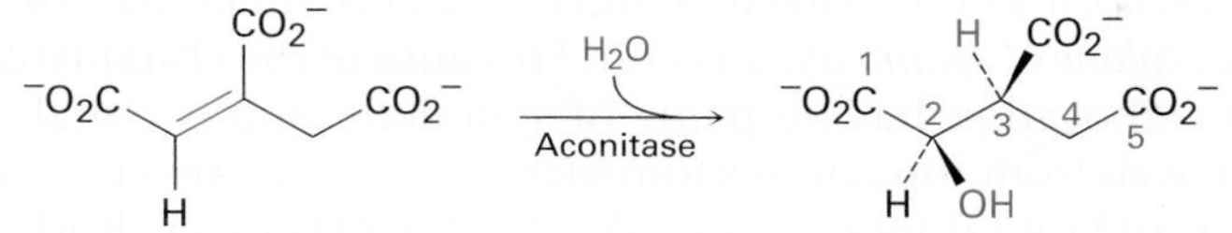

cis-Aconitate
(achiral)

(2R,3S)-Isocitrate

Even though *cis*-aconitate is achiral, only the (2R,3S) enantiomer of the product is formed. As discussed in **Sections 5.11 and 5.12**, *cis*-aconitate is a prochiral molecule, which is held in a chiral environment by the aconitase enzyme during the reaction. In that chiral environment, the two faces of the double bond are chemically distinct, and addition occurs on only the *Re* face at C2.

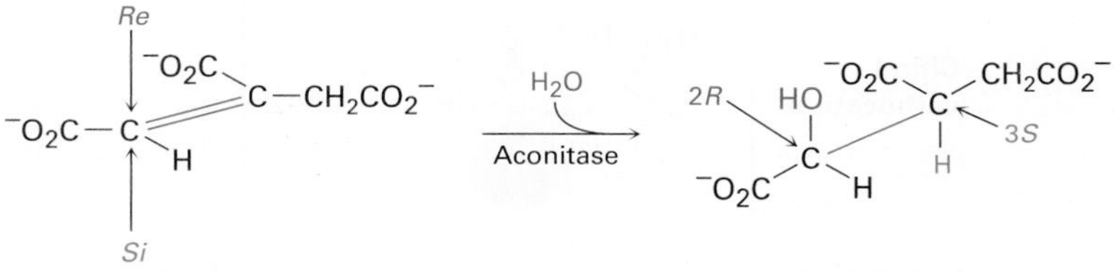

cis-Aconitate

(2R,3S)-Isocitrate

8.13 Reaction Stereochemistry: Addition of H_2O to a Chiral Alkene

The reaction discussed in the previous section involves an addition to an achiral reactant and forms an optically inactive, racemic mixture of two enantiomeric products. What would happen, though, if we were to carry out the reaction on a *single* enantiomer of a *chiral* reactant? For example, what stereochemical result would be obtained from addition of H_2O to a chiral alkene, such as (R)-4-methyl-1-hexene? The product of the reaction, 4-methyl-2-hexanol, has two chirality centers and so has four possible stereoisomers.

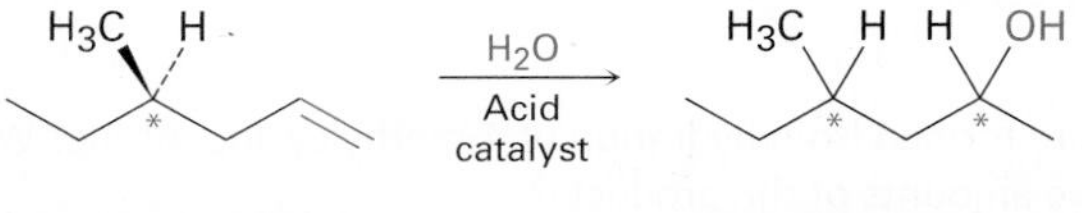

(R)-4-Methyl-1-hexene
(chiral)

4-Methyl-2-hexanol
(chiral)

Let's think about the two chirality centers separately. What about the configuration at C4, the methyl-bearing carbon atom? Since C4 has the *R* configuration in the starting material and this chirality center is unaffected by the reaction, its configuration is unchanged. Thus, the configuration at C4 in the product remains *R* (assuming that the relative rankings of the four attached groups are not changed by the reaction).

What about the configuration at C2, the newly formed chirality center? As shown in **Figure 8.12**, the stereochemistry at C2 is established by reaction of H_2O with a carbocation intermediate in the usual manner. *But this carbocation does not have a plane of symmetry;* it is chiral because of the chirality center at C4. Because the carbocation has no plane of symmetry and is chiral, it does not react equally well from top and bottom faces. One of the two faces is likely, for steric reasons, to be a bit more accessible than the other face, leading to a mixture of *R* and *S* products in some ratio other than 50:50. Thus, two diastereomeric products, (2*R*,4*R*)-4-methyl-2-hexanol and (2*S*,4*R*)-4-methyl-2-hexanol, are formed in unequal amounts, and the mixture is optically active.

Figure 8.12 Stereochemistry of the acid-catalyzed addition of H_2O to the chiral alkene, (*R*)-4-methyl-1-hexene. A mixture of diastereomeric 2*R*,4*R* and 2*S*,4*R* products is formed in unequal amounts because reaction of the chiral carbocation intermediate is not equally likely from top and bottom. The product mixture is optically active.

(2*S*,4*R*)-4-Methyl-2-hexanol   (2*R*,4*R*)-4-Methyl-2-hexanol

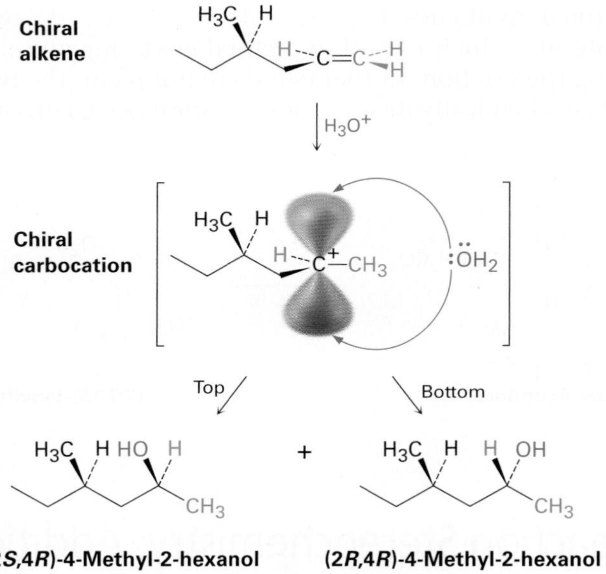

As a general rule, the formation of a new chirality center by the reaction of a chiral reactant leads to unequal amounts of diastereomeric products. If the chiral reactant is optically active because only one enantiomer is used rather than a racemic mixture, then the products are also optically active.

Problem 8.20
What products are formed from acid-catalyzed hydration of racemic (±)-4-methyl-1-hexene? What can you say about the relative amounts of the products? Is the product mixture optically active?

Problem 8.21
What products are formed from hydration of 4-methylcyclopentene? What can you say about the relative amounts of the products?

A DEEPER LOOK: Terpenes: Naturally Occurring Alkenes

The wonderful fragrance of leaves from the California bay laurel is due primarily to myrcene, a simple terpene.

Ever since its discovery in Persia around 1000 A.D., it has been known that *steam distillation,* the codistillation of plant materials with water, produces a fragrant mixture of liquids called *essential oils.* The resulting oils have long been used as medicines, spices, and perfumes, and their investigation played a major role in the emergence of organic chemistry as a science during the 19th century.

Chemically, plant essential oils consist largely of mixtures of compounds called *terpenoids*—small organic molecules with an immense diversity of structure. More than 35,000 different terpenoids are known. Some are open-chain molecules, and others contain rings; some are hydrocarbons, and others contain oxygen. Hydrocarbon terpenoids, in particular, are known as *terpenes,* and all contain double bonds. For example:

Myrcene
(oil of bay)

α-Pinene
(turpentine)

Humulene
(oil of hops)

β-Santalene
(sandalwood oil)

Regardless of their apparent structural differences, all terpenoids are related. According to a formalism called the *isoprene rule,* they can be thought of as arising from head-to-tail joining of 5-carbon isoprene units (2-methyl-1,3-butadiene). Carbon 1 is the head of the isoprene unit, and carbon 4 is the tail. For example, myrcene contains two isoprene units joined head to tail, forming an 8-carbon chain with two 1-carbon branches. α-Pinene similarly contains two isoprene units assembled into a more complex cyclic structure, and humulene contains three isoprene units. See if you can identify the isoprene units in α-pinene, humulene, and β-santalene.

Isoprene → **Myrcene**

(continued)

Terpenes (and terpenoids) are further classified according to the number of 5-carbon units they contain. Thus, *monoterpenes* are 10-carbon substances derived from two isoprene units, *sesquiterpenes* are 15-carbon molecules derived from three isoprene units, *diterpenes* are 20-carbon substances derived from four isoprene units, and so on. Monoterpenes and sesquiterpenes are found primarily in plants, but the higher terpenoids occur in both plants and animals, and many have important biological roles. The triterpenoid lanosterol, for instance, is the biological precursor from which all steroid hormones are made.

Lanosterol (a triterpene, C_{30})

Isoprene itself is not the true biological precursor of terpenoids. Nature instead uses two "isoprene equivalents"—isopentenyl diphosphate and dimethylallyl diphosphate—which are themselves made by two different routes depending on the organism. Lanosterol, in particular, is biosynthesized from acetic acid by a complex pathway that has been worked out in great detail. We'll look at the subject more closely in **Sections 27.5 and 27.7**.

Isopentenyl diphosphate   **Dimethylallyl diphosphate**

Summary

Key words

anti stereochemistry, 265
bromonium ion, 265
carbene, 287
chain-growth polymer, 291
epoxide, 281
glycol, 282
halohydrin, 267
hydroboration, 272
hydrogenation, 276
hydroxylation, 282
monomer, 289
oxidation, 281
oxymercuration–demercuration, 271
ozonide, 284
polymer, 289

With the background needed to understand organic reactions now covered, this chapter has begun the systematic description of major functional groups.

Alkenes are generally prepared by an *elimination reaction*, such as *dehydrohalogenation*, the elimination of HX from an alkyl halide, or *dehydration*, the elimination of water from an alcohol. The flip side of that elimination reaction to prepare alkenes is the addition of various substances to the alkene double bond to give saturated products.

HCl, HBr, and HI add to alkenes by a two-step electrophilic addition mechanism. Initial reaction of the nucleophilic double bond with H$^+$ gives a carbocation intermediate, which then reacts with halide ion. Bromine and chlorine add to alkenes via three-membered-ring **bromonium ion** or chloronium ion intermediates to give addition products having **anti stereochemistry**. If water is present during the halogen addition reaction, a **halohydrin** is formed.

Hydration of an alkene—the addition of water—is carried out by either of two procedures, depending on the product desired. **Oxymercuration–demercuration** involves electrophilic addition of Hg^{2+} to an alkene, followed by trapping of the cation intermediate with water and subsequent treatment with NaBH$_4$. **Hydroboration** involves addition of borane (BH$_3$) followed by oxidation of the intermediate organoborane with alkaline H_2O_2. The two hydration methods are complementary: oxymercuration–demercuration gives the product of Markovnikov addition, whereas hydroboration–oxidation gives the product with non-Markovnikov **syn stereochemistry**.

Alkenes are **reduced** by addition of H_2 in the presence of a catalyst such as platinum or palladium to yield alkanes, a process called catalytic **hydrogenation**. Alkenes are also **oxidized** by reaction with a peroxyacid to give **epoxides**, which can be converted into trans-1,2-diols by acid-catalyzed hydrolysis. The corresponding cis-1,2-diols can be made directly from alkenes by **hydroxylation** with OsO_4. Alkenes can also be cleaved to produce carbonyl compounds by reaction with ozone, followed by reduction with zinc metal. In addition, alkenes react with divalent substances called **carbenes, R_2C:**, to give cyclopropanes. Nonhalogenated cyclopropanes are best prepared by treatment of the alkene with CH_2I_2 and zinc–copper, a process called the **Simmons–Smith reaction**.

Alkene **polymers**—large molecules resulting from repetitive bonding together of many hundreds or thousands of small **monomer** units—are formed by chain-reaction polymerization of simple alkenes. Polyethylene, polypropylene, and polystyrene are examples. As a general rule, radical addition reactions are not common in the laboratory but occur much more frequently in biological pathways.

Many reactions give chiral products. If the reactants are optically inactive, the products are also optically inactive. If one or both of the reactants is optically active, the products can also be optically active.

Key words—cont'd
reduction, 277
Simmons–Smith reaction, 288
stereospecific, 288
syn stereochemistry, 273

Learning Reactions

What's seven times nine? Sixty-three, of course. You didn't have to stop and figure it out; you knew the answer immediately because you long ago learned the multiplication tables. Learning the reactions of organic chemistry requires the same approach: reactions have to be learned for immediate recall if they are to be useful.

Different people take different approaches to learning reactions. Some people make flashcards; others find studying with friends to be helpful. To help guide your study, most chapters in this book end with a summary of the reactions just presented. In addition, the accompanying *Study Guide and Solutions Manual* has several appendixes that organize organic reactions from other viewpoints. Fundamentally, though, there are no shortcuts. Learning organic chemistry does take effort.

Summary of Reactions

Note: No stereochemistry is implied unless specifically indicated with wedged, solid, and dashed lines.

1. Addition reactions of alkenes
    (a) Addition of HCl, HBr, and HI (Sections 7.7 and 7.8)
        Markovnikov regiochemistry occurs, with H adding to the less highly substituted alkene carbon and halogen adding to the more highly substituted carbon.

$$\diagdown C=C \diagup \xrightarrow[\text{Ether}]{HX} \diagdown \overset{H}{C}-\overset{X}{C} \diagup$$

(continued)

(b) Addition of halogens Cl_2 and Br_2 (Section 8.2)
Anti addition is observed through a halonium ion intermediate.

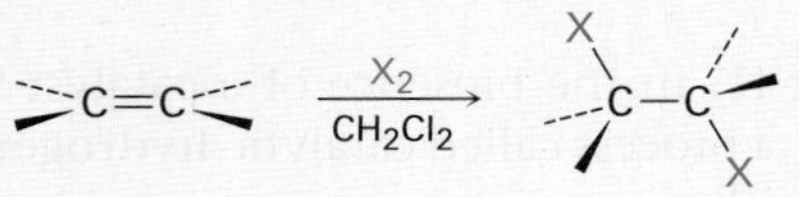

$$\text{C=C} \xrightarrow{X_2, CH_2Cl_2} \text{X-C-C-X}$$

(c) Halohydrin formation (Section 8.3)
Markovnikov regiochemistry and anti stereochemistry occur.

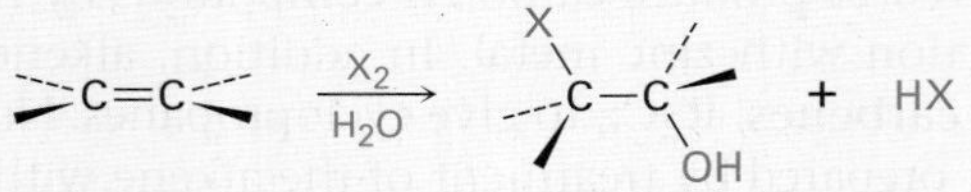

$$\text{C=C} \xrightarrow{X_2, H_2O} \text{X-C-C-OH} + HX$$

(d) Addition of water by oxymercuration–demercuration (Section 8.4)
Markovnikov regiochemistry occurs.

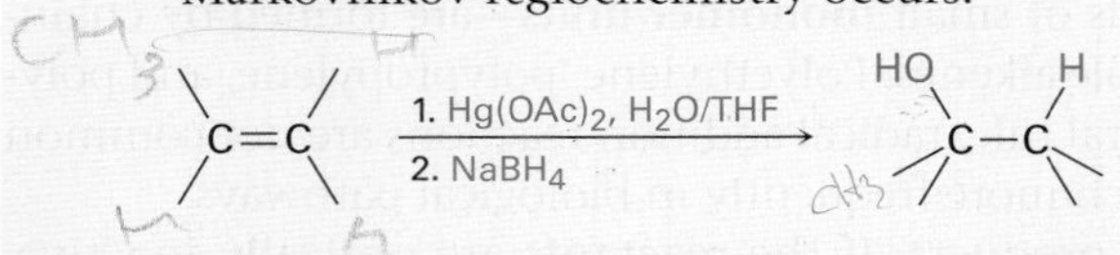

$$\text{C=C} \xrightarrow[\text{2. NaBH}_4]{\text{1. Hg(OAc)}_2, \text{H}_2\text{O/THF}} \text{HO-C-C-H}$$

(e) Addition of water by hydroboration–oxidation (Section 8.5)
Non-Markovnikov syn addition occurs.

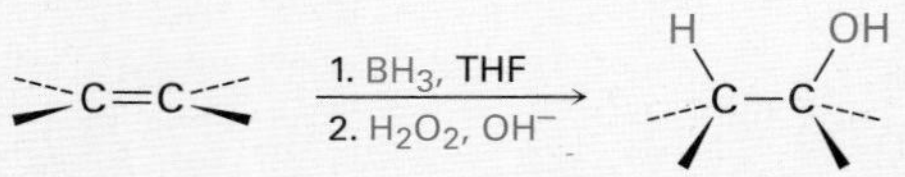

$$\text{C=C} \xrightarrow[\text{2. H}_2\text{O}_2, \text{OH}^-]{\text{1. BH}_3, \text{THF}} \text{H-C-C-OH}$$

(f) Catalytic hydrogenation (Section 8.6)
Syn addition occurs.

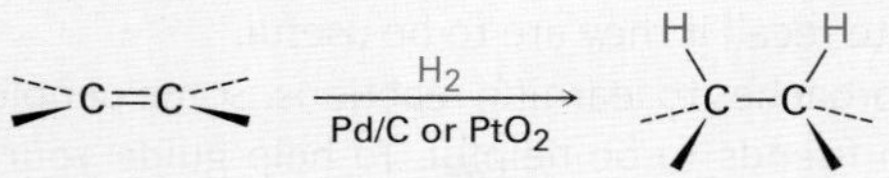

$$\text{C=C} \xrightarrow[\text{Pd/C or PtO}_2]{\text{H}_2} \text{H-C-C-H}$$

(g) Epoxidation with a peroxyacid (Section 8.7)
Syn addition occurs.

$$\text{C=C} \xrightarrow{\text{RCOOH}} \text{C-O-C (epoxide)}$$

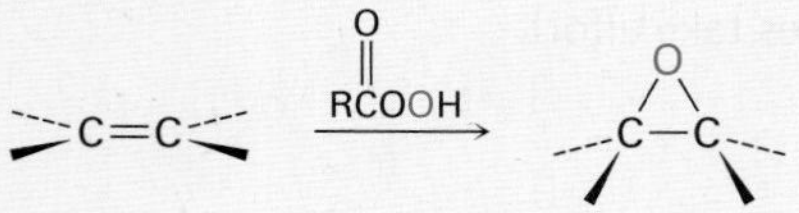

(h) Hydroxylation with OsO_4 (Section 8.7)
Syn addition occurs.

$$\text{C=C} \xrightarrow[\substack{\text{2. NaHSO}_3, \text{H}_2\text{O} \\ \text{or OsO}_4, \text{NMO}}]{\text{1. OsO}_4} \text{HO-C-C-OH}$$

(i) Addition of carbenes to yield cyclopropanes (Section 8.9)

 (1) Dichlorocarbene addition

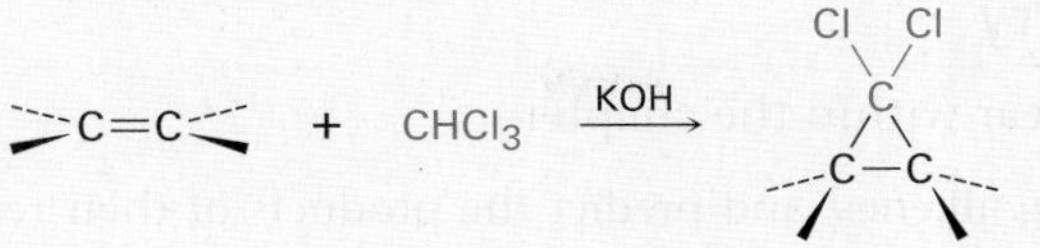

$$\text{C=C} + \text{CHCl}_3 \xrightarrow{\text{KOH}} \text{cyclopropane with CCl}_2$$

 (2) Simmons–Smith reaction

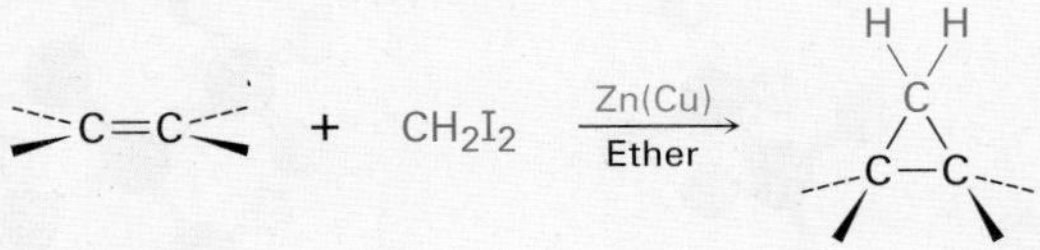

$$\text{C=C} + \text{CH}_2\text{I}_2 \xrightarrow[\text{Ether}]{\text{Zn(Cu)}} \text{cyclopropane with CH}_2$$

2. Hydroxylation by acid-catalyzed epoxide hydrolysis (Section 8.7)

 Anti stereochemistry occurs.

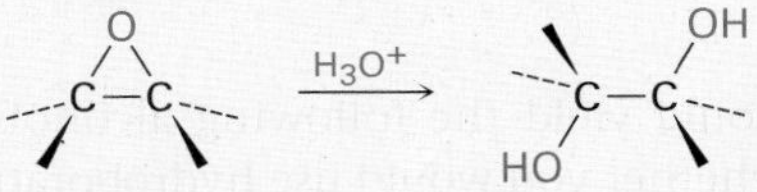

$$\text{epoxide} \xrightarrow{\text{H}_3\text{O}^+} \text{trans-1,2-diol}$$

3. Oxidative cleavage of alkenes (Section 8.8)
 (a) Reaction with ozone followed by zinc in acetic acid

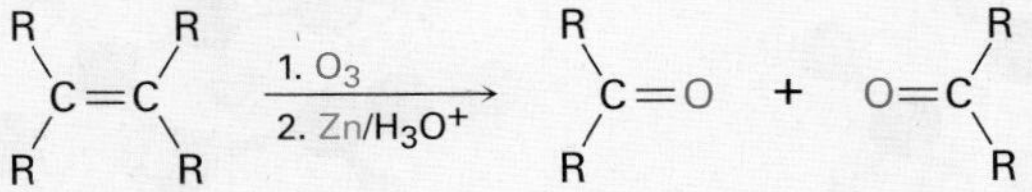

$$\underset{R}{\overset{R}{>}}\!\!C\!=\!C\!\!\underset{R}{\overset{R}{<}} \xrightarrow[\text{2. Zn/H}_3\text{O}^+]{\text{1. O}_3} \underset{R}{\overset{R}{>}}\!\!C\!=\!O \;+\; O\!=\!C\!\!\underset{R}{\overset{R}{<}}$$

 (b) Reaction with KMnO$_4$ in acidic solution

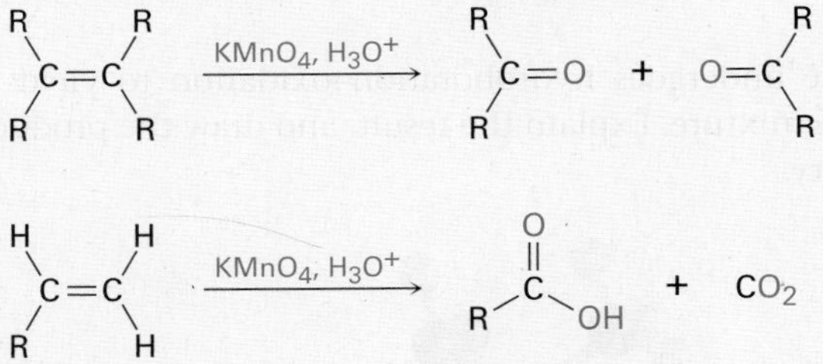

$$\underset{R}{\overset{R}{>}}\!\!C\!=\!C\!\!\underset{R}{\overset{R}{<}} \xrightarrow{\text{KMnO}_4,\, \text{H}_3\text{O}^+} \underset{R}{\overset{R}{>}}\!\!C\!=\!O \;+\; O\!=\!C\!\!\underset{R}{\overset{R}{<}}$$

$$\underset{R}{\overset{H}{>}}\!\!C\!=\!C\!\!\underset{H}{\overset{R}{<}} \xrightarrow{\text{KMnO}_4,\, \text{H}_3\text{O}^+} R\!-\!\underset{\text{OH}}{\overset{O}{\|C}} \;+\; \text{CO}_2$$

4. Cleavage of 1,2-diols (Section 8.8)

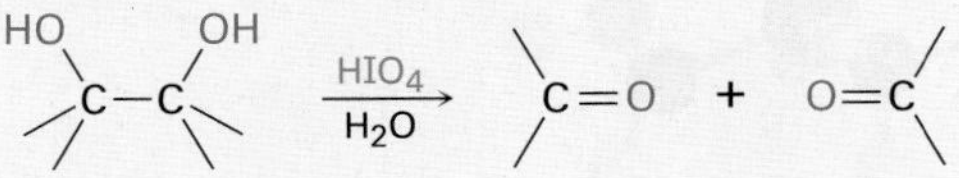

$$\text{HO-C-C-OH} \xrightarrow[\text{H}_2\text{O}]{\text{HIO}_4} \text{C=O} \;+\; O\!=\!C$$

Exercises

OWL Interactive versions of these problems are assignable in OWL for Organic Chemistry.

Visualizing Chemistry

(Problems 8.1–8.21 appear within the chapter.)

8.22 Name the following alkenes, and predict the products of their reaction with (1) *meta*-chloroperoxybenzoic acid, (2) $KMnO_4$ in aqueous acid, and (3) O_3, followed by Zn in acetic acid:

(a)   (b)

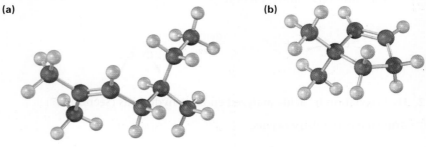

8.23 Draw the structures of alkenes that would yield the following alcohols on hydration (red = O). Tell in each case whether you would use hydroboration–oxidation or oxymercuration–demercuration.

(a)   (b)

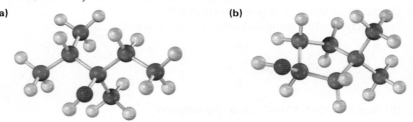

8.24 The following alkene undergoes hydroboration–oxidation to yield a single product rather than a mixture. Explain the result, and draw the product showing its stereochemistry.

8.25 From what alkene was the following 1,2-diol made, and what method was used, epoxide hydrolysis or OsO$_4$?

Additional Problems

Reactions of Alkenes

8.26 Predict the products of the following reactions (the aromatic ring is unreactive in all cases). Indicate regiochemistry when relevant.

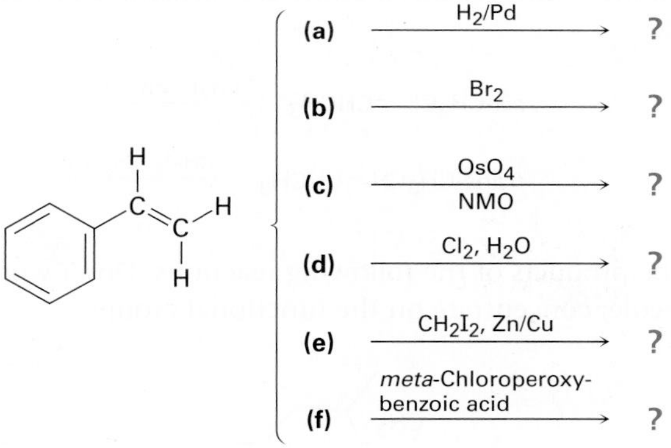

(a) H$_2$/Pd ⟶ ?
(b) Br$_2$ ⟶ ?
(c) OsO$_4$, NMO ⟶ ?
(d) Cl$_2$, H$_2$O ⟶ ?
(e) CH$_2$I$_2$, Zn/Cu ⟶ ?
(f) *meta*-Chloroperoxybenzoic acid ⟶ ?

8.27 Suggest structures for alkenes that give the following reaction products. There may be more than one answer for some cases.

(a) ? —H$_2$/Pd→ CH$_3$CHCH$_2$CH$_2$CH$_2$CH$_3$ with CH$_3$ substituent

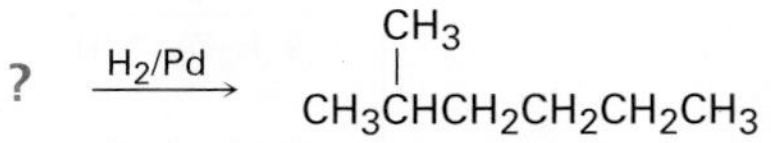

(b) ? —H$_2$/Pd→ cyclohexane with two CH$_3$ groups

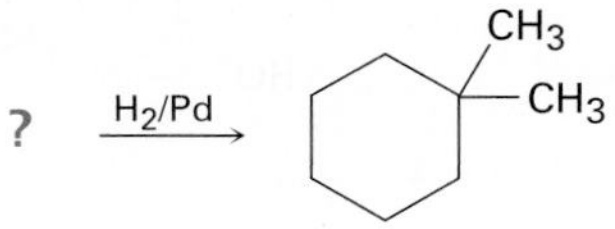

(c) ? —Br$_2$→ CH$_3$CHCHCH$_2$CHCH$_3$ with Br, CH$_3$, Br substituents

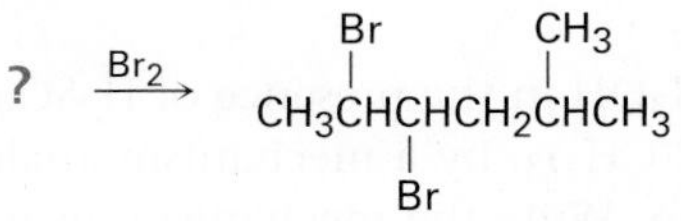

(d) ? —HCl→ CH$_3$CHCHCH$_2$CH$_2$CH$_3$ with Cl and CH$_3$ substituents

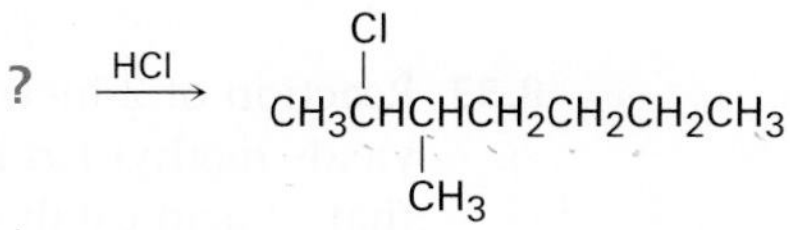

(e) ? —1. Hg(OAc)$_2$, H$_2$O; 2. NaBH$_4$→ CH$_3$CH$_2$CH$_2$CHCH$_3$ with OH

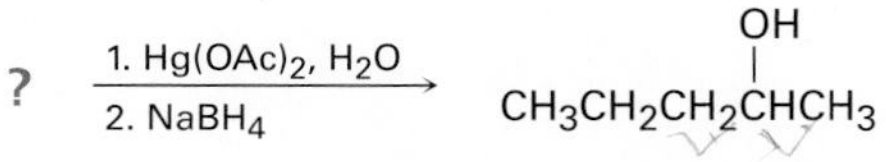

(f) ? —CH$_2$I$_2$, Zn/Cu→ cyclohexane-spiro-cyclopropane

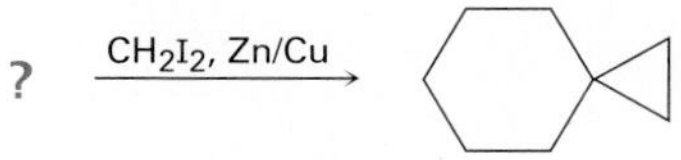

8.28 Predict the products of the following reactions, showing both regiochemistry and stereochemistry where appropriate:

(a) 1-methyl-cyclohexene (with H shown) → 1. O$_3$; 2. Zn, H$_3$O$^+$ → ?

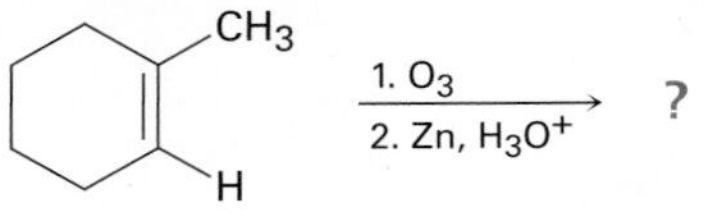

(b) cyclohexene → KMnO$_4$, H$_3$O$^+$ → ?

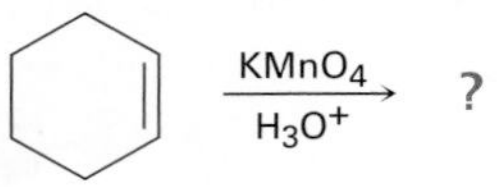

(c) 1-methylcyclohexene → 1. BH$_3$; 2. H$_2$O$_2$, $^-$OH → ?

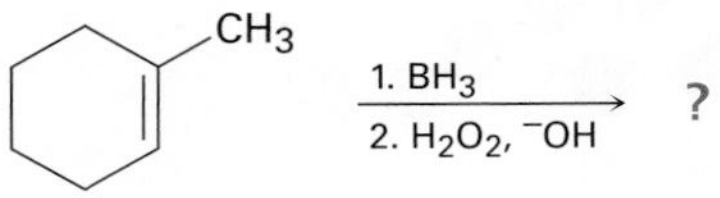

(d) 1-methylcyclohexene → 1. Hg(OAc)$_2$, H$_2$O; 2. NaBH$_4$ → ?

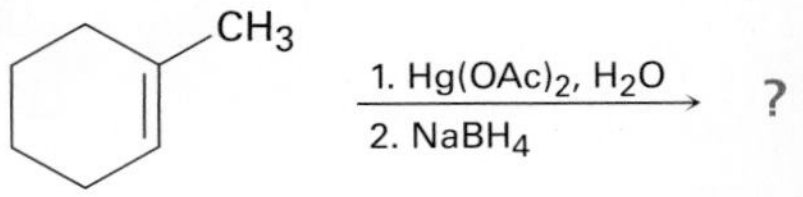

8.29 Which reaction would you expect to be faster, addition of HBr to cyclohexene or to 1-methylcyclohexene? Explain.

8.30 What product will result from hydroboration–oxidation of 1-methylcyclopentene with deuterated borane, BD$_3$? Show both the stereochemistry (spatial arrangement) and the regiochemistry (orientation) of the product.

8.31 The cis and trans isomers of 2-butene give different cyclopropane products in the Simmons–Smith reaction. Show the structures of both, and explain the difference.

cis-CH$_3$CH=CHCH$_3$ $\xrightarrow{\text{CH}_2\text{I}_2,\ \text{Zn(Cu)}}$?

trans-CH$_3$CH=CHCH$_3$ $\xrightarrow{\text{CH}_2\text{I}_2,\ \text{Zn(Cu)}}$?

8.32 Predict the products of the following reactions. Don't worry about the size of the molecule; concentrate on the functional groups.

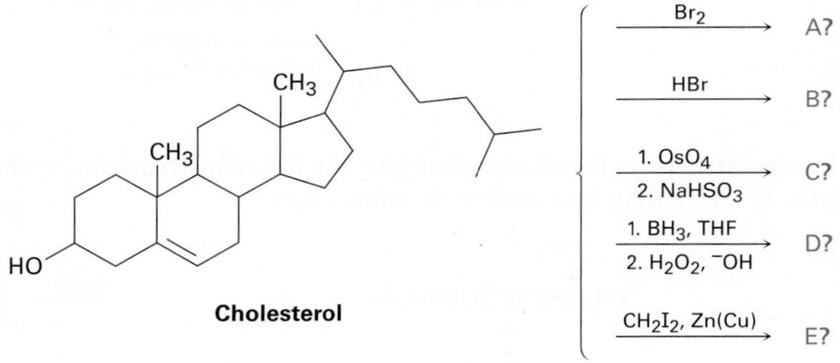

Cholesterol:
- $\xrightarrow{\text{Br}_2}$ A?
- $\xrightarrow{\text{HBr}}$ B?
- $\xrightarrow{\text{1. OsO}_4;\ \text{2. NaHSO}_3}$ C?
- $\xrightarrow{\text{1. BH}_3,\ \text{THF};\ \text{2. H}_2\text{O}_2,\ ^-\text{OH}}$ D?
- $\xrightarrow{\text{CH}_2\text{I}_2,\ \text{Zn(Cu)}}$ E?

8.33 Reaction of 2-methylpropene with CH$_3$OH in the presence of H$_2$SO$_4$ catalyst yields methyl *tert*-butyl ether, CH$_3$OC(CH$_3$)$_3$, by a mechanism analogous to that of acid-catalyzed alkene hydration. Write the mechanism, using curved arrows for each step.

8.34 Addition of HCl to 1-methoxycyclohexene yields 1-chloro-1-methoxycyclohexane as the sole product. Use resonance structures of the carbocation intermediate to explain why none of the other regioisomer is formed.

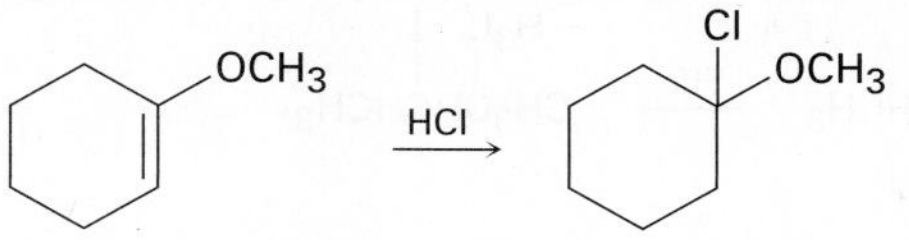

1-Methoxycyclohexene → (HCl) → 1-Chloro-1-methoxycyclohexane

Synthesis Using Alkenes

8.35 How would you carry out the following transformations? Tell the reagents you would use in each case.

(a) cyclopentene → cis-1,2-cyclopentanediol (H, OH on same face)

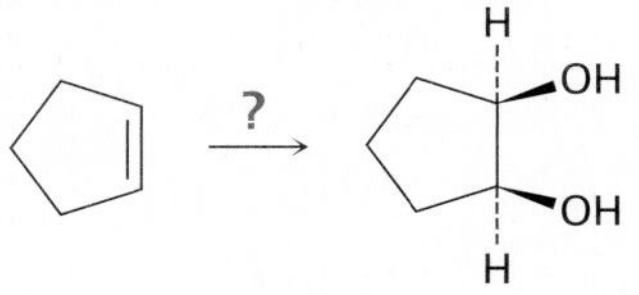

(b) cyclopentene → cyclopentanol

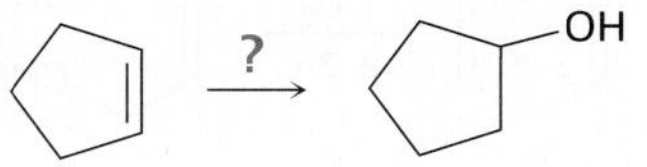

(c) cyclopentene → 1,1-dichlorocyclopropane-fused bicyclic (dichlorocyclopropanation)

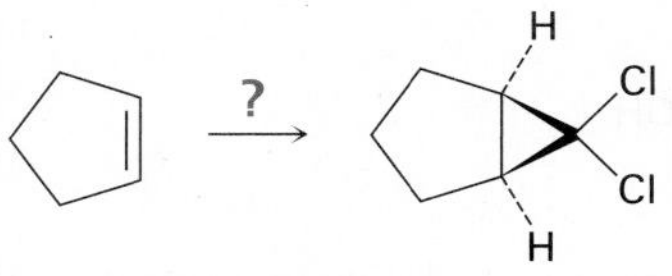

(d) 1-methylcyclohexanol → 1-methylcyclohexene

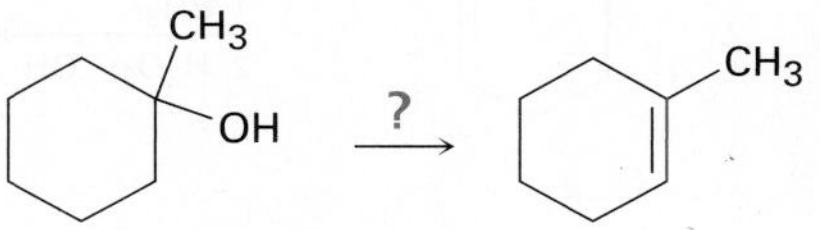

(e) $CH_3CH=CHCH(CH_3)CH_3$ → CH_3CHO + $(CH_3)_2CHCHO$... i.e. $CH_3CH=CHCH(CH_3)$ → $CH_3CHO + CH_3CH(CH_3)CHO$

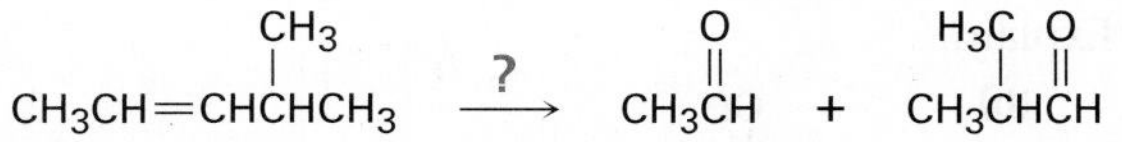

(f) $CH_3C(CH_3)=CH_2$ → $CH_3CH(CH_3)CH_2OH$

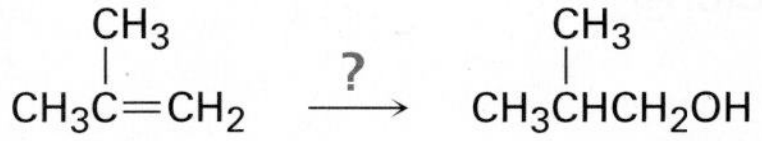

8.36 Draw the structure of an alkene that yields only acetone, $(CH_3)_2C=O$, on ozonolysis followed by treatment with Zn.

8.37 Show the structures of alkenes that give the following products on oxidative cleavage with $KMnO_4$ in acidic solution:

(a) $CH_3CH_2CO_2H$ + CO_2

(b) $(CH_3)_2C=O$ + $CH_3CH_2CH_2CO_2H$

(c) 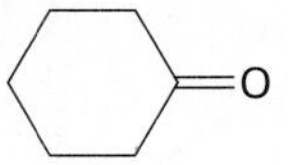 cyclohexanone (=O) + $(CH_3)_2C=O$

(d) $CH_3CH_2CCH_2CH_2CH_2CH_2CO_2H$ (with C=O)

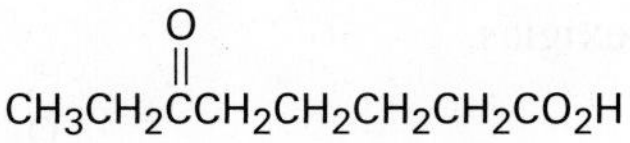

8.38 In planning the synthesis of one compound from another, it's just as important to know what *not* to do as to know what to do. The following reactions all have serious drawbacks to them. Explain the potential problems of each.

(a) CH₃C(CH₃)=CHCH₃ —HI→ CH₃C(CH₃)(I)CH(H)CH₃ — shown as (CH₃)₂C=CHCH₃ + HI → CH₃CH(I)CH(CH₃)CH₃ (with methyl and I on adjacent carbons)

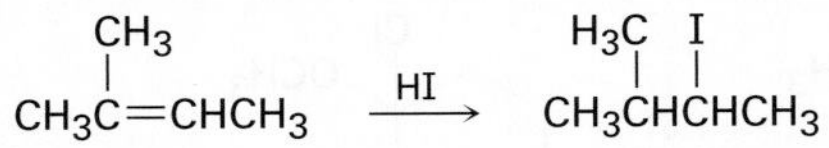

(b) cyclopentene —1. OsO₄; 2. NaHSO₃→ trans-1,2-cyclopentanediol

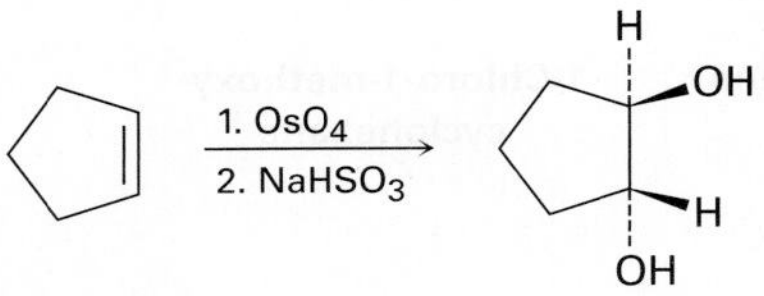

(c) 1,3-cyclohexadiene —1. O₃; 2. Zn→ cis-OHC-CH=CH-CH₂-CH₂-CHO (with both CHO groups on a cis alkene)

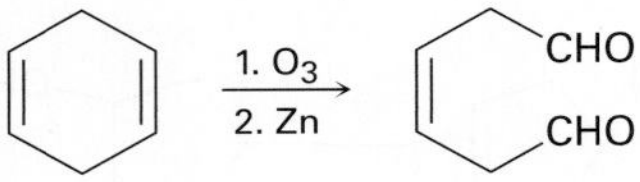

(d) 1-methylcyclohexene —1. BH₃; 2. H₂O₂, ⁻OH→ trans-1-methyl-2-hydroxycyclohexane

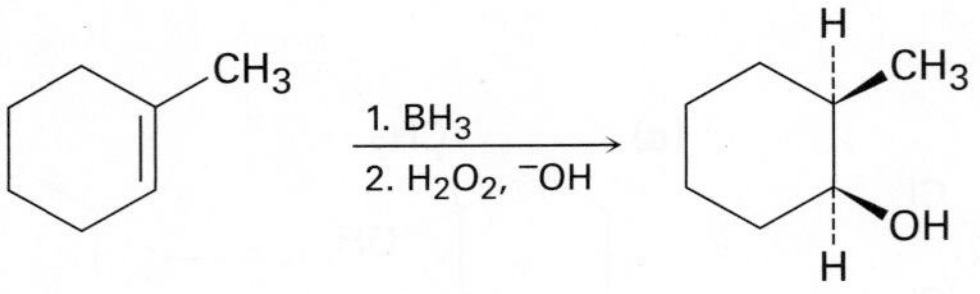

8.39 Which of the following alcohols could *not* be made selectively by hydroboration–oxidation of an alkene? Explain.

(a) CH₃CH₂CH₂CH(OH)CH₃

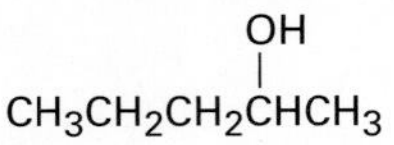

(b) (CH₃)₂CHCH(OH)(CH₃)₂

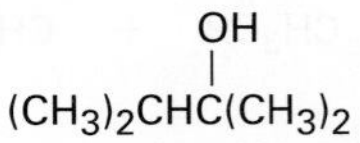

(c) trans-1-methyl-2-hydroxycyclohexane

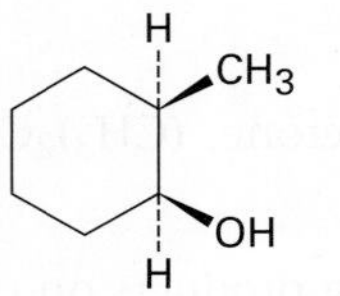

(d) cis-1-methyl-2-hydroxycyclohexane

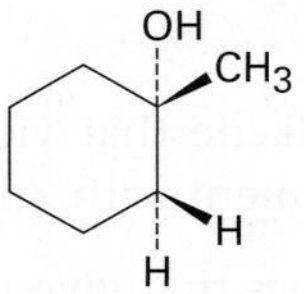

Polymers

8.40 Plexiglas, a clear plastic used to make many molded articles, is made by polymerization of methyl methacrylate. Draw a representative segment of Plexiglas.

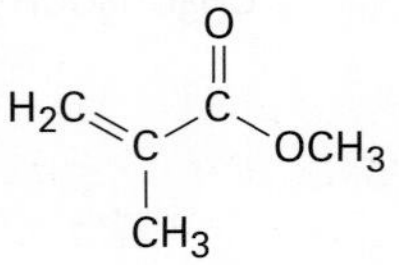

H₂C=C(CH₃)C(=O)OCH₃   **Methyl methacrylate**

8.41 Poly(vinyl pyrrolidone), prepared from *N*-vinylpyrrolidone, is used both in cosmetics and as a synthetic blood substitute. Draw a representative segment of the polymer.

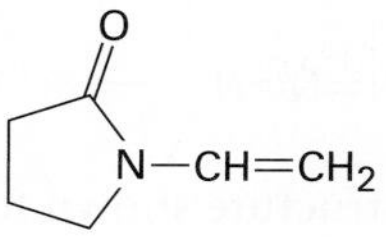

N-Vinylpyrrolidone

8.42 When a single alkene monomer, such as ethylene, is polymerized, the product is a *homopolymer*. If a mixture of two alkene monomers is polymerized, however, a *copolymer* often results. The following structure represents a segment of a copolymer called *Saran*. What two monomers were copolymerized to make Saran?

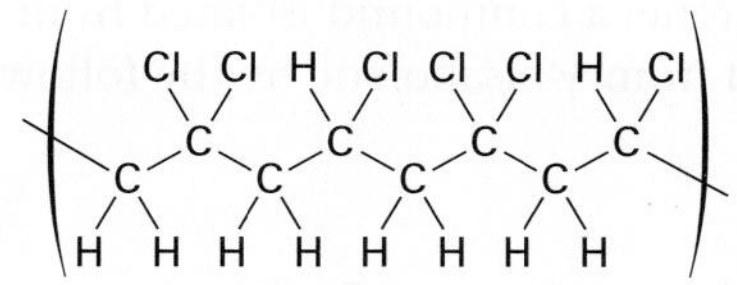

Saran

General Problems

8.43 Compound **A** has the formula $C_{10}H_{16}$. On catalytic hydrogenation over palladium, it reacts with only 1 molar equivalent of H_2. Compound **A** also undergoes reaction with ozone, followed by zinc treatment, to yield a symmetrical diketone, **B** ($C_{10}H_{16}O_2$).
(a) How many rings does **A** have?
(b) What are the structures of **A** and **B**?
(c) Write the reactions.

8.44 An unknown hydrocarbon **A** with the formula C_6H_{12} reacts with 1 molar equivalent of H_2 over a palladium catalyst. Hydrocarbon **A** also reacts with OsO_4 to give diol **B**. When oxidized with $KMnO_4$ in acidic solution, **A** gives two fragments. One fragment is propanoic acid, $CH_3CH_2CO_2H$, and the other fragment is ketone **C**. What are the structures of **A**, **B**, and **C**? Write all reactions, and show your reasoning.

8.45 Using an oxidative cleavage reaction, explain how you would distinguish between the following two isomeric dienes:

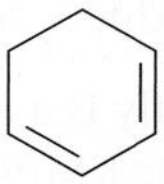

and

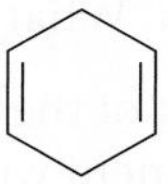

8.46 Compound **A**, $C_{10}H_{18}O$, undergoes reaction with dilute H_2SO_4 at 50 °C to yield a mixture of two alkenes, $C_{10}H_{16}$. The major alkene product, **B**, gives only cyclopentanone after ozone treatment followed by reduction with zinc in acetic acid. Identify **A** and **B**, and write the reactions.

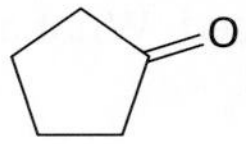

Cyclopentanone

8.47 Iodine azide, IN₃, adds to alkenes by an electrophilic mechanism similar to that of bromine. If a monosubstituted alkene such as 1-butene is used, only one product results:

$$CH_3CH_2CH=CH_2 \;+\; I-N=N=N \;\longrightarrow\; CH_3CH_2\underset{\underset{N=N=N}{|}}{C}HCH_2I$$

(a) Add lone-pair electrons to the structure shown for IN₃, and draw a second resonance form for the molecule.
(b) Calculate formal charges for the atoms in both resonance structures you drew for IN₃ in part (a).
(c) In light of the result observed when IN₃ adds to 1-butene, what is the polarity of the I–N₃ bond? Propose a mechanism for the reaction using curved arrows to show the electron flow in each step.

8.48 10-Bromo-α-chamigrene, a compound isolated from marine algae, is thought to be biosynthesized from γ-bisabolene by the following route:

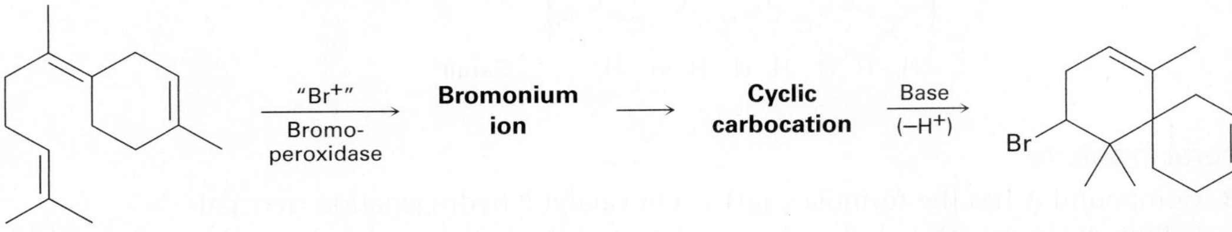

γ-Bisabolene $\xrightarrow[\text{Bromoperoxidase}]{\text{"Br}^+\text{"}}$ **Bromonium ion** $\longrightarrow$ **Cyclic carbocation** $\xrightarrow[(-H^+)]{\text{Base}}$ 10-Bromo-α-chamigrene

Draw the structures of the intermediate bromonium and cyclic carbocation, and propose mechanisms for all three steps.

8.49 Draw the structure of a hydrocarbon that absorbs 2 molar equivalents of H₂ on catalytic hydrogenation and gives only butanedial on ozonolysis.

$$\underset{\text{Butanedial}}{HCCH_2CH_2CH}$$ (with two C=O groups)

8.50 Simmons–Smith reaction of cyclohexene with diiodomethane gives a single cyclopropane product, but the analogous reaction of cyclohexene with 1,1-diiodoethane gives (in low yield) a mixture of two isomeric methylcyclopropane products. What are the two products, and how do they differ?

8.51 The sex attractant of the common housefly is a hydrocarbon with the formula $C_{23}H_{46}$. On treatment with aqueous acidic KMnO₄, two products are obtained, $CH_3(CH_2)_{12}CO_2H$ and $CH_3(CH_2)_7CO_2H$. Propose a structure.

8.52 Compound **A** has the formula C_8H_8. It reacts rapidly with KMnO₄ to give CO₂ and a carboxylic acid, **B** ($C_7H_6O_2$), but reacts with only 1 molar equivalent of H₂ on catalytic hydrogenation over a palladium catalyst. On hydrogenation under conditions that reduce aromatic rings, 4 equivalents of H₂ are taken up and hydrocarbon **C** (C_8H_{16}) is produced. What are the structures of **A**, **B**, and **C**? Write the reactions.

8.53 Isolated from marine algae, prelaureatin is thought to be biosynthesized from laurediol by the following route. Propose a mechanism.

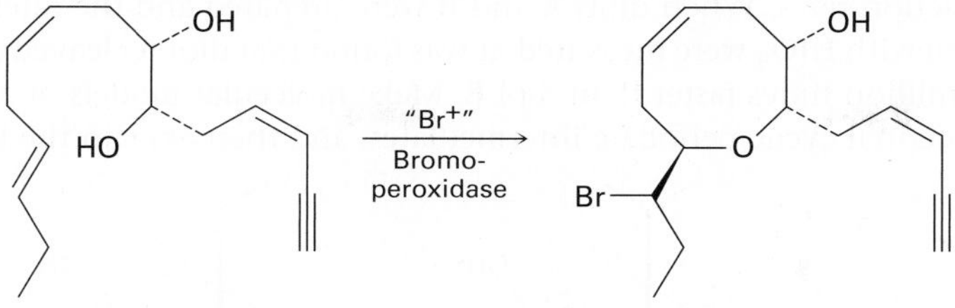

Laurediol → "Br⁺" / Bromoperoxidase → **Prelaureatin**

8.54 How would you distinguish between the following pairs of compounds using simple chemical tests? Tell what you would do and what you would see.
(a) Cyclopentene and cyclopentane (b) 2-Hexene and benzene

8.55 Dichlorocarbene can be generated by heating sodium trichloroacetate. Propose a mechanism for the reaction, and use curved arrows to indicate the movement of electrons in each step. What relationship does your mechanism bear to the base-induced elimination of HCl from chloroform?

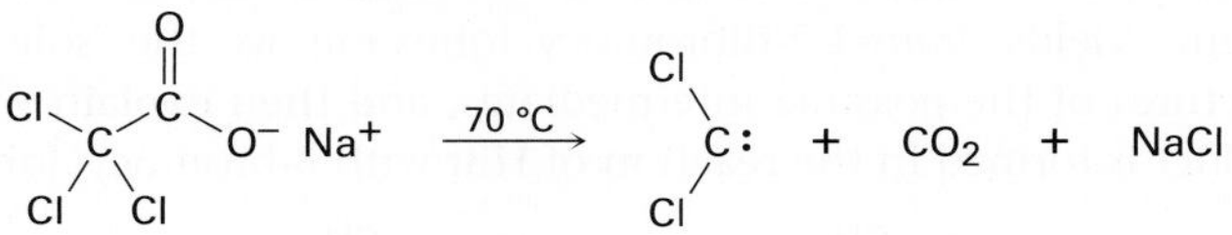

$$Cl_3C-CO_2^- \;Na^+ \xrightarrow{70\,°C} :CCl_2 + CO_2 + NaCl$$

8.56 α-Terpinene, $C_{10}H_{16}$, is a pleasant-smelling hydrocarbon that has been isolated from oil of marjoram. On hydrogenation over a palladium catalyst, α-terpinene reacts with 2 molar equivalents of H_2 to yield a hydrocarbon, $C_{10}H_{20}$. On ozonolysis, followed by reduction with zinc and acetic acid, α-terpinene yields two products, glyoxal and 6-methyl-2,5-heptanedione.

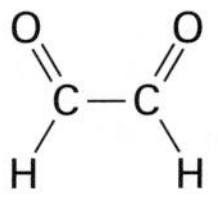

Glyoxal: OHC–CHO

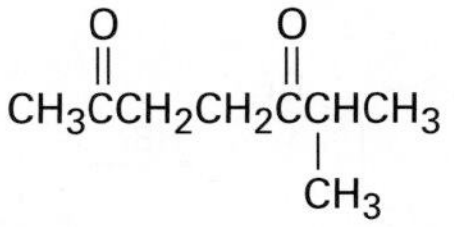

6-Methyl-2,5-heptanedione: $CH_3CCH_2CH_2CCHCH_3$ with CH_3 branch

(a) How many degrees of unsaturation does α-terpinene have?
(b) How many double bonds and how many rings does it have?
(c) Propose a structure for α-terpinene.

8.57 Evidence that cleavage of 1,2-diols by HIO_4 occurs through a five-membered cyclic periodate intermediate is based on *kinetic data*—the measurement of reaction rates. When diols **A** and **B** were prepared and the rates of their reaction with HIO_4 were measured, it was found that diol **A** cleaved approximately 1 million times faster than diol **B**. Make molecular models of **A** and **B** and of potential cyclic periodate intermediates, and then explain the kinetic results.

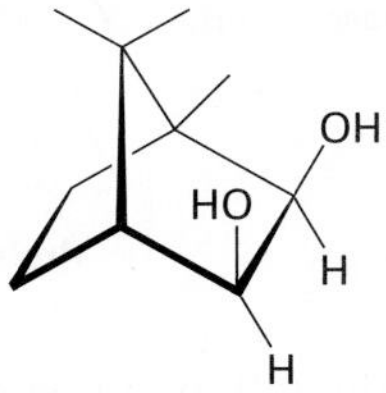

A
(*cis* diol)

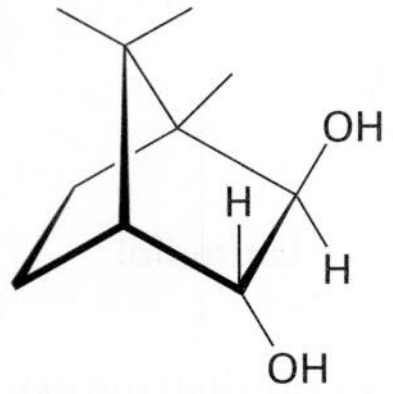

B
(*trans* diol)

8.58 Reaction of HBr with 3-methylcyclohexene yields a mixture of four products: *cis*- and *trans*-1-bromo-3-methylcyclohexane and *cis*- and *trans*-1-bromo-2-methylcyclohexane. The analogous reaction of HBr with 3-bromocyclohexene yields *trans*-1,2-dibromocyclohexane as the sole product. Draw structures of the possible intermediates, and then explain why only a single product is formed in the reaction of HBr with 3-bromocyclohexene.

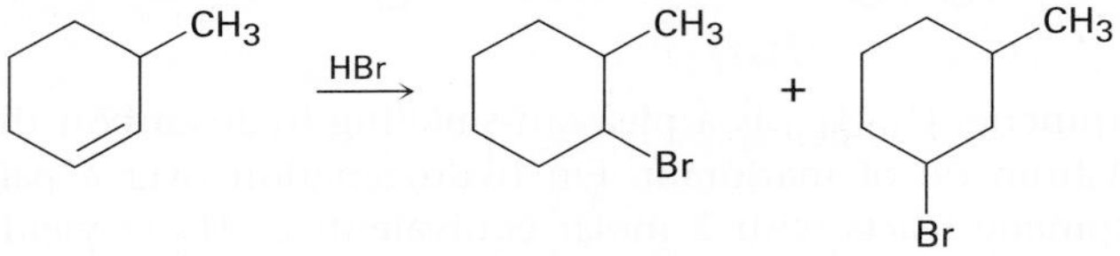

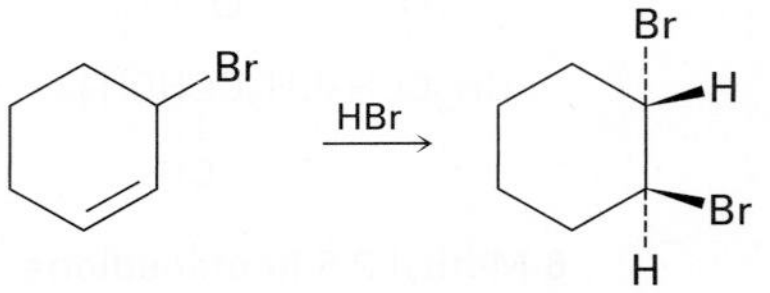

8.59 Reaction of cyclohexene with mercury(II) acetate in CH_3OH rather than H_2O, followed by treatment with $NaBH_4$, yields cyclohexyl methyl ether rather than cyclohexanol. Suggest a mechanism.

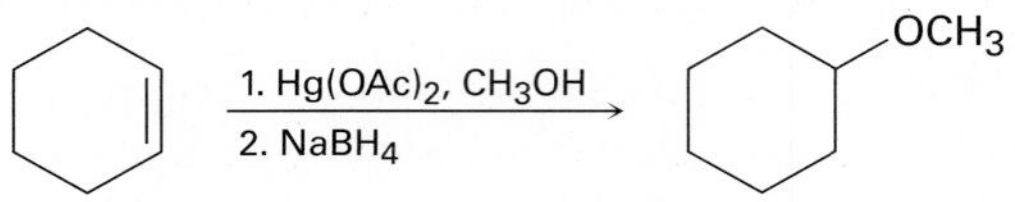

Cyclohexene → 1. $Hg(OAc)_2$, CH_3OH 2. $NaBH_4$ → **Cyclohexyl methyl ether**

8.60 Use your general knowledge of alkene chemistry to suggest a mechanism for the following reaction.

[Structure: bicyclic diene with CO₂CH₃ group] —Hg(OAc)₂→ [Structure: decalin-type product with CO₂CH₃ and AcO—Hg groups]

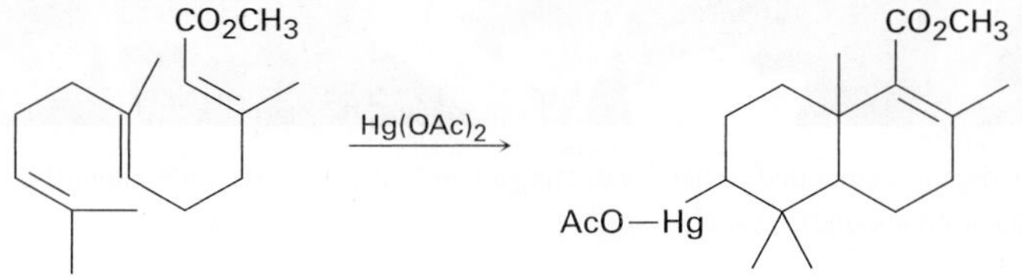

8.61 Treatment of 4-penten-1-ol with aqueous Br₂ yields a cyclic bromo ether rather than the expected bromohydrin, Suggest a mechanism, using curved arrows to show electron movement.

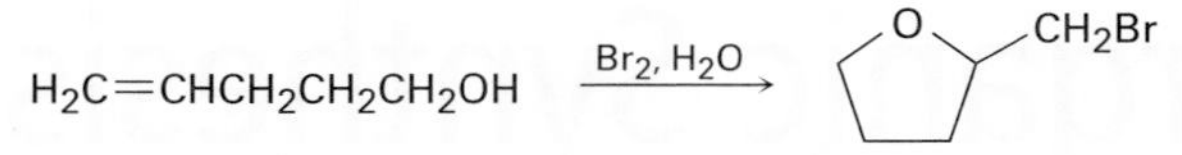

$H_2C=CHCH_2CH_2CH_2OH$ —Br₂, H₂O→ [tetrahydrofuran with CH₂Br substituent]

4-Penten-1-ol   **2-(Bromomethyl)tetrahydrofuran**

8.62 Hydroboration of 2-methyl-2-pentene at 25 °C followed by oxidation with alkaline H_2O_2 yields 2-methyl-3-pentanol, but hydroboration at 160 °C followed by oxidation yields 4-methyl-1-pentanol. Suggest a mechanism.

$$\begin{array}{c}CH_3\\|\\CH_3C=CHCH_2CH_3\end{array}$$

2-Methyl-2-pentene

1. BH₃, THF, 25 °C
2. H₂O₂, OH⁻

→ $CH_3\underset{|}{\overset{H_3C\ OH}{\underset{|}{C}H}}CHCH_2CH_3$

2-Methyl-3-pentanol

1. BH₃, THF, 160 °C
2. H₂O₂, OH⁻

→ $CH_3\underset{|}{\overset{CH_3}{C}H}CH_2CH_2CH_2OH$

4-Methyl-1-pentanol

8.63 We'll see in the next chapter that alkynes undergo many of the same reactions that alkenes do. What product might you expect from each of the following reactions?

$CH_3\underset{|}{\overset{CH_3}{C}H}CH_2CH_2C≡CH$

(a) —1 equiv Br₂→ ?
(b) —2 equiv H₂, Pd/C→ ?
(c) —1 equiv HBr→ ?

8.64 Hydroxylation of *cis*-2-butene with OsO₄ yields a different product than hydroxylation of *trans*-2-butene. Draw the structure, show the stereochemistry of each product, and explain the difference between them.

8.65 Compound **A**, $C_{11}H_{16}O$, was found to be an optically active alcohol. Despite its apparent unsaturation, no hydrogen was absorbed on catalytic reduction over a palladium catalyst. On treatment of **A** with dilute sulfuric acid, dehydration occurred and an optically inactive alkene **B**, $C_{11}H_{14}$, was produced as the major product. Alkene **B**, on ozonolysis, gave two products. One product was identified as propanal, CH_3CH_2CHO. Compound **C**, the other product, was shown to be a ketone, C_8H_8O. How many degrees of unsaturation does **A** have? Write the reactions, and identify **A**, **B**, and **C**.

9

Synthesizing organic compounds is like conducting an orchestra. When in tune, chemists can create highly complex organic compounds. © Olaf Doering/Alamy

Alkynes: An Introduction to Organic Synthesis

9.1 Naming Alkynes
9.2 Preparation of Alkynes: Elimination Reactions of Dihalides
9.3 Reactions of Alkynes: Addition of HX and X_2
9.4 Hydration of Alkynes
9.5 Reduction of Alkynes
9.6 Oxidative Cleavage of Alkynes
9.7 Alkyne Acidity: Formation of Acetylide Anions
9.8 Alkylation of Acetylide Anions
9.9 An Introduction to Organic Synthesis

A Deeper Look—The Art of Organic Synthesis

An **alkyne** is a hydrocarbon that contains a carbon–carbon triple bond. Acetylene, H—C≡C—H, the simplest alkyne, was once widely used in industry as the starting material for the preparation of acetaldehyde, acetic acid, vinyl chloride, and other high-volume chemicals, but more efficient routes to these substances using ethylene as starting material are now available. Acetylene is still used in the preparation of acrylic polymers, but is probably best known as the gas burned in high-temperature oxy–acetylene welding torches.

In addition to simple alkynes with one triple bond, research is also being carried out on *polyynes*—linear carbon chains of *sp*-hybridized carbon atoms. Polyynes with up to eight triple bonds have been detected in interstellar space, and evidence has been presented for the existence of *carbyne,* an allotrope of carbon consisting of repeating triple bonds in long chains of indefinite length. The electronic properties of polyynes are being explored for potential use in nanotechnology applications.

H—C≡C—C≡C—C≡C—C≡C—C≡C—C≡C—C≡C—H

A polyyne detected in interstellar space

Why This Chapter? Alkynes are less common than alkenes, both in the laboratory and in living organisms, so we won't cover them in great detail. The real importance of this chapter is that we'll use alkyne chemistry as a vehicle to begin looking at some of the general strategies used in organic synthesis—the construction of complex molecules in the laboratory. Without the ability to design and synthesize new molecules in the laboratory, many of the medicines we take for granted would not exist and few new ones would be made.

9.1 Naming Alkynes

Alkyne nomenclature follows the general rules for hydrocarbons discussed in **Sections 3.4 and 7.3**. The suffix *-yne* is used, and the position of the triple bond is indicated by giving the number of the first alkyne carbon in the chain.

⏻WL Sign in to OWL for Organic Chemistry at **www.cengage.com/owl** to view tutorials and simulations, develop problem-solving skills, and complete online homework assigned by your professor.

9.1 | Naming Alkynes

Numbering the main chain begins at the end nearer the triple bond so that the triple bond receives as low a number as possible.

$$\underset{876543\,21}{CH_3CH_2\underset{\underset{CH_3}{|}}{CH}CH_2C\equiv CCH_2CH_3}$$ Begin numbering at the end nearer the triple bond.

6-Methyl-3-octyne

(New: **6-Methyloct-3-yne**)

Compounds with more than one triple bond are called diynes, triynes, and so forth; compounds containing both double and triple bonds are called enynes (not ynenes). Numbering of an enyne chain starts from the end nearer the first multiple bond, whether double or triple. When there is a choice in numbering, double bonds receive lower numbers than triple bonds. For example:

$$\underset{7654321}{HC\equiv CCH_2CH_2CH_2CH=CH_2}$$
$$\underset{12\,345678\,9}{HC\equiv CCH_2\underset{\underset{CH_3}{|}}{CH}CH_2CH_2CH=CHCH_3}$$

1-Hepten-6-yne   **4-Methyl-7-nonen-1-yne**

(New: **Hept-1-en-6-yne**)   (New: **4-Methylnon-7-en-1-yne**)

As with alkyl and alkenyl substituents derived from alkanes and alkenes, respectively, alkynyl groups are also possible.

CH₃CH₂CH₂CH₂—   CH₃CH₂CH=CH—   CH₃CH₂C≡C—

Butyl   **1-Butenyl**   **1-Butynyl**
(an alkyl group)   (a vinylic group)   (an alkynyl group)

                      (New: **But-1-enyl**)   (New: **But-1-ynyl**)

Problem 9.1
Name the following compounds:

(a) $$CH_3\underset{\underset{CH_3}{|}}{CH}C\equiv C\underset{\underset{CH_3}{|}}{CH}CH_3$$

(b) $$HC\equiv C\underset{\underset{CH_3}{|}}{\overset{\overset{CH_3}{|}}{C}}CH_3$$

(c) $$CH_3CH_2\underset{\underset{CH_3}{|}}{\overset{\overset{CH_3}{|}}{C}}C\equiv CCH_2CH_2CH_3$$

(d) $$CH_3CH_2\underset{\underset{CH_3}{|}}{\overset{\overset{CH_3}{|}}{C}}C\equiv C\underset{\underset{CH_3}{|}}{CH}CH_3$$

(e)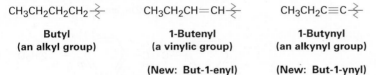

(f) CH₃CH=CHCH=CHC≡CCH₃

Problem 9.2
There are seven isomeric alkynes with the formula C_6H_{10}. Draw and name them.

9.2 Preparation of Alkynes: Elimination Reactions of Dihalides

Alkynes can be prepared by the elimination of HX from alkyl halides in much the same manner as alkenes (Section 8.1). Treatment of a 1,2-dihaloalkane (a *vicinal* dihalide) with an excess amount of a strong base such as KOH or $NaNH_2$ results in a twofold elimination of HX and formation of an alkyne. As with the elimination of HX to form an alkene, we'll defer a full discussion of this topic and the relevant reaction mechanisms until Chapter 11.

The starting vicinal dihalides are themselves readily available by addition of Br_2 or Cl_2 to alkenes. Thus, the overall halogenation/dehydrohalogenation sequence makes it possible to go from an alkene to an alkyne. For example, diphenylethylene is converted into diphenylacetylene by reaction with Br_2 and subsequent base treatment.

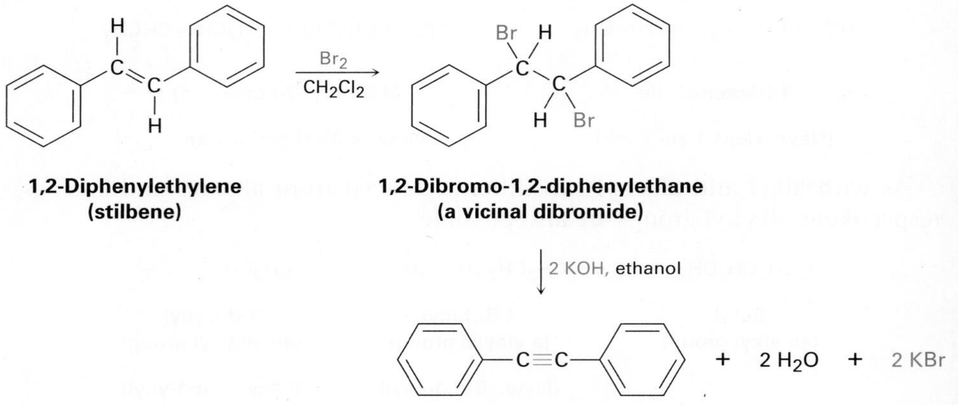

1,2-Diphenylethylene (stilbene) $\xrightarrow{Br_2, CH_2Cl_2}$ **1,2-Dibromo-1,2-diphenylethane (a vicinal dibromide)** $\xrightarrow{2\text{ KOH, ethanol}}$ **Diphenylacetylene (85%)** + $2 H_2O$ + $2 KBr$

The twofold dehydrohalogenation takes place through a vinylic halide intermediate, which suggests that vinylic halides themselves should give alkynes when treated with strong base. (*Remember:* A *vinylic* substituent is one that is attached to a double-bond carbon.) This is indeed the case. For example:

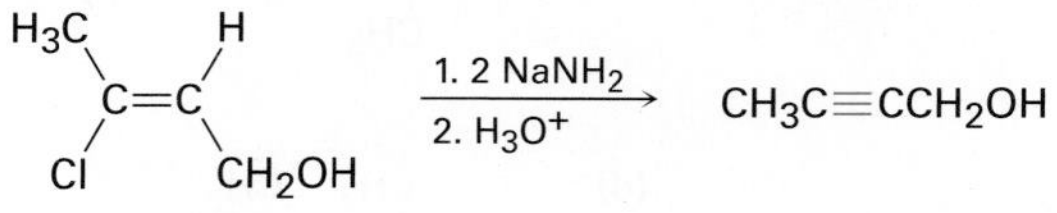

(Z)-3-Chloro-2-buten-1-ol $\xrightarrow{\text{1. 2 NaNH}_2 \\ \text{2. H}_3\text{O}^+}$ $CH_3C\equiv CCH_2OH$ **2-Butyn-1-ol**

9.3 Reactions of Alkynes: Addition of HX and X_2

You might recall from Section 1.9 that a carbon–carbon triple bond results from the interaction of two *sp*-hybridized carbon atoms. The two *sp* hybrid orbitals of carbon lie at an angle of 180° to each other along an axis perpendicular to the axes of the two unhybridized $2p_y$ and $2p_z$ orbitals. When two *sp*-hybridized carbons approach each other, one *sp–sp* σ bond and two *p–p* π bonds are formed.

The two remaining *sp* orbitals form bonds to other atoms at an angle of 180° from the carbon–carbon bond. Thus, acetylene is a linear molecule with H—C≡C bond angles of 180° **(Figure 9.1)**. The length of the C≡C bond is 120 pm, and its strength is approximately 965 kJ/mol (231 kcal/mol), making it the shortest and strongest known carbon–carbon bond.

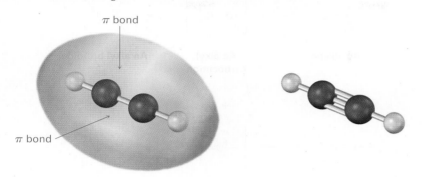

Figure 9.1 The structure of acetylene, H—C≡C—H. The H—C≡C bond angles are 180°, and the C≡C bond length is 120 pm. The electrostatic potential map shows that the π bonds create a **negative belt** around the molecule.

As a general rule, electrophiles undergo addition reactions with alkynes much as they do with alkenes. Take the reaction of alkynes with HX, for instance. The reaction often can be stopped after addition of 1 equivalent of HX, but reaction with an excess of HX leads to a dihalide product. For example, reaction of 1-hexyne with 2 equivalents of HBr yields 2,2-dibromohexane. As the following examples indicate, the regiochemistry of addition follows Markovnikov's rule, with halogen adding to the more highly substituted side of the alkyne bond and hydrogen adding to the less highly substituted side. Trans stereochemistry of H and X normally, although not always, results in the product.

$CH_3CH_2CH_2CH_2C \equiv CH$ $\xrightarrow{HBr}$ 2-Bromo-1-hexene $\xrightarrow{HBr}$ 2,2-Dibromohexane

1-Hexyne

$CH_3CH_2C \equiv CCH_2CH_3$ $\xrightarrow[CH_3CO_2H]{HCl}$ (Z)-3-Chloro-3-hexene $\xrightarrow[CH_3CO_2H]{HCl}$ 3,3-Dichlorohexane

3-Hexyne

Bromine and chlorine also add to alkynes to give addition products, and trans stereochemistry again results.

$CH_3CH_2C \equiv CH$ $\xrightarrow[CH_2Cl_2]{Br_2}$ (E)-1,2-Dibromo-1-butene $\xrightarrow[CH_2Cl_2]{Br_2}$ 1,1,2,2-Tetrabromobutane

1-Butyne

The mechanism of alkyne additions is similar but not identical to that of alkene additions. When an electrophile such as HBr adds to an alkene, the reaction takes place in two steps and involves an alkyl carbocation intermediate

(Sections 7.7 and 7.8). If HBr were to add by the same mechanism to an *alkyne*, an analogous *vinylic* carbocation would be formed as the intermediate.

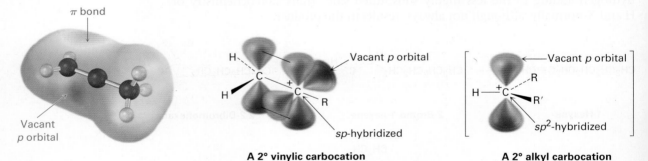

A vinylic carbocation has an *sp*-hybridized carbon and generally forms less readily than an alkyl carbocation (**Figure 9.2**). As a rule, a secondary vinylic carbocation forms about as readily as a primary alkyl carbocation, but a primary vinylic carbocation is so difficult to form that there is no clear evidence it even exists. Thus, many alkyne additions occur through more complex mechanistic pathways.

Figure 9.2 The structure of a secondary vinylic carbocation. The cationic carbon atom is *sp*-hybridized and has a vacant *p* orbital perpendicular to the plane of the π bond orbitals. Only one R group is attached to the positively charged carbon rather than two, as in a secondary alkyl carbocation. The electrostatic potential map shows that the **most positive regions** coincide with lobes of the vacant *p* orbital and are perpendicular to the **most negative regions** associated with the π bond.

Problem 9.3
What products would you expect from the following reactions?
(a) $CH_3CH_2CH_2C\equiv CH\ +\ 2\ Cl_2\ \longrightarrow\ ?$

(b) [cyclopentyl]—C≡CH + 1 HBr ⟶ ?

(c) $CH_3CH_2CH_2CH_2C\equiv CCH_3\ +\ 1\ HBr\ \longrightarrow\ ?$

9.4 Hydration of Alkynes

Like alkenes **(Sections 8.4 and 8.5)**, alkynes can be hydrated by either of two methods. Direct addition of water catalyzed by mercury(II) ion yields the Markovnikov product, and indirect addition of water by a hydroboration–oxidation sequence yields the non-Markovnikov product.

Mercury(II)-Catalyzed Hydration of Alkynes

Alkynes don't react directly with aqueous acid but will undergo hydration readily in the presence of mercury(II) sulfate as a Lewis acid catalyst. The reaction occurs with Markovnikov regiochemistry, so the —OH group adds to the more highly substituted carbon and the —H attaches to the less highly substituted one.

$$CH_3CH_2CH_2CH_2C\equiv CH \xrightarrow[HgSO_4]{H_2O,\ H_2SO_4} \left[CH_3CH_2CH_2CH_2\underset{OH}{C}=CH_2\right] \longrightarrow CH_3CH_2CH_2CH_2\underset{H\ H}{\overset{O}{C}}\underset{}{C}H$$

1-Hexyne   **An enol**   **2-Hexanone (78%)**

Interestingly, the product actually isolated from alkyne hydration is not the vinylic alcohol, or **enol** (ene + ol), but is instead a ketone. Although the enol is an intermediate in the reaction, it immediately rearranges to a ketone by a process called *keto–enol tautomerism*. The individual keto and enol forms are said to be **tautomers**, a word used to describe two isomers that under spontaneous interconversion accompanied by the change in position of a hydrogen. With few exceptions, the keto–enol tautomeric equilibrium lies on the side of the ketone; enols are almost never isolated. We'll look more closely at this equilibrium in **Section 22.1**.

Enol tautomer   **Keto tautomer**
(less favored)   **(more favored)**

As shown in **Figure 9.3**, the mechanism of the mercury(II)-catalyzed alkyne hydration reaction is analogous to the oxymercuration reaction of alkenes **(Section 8.4)**. Electrophilic addition of mercury(II) ion to the alkyne gives a vinylic cation, which reacts with water and loses a proton to yield a mercury-containing enol intermediate. In contrast with alkene oxymercuration, however, no treatment with NaBH$_4$ is necessary to remove the mercury. The acidic reaction conditions alone are sufficient to effect replacement of mercury by hydrogen. Tautomerization then gives the ketone.

Figure 9.3 | MECHANISM

Mechanism of the mercury(II)-catalyzed hydration of an alkyne to yield a ketone. The reaction occurs through initial formation of an intermediate enol, which tautomerizes to the ketone.

1 The alkyne uses a pair of electrons to attack the electrophilic mercury(II) ion, yielding a mercury-containing vinylic carbocation intermediate.

2 Nucleophilic attack of water on the carbocation forms a C–O bond and yields a protonated mercury-containing enol.

3 Abstraction of H⁺ from the protonated enol by water gives an organomercury compound.

4 Replacement of Hg^{2+} by H⁺ occurs to give a neutral enol.

5 The enol undergoes tautomerization to give the final ketone product.

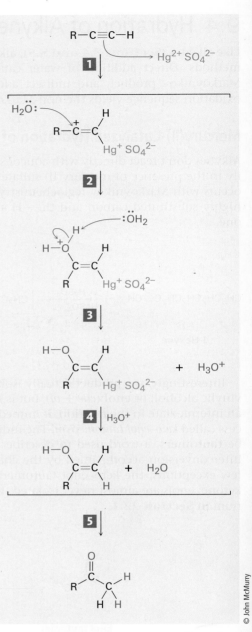

A mixture of both possible ketones results when an unsymmetrically substituted internal alkyne (RC≡CR′) is hydrated. The reaction is therefore most useful when applied to a terminal alkyne (RC≡CH) because only a methyl ketone is formed.

An internal alkyne

$$R-C\equiv C-R' \xrightarrow[HgSO_4]{H_3O^+} R-\overset{O}{\underset{}{C}}-CH_2R' \;+\; RCH_2-\overset{O}{\underset{}{C}}-R'$$

Mixture

9.4 | Hydration of Alkynes

A terminal alkyne

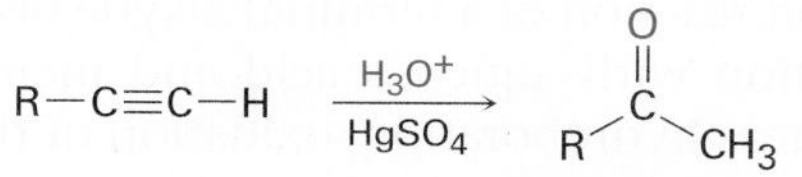

$$R-C\equiv C-H \xrightarrow[HgSO_4]{H_3O^+} R-\underset{\underset{\textbf{A methyl ketone}}{}}{\overset{O}{\underset{\|}{C}}}-CH_3$$

Problem 9.4
What product would you obtain by hydration of the following alkynes?
(a) $CH_3CH_2CH_2C\equiv CCH_2CH_2CH_3$
(b) $CH_3CH(CH_3)CH_2C\equiv CCH_2CH_2CH_3$

Problem 9.5
What alkynes would you start with to prepare the following ketones?
(a) $CH_3CH_2CH_2\overset{O}{\underset{\|}{C}}CH_3$
(b) $CH_3CH_2\overset{O}{\underset{\|}{C}}CH_2CH_3$

Hydroboration–Oxidation of Alkynes

Borane adds rapidly to an alkyne just as it does to an alkene, and the resulting vinylic borane can be oxidized by H_2O_2 to yield an enol. Tautomerization then gives either a ketone or an aldehyde, depending on the structure of the alkyne reactant. Hydroboration–oxidation of an internal alkyne such as 3-hexyne gives a ketone, and hydroboration–oxidation of a terminal alkyne gives an aldehyde. Note that the relatively unhindered terminal alkyne undergoes two additions, giving a doubly hydroborated intermediate. Oxidation with H_2O_2 at pH 8 then replaces both boron atoms by oxygen and generates the aldehyde.

An internal alkyne

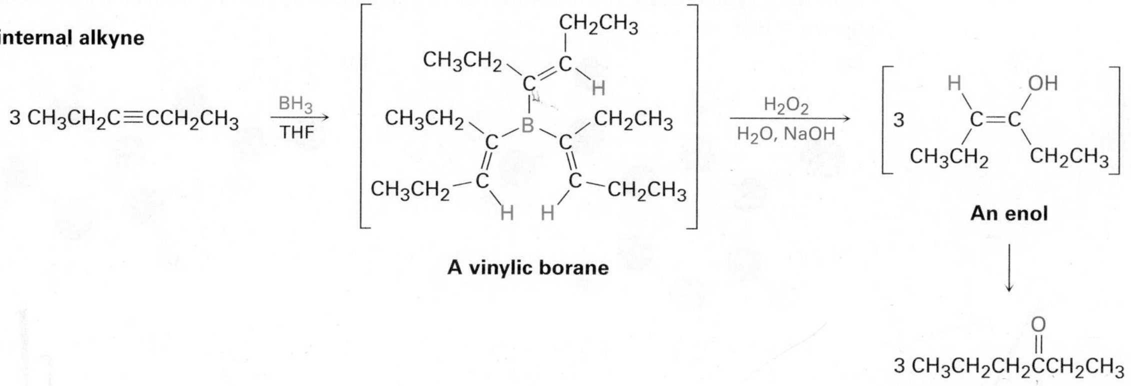

$$3\ CH_3CH_2C\equiv CCH_2CH_3 \xrightarrow[THF]{BH_3} \text{A vinylic borane} \xrightarrow[H_2O,\ NaOH]{H_2O_2} 3\ \text{An enol} \longrightarrow 3\ CH_3CH_2CH_2\overset{O}{\underset{\|}{C}}CH_2CH_3$$

3-Hexanone

A terminal alkyne

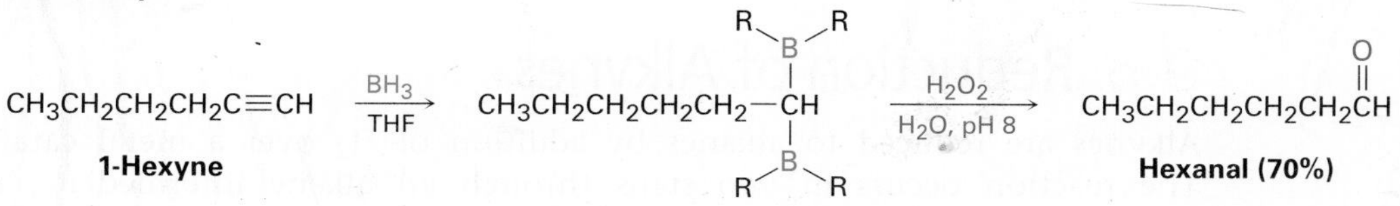

$$CH_3CH_2CH_2CH_2C\equiv CH \xrightarrow[THF]{BH_3} CH_3CH_2CH_2CH_2CH_2-CH(BR_2)_2 \xrightarrow[H_2O,\ pH\ 8]{H_2O_2} CH_3CH_2CH_2CH_2CH_2CHO$$

1-Hexyne   **Hexanal (70%)**

CHAPTER 9 | Alkynes: An Introduction to Organic Synthesis

The hydroboration–oxidation sequence is complementary to the direct, mercury(II)-catalyzed hydration reaction of a terminal alkyne because different products result. Direct hydration with aqueous acid and mercury(II) sulfate leads to a methyl ketone, whereas hydroboration–oxidation of the same terminal alkyne leads to an aldehyde.

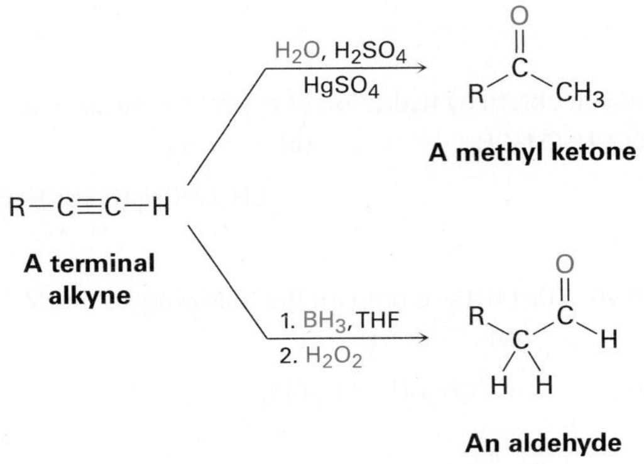

Problem 9.6
What alkyne would you start with to prepare each of the following compounds by a hydroboration–oxidation reaction?

(a) C₆H₅—CH₂CHO

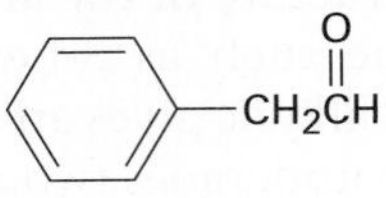

(b) CH₃CHCH₂C(O)CHCH₃ with CH₃ groups

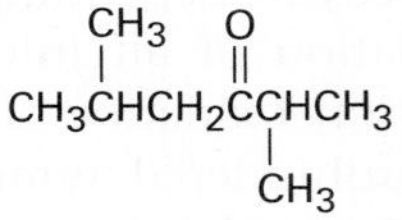

Problem 9.7
How would you prepare the following carbonyl compounds starting from an alkyne (reddish brown = Br)?

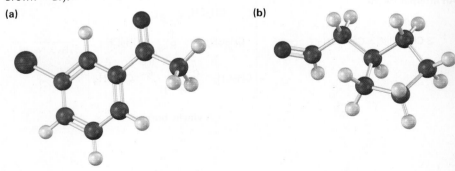

(a)   (b)

9.5 Reduction of Alkynes

Alkynes are reduced to alkanes by addition of H₂ over a metal catalyst. The reaction occurs in two steps through an alkene intermediate, and measurements show that the first step in the reaction is more exothermic than the second step.

$$HC \equiv CH \xrightarrow[\text{Catalyst}]{H_2} H_2C=CH_2 \qquad \Delta H°_{hydrog} = -176 \text{ kJ/mol } (-42 \text{ kcal/mol})$$

$$H_2C=CH_2 \xrightarrow[\text{Catalyst}]{H_2} CH_3-CH_3 \qquad \Delta H°_{hydrog} = -137 \text{ kJ/mol } (-33 \text{ kcal/mol})$$

Complete reduction to the alkane occurs when palladium on carbon (Pd/C) is used as catalyst, but hydrogenation can be stopped at the alkene stage if the less active *Lindlar catalyst* is used. The Lindlar catalyst is a finely divided palladium metal that has been precipitated onto a calcium carbonate support and then deactivated by treatment with lead acetate and quinoline, an aromatic amine. The hydrogenation occurs with syn stereochemistry **(Section 8.5)**, giving a cis alkene product.

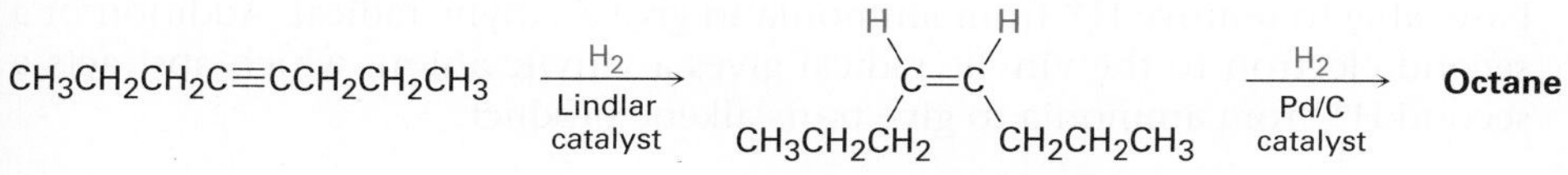

CH₃CH₂CH₂C≡CCH₂CH₂CH₃ →(H₂, Lindlar catalyst)→ cis-4-Octene →(H₂, Pd/C catalyst)→ Octane

4-Octyne

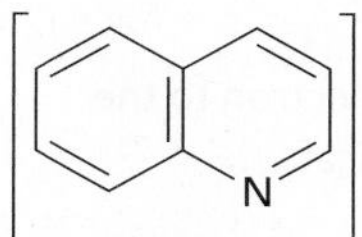

Quinoline

The alkyne hydrogenation reaction has been explored extensively by the Hoffmann–LaRoche pharmaceutical company, where it is used in the commercial synthesis of vitamin A. The cis isomer of vitamin A produced initially on hydrogenation is converted to the trans isomer by heating.

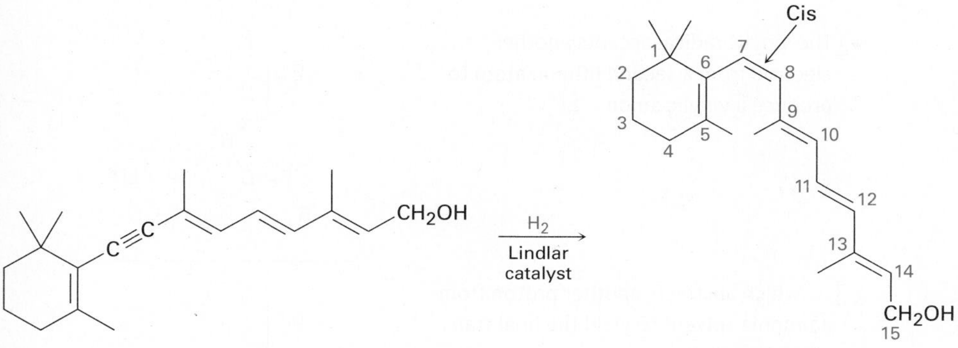

7-*cis*-Retinol
(7-*cis*-vitamin A; vitamin A has a trans double bond at C7)

An alternative method for the conversion of an alkyne to an alkene uses sodium or lithium metal as the reducing agent in liquid ammonia as solvent. This method is complementary to the Lindlar reduction because it produces trans rather than cis alkenes. For example, 5-decyne gives *trans*-5-decene on treatment with lithium in liquid ammonia.

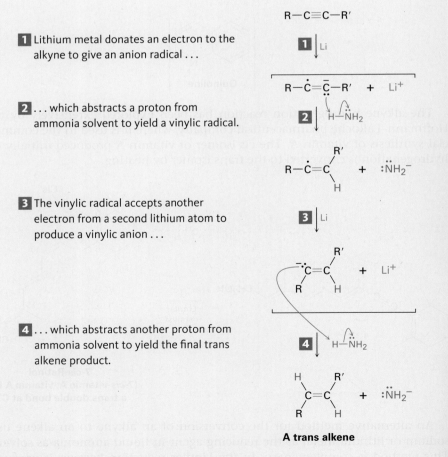

Alkali metals dissolve in liquid ammonia at $-33\ °C$ to produce a deep blue solution containing the metal cation and ammonia-solvated electrons. When an alkyne is then added to the solution, reduction occurs by the mechanism shown in **Figure 9.4**. An electron first adds to the triple bond to yield an intermediate anion radical—a species that is both an anion (has a negative charge) and a radical (has an odd number of electrons). This anion radical is a strong base, able to remove H^+ from ammonia to give a vinylic radical. Addition of a second electron to the vinylic radical gives a vinylic anion, which abstracts a second H^+ from ammonia to give trans alkene product.

Figure 9.4 | MECHANISM

Mechanism of the lithium/ammonia reduction of an alkyne to produce a trans alkene.

1 Lithium metal donates an electron to the alkyne to give an anion radical . . .

2 . . . which abstracts a proton from ammonia solvent to yield a vinylic radical.

3 The vinylic radical accepts another electron from a second lithium atom to produce a vinylic anion . . .

4 . . . which abstracts another proton from ammonia solvent to yield the final trans alkene product.

Trans stereochemistry of the alkene product is established during the second reduction step (**3**) when the less hindered trans vinylic anion is formed from the vinylic radical. Vinylic radicals undergo rapid cis–trans equilibration, but vinylic anions equilibrate much less rapidly. Thus, the more stable trans vinylic anion is formed rather than the less stable cis anion and is then protonated without equilibration.

Problem 9.8
Using any alkyne needed, how would you prepare the following alkenes?
(a) *trans*-2-Octene   (b) *cis*-3-Heptene   (c) 3-Methyl-1-pentene

---

9.6 Oxidative Cleavage of Alkynes

Alkynes, like alkenes, can be cleaved by reaction with powerful oxidizing agents such as ozone or $KMnO_4$, although the reaction is of little value and we mention it only for completeness. A triple bond is generally less reactive than a double bond, and yields of cleavage products are sometimes low. The products obtained from cleavage of an internal alkyne are carboxylic acids; from a terminal alkyne, CO_2 is formed as one product.

An internal alkyne

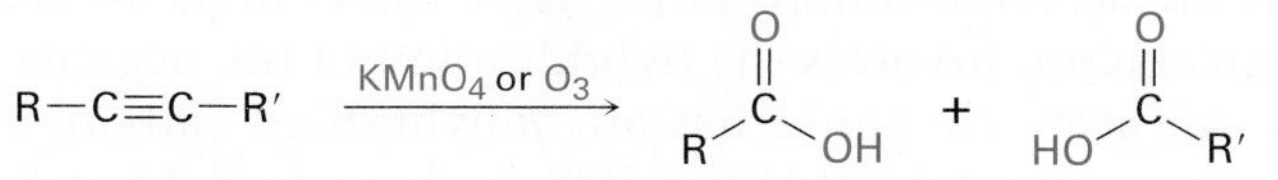

$$R-C\equiv C-R' \xrightarrow{KMnO_4 \text{ or } O_3} R-\underset{O}{\overset{O}{\|}}C-OH + HO-\underset{}{\overset{O}{\|}}C-R'$$

A terminal alkyne

$$R-C\equiv C-H \xrightarrow{KMnO_4 \text{ or } O_3} R-\underset{}{\overset{O}{\|}}C-OH + O=C=O$$

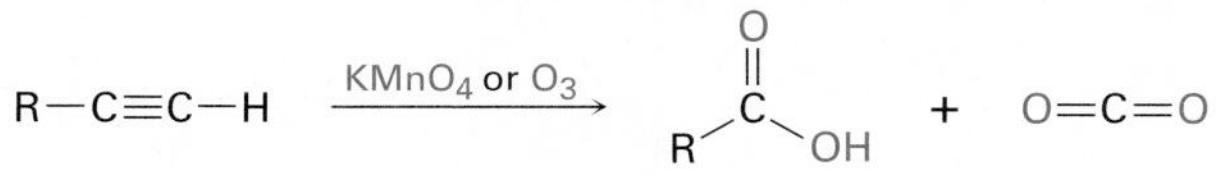

9.7 Alkyne Acidity: Formation of Acetylide Anions

The most striking difference between alkenes and alkynes is that terminal alkynes are relatively acidic. When a terminal alkyne is treated with a strong base, such as sodium amide, $Na^+ \: ^-NH_2$, the terminal hydrogen is removed and an **acetylide anion** is formed.

$$R-C\equiv C-H \xrightarrow{:NH_2 \: Na^+} R-C\equiv C:^- \: Na^+ \: + \: :NH_3$$

A terminal alkyne   **An acetylide anion**

According to the Brønsted–Lowry definition **(Section 2.7)**, an acid is a substance that donates H^+. Although we usually think of oxyacids (H_2SO_4, HNO_3) or halogen acids (HCl, HBr) in this context, any compound containing a hydrogen atom can be an acid under the right circumstances. By measuring dissociation constants of different acids and expressing the results as pK_a values, an acidity order can be established. Recall from **Section 2.8** that a lower pK_a corresponds to a stronger acid and a higher pK_a corresponds to a weaker acid.

Where do hydrocarbons lie on the acidity scale? As the data in Table 9.1 show, both methane (p$K_a \approx 60$) and ethylene (p$K_a = 44$) are very weak acids and thus do not react with any of the common bases. Acetylene, however, has p$K_a = 25$ and can be deprotonated by the conjugate base of any acid whose pK_a is greater than 25. Amide ion (NH_2^-), for example, the conjugate base of ammonia (p$K_a = 35$), is often used to deprotonate terminal alkynes.

Table 9.1 Acidity of Simple Hydrocarbons

Family	Example	K_a	pK_a	
Alkyne	HC≡CH	10^{-25}	25	Stronger acid
Alkene	$H_2C=CH_2$	10^{-44}	44	↑
Alkane	CH_4	10^{-60}	60	Weaker acid

Why are terminal alkynes more acidic than alkenes or alkanes? In other words, why are acetylide anions more stable than vinylic or alkyl anions? The simplest explanation involves the hybridization of the negatively charged carbon atom. An acetylide anion has an *sp*-hybridized carbon, so the negative charge resides in an orbital that has 50% "*s* character." A vinylic anion has an *sp*^{2}-hybridized carbon with 33% *s* character, and an alkyl anion (*sp*^{3}) has only 25% *s* character. Because *s* orbitals are nearer the positive nucleus and lower in energy than *p* orbitals, the negative charge is stabilized to a greater extent in an orbital with higher *s* character **(Figure 9.5)**.

Figure 9.5 A comparison of alkyl, vinylic, and acetylide anions. The acetylide anion, with *sp* hybridization, has more *s* character and is more stable. Electrostatic potential maps show that placing the negative charge closer to the carbon nucleus makes carbon appear less negative (red).

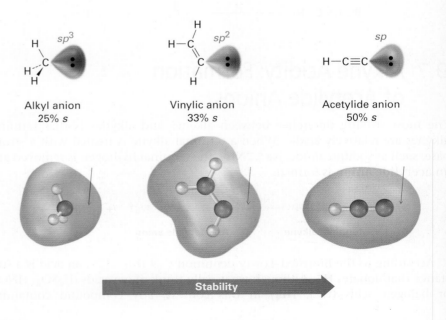

Problem 9.9
The pK_a of acetone, CH_3COCH_3, is 19.3. Which of the following bases is strong enough to deprotonate acetone?
(a) KOH (pK_a of H_2O = 15.7)
(b) Na^+ $^-C{\equiv}CH$ (pK_a of C_2H_2 = 25)
(c) $NaHCO_3$ (pK_a of H_2CO_3 = 6.4)
(d) $NaOCH_3$ (pK_a of CH_3OH = 15.6)

9.8 Alkylation of Acetylide Anions

The negative charge and unshared electron pair on carbon make an acetylide anion strongly nucleophilic. As a result, an acetylide anion can react with electrophiles, such as alkyl halides, in a process that replaces the halide and yields a new alkyne product.

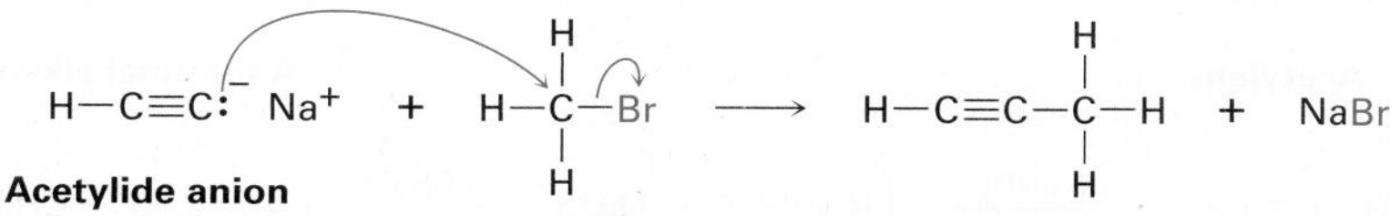

$$H-C{\equiv}C:^- \; Na^+ \; + \; H_3C-Br \; \longrightarrow \; H-C{\equiv}C-CH_3 \; + \; NaBr$$

Acetylide anion → **Propyne**

We won't study the details of this substitution reaction until Chapter 11 but for now can picture it as happening by the pathway shown in **Figure 9.6**. The nucleophilic acetylide ion uses an electron pair to form a bond to the positively polarized, electrophilic carbon atom of bromomethane. As the new C–C bond forms, Br⁻ departs, taking with it the electron pair from the former C–Br bond and yielding propyne as product. We call such a reaction an **alkylation** because a new alkyl group has become attached to the starting alkyne.

Figure 9.6 MECHANISM

A mechanism for the alkylation reaction of acetylide anion with bromomethane to give propyne.

1 The nucleophilic acetylide anion uses its electron lone pair to form a bond to the positively polarized, electrophilic carbon atom of bromomethane. As the new C–C bond begins to form, the C–Br bond begins to break in the transition state.

2 The new C–C bond is fully formed and the old C–Br bond is fully broken at the end of the reaction.

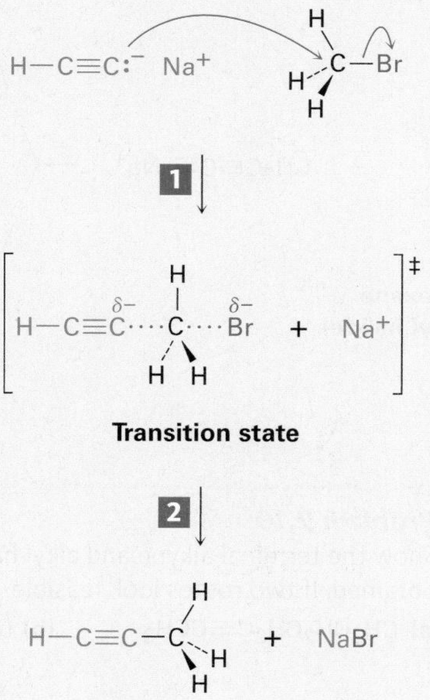

© John McMurry

Alkyne alkylation is not limited to acetylene itself. Any terminal alkyne can be converted into its corresponding anion and then alkylated by treatment with an alkyl halide, yielding an internal alkyne. For example, conversion of 1-hexyne into its anion, followed by reaction with 1-bromobutane, yields 5-decyne.

$$CH_3CH_2CH_2CH_2C{\equiv}CH \xrightarrow[\text{2. } CH_3CH_2CH_2CH_2Br]{\text{1. } NaNH_2,\ NH_3} CH_3CH_2CH_2CH_2C{\equiv}CCH_2CH_2CH_2CH_3$$

1-Hexyne   **5-Decyne (76%)**

Because of its generality, acetylide alkylation is a good method for preparing substituted alkynes from simpler precursors. A terminal alkyne can be prepared by alkylation of acetylene itself, and an internal alkyne can be prepared by further alkylation of a terminal alkyne.

$$H-C{\equiv}C-H \xrightarrow{NaNH_2} [H-C{\equiv}C{:}^-\ Na^+] \xrightarrow{RCH_2Br} H-C{\equiv}C-CH_2R$$

Acetylene   **A terminal alkyne**

$$R-C{\equiv}C-H \xrightarrow{NaNH_2} [R-C{\equiv}C{:}^-\ Na^+] \xrightarrow{R'CH_2Br} R-C{\equiv}C-CH_2R'$$

A terminal alkyne   **An internal alkyne**

The alkylation reaction is limited to the use of primary alkyl bromides and alkyl iodides because acetylide ions are sufficiently strong bases to cause elimination instead of substitution when they react with secondary and tertiary alkyl halides. For example, reaction of bromocyclohexane with propyne anion yields the elimination product cyclohexene rather than the substitution product 1-propynylcyclohexane.

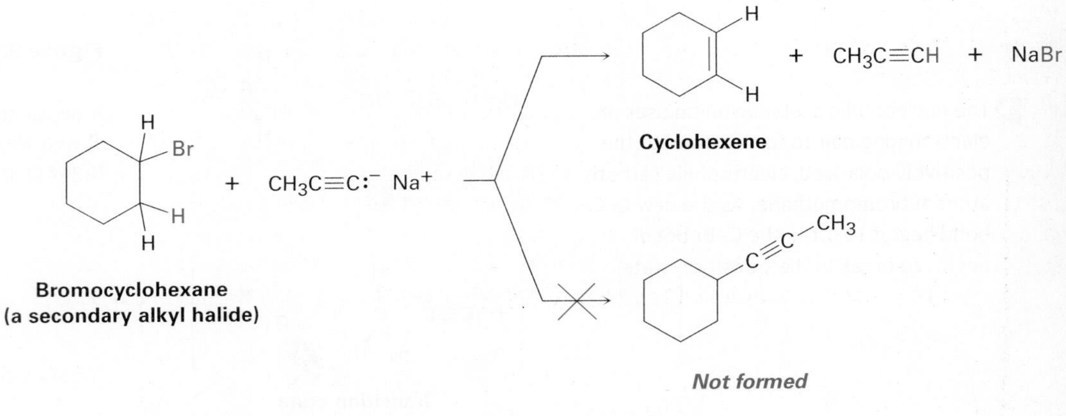

Bromocyclohexane
(a secondary alkyl halide)

$+\ CH_3C{\equiv}C{:}^-\ Na^+$

Cyclohexene $+\ CH_3C{\equiv}CH\ +\ NaBr$

Not formed

Problem 9.10
Show the terminal alkyne and alkyl halide from which the following products can be obtained. If two routes look feasible, list both.

(a) $CH_3CH_2CH_2C{\equiv}CCH_3$ (b) $(CH_3)_2CHC{\equiv}CCH_2CH_3$ (c) cyclohexyl$-C{\equiv}CCH_3$

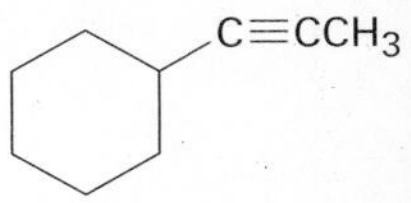

Problem 9.11
How would you prepare *cis*-2-butene starting from propyne, an alkyl halide, and any other reagents needed? This problem can't be worked in a single step. You'll have to carry out more than one reaction.

9.9 An Introduction to Organic Synthesis

There are many reasons for carrying out the laboratory synthesis of an organic compound. In the pharmaceutical industry, new molecules are designed and synthesized in the hope that some might be useful new drugs. In the chemical industry, syntheses are done to devise more economical routes to known compounds. In academic laboratories, the synthesis of extremely complex molecules is sometimes done just for the intellectual challenge involved in mastering so difficult a subject. The successful synthesis route is a highly creative work that is sometimes described by such subjective terms as *elegant* or *beautiful*.

In this book, too, we will often devise syntheses of molecules from simpler precursors, but our purpose is to learn. The ability to plan a successful multistep synthetic sequence requires a working knowledge of the uses and limitations of many different organic reactions. Furthermore, it requires the practical ability to fit together the steps in a sequence such that each reaction does only what is desired without causing changes elsewhere in the molecule. Planning a synthesis makes you approach a chemical problem in a logical way, draw on your knowledge of chemical reactions, and organize that knowledge into a workable plan—it helps you learn organic chemistry.

There's no secret to planning an organic synthesis: all it takes is a knowledge of the different reactions and some practice. The only real trick is to *work backward* in what is often called a **retrosynthetic** direction. Don't look at a potential starting material and ask yourself what reactions it might undergo. Instead, look at the final product and ask, "What was the immediate precursor of that product?" For example, if the final product is an alkyl halide, the immediate precursor might be an alkene, to which you could add HX. If the final product is a cis alkene, the immediate precursor might be an alkyne, which you could hydrogenate using the Lindlar catalyst. Having found an immediate precursor, work backward again, one step at a time, until you get back to the starting material. You have to keep the starting material in mind, of course, so that you can work back to it, but you don't want that starting material to be your main focus.

Let's work several examples of increasing complexity.

Worked Example 9.1
Devising a Synthesis Route

Synthesize *cis*-2-hexene from 1-pentyne and an alkyl halide. More than one step is needed.

$$CH_3CH_2CH_2C\equiv CH \;+\; RX \;\longrightarrow\; \underset{\text{cis-2-Hexene}}{\overset{CH_3CH_2CH_2 \quad CH_3}{\underset{H \qquad\quad H}{C=C}}}$$

1-Pentyne   Alkyl halide

Strategy

When undertaking any synthesis problem, you should look at the product, identify the functional groups it contains, and then ask yourself how those functional groups can be prepared. Always work retrosynthetically, one step at a time.

The product in this case is a cis-disubstituted alkene, so the first question is, "What is an immediate precursor of a cis-disubstituted alkene?" We know that an alkene can be prepared from an alkyne by reduction and that the right choice of experimental conditions will allow us to prepare either a trans-disubstituted alkene (using lithium in liquid ammonia) or a cis-disubstituted alkene (using catalytic hydrogenation over the Lindlar catalyst). Thus, reduction of 2-hexyne by catalytic hydrogenation using the Lindlar catalyst should yield cis-2-hexene.

$$CH_3CH_2CH_2C\equiv CCH_3 \xrightarrow[\text{Lindlar catalyst}]{H_2} \begin{array}{c} CH_3CH_2CH_2 \quad CH_3 \\ \diagdown \quad \diagup \\ C=C \\ \diagup \quad \diagdown \\ H \quad\quad H \end{array}$$

2-Hexyne → **cis-2-Hexene**

Next ask, "What is an immediate precursor of 2-hexyne?" We've seen that an internal alkyne can be prepared by alkylation of a terminal alkyne anion. In the present instance, we're told to start with 1-pentyne and an alkyl halide. Thus, alkylation of the anion of 1-pentyne with iodomethane should yield 2-hexyne.

$$CH_3CH_2CH_2C\equiv CH + NaNH_2 \xrightarrow{\text{In } NH_3} CH_3CH_2CH_2C\equiv C:^- \ Na^+$$

1-Pentyne

$$CH_3CH_2CH_2C\equiv C:^- \ Na^+ + CH_3I \xrightarrow{\text{In THF}} CH_3CH_2CH_2C\equiv CCH_3$$

2-Hexyne

Solution

cis-2-Hexene can be synthesized from the given starting materials in three steps.

$$CH_3CH_2CH_2C\equiv CH \xrightarrow[\text{2. } CH_3I, THF]{\text{1. } NaNH_2, NH_3} CH_3CH_2CH_2C\equiv CCH_3 \xrightarrow[\text{Lindlar catalyst}]{H_2} \begin{array}{c} CH_3CH_2CH_2 \quad CH_3 \\ \diagdown \quad \diagup \\ C=C \\ \diagup \quad \diagdown \\ H \quad\quad H \end{array}$$

1-Pentyne → **2-Hexyne** → **cis-2-Hexene**

Worked Example 9.2 Devising a Synthesis Route

Synthesize 2-bromopentane from acetylene and an alkyl halide. More than one step is needed.

$$HC\equiv CH + RX \longrightarrow CH_3CH_2CH_2\overset{\overset{\displaystyle Br}{|}}{C}HCH_3$$

Acetylene **Alkyl halide** **2-Bromopentane**

9.9 | An Introduction to Organic Synthesis

Strategy

Identify the functional group in the product (an alkyl bromide) and work the problem retrosynthetically. What is an immediate precursor of an alkyl bromide? Perhaps an alkene plus HBr. Of the two possibilities, Markovnikov addition of HBr to 1-pentene looks like a better choice than addition to 2-pentene because the latter reaction would give a mixture of isomers.

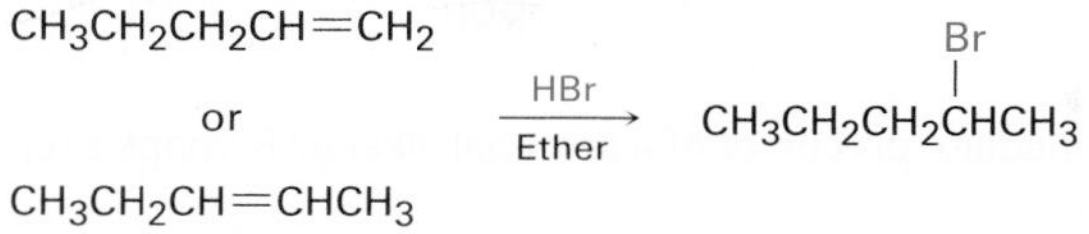

$$CH_3CH_2CH_2CH=CH_2$$
or
$$CH_3CH_2CH=CHCH_3$$
$$\xrightarrow[\text{Ether}]{\text{HBr}} CH_3CH_2CH_2\overset{\overset{\displaystyle Br}{|}}{C}HCH_3$$

What is an immediate precursor of an alkene? Perhaps an alkyne, which could be reduced.

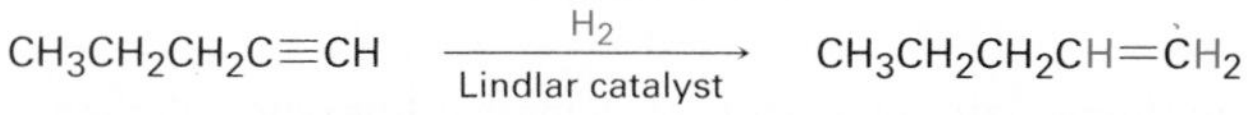

$$CH_3CH_2CH_2C\equiv CH \xrightarrow[\text{Lindlar catalyst}]{H_2} CH_3CH_2CH_2CH=CH_2$$

What is an immediate precursor of a terminal alkyne? Perhaps sodium acetylide and an alkyl halide.

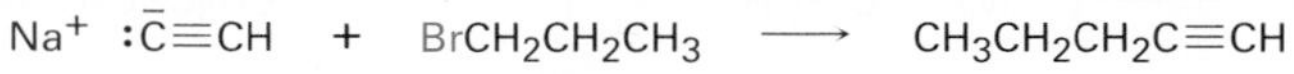

$$Na^+ :\bar{C}\equiv CH + BrCH_2CH_2CH_3 \longrightarrow CH_3CH_2CH_2C\equiv CH$$

Solution

The desired product can be synthesized in four steps from acetylene and 1-bromopropane.

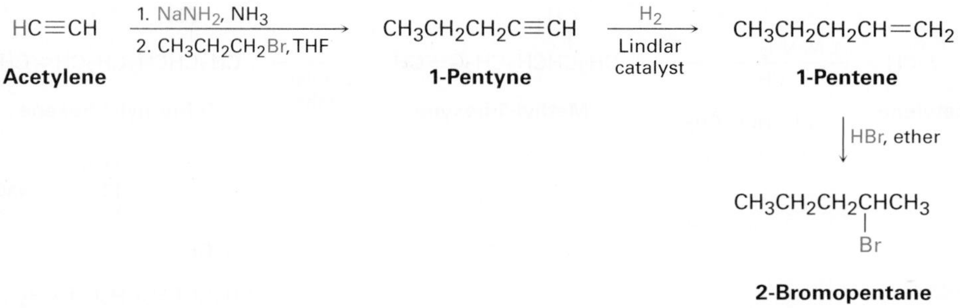

$$HC\equiv CH \xrightarrow[\text{2. }CH_3CH_2CH_2Br, THF]{\text{1. }NaNH_2,\ NH_3} CH_3CH_2CH_2C\equiv CH \xrightarrow[\substack{\text{Lindlar}\\\text{catalyst}}]{H_2} CH_3CH_2CH_2CH=CH_2$$

Acetylene   **1-Pentyne**   **1-Pentene**

$$\downarrow \text{HBr, ether}$$

$$CH_3CH_2CH_2\overset{\overset{\displaystyle Br}{|}}{C}HCH_3$$

2-Bromopentane

Worked Example 9.3
Devising a Synthesis Route

Synthesize 5-methyl-1-hexanol (5-methyl-1-hydroxyhexane) from acetylene and an alkyl halide.

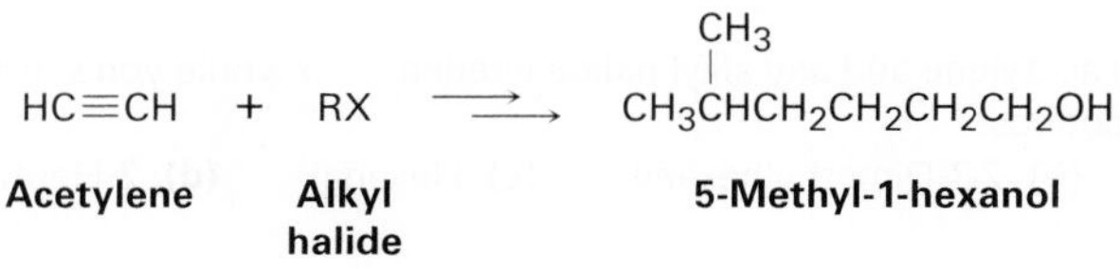

$$HC\equiv CH + RX \longrightarrow CH_3\overset{\overset{\displaystyle CH_3}{|}}{C}HCH_2CH_2CH_2CH_2OH$$

Acetylene   **Alkyl halide**   **5-Methyl-1-hexanol**

Strategy

What is an immediate precursor of a primary alcohol? Perhaps a terminal alkene, which could be hydrated with non-Markovnikov regiochemistry by reaction with borane followed by oxidation with H_2O_2.

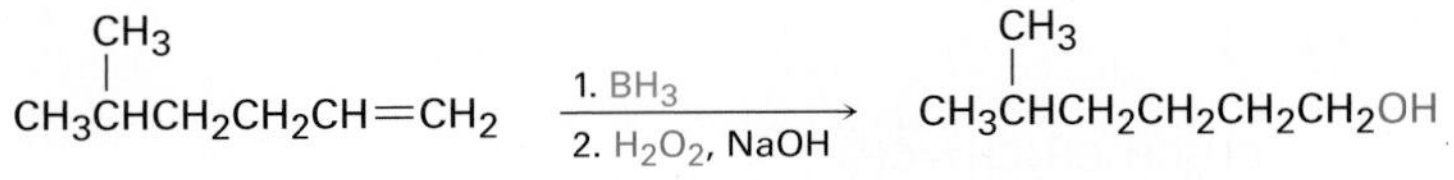

$$CH_3CHCH_2CH_2CH=CH_2 \xrightarrow[\text{2. } H_2O_2,\text{ NaOH}]{\text{1. } BH_3} CH_3CHCH_2CH_2CH_2CH_2OH$$
(with CH_3 substituent on the carbon bearing H)

What is an immediate precursor of a terminal alkene? Perhaps a terminal alkyne, which could be reduced.

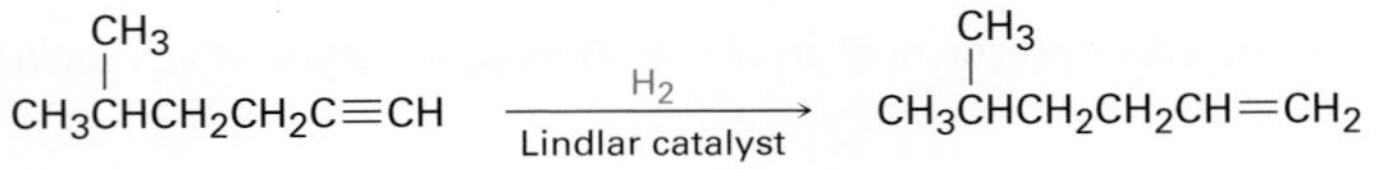

$$CH_3CHCH_2CH_2C{\equiv}CH \xrightarrow[\text{Lindlar catalyst}]{H_2} CH_3CHCH_2CH_2CH=CH_2$$

What is an immediate precursor of 5-methyl-1-hexyne? Perhaps acetylene and 1-bromo-3-methylbutane.

$$HC{\equiv}CH \xrightarrow{NaNH_2} Na^+ \; {}^-C{\equiv}CH \xrightarrow{CH_3CHCH_2CH_2Br} CH_3CHCH_2CH_2C{\equiv}CH$$

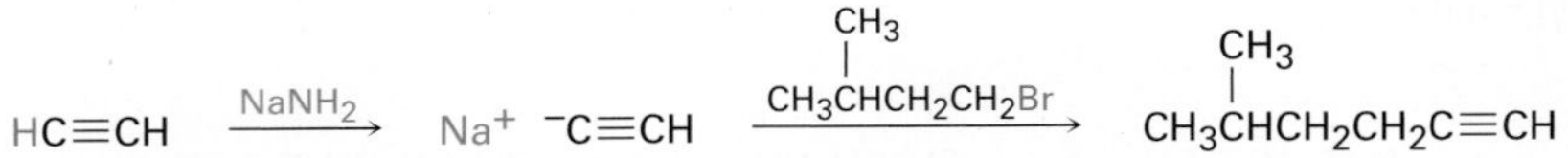

Solution

The synthesis can be completed in four steps from acetylene and 1-bromo-3-methylbutane:

$$HC{\equiv}CH \xrightarrow[\text{2. } CH_3CHCH_2CH_2Br]{\text{1. } NaNH_2} CH_3CHCH_2CH_2C{\equiv}CH \xrightarrow[\text{Lindlar catalyst}]{H_2} CH_3CHCH_2CH_2CH=CH_2$$

Acetylene → **5-Methyl-1-hexyne** → **5-Methyl-1-hexene**

$$\xrightarrow[\text{2. } H_2O_2,\text{ NaOH}]{\text{1. } BH_3} CH_3CHCH_2CH_2CH_2CH_2OH$$

5-Methyl-1-hexanol

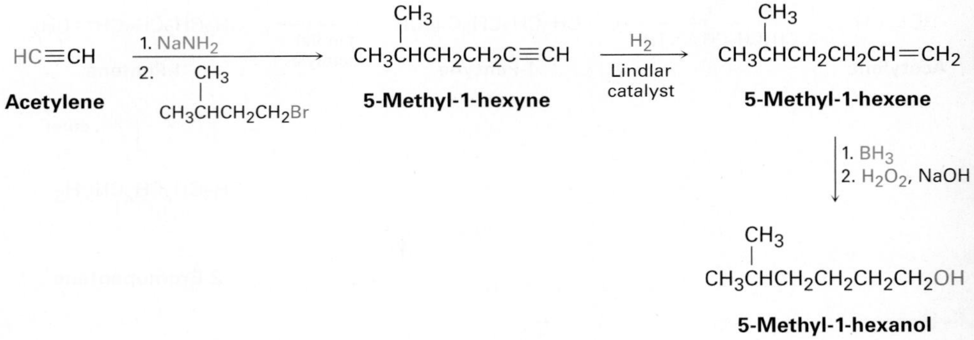

Problem 9.12
Beginning with 4-octyne as your only source of carbon, and using any inorganic reagents necessary, how would you synthesize the following compounds?
(a) *cis*-4-Octene (b) Butanal (c) 4-Bromooctane
(d) 4-Octanol (e) 4,5-Dichlorooctane (f) Butanoic acid

Problem 9.13
Beginning with acetylene and any alkyl halide needed, how would you synthesize the following compounds?
(a) Decane (b) 2,2-Dimethylhexane (c) Hexanal (d) 2-Heptanone

A DEEPER LOOK: The Art of Organic Synthesis

If you think some of the synthesis problems at the end of this chapter are hard, try devising a synthesis of vitamin B_{12} starting only from simple substances you can buy in a chemical catalog. This extraordinary achievement was reported in 1973 as the culmination of a collaborative effort headed by Robert B. Woodward of Harvard University and Albert Eschenmoser of the Swiss Federal Institute of Technology in Zürich. More than 100 graduate students and postdoctoral associates contributed to the work, which took more than a decade to complete.

Vitamin B_{12} has been synthesized from scratch in the laboratory, but the bacteria growing on sludge from municipal sewage plants do a much better job.

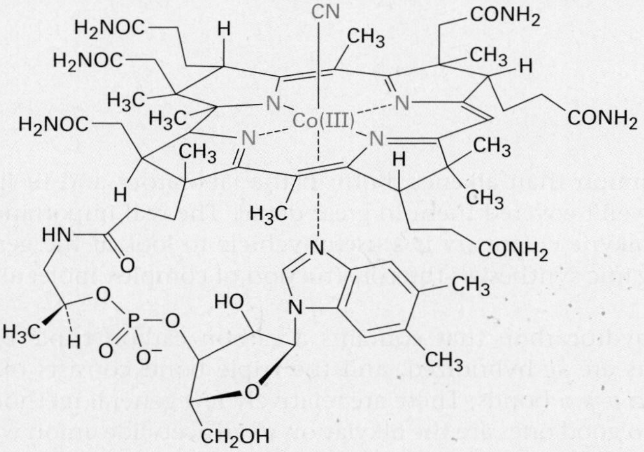

Vitamin B_{12}

Why put such extraordinary effort into the laboratory synthesis of a molecule so easily obtained from natural sources? There are many reasons. On a basic human level, a chemist might be motivated primarily by the challenge, much as a climber might be challenged by the ascent of a difficult peak. Beyond the pure challenge, the completion of a difficult synthesis is also valuable for the way in which it establishes new standards and raises the field to a new level. If vitamin B_{12} can be made, then why can't any molecule found in nature be made? Indeed, the decades that have passed since the work of Woodward and Eschenmoser have seen the laboratory synthesis of many enormously complex and valuable substances. Sometimes these substances—for instance, the anticancer compound paclitaxel, trade named Taxol—are not easily available in nature, so laboratory synthesis is the only method for obtaining larger quantities.

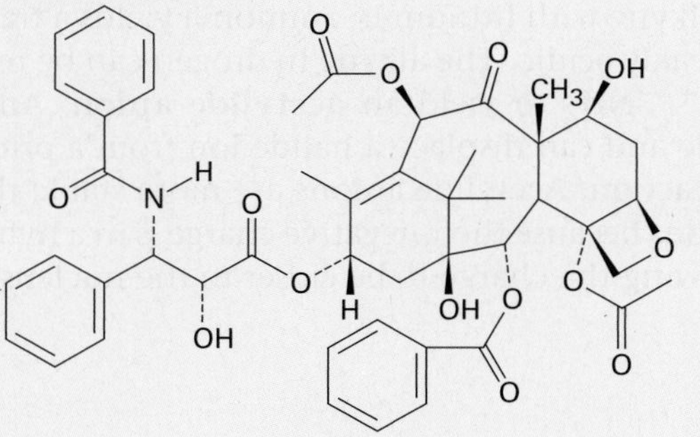

Paclitaxel (Taxol)

(continued)

But perhaps the most important reason for undertaking a complex synthesis is that, in so doing, new reactions and new chemistry are discovered. It invariably happens in synthesis that a point is reached at which the planned route fails. At such a time, the only alternatives are to quit or to devise a way around the difficulty. New reactions and new principles come from such situations, and it is in this way that the science of organic chemistry grows richer. In the synthesis of vitamin B_{12}, for example, unexpected findings emerged that led to the understanding of an entire new class of reactions—the *pericyclic* reactions that are the subject of Chapter 30 in this book. From synthesizing vitamin B_{12} to understanding pericyclic reactions—no one could have possibly predicted such a link at the beginning of the synthesis, but that is the way of science.

Summary

Key words
acetylide anion, 325
alkylation, 327
alkyne (RC≡CR), 314
enol, 319
retrosynthetic, 329
tautomer, 319

Alkynes are less common than alkenes, both in the laboratory and in living organisms, so we haven't covered them in great detail. The real importance of this chapter is that alkyne chemistry is a useful vehicle to look at the general strategies used in organic synthesis—the construction of complex molecules in the laboratory.

An **alkyne** is a hydrocarbon that contains a carbon–carbon triple bond. Alkyne carbon atoms are *sp*-hybridized, and the triple bond consists of one *sp–sp* σ bond and two *p–p* π bonds. There are relatively few general methods of alkyne synthesis. Two good ones are the alkylation of an acetylide anion with a primary alkyl halide and the twofold elimination of HX from a vicinal dihalide.

The chemistry of alkynes is dominated by electrophilic addition reactions, similar to those of alkenes. Alkynes react with HBr and HCl to yield vinylic halides and with Br_2 and Cl_2 to yield 1,2-dihalides (vicinal dihalides). Alkynes can be hydrated by reaction with aqueous sulfuric acid in the presence of mercury(II) catalyst. The reaction leads to an intermediate **enol** that immediately **tautomerizes** to yield a ketone. Because the addition reaction occurs with Markovnikov regiochemistry, a methyl ketone is produced from a terminal alkyne. Alternatively, hydroboration–oxidation of a terminal alkyne yields an aldehyde.

Alkynes can be reduced to yield alkenes and alkanes. Complete reduction of the triple bond over a palladium hydrogenation catalyst yields an alkane; partial reduction by catalytic hydrogenation over a Lindlar catalyst yields a cis alkene. Reduction of the alkyne with lithium in ammonia yields a trans alkene.

Terminal alkynes are weakly acidic. The alkyne hydrogen can be removed by a strong base such as Na^+ $^-NH_2$ to yield an **acetylide anion**. An acetylide anion acts as a nucleophile and can displace a halide ion from a primary alkyl halide in an **alkylation** reaction. Acetylide anions are more stable than either alkyl anions or vinylic anions because their negative charge is in a hybrid orbital with 50% *s* character, allowing the charge to be closer to the nucleus.

Summary of Reactions

1. Preparation of alkynes
   (a) Dehydrohalogenation of vicinal dihalides (Section 9.2)

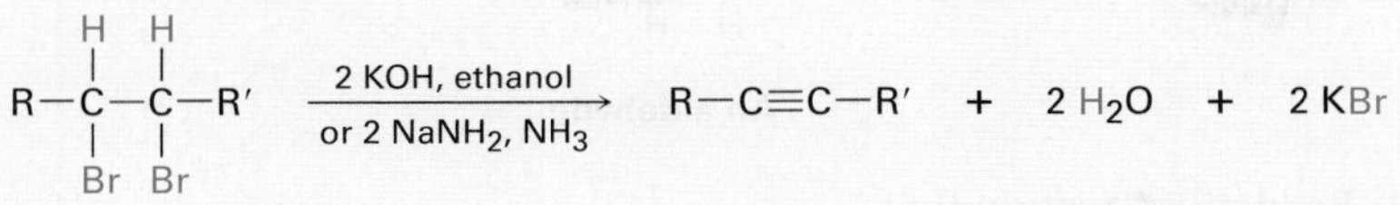

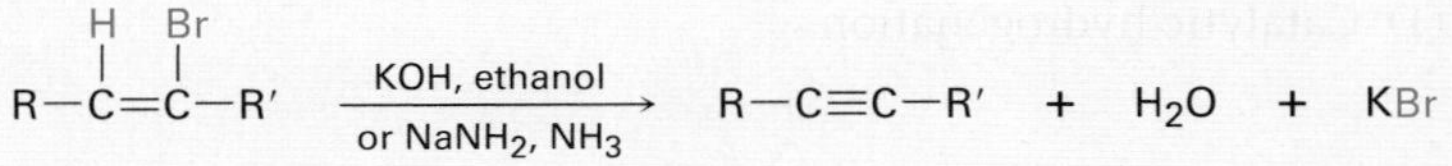

   $$R-\underset{\underset{Br}{|}}{\overset{\overset{H}{|}}{C}}-\underset{\underset{Br}{|}}{\overset{\overset{H}{|}}{C}}-R' \xrightarrow[\text{or 2 NaNH}_2,\text{ NH}_3]{\text{2 KOH, ethanol}} R-C\equiv C-R' + 2\,H_2O + 2\,KBr$$

   $$R-\underset{\underset{}{}}{\overset{\overset{H}{|}}{C}}=\underset{\underset{}{}}{\overset{\overset{Br}{|}}{C}}-R' \xrightarrow[\text{or NaNH}_2,\text{ NH}_3]{\text{KOH, ethanol}} R-C\equiv C-R' + H_2O + KBr$$

   (b) Alkylation of acetylide anions (Section 9.8)

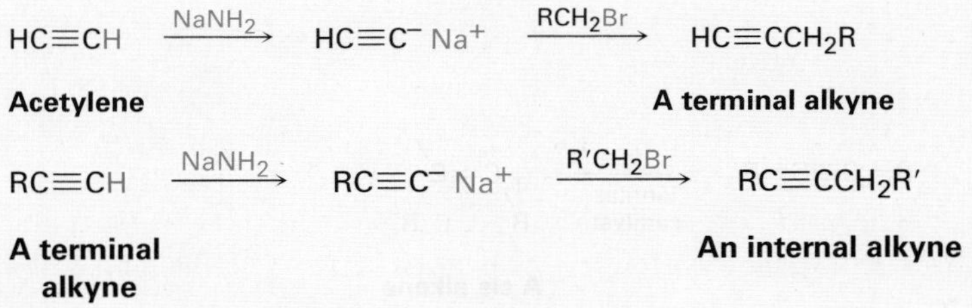

   $$HC\equiv CH \xrightarrow{\text{NaNH}_2} HC\equiv C^-\,Na^+ \xrightarrow{RCH_2Br} HC\equiv CCH_2R$$

   Acetylene   **A terminal alkyne**

   $$RC\equiv CH \xrightarrow{\text{NaNH}_2} RC\equiv C^-\,Na^+ \xrightarrow{R'CH_2Br} RC\equiv CCH_2R'$$

   A terminal alkyne   **An internal alkyne**

2. Reactions of alkynes
   (a) Addition of HCl and HBr (Section 9.3)

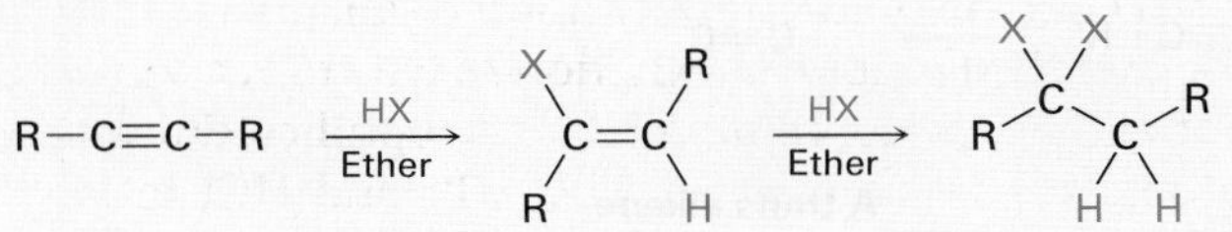

   $$R-C\equiv C-R \xrightarrow[\text{Ether}]{HX} \underset{R}{\overset{X}{C}}=\underset{H}{\overset{R}{C}} \xrightarrow[\text{Ether}]{HX} R-\underset{H}{\overset{X}{C}}-\underset{H}{\overset{X}{C}}-R$$

   (b) Addition of Cl_2 and Br_2 (Section 9.3)

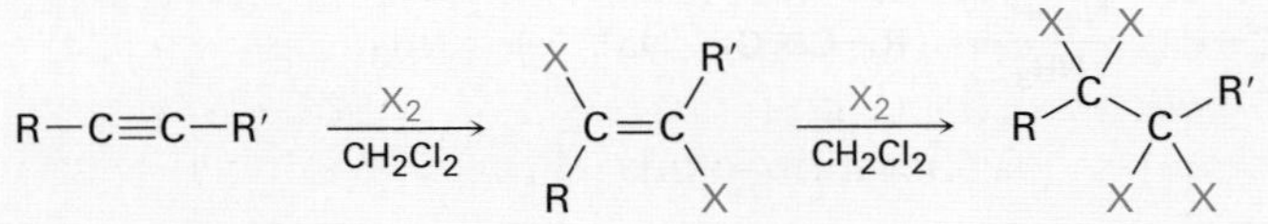

   $$R-C\equiv C-R' \xrightarrow[\text{CH}_2\text{Cl}_2]{X_2} \underset{R}{\overset{X}{C}}=\underset{X}{\overset{R'}{C}} \xrightarrow[\text{CH}_2\text{Cl}_2]{X_2} R-\underset{X}{\overset{X}{C}}-\underset{X}{\overset{X}{C}}-R'$$

   (c) Hydration (Section 9.4)

   (1) Mercuric sulfate catalyzed

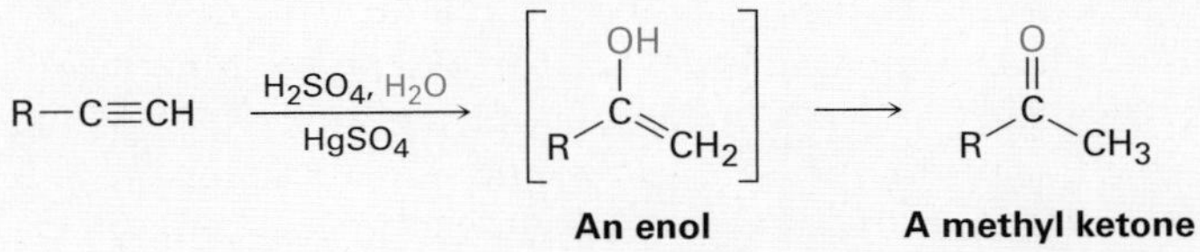

   $$R-C\equiv CH \xrightarrow[\text{HgSO}_4]{H_2SO_4,\,H_2O} \left[\underset{R}{\overset{OH}{C}}=CH_2\right] \longrightarrow R-\overset{\overset{O}{\|}}{C}-CH_3$$

                                                   An enol   **A methyl ketone**

   (continued)

(2) Hydroboration–oxidation

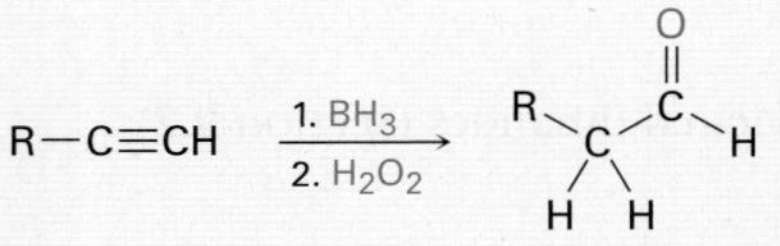

$$R-C\equiv CH \xrightarrow[\text{2. }H_2O_2]{\text{1. }BH_3} R-\underset{\underset{H}{|}}{\overset{\overset{H}{|}}{C}}-\overset{\overset{O}{\|}}{C}-H$$

An aldehyde

(d) Reduction (Section 9.5)

(1) Catalytic hydrogenation

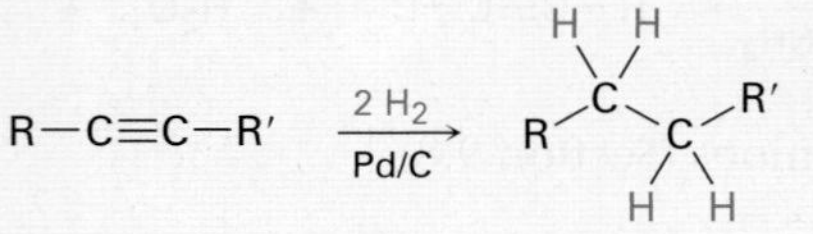

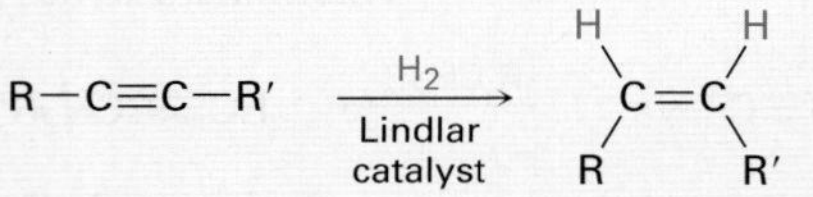

$$R-C\equiv C-R' \xrightarrow[\text{Pd/C}]{2\ H_2} R-\underset{\underset{H}{|}}{\overset{\overset{H}{|}}{C}}-\underset{\underset{H}{|}}{\overset{\overset{H}{|}}{C}}-R'$$

$$R-C\equiv C-R' \xrightarrow[\substack{\text{Lindlar}\\\text{catalyst}}]{H_2} \underset{R}{\overset{H}{\diagdown}}C=C\underset{R'}{\overset{H}{\diagup}}$$

A cis alkene

(2) Lithium in liquid ammonia

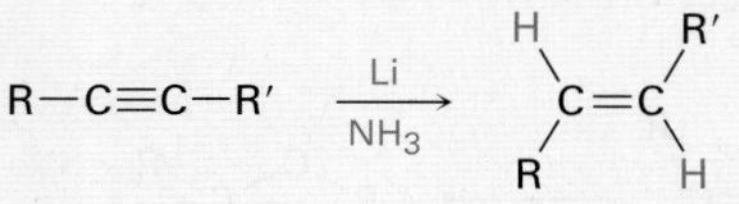

$$R-C\equiv C-R' \xrightarrow[NH_3]{Li} \underset{R}{\overset{H}{\diagdown}}C=C\underset{H}{\overset{R'}{\diagup}}$$

A trans alkene

(e) Conversion into acetylide anions (Section 9.7)

$$R-C\equiv C-H \xrightarrow[NH_3]{NaNH_2} R-C\equiv C:^-\ Na^+\ +\ NH_3$$

Exercises

Visualizing Chemistry

(Problems 9.1–9.13 appear within the chapter.)

OWL Interactive versions of these problems are assignable in OWL for Organic Chemistry.

9.14 Name the following alkynes, and predict the products of their reaction with **(1)** H_2 in the presence of a Lindlar catalyst and **(2)** H_3O^+ in the presence of $HgSO_4$:

(a)

(b)

9.15 From what alkyne might each of the following substances have been made? (Green = Cl.)

(a)

(b)

9.16 How would you prepare the following substances, starting from any compounds having four carbons or fewer?

(a)

(b)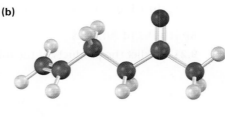

9.17 The following cycloalkyne is too unstable to exist. Explain.

Additional Problems

Naming Alkynes

9.18 Give IUPAC names for the following compounds:

(a) CH₃CH₂C≡CCCH₃ with two CH₃ groups on the carbon bearing C

    $$\text{CH}_3\text{CH}_2\text{C}\equiv\text{CC}(\text{CH}_3)_2\text{CH}_3$$

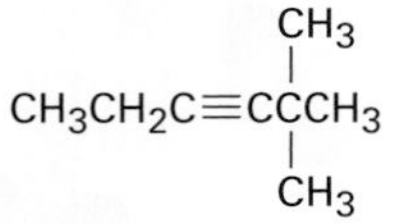

(b) $CH_3C\equiv CCH_2C\equiv CCH_2CH_3$

(c) $CH_3CH=C(CH_3)C\equiv CC(CH_3)HCH_3$

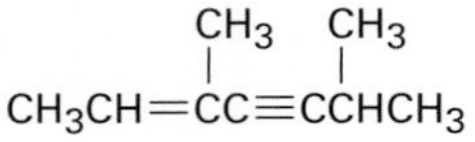

(d) $HC\equiv CC(CH_3)CH_2C\equiv CH$ with CH₃

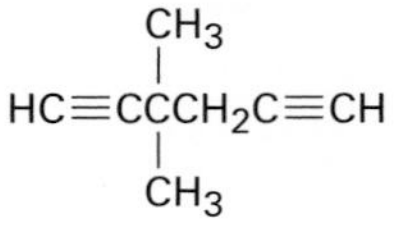

(e) $H_2C=CHCH=CHC\equiv CH$

(f) $CH_3CH_2CH(CH_2CH_3)C\equiv CC(CH_3)(CH_2CH_3)CH_2CH_3$...

Structures as drawn:

(a)
$$\begin{array}{c} \text{CH}_3 \\ | \\ \text{CH}_3\text{CH}_2\text{C}\equiv\text{CCCH}_3 \\ | \\ \text{CH}_3 \end{array}$$

(b) $CH_3C\equiv CCH_2C\equiv CCH_2CH_3$

(c)
$$\begin{array}{c} \text{CH}_3 \quad \text{CH}_3 \\ | \quad\quad | \\ \text{CH}_3\text{CH}=\text{CC}\equiv\text{CCHCH}_3 \end{array}$$

(d)
$$\begin{array}{c} \text{CH}_3 \\ | \\ \text{HC}\equiv\text{CCCH}_2\text{C}\equiv\text{CH} \\ | \\ \text{CH}_3 \end{array}$$

(e) $H_2C=CHCH=CHC\equiv CH$

(f)
$$\begin{array}{c} \text{CH}_2\text{CH}_3 \\ | \\ \text{CH}_3\text{CH}_2\text{CHC}\equiv\text{CCHCHCH}_3 \\ | \quad\quad\quad\quad\quad\quad\quad | \\ \text{CH}_2\text{CH}_3 \quad\quad \text{CH}_3 \end{array}$$

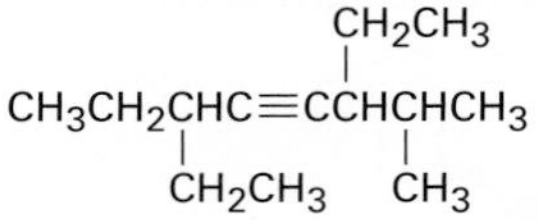

9.19 Draw structures corresponding to the following names:
 (a) 3,3-Dimethyl-4-octyne
 (b) 3-Ethyl-5-methyl-1,6,8-decatriyne
 (c) 2,2,5,5-Tetramethyl-3-hexyne
 (d) 3,4-Dimethylcyclodecyne
 (e) 3,5-Heptadien-1-yne
 (f) 3-Chloro-4,4-dimethyl-1-nonen-6-yne
 (g) 3-sec-Butyl-1-heptyne
 (h) 5-tert-Butyl-2-methyl-3-octyne

9.20 The following two hydrocarbons have been isolated from various plants in the sunflower family. Name them according to IUPAC rules.
 (a) $CH_3CH=CHC\equiv CC\equiv CCH=CHCH=CH_2$ (all trans)
 (b) $CH_3C\equiv CC\equiv CC\equiv CC\equiv CCH=CH_2$

Reactions of Alkynes

9.21 Predict the products of the following reactions:

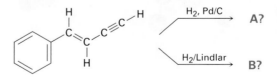

9.22 Predict the products from reaction of 1-hexyne with the following reagents:
 (a) 1 equiv HBr
 (b) 1 equiv Cl₂
 (c) H₂, Lindlar catalyst
 (d) NaNH₂ in NH₃, then CH₃Br
 (e) H₂O, H₂SO₄, HgSO₄
 (f) 2 equiv HCl

9.23 Predict the products from reaction of 5-decyne with the following reagents:
(a) H_2, Lindlar catalyst
(b) Li in NH_3
(c) 1 equiv Br_2
(d) BH_3 in THF, then H_2O_2, OH^-
(e) H_2O, H_2SO_4, $HgSO_4$
(f) Excess H_2, Pd/C catalyst

9.24 Predict the products from reaction of 2-hexyne with the following reagents:
(a) 2 equiv Br_2
(b) 1 equiv HBr
(c) Excess HBr
(d) Li in NH_3
(e) H_2O, H_2SO_4, $HgSO_4$

9.25 Propose structures for hydrocarbons that give the following products on oxidative cleavage by $KMnO_4$ or O_3:
(a) CO_2 + $CH_3(CH_2)_5CO_2H$
(b) CH_3CO_2H + (benzoic acid, $C_6H_5CO_2H$)
(c) $HO_2C(CH_2)_8CO_2H$
(d) CH_3CHO + $CH_3COCH_2CH_2CO_2H$ + CO_2
(e) $HCOCH_2CH_2CH_2CH_2COCO_2H$ + CO_2

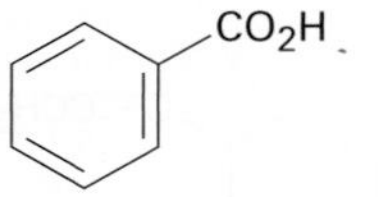

9.26 Identify the reagents **a–c** in the following scheme:

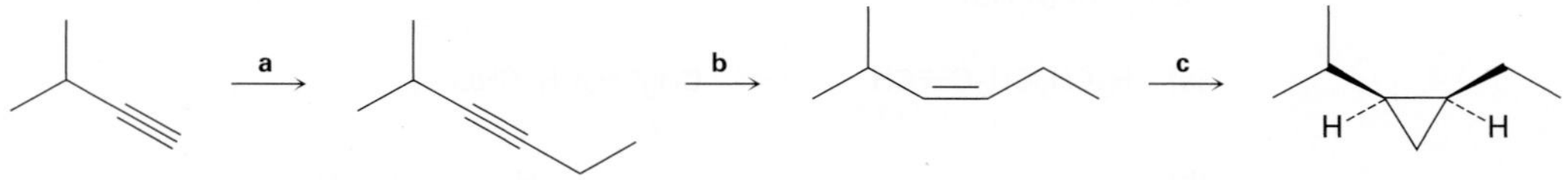

Organic Synthesis

9.27 How would you carry out the following conversions? More than one step may be needed in some instances.

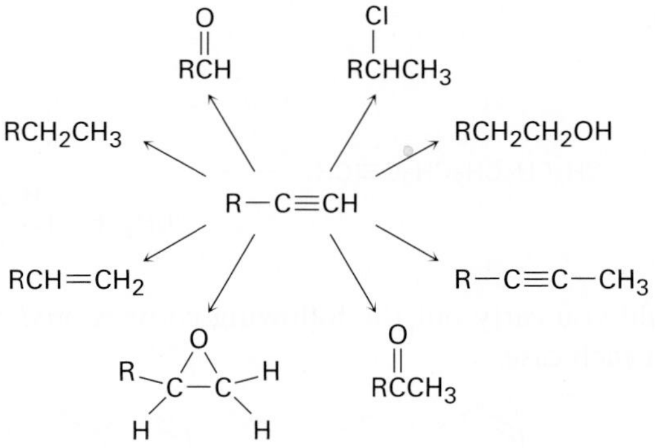

9.28 How would you carry out the following reactions?

(a) CH₃CH₂C≡CH $\xrightarrow{?}$ CH₃CH₂C(=O)CH₃

(b) CH₃CH₂C≡CH $\xrightarrow{?}$ CH₃CH₂CH₂CHO

(c) Ph–C≡CH $\xrightarrow{?}$ Ph–C≡C–CH₃

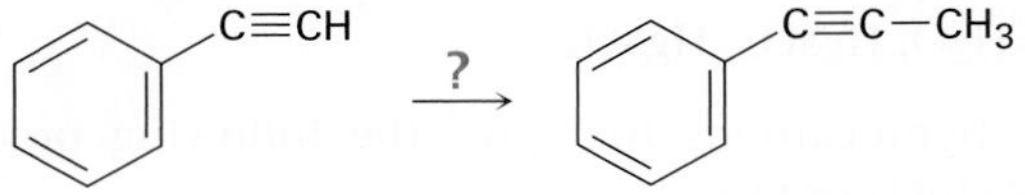

(d) Ph–C≡CCH₃ $\xrightarrow{?}$ cis-Ph–CH=CH–CH₃ (with H, H, CH₃ shown)

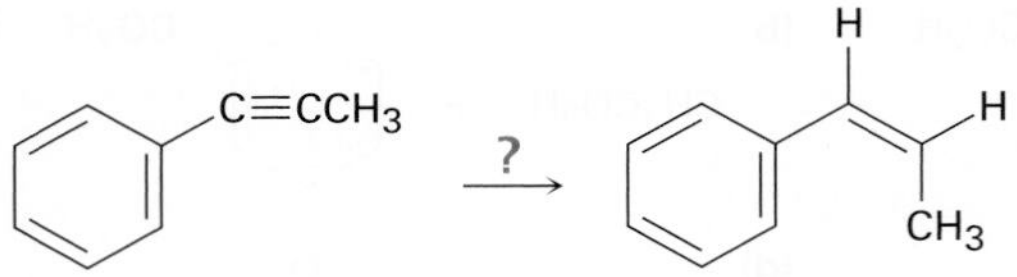

(e) CH₃CH₂C≡CH $\xrightarrow{?}$ CH₃CH₂CO₂H

(f) CH₃CH₂CH₂CH₂CH=CH₂ $\xrightarrow[\text{(2 steps)}]{?}$ CH₃CH₂CH₂CH₂C≡CH

9.29 Each of the following syntheses requires more than one step. How would you carry them out?

(a) CH₃CH₂CH₂C≡CH $\xrightarrow{?}$ CH₃CH₂CH₂CHO

(b) (CH₃)₂CHCH₂C≡CH $\xrightarrow{?}$ (CH₃)₂CHCH₂–CH=CH–CH₂CH₃ (cis alkene)

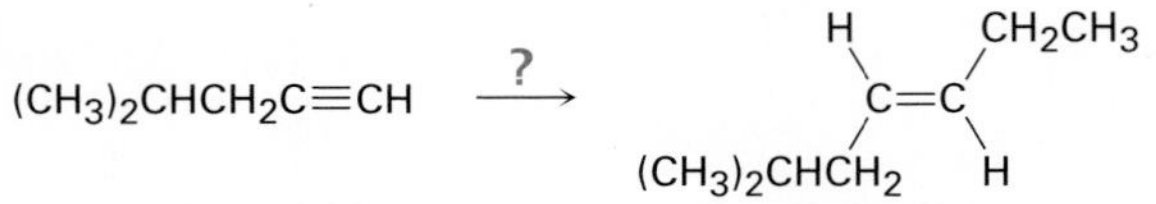

9.30 How would you carry out the following transformation? More than one step is needed.

CH₃CH₂CH₂CH₂C≡CH $\xrightarrow{?}$ cyclopropane with CH₃CH₂CH₂CH₂ and CH₃ substituents

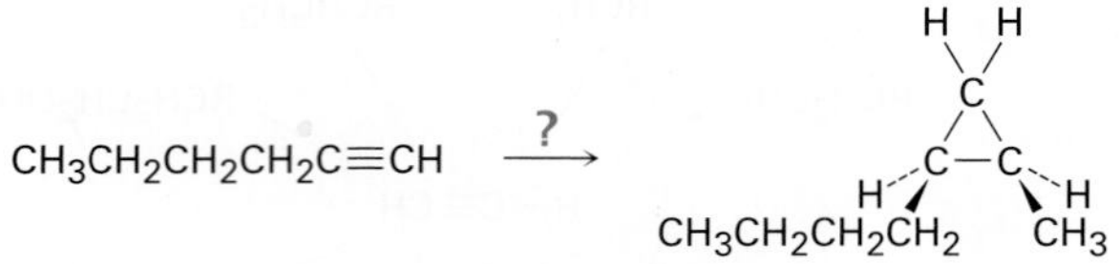

9.31 How would you carry out the following conversions? More than one step is needed in each case.

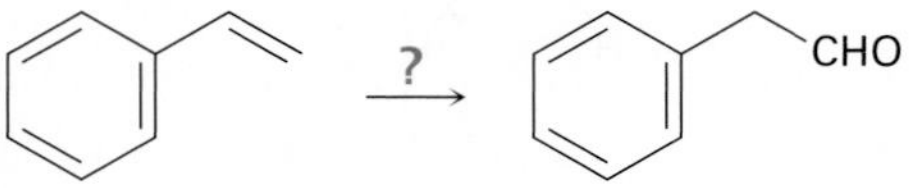

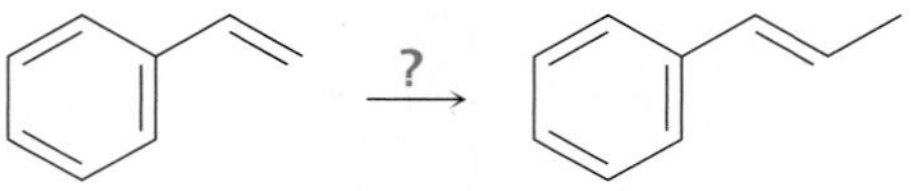

Ph–CH=CH₂ $\xrightarrow{?}$ Ph–CH₂–CHO

Ph–CH=CH₂ $\xrightarrow{?}$ Ph–CH=CH–CH₃

9.32 Synthesize the following compounds using 1-butyne as the only source of carbon, along with any inorganic reagents you need. More than one step may be needed.
  (a) 1,1,2,2-Tetrachlorobutane
  (b) 1,1-Dichloro-2-ethylcyclopropane

9.33 How would you synthesize the following compounds from acetylene and any alkyl halides with four or fewer carbons? More than one step may be needed.
  (a) $CH_3CH_2CH_2C{\equiv}CH$
  (b) $CH_3CH_2C{\equiv}CCH_2CH_3$

  (c) $CH_3\underset{\underset{CH_3}{|}}{C}HCH_2CH{=}CH_2$

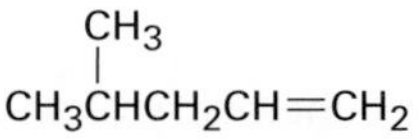

  (d) $CH_3CH_2CH_2\overset{\overset{O}{\|}}{C}CH_2CH_2CH_2CH_3$

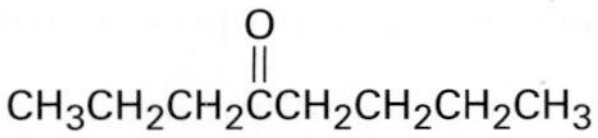

  (e) $CH_3CH_2CH_2CH_2CH_2CHO$

9.34 How would you carry out the following reactions to introduce deuterium into organic molecules?

  (a) $CH_3CH_2C{\equiv}CCH_2CH_3 \xrightarrow{?}$ (Z)-3,4-dideutero-3-hexene with D, D on same side; C₂H₅, C₂H₅

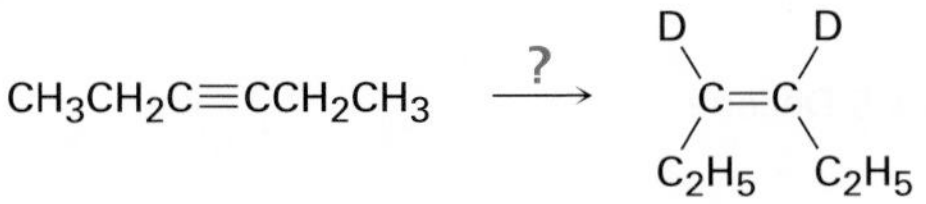

  (b) $CH_3CH_2C{\equiv}CCH_2CH_3 \xrightarrow{?}$ (E)-isomer with D and C₂H₅ on opposite corners

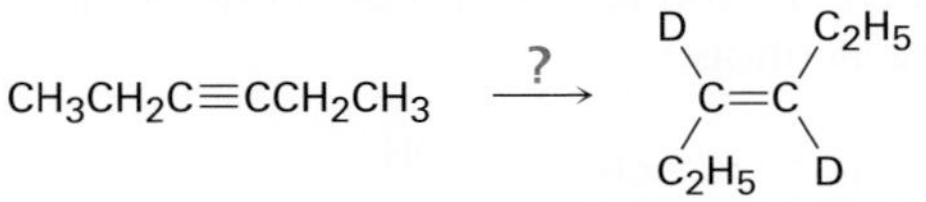

  (c) $CH_3CH_2CH_2C{\equiv}CH \xrightarrow{?} CH_3CH_2CH_2C{\equiv}CD$

  (d) Ph–C≡CH $\xrightarrow{?}$ Ph–CD=CD₂

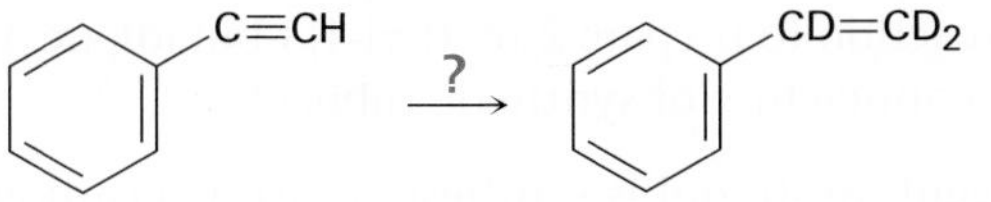

9.35 How would you prepare cyclodecyne starting from acetylene and any alkyl halide needed?

9.36 The sex attractant given off by the common housefly is an alkene named *muscalure*. Propose a synthesis of muscalure starting from acetylene and any alkyl halides needed. What is the IUPAC name for muscalure?

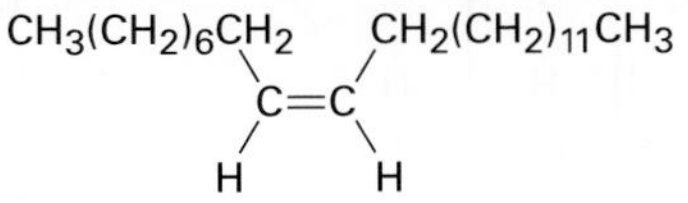

$$\underset{HH}{\underset{||}{CH_3(CH_2)_6CH_2}}C{=}C\underset{}{CH_2(CH_2)_{11}CH_3} \quad \textbf{Muscalure}$$

General Problems

9.37 A hydrocarbon of unknown structure has the formula C_8H_{10}. On catalytic hydrogenation over the Lindlar catalyst, 1 equivalent of H_2 is absorbed. On hydrogenation over a palladium catalyst, 3 equivalents of H_2 are absorbed.
  (a) How many degrees of unsaturation are present in the unknown?
  (b) How many triple bonds are present?
  (c) How many double bonds are present?
  (d) How many rings are present?
  (e) Draw a structure that fits the data.

9.38 Compound **A** (C_9H_{12}) absorbed 3 equivalents of H_2 on catalytic reduction over a palladium catalyst to give **B** (C_9H_{18}). On ozonolysis, compound **A** gave, among other things, a ketone that was identified as cyclohexanone. On treatment with $NaNH_2$ in NH_3, followed by addition of iodomethane, compound **A** gave a new hydrocarbon, **C** ($C_{10}H_{14}$). What are the structures of **A**, **B**, and **C**?

9.39 Hydrocarbon **A** has the formula $C_{12}H_8$. It absorbs 8 equivalents of H_2 on catalytic reduction over a palladium catalyst. On ozonolysis, only two products are formed: oxalic acid (HO_2CCO_2H) and succinic acid ($HO_2CCH_2CH_2CO_2H$). Write the reactions, and propose a structure for **A**.

9.40 Occasionally, a chemist might need to *invert* the stereochemistry of an alkene—that is, to convert a cis alkene to a trans alkene, or vice versa. There is no one-step method for doing an alkene inversion, but the transformation can be carried out by combining several reactions in the proper sequence. How would you carry out the following reactions?

(a) *trans*-5-Decene $\xrightarrow{?}$ *cis*-5-Decene

(b) *cis*-5-Decene $\xrightarrow{?}$ *trans*-5-Decene

9.41 Organometallic reagents such as sodium acetylide undergo an addition reaction with ketones, giving alcohols:

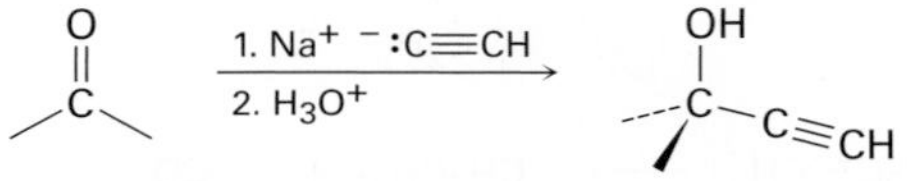

How might you use this reaction to prepare 2-methyl-1,3-butadiene, the starting material used in the manufacture of synthetic rubber?

9.42 The oral contraceptive agent Mestranol is synthesized using a carbonyl addition reaction like that shown in Problem 9.41. Draw the structure of the ketone needed.

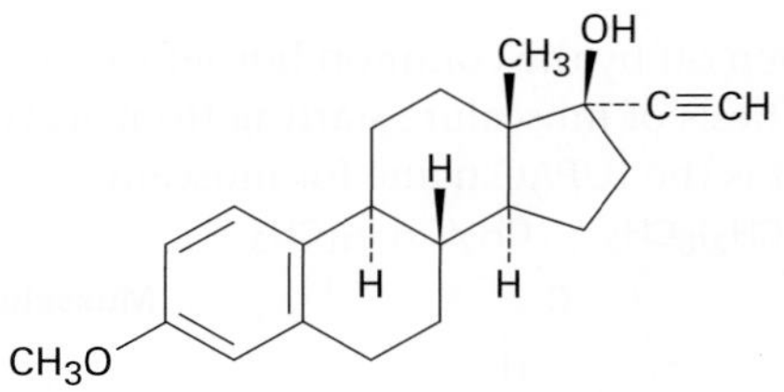

Mestranol

9.43 1-Octen-3-ol, a potent mosquito attractant commonly used in mosquito traps, can be prepared in two steps from hexanal, $CH_3CH_2CH_2CH_2CH_2CHO$. The first step is an acetylide-addition reaction like that described in Problem 9.41. What is the structure of the product from the first step, and how can it be converted into 1-octen-3-ol?

$CH_3CH_2CH_2CH_2CH_2\overset{OH}{\underset{|}{C}}HCH=CH_2$ **1-Octen-3-ol**

9.44 Erythrogenic acid, $C_{18}H_{26}O_2$, is an acetylenic fatty acid that turns a vivid red on exposure to light. On catalytic hydrogenation over a palladium catalyst, 5 equivalents of H_2 are absorbed, and stearic acid, $CH_3(CH_2)_{16}CO_2H$, is produced. Ozonolysis of erythrogenic acid gives four products: formaldehyde, CH_2O; oxalic acid, HO_2CCO_2H; azelaic acid, $HO_2C(CH_2)_7CO_2H$; and the aldehyde acid $OHC(CH_2)_4CO_2H$. Draw two possible structures for erythrogenic acid, and suggest a way to tell them apart by carrying out some simple reactions.

9.45 Hydrocarbon **A** has the formula C_9H_{12} and absorbs 3 equivalents of H_2 to yield **B**, C_9H_{18}, when hydrogenated over a Pd/C catalyst. On treatment of **A** with aqueous H_2SO_4 in the presence of mercury(II), two isomeric ketones, **C** and **D**, are produced. Oxidation of **A** with $KMnO_4$ gives a mixture of acetic acid (CH_3CO_2H) and the tricarboxylic acid **E**. Propose structures for compounds **A–D**, and write the reactions.

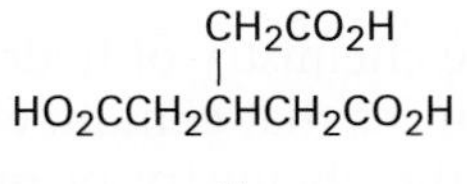

$$HO_2CCH_2CHCH_2CO_2H$$
$$|$$
$$CH_2CO_2H$$

E

9.46 Terminal alkynes react with Br_2 and water to yield bromo ketones. For example:

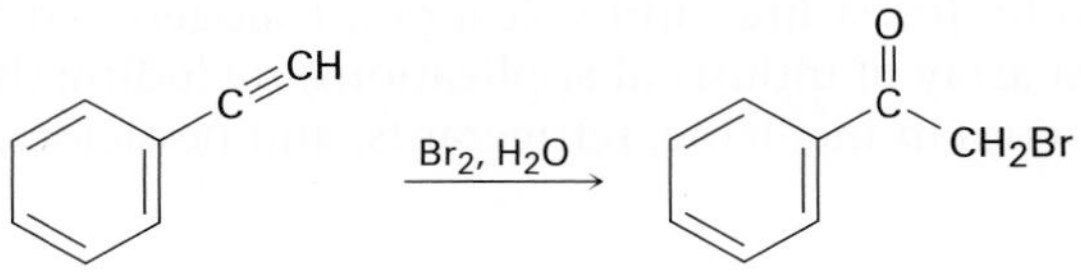

PhC≡CH $\xrightarrow{Br_2, H_2O}$ PhC(=O)CH$_2$Br

Propose a mechanism for the reaction. To what reaction of alkenes is the process analogous?

9.47 A *cumulene* is a compound with three adjacent double bonds. Draw an orbital picture of a cumulene. What kind of hybridization do the two central carbon atoms have? What is the geometric relationship of the substituents on one end to the substituents on the other end? What kind of isomerism is possible? Make a model to help see the answer.

$$R_2C=C=C=CR_2$$

A cumulene

9.48 Reaction of acetone with D_3O^+ yields hexadeuterioacetone. That is, all the hydrogens in acetone are exchanged for deuterium. Review the mechanism of mercuric ion–catalyzed alkyne hydration, and then propose a mechanism for this deuterium incorporation.

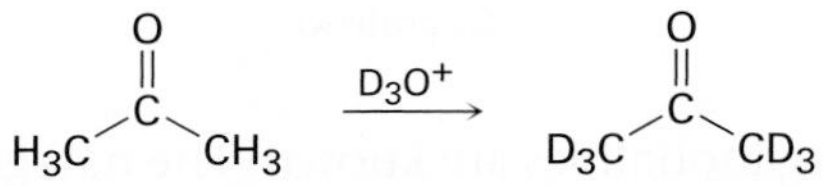

$H_3C-C(=O)-CH_3$ $\xrightarrow{D_3O^+}$ $D_3C-C(=O)-CD_3$

Acetone   **Hexadeuterioacetone**

10

The gases released during volcanic eruptions contain large amounts of organohalides, including chloromethane, chloroform, dichlorodifluoromethane, and many others.

Image copyright Vulkanette, 2010. Used under license from Shutterstock.com

Organohalides

10.1 Names and Properties of Alkyl Halides
10.2 Preparing Alkyl Halides from Alkanes: Radical Halogenation
10.3 Preparing Alkyl Halides from Alkenes: Allylic Bromination
10.4 Stability of the Allyl Radical: Resonance Revisited
10.5 Preparing Alkyl Halides from Alcohols
10.6 Reactions of Alkyl Halides: Grignard Reagents
10.7 Organometallic Coupling Reactions
10.8 Oxidation and Reduction in Organic Chemistry

A Deeper Look—Naturally Occurring Organohalides

OWL Sign in to OWL for Organic Chemistry at **www.cengage.com/owl** to view tutorials and simulations, develop problem-solving skills, and complete online homework assigned by your professor.

Now that we've covered the chemistry of hydrocarbons, it's time to start looking at more complex substances that contain elements in addition to C and H. We'll begin by discussing the chemistry of **organohalides**, compounds that contain one or more halogen atoms.

Halogen-substituted organic compounds are widespread in nature, and more than 5000 organohalides have been found in algae and various other marine organisms. Chloromethane, for example, is released in large amounts by ocean kelp, as well as by forest fires and volcanoes. Halogen-containing compounds also have a vast array of industrial applications, including their use as solvents, inhaled anesthetics in medicine, refrigerants, and pesticides.

Trichloroethylene (a solvent)

Halothane (an inhaled anesthetic)

Dichlorodifluoromethane (a refrigerant)

Bromomethane (a fumigant)

Still other halo-substituted compounds are used as medicines and food additives. The nonnutritive sweetener sucralose, marketed as Splenda, contains four chlorine atoms, for instance. Sucralose is about 600 times as sweet as sucrose, so only 1 mg is equivalent to an entire teaspoon of table sugar.

Sucralose

A large variety of organohalides are known. The halogen might be bonded to an alkynyl group (C≡C–X), a vinylic group (C=C–X), an aromatic ring (Ar–X), or an alkyl group. We'll be concerned in this chapter, however, primarily with **alkyl halides**, compounds with a halogen atom bonded to a saturated, sp^3-hybridized carbon atom.

Why This Chapter? Alkyl halides are encountered less frequently than their oxygen-containing relatives and are not often involved in the biochemical pathways of terrestrial organisms, but some of the *kinds* of reactions they undergo—nucleophilic substitutions and eliminations—*are* encountered frequently. Thus, alkyl halide chemistry acts as a relatively simple model for many mechanistically similar but structurally more complex reactions found in biomolecules. We'll begin in this chapter with a look at how to name and prepare alkyl halides, and we'll see several of their reactions. Then in the next chapter, we'll make a detailed study of the substitution and elimination reactions of alkyl halides—two of the most important and well-studied reaction types in organic chemistry.

10.1 Names and Properties of Alkyl Halides

Although commonly called *alkyl halides,* halogen-substituted alkanes are named systematically as *haloalkanes* (Section 3.4), treating the halogen as a substituent on a parent alkane chain. There are three steps:

STEP 1
Find the longest chain, and name it as the parent. If a double or triple bond is present, the parent chain must contain it.

STEP 2
Number the carbons of the parent chain beginning at the end nearer the first substituent, whether alkyl or halo. Assign each substituent a number according to its position on the chain.

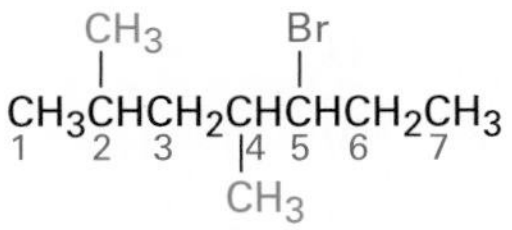

$$\underset{\text{5-Bromo-2,4-dimethyl}\textbf{heptane}}{\underset{1\ \ \ 2\ \ \ 3\ \ \ \ 4\ 5\ \ 6\ \ \ 7}{CH_3\overset{\overset{CH_3}{|}}{C}HCH_2\overset{\overset{Br}{|}}{C}H\underset{\underset{CH_3}{|}}{C}HCH_2CH_3}} \qquad \underset{\text{2-Bromo-4,5-dimethyl}\textbf{heptane}}{\underset{1\ \ \ 2\ \ \ 3\ \ \ \ 4\ 5\ \ 6\ \ \ 7}{CH_3\overset{\overset{Br}{|}}{C}HCH_2\overset{\overset{CH_3}{|}}{C}H\underset{\underset{CH_3}{|}}{C}HCH_2CH_3}}$$

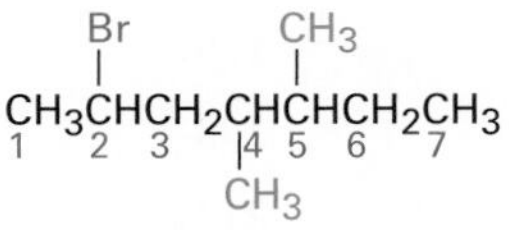

If different halogens are present, number all and list them in alphabetical order when writing the name.

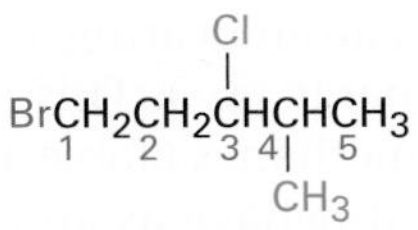

$$\underset{\text{1-Bromo-3-chloro-4-methyl}\textbf{pentane}}{\underset{1\ \ \ \ \ 2\ \ \ \ \ 3\ \ 4\ \ \ \ 5}{BrCH_2CH_2\overset{\overset{Cl}{|}}{C}H\underset{\underset{CH_3}{|}}{C}HCH_3}}$$

STEP 3
If the parent chain can be properly numbered from either end by step 2, begin at the end nearer the substituent that has alphabetical precedence.

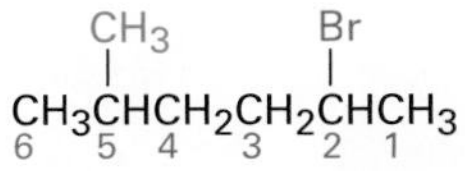

$$\underset{\begin{array}{c}\textbf{2-Bromo-5-methylhexane}\\(\textit{Not}\ \textbf{5-bromo-2-methylhexane})\end{array}}{\underset{6\ \ \ \ 5\ \ \ 4\ \ \ \ 3\ \ \ 2\ \ \ 1}{CH_3\overset{\overset{CH_3}{|}}{C}HCH_2CH_2\overset{\overset{Br}{|}}{C}HCH_3}}$$

In addition to their systematic names, many simple alkyl halides are also named by identifying first the alkyl group and then the halogen. For example, CH₃I can be called either iodomethane or methyl iodide. Such names are well entrenched in the chemical literature and in daily usage, but they won't be used in this book.

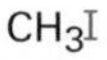

CH₃I

Iodomethane
(or methyl iodide)

CH₃CHCH₃ with Cl

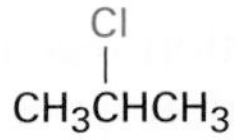

2-Chloropropane
(or isopropyl chloride)

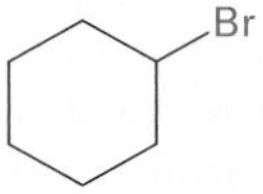

Cyclohexane with Br

Bromocyclohexane
(or cyclohexyl bromide)

Halogens increase in size going down the periodic table, so the lengths of the corresponding carbon–halogen bonds increase accordingly (Table 10.1). In addition, C–X bond strengths decrease going down the periodic table. As we've been doing thus far, we'll continue to use the abbreviation X to represent any of the halogens F, Cl, Br, or I.

Table 10.1 A Comparison of the Halomethanes

Halomethane	Bond length (pm)	Bond strength (kJ/mol)	Bond strength (kcal/mol)	Dipole moment (D)
CH₃F	139	460	110	1.85
CH₃Cl	178	350	84	1.87
CH₃Br	193	294	70	1.81
CH₃I	214	239	57	1.62

In our discussion of bond polarity in functional groups in **Section 6.4**, we noted that halogens are more electronegative than carbon. The C–X bond is therefore polar, with the carbon atom bearing a slight positive charge ($\delta+$) and the halogen a slight negative charge ($\delta-$). This polarity results in a substantial dipole moment for all the halomethanes (Table 10.1) and implies that the alkyl halide C–X carbon atom should behave as an electrophile in polar reactions. We'll soon see that this is indeed the case.

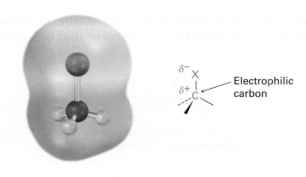

10.2 | Preparing Alkyl Halides from Alkanes: Radical Halogenation

Problem 10.1
Give IUPAC names for the following alkyl halides:

(a) CH₃CH₂CH₂CH₂I

(b) 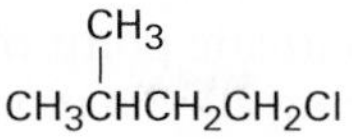 CH₃CHCH₂CH₂Cl
         |
         CH₃

(c) 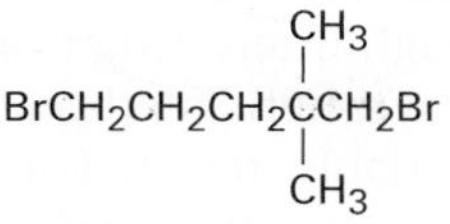 BrCH₂CH₂CH₂CCH₂Br
                |
          CH₃—C—CH₃
                |
                CH₃

(d) 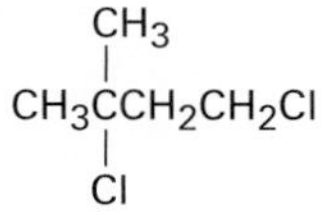 CH₃CCH₂CH₂Cl
         |
         CH₃ (top), Cl (bottom)

(e) 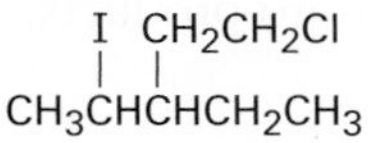 CH₃CHCHCH₂CH₃
         | |
         I CH₂CH₂Cl

(f) 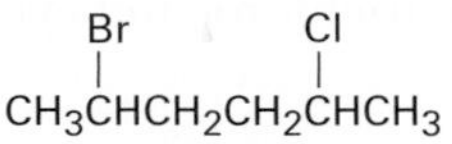 CH₃CHCH₂CH₂CHCH₃
         | |
         Br   Cl

Problem 10.2
Draw structures corresponding to the following IUPAC names:
(a) 2-Chloro-3,3-dimethylhexane
(b) 3,3-Dichloro-2-methylhexane
(c) 3-Bromo-3-ethylpentane
(d) 1,1-Dibromo-4-isopropylcyclohexane
(e) 4-*sec*-Butyl-2-chlorononane
(f) 1,1-Dibromo-4-*tert*-butylcyclohexane

10.2 Preparing Alkyl Halides from Alkanes: Radical Halogenation

Simple alkyl halides can sometimes be prepared by reaction of an alkane with Cl_2 or Br_2 in the presence of light through a radical chain-reaction pathway (**Section 6.3**). The mechanism is shown in **Figure 10.1** for chlorination.

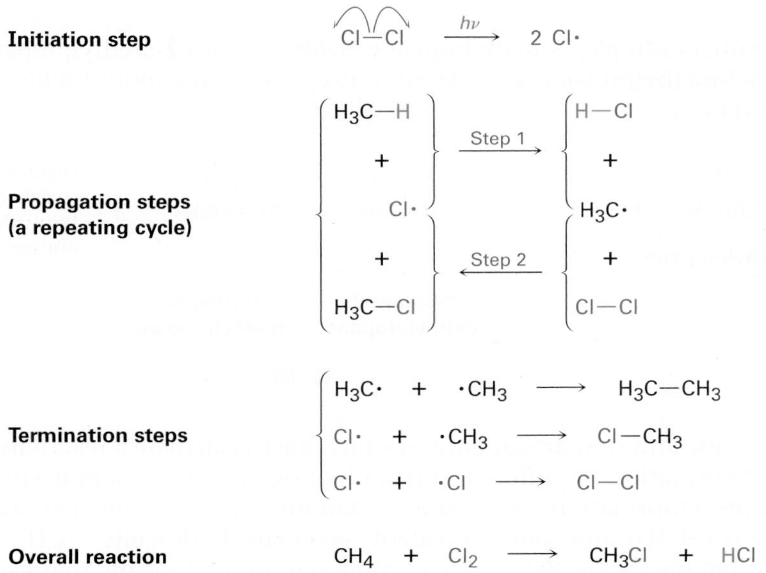

Initiation step

Cl—Cl $\xrightarrow{h\nu}$ 2 Cl·

Propagation steps (a repeating cycle)

$H_3C—H$ + Cl· $\xrightarrow{\text{Step 1}}$ H—Cl + H_3C·

H_3C· + Cl—Cl $\xrightarrow{\text{Step 2}}$ H_3C—Cl + Cl·

Termination steps

H_3C· + ·CH_3 ⟶ H_3C—CH_3
Cl· + ·CH_3 ⟶ Cl—CH_3
Cl· + ·Cl ⟶ Cl—Cl

Overall reaction

CH_4 + Cl_2 ⟶ CH_3Cl + HCl

Figure 10.1 Mechanism of the radical chlorination of methane. Three kinds of steps are required: initiation, propagation, and termination. The propagation steps are a repeating cycle, with Cl· a reactant in step 1 and a product in step 2, and with ·CH_3 a product in step 1 and a reactant in step 2. (The symbol $h\nu$ shown in the initiation step is the standard way of indicating irradiation with light.)

Recall from **Section 6.3** that radical substitution reactions require three kinds of steps: *initiation*, *propagation*, and *termination*. Once an initiation step has started the process by producing radicals, the reaction continues in a self-sustaining cycle. The cycle requires two repeating propagation steps in which a radical, the halogen, and the alkane yield alkyl halide product plus more radical to carry on the chain. The chain is occasionally terminated by the combination of two radicals.

Although interesting from a mechanistic point of view, alkane halogenation is a poor synthetic method for preparing alkyl halides because mixtures of products invariably result. For example, chlorination of methane does not stop cleanly at the monochlorinated stage but continues to give a mixture of dichloro, trichloro, and even tetrachloro products.

$$CH_4 + Cl_2 \xrightarrow{h\nu} CH_3Cl + HCl$$
$$\xrightarrow{Cl_2} CH_2Cl_2 + HCl$$
$$\xrightarrow{Cl_2} CHCl_3 + HCl$$
$$\xrightarrow{Cl_2} CCl_4 + HCl$$

The situation is even worse for chlorination of alkanes that have more than one sort of hydrogen. For example, chlorination of butane gives two monochlorinated products in a 30:70 ratio in addition to dichlorobutane, trichlorobutane, and so on.

$$CH_3CH_2CH_2CH_3 + Cl_2 \xrightarrow{h\nu} \underbrace{CH_3CH_2CH_2CH_2Cl + CH_3CH_2CHClCH_3}_{30\,:\,70} + \text{Dichloro-, trichloro-, tetrachloro-, and so on}$$

Butane   **1-Chlorobutane**   **2-Chlorobutane**

As another example, 2-methylpropane yields 2-chloro-2-methylpropane and 1-chloro-2-methylpropane in a 35:65 ratio, along with more highly chlorinated products.

$$(CH_3)_2CHCH_3 + Cl_2 \xrightarrow{h\nu} \underbrace{(CH_3)_3CCl + (CH_3)_2CHCH_2Cl}_{35\,:\,65} + \text{Dichloro-, trichloro-, tetrachloro-, and so on}$$

2-Methylpropane   **2-Chloro-2-methylpropane**   **1-Chloro-2-methylpropane**

From these and similar reactions, it's possible to calculate a reactivity order toward chlorination for different sorts of hydrogen atoms in a molecule. Take the butane chlorination, for instance. Butane has six equivalent primary hydrogens (–CH$_3$) and four equivalent secondary hydrogens (–CH$_2$–). The fact that butane yields 30% of 1-chlorobutane product means that each one of the six primary hydrogens is responsible for $30\% \div 6 = 5\%$ of the product. Similarly, the fact that 70% of 2-chlorobutane is formed means that each of the four secondary hydrogens is responsible for $70\% \div 4 = 17.5\%$ of the product. Thus, a secondary hydrogen reacts $17.5\% \div 5\% = 3.5$ times as often as a primary hydrogen.

A similar calculation for the chlorination of 2-methylpropane indicates that each of the nine primary hydrogens accounts for 65% ÷ 9 = 7.2% of the product, while the single tertiary hydrogen (R₃CH) accounts for 35% of the product. Thus, a tertiary hydrogen is 35 ÷ 7.2 = 5 times as reactive as a primary hydrogen toward chlorination.

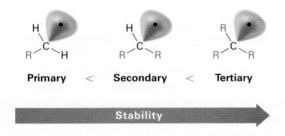

| Primary | < | Secondary | < | Tertiary |
| 1.0 | | 3.5 | | 5.0 |

Reactivity →

The observed reactivity order of alkane hydrogens toward radical chlorination can be explained by looking at the bond dissociation energies given previously in Table 6.3 on page 204. The data show that a tertiary C–H bond (400 kJ/mol; 96 kcal/mol) is weaker than a secondary C–H bond (410 kJ/mol; 98 kcal/mol), which is in turn weaker than a primary C–H bond (421 kJ/mol; 101 kcal/mol). Since less energy is needed to break a tertiary C–H bond than to break a primary or secondary C–H bond, the resultant tertiary radical is more stable than a primary or secondary radical.

Primary < Secondary < Tertiary

Stability →

Problem 10.3
Draw and name all monochloro products you would expect to obtain from radical chlorination of 2-methylpentane. Which, if any, are chiral?

Problem 10.4
Taking the relative reactivities of 1°, 2°, and 3° hydrogen atoms into account, what product(s) would you expect to obtain from monochlorination of 2-methylbutane? What would the approximate percentage of each product be? (Don't forget to take into account the number of each sort of hydrogen.)

10.3 Preparing Alkyl Halides from Alkenes: Allylic Bromination

We've already seen several methods for preparing alkyl halides from alkenes, including the reactions of HX and X₂ with alkenes in electrophilic addition reactions (**Sections 7.7 and 8.2**). The hydrogen halides HCl, HBr, and HI react with alkenes by a polar mechanism to give the product of Markovnikov addition. Bromine and chlorine undergo anti addition through halonium ion intermediates to give 1,2-dihalogenated products.

X = Cl or Br

X = Cl, Br, or I

Another method for preparing alkyl halides from alkenes is by reaction with N-bromosuccinimide (abbreviated NBS) in the presence of light to give products resulting from substitution of hydrogen by bromine at the position next to the double bond—the **allylic** position. Cyclohexene, for example, gives 3-bromocyclohexene.

Cyclohexene

3-Bromocyclohexene (85%)

This allylic bromination with NBS is analogous to the alkane chlorination reaction discussed in the previous section and occurs by a radical chain reaction pathway **(Figure 10.2)**. As in alkane halogenation, a Br· radical abstracts an allylic hydrogen atom, forming an allylic radical plus HBr. The HBr then reacts with NBS to form Br_2, which in turn reacts with the allylic radical to yield the brominated product and a Br· radical that cycles back into the first step and carries on the chain.

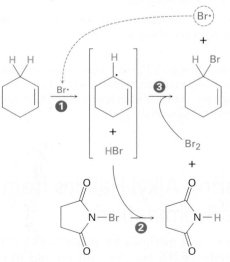

Figure 10.2 Mechanism of allylic bromination of an alkene with NBS. The process is a radical chain reaction in which (**1**) a Br· radical abstracts an allylic hydrogen atom of the alkene and gives an allylic radical plus HBr. (**2**) The HBr then reacts with NBS to form Br_2, which (**3**) reacts with the allylic radical to yield the bromoalkene product and a Br· radical that carries on the chain.

Why does bromination with NBS occur exclusively at an allylic position rather than elsewhere in the molecule? The answer, once again, is found by looking at bond dissociation energies to see the relative stabilities of various kinds of radicals. Although a typical secondary alkyl C–H bond has a strength of about 410 kJ/mol (98 kcal/mol) and a typical vinylic C–H bond has a strength of 465 kJ/mol (111 kcal/mol), an *allylic* C–H bond has a strength of only about 370 kJ/mol (88 kcal/mol). An allylic radical is therefore more stable than a typical alkyl radical with the same substitution by about 40 kJ/mol (9 kcal/mol).

Allylic
370 kJ/mol (88 kcal/mol)

Alkyl
410 kJ/mol (98 kcal/mol)

Vinylic
465 kJ/mol (111 kcal/mol)

We can thus expand the stability ordering to include vinylic and allylic radicals.

Vinylic < Methyl < Primary < Secondary < Tertiary < Allylic

Stability

10.4 Stability of the Allyl Radical: Resonance Revisited

To see why an allylic radical is so stable, look at the orbital picture in **Figure 10.3**. The radical carbon atom with an unpaired electron can adopt sp^2 hybridization, placing the unpaired electron in a *p* orbital and giving a structure that is electronically symmetrical. The *p* orbital on the central carbon can therefore overlap equally well with a *p* orbital on either of the two neighboring carbons.

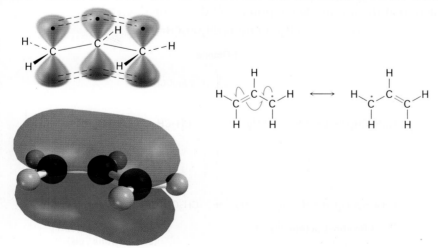

Figure 10.3 An orbital view of the allyl radical. The *p* orbital on the central carbon can overlap equally well with a *p* orbital on either neighboring carbon, giving rise to two equivalent resonance structures.

Because the allyl radical is electronically symmetrical, it has two resonance forms—one with the unpaired electron on the left and the double bond on the right and another with the unpaired electron on the right and the double bond on the left. Neither structure is correct by itself; the true structure of the allyl radical is a resonance hybrid of the two. (You might want to review **Sections 2.4–2.6** to brush up on resonance.) As noted in **Section 2.5**, the greater the number of resonance forms, the greater the stability of a compound because bonding electrons are attracted to more nuclei. An allyl radical, with two resonance forms, is therefore more stable than a typical alkyl radical, which has only a single structure.

In molecular orbital terms, the stability of the allyl radical is due to the fact that the unpaired electron is **delocalized**, or spread out, over an extended π orbital network rather than localized at only one site, as shown by the computer-generated MO in Figure 10.3. This delocalization is particularly apparent in the so-called spin density surface in **Figure 10.4**, which shows the calculated location of the unpaired electron. The two terminal carbons share the unpaired electron equally.

Figure 10.4 The spin density surface of the allyl radical locates the position of the unpaired electron and shows that it is equally shared between the two terminal carbons.

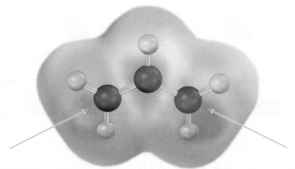

In addition to its effect on stability, delocalization of the unpaired electron in the allyl radical has other chemical consequences. Because the unpaired electron is delocalized over both ends of the π orbital system, reaction with Br_2 can occur at either end. As a result, allylic bromination of an unsymmetrical alkene often leads to a mixture of products. For example, bromination of 1-octene gives a mixture of 3-bromo-1-octene and 1-bromo-2-octene. The two products are not formed in equal amounts, however, because the intermediate allylic radical is not symmetrical and reaction at the two ends is not equally likely. Reaction at the less hindered, primary end is favored.

$$CH_3CH_2CH_2CH_2CH_2CH_2CH=CH_2$$

1-Octene

↓ NBS, $h\nu$, CCl_4

$$[CH_3CH_2CH_2CH_2CH_2\dot{C}HCH=CH_2 \longleftrightarrow CH_3CH_2CH_2CH_2CH_2CH=CH\dot{C}H_2]$$

↓

$$\underset{\textbf{3-Bromo-1-octene (17\%)}}{CH_3CH_2CH_2CH_2CH_2\overset{Br}{\overset{|}{C}}HCH=CH_2} \quad + \quad \underset{\substack{\textbf{1-Bromo-2-octene (83\%)} \\ \textbf{(53 : 47 trans : cis)}}}{CH_3CH_2CH_2CH_2CH_2CH=CHCH_2Br}$$

10.4 | Stability of the Allyl Radical: Resonance Revisited

The products of allylic bromination reactions are useful for conversion into dienes by dehydrohalogenation with base. Cyclohexene can be converted into 1,3-cyclohexadiene, for example.

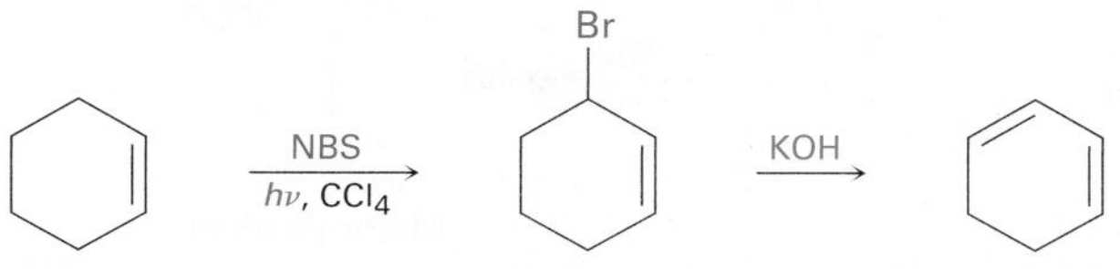

Cyclohexene → (NBS, hν, CCl₄) → 3-Bromocyclohexene → (KOH) → 1,3-Cyclohexadiene

Worked Example 10.1
Predicting the Product of an Allylic Bromination Reaction

What products would you expect from reaction of 4,4-dimethylcyclohexene with NBS?

Strategy
Draw the alkene reactant, and identify the allylic positions. In this case, there are two different allylic positions; we'll label them **A** and **B**. Now abstract an allylic hydrogen from each position to generate the two corresponding allylic radicals. Each of the two allylic radicals can add a Br atom at either end (**A** or **A'**; **B** or **B'**), to give a mixture of up to four products. Draw and name the products. In the present instance, the "two" products from reaction at position **B** are identical, so only three products are formed in this reaction.

Solution

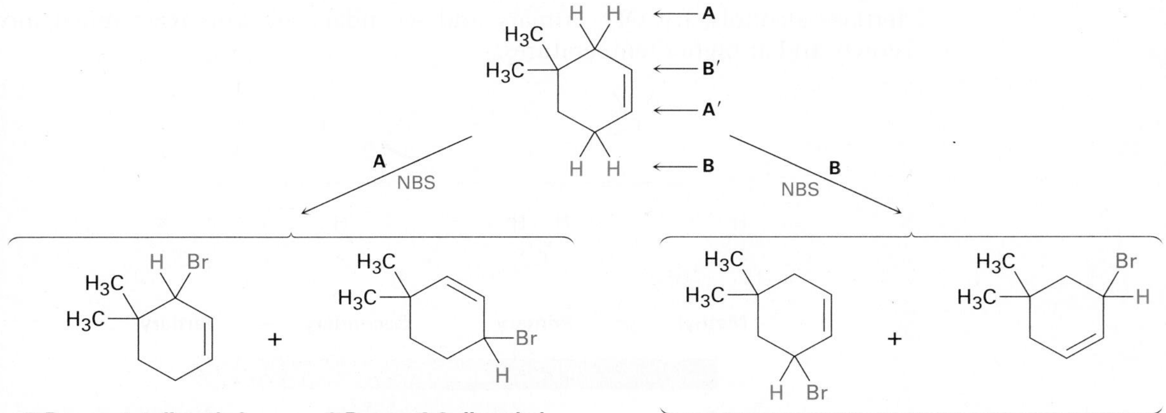

3-Bromo-4,4-dimethyl-cyclohexene

6-Bromo-3,3-dimethyl-cyclohexene

3-Bromo-5,5-dimethyl-cyclohexene

Problem 10.5
Draw three resonance forms for the cyclohexadienyl radical.

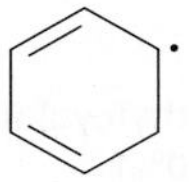

Cyclohexadienyl radical

Problem 10.6
The major product of the reaction of methylenecyclohexane with N-bromosuccinimide is 1-(bromomethyl)cyclohexene. Explain.

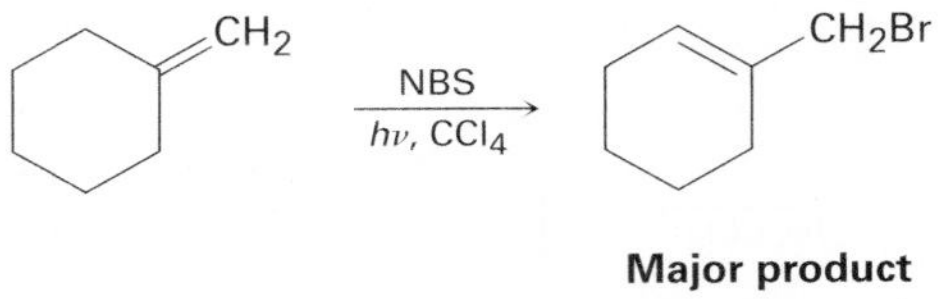

$$\text{methylenecyclohexane} \xrightarrow[h\nu,\ CCl_4]{NBS} \text{1-(bromomethyl)cyclohexene (Major product)}$$

Problem 10.7
What products would you expect from reaction of the following alkenes with NBS? If more than one product is formed, show the structures of all.

(a) 4-methylcyclohexene

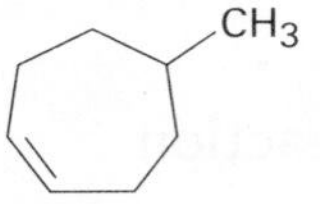

(b) $CH_3CHCH=CHCH_2CH_3$ with CH_3 substituent

10.5 Preparing Alkyl Halides from Alcohols

The most generally useful method for preparing alkyl halides is to make them from alcohols, which themselves can be obtained from carbonyl compounds as we'll see in **Sections 17.4 and 17.5**. Because of the importance of the process, many different methods have been developed to transform alcohols into alkyl halides. The simplest method is to treat the alcohol with HCl, HBr, or HI. For reasons that will be discussed in **Section 11.5**, the reaction works best with tertiary alcohols, R_3COH. Primary and secondary alcohols react much more slowly and at higher temperatures.

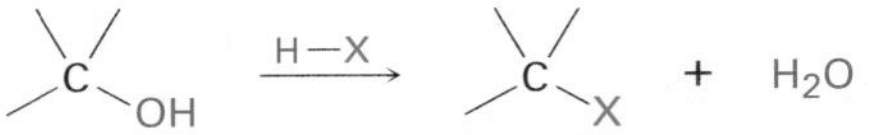

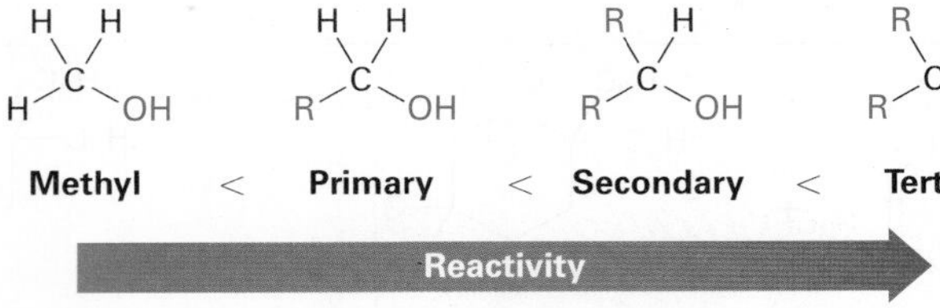

$$\text{R}_3\text{C-OH} \xrightarrow{H-X} \text{R}_3\text{C-X} + H_2O$$

Methyl < Primary < Secondary < Tertiary

Reactivity →

The reaction of HX with a tertiary alcohol is so rapid that it's often carried out simply by bubbling the pure HCl or HBr gas into a cold ether solution of the alcohol. 1-Methylcyclohexanol, for example, is converted into 1-chloro-1-methylcyclohexane by treating with HCl.

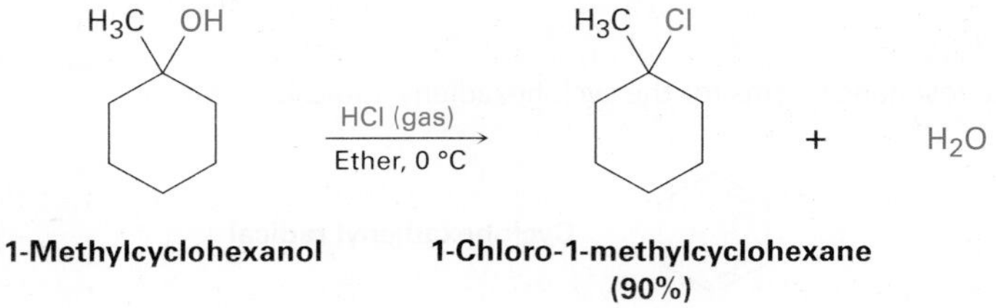

$$\text{1-Methylcyclohexanol} \xrightarrow[\text{Ether, 0 °C}]{\text{HCl (gas)}} \text{1-Chloro-1-methylcyclohexane (90\%)} + H_2O$$

Primary and secondary alcohols are best converted into alkyl halides by treatment with either thionyl chloride (SOCl$_2$) or phosphorus tribromide (PBr$_3$). These reactions, which normally take place readily under mild conditions, are less acidic and less likely to cause acid-catalyzed rearrangements than the HX method.

Benzoin → (86%) + SO$_2$ + HCl (via SOCl$_2$, Pyridine)

3 CH$_3$CH$_2$CHCH$_3$ (OH) — PBr$_3$, Ether, 35 °C → 3 CH$_3$CH$_2$CHCH$_3$ (Br) + H$_3$PO$_3$

2-Butanol → 2-Bromobutane (86%)

As the preceding examples indicate, the yields of these SOCl$_2$ and PBr$_3$ reactions are generally high and other functional groups such as ethers, carbonyls, and aromatic rings don't usually interfere. We'll look at the mechanisms of these and other related substitution reactions in **Section 11.3**.

Alkyl fluorides can also be prepared from alcohols. Numerous alternative reagents are used for the reaction, including diethylaminosulfur trifluoride [(CH$_3$CH$_2$)$_2$NSF$_3$] and HF in pyridine solvent.

Cyclohexanol → Fluorocyclohexane (99%) (via HF, Pyridine)

Problem 10.8
How would you prepare the following alkyl halides from the corresponding alcohols?

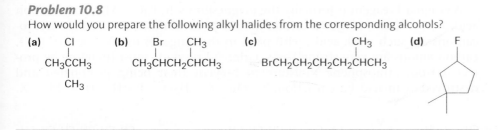

(a) CH$_3$CCH$_3$ with Cl and CH$_3$ substituents
(b) CH$_3$CHCH$_2$CHCH$_3$ with Br and CH$_3$
(c) BrCH$_2$CH$_2$CH$_2$CH$_2$CHCH$_3$ with CH$_3$
(d) fluorocyclopentane with two methyl groups

10.6 Reactions of Alkyl Halides: Grignard Reagents

Alkyl halides, RX, react with magnesium metal in ether or tetrahydrofuran (THF) solvent to yield alkylmagnesium halides, RMgX. The products, called **Grignard reagents** after their discoverer, Victor Grignard, are examples of *organometallic* compounds because they contain a carbon–metal bond. In addition to alkyl halides, Grignard reagents can also be made from alkenyl (vinylic) and aryl (aromatic) halides. The halogen can be Cl, Br, or I, although chlorides are less reactive than bromides and iodides. Organofluorides rarely react with magnesium.

$$\left.\begin{array}{l}1° \text{ alkyl} \\ 2° \text{ alkyl} \\ 3° \text{ alkyl} \\ \text{alkenyl} \\ \text{aryl}\end{array}\right\} \longrightarrow \text{R—X} \longleftarrow \left\{\begin{array}{l}\text{Cl} \\ \text{Br} \\ \text{I}\end{array}\right.$$

$$\downarrow \text{Mg} \quad \text{Ether or THF}$$

$$\text{R—Mg—X}$$

As you might expect from the discussion of electronegativity and bond polarity in **Section 6.4**, the carbon–magnesium bond is polarized, making the carbon atom of Grignard reagents both nucleophilic and basic. An electrostatic potential map of methylmagnesium iodide, for instance, indicates the electron-rich (red) character of the carbon bonded to magnesium.

Iodomethane $\xrightarrow{\text{Mg, Ether}}$ **Methylmagnesium iodide** — Basic and nucleophilic

A Grignard reagent is formally the magnesium salt, $R_3C^-{}^+MgX$, of a carbon acid, R_3C-H, and is thus a carbon anion, or **carbanion**. But because hydrocarbons are such weak acids, with pK_a's in the range 44 to 60 **(Section 9.7)**, carbon anions are very strong bases. Grignard reagents must therefore be protected from atmospheric moisture to prevent their being protonated and destroyed in an acid–base reaction: $R-Mg-X + H_2O \rightarrow R-H + HO-Mg-X$.

$$CH_3CH_2CH_2CH_2CH_2CH_2Br \xrightarrow[\text{Ether}]{\text{Mg}} CH_3CH_2CH_2CH_2CH_2CH_2MgBr \xrightarrow{H_2O} CH_3CH_2CH_2CH_2CH_2CH_3$$

1-Bromohexane   **1-Hexylmagnesium bromide**   **Hexane**

Grignard reagents themselves don't occur in living organisms, but they are useful carbon-based nucleophiles in several important laboratory reactions, which we'll look at in detail in Chapter 17. In addition, they act as a simple model for other, more complex carbon-based nucleophiles that *are* important in biological chemistry. We'll see many examples in Chapter 29.

Problem 10.9

How strong a base would you expect a Grignard reagent to be? Look at Table 9.1 on page 326, and predict whether the following reactions will occur as written. (The pK_a of NH_3 is 35.)

(a) $CH_3MgBr + H-C\equiv C-H \rightarrow CH_4 + H-C\equiv C-MgBr$
(b) $CH_3MgBr + NH_3 \rightarrow CH_4 + H_2N-MgBr$

Problem 10.10

How might you replace a halogen substituent by a deuterium atom if you wanted to prepare a deuterated compound?

$$\underset{CH_3CHCH_2CH_3}{\overset{Br}{|}} \xrightarrow{?} \underset{CH_3CHCH_2CH_3}{\overset{D}{|}}$$

10.7 Organometallic Coupling Reactions

Many other kinds of organometallic compounds can be prepared in a manner similar to that of Grignard reagents. For instance, alkyllithium reagents, RLi, can be prepared by the reaction of an alkyl halide with lithium metal. Alkyllithiums are both nucleophiles and strong bases, and their chemistry is similar in many respects to that of alkylmagnesium halides.

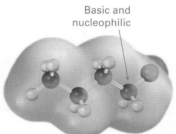

Basic and nucleophilic

$$CH_3CH_2CH_2CH_2Br \xrightarrow[\text{Pentane}]{2\ Li} CH_3CH_2CH_2CH_2Li + LiBr$$

1-Bromobutane → **Butyllithium**

One particularly valuable reaction of alkyllithiums is in making lithium diorganocopper compounds, R_2CuLi, by reaction with copper(I) iodide in diethyl ether as solvent. Called **Gilman reagents**, lithium diorganocopper compounds are useful because they undergo a *coupling* reaction with organochlorides, bromides, and iodides (but not fluorides). One of the alkyl groups from the Gilman reagent replaces the halogen of the organohalide, forming a new carbon–carbon bond and yielding a hydrocarbon product. Lithium dimethylcopper, for instance, reacts with 1-iododecane to give undecane in 90% yield.

$$2\ CH_3Li + CuI \xrightarrow{\text{Ether}} (CH_3)_2Cu^-\ Li^+ + LiI$$

Methyllithium → **Lithium dimethylcopper (a Gilman reagent)**

$$(CH_3)_2CuLi + CH_3(CH_2)_8CH_2I \xrightarrow[0\ °C]{\text{Ether}} CH_3(CH_2)_8CH_2CH_3 + LiI + CH_3Cu$$

Lithium dimethylcopper **1-Iododecane** → **Undecane (90%)**

This organometallic coupling reaction is useful in organic synthesis because it forms carbon–carbon bonds, thereby making possible the preparation of larger molecules from smaller ones. As the following examples indicate, the coupling reaction can be carried out on aryl and vinylic halides as well as on alkyl halides.

$$\text{n-C}_7\text{H}_{15}\text{CH=CHI (trans-1-Iodo-1-nonene)} + (n\text{-C}_4\text{H}_9)_2\text{CuLi} \longrightarrow \text{n-C}_7\text{H}_{15}\text{CH=CH-C}_4\text{H}_9\text{-n (trans-5-Tridecene (71\%))} + n\text{-C}_4\text{H}_9\text{Cu} + \text{LiI}$$

trans-1-Iodo-1-nonene → **trans-5-Tridecene (71%)**

Iodobenzene + $(CH_3)_2$CuLi ⟶ Toluene (91%) + CH_3Cu + LiI

An organocopper coupling reaction is carried out commercially to synthesize muscalure, (9Z)-tricosene, the sex attractant secreted by the common housefly. Minute amounts of muscalure greatly increase the lure of insecticide-treated fly bait and provide an effective and species-specific means of insect control.

cis-1-Bromo-9-octadecene $\xrightarrow{[CH_3(CH_2)_4]_2\text{CuLi}}$ Muscalure (9Z-tricosene)

The mechanism of the coupling reaction involves initial formation of a triorganocopper intermediate, followed by coupling and loss of RCu. The coupling is not a typical polar nucleophilic substitution reaction of the sort considered in the next chapter.

$$R-X + [R'-Cu-R']^- \text{Li}^+ \longrightarrow \left[\begin{array}{c} R \\ | \\ R'-Cu-R' \end{array}\right] \longrightarrow R-R' + R'-Cu$$

In addition to the coupling reaction of diorganocopper reagents with organohalides, related processes also occur with other organometallic reagents, particularly organopalladium compounds. One of the most commonly used procedures is the coupling reaction of an aromatic or vinyl substituted boronic acid [R—B(OH)$_2$] with an aromatic or vinyl substituted organohalide in the presence of a base and a palladium catalyst. The reaction is less general than the diorganocopper reaction because it does not work with alkyl substrates, but it is preferred when possible because it uses only a catalytic amount of metal rather than a full equivalent and because palladium compounds are less toxic than copper compounds. For example:

Called the *Suzuki–Miyaura reaction*, the process is particularly useful for preparing so-called biaryl compounds, which have two aromatic rings joined together. A large number of commonly used drugs fit this description, so the Suzuki–Miyaura reaction is much-used in the pharmaceutical industry. As an example, valsartan, marketed as Diovan, is a widely prescribed antihypertensive agent whose synthesis begins with a Suzuki–Miyaura coupling of *ortho*-chlorobenzonitrile with *para*-methylbenzeneboronic acid.

Shown in a simplified form in **Figure 10.5**, the mechanism of the Suzuki–Miyaura reaction involves initial reaction of the aromatic halide with the palladium catalyst to form an organopalladium intermediate, followed by reaction of that intermediate with the aromatic boronic acid. The resultant diorganopalladium complex then decomposes to the coupled biaryl product plus regenerated catalyst.

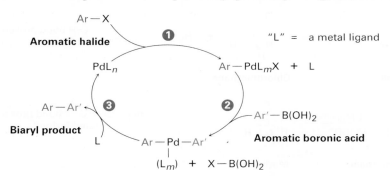

Figure 10.5 Mechanism of the Suzuki–Miyaura coupling reaction of an aromatic boronic acid with an aromatic halide to give a biaryl. The reaction takes place by (**1**) reaction of the aromatic halide, ArX, with the catalyst to form an organopalladium intermediate, followed by (**2**) reaction with the aromatic boronic acid. (**3**) Subsequent decomposition of the diarylpalladium intermediate gives the biaryl product.

Problem 10.11
How would you carry out the following transformations using an organocopper coupling reaction? More than one step is required in each case.

(a) cyclohexene → 3-methylcyclohexene (with CH₃ substituent)

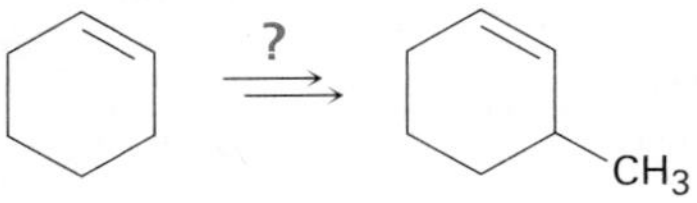

(b) $CH_3CH_2CH_2CH_2Br \xrightarrow{?} CH_3CH_2CH_2CH_2CH_2CH_2CH_2CH_3$

(c) $CH_3CH_2CH_2CH=CH_2 \xrightarrow{?} CH_3CH_2CH_2CH_2CH_2CH_2CH_2CH_2CH_3$

10.8 Oxidation and Reduction in Organic Chemistry

We've pointed out on several occasions that some of the reactions discussed in this and earlier chapters are either oxidations or reductions. As noted in **Section 8.7**, an organic oxidation results in a loss of electron density by carbon, caused either by bond formation between carbon and a more electronegative atom (usually O, N, or a halogen) or by bond-breaking between carbon and a less electronegative atom (usually H). Conversely, an organic reduction results in a gain of electron density by carbon, caused either by bond formation between carbon and a less electronegative atom or by bond-breaking between carbon and a more electronegative atom **(Section 8.6)**.

Oxidation Decreases electron density on carbon by:
– forming one of these: C–O   C–N   C–X
– or breaking this: C–H

Reduction Increases electron density on carbon by:
– forming this: C–H
– or breaking one of these: C–O   C–N   C–X

Based on these definitions, the chlorination reaction of methane to yield chloromethane is an oxidation because a C–H bond is broken and a C–Cl bond is formed. The conversion of an alkyl chloride to an alkane via a Grignard reagent followed by protonation is a reduction, however, because a C–Cl bond is broken and a C–H bond is formed.

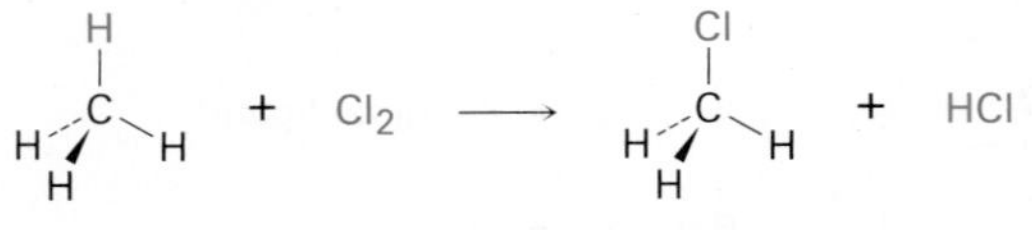

Methane + Cl₂ → Chloromethane + HCl   Oxidation: C–H bond broken and C–Cl bond formed

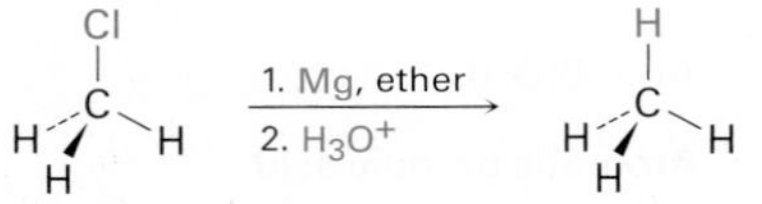

Chloromethane (1. Mg, ether; 2. H₃O⁺) → Methane   Reduction: C–Cl bond broken and C–H bond formed

10.8 | Oxidation and Reduction in Organic Chemistry

As other examples, the reaction of an alkene with Br_2 to yield a 1,2-dibromide is an oxidation because two C—Br bonds are formed, but the reaction of an alkene with HBr to yield an alkyl bromide is neither an oxidation nor a reduction because both a C—H and a C—Br bond are formed.

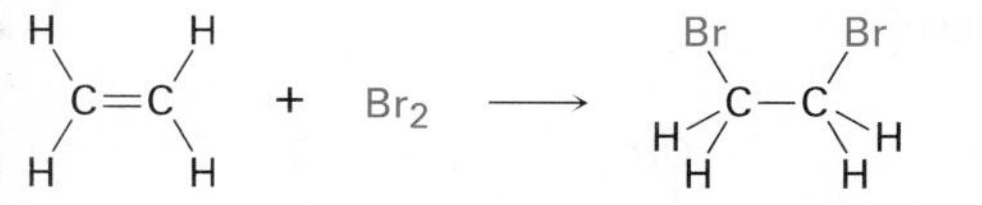

Ethylene + Br_2 → **1,2-Dibromoethane**   Oxidation: Two new bonds formed between carbon and a more electronegative element

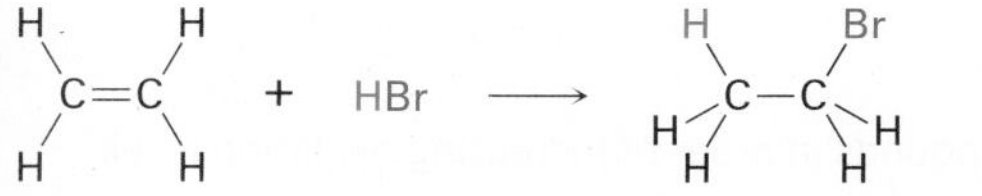

Ethylene + HBr → **Bromoethane**   Neither oxidation nor reduction: One new C—H bond and one new C—Br bond formed

A list of compounds of increasing oxidation level is shown in **Figure 10.6**. Alkanes are at the lowest oxidation level because they have the maximum possible number of C—H bonds per carbon, and CO_2 is at the highest level because it has the maximum possible number of C—O bonds per carbon. Any reaction that converts a compound from a lower level to a higher level is an oxidation, any reaction that converts a compound from a higher level to a lower level is a reduction, and any reaction that doesn't change the level is neither an oxidation nor a reduction.

CH_3CH_3	$H_2C=CH_2$	$HC\equiv CH$	
CH_3OH	$H_2C=O$	HCO_2H	CO_2
CH_3Cl	CH_2Cl_2	$CHCl_3$	CCl_4
CH_3NH_2	$H_2C=NH$	$HC\equiv N$	

Low oxidation level ⟶ High oxidation level

Figure 10.6 Oxidation levels of some common types of compounds.

Worked Example 10.2 shows how to compare the oxidation levels of different compounds with the same number of carbon atoms.

Worked Example 10.2

Comparing Oxidation Levels

Rank the following compounds in order of increasing oxidation level:

$CH_3CH=CH_2$ $CH_3\overset{OH}{\underset{|}{C}}HCH_3$ $CH_3\overset{O}{\underset{\|}{C}}CH_3$ $CH_3CH_2CH_3$

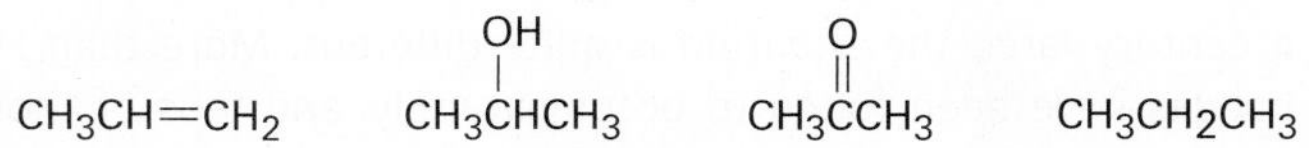

Strategy
Compounds that have the same number of carbon atoms can be compared by adding the number of C—O, C—N, and C—X bonds in each and then subtracting the number of C—H bonds. The larger the resultant value, the higher the oxidation level.

Solution

The first compound (propene) has six C–H bonds, giving an oxidation level of −6; the second (2-propanol) has one C–O bond and seven C–H bonds, giving an oxidation level of −6; the third (acetone) has two C–O bonds and six C–H bonds, giving an oxidation level of −4; and the fourth (propane) has eight C–H bonds, giving an oxidation level of −8. Thus, the order of increasing oxidation level is

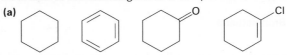

Problem 10.12
Rank each of the following series of compounds in order of increasing oxidation level:

(a) [cyclohexane] [benzene] [cyclohexanone] [1-chlorocyclohexene]

(b) CH₃CN   CH₃CH₂NH₂   H₂NCH₂CH₂NH₂

Problem 10.13
Tell whether each of the following reactions is an oxidation, a reduction, or neither.

(a) CH₃CH₂CHO —NaBH₄/H₂O→ CH₃CH₂CH₂OH

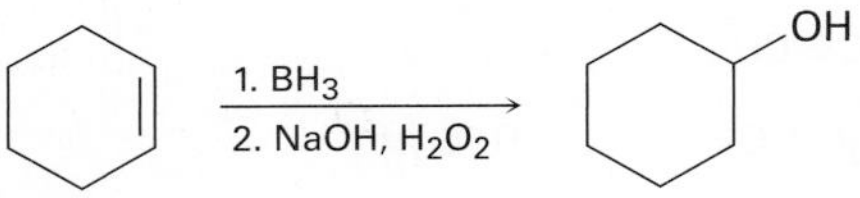

(b) cyclohexene —1. BH₃; 2. NaOH, H₂O₂→ cyclohexanol

Naturally Occurring Organohalides
A DEEPER LOOK

As recently as 1970, only about 30 naturally occurring organohalides were known. It was simply assumed that chloroform, halogenated phenols, chlorinated aromatic compounds called PCBs, and other such substances found in the environment were industrial pollutants. Now, a bit more than a third of a century later, the situation is quite different. More than 5000 organohalides have been found to occur naturally, and tens of thousands more surely exist. From a simple compound like chloromethane to an extremely complex one like the antibiotic vancomycin, a remarkably diverse range of organohalides exists in plants, bacteria, and animals. Many even have valuable physiological activity. The pentahalogenated alkene halomon, for

Marine corals secrete organohalogen compounds that act as a feeding deterrent to fish.

(continued)

instance, has been isolated from the red alga *Portieria hornemannii* and found to have anticancer activity against several human tumor cell lines.

Halomon

Some naturally occurring organohalides are produced in massive quantities. Forest fires, volcanoes, and marine kelp release up to *5 million tons* of CH_3Cl per year, for example, while annual industrial emissions total about 26,000 tons. Termites are thought to release as much as 10^8 kg of chloroform per year. A detailed examination of the Okinawan acorn worm *Ptychodera flava* found that the 64 million worms living in a 1 km^2 study area excreted nearly 8000 pounds per year of bromophenols and bromoindoles, compounds previously thought to be nonnatural pollutants.

Why do organisms produce organohalides, many of which are undoubtedly toxic? The answer seems to be that many organisms use organohalogen compounds for self-defense, either as feeding deterrents, as irritants to predators, or as natural pesticides. Marine sponges, coral, and sea hares, for example, release foul-tasting organohalides that deter fish, starfish, and other predators from eating them. Even humans appear to produce halogenated compounds as part of their defense against infection. The human immune system contains a peroxidase enzyme capable of carrying out halogenation reactions on fungi and bacteria, thereby killing the pathogen. And most remarkable of all, even free chlorine—Cl_2—has been found to be present in humans.

Much remains to be learned—only a few hundred of the more than 500,000 known species of marine organisms have been examined—but it is clear that organohalides are an integral part of the world around us.

Summary

Alkyl halides are not often found in terrestrial organisms, but the kinds of reactions they undergo are among the most important and well-studied reaction types in organic chemistry. In this chapter, we saw how to name and prepare alkyl halides, and we'll soon make a detailed study of their substitution and elimination reactions.

Simple alkyl halides can be prepared by radical halogenation of alkanes, but mixtures of products usually result. The reactivity order of alkanes toward halogenation is identical to the stability order of radicals: $R_3C\cdot > R_2CH\cdot > RCH_2\cdot$. Alkyl halides can also be prepared from alkenes by reaction with *N*-bromosuccinimide (NBS) to give the product of **allylic** bromination. The NBS bromination of alkenes takes place through an intermediate allylic radical, which is stabilized by resonance.

Alcohols react with HX to form alkyl halides, but the reaction works well only for tertiary alcohols, R_3COH. Primary and secondary alkyl halides are normally

Key words

alkyl halide, 344
allylic, 350
carbanion, 356
delocalized, 352
Gilman reagent (LiR_2Cu), 357
Grignard reagent (RMgX), 355
organohalide, 344

prepared from alcohols using either $SOCl_2$, PBr_3, or HF in pyridine. Alkyl halides react with magnesium in ether solution to form organomagnesium halides, called **Grignard reagents (RMgX)**, which are both nucleophilic and strongly basic.

Alkyl halides also react with lithium metal to form organolithium reagents, RLi. In the presence of CuI, these form diorganocoppers, or **Gilman reagents (LiR$_2$Cu)**. Gilman reagents react with organohalides to yield coupled hydrocarbon products.

Summary of Reactions

1. Preparation of alkyl halides
   (a) From alkenes by allylic bromination (Section 10.3)

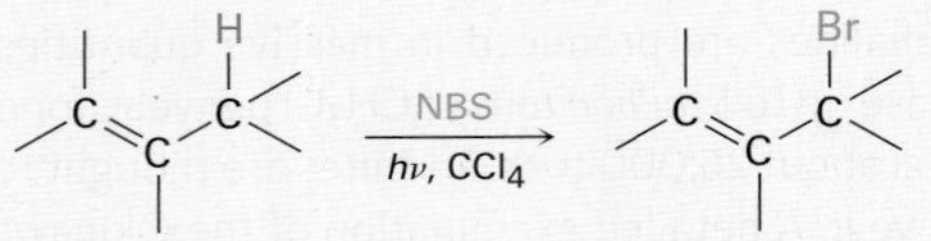

   $$\underset{}{\overset{H}{\underset{}{>C=C-C<}}} \xrightarrow[h\nu,\ CCl_4]{NBS} \underset{}{\overset{Br}{\underset{}{>C=C-C<}}}$$

   (b) From alcohols (Section 10.5)
      (1) Reaction with HX

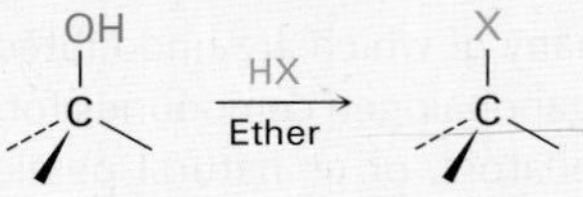

      $$\overset{OH}{\underset{}{>C<}} \xrightarrow[Ether]{HX} \overset{X}{\underset{}{>C<}}$$

      Reactivity order: 3° > 2° > 1°

      (2) Reaction of 1° and 2° alcohols with $SOCl_2$

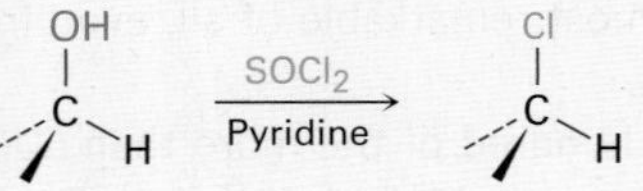

      $$\overset{OH}{\underset{H}{>C<}} \xrightarrow[Pyridine]{SOCl_2} \overset{Cl}{\underset{H}{>C<}}$$

      (3) Reaction of 1° and 2° alcohols with PBr_3

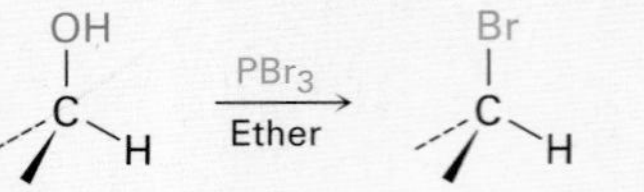

      $$\overset{OH}{\underset{H}{>C<}} \xrightarrow[Ether]{PBr_3} \overset{Br}{\underset{H}{>C<}}$$

      (4) Reaction of 1° and 2° alcohols with HF–pyridine

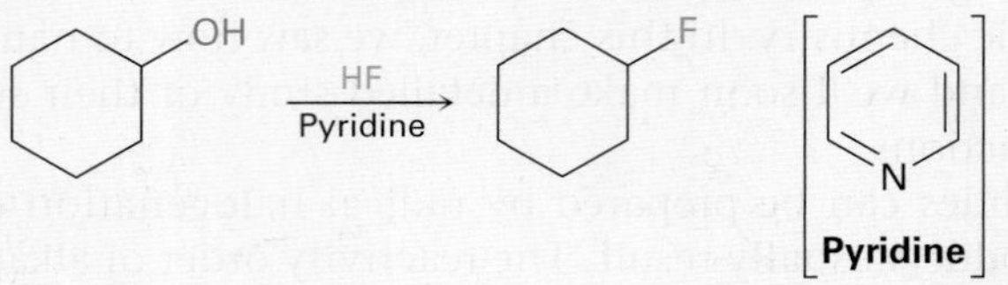

      cyclohexanol $\xrightarrow[Pyridine]{HF}$ fluorocyclohexane [Pyridine]

2. Reactions of alkyl halides
   (a) Formation of Grignard (organomagnesium) reagents (Section 10.6)

   $$R-X \xrightarrow[Ether]{Mg} R-Mg-X$$

(continued)

(b) Formation of Gilman (diorganocopper) reagents (Section 10.7)

$$R-X \xrightarrow[\text{Pentane}]{2\ Li} R-Li + LiX$$

$$2\ R-Li + CuI \xrightarrow{\text{In ether}} [R-Cu-R]^- Li^+ + LiI$$

(c) Organometallic coupling (Section 10.7)
   (1) Diorganocopper reaction

$$R_2CuLi + R'-X \xrightarrow{\text{In ether}} R-R' + RCu + LiX$$

   (2) Palladium-catalyzed Suzuki–Miyaura reaction

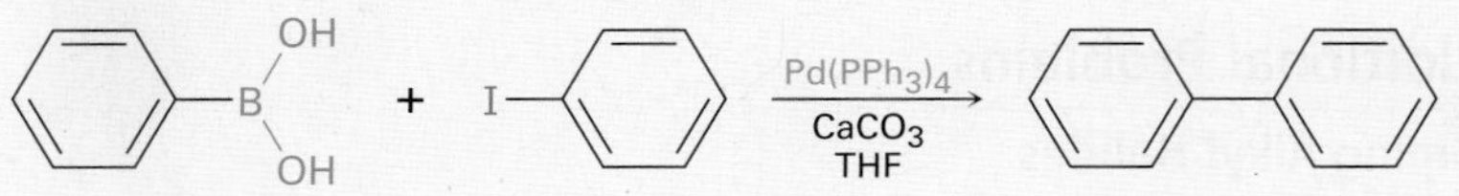

Exercises

Visualizing Chemistry

(Problems 10.1–10.13 appear within the chapter.)

OWL Interactive versions of these problems are assignable in OWL for Organic Chemistry.

10.14 Give IUPAC names for the following alkyl halides (green = Cl):

(a)  (b)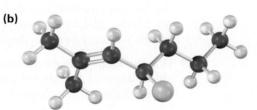

10.15 Show the product(s) of reaction of the following alkenes with NBS:

(a) 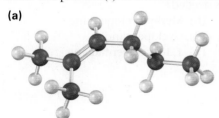 (b)

10.16 The following alkyl bromide can be prepared by reaction of the alcohol (S)-2-pentanol with PBr₃. Name the compound, assign (R) or (S) stereochemistry, and tell whether the reaction of the alcohol occurs with retention of the same stereochemistry or with a change in stereochemistry (reddish brown = Br).

Additional Problems

Naming Alkyl Halides

10.17 Name the following alkyl halides:

(a)
H₃C Br Br   CH₃
 | | | |
CH₃CHCHCHCH₂CHCH₃

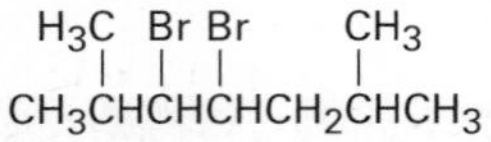

(b)
        I
        |
CH₃CH=CHCH₂CHCH₃

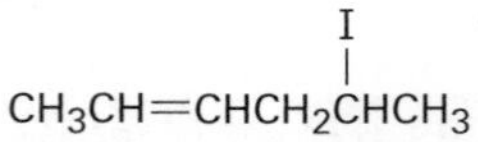

(c)
   Br Cl CH₃
    | | |
CH₃CCH₂CHCHCH₃
    |
    CH₃

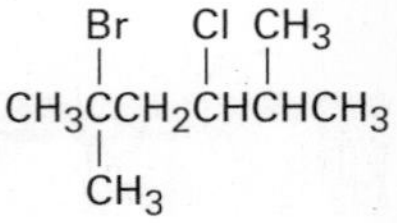

(d)
      CH₂Br
       |
CH₃CH₂CHCH₂CH₂CH₃

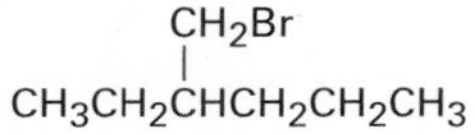

(e) ClCH₂CH₂CH₂C≡CCH₂Br

10.18 Draw structures corresponding to the following IUPAC names:
(a) 2,3-Dichloro-4-methylhexane
(b) 4-Bromo-4-ethyl-2-methylhexane
(c) 3-Iodo-2,2,4,4-tetramethylpentane
(d) *cis*-1-Bromo-2-ethylcyclopentane

10.19 Draw and name the monochlorination products you might obtain by radical chlorination of 2-methylbutane. Which of the products are chiral? Are any of the products optically active?

Synthesizing Alkyl Halides

10.20 How would you prepare the following compounds, starting with cyclopentene and any other reagents needed?
(a) Chlorocyclopentane   (b) Methylcyclopentane
(c) 3-Bromocyclopentene   (d) Cyclopentanol
(e) Cyclopentylcyclopentane   (f) 1,3-Cyclopentadiene

10.21 Predict the product(s) of the following reactions:

(a) 1-methyl-1-cyclohexanol $\xrightarrow{\text{HBr, Ether}}$?

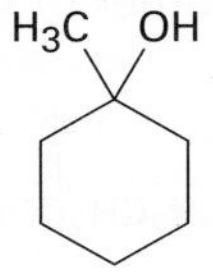

(b) $CH_3CH_2CH_2CH_2OH \xrightarrow{SOCl_2}$?

(c) octahydronaphthalene (with double bond at ring junction) $\xrightarrow{\text{NBS, } h\nu, CCl_4}$?

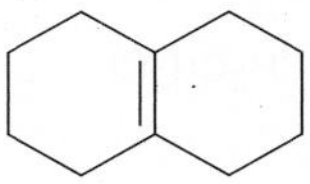

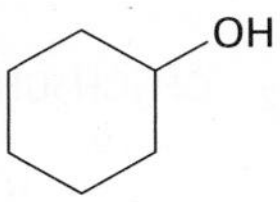

(d) cyclohexanol with adjacent carbon $\xrightarrow{PBr_3, \text{Ether}}$?

(e) $CH_3CH_2CHBrCH_3 \xrightarrow{\text{Mg, Ether}}$ A? $\xrightarrow{H_2O}$ B?

(f) $CH_3CH_2CH_2CH_2Br \xrightarrow{\text{Li, Pentane}}$ A? $\xrightarrow{\text{CuI}}$ B?

(g) $CH_3CH_2CH_2CH_2Br$ + $(CH_3)_2CuLi \xrightarrow{\text{Ether}}$?

10.22 A chemist requires a large amount of 1-bromo-2-pentene as starting material for a synthesis and decides to carry out an NBS allylic bromination reaction. What is wrong with the following synthesis plan? What side products would form in addition to the desired product?

$$CH_3CH_2CH=CHCH_3 \xrightarrow[h\nu, CCl_4]{NBS} CH_3CH_2CH=CHCH_2Br$$

10.23 What product(s) would you expect from the reaction of 1-methylcyclohexene with NBS? Would you use this reaction as part of a synthesis?

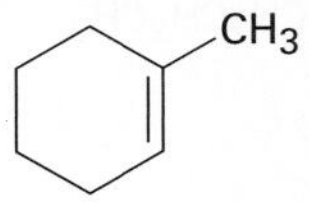

1-methylcyclohexene $\xrightarrow{\text{NBS, } h\nu, CCl_4}$?

10.24 What product(s) would you expect from the reaction of 1,4-hexadiene with NBS? What is the structure of the most stable radical intermediate?

10.25 What product would you expect from the reaction of 1-phenyl-2-butene with NBS? Explain.

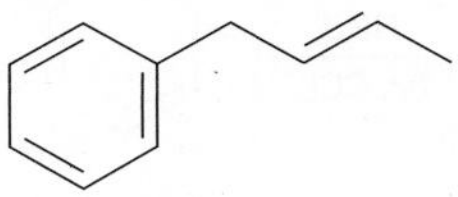

1-Phenyl-2-butene

Oxidation and Reduction

10.26 Rank the compounds in each of the following series in order of increasing oxidation level:

(a)

$CH_3CH=CHCH_3$ $CH_3CH_2CH=CH_2$ $CH_3CH_2CH_2\overset{\overset{O}{\|}}{C}H$ $CH_3CH_2CH_2\overset{\overset{O}{\|}}{C}OH$

(b)

$CH_3CH_2CH_2NH_2$ $CH_3CH_2CH_2Br$ $CH_3\overset{\overset{O}{\|}}{C}CH_2Cl$ $BrCH_2CH_2CH_2Cl$

10.27 Which of the following compounds have the same oxidation level, and which have different levels?

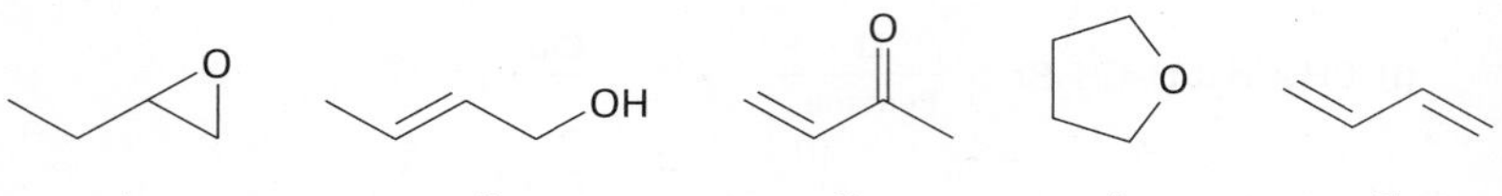

1   2   3   4   5

10.28 Tell whether each of the following reactions is an oxidation, a reduction, or neither:

(a)

$CH_3CH_2OH \xrightarrow{CrO_3} CH_3\overset{\overset{O}{\|}}{C}H$

(b)

$H_2C=CH\overset{\overset{O}{\|}}{C}CH_3 + NH_3 \longrightarrow H_2NCH_2CH_2\overset{\overset{O}{\|}}{C}CH_3$

(c)

$CH_3CH_2\overset{\overset{Br}{|}}{C}HCH_3 \xrightarrow[\text{2. }H_2O]{\text{1. Mg}} CH_3CH_2CH_2CH_3$

General Problems

10.29 Alkylbenzenes such as toluene (methylbenzene) react with NBS to give products in which bromine substitution has occurred at the position next to the aromatic ring (the *benzylic* position). Explain, based on the bond dissociation energies in Table 6.3 on page 204.

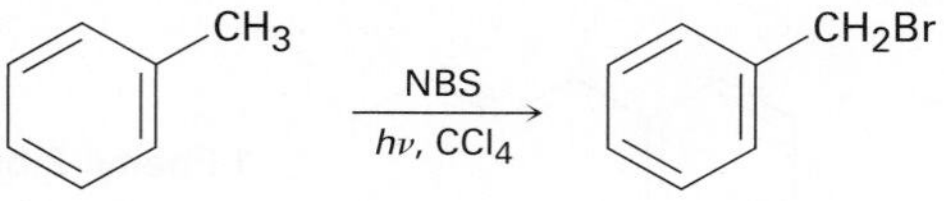

$PhCH_3 \xrightarrow[h\nu, CCl_4]{NBS} PhCH_2Br$

10.30 Draw resonance structures for the benzyl radical, $C_6H_5CH_2\cdot$, the intermediate produced in the NBS bromination reaction of toluene (Problem 10.29).

10.31 Draw resonance structures for the following species:

(a) $CH_3CH=CHCH=CHCH=\overset{+}{C}HCH_2$ (b) 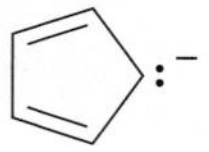   (c) $CH_3C\equiv\overset{+}{N}-\overset{..}{\underset{..}{O}}:^-$

10.32 (S)-3-Methylhexane undergoes radical bromination to yield optically inactive 3-bromo-3-methylhexane as the major product. Is the product chiral? What conclusions can you draw about the radical intermediate?

10.33 Assume that you have carried out a radical chlorination reaction on (R)-2-chloropentane and have isolated (in low yield) 2,4-dichloropentane. How many stereoisomers of the product are formed and in what ratio? Are any of the isomers optically active? (See Problem 10.32.)

10.34 How would you carry out the following syntheses?

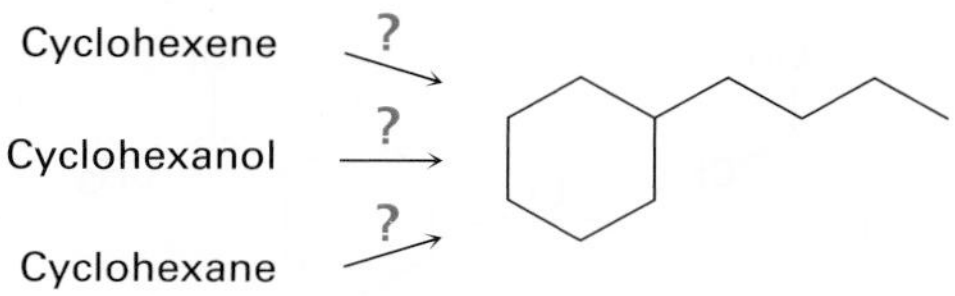

Cyclohexene $\xrightarrow{?}$
Cyclohexanol $\xrightarrow{?}$ butylcyclohexane
Cyclohexane $\xrightarrow{?}$

10.35 The syntheses shown here are unlikely to occur as written. What is wrong with each?

(a) $CH_3CH_2CH_2F \xrightarrow[2.\ H_3O^+]{1.\ Mg} CH_3CH_2CH_3$

(b) 2-methyl-1-methylenecyclohexane $\xrightarrow[h\nu,\ CCl_4]{NBS}$ 1-bromo-2-methyl-6-methylenecyclohexane

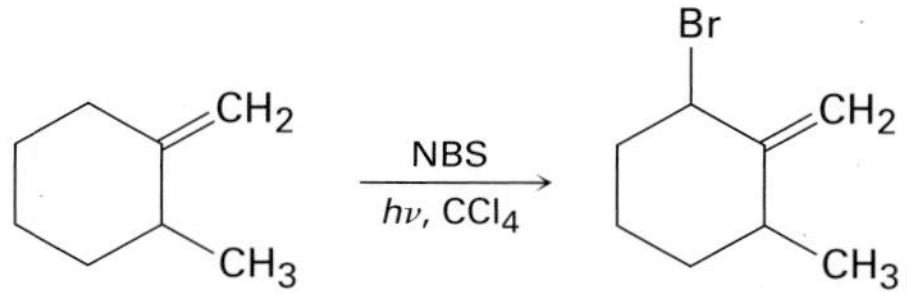

(c) fluorocyclohexane $\xrightarrow[Ether]{(CH_3)_2CuLi}$ methylcyclohexane

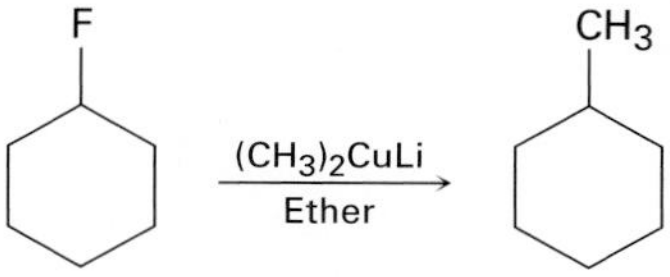

10.36 Why do you suppose it's not possible to prepare a Grignard reagent from a bromo alcohol such as 4-bromo-1-pentanol? Give another example of a molecule that is unlikely to form a Grignard reagent.

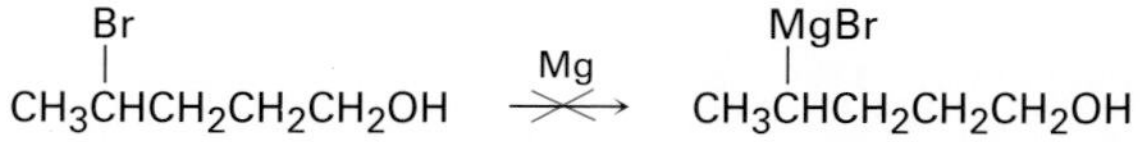

$$CH_3\overset{Br}{\underset{|}{C}}HCH_2CH_2CH_2OH \xrightarrow{Mg} \not\rightarrow CH_3\overset{MgBr}{\underset{|}{C}}HCH_2CH_2CH_2OH$$

10.37 Addition of HBr to a double bond with an ether (−OR) substituent occurs regiospecifically to give a product in which the −Br and −OR are bonded to the same carbon. Draw the two possible carbocation intermediates in this electrophilic addition reaction, and explain using resonance why the observed product is formed.

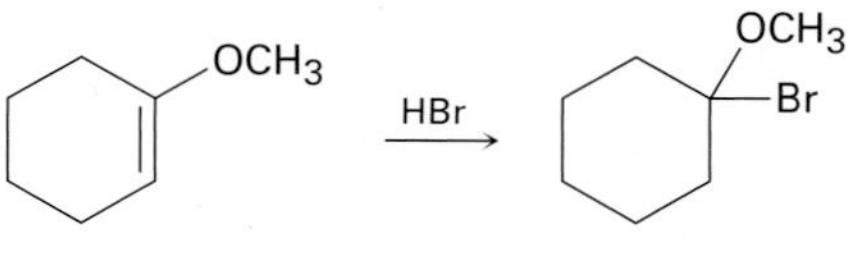

1-methoxycyclohexene $\xrightarrow{HBr}$ 1-bromo-1-methoxycyclohexane

10.38 Alkyl halides can be reduced to alkanes by a radical reaction with tributyltin hydride, $(C_4H_9)_3SnH$, in the presence of light ($h\nu$). Propose a radical chain mechanism by which the reaction might occur. The initiation step is the light-induced homolytic cleavage of the Sn–H bond to yield a tributyltin radical.

$$R-X \ + \ (C_4H_9)_3SnH \ \xrightarrow{h\nu} \ R-H \ + \ (C_4H_9)_3SnX$$

10.39 Identify the reagents **a–c** in the following scheme:

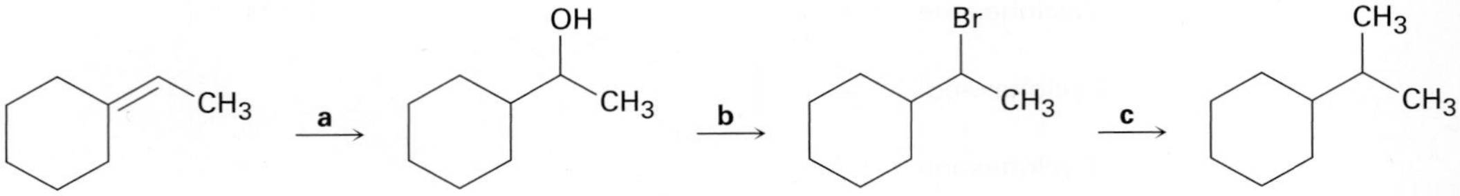

10.40 Tertiary alkyl halides, R_3CX, undergo spontaneous dissociation to yield a carbocation, R_3C^+, plus halide ion. Which do you think reacts faster, $(CH_3)_3CBr$ or $H_2C=CHC(CH_3)_2Br$? Explain.

10.41 In light of the fact that tertiary alkyl halides undergo spontaneous dissociation to yield a carbocation plus halide ion (Problem 10.40), propose a mechanism for the following reaction.

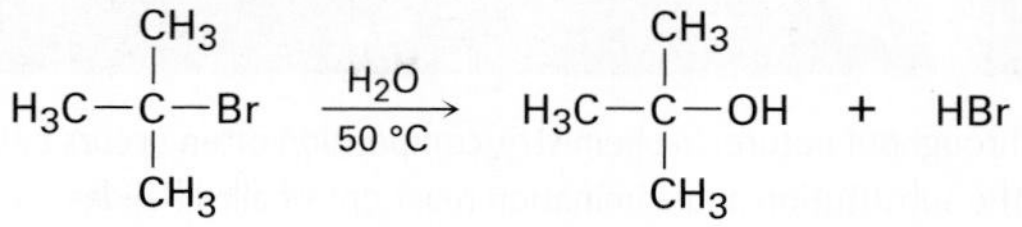

$$H_3C-\underset{\underset{CH_3}{|}}{\overset{\overset{CH_3}{|}}{C}}-Br \xrightarrow[50\ °C]{H_2O} H_3C-\underset{\underset{CH_3}{|}}{\overset{\overset{CH_3}{|}}{C}}-OH + HBr$$

10.42 Carboxylic acids (RCO_2H; $pK_a \approx 5$) are approximately 10^{11} times more acidic than alcohols (ROH; $pK_a \approx 16$). In other words, a carboxylate ion (RCO_2^-) is more stable than an alkoxide ion (RO^-). Explain, using resonance.

10.43 How might you use a Suzuki–Miyaura coupling to prepare the following biaryl compound? Show the two potential reaction partners.

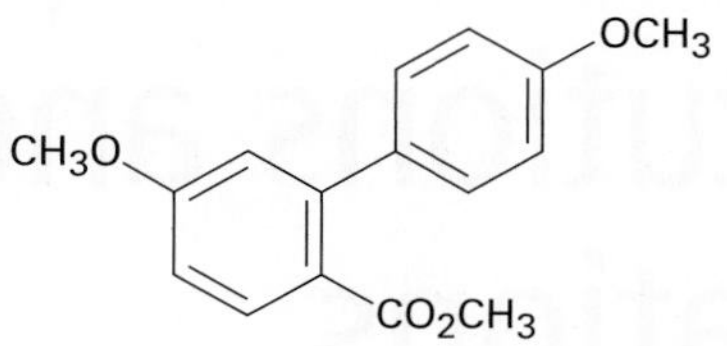

11

Competition occurs throughout nature. In chemistry, competition often occurs between alternative reaction pathways, such as in the substitution and elimination reactions of alkyl halides. Cheryl Ann Quigley/Shutterstock

Reactions of Alkyl Halides: Nucleophilic Substitutions and Eliminations

11.1 The Discovery of Nucleophilic Substitution Reactions
11.2 The S_N2 Reaction
11.3 Characteristics of the S_N2 Reaction
11.4 The S_N1 Reaction
11.5 Characteristics of the S_N1 Reaction
11.6 Biological Substitution Reactions
11.7 Elimination Reactions: Zaitsev's Rule
11.8 The E2 Reaction and the Deuterium Isotope Effect
11.9 The E2 Reaction and Cyclohexane Conformation
11.10 The E1 and E1cB Reactions
11.11 Biological Elimination Reactions
11.12 A Summary of Reactivity: S_N1, S_N2, E1, E1cB, and E2

A Deeper Look— Green Chemistry

OWL Sign in to OWL for Organic Chemistry at www.cengage.com/owl to view tutorials and simulations, develop problem-solving skills, and complete online homework assigned by your professor.

We saw in the preceding chapter that the carbon–halogen bond in an alkyl halide is polar and that the carbon atom is electron-poor. Thus, alkyl halides are electrophiles, and much of their chemistry involves polar reactions with nucleophiles and bases. Alkyl halides do one of two things when they react with a nucleophile/base, such as hydroxide ion: either they undergo *substitution* of the X group by the nucleophile, or they undergo *elimination* of HX to yield an alkene.

Why This Chapter? Nucleophilic substitution and base-induced elimination are two of the most widely occurring and versatile reaction types in organic chemistry, both in the laboratory and in biological pathways. We'll look at them closely in this chapter to see how they occur, what their characteristics are, and how they can be used. We'll begin with substitution reactions.

11.1 The Discovery of Nucleophilic Substitution Reactions

The discovery of the nucleophilic substitution reaction of alkyl halides dates back to work carried out in 1896 by the German chemist Paul Walden. Walden found that the pure enantiomeric (+)- and (−)-malic acids could be interconverted through a series of simple substitution reactions. When Walden treated (−)-malic acid with PCl₅, he isolated (+)-chlorosuccinic acid. This, on treatment with wet Ag₂O, gave (+)-malic acid. Similarly, reaction of (+)-malic acid with PCl₅ gave (−)-chlorosuccinic acid, which was converted into (−)-malic acid when treated with wet Ag₂O. The full cycle of reactions is shown in **Figure 11.1**.

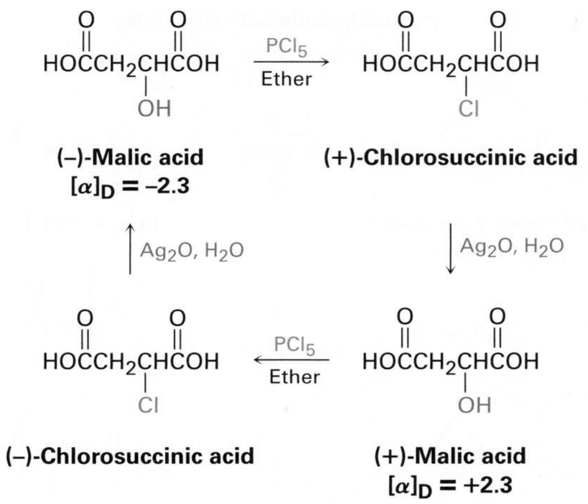

Figure 11.1 Walden's cycle of reactions interconverting (+)- and (−)-malic acids.

$$\text{HOCCH}_2\text{CHCOH} \atop \text{OH} \xrightarrow[\text{Ether}]{\text{PCl}_5} \text{HOCCH}_2\text{CHCOH} \atop \text{Cl}$$

(−)-Malic acid
$[\alpha]_D = -2.3$

(+)-Chlorosuccinic acid

↑ Ag₂O, H₂O   ↓ Ag₂O, H₂O

$$\text{HOCCH}_2\text{CHCOH} \atop \text{Cl} \xleftarrow[\text{Ether}]{\text{PCl}_5} \text{HOCCH}_2\text{CHCOH} \atop \text{OH}$$

(−)-Chlorosuccinic acid

(+)-Malic acid
$[\alpha]_D = +2.3$

At the time, the results were astonishing. The eminent chemist Emil Fischer called Walden's discovery "the most remarkable observation made in the field of optical activity since the fundamental observations of Pasteur." Because (−)-malic acid was converted into (+)-malic acid, *some reactions in the cycle must have occurred with a change, or inversion, in configuration at the chirality center.* But which ones, and how? (Remember from **Section 5.5** that the direction of light rotation and the configuration of a chirality center aren't directly related. You can't tell by looking at the sign of rotation whether a change in configuration has occurred during a reaction.)

Today, we refer to the transformations taking place in Walden's cycle as **nucleophilic substitution reactions** because each step involves the substitution of one nucleophile (chloride ion, Cl⁻, or hydroxide ion, HO⁻) by another. Nucleophilic substitution reactions are one of the most common and versatile reaction types in organic chemistry.

$$R-X + Nu:^- \longrightarrow R-Nu + X:^-$$

Following the work of Walden, further investigations were undertaken during the 1920s and 1930s to clarify the mechanism of nucleophilic substitution reactions and to find out how inversions of configuration occur. Among the first series studied was one that interconverted the two enantiomers of 1-phenyl-2-propanol (**Figure 11.2**). Although this particular series of reactions involves nucleophilic substitution of an alkyl *p*-toluenesulfonate (called a *tosylate*) rather than an alkyl halide, exactly the same type of reaction is involved as that studied by Walden. For all practical purposes, the entire tosylate group acts as if it were simply a halogen substituent. (In fact, when you see a tosylate

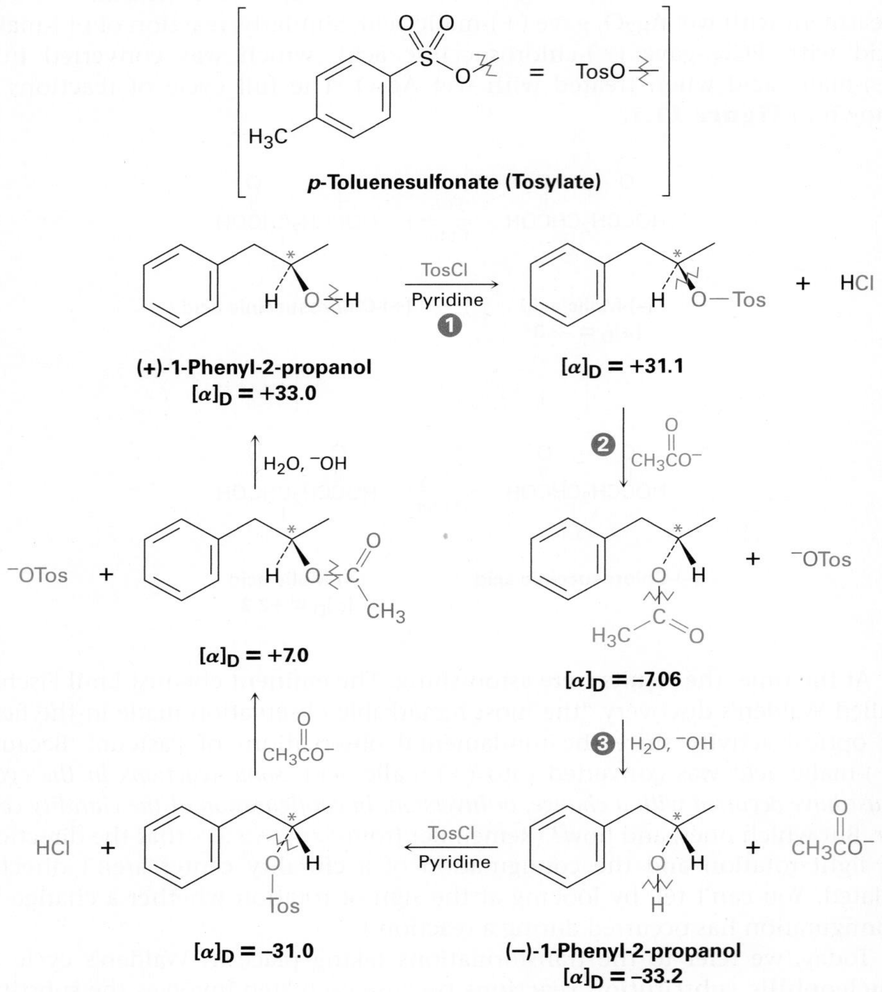

Figure 11.2 A Walden cycle interconverting (+) and (−) enantiomers of 1-phenyl-2-propanol. Chirality centers are marked by asterisks, and the bonds broken in each reaction are indicated by red wavy lines. The inversion of chirality occurs in step ❷, where acetate ion substitutes for tosylate ion.

p-Toluenesulfonate (Tosylate)

(+)-1-Phenyl-2-propanol
$[\alpha]_D = +33.0$

$[\alpha]_D = +31.1$

$[\alpha]_D = +7.0$

$[\alpha]_D = -7.06$

$[\alpha]_D = -31.0$

(−)-1-Phenyl-2-propanol
$[\alpha]_D = -33.2$

In the three-step reaction sequence shown in Figure 11.2, (+)-1-phenyl-2-propanol is interconverted with its (−) enantiomer, so at least one of the three steps must involve an inversion of configuration at the chirality center. Step 1, formation of a tosylate, occurs by breaking the O–H bond of the alcohol rather than the C–O bond to the chiral carbon, so the configuration around carbon is unchanged. Similarly, step 3, hydroxide-ion cleavage of the acetate, takes place without breaking the C–O bond at the chirality center. *The inversion of stereochemical configuration must therefore take place in step 2, the nucleophilic substitution of tosylate ion by acetate ion.*

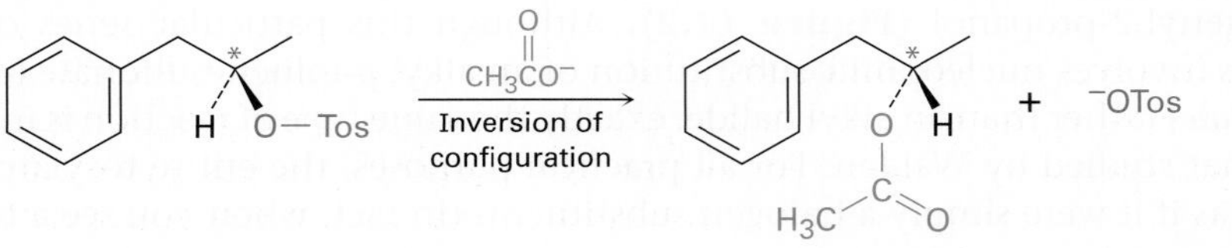

From this and nearly a dozen other series of similar reactions, workers concluded that the nucleophilic substitution reaction of a primary or secondary alkyl halide or tosylate always proceeds with inversion of configuration. (Tertiary alkyl halides and tosylates, as we'll see shortly, give different stereochemical results and react by a different mechanism.)

> **Worked Example 11.1**
>
> **Predicting the Stereochemistry of a Nucleophilic Substitution Reaction**
>
> What product would you expect from a nucleophilic substitution reaction of (R)-1-bromo-1-phenylethane with cyanide ion, $^-C{\equiv}N$, as nucleophile? Show the stereochemistry of both reactant and product, assuming that inversion of configuration occurs. 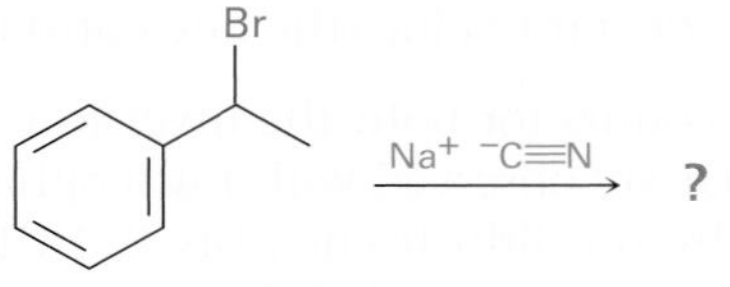
>
> **Strategy**
> Draw the R enantiomer of the reactant, and then change the configuration of the chirality center while replacing the ^-Br with a ^-CN.
>
> **Solution** 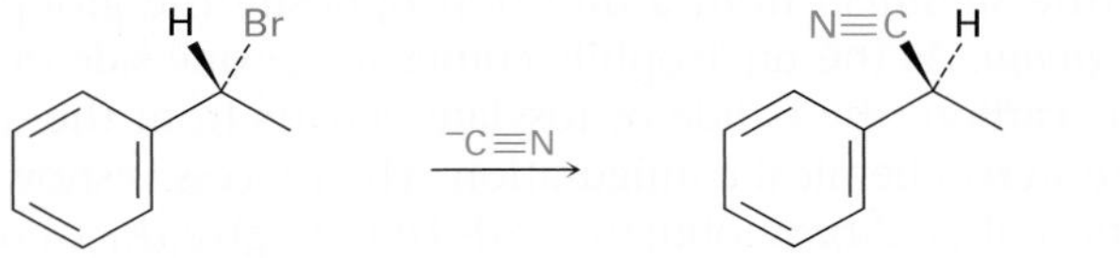
>
> (R)-1-Bromo-1-phenylethane → (S)-2-Phenylpropanenitrile

> **Problem 11.1**
> What product would you expect to obtain from a nucleophilic substitution reaction of (S)-2-bromohexane with acetate ion, $CH_3CO_2^-$? Assume that inversion of configuration occurs, and show the stereochemistry of both reactant and product.

11.2 The S$_N$2 Reaction

In every chemical reaction, there is a direct relationship between the rate at which the reaction occurs and the concentrations of the reactants. When we measure this relationship, we measure the **kinetics** of the reaction. For example, let's look at the kinetics of a simple nucleophilic substitution—the reaction of CH_3Br with OH^- to yield CH_3OH plus Br^-.

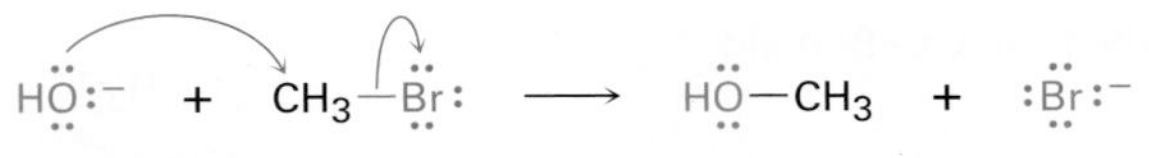

$$HO{:}^- + CH_3{-}Br{:} \longrightarrow HO{-}CH_3 + {:}Br{:}^-$$

At a given temperature, solvent, and concentration of reactants, the substitution occurs at a certain rate. If we double the concentration of OH^-, the frequency of encounter between the reaction partners doubles and we find that the reaction rate also doubles. Similarly, if we double the concentration of

CH₃Br, the reaction rate again doubles. We call such a reaction, in which the rate is linearly dependent on the concentrations of two species, a **second-order reaction**. Mathematically, we can express this second-order dependence of the nucleophilic substitution reaction by setting up a *rate equation*. As either [RX] or [⁻OH] changes, the rate of the reaction changes proportionately.

$$\text{Reaction rate} = \text{Rate of disappearance of reactant}$$

$$= k \times [\text{RX}] \times [^-\text{OH}]$$

where   [RX] = CH₃Br concentration in molarity
[⁻OH] = ⁻OH concentration in molarity
k = A constant value (the rate constant)

A mechanism that accounts for both the inversion of configuration and the second-order kinetics that are observed with nucleophilic substitution reactions was suggested in 1937 by the British chemists E. D. Hughes and Christopher Ingold, who formulated what they called the **S$_N$2 reaction**—short for *substitution, nucleophilic, bimolecular*. (*Bimolecular* means that two molecules, nucleophile and alkyl halide, take part in the step whose kinetics are measured.)

The essential feature of the S$_N$2 mechanism is that it takes place in a single step without intermediates when the incoming nucleophile reacts with the alkyl halide or tosylate (the *substrate*) from a direction opposite the group that is displaced (the *leaving group*). As the nucleophile comes in on one side of the substrate and bonds to the carbon, the halide or tosylate departs from the other side, thereby inverting the stereochemical configuration. The process is shown in **Figure 11.3** for the reaction of (S)-2-bromobutane with HO⁻ to give (R)-2-butanol.

Figure 11.3 | MECHANISM

The mechanism of the S$_N$2 reaction. The reaction takes place in a single step when the incoming nucleophile approaches from a direction 180° away from the leaving halide ion, thereby inverting the stereochemistry at carbon.

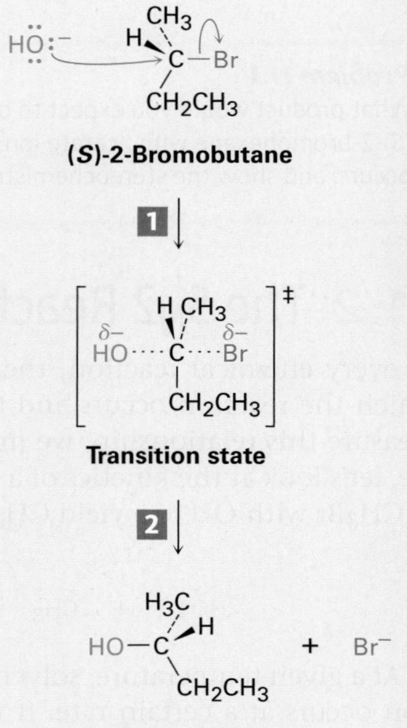

1 The nucleophile –OH uses its lone-pair electrons to attack the alkyl halide carbon 180° away from the departing halogen. This leads to a transition state with a partially formed C–OH bond and a partially broken C–Br bond.

2 The stereochemistry at carbon is inverted as the C–OH bond forms fully and the bromide ion departs with the electron pair from the former C–Br bond.

(S)-2-Bromobutane

Transition state

(R)-2-Butanol

© John McMurry

As shown in Figure 11.3, the S$_N$2 reaction occurs when an electron pair on the nucleophile Nu:⁻ forces out the group X:⁻, which takes with it the electron pair from the former C−X bond. This occurs through a transition state in which the new Nu−C bond is partially forming at the same time that the old C−X bond is partially breaking and in which the negative charge is shared by both the incoming nucleophile and the outgoing halide ion. The transition state for this inversion has the remaining three bonds to carbon in a planar arrangement **(Figure 11.4)**.

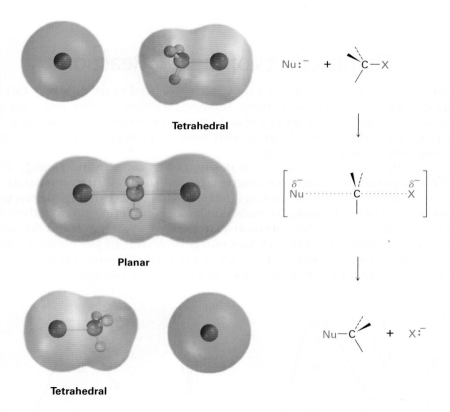

Figure 11.4 The transition state of an S$_N$2 reaction has a planar arrangement of the carbon atom and the remaining three groups. Electrostatic potential maps show that **negative charge** is delocalized in the transition state.

The mechanism proposed by Hughes and Ingold is fully consistent with experimental results, explaining both stereochemical and kinetic data. Thus, the requirement for backside approach of the entering nucleophile from a direction 180° away from the departing X group causes the stereochemistry of the substrate to invert, much like an umbrella turning inside out in the wind. The Hughes–Ingold mechanism also explains why second-order kinetics are found: the S$_N$2 reaction occurs in a single step that involves both alkyl halide and nucleophile. Two molecules are involved in the step whose rate is measured.

Problem 11.2
What product would you expect to obtain from S$_N$2 reaction of OH⁻ with (*R*)-2-bromobutane? Show the stereochemistry of both reactant and product.

Problem 11.3
Assign configuration to the following substance, and draw the structure of the product that would result on nucleophilic substitution reaction with HS⁻ (reddish brown = Br):

11.3 Characteristics of the S$_N$2 Reaction

Key IDEAS

Test your knowledge of Key Ideas by answering end-of-chapter exercises marked with ▲.

Now that we know how S$_N$2 reactions occur, we need to see how they can be used and what variables affect them. Some S$_N$2 reactions are fast, and some are slow; some take place in high yield and others in low yield. Understanding the factors involved can be of tremendous value. Let's begin by recalling a few things about reaction rates in general.

The rate of a chemical reaction is determined by the activation energy $\Delta G^{\ddagger}$, the energy difference between reactant ground state and transition state. A change in reaction conditions can affect $\Delta G^{\ddagger}$ either by changing the reactant energy level or by changing the transition-state energy level. Lowering the reactant energy or raising the transition-state energy increases $\Delta G^{\ddagger}$ and decreases the reaction rate; raising the reactant energy or decreasing the transition-state energy decreases $\Delta G^{\ddagger}$ and increases the reaction rate **(Figure 11.5)**. We'll see examples of all these effects as we look at S$_N$2 reaction variables.

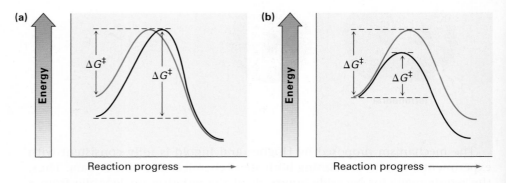

Figure 11.5 The effects of changes in reactant and transition-state energy levels on reaction rate. **(a)** A higher reactant energy level (red curve) corresponds to a faster reaction (smaller $\Delta G^{\ddagger}$). **(b)** A higher transition-state energy level (red curve) corresponds to a slower reaction (larger $\Delta G^{\ddagger}$).

The Substrate: Steric Effects in the S$_N$2 Reaction

The first S$_N$2 reaction variable to look at is the structure of the substrate. Because the S$_N$2 transition state involves partial bond formation between the incoming nucleophile and the alkyl halide carbon atom, it seems reasonable that a hindered, bulky substrate should prevent easy approach of the nucleophile, making bond formation difficult. In other words, the transition state for reaction of a sterically hindered substrate, whose carbon atom is "shielded" from approach of the incoming nucleophile, is higher in energy and forms more slowly than the corresponding transition state for a less hindered substrate (**Figure 11.6**).

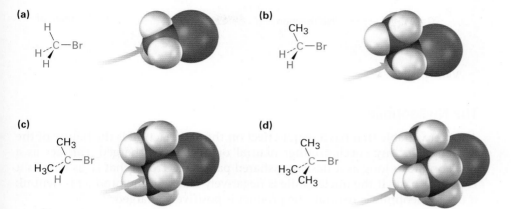

Figure 11.6 Steric hindrance to the S$_N$2 reaction. As the models indicate, the carbon atom in **(a)** bromomethane is readily accessible, resulting in a fast S$_N$2 reaction. The carbon atoms in **(b)** bromoethane (primary), **(c)** 2-bromopropane (secondary), and **(d)** 2-bromo-2-methylpropane (tertiary) are successively more hindered, resulting in successively slower S$_N$2 reactions.

As Figure 11.6 shows, the difficulty of nucleophile approach increases as the three substituents bonded to the halo-substituted carbon atom increase in size. Methyl halides are by far the most reactive substrates in S$_N$2 reactions, followed by primary alkyl halides such as ethyl and propyl. Alkyl branching at the reacting center, as in isopropyl halides (2°), slows the reaction greatly, and further branching, as in *tert*-butyl halides (3°), effectively halts the reaction. Even branching one carbon removed from the reacting center, as in 2,2-dimethylpropyl *(neopentyl)* halides, greatly slows nucleophilic displacement. As a result, S$_N$2 reactions occur only at relatively unhindered sites and are normally useful only with methyl halides, primary halides, and a few simple secondary halides. Relative reactivities for some different substrates are as follows:

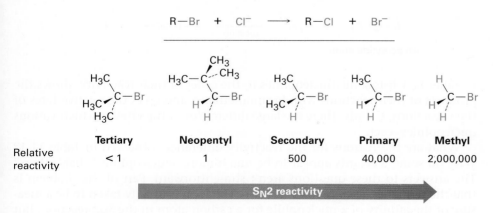

Vinylic halides (R$_2$C=CRX) and aryl halides are not shown on this reactivity list because they are unreactive toward S$_N$2 displacement. This lack of reactivity is due to steric factors: the incoming nucleophile would have to approach in the plane of the carbon–carbon double bond and burrow through part of the molecule to carry out a backside displacement.

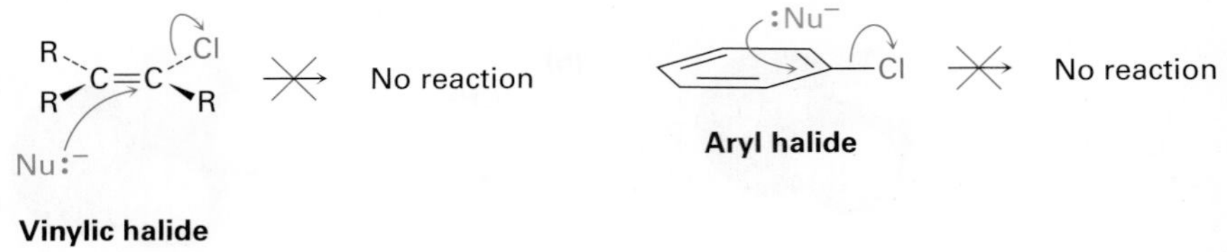

Vinylic halide → No reaction

Aryl halide → No reaction

The Nucleophile

Another variable that has a major effect on the S_N2 reaction is the nature of the nucleophile. Any species, either neutral or negatively charged, can act as a nucleophile as long as it has an unshared pair of electrons; that is, as long as it is a Lewis base. If the nucleophile is negatively charged, the product is neutral; if the nucleophile is neutral, the product is positively charged.

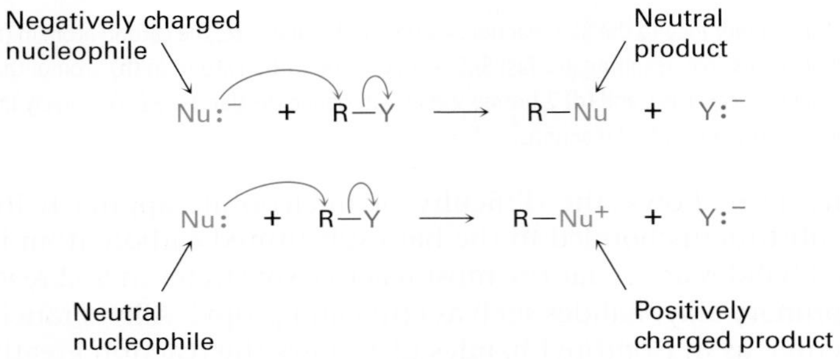

Negatively charged nucleophile: $Nu:^- + R-Y \longrightarrow R-Nu + Y:^-$ (Neutral product)

Neutral nucleophile: $Nu: + R-Y \longrightarrow R-Nu^+ + Y:^-$ (Positively charged product)

A wide array of substances can be prepared using nucleophilic substitution reactions. In fact, we've already seen examples in previous chapters. The reaction of an acetylide anion with an alkyl halide discussed in **Section 9.8**, for instance, is an S_N2 reaction in which the acetylide nucleophile displaces a halide leaving group.

$$R-C\equiv C:^- + CH_3Br \xrightarrow[\text{reaction}]{S_N2} R-C\equiv C-CH_3 + Br^-$$

An acetylide anion

Table 11.1 lists some nucleophiles in the order of their reactivity, shows the products of their reactions with bromomethane, and gives the relative rates of their reactions. Clearly, there are large differences in the rates at which various nucleophiles react.

What are the reasons for the reactivity differences observed in Table 11.1? Why do some reactants appear to be much more "nucleophilic" than others? The answers to these questions aren't straightforward. Part of the problem is that the term *nucleophilicity* is imprecise. The term is usually taken to be a measure of the affinity of a nucleophile for a carbon atom in the S_N2 reaction, but the reactivity of a given nucleophile can change from one reaction to the next. The exact nucleophilicity of a species in a given reaction depends on the substrate, the solvent, and even the reactant concentrations. Detailed explanations

Table 11.1 Some S$_N$2 Reactions with Bromomethane

$$\text{Nu:}^- + CH_3Br \rightarrow CH_3Nu + Br^-$$

Nucleophile		Product		Relative rate of reaction
Formula	Name	Formula	Name	
H_2O	Water	$CH_3OH_2^+$	Methylhydronium ion	1
$CH_3CO_2^-$	Acetate	$CH_3CO_2CH_3$	Methyl acetate	500
NH_3	Ammonia	$CH_3NH_3^+$	Methylammonium ion	700
Cl^-	Chloride	CH_3Cl	Chloromethane	1,000
HO^-	Hydroxide	CH_3OH	Methanol	10,000
CH_3O^-	Methoxide	CH_3OCH_3	Dimethyl ether	25,000
I^-	Iodide	CH_3I	Iodomethane	100,000
^-CN	Cyanide	CH_3CN	Acetonitrile	125,000
HS^-	Hydrosulfide	CH_3SH	Methanethiol	125,000

for the observed nucleophilicities aren't always simple, but some trends can be detected in the data of Table 11.1.

- **Nucleophilicity roughly parallels basicity** when comparing nucleophiles that have the same reacting atom. Thus, OH$^-$ is both more basic and more nucleophilic than acetate ion, CH$_3$CO$_2^-$, which in turn is more basic and more nucleophilic than H$_2$O. Since "nucleophilicity" is usually taken as the affinity of a Lewis base for a carbon atom in the S$_N$2 reaction and "basicity" is the affinity of a base for a proton, it's easy to see why there might be a correlation between the two kinds of behavior.

- **Nucleophilicity usually increases going down a column of the periodic table.** Thus, HS$^-$ is more nucleophilic than HO$^-$, and the halide reactivity order is I$^-$ > Br$^-$ > Cl$^-$. Going down the periodic table, elements have their valence electrons in successively larger shells where they are successively farther from the nucleus, less tightly held, and consequently more reactive. The matter is complex, though, and the nucleophilicity order can change depending on the solvent.

- **Negatively charged nucleophiles are usually more reactive than neutral ones.** As a result, S$_N$2 reactions are often carried out under basic conditions rather than neutral or acidic conditions.

Problem 11.4
What product would you expect from S$_N$2 reaction of 1-bromobutane with each of the following?
(a) NaI   (b) KOH   (c) H—C≡C—Li   (d) NH$_3$

Problem 11.5
Which substance in each of the following pairs is more reactive as a nucleophile? Explain.
(a) (CH$_3$)$_2$N$^-$ or (CH$_3$)$_2$NH   (b) (CH$_3$)$_3$B or (CH$_3$)$_3$N   (c) H$_2$O or H$_2$S

The Leaving Group

Still another variable that can affect the S_N2 reaction is the nature of the group displaced by the incoming nucleophile. Because the leaving group is expelled with a negative charge in most S_N2 reactions, the best leaving groups are those that best stabilize the negative charge in the transition state. The greater the extent of charge stabilization by the leaving group, the lower the energy of the transition state and the more rapid the reaction. But as we saw in **Section 2.8**, those groups that best stabilize a negative charge are also the weakest bases. Thus, weak bases such as Cl^-, Br^-, and tosylate ion make good leaving groups, while strong bases such as OH^- and NH_2^- make poor leaving groups.

Relative reactivity	OH^-, NH_2^-, OR^-	F^-	Cl^-	Br^-	I^-	$TosO^-$
	<<1	1	200	10,000	30,000	60,000

Leaving group reactivity →

It's just as important to know which are poor leaving groups as to know which are good, and the preceding data clearly indicate that F^-, HO^-, RO^-, and H_2N^- are not displaced by nucleophiles. In other words, alkyl fluorides, alcohols, ethers, and amines do not typically undergo S_N2 reactions. To carry out an S_N2 reaction with an alcohol, it's necessary to convert the ^-OH into a better leaving group. This, in fact, is just what happens when a primary or secondary alcohol is converted into either an alkyl chloride by reaction with $SOCl_2$ or an alkyl bromide by reaction with PBr_3 **(Section 10.5)**.

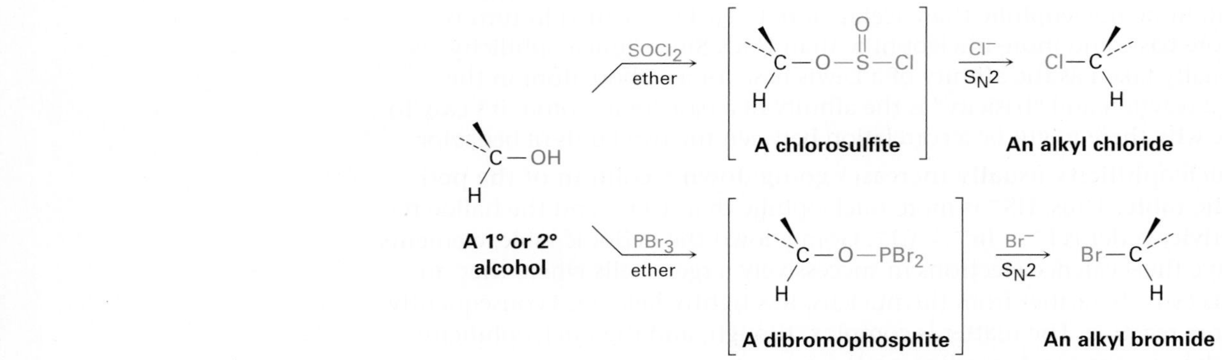

Alternatively, an alcohol can be made more reactive toward nucleophilic substitution by treating it with *para*-toluenesulfonyl chloride to form a tosylate. As noted previously, tosylates are even more reactive than halides in nucleophilic substitutions. Note that tosylate formation does not change the configuration of the oxygen-bearing carbon because the C–O bond is not broken.

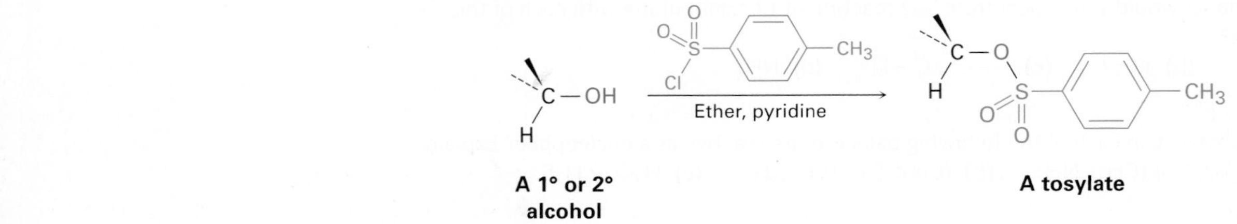

The one general exception to the rule that ethers don't typically undergo S_N2 reactions occurs with epoxides, the three-membered cyclic ethers that we saw in **Section 8.7**. Epoxides, because of the angle strain in the three-membered ring, are much more reactive than other ethers. They react with aqueous acid to give 1,2-diols, as we saw in **Section 8.7**, and they react readily with many other nucleophiles as well. Propene oxide, for instance, reacts with HCl to give 1-chloro-2-propanol by S_N2 backside attack on the less hindered primary carbon atom. We'll look at the process in more detail in **Section 18.6**.

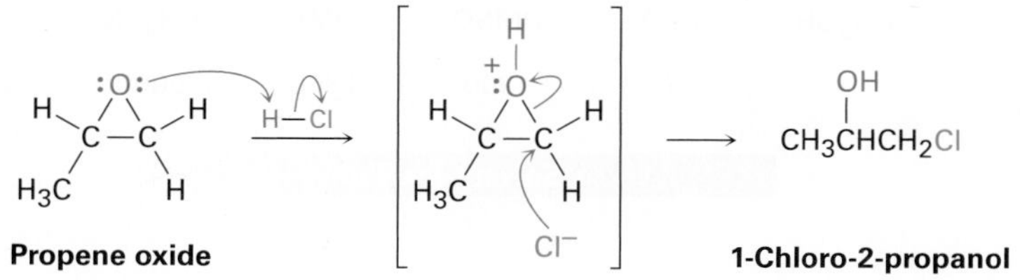

Problem 11.6

Rank the following compounds in order of their expected reactivity toward S_N2 reaction:

CH_3Br, CH_3OTos, $(CH_3)_3CCl$, $(CH_3)_2CHCl$

The Solvent

The rates of S_N2 reactions are strongly affected by the solvent. Protic solvents—those that contain an –OH or –NH group—are generally the worst for S_N2 reactions, while polar aprotic solvents, which are polar but don't have an –OH or –NH group, are the best.

Protic solvents, such as methanol and ethanol, slow down S_N2 reactions by **solvation** of the reactant nucleophile. The solvent molecules hydrogen bond to the nucleophile and form a cage around it, thereby lowering its energy and reactivity.

A solvated anion (reduced nucleophilicity due to enhanced ground-state stability)

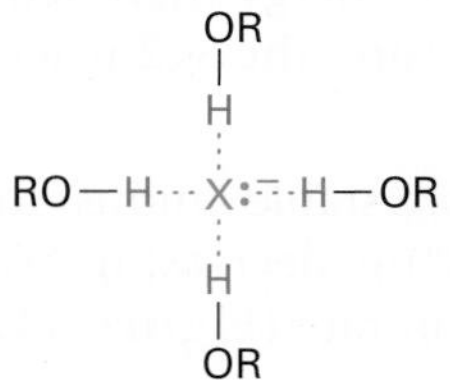

In contrast with protic solvents, which decrease the rates of S_N2 reactions by lowering the ground-state energy of the nucleophile, polar aprotic solvents increase the rates of S_N2 reactions by raising the ground-state energy of the nucleophile. Acetonitrile (CH_3CN), dimethylformamide [$(CH_3)_2NCHO$, abbreviated DMF], dimethyl sulfoxide [$(CH_3)_2SO$, abbreviated DMSO], and hexamethylphosphoramide {[(CH$_3$)$_2$N]$_3$PO, abbreviated HMPA} are particularly useful. These solvents can dissolve many salts because of their high polarity, but they tend to solvate metal cations rather than nucleophilic anions. As a result, the bare unsolvated anions have a greater nucleophilicity and S$_N$2 reactions take place at correspondingly faster rates. For instance, a rate increase of 200,000 has been observed on changing from methanol to HMPA for the reaction of azide ion with 1-bromobutane.

$$CH_3CH_2CH_2CH_2-Br + N_3^- \longrightarrow CH_3CH_2CH_2CH_2-N_3 + Br^-$$

Solvent	CH$_3$OH	H$_2$O	DMSO	DMF	CH$_3$CN	HMPA
Relative reactivity	1	7	1300	2800	5000	200,000

→ Solvent reactivity

Problem 11.7
Organic solvents like benzene, ether, and chloroform are neither protic nor strongly polar. What effect would you expect these solvents to have on the reactivity of a nucleophile in S$_N$2 reactions?

A Summary of S$_N$2 Reaction Characteristics

The effects on S$_N$2 reactions of the four variables—substrate structure, nucleophile, leaving group, and solvent—are summarized in the following statements and in the energy diagrams of **Figure 11.7**:

Substrate Steric hindrance raises the energy of the S$_N$2 transition state, increasing $\Delta G^\ddagger$ and decreasing the reaction rate **(Figure 11.7a)**. As a result, S$_N$2 reactions are best for methyl and primary substrates. Secondary substrates react slowly, and tertiary substrates do not react by an S$_N$2 mechanism.

Nucleophile Basic, negatively charged nucleophiles are less stable and have a higher ground-state energy than neutral ones, decreasing $\Delta G^\ddagger$ and increasing the S$_N$2 reaction rate **(Figure 11.7b)**.

Leaving group Good leaving groups (more stable anions) lower the energy of the transition state, decreasing $\Delta G^\ddagger$ and increasing the S$_N$2 reaction rate **(Figure 11.7c)**.

Solvent Protic solvents solvate the nucleophile, thereby lowering its ground-state energy, increasing $\Delta G^\ddagger$, and decreasing the S$_N$2 reaction rate. Polar aprotic solvents surround the accompanying cation but not the nucleophilic anion, thereby raising the groundstate energy of the nucleophile, decreasing $\Delta G^\ddagger$, and increasing the reaction rate **(Figure 11.7d)**.

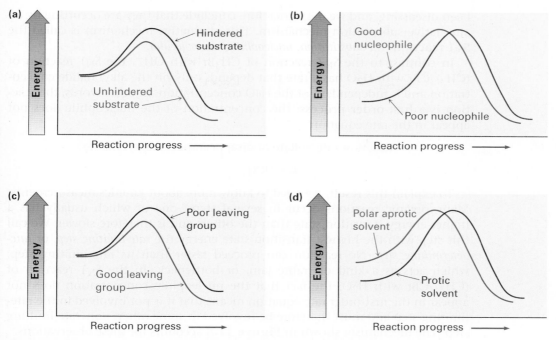

Figure 11.7 Energy diagrams showing the effects of **(a)** substrate, **(b)** nucleophile, **(c)** leaving group, and **(d)** solvent on S_N2 reaction rates. Substrate and leaving group effects are felt primarily in the transition state. Nucleophile and solvent effects are felt primarily in the reactant ground state.

11.4 The S_N1 Reaction

Most nucleophilic substitutions take place by the S_N2 pathway just discussed. The reaction is favored when carried out with an unhindered substrate and a negatively charged nucleophile in a polar aprotic solvent, but is disfavored when carried out with a hindered substrate and a neutral nucleophile in a protic solvent. You might therefore expect the reaction of a tertiary substrate (hindered) with water (neutral, protic) to be among the slowest of substitution reactions. Remarkably, however, the opposite is true. The reaction of the tertiary halide $(CH_3)_3CBr$ with H_2O to give the alcohol 2-methyl-2-propanol is more than 1 million times as fast as the corresponding reaction of CH_3Br to give methanol.

What's going on here? Clearly, a nucleophilic substitution reaction is occurring—a halogen is replacing a hydroxyl group—yet the reactivity order seems backward. These reactions can't be taking place by the S_N2 mechanism we've been discussing, and we must therefore conclude that they are occurring by an alternative substitution mechanism. This alternative mechanism is called the **S$_N$1 reaction**, for *substitution, nucleophilic, unimolecular*.

In contrast to the S$_N$2 reaction of CH$_3$Br with OH$^-$, the S$_N$1 reaction of (CH$_3$)$_3$CBr with H$_2$O has a rate that depends only on the alkyl halide concentration and is independent of the H$_2$O concentration. In other words, the reaction is a **first-order process**; the concentration of the nucleophile does not appear in the rate equation.

$$\text{Reaction rate} = \text{Rate of disappearance of alkyl halide}$$
$$= k \times [\text{RX}]$$

To explain this result, we need to know more about kinetics measurements. Many organic reactions occur in several steps, one of which usually has a higher-energy transition state than the others and is therefore slower. We call this step with the highest transition-state energy the *rate-limiting step*, or *rate-determining step*. No reaction can proceed faster than its rate-limiting step, which acts as a kind of traffic jam, or bottleneck. In the S$_N$1 reaction of (CH$_3$)$_3$CBr with H$_2$O, the fact that the nucleophile concentration does not appear in the first-order rate equation means that it is not involved in the rate-limiting step and must therefore be involved in some other, non–rate-limiting step. The mechanism shown in **Figure 11.8** accounts for these observations.

Figure 11.8 | MECHANISM

The mechanism of the S$_N$1 reaction of 2-bromo-2-methylpropane with H$_2$O involves three steps. Step **1**—the spontaneous, unimolecular dissociation of the alkyl bromide to yield a carbocation—is rate-limiting.

1 Spontaneous dissociation of the alkyl bromide occurs in a slow, rate-limiting step to generate a carbocation intermediate plus bromide ion.

2 The carbocation intermediate reacts with water as nucleophile in a fast step to yield protonated alcohol as product.

3 Loss of a proton from the protonated alcohol intermediate then gives the neutral alcohol product.

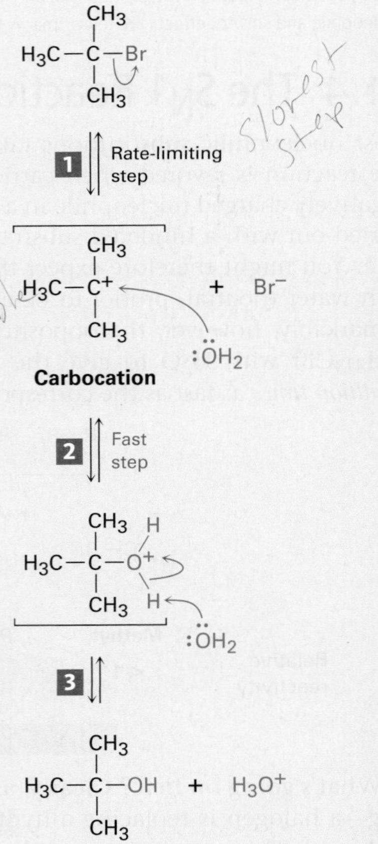

Unlike what happens in an S_N2 reaction, where the leaving group is displaced at the same time the incoming nucleophile approaches, an S_N1 reaction takes place by loss of the leaving group *before* the nucleophile approaches. 2-Bromo-2-methylpropane spontaneously dissociates to the *tert*-butyl carbocation plus Br^- in a slow rate-limiting step, and the intermediate carbocation is then immediately trapped by the nucleophile water in a faster second step. Water is not a reactant in the step whose rate is measured. The energy diagram is shown in **Figure 11.9**.

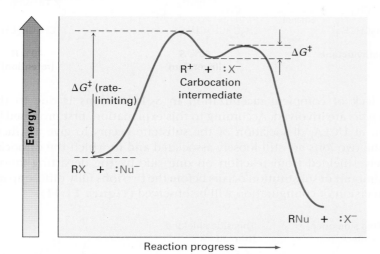

Figure 11.9 An energy diagram for an S_N1 reaction. The rate-limiting step is the spontaneous dissociation of the alkyl halide to give a carbocation intermediate. Reaction of the carbocation with a nucleophile then occurs in a second, faster step.

Because an S_N1 reaction occurs through a carbocation intermediate, its stereochemical outcome is different from that of an S_N2 reaction. Carbocations, as we've seen, are planar, sp^2-hybridized, and achiral. Thus, if we carry out an S_N1 reaction on one enantiomer of a chiral reactant and go through an achiral carbocation intermediate, the product must lose its optically activity **(Section 8.12)**. That is, the symmetrical intermediate carbocation can react with a nucleophile equally well from either side, leading to a racemic, 50:50 mixture of enantiomers **(Figure 11.10)**.

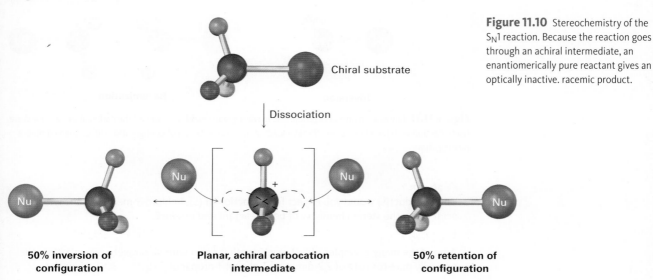

Figure 11.10 Stereochemistry of the S_N1 reaction. Because the reaction goes through an achiral intermediate, an enantiomerically pure reactant gives an optically inactive. racemic product.

The conclusion that S_N1 reactions on enantiomerically pure substrates should give racemic products is nearly, but not exactly, what is found. In fact, few S_N1 displacements occur with complete racemization. Most give a minor (0–20%) excess of inversion. The reaction of (R)-6-chloro-2,6-dimethyloctane with H_2O, for example, leads to an alcohol product that is approximately 80% racemized and 20% inverted (80% R,S + 20% S is equivalent to 40% R + 60% S).

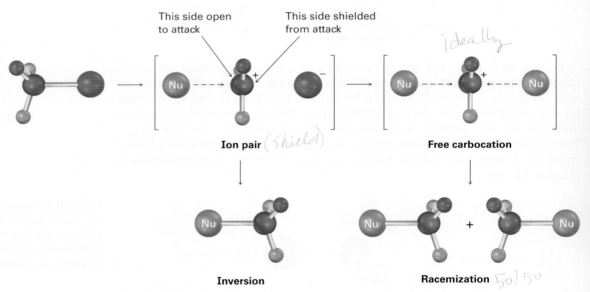

This lack of complete racemization in S_N1 reactions is due to the fact that *ion pairs* are involved. According to this explanation, first proposed by Saul Winstein at UCLA, dissociation of the substrate occurs to give a structure in which the two ions are still loosely associated and in which the carbocation is effectively shielded from reaction on one side by the departing anion. If a certain amount of substitution occurs before the two ions fully diffuse apart, then a net inversion of configuration will be observed **(Figure 11.11)**.

Figure 11.11 Ion-pairs in an S_N1 reaction. The leaving group shields one side of the carbocation intermediate from reaction with the nucleophile, thereby leading to some inversion of configuration rather than complete racemization.

Problem 11.8
What product(s) would you expect from reaction of (S)-3-chloro-3-methyloctane with acetic acid? Show the stereochemistry of both reactant and product.

Problem 11.9
Among the many examples of S_N1 reactions that occur with incomplete racemization, the optically pure tosylate of 2,2-dimethyl-1-phenyl-1-propanol ($[\alpha]_D = -30.3$) gives the

corresponding acetate ($[\alpha]_D = +5.3$) when heated in acetic acid. If complete inversion had occurred, the optically pure acetate would have had $[\alpha]_D = +53.6$. What percentage racemization and what percentage inversion occurred in this reaction?

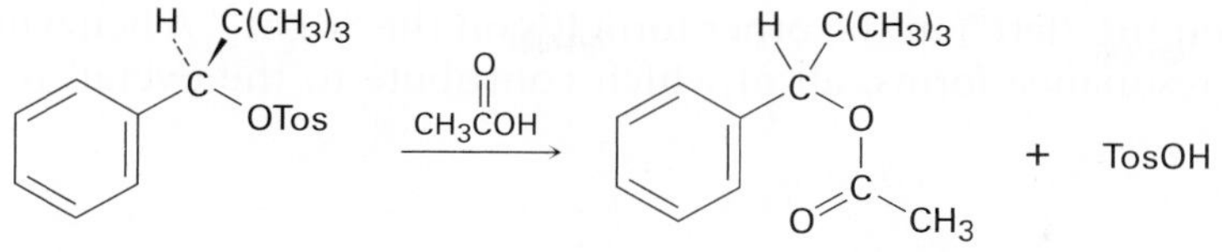

$[\alpha]_D = -30.3$

Observed $[\alpha]_D = +5.3$
(optically pure $[\alpha]_D = +53.6$)

Problem 11.10
Assign configuration to the following substrate, and show the stereochemistry and identity of the product you would obtain by S_N1 reaction with water (reddish brown = Br):

11.5 Characteristics of the S_N1 Reaction

Just as the S_N2 reaction is strongly influenced by the structure of the substrate, the leaving group, the nucleophile, and the solvent, the S_N1 reaction is similarly influenced. Factors that lower $\Delta G^{\ddagger}$, either by lowering the energy level of the transition state or by raising the energy level of the ground state, favor faster S_N1 reactions. Conversely, factors that raise $\Delta G^{\ddagger}$, either by raising the energy level of the transition state or by lowering the energy level of the reactant, slow down the S_N1 reaction.

Key IDEAS

Test your knowledge of Key Ideas by answering end-of-chapter exercises marked with ▲.

The Substrate

According to the Hammond postulate **(Section 7.10)**, any factor that stabilizes a high-energy intermediate also stabilizes the transition state leading to that intermediate. Since the rate-limiting step in an S_N1 reaction is the spontaneous, unimolecular dissociation of the substrate to yield a carbocation, the reaction is favored whenever a stabilized carbocation intermediate is formed. The more stable the carbocation intermediate, the faster the S_N1 reaction.

We saw in **Section 7.9** that the stability order of alkyl carbocations is $3° > 2° > 1° > -CH_3$. To this list we must also add the resonance-stabilized allyl and benzyl cations. Just as allylic radicals are unusually stable because the unpaired electron can be delocalized over an extended π orbital system **(Section 10.4)**, so allylic and benzylic carbocations are unusually stable. (The word **benzylic** means "next to an aromatic ring.") As **Figure 11.12** indicates, an allylic cation has two resonance forms. In one form the double bond is on the "left"; in the other form it's on the "right." A benzylic cation has five resonance forms, all of which contribute to the overall resonance hybrid.

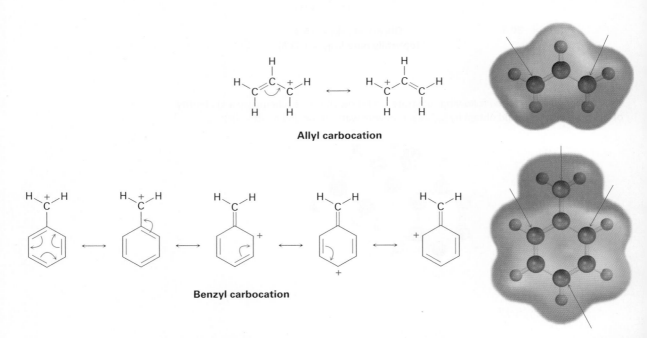

Figure 11.12 Resonance forms of allylic and benzylic carbocations. The positive charge is delocalized over the π system in both. **Electron-poor atoms** are indicated by blue arrows.

Because of resonance stabilization, a primary allylic or benzylic carbocation is about as stable as a secondary alkyl carbocation and a secondary allylic or benzylic carbocation is about as stable as a tertiary alkyl carbocation. This stability order of carbocations is the same as the order of S_N1 reactivity for alkyl halides and tosylates.

Methyl < Primary < Allylic ≈ Benzylic ≈ Secondary < Tertiary

Carbocation stability →

We should also note parenthetically that primary allylic and benzylic substrates are particularly reactive in S_N2 reactions as well as in S_N1 reactions. Allylic and benzylic C–X bonds are about 50 kJ/mol (12 kcal/mol) weaker than the corresponding saturated bonds and are therefore more easily broken.

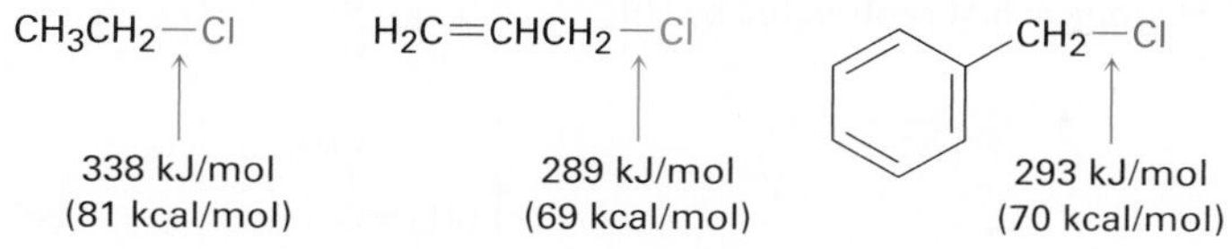

$$CH_3CH_2{-}Cl \quad\quad H_2C{=}CHCH_2{-}Cl \quad\quad C_6H_5CH_2{-}Cl$$

$$338 \text{ kJ/mol} \quad\quad 289 \text{ kJ/mol} \quad\quad 293 \text{ kJ/mol}$$
$$(81 \text{ kcal/mol}) \quad\quad (69 \text{ kcal/mol}) \quad\quad (70 \text{ kcal/mol})$$

Problem 11.11
Rank the following substances in order of their expected S_N1 reactivity:

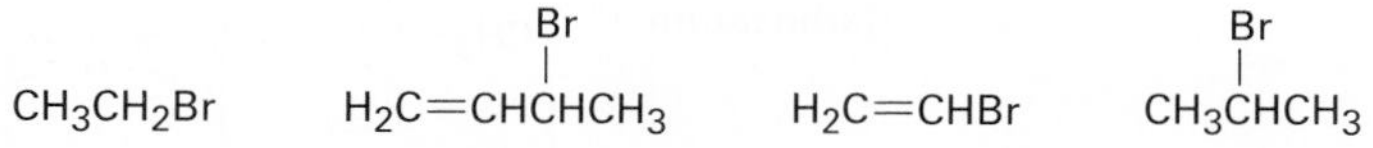

$$CH_3CH_2Br \quad\quad H_2C{=}CHCHBrCH_3 \quad\quad H_2C{=}CHBr \quad\quad CH_3CHBrCH_3$$

Problem 11.12
3-Bromo-1-butene and 1-bromo-2-butene undergo S_N1 reaction at nearly the same rate even though one is a secondary halide and the other is primary. Explain.

The Leaving Group

We said during the discussion of S_N2 reactivity that the best leaving groups are those that are most stable; that is, those that are the conjugate bases of strong acids. An identical reactivity order is found for the S_N1 reaction because the leaving group is directly involved in the rate-limiting step. Thus, the S_N1 reactivity order is

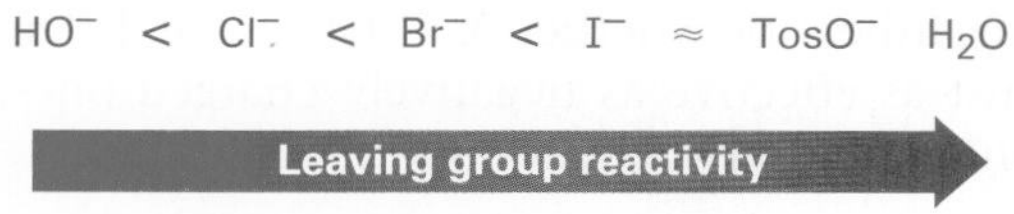

$$HO^- < Cl^- < Br^- < I^- \approx TosO^- \quad H_2O$$

Leaving group reactivity →

Note that in the S_N1 reaction, which is often carried out under acidic conditions, neutral water is sometimes the leaving group. This occurs, for example, when an alkyl halide is prepared from a tertiary alcohol by reaction with HBr or HCl **(Section 10.5)**. As shown in **Figure 11.13**, the alcohol is first protonated and then spontaneously loses H_2O to generate a carbocation, which reacts with halide ion to give the alkyl halide. Knowing that an S_N1 reaction is involved in the conversion of alcohols to alkyl halides explains why the reaction works well only for tertiary alcohols. Tertiary alcohols react fastest because they give the most stable carbocation intermediates.

Figure 11.13 MECHANISM

The mechanism of the S_N1 reaction of a tertiary alcohol with HBr to yield an alkyl halide. Neutral water is the leaving group (step **2**).

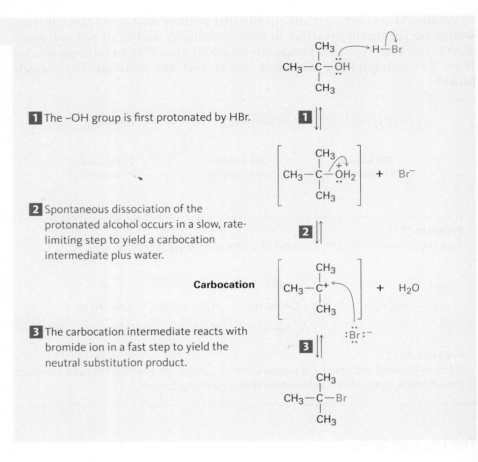

1 The –OH group is first protonated by HBr.

2 Spontaneous dissociation of the protonated alcohol occurs in a slow, rate-limiting step to yield a carbocation intermediate plus water.

3 The carbocation intermediate reacts with bromide ion in a fast step to yield the neutral substitution product.

© John McMurry

The Nucleophile

The nature of the nucleophile plays a major role in the S_N2 reaction but does not affect an S_N1 reaction. Because the S_N1 reaction occurs through a rate-limiting step in which the added nucleophile has no part, the nucleophile can't affect the reaction rate. The reaction of 2-methyl-2-propanol with HX, for instance, occurs at the same rate regardless of whether X is Cl, Br, or I. Furthermore, neutral nucleophiles are just as effective as negatively charged ones, so S_N1 reactions frequently occur under neutral or acidic conditions.

2-Methyl-2-propanol   (Same rate for X = Cl, Br, I)

The Solvent

What about the solvent? Do solvents have the same effect in S_N1 reactions that they have in S_N2 reactions? The answer is both yes and no. Yes, solvents have a large effect on S_N1 reactions, but no, the reasons for the effects on S_N1 and S_N2 reactions are not the same. Solvent effects in the S_N2 reaction are due largely to stabilization or destabilization of the nucleophile *reactant,* while solvent effects in the S_N1 reaction are due largely to stabilization or destabilization of the *transition state.*

The Hammond postulate says that any factor stabilizing the intermediate carbocation should increase the rate of an S_N1 reaction. Solvation of the carbocation—the interaction of the ion with solvent molecules—has just such an effect. Solvent molecules orient around the carbocation so that the electron-rich ends of the solvent dipoles face the positive charge **(Figure 11.14)**, thereby lowering the energy of the ion and favoring its formation.

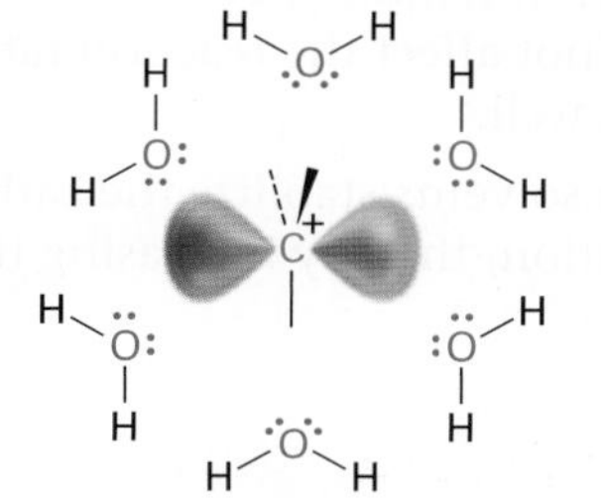

Figure 11.14 Solvation of a carbocation by water. The electron-rich oxygen atoms of solvent molecules orient around the positively charged carbocation and thereby stabilize it.

The properties of a solvent that contribute to its ability to stabilize ions by solvation are related to the solvent's polarity. S_N1 reactions take place much more rapidly in strongly polar solvents, such as water and methanol, than in less polar solvents, such as ether and chloroform. In the reaction of 2-chloro-2-methylpropane, for example, a rate increase of 100,000 is observed on going from ethanol (less polar) to water (more polar). The rate increases on going from a hydrocarbon solvent to water are so large they can't be measured accurately.

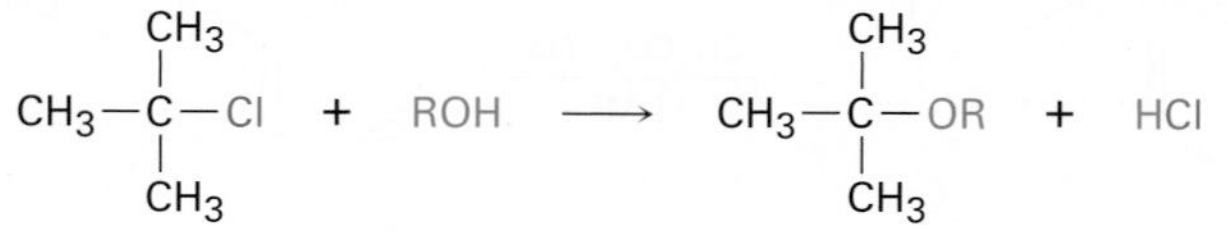

$$CH_3-C(CH_3)_2-Cl + ROH \longrightarrow CH_3-C(CH_3)_2-OR + HCl$$

	Ethanol	40% Water/60% Ethanol	80% Water/20% Ethanol	Water
Relative reactivity	1	100	14,000	100,000

Solvent reactivity →

It should be emphasized again that both the S_N1 and the S_N2 reaction show solvent effects, but that they do so for different reasons. S_N2 reactions are *disfavored* in protic solvents because the *ground-state energy* of the nucleophile is lowered by solvation. S_N1 reactions are *favored* in protic solvents because the *transition-state energy* leading to carbocation intermediate is lowered by solvation.

A Summary of S$_N$1 Reaction Characteristics

The effects on S$_N$1 reactions of the four variables—substrate, leaving group, nucleophile, and solvent—are summarized in the following statements:

Substrate The best substrates yield the most stable carbocations. As a result, S$_N$1 reactions are best for tertiary, allylic, and benzylic halides.

Leaving group Good leaving groups increase the reaction rate by lowering the energy level of the transition state for carbocation formation.

Nucleophile The nucleophile must be nonbasic to prevent a competitive elimination of HX **(Section 11.7)**, but otherwise does not affect the reaction rate. Neutral nucleophiles work well.

Solvent Polar solvents stabilize the carbocation intermediate by solvation, thereby increasing the reaction rate.

Worked Example 11.2

Predicting the Mechanism of a Nucleophilic Substitution Reaction

Predict whether each of the following substitution reactions is likely to be S$_N$1 or S$_N$2:

(a) [benzylic secondary chloride] $\xrightarrow[\text{CH}_3\text{CO}_2\text{H, H}_2\text{O}]{\text{CH}_3\text{CO}_2^- \text{ Na}^+}$ [benzylic secondary acetate, OAc]

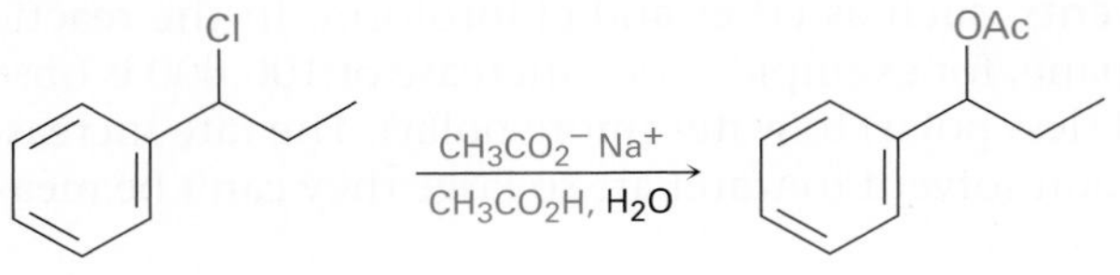

(b) PhCH$_2$CH$_2$Br $\xrightarrow[\text{DMF}]{\text{CH}_3\text{CO}_2^- \text{ Na}^+}$ PhCH$_2$CH$_2$OAc

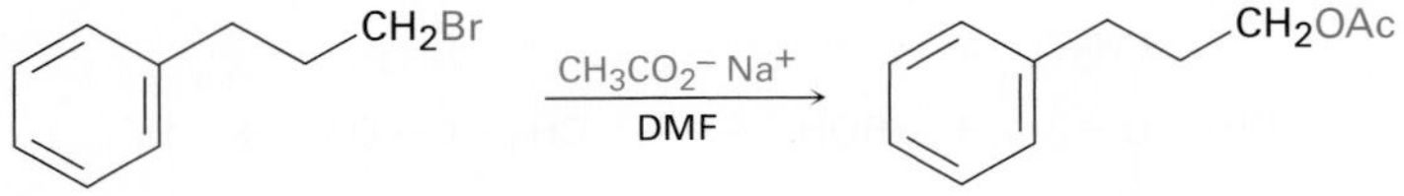

Strategy
Look at the substrate, leaving group, nucleophile, and solvent. Then decide from the summaries at the ends of **Sections 11.3 and 11.5** whether an S$_N$1 or an S$_N$2 reaction is favored. S$_N$1 reactions are favored by tertiary, allylic, or benzylic substrates, by good leaving groups, by nonbasic nucleophiles, and by protic solvents. S$_N$2 reactions are favored by primary substrates, by good leaving groups, by good nucleophiles, and by polar aprotic solvents.

Solution
(a) This is likely to be an S$_N$1 reaction because the substrate is secondary and benzylic, the nucleophile is weakly basic, and the solvent is protic.
(b) This is likely to be an S$_N$2 reaction because the substrate is primary, the nucleophile is a reasonably good one, and the solvent is polar aprotic.

Problem 11.13
Predict whether each of the following substitution reactions is likely to be S_N1 or S_N2:

(a) Cyclohexenyl-CH(OH)CH$_3$ $\xrightarrow[\text{CH}_3\text{OH}]{\text{HCl}}$ Cyclohexenyl-CH(Cl)CH$_3$

(b) $H_2C=C(CH_3)CH_2Br$ $\xrightarrow[\text{CH}_3\text{CN}]{\text{Na}^+ \ ^-\text{SCH}_3}$ $H_2C=C(CH_3)CH_2SCH_3$

11.6 Biological Substitution Reactions

Both S_N1 and S_N2 reactions are well-known in biological chemistry, particularly in the pathways for biosynthesis of the many thousands of plant-derived substances called *terpenoids,* which we'll discuss in **Section 27.5**. Unlike what typically happens in the laboratory, however, the substrate in a biological substitution reaction is usually an organodiphosphate rather than an alkyl halide. Thus, the leaving group is the diphosphate ion, abbreviated PP_i, rather than a halide ion. In fact, it's useful to think of the diphosphate group as the "biological equivalent" of a halogen. The dissociation of an organodiphosphate in a biological reaction is typically assisted by complexation to a divalent metal cation such as Mg^{2+} to help neutralize charge and make the diphosphate a better leaving group.

$$R-Cl \xrightarrow{\text{Dissociation}} R^+ + Cl^-$$
An alkyl chloride

$$R-OPOPO^-\text{(with }O^-, O^-, Mg^{2+}\text{)} \xrightarrow{\text{Dissociation}} R^+ + {}^-OPOPO^-\text{(with }O^-, O^-, Mg^{2+}\text{)} \ (PP_i)$$
An organodiphosphate   **Diphosphate ion**

Two S_N1 reactions occur during the biosynthesis of geraniol, a fragrant alcohol found in roses and used in perfumery. Geraniol biosynthesis begins with dissociation of dimethylallyl diphosphate to give an allylic carbocation, which reacts with isopentenyl diphosphate **(Figure 11.15)**. From the viewpoint of isopentenyl diphosphate, the reaction is an electrophilic alkene addition, but from the viewpoint of dimethylallyl diphosphate, the process is an S_N1 reaction in which the carbocation intermediate reacts with a double bond as the nucleophile.

Following this initial S_N1 reaction, loss of the *pro-R* hydrogen gives geranyl diphosphate, itself an allylic diphosphate that dissociates a second time. Reaction of the geranyl carbocation with water in a second S_N1 reaction, followed by loss of a proton, then yields geraniol.

Figure 11.15 Biosynthesis of geraniol from dimethylallyl diphosphate. Two S_N1 reactions occur, both with diphosphate ion as the leaving group.

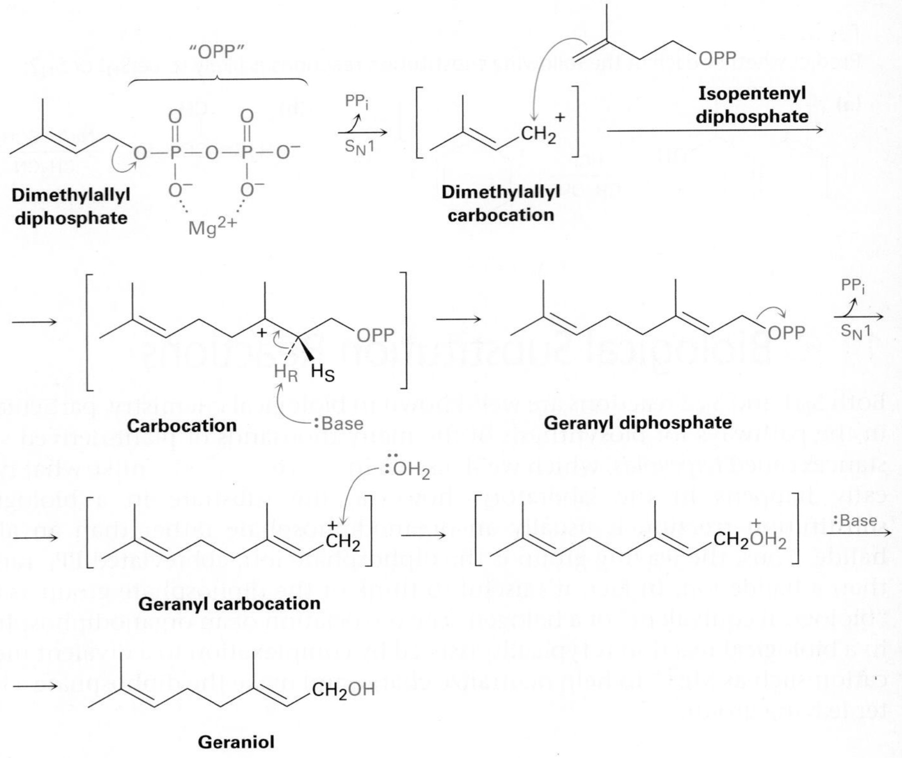

S_N2 reactions are involved in almost all biological methylations, which transfer a $-CH_3$ group from an electrophilic donor to a nucleophile. The donor is *S*-adenosylmethionine (abbreviated SAM), which contains a positively charged sulfur (a sulfonium ion, Section 5.12), and the leaving group is the neutral *S*-adenosylhomocysteine molecule. In the biosynthesis of epinephrine (adrenaline) from norepinephrine, for instance, the nucleophilic nitrogen atom of norepinephrine attacks the electrophilic methyl carbon atom of *S*-adenosylmethionine in an S_N2 reaction, displacing *S*-adenosylhomocysteine (**Figure 11.16**). In effect, *S*-adenosylmethionine is simply a biological equivalent of CH_3Cl.

Figure 11.16 The biosynthesis of epinephrine from norepinephrine occurs by an S_N2 reaction with *S*-adenosylmethionine.

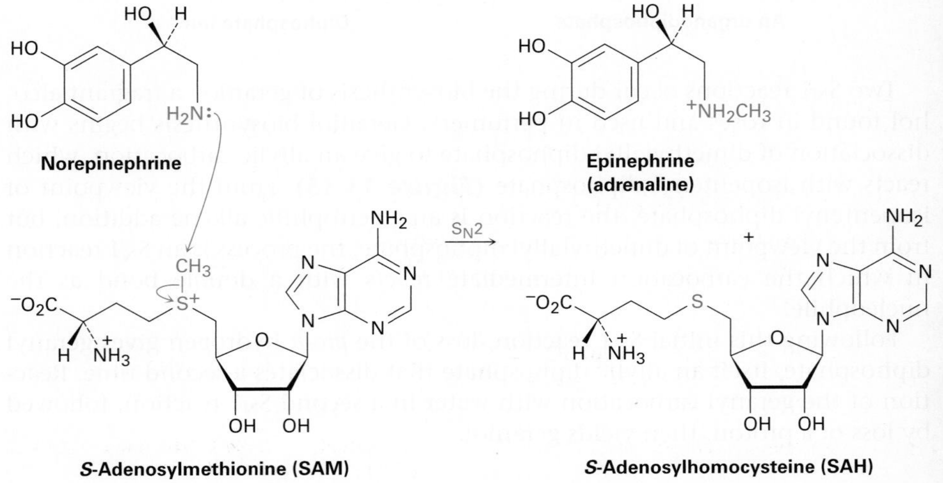

Problem 11.14
Review the mechanism of geraniol biosynthesis shown in Figure 11.15, and propose a mechanism for the biosynthesis of limonene from linalyl diphosphate.

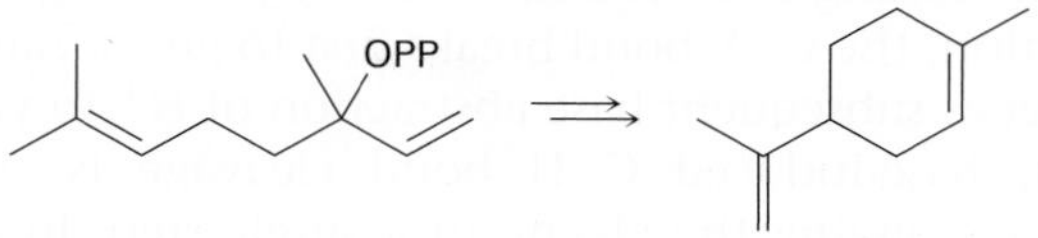

Linalyl diphosphate → **Limonene**

11.7 Elimination Reactions: Zaitsev's Rule

We said at the beginning of this chapter that two kinds of reactions can take place when a nucleophile/Lewis base reacts with an alkyl halide. The nucleophile can either substitute for the halide by reaction at carbon or can cause elimination of HX by reaction at a neighboring hydrogen:

Substitution

$$\text{C-C(Br)} + \text{OH}^- \longrightarrow \text{C-C(OH)} + \text{Br}^-$$

Elimination

$$\text{C-C(Br)} + \text{OH}^- \longrightarrow \text{C=C} + \text{H}_2\text{O} + \text{Br}^-$$

> **Key IDEAS**
>
> Test your knowledge of Key Ideas by answering end-of-chapter exercises marked with ▲.

Elimination reactions are more complex than substitution reactions for several reasons. One is the problem of regiochemistry. What products result by loss of HX from an unsymmetrical halide? In fact, elimination reactions almost always give mixtures of alkene products, and the best we can usually do is to predict which will be the major product.

According to **Zaitsev's rule**, formulated in 1875 by the Russian chemist Alexander Zaitsev, base-induced elimination reactions generally (although not always) give the more stable alkene product—that is, the alkene with more alkyl substituents on the double-bond carbons. In the following two cases, for example, the more highly substituted alkene product predominates.

> **ZAITSEV'S RULE**
>
> In the elimination of HX from an alkyl halide, the more highly substituted alkene product predominates.

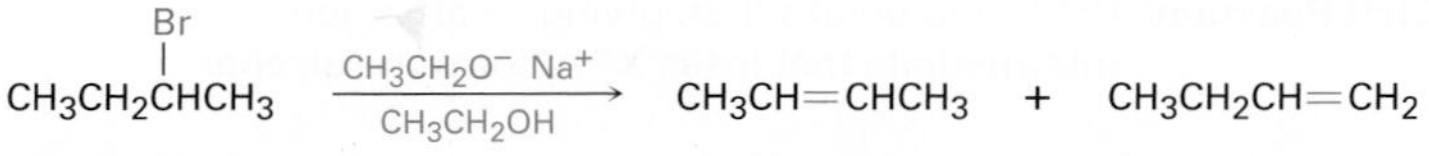

$$\underset{\text{2-Bromobutane}}{\text{CH}_3\text{CH}_2\text{CHBrCH}_3} \xrightarrow[\text{CH}_3\text{CH}_2\text{OH}]{\text{CH}_3\text{CH}_2\text{O}^-\text{ Na}^+} \underset{\substack{\text{2-Butene}\\(81\%)}}{\text{CH}_3\text{CH}=\text{CHCH}_3} + \underset{\substack{\text{1-Butene}\\(19\%)}}{\text{CH}_3\text{CH}_2\text{CH}=\text{CH}_2}$$

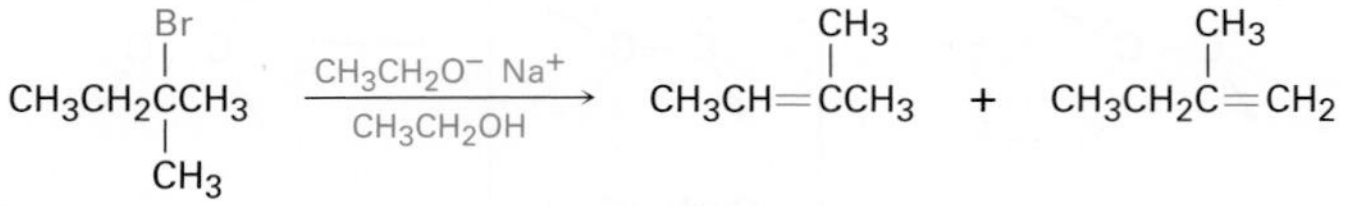

$$\underset{\text{2-Bromo-2-methylbutane}}{\text{CH}_3\text{CH}_2\text{C(Br)(CH}_3\text{)CH}_3} \xrightarrow[\text{CH}_3\text{CH}_2\text{OH}]{\text{CH}_3\text{CH}_2\text{O}^-\text{ Na}^+} \underset{\substack{\text{2-Methyl-2-butene}\\(70\%)}}{\text{CH}_3\text{CH}=\text{C(CH}_3\text{)CH}_3} + \underset{\substack{\text{2-Methyl-1-butene}\\(30\%)}}{\text{CH}_3\text{CH}_2\text{C(CH}_3\text{)}=\text{CH}_2}$$

Another factor that complicates a study of elimination reactions is that they can take place by different mechanisms, just as substitutions can. We'll consider three of the most common mechanisms—the E1, E2, and E1cB reactions—which differ in the timing of C–H and C–X bond-breaking.

In the E1 reaction, the C–X bond breaks first to give a carbocation intermediate that undergoes subsequent base abstraction of H^+ to yield the alkene. In the E2 reaction, base-induced C–H bond cleavage is simultaneous with C–X bond cleavage, giving the alkene in a single step. In the E1cB reaction (cB for "conjugate base"), base abstraction of the proton occurs first, giving a carbanion ($R{:}^-$) intermediate. This anion, the conjugate base of the reactant "acid," then undergoes loss of X^- in a subsequent step to give the alkene. All three mechanisms occur frequently in the laboratory, but the E1cB mechanism predominates in biological pathways.

E1 Reaction: C–X bond breaks first to give a carbocation intermediate, followed by base removal of a proton to yield the alkene.

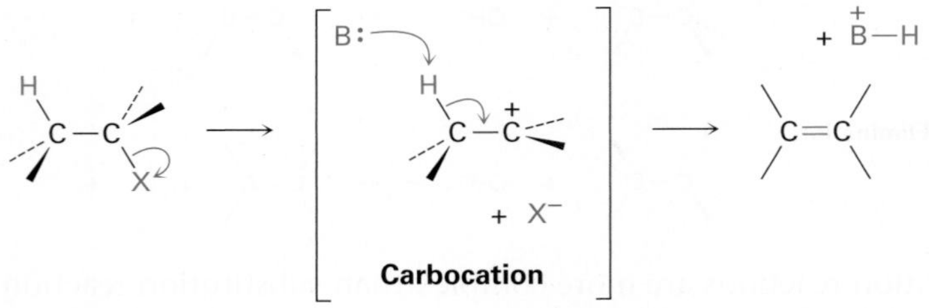

E2 Reaction: C–H and C–X bonds break simultaneously, giving the alkene in a single step without intermediates.

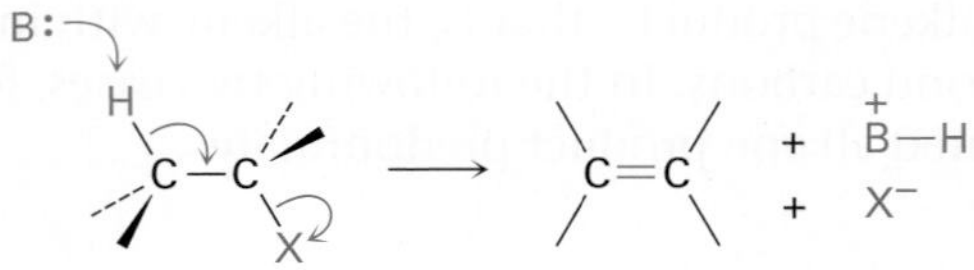

E1cB Reaction: C–H bond breaks first, giving a carbanion intermediate that loses X^- to form the alkene.

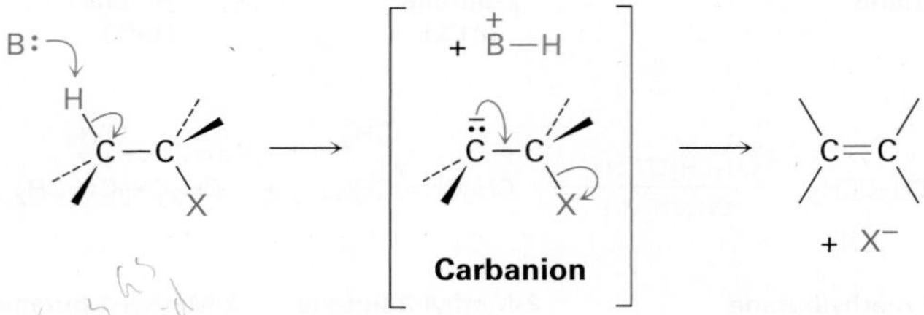

Worked Example 11.3

Predicting the Product of an Elimination Reaction

What product would you expect from reaction of 1-chloro-1-methylcyclohexane with KOH in ethanol?

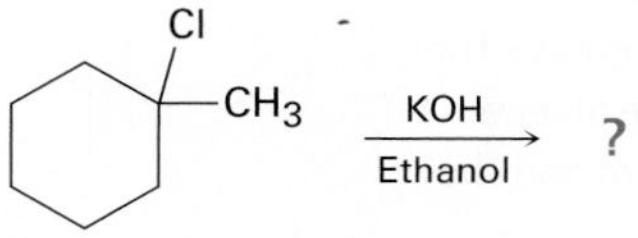

[Structure: 1-chloro-1-methylcyclohexane + KOH/Ethanol → ?]

Strategy

Treatment of an alkyl halide with a strong base such as KOH yields an alkene. To find the products in a specific case, locate the hydrogen atoms on each carbon next to the leaving group, and then generate the potential alkene products by removing HX in as many ways as possible. The major product will be the one that has the most highly substituted double bond—in this case, 1-methylcyclohexene.

Solution

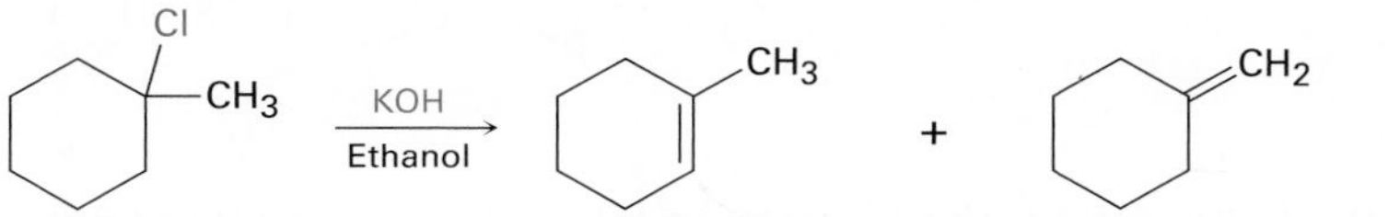

[Structures showing 1-Chloro-1-methylcyclohexane + KOH/Ethanol → 1-Methylcyclohexene (major) + Methylenecyclohexane (minor)]

Problem 11.15

Ignoring double-bond stereochemistry, what products would you expect from elimination reactions of the following alkyl halides? Which product will be the major product in each case?

(a) $CH_3CH_2CHCHCH_3$ with Br and CH_3 substituents

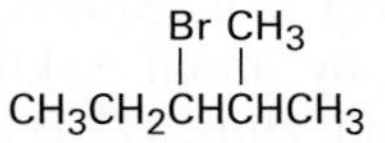

(b) $CH_3CHCH_2-C-CHCH_3$ with CH_3, Cl, and CH_3 substituents

(c) Cyclohexane with Br and $CHCH_3$ substituents

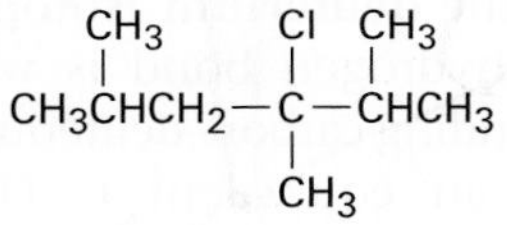

[handwritten: apply Zaitsev's rule, mult product formed, major product is most stable alkene]

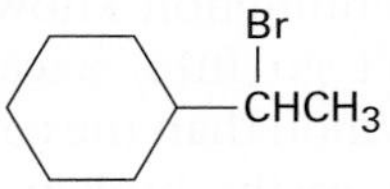

Problem 11.16

What alkyl halides might the following alkenes have been made from?

(a) $CH_3CHCH_2CH_2CHCH=CH_2$ with two CH_3 groups

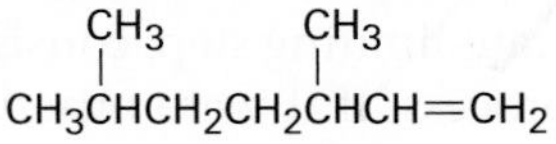

(b) Cyclopentene with two CH_3 substituents

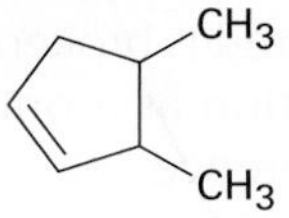

11.8 The E2 Reaction and the Deuterium Isotope Effect

The **E2 reaction** (for *elimination, bimolecular*) occurs when an alkyl halide is treated with a strong base, such as hydroxide ion or alkoxide ion (RO^-). It is the most commonly occurring pathway for elimination and can be formulated as shown in **Figure 11.17**.

Figure 11.17 | MECHANISM

Mechanism of the E2 reaction of an alkyl halide. The reaction takes place in a single step through a transition state in which the double bond begins to form at the same time the H and X groups are leaving.

1. Base (B:) attacks a neighboring hydrogen and begins to remove the H at the same time as the alkene double bond starts to form and the X group starts to leave.

2. Neutral alkene is produced when the C–H bond is fully broken and the X group has departed with the C–X bond electron pair.

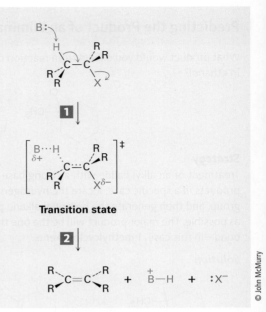

Transition state

© John McMurry

Like the S_N2 reaction, the E2 reaction takes place in one step without intermediates. As the base begins to abstract H^+ from a carbon next to the leaving group, the C–H bond begins to break, a C=C bond begins to form, and the leaving group begins to depart, taking with it the electron pair from the C–X bond. Among the pieces of evidence supporting this mechanism is that E2 reactions show second-order kinetics and follow the rate law: rate = k × [RX] × [Base]. That is, both base and alkyl halide take part in the rate-limiting step.

A second piece of evidence in support of the E2 mechanism is provided by a phenomenon known as the **deuterium isotope effect**. For reasons that we won't go into, a carbon–hydrogen bond is weaker by about 5 kJ/mol (1.2 kcal/mol) than the corresponding carbon–deuterium bond. Thus, a C–H bond is more easily broken than an equivalent C–D bond, and the rate of C–H bond cleavage is faster. For instance, the base-induced elimination of HBr from 1-bromo-2-phenylethane proceeds 7.11 times as fast as the corresponding elimination of DBr from 1-bromo-2,2-dideuterio-2-phenylethane. This result tells us that the C–H (or C–D) bond is broken in the rate-limiting step, consistent with our picture of the E2 reaction as a one-step process. If it were otherwise, we couldn't measure a rate difference.

(H)—Faster reaction
(D)—Slower reaction

Yet a third piece of mechanistic evidence involves the stereochemistry of E2 eliminations. As shown by a large number of experiments, E2 reactions occur with *periplanar* geometry, meaning that all four reacting atoms—the hydrogen, the two carbons, and the leaving group—lie in the same plane. Two such geometries are possible: **syn periplanar** geometry, in which the H and the X are on the same side of the molecule, and **anti periplanar** geometry, in which the H and the X are on opposite sides of the molecule. Of the two, anti periplanar geometry is energetically preferred because it allows the substituents on the two carbons to adopt a staggered relationship, whereas syn geometry requires that the substituents be eclipsed.

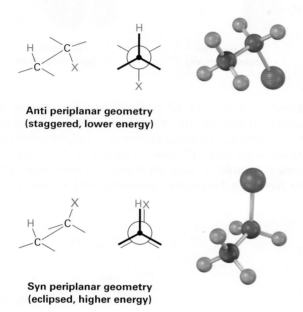

**Anti periplanar geometry
(staggered, lower energy)**

**Syn periplanar geometry
(eclipsed, higher energy)**

What's so special about periplanar geometry? Because the sp^3 σ orbitals in the reactant C–H and C–X bonds must overlap and become p π orbitals in the alkene product, there must also be some overlap in the transition state. This can occur most easily if all the orbitals are in the same plane to begin with—that is, if they're periplanar **(Figure 11.18)**.

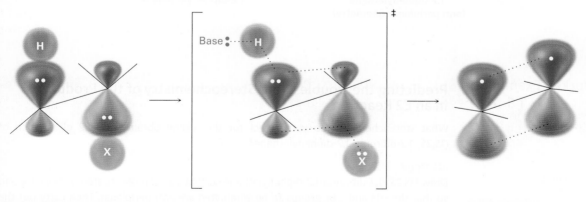

Anti periplanar reactant   **Anti transition state**   **Alkene product**

Figure 11.18 The transition state for the E2 reaction of an alkyl halide with base. Overlap of the developing *p* orbitals in the transition state requires periplanar geometry of the reactant.

You can think of E2 elimination reactions with periplanar geometry as being similar to S_N2 reactions with 180° geometry. In an S_N2 reaction, an electron pair from the incoming nucleophile pushes out the leaving group on the opposite side of the molecule. In an E2 reaction, an electron pair from a neighboring C–H bond pushes out the leaving group on the opposite side of the molecule.

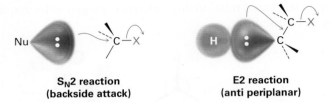

S_N2 reaction
(backside attack)

E2 reaction
(anti periplanar)

Anti periplanar geometry for E2 eliminations has specific stereochemical consequences that provide strong evidence for the proposed mechanism. To take just one example, *meso*-1,2-dibromo-1,2-diphenylethane undergoes E2 elimination on treatment with base to give only the *E* alkene. None of the isomeric *Z* alkene is formed because the transition state leading to the *Z* alkene would have to have syn periplanar geometry and would thus be higher in energy.

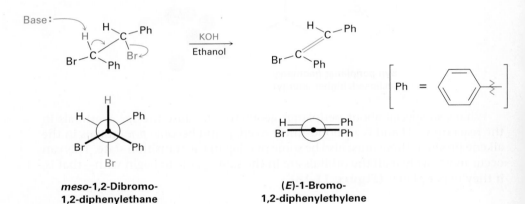

meso-1,2-Dibromo-
1,2-diphenylethane
(anti periplanar geometry)

(*E*)-1-Bromo-
1,2-diphenylethylene

Worked Example 11.4

Predicting the Double-Bond Stereochemistry of the Product in an E2 Reaction

What stereochemistry do you expect for the alkene obtained by E2 elimination of (1*S*,2*S*)-1,2-dibromo-1,2-diphenylethane?

Strategy

Draw (1*S*,2*S*)-1,2-dibromo-1,2-diphenylethane so that you can see its stereochemistry and so that the –H and –Br groups to be eliminated are anti periplanar. Then carry out the elimination while keeping all substituents in approximately their same positions, and see what alkene results.

Solution

Anti periplanar elimination of HBr gives (Z)-1-bromo-1,2-diphenylethylene.

Problem 11.17
What stereochemistry do you expect for the alkene obtained by E2 elimination of (1R,2R)-1,2-dibromo-1,2-diphenylethane? Draw a Newman projection of the reacting conformation.

Problem 11.18
What stereochemistry do you expect for the trisubstituted alkene obtained by E2 elimination of the following alkyl halide on treatment with KOH? (Reddish brown = Br.)

11.9 The E2 Reaction and Cyclohexane Conformation

Anti periplanar geometry for E2 reactions is particularly important in cyclohexane rings, where chair geometry forces a rigid relationship between the substituents on neighboring carbon atoms **(Section 4.8)**. The anti periplanar requirement for E2 reactions overrides Zaitsev's rule and can be met in cyclohexanes only if the hydrogen and the leaving group are trans diaxial **(Figure 11.19)**. If either the leaving group or the hydrogen is equatorial, E2 elimination can't occur.

Axial chlorine: H and Cl are anti periplanar

Equatorial chlorine: H and Cl are not anti periplanar

Figure 11.19 The geometric requirement for an E2 reaction in a substituted cyclohexane. The leaving group and the hydrogen must both be axial for anti periplanar elimination to occur.

The elimination of HCl from the isomeric menthyl and neomenthyl chlorides shown in **Figure 11.20** gives a good illustration of this trans-diaxial requirement. Neomenthyl chloride undergoes elimination of HCl on reaction with ethoxide ion 200 times as fast as menthyl chloride. Furthermore, neomenthyl chloride yields 3-menthene as the major alkene product, whereas menthyl chloride yields 2-menthene.

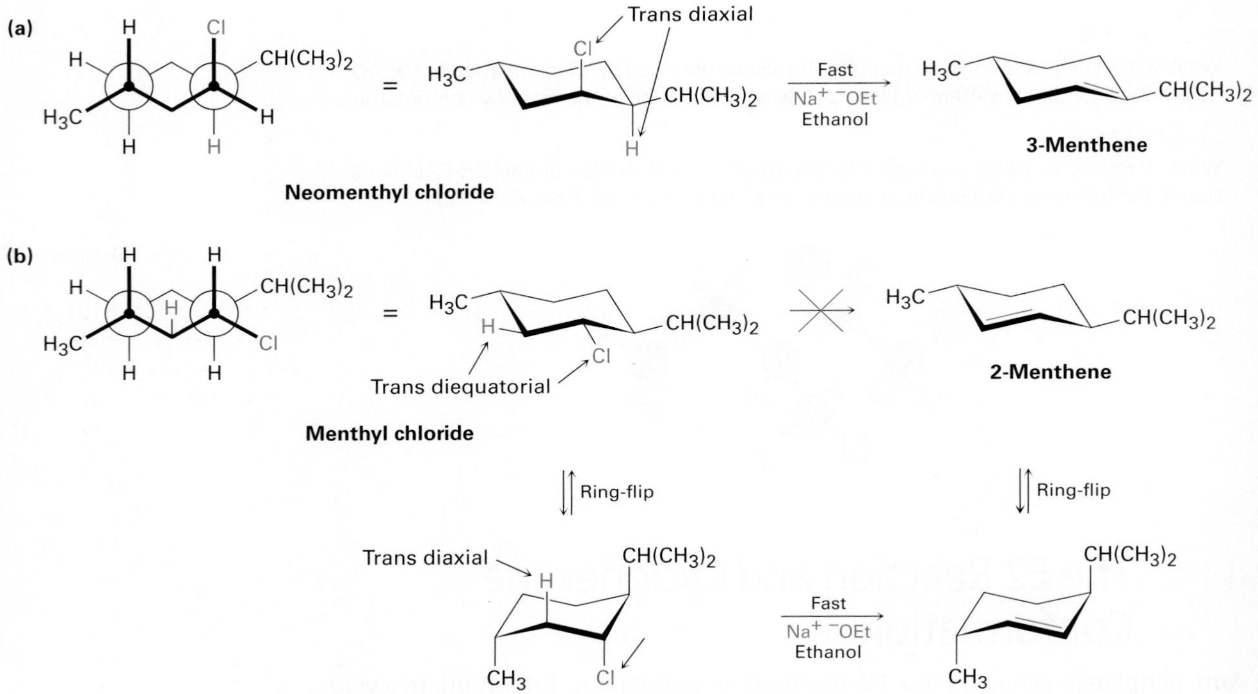

Figure 11.20 Dehydrochlorination of menthyl and neomenthyl chlorides. **(a)** Neomenthyl chloride loses HCl directly from its more stable conformation, but **(b)** menthyl chloride must first ring-flip to a higher energy conformation before HCl loss can occur. The abbreviation "Et" represents an ethyl group.

The difference in reactivity between the isomeric menthyl chlorides is due to the difference in their conformations. Neomenthyl chloride has the conformation shown in Figure 11.20a, with the methyl and isopropyl groups equatorial and the chlorine axial—a perfect geometry for E2 elimination. Loss of the hydrogen atom at C4 occurs easily to yield the more substituted alkene product, 3-menthene, as predicted by Zaitsev's rule.

Menthyl chloride, by contrast, has a conformation in which all three substituents are equatorial (Figure 11.20b). To achieve the necessary geometry for elimination, menthyl chloride must first ring-flip to a higher-energy chair conformation, in which all three substituents are axial. E2 elimination then occurs with loss of the only trans-diaxial hydrogen available, leading to the non-Zaitsev product 2-menthene. The net effect of the simple change in chlorine stereochemistry is a 200-fold change in reaction rate and a complete change of product. The chemistry of the molecule is controlled by its conformation.

Problem 11.19
Which isomer would you expect to undergo E2 elimination faster, *trans*-1-bromo-4-*tert*-butylcyclohexane or *cis*-1-bromo-4-*tert*-butylcyclohexane? Draw each molecule in its more stable chair conformation, and explain your answer.

11.10 The E1 and E1cB Reactions

The E1 Reaction

Just as the E2 reaction is analogous to the S_N2 reaction, the S_N1 reaction has a close analog called the **E1 reaction** (for *elimination, unimolecular*). The E1 reaction can be formulated as shown in **Figure 11.21** for the elimination of HCl from 2-chloro-2-methylpropane.

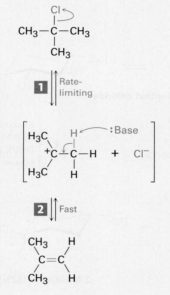

1 Spontaneous dissociation of the tertiary alkyl chloride yields an intermediate carbocation in a slow, rate-limiting step.

2 Loss of a neighboring H⁺ in a fast step yields the neutral alkene product. The electron pair from the C–H bond goes to form the alkene π bond.

Figure 11.21 | MECHANISM

Mechanism of the E1 reaction. Two steps are involved, the first of which is rate-limiting, and a carbocation intermediate is present.

© John McMurry

E1 eliminations begin with the same unimolecular dissociation to give a carbocation that we saw in the S_N1 reaction, but the dissociation is followed by loss of H⁺ from the adjacent carbon rather than by substitution. In fact, the E1 and S_N1 reactions normally occur together whenever an alkyl halide is treated in a protic solvent with a nonbasic nucleophile. Thus, the best E1 substrates are also the best S_N1 substrates, and mixtures of substitution and elimination products are usually obtained. For example, when 2-chloro-2-methylpropane is warmed to 65 °C in 80% aqueous ethanol, a 64:36 mixture of 2-methyl-2-propanol (S_N1) and 2-methylpropene (E1) results.

2-Chloro-2-methylpropane → (H₂O, ethanol, 65 °C) → 2-Methyl-2-propanol (64%) + 2-Methylpropene (36%)

Much evidence has been obtained in support of the E1 mechanism. For example, E1 reactions show first-order kinetics, consistent with a rate-limiting, unimolecular dissociation process. Furthermore, E1 reactions show no deuterium isotope effect because rupture of the C–H (or C–D) bond occurs after the rate-limiting step rather than during it. Thus, we can't measure a rate difference between a deuterated and nondeuterated substrate.

A final piece of evidence involves the stereochemistry of elimination. Unlike the E2 reaction, where anti periplanar geometry is required, there is no geometric requirement on the E1 reaction because the halide and the hydrogen are lost in separate steps. We might therefore expect to obtain the more stable (Zaitsev's rule) product from E1 reaction, which is just what we find. To return to a familiar example, menthyl chloride loses HCl under E1 conditions in a polar solvent to give a mixture of alkenes in which the Zaitsev product, 3-menthene, predominates **(Figure 11.22)**.

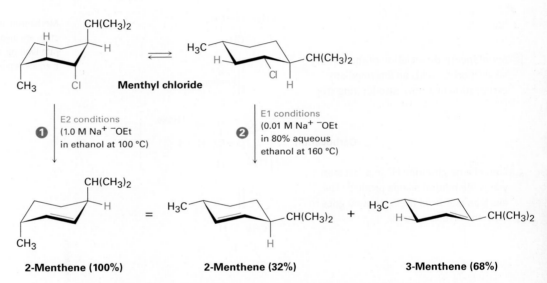

Figure 11.22 Elimination reactions of menthyl chloride. E2 conditions (❶, strong base in 100% ethanol) lead to 2-menthene through an anti periplanar elimination, whereas E1 conditions (❷, dilute base in 80% aqueous ethanol) lead to a mixture of 2-menthene and 3-menthene.

The E1cB Reaction

In contrast to the E1 reaction, which involves a carbocation intermediate, the **E1cB reaction** takes place through a carbanion intermediate. Base-induced abstraction of a proton in a slow, rate-limiting step gives an anion, which expels a leaving group on the adjacent carbon. The reaction is particularly common in substrates that have a poor leaving group, such as –OH, two carbons removed from a carbonyl group, HO—C—CH—C=O. The poor leaving group disfavors the alternative E1 and E2 possibilities, and the carbonyl group makes the adjacent hydrogen unusually acidic by resonance stabilization of the anion intermediate. We'll look at this acidifying effect of a carbonyl group in **Section 22.5**.

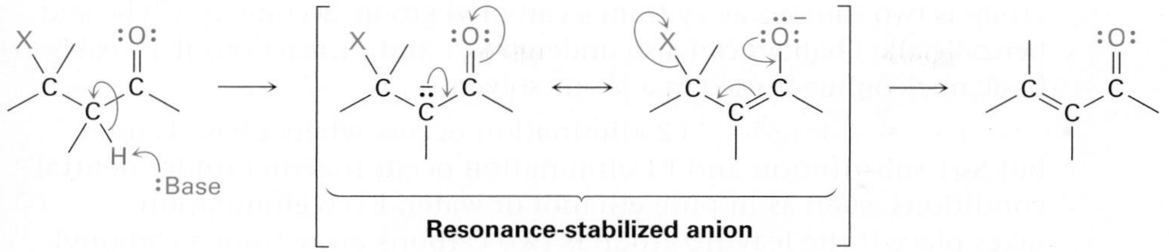

11.11 Biological Elimination Reactions

All three elimination reactions—E2, E1, and E1cB—occur in biological pathways, but the E1cB mechanism is particularly common. The substrate is usually an alcohol rather than an alkyl halide, and the H atom removed is usually adjacent to a carbonyl group, just as in laboratory reactions. Thus, 3-hydroxy carbonyl compounds are frequently converted to unsaturated carbonyl compounds by elimination reactions. A typical example occurs during the biosynthesis of fats when a 3-hydroxybutyryl thioester is dehydrated to the corresponding unsaturated (crotonyl) thioester. The base in this reaction is a histidine amino acid in the enzyme, and loss of the −OH group is assisted by simultaneous protonation.

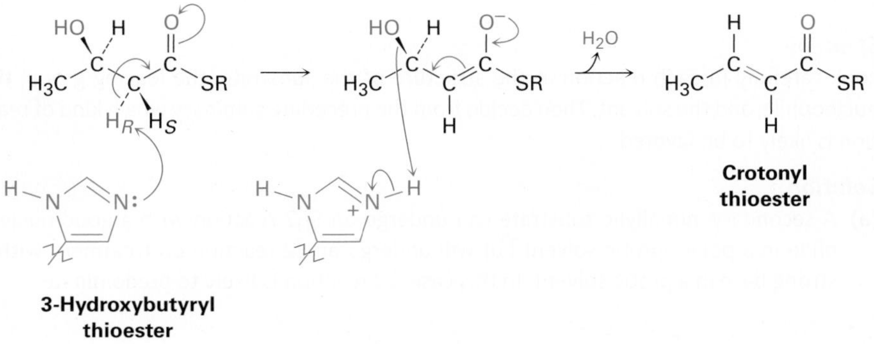

11.12 A Summary of Reactivity: S_N1, S_N2, E1, E1cB, and E2

S_N1, S_N2, E1, E1cB, E2—how can you keep it all straight and predict what will happen in any given case? Will substitution or elimination occur? Will the reaction be bimolecular or unimolecular? There are no rigid answers to these questions, but it's possible to recognize some trends and make some generalizations.

- **Primary alkyl halides** S_N2 substitution occurs if a good nucleophile is used, E2 elimination occurs if a strong, sterically hindered base is used, and E1cB elimination occurs if the leaving group is two carbons away from a carbonyl group.

- **Secondary alkyl halides** S_N2 substitution occurs if a weakly basic nucleophile is used in a polar aprotic solvent, E2 elimination predominates if a strong base is used, and E1cB elimination takes place if the leaving group is two carbons away from a carbonyl group. Secondary allylic and benzylic alkyl halides can also undergo S_N1 and E1 reactions if a weakly basic nucleophile is used in a protic solvent.
- **Tertiary alkyl halides** E2 elimination occurs when a base is used, but S_N1 substitution and E1 elimination occur together under neutral conditions, such as in pure ethanol or water. E1cB elimination takes place if the leaving group is two carbons away from a carbonyl group.

Worked Example 11.5 — Predicting the Product and Mechanism of Reactions

Tell whether each of the following reactions is likely to be S_N1, S_N2, E1, E1cB, or E2, and predict the product of each:

(a) Cyclopentyl chloride + Na⁺ ⁻OCH₃ / CH₃OH → ?

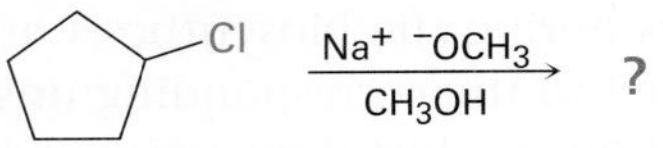

(b) 1-bromo-1-phenylethane + HCO₂H / H₂O → ?

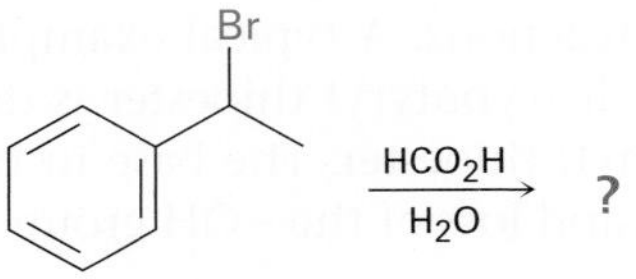

Strategy

Look carefully in each reaction at the structure of the substrate, the leaving group, the nucleophile, and the solvent. Then decide from the preceding summary which kind of reaction is likely to be favored.

Solution

(a) A secondary, nonallylic substrate can undergo an S_N2 reaction with a good nucleophile in a polar aprotic solvent but will undergo an E2 reaction on treatment with a strong base in a protic solvent. In this case, E2 reaction is likely to predominate.

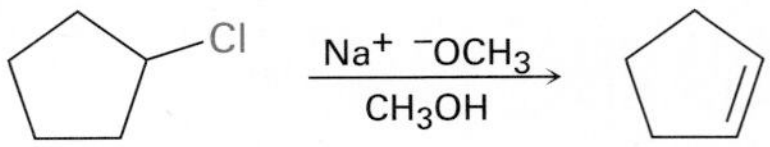

Cyclopentyl chloride + Na⁺ ⁻OCH₃ / CH₃OH → cyclopentene **E2 reaction**

(b) A secondary benzylic substrate can undergo an S_N2 reaction on treatment with a nonbasic nucleophile in a polar aprotic solvent and will undergo an E2 reaction on treatment with a base. Under protic conditions, such as aqueous formic acid (HCO₂H), an S_N1 reaction is likely, along with some E1 reaction.

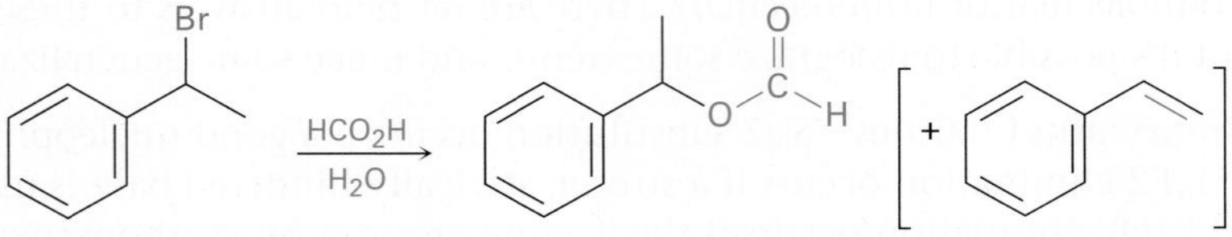

1-bromo-1-phenylethane + HCO₂H / H₂O → 1-phenylethyl formate (**S_N1**) + styrene (**E1**)

Problem 11.20

Tell whether each of the following reactions is likely to be S_N1, S_N2, E1, E1cB, or E2:

(a) $CH_3CH_2CH_2CH_2Br \xrightarrow[THF]{NaN_3} CH_3CH_2CH_2CH_2N=N=N$

(b) $CH_3CH_2\overset{Cl}{\underset{|}{C}}HCH_2CH_3 \xrightarrow[Ethanol]{KOH} CH_3CH_2CH=CHCH_3$

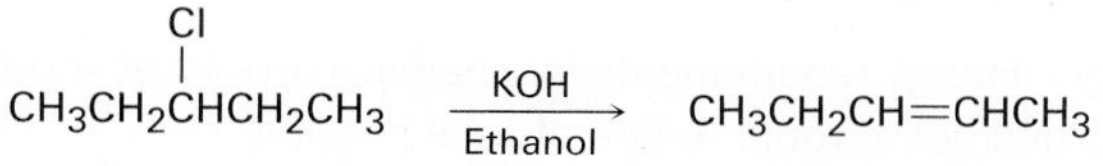

(c) [cyclohexane with Cl and CH₃] $\xrightarrow{CH_3CO_2H}$ [cyclohexane with OCOCH₃ and CH₃]

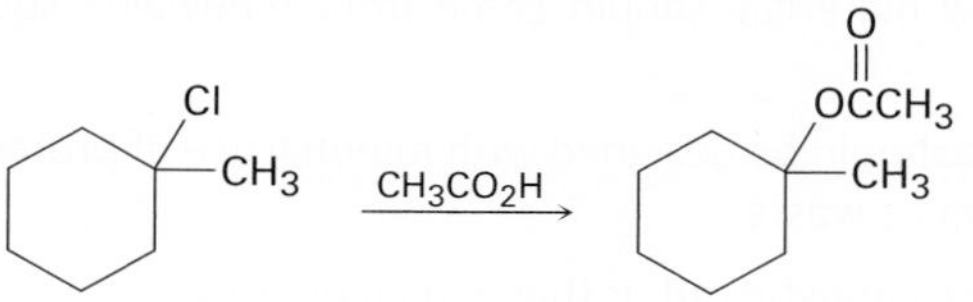

(d) [2-(1-hydroxycyclohexyl)cyclohexanone] $\xrightarrow[Ethanol]{NaOH}$ [2-cyclohexylidenecyclohexanone]

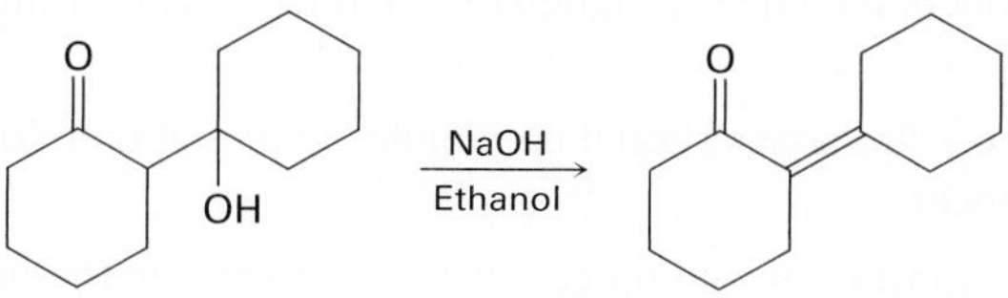

A DEEPER LOOK

Green Chemistry

Organic chemistry in the 20th century changed the world, giving us new medicines, insecticides, adhesives, textiles, dyes, building materials, composites, and all manner of polymers. But these advances did not come without a cost: every chemical process produces wastes that must be dealt with, including reaction solvents and toxic by-products that might evaporate into the air or be leached into groundwater if not disposed of properly. Even apparently harmless by-products must be safely buried or otherwise sequestered. As always, there's no such thing as a free lunch; with the good also comes the bad.

It may never be possible to make organic chemistry completely benign, but awareness of the environmental problems caused by many chemical processes has grown dramatically in recent years, giving rise to a movement called *green chemistry*. Green chemistry is the design and implementation of chemical products and processes that reduce waste and attempt to eliminate the generation of hazardous substances. There are 12 principles of green chemistry:

Let's hope disasters like this are never repeated.

Prevent waste – Waste should be prevented rather than treated or cleaned up after it has been created.

Maximize atom economy – Synthetic methods should maximize the incorporation of all materials used in a process into the final product so that waste is minimized.

Use less hazardous processes – Synthetic methods should use reactants and generate wastes with minimal toxicity to health and the environment.

(continued)

(continued)

Design safer chemicals – Chemical products should be designed to have minimal toxicity.

Use safer solvents – Minimal use should be made of solvents, separation agents, and other auxiliary substances in a reaction.

Design for energy efficiency – Energy requirements for chemical processes should be minimized, with reactions carried out at room temperature if possible.

Use renewable feedstocks – Raw materials should come from renewable sources when feasible.

Minimize derivatives – Syntheses should be designed with minimal use of protecting groups to avoid extra steps and reduce waste.

Use catalysis – Reactions should be catalytic rather than stoichiometric.

Design for degradation – Products should be designed to be biodegradable at the end of their useful lifetimes.

Monitor pollution in real time – Processes should be monitored in real time for the formation of hazardous substances.

Prevent accidents – Chemical substances and processes should minimize the potential for fires, explosions, or other accidents.

The foregoing 12 principles won't all be met in most real-world applications, but they provide a worthy goal to aim for and they can make chemists think more carefully about the environmental implications of their work. Real success stories are already occurring, and more are in progress. Approximately 7 million pounds per year of ibuprofen (6 billion tablets!) is now made by a "green" process that produces approximately 99% less waste than the process it replaces. Only three steps are needed, the anhydrous HF solvent used in the first step is recovered and reused, and the second and third steps are catalytic.

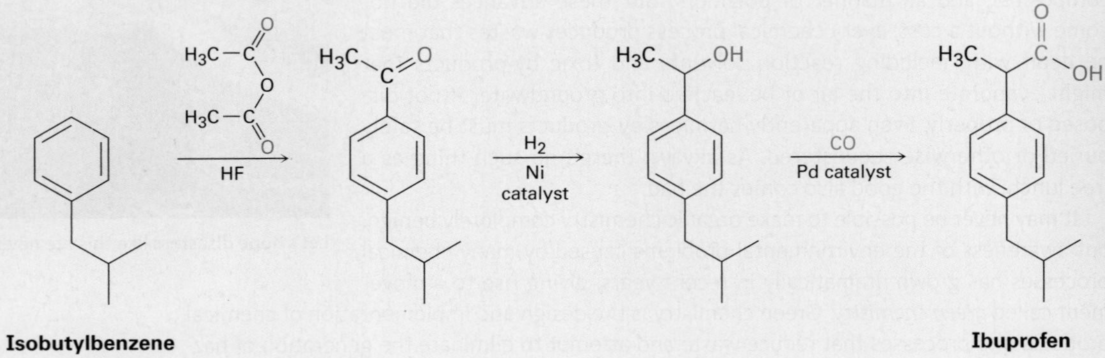

Isobutylbenzene → **Ibuprofen**

Summary

The reaction of an alkyl halide or tosylate with a nucleophile/base results either in *substitution* or in *elimination*. The resultant nucleophilic substitution and base-induced elimination reactions are two of the most widely occurring and versatile reaction types in organic chemistry, both in the laboratory and in biological pathways.

Nucleophilic substitutions are of two types: **S_N2 reactions** and **S_N1 reactions**. In the S_N2 reaction, the entering nucleophile approaches the halide from a direction 180° away from the leaving group, resulting in an umbrella-like inversion of configuration at the carbon atom. The reaction is kinetically **second-order** and is strongly inhibited by increasing steric bulk of the reactants. Thus, S_N2 reactions are favored for primary and secondary substrates.

In the S_N1 reaction, the substrate spontaneously dissociates to a carbocation in a slow **rate-limiting step**, followed by a rapid reaction with the nucleophile. As a result, S_N1 reactions are kinetically **first-order** and take place with substantial racemization of configuration at the carbon atom. They are most favored for tertiary substrates. Both S_N1 and S_N2 reactions occur in biological pathways, although the leaving group is typically a diphosphate ion rather than a halide.

Eliminations of alkyl halides to yield alkenes occur by three mechanisms: **E2 reactions**, **E1 reactions**, and **E1cB reactions**, which differ in the timing of C–H and C–X bond-breaking. In the E2 reaction, C–H and C–X bond-breaking occur simultaneously when a base abstracts H$^+$ from one carbon at the same time the leaving group departs from the neighboring carbon. The reaction takes place preferentially through an **anti periplanar** transition state in which the four reacting atoms—hydrogen, two carbons, and leaving group—are in the same plane. The reaction shows second-order kinetics and a **deuterium isotope effect**, and occurs when a secondary or tertiary substrate is treated with a strong base. These elimination reactions usually give a mixture of alkene products in which the more highly substituted alkene predominates (**Zaitsev's rule**).

In the E1 reaction, C–X bond-breaking occurs first. The substrate dissociates to yield a carbocation in the slow rate-limiting step before losing H$^+$ from an adjacent carbon in a second step. The reaction shows first-order kinetics and no deuterium isotope effect and occurs when a tertiary substrate reacts in polar, nonbasic solution.

In the E1cB reaction, C–H bond-breaking occurs first. A base abstracts a proton to give a carbanion, followed by loss of the leaving group from the adjacent carbon in a second step. The reaction is favored when the leaving group is two carbons removed from a carbonyl, which stabilizes the intermediate anion by resonance. Biological elimination reactions typically occur by this E1cB mechanism.

In general, substrates react in the following way:

RCH$_2$X (primary) ⟶ Mostly S_N2 substitution

R$_2$CHX (secondary) ⟶ S_N2 substitution with nonbasic nucleophiles
E2 elimination with strong bases

R$_3$CX (tertiary) ⟶ Mostly E2 elimination
(S_N1 substitution and E1 elimination in nonbasic solvents)

Key words

anti periplanar, 401
benzylic, 390
deuterium isotope effect, 400
E1 reaction, 405
E1cB reaction, 406
E2 reaction, 399
first-order reaction, 386
kinetics, 375
nucleophilic substitution reaction, 373
second-order reaction, 376
S_N1 reaction, 386
S_N2 reaction, 376
solvation, 383
syn periplanar, 401
Zaitsev's rule, 397

Summary of Reactions

1. Nucleophilic substitutions
   (a) S_N1 reaction of 3°, allylic, and benzylic halides (Sections 11.4 and 11.5)

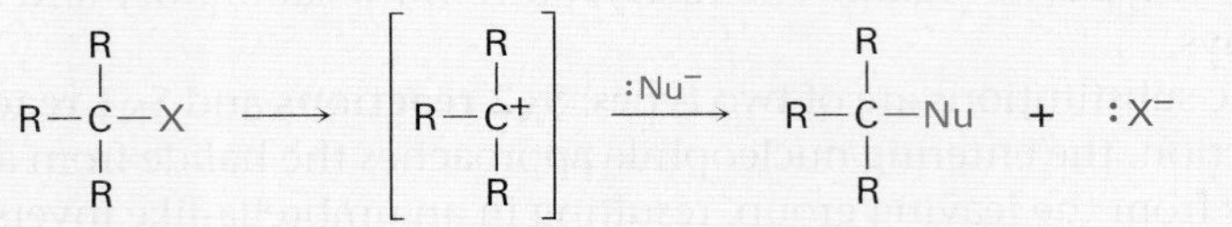

   $$R-\underset{R}{\underset{|}{\overset{R}{\overset{|}{C}}}}-X \longrightarrow \left[R-\underset{R}{\underset{|}{\overset{R}{\overset{|}{C^+}}}}\right] \xrightarrow{:Nu^-} R-\underset{R}{\underset{|}{\overset{R}{\overset{|}{C}}}}-Nu + :X^-$$

   (b) S_N2 reaction of 1° and simple 2° halides (Sections 11.2 and 11.3)

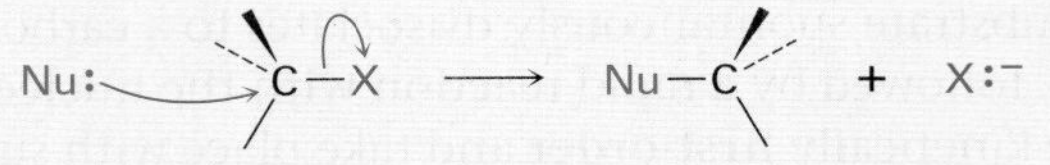

2. Eliminations
   (a) E1 reaction (Section 11.10)

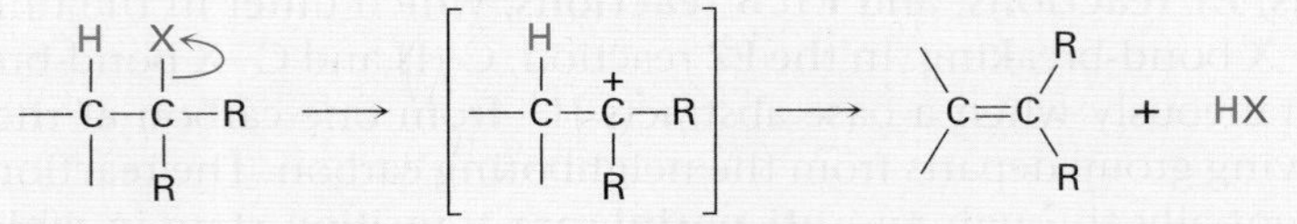

   (b) E1cB reaction (Section 11.10)

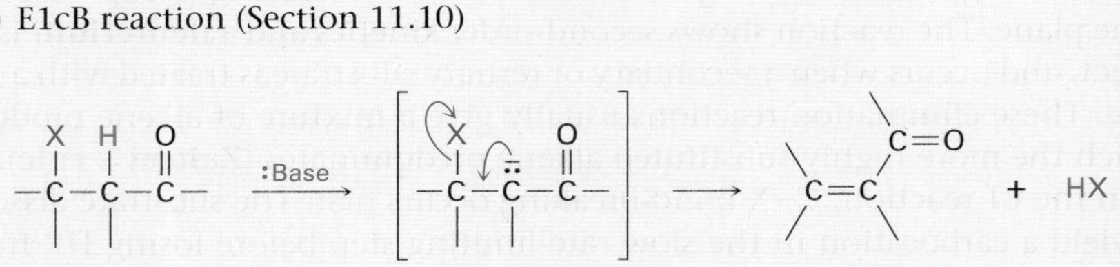

   (c) E2 reaction (Section 11.8)

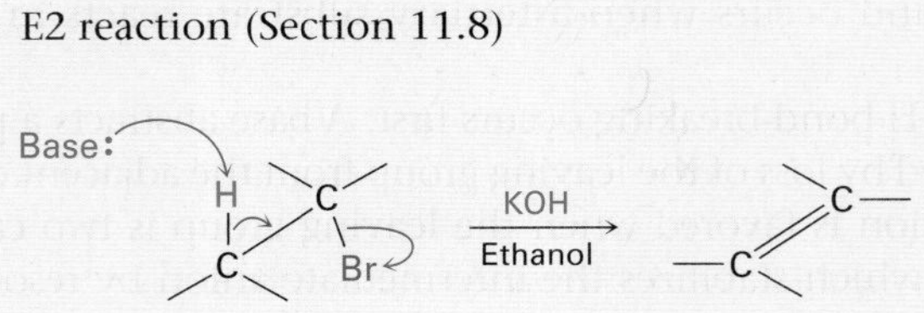

   $$\xrightarrow{\text{KOH}}_{\text{Ethanol}}$$

Exercises

Visualizing Chemistry

(Problems 11.1–11.20 appear within the chapter.)

 Interactive versions of these problems are assignable in OWL for Organic Chemistry.

▲ denotes problems linked to the Key Ideas in this chapter.

11.21 Write the product you would expect from reaction of each of the following alkyl halides with (1) Na^+ $^-SCH_3$ and (2) Na^+ ^-OH (green = Cl):

(a)  (b) 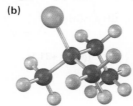 (c)

11.22 From what alkyl bromide was the following alkyl acetate made by S_N2 reaction? Write the reaction, showing all stereochemistry.

11.23 Assign R or S configuration to the following molecule, write the product you would expect from S_N2 reaction with NaCN, and assign R or S configuration to the product (green = Cl):

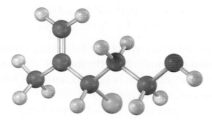

11.24 Draw the structure and assign Z or E stereochemistry to the product you expect from E2 reaction of the following molecule with NaOH (green = Cl):

▲ Problems linked to Key Ideas in this chapter

Additional Problems

Nucleophilic Substitution Reactions

11.25 Draw all isomers of C$_4$H$_9$Br, name them, and arrange them in order of decreasing reactivity in the S$_N$2 reaction.

11.26 The following Walden cycle has been carried out. Explain the results, and indicate where Walden inversion is occurring.

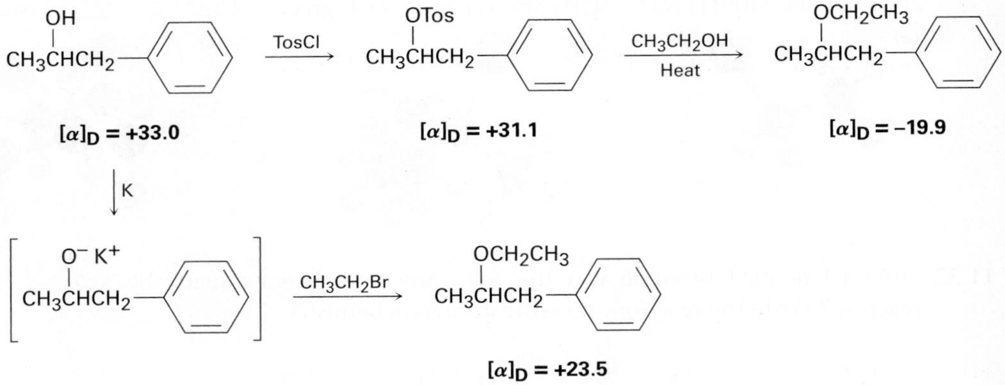

[Walden cycle diagram: CH$_3$CHCH$_2$(Ph)-OH ($[\alpha]_D = +33.0$) → TosCl → CH$_3$CHCH$_2$(Ph)-OTos ($[\alpha]_D = +31.1$) → CH$_3$CH$_2$OH, Heat → CH$_3$CHCH$_2$(Ph)-OCH$_2$CH$_3$ ($[\alpha]_D = -19.9$); and starting alcohol → K → [CH$_3$CHCH$_2$(Ph)-O$^-$ K$^+$] → CH$_3$CH$_2$Br → CH$_3$CHCH$_2$(Ph)-OCH$_2$CH$_3$ ($[\alpha]_D = +23.5$)]

11.27 ▲ Which compound in each of the following pairs will react faster in an S$_N$2 reaction with OH$^-$?
(a) CH$_3$Br or CH$_3$I
(b) CH$_3$CH$_2$I in ethanol or in dimethyl sulfoxide
(c) (CH$_3$)$_3$CCl or CH$_3$Cl
(d) H$_2$C=CHBr or H$_2$C=CHCH$_2$Br

11.28 Which reactant in each of the following pairs is more nucleophilic? Explain.
(a) $^-$NH$_2$ or NH$_3$ (b) H$_2$O or CH$_3$CO$_2^-$
(c) BF$_3$ or F$^-$ (d) (CH$_3$)$_3$P or (CH$_3$)$_3$N
(e) I$^-$ or Cl$^-$ (f) $^-$C≡N or $^-$OCH$_3$

11.29 What effect would you expect the following changes to have on the rate of the S$_N$2 reaction of 1-iodo-2-methylbutane with cyanide ion?
(a) The CN$^-$ concentration is halved, and the 1-iodo-2-methylbutane concentration is doubled.
(b) Both the CN$^-$ and the 1-iodo-2-methylbutane concentrations are tripled.

11.30 What effect would you expect the following changes to have on the rate of the reaction of ethanol with 2-iodo-2-methylbutane?
(a) The concentration of the halide is tripled.
(b) The concentration of the ethanol is halved by adding diethyl ether as an inert solvent.

▲ Problems linked to Key Ideas in this chapter

11.31 How might you prepare each of the following molecules using a nucleophilic substitution reaction at some step?

(a) CH$_3$C≡CCH(CH$_3$)CH$_3$
(b) CH$_3$—O—C(CH$_3$)$_3$
(c) CH$_3$CH$_2$CH$_2$CH$_2$CN
(d) CH$_3$CH$_2$CH$_2$NH$_2$

11.32 ▲ Which reaction in each of the following pairs would you expect to be faster?
(a) The S$_N$2 displacement by I$^-$ on CH$_3$Cl or on CH$_3$OTos
(b) The S$_N$2 displacement by CH$_3$CO$_2$$^-$ on bromoethane or on bromocyclohexane
(c) The S$_N$2 displacement on 2-bromopropane by CH$_3$CH$_2$O$^-$ or by CN$^-$
(d) The S$_N$2 displacement by HC≡C$^-$ on bromomethane in benzene or in acetonitrile

11.33 Predict the product and give the stereochemistry resulting from reaction of each of the following nucleophiles with (R)-2-bromooctane:
(a) $^-$CN (b) CH$_3$CO$_2$$^-$ (c) CH$_3$S$^-$

11.34 (R)-2-Bromooctane undergoes racemization to give (±)-2-bromooctane when treated with NaBr in dimethyl sulfoxide. Explain.

Elimination Reactions

11.35 Propose structures for compounds that fit the following descriptions:
(a) An alkyl halide that gives a mixture of three alkenes on E2 reaction
(b) An organohalide that will not undergo nucleophilic substitution
(c) An alkyl halide that gives the non-Zaitsev product on E2 reaction
(d) An alcohol that reacts rapidly with HCl at 0 °C

11.36 What products would you expect from the reaction of 1-bromopropane with each of the following?
(a) NaNH$_2$ (b) KOC(CH$_3$)$_3$ (c) NaI
(d) NaCN (e) NaC≡CH (f) Mg, then H$_2$O

11.37 1-Chloro-1,2-diphenylethane can undergo E2 elimination to give either cis- or trans-1,2-diphenylethylene (stilbene). Draw Newman projections of the reactive conformations leading to both possible products, and suggest a reason why the trans alkene is the major product.

Ph—CHCl—CH$_2$—Ph $\xrightarrow{^-OCH_3}$ Ph—CH=CH—Ph

1-Chloro-1,2-diphenylethane → trans-1,2-Diphenylethylene

11.38 Predict the major alkene product of the following E1 reaction:

(CH$_3$)$_2$CHCH(Br)CH$_2$CH$_3$ $\xrightarrow[\text{Heat}]{\text{HOAc}}$?

▲ Problems linked to Key Ideas in this chapter

11.39 There are eight diastereomers of 1,2,3,4,5,6-hexachlorocyclohexane. Draw each in its more stable chair conformation. One isomer loses HCl in an E2 reaction nearly 1000 times more slowly than the others. Which isomer reacts so slowly, and why?

General Problems

11.40 The reactions shown below are unlikely to occur as written. Tell what is wrong with each, and predict the actual product.

(a)
$$\text{CH}_3\text{CHBrCH}_2\text{CH}_3 \xrightarrow[\text{(CH}_3)_3\text{COH}]{\text{K}^+ {}^-\text{OC(CH}_3)_3} \text{CH}_3\text{CH(OC(CH}_3)_3)\text{CH}_2\text{CH}_3$$

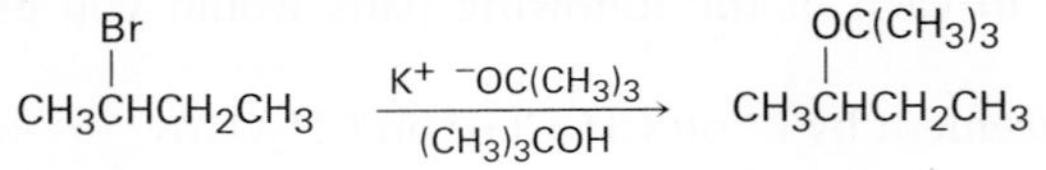

(b) cyclohexyl-F $\xrightarrow{\text{Na}^+ {}^-\text{OH}}$ cyclohexyl-OH

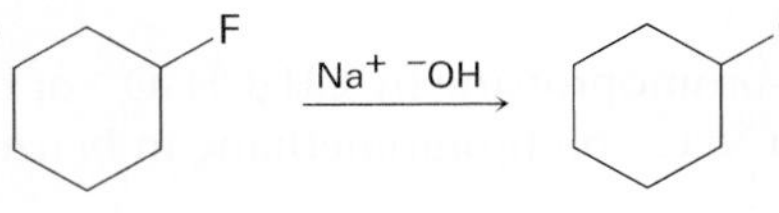

(c) 1-methylcyclohexan-1-ol $\xrightarrow[\text{Pyridine (a base)}]{\text{SOCl}_2}$ 1-chloro-1-methylcyclohexane

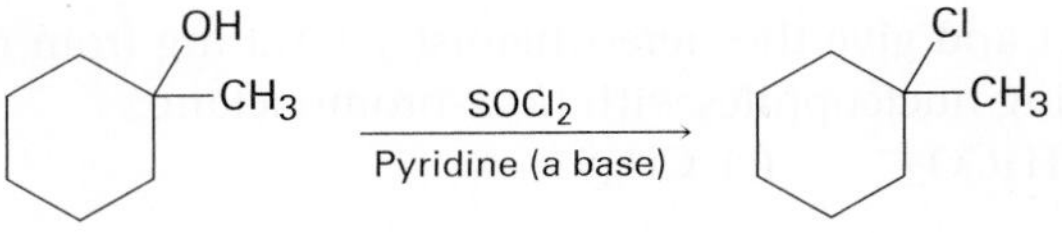

11.41 ▲ Order each of the following sets of compounds with respect to S_N1 reactivity:

(a) (CH₃)₃CCl, PhC(CH₃)₂Cl, CH₃CH₂CH(NH₂)CH₃

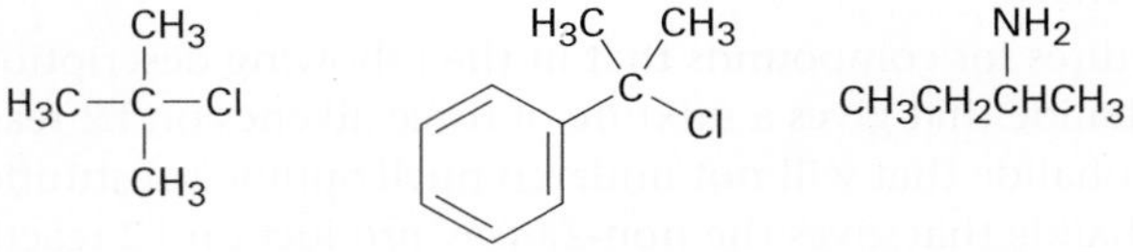

(b) (CH₃)₃CCl, (CH₃)₃CBr, (CH₃)₃COH

(c) PhCH₂Br, PhCHBrCH₃, (Ph)₃CBr

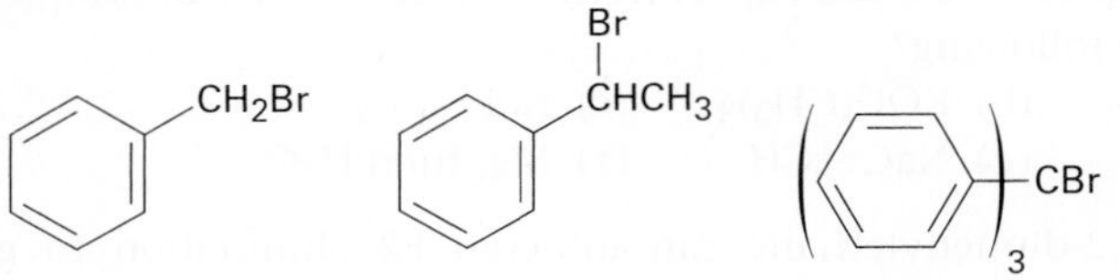

11.42 ▲ Order each of the following sets of compounds with respect to S_N2 reactivity:

(a) (CH₃)₃CCl, CH₃CH₂CH₂Cl, CH₃CH₂CHClCH₃

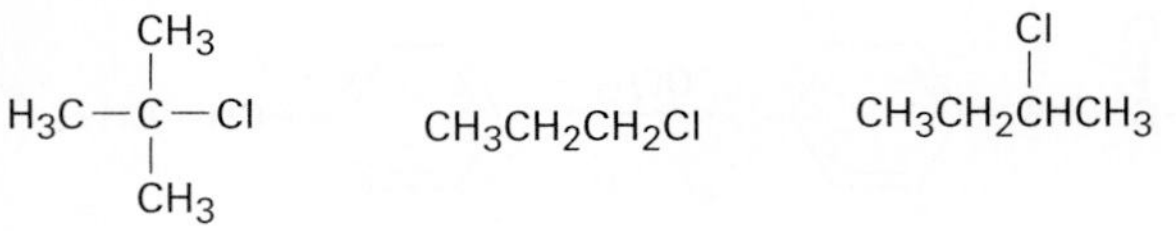

(b) CH₃CH(CH₃)CHBrCH₃, CH₃CH(CH₃)CH₂Br, (CH₃)₃CCH₂Br

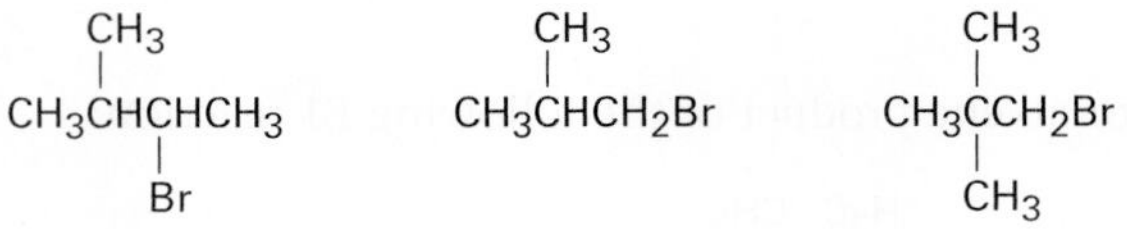

(c) CH₃CH₂CH₂OCH₃, CH₃CH₂CH₂OTos, CH₃CH₂CH₂Br

▲ Problems linked to Key Ideas in this chapter

11.43 Reaction of the following *S* tosylate with cyanide ion yields a nitrile product that also has *S* stereochemistry. Explain.

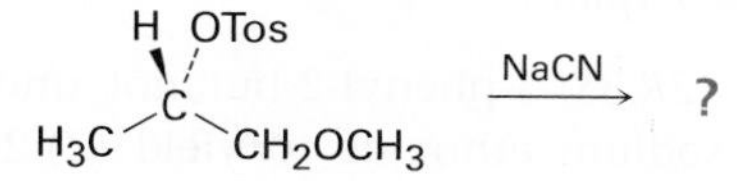

$$\underset{\textbf{(\textit{S} stereochemistry)}}{\overset{H_3C\diagup \overset{H\;\;OTos}{\underset{|\;\;\;/}{C}}\diagdown CH_2OCH_3}{}} \xrightarrow{\text{NaCN}} ?$$

11.44 Ethers can often be prepared by S_N2 reaction of alkoxide ions, RO⁻, with alkyl halides. Suppose you wanted to prepare cyclohexyl methyl ether. Which of the two possible routes shown below would you choose? Explain.

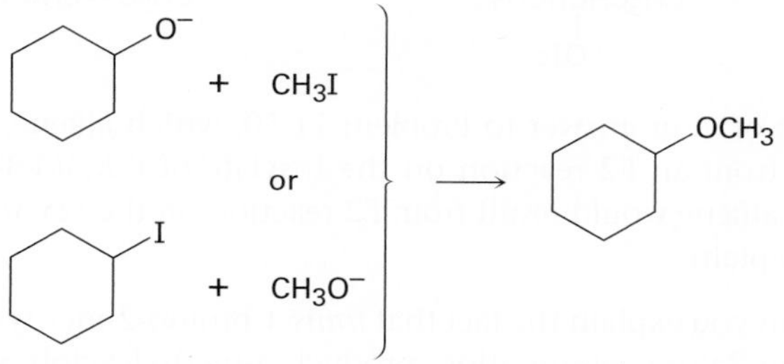

11.45 We saw in Section 8.7 that bromohydrins are converted into epoxides when treated with base. Propose a mechanism, using curved arrows to show the electron flow.

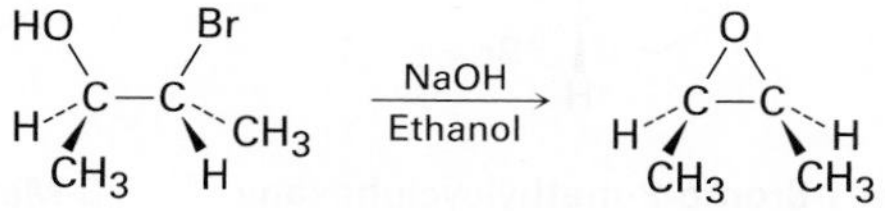

11.46 Show the stereochemistry of the epoxide (see Problem 11.45) you would obtain by formation of a bromohydrin from *trans*-2-butene, followed by treatment with base.

11.47 In light of your answer to Problem 11.45, what product might you expect from treatment of 4-bromo-1-butanol with base?

$$\text{BrCH}_2\text{CH}_2\text{CH}_2\text{CH}_2\text{OH} \xrightarrow{\text{Base}} ?$$

11.48 ▲ The following tertiary alkyl bromide does not undergo a nucleophilic substitution reaction by either S_N1 or S_N2 mechanisms. Explain.

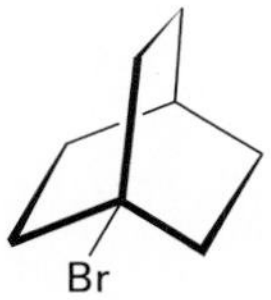

▲ Problems linked to Key Ideas in this chapter

11.49 In addition to not undergoing substitution reactions, the alkyl bromide shown in Problem 11.48 also fails to undergo an elimination reaction when treated with base. Explain.

11.50 The tosylate of (2R,3S)-3-phenyl-2-butanol undergoes E2 elimination on treatment with sodium ethoxide to yield (Z)-2-phenyl-2-butene. Explain, using Newman projections.

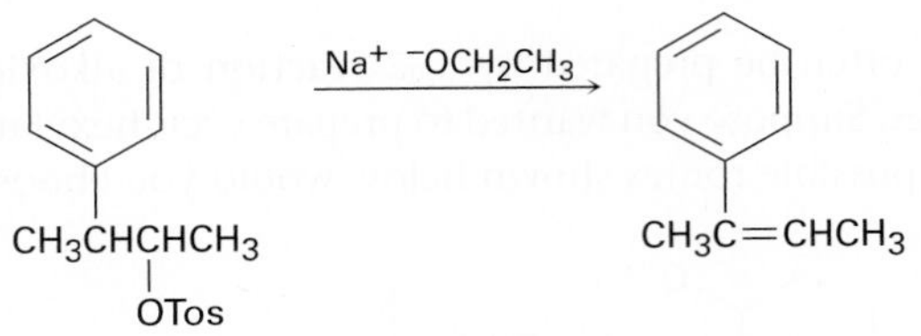

$$\text{CH}_3\text{CHCHCH}_3 \text{ (OTos)} \xrightarrow{\text{Na}^+ \text{ }^-\text{OCH}_2\text{CH}_3} \text{CH}_3\text{C}=\text{CHCH}_3$$

11.51 In light of your answer to Problem 11.50, which alkene, E or Z, would you expect from an E2 reaction on the tosylate of (2R,3R)-3-phenyl-2-butanol? Which alkene would result from E2 reaction on the (2S,3R) and (2S,3S) tosylates? Explain.

11.52 How can you explain the fact that *trans*-1-bromo-2-methylcyclohexane yields the non-Zaitsev elimination product 3-methylcyclohexene on treatment with base?

trans-1-Bromo-2-methylcyclohexane $\xrightarrow{\text{KOH}}$ 3-Methylcyclohexene

11.53 Predict the product(s) of the following reaction, indicating stereochemistry where necessary:

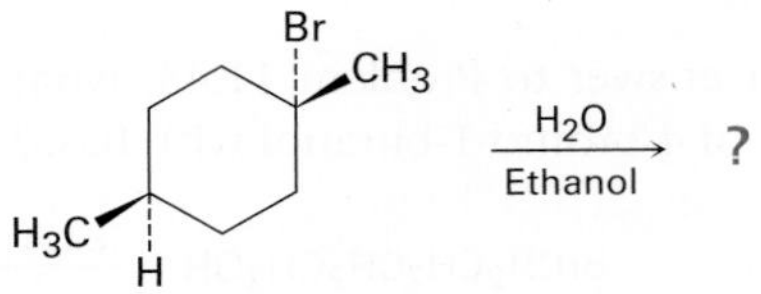

$\xrightarrow[\text{Ethanol}]{\text{H}_2\text{O}}$?

▲ Problems linked to Key Ideas in this chapter

11.54 Metabolism of *S*-adenosylhomocysteine (Section 11.6) involves the following sequence. Propose a mechanism for the second step.

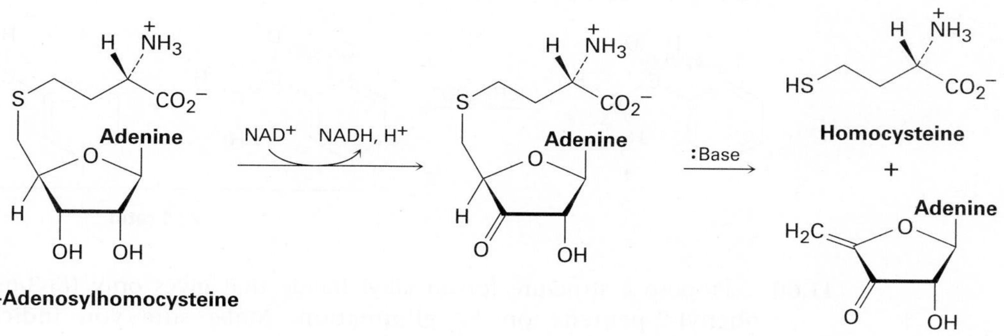

11.55 Reaction of iodoethane with CN^- yields a small amount of *isonitrile*, $CH_3CH_2N{\equiv}C$, along with the nitrile $CH_3CH_2C{\equiv}N$ as the major product. Write electron-dot structures for both products, assign formal charges as necessary, and propose mechanisms to account for their formation.

11.56 ▲ Alkynes can be made by dehydrohalogenation of vinylic halides in a reaction that is essentially an E2 process. In studying the stereochemistry of this elimination, it was found that (*Z*)-2-chloro-2-butenedioic acid reacts 50 times as fast as the corresponding *E* isomer. What conclusion can you draw about the stereochemistry of eliminations in vinylic halides? How does this result compare with eliminations of alkyl halides?

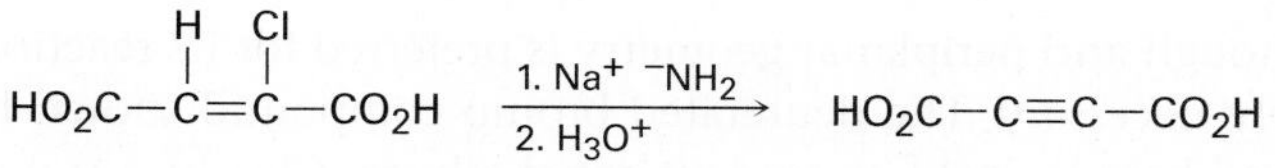

$$HO_2C-\underset{H}{\overset{|}{C}}=\underset{Cl}{\overset{|}{C}}-CO_2H \quad \xrightarrow[\text{2. } H_3O^+]{\text{1. Na}^+ \ ^-NH_2} \quad HO_2C-C{\equiv}C-CO_2H$$

11.57 (*S*)-2-Butanol slowly racemizes on standing in dilute sulfuric acid. Explain.

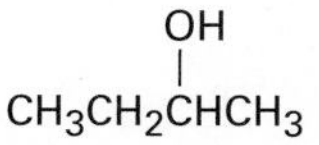

$$CH_3CH_2\overset{\overset{OH}{|}}{C}HCH_3 \quad \text{2-Butanol}$$

11.58 Reaction of HBr with (*R*)-3-methyl-3-hexanol leads to racemic 3-bromo-3-methylhexane. Explain.

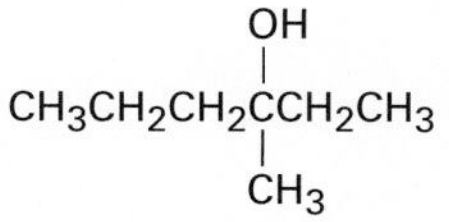

$$CH_3CH_2CH_2\underset{\underset{CH_3}{|}}{\overset{\overset{OH}{|}}{C}}CH_2CH_3 \quad \text{3-Methyl-3-hexanol}$$

▲ Problems linked to Key Ideas in this chapter

11.59 Treatment of 1-bromo-2-deuterio-2-phenylethane with strong base leads to a mixture of deuterated and nondeuterated phenylethylenes in an approximately 7:1 ratio. Explain.

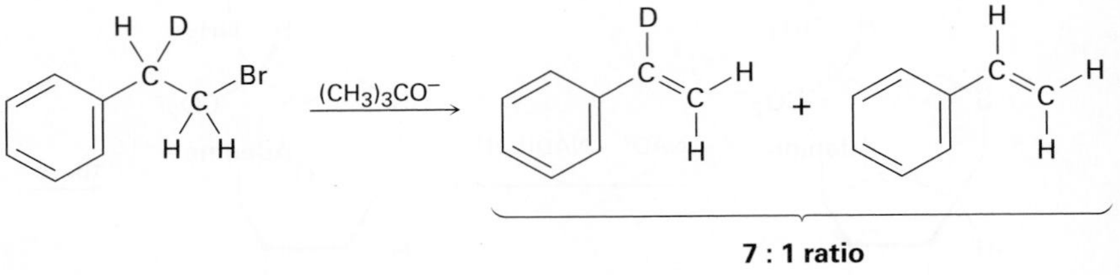

11.60 ▲ Propose a structure for an alkyl halide that gives only (*E*)-3-methyl-2-phenyl-2-pentene on E2 elimination. Make sure you indicate the stereochemistry.

11.61 One step in the urea cycle for ridding the body of ammonia is the conversion of argininosuccinate to the amino acid arginine plus fumarate. Propose a mechanism for the reaction, and show the structure of arginine.

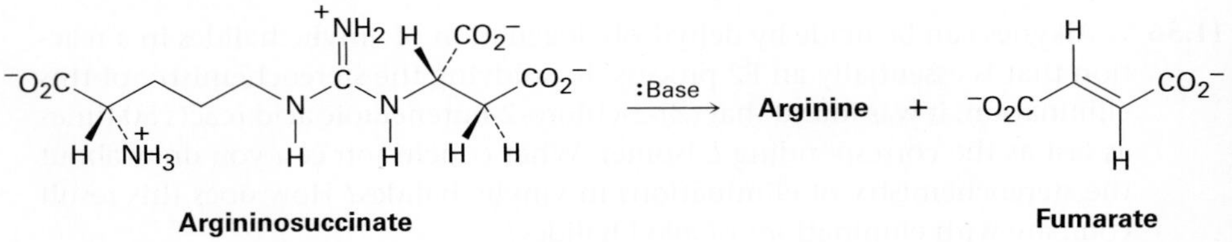

11.62 Although anti periplanar geometry is preferred for E2 reactions, it isn't absolutely necessary. The deuterated bromo compound shown here reacts with strong base to yield an undeuterated alkene. Clearly, a syn elimination has occurred. Make a molecular model of the reactant, and explain the result.

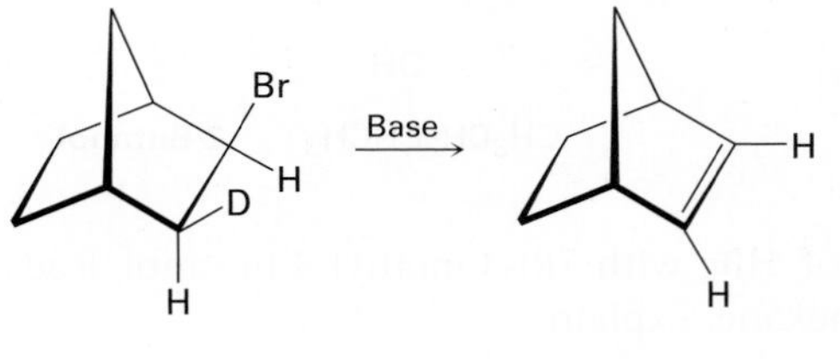

▲ Problems linked to Key Ideas in this chapter

11.63 In light of your answer to Problem 11.62, explain why one of the following isomers undergoes E2 reaction approximately 100 times as fast as the other. Which isomer is more reactive, and why?

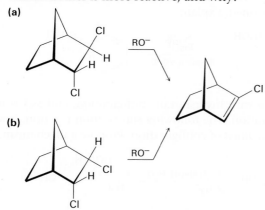

11.64 Methyl esters (RCO$_2$CH$_3$) undergo a cleavage reaction to yield carboxylate ions plus iodomethane on heating with LiI in dimethylformamide:

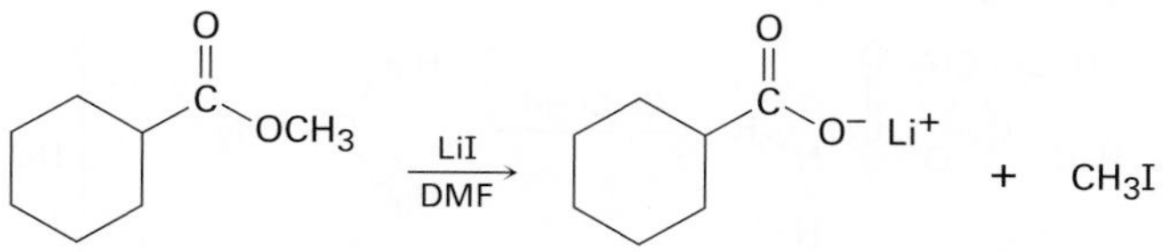

The following evidence has been obtained: (1) The reaction occurs much faster in DMF than in ethanol. (2) The corresponding ethyl ester (RCO$_2$CH$_2$CH$_3$) cleaves approximately 10 times more slowly than the methyl ester. Propose a mechanism for the reaction. What other kinds of experimental evidence could you gather to support your hypothesis?

11.65 The reaction of 1-chlorooctane with CH$_3$CO$_2^-$ to give octyl acetate is greatly accelerated by adding a small quantity of iodide ion. Explain.

11.66 Compound **X** is optically inactive and has the formula C$_{16}$H$_{16}$Br$_2$. On treatment with strong base, **X** gives hydrocarbon **Y**, C$_{16}$H$_{14}$. Compound **Y** absorbs 2 equivalents of hydrogen when reduced over a palladium catalyst and reacts with ozone to give two fragments. One fragment, **Z**, is an aldehyde with formula C$_7$H$_6$O. The other fragment is glyoxal, (CHO)$_2$. Write the reactions involved, and suggest structures for **X**, **Y**, and **Z**. What is the stereochemistry of **X**?

▲ Problems linked to Key Ideas in this chapter

11.67 When a primary alcohol is treated with *p*-toluenesulfonyl chloride at room temperature in the presence of an organic base such as pyridine, a tosylate is formed. When the same reaction is carried out at higher temperature, an alkyl chloride is often formed. Explain.

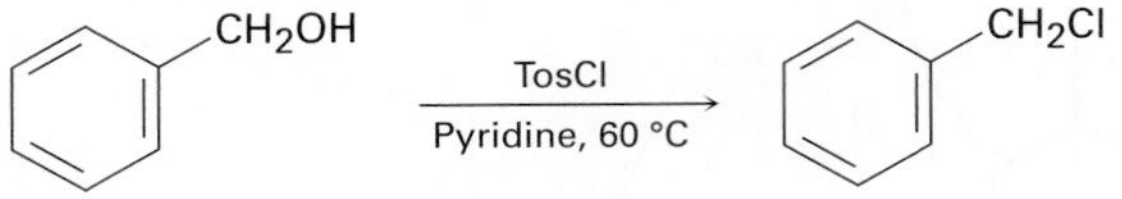

11.68 S_N2 reactions take place with inversion of configuration, and S_N1 reactions take place with racemization. The following substitution reaction, however, occurs with complete *retention* of configuration. Propose a mechanism.

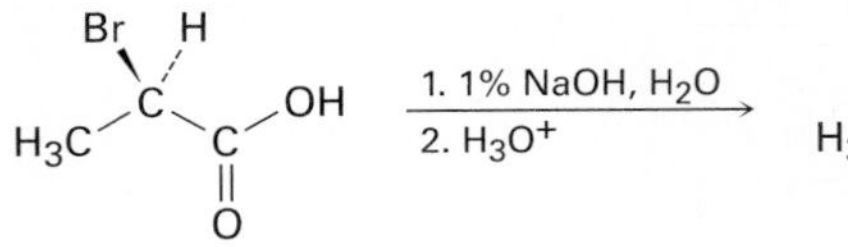

11.69 Propose a mechanism for the following reaction, an important step in the laboratory synthesis of proteins:

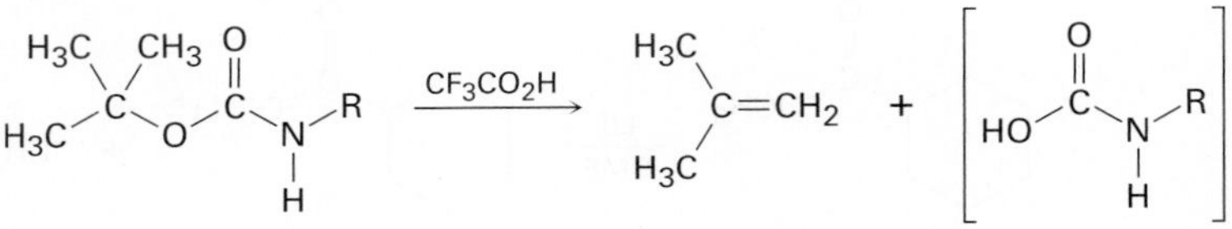

11.70 The amino acid methionine is formed by a methylation reaction of homocysteine with *N*-methyltetrahydrofolate. The stereochemistry of the reaction has been probed by carrying out the transformation using a donor with a "chiral methyl group" that contains protium (H), deuterium (D), and tritium (T) isotopes of hydrogen. Does the methylation reaction occur with inversion or retention of configuration?

▲ Problems linked to Key Ideas in this chapter

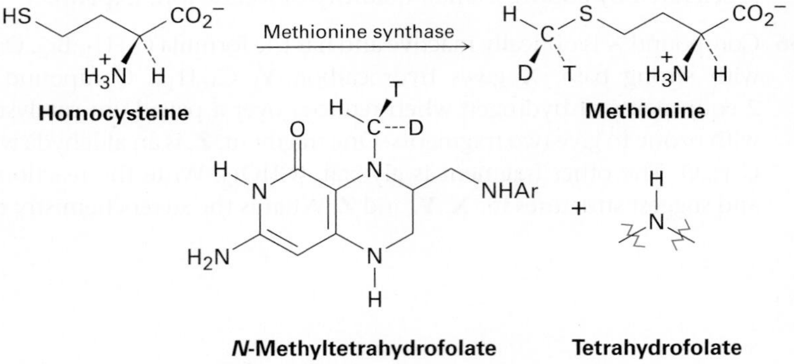

11.71 Amines are converted into alkenes by a two-step process called the *Hofmann elimination*. S$_N$2 reaction of the amine with an excess of CH$_3$I in the first step yields an intermediate that undergoes E2 reaction when treated with silver oxide as base. Pentylamine, for example, yields 1-pentene. Propose a structure for the intermediate, and explain why it undergoes ready elimination.

$$CH_3CH_2CH_2CH_2CH_2NH_2 \xrightarrow[\text{2. Ag}_2\text{O, H}_2\text{O}]{\text{1. Excess CH}_3\text{I}} CH_3CH_2CH_2CH=CH_2$$

11.72 The antipsychotic drug flupentixol is prepared by the following scheme:

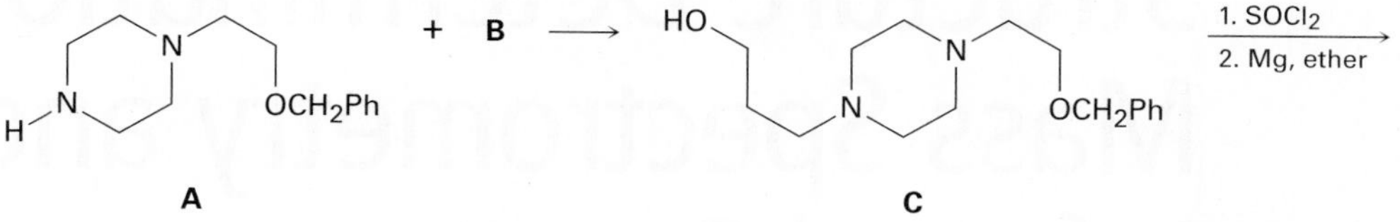

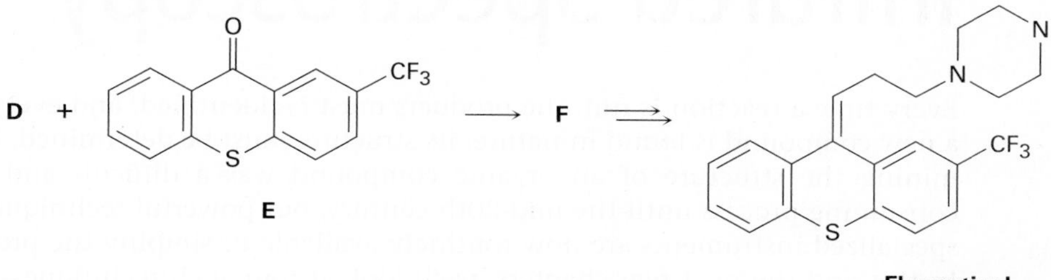

(a) What alkyl chloride **B** reacts with amine **A** to form **C**?
(b) Compound **C** is treated with SOCl$_2$, and the product is allowed to react with magnesium metal to give a Grignard reagent **D**. What is the structure of **D**?
(c) We'll see in Section 19.7 that Grignard reagents add to ketones, such as **E**, to give tertiary alcohols, such as **F**. Because of the newly formed chirality center, compound **F** exists as a pair of enantiomers. Draw both, and assign *R,S* configuration.
(d) Two stereoisomers of flupentixol are subsequently formed from **F**, but only one is shown. Draw the other isomer, and identify the type of stereoisomerism.

▲ Problems linked to Key Ideas in this chapter

12

More than a thousand different chemical compounds have been isolated from coffee. Their structures were determined using various spectroscopic techniques. © webphotographeer/iStockphoto

Structure Determination: Mass Spectrometry and Infrared Spectroscopy

12.1 Mass Spectrometry of Small Molecules: Magnetic-Sector Instruments
12.2 Interpreting Mass Spectra
12.3 Mass Spectrometry of Some Common Functional Groups
12.4 Mass Spectrometry in Biological Chemistry: Time-of-Flight (TOF) Instruments
12.5 Spectroscopy and the Electromagnetic Spectrum
12.6 Infrared Spectroscopy
12.7 Interpreting Infrared Spectra
12.8 Infrared Spectra of Some Common Functional Groups
  A Deeper Look—
  X-Ray Crystallography

Every time a reaction is run, the products must be identified, and every time a new compound is found in nature, its structure must be determined. Determining the structure of an organic compound was a difficult and time-consuming process until the mid-20th century, but powerful techniques and specialized instruments are now routinely available to simplify the problem. In this and the next two chapters, we'll look at four such techniques—mass spectrometry (MS), infrared (IR) spectroscopy, ultraviolet spectroscopy (UV), and nuclear magnetic resonance spectroscopy (NMR)—and we'll see the kind of information that can be obtained from each.

Mass spectrometry	What is the size and formula?
Infrared spectroscopy	What functional groups are present?
Ultraviolet spectroscopy	Is a conjugated π electron system present?
Nuclear magnetic resonance spectroscopy	What is the carbon–hydrogen framework?

Why This Chapter? Finding the structures of new molecules, whether small ones synthesized in the laboratory or large proteins and nucleic acids found in living organisms, is central to progress in chemistry and biochemistry. We can only scratch the surface of structure determination in this book, but after reading this and the following two chapters, you should have a good idea of the range of structural techniques available and of how and when each is used.

12.1 Mass Spectrometry of Small Molecules: Magnetic-Sector Instruments

At its simplest, **mass spectrometry (MS)** is a technique for measuring the mass, and therefore the molecular weight (MW), of a molecule. In addition,

OWL Sign in to OWL for Organic Chemistry at **www.cengage.com/owl** to view tutorials and simulations, develop problem-solving skills, and complete online homework assigned by your professor.

it's often possible to gain structural information about a molecule by measuring the masses of the fragments produced when molecules are broken apart.

More than 20 different kinds of commercial mass spectrometers are available depending on the intended application, but all have three basic parts: an *ionization source* in which sample molecules are given an electrical charge, a *mass analyzer* in which ions are separated by their mass-to-charge ratio, and a *detector* in which the separated ions are observed and counted.

Sample		Display
Ionization source	**Mass analyzer**	**Detector**
Electron impact (EI), or Electrospray ionization (ESI), or Matrix-assisted laser desorption ionization (MALDI)	Magnetic sector, or Time-of-flight (TOF), or Quadrupole (Q)	Photomultiplier, or Electron multiplier, or Micro-channel plate

Among the most common mass spectrometers used for routine purposes in the laboratory is the electron-impact, magnetic-sector instrument shown schematically in **Figure 12.1**. A small amount of sample is vaporized into the ionization source, where it is bombarded by a stream of high-energy electrons. The energy of the electron beam can be varied but is commonly around 70 electron volts (eV), or 6700 kJ/mol. When a high-energy electron strikes an organic molecule, it dislodges a valence electron from the molecule, producing a *cation radical*—*cation* because the molecule has lost an electron and now has a positive charge; *radical* because the molecule now has an odd number of electrons.

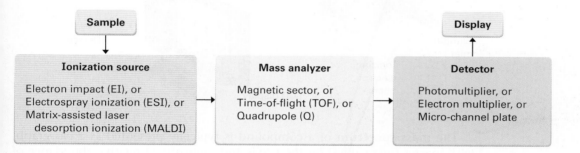

$$RH \xrightarrow{e^-} RH^{+\cdot} + e^-$$

Organic molecule → **Cation radical**

Electron bombardment transfers so much energy that most of the cation radicals fragment after formation. They fly apart into smaller pieces, some of which retain the positive charge and some of which are neutral. The fragments then flow through a curved pipe in a strong magnetic field, which deflects them into different paths according to their mass-to-charge ratio (m/z). Neutral fragments are not deflected by the magnetic field and are lost on the walls of the pipe, but positively charged fragments are sorted by the mass spectrometer onto a detector, which records them as peaks at the various m/z ratios. Since the number of charges z on each ion is usually 1, the value of m/z for each ion is simply its mass m. Masses up to approximately 2500 atomic mass units (amu) can be analyzed.

Figure 12.1 A representation of an electron-ionization, magnetic-sector mass spectrometer. Molecules are ionized by collision with high-energy electrons, causing some of the molecules to fragment. Passage of the charged fragments through a magnetic field then sorts them according to their mass.

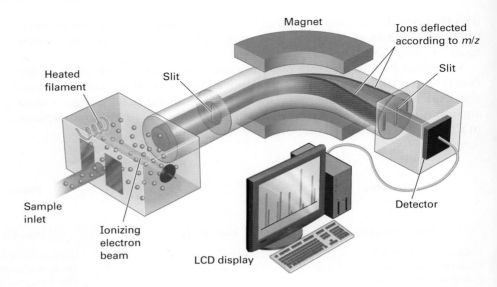

The **mass spectrum** of a compound is typically presented as a bar graph, with masses (m/z values) on the x axis and intensity, or relative abundance of ions of a given m/z striking the detector, on the y axis. The tallest peak, assigned an intensity of 100%, is called the **base peak**, and the peak that corresponds to the unfragmented cation radical is called the **parent peak**, or the *molecular ion* (M^+). **Figure 12.2** shows the mass spectrum of propane.

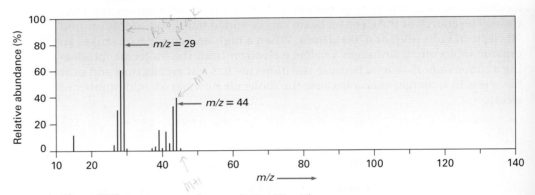

Figure 12.2 Mass spectrum of propane (C_3H_8; MW = 44).

Mass spectral fragmentation patterns are usually complex, and the molecular ion is often not the base peak. The mass spectrum of propane in Figure 12.2, for instance, shows a molecular ion at $m/z = 44$ that is only about 30% as high as the base peak at $m/z = 29$. In addition, many other fragment ions are present.

12.2 Interpreting Mass Spectra

What kinds of information can we get from a mass spectrum? The most obvious information is the molecular weight of the sample, which in itself can be invaluable. If we were given samples of hexane (MW = 86), 1-hexene (MW = 84), and 1-hexyne (MW = 82), for example, mass spectrometry would easily distinguish them.

Some instruments, called *double-focusing mass spectrometers*, have such high resolution that they provide exact mass measurements accurate to 5 ppm, or about 0.0005 amu, making it possible to distinguish between two formulas with the same nominal mass. For example, both C_5H_{12} and C_4H_8O have MW=72, but they differ slightly beyond the decimal point: C_5H_{12} has an exact mass of 72.0939 amu, whereas C_4H_8O has an exact mass of 72.0575 amu. A high-resolution instrument can easily distinguish between them. Note, however, that exact mass measurements refer to molecules with specific isotopic compositions. Thus, the sum of the exact atomic masses of the specific isotopes in a molecule is measured—1.00783 amu for ^1H, 12.00000 amu for ^{12}C, 14.00307 amu for ^{14}N, 15.99491 amu for ^{16}O, and so on—rather than the sum of the average atomic masses of elements as found on a periodic table.

Unfortunately, not every compound shows a molecular ion in its electron-impact mass spectrum. Although M^+ is usually easy to identify if it's abundant, some compounds, such as 2,2-dimethylpropane, fragment so easily that no molecular ion is observed **(Figure 12.3)**. In such cases, alternative "soft" ionization methods that do not use electron bombardment can prevent or minimize fragmentation.

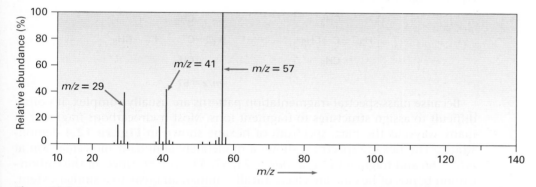

Figure 12.3 Mass spectrum of 2,2-dimethylpropane (C_5H_{12}; MW = 72). No molecular ion is observed when electron-impact ionization is used. What do you think is the formula and structure of the M^+ peak at m/z = 57?

Knowing the molecular weight makes it possible to narrow greatly the choices of molecular formula. For example, if the mass spectrum of an unknown compound shows a molecular ion at m/z = 110, the molecular formula is likely to be C_8H_{14}, $C_7H_{10}O$, $C_6H_6O_2$, or $C_6H_{10}N_2$. There are always a number of molecular formulas possible for all but the lowest molecular weights, and a computer can easily generate a list of choices.

A further point about mass spectrometry, noticeable in the spectra of both propane (Figure 12.2) and 2,2-dimethylpropane (Figure 12.3), is that the peak for the molecular ion is not at the highest m/z value. There is also a small peak at M+1 because of the presence of different isotopes in the molecules. Although ^{12}C is the most abundant carbon isotope, a small amount (1.10% natural abundance) of ^{13}C is also present. Thus, a certain percentage of the molecules analyzed in the mass spectrometer are likely to contain a ^{13}C atom, giving rise to the observed M+1 peak. In addition, a small amount of ^2H (deuterium; 0.015% natural abundance) is present, making a further contribution to the M+1 peak.

Mass spectrometry would be useful even if molecular weight and formula were the only information that could be obtained, but in fact we can get much more. For one thing, the mass spectrum of a compound serves as a kind of "molecular fingerprint." Each organic compound fragments in a unique way depending on its structure, and the likelihood of two compounds having identical mass spectra is small. Thus, it's sometimes possible to identify an unknown by computer-based matching of its mass spectrum to one of the more than 592,000 spectra recorded in a data base called the *Registry of Mass Spectral Data*.

It's also possible to derive structural information about a molecule by interpreting its fragmentation pattern. Fragmentation occurs when the high-energy cation radical flies apart by spontaneous cleavage of a chemical bond. One of the two fragments retains the positive charge and is a carbocation, while the other fragment is a neutral radical.

Not surprisingly, the positive charge often remains with the fragment that is best able to stabilize it. In other words, a relatively stable carbocation is often formed during fragmentation. For example, 2,2-dimethylpropane tends to fragment in such a way that the positive charge remains with the *tert*-butyl group. 2,2-Dimethylpropane therefore has a base peak at $m/z = 57$, corresponding to $C_4H_9^+$ (Figure 12.3).

$$\begin{bmatrix} & CH_3 & \\ H_3C-&\underset{\underset{CH_3}{|}}{\overset{\overset{CH_3}{|}}{C}}&-CH_3 \end{bmatrix}^{+\cdot} \longrightarrow H_3C-\underset{\underset{CH_3}{|}}{\overset{\overset{CH_3}{|}}{C^+}} + \cdot CH_3$$

$$m/z = 57$$

Because mass-spectral fragmentation patterns are usually complex, it's often difficult to assign structures to fragment ions. Most hydrocarbons fragment in many ways, as the mass spectrum of hexane shown in **Figure 12.4** demonstrates. The hexane spectrum shows a moderately abundant molecular ion at $m/z = 86$ and fragment ions at $m/z = 71, 57, 43$, and 29. Since all the carbon–carbon bonds of hexane are electronically similar, all break to a similar extent, giving rise to the observed mixture of ions.

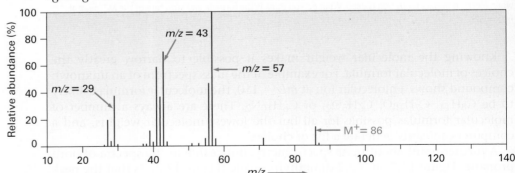

Figure 12.4 Mass spectrum of hexane (C_6H_{14}; MW = 86). The base peak is at $m/z = 57$, and numerous other ions are present.

Figure 12.5 shows how the hexane fragments might arise. The loss of a methyl radical from the hexane cation radical ($M^+ = 86$) gives rise to a fragment of mass 71; the loss of an ethyl radical accounts for a fragment of mass 57; the loss of a propyl radical accounts for a fragment of mass 43; and the loss of a butyl radical accounts for a fragment of mass 29. With practice, it's sometimes possible to analyze the fragmentation pattern of an unknown compound and work backward to a structure that is compatible with the data.

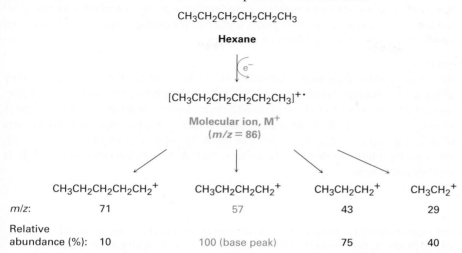

Figure 12.5 Fragmentation of hexane in a mass spectrometer.

We'll see in the next section and in later chapters that specific functional groups, such as alcohols, ketones, aldehydes, and amines, show specific kinds of mass spectral fragmentations that can be interpreted to provide structural information.

Using Mass Spectra to Identify Compounds

Worked Example 12.1

Assume that you have two unlabeled samples, one of methylcyclohexane and the other of ethylcyclopentane. How could you use mass spectrometry to tell them apart? The mass spectra of both are shown in **Figure 12.6**.

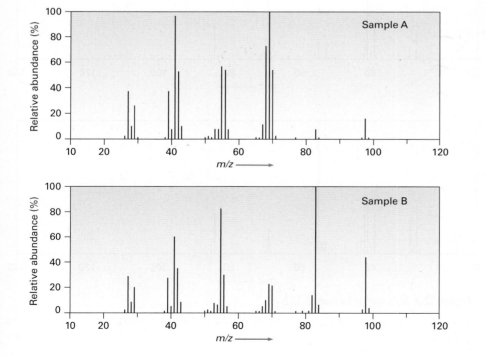

Figure 12.6 Mass spectra of unlabeled samples **A** and **B** for Worked Example 12.1.

Strategy

Look at the possible structures and decide on how they differ. Then think about how any of these differences in structure might give rise to differences in mass spectra. Methylcyclohexane, for instance, has a —CH₃ group, and ethylcyclopentane has a —CH₂CH₃ group, which should affect the fragmentation patterns.

Solution

Both mass spectra show molecular ions at $M^+ = 98$, corresponding to C_7H_{14}, but they differ in their fragmentation patterns. Sample **A** has its base peak at $m/z = 69$, corresponding to the loss of a CH_2CH_3 group (29 mass units), but **B** has a rather small peak at $m/z = 69$. Sample **B** shows a base peak at $m/z = 83$, corresponding to the loss of a CH_3 group (15 mass units), but sample **A** has only a small peak at $m/z = 83$. We can therefore be reasonably certain that **A** is ethylcyclopentane and **B** is methylcyclohexane.

Problem 12.1

The male sex hormone testosterone contains only C, H, and O and has a mass of 288.2089 amu as determined by high-resolution mass spectrometry. What is the likely molecular formula of testosterone?

Problem 12.2

Two mass spectra are shown in **Figure 12.7**. One spectrum is that of 2-methyl-2-pentene; the other is of 2-hexene. Which is which? Explain.

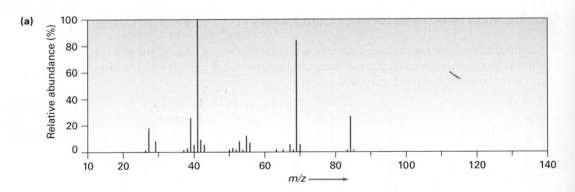

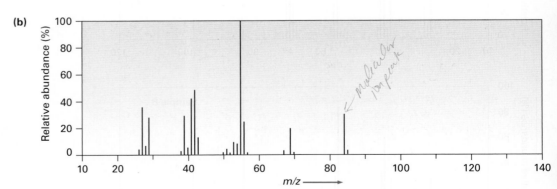

Figure 12.7 Mass spectra for Problem 12.2.

12.3 Mass Spectrometry of Some Common Functional Groups

As each functional group is discussed in future chapters, mass-spectral fragmentations characteristic of that group will be described. As a preview, though, we'll point out some distinguishing features of several common functional groups.

Alcohols

Alcohols undergo fragmentation in the mass spectrometer by two pathways: *alpha cleavage* and *dehydration*. In the α-cleavage pathway, a C–C bond nearest the hydroxyl group is broken, yielding a neutral radical plus a resonance-stabilized, oxygen-containing cation.

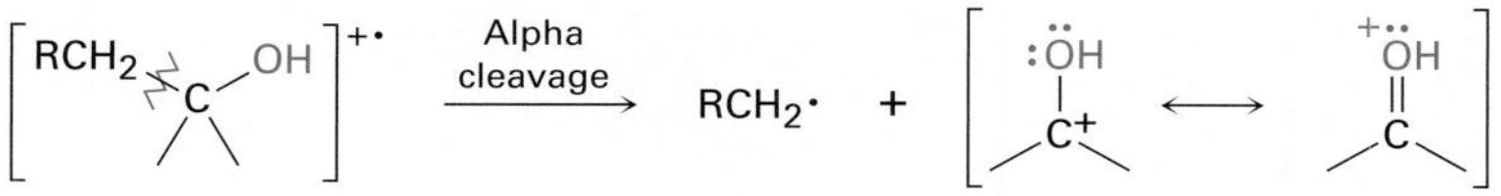

In the dehydration pathway, water is eliminated, yielding an alkene radical cation with a mass 18 units less than M^+.

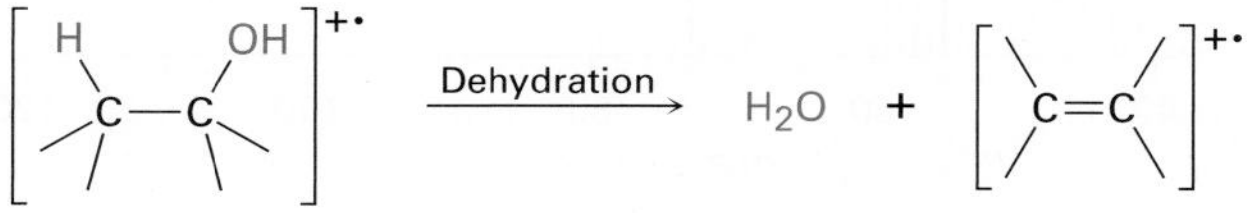

Amines

Aliphatic amines undergo a characteristic α cleavage in the mass spectrometer, similar to that observed for alcohols. A C–C bond nearest the nitrogen atom is broken, yielding an alkyl radical and a resonance-stabilized, nitrogen-containing cation.

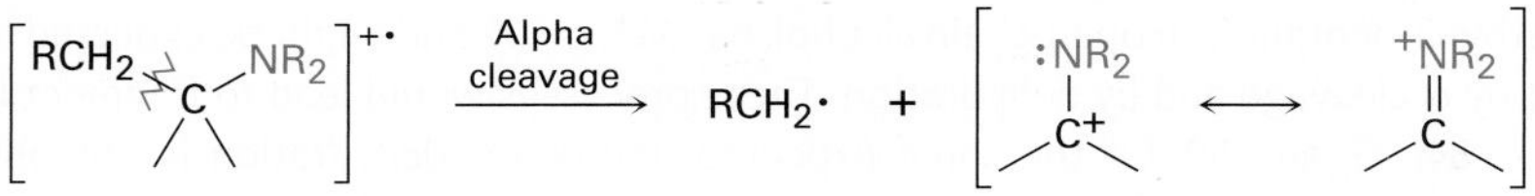

Carbonyl Compounds

Ketones and aldehydes that have a hydrogen on a carbon three atoms away from the carbonyl group undergo a characteristic mass-spectral cleavage called the *McLafferty rearrangement*. The hydrogen atom is transferred to the carbonyl oxygen, a C–C bond is broken, and a neutral alkene fragment is produced. The charge remains with the oxygen-containing fragment.

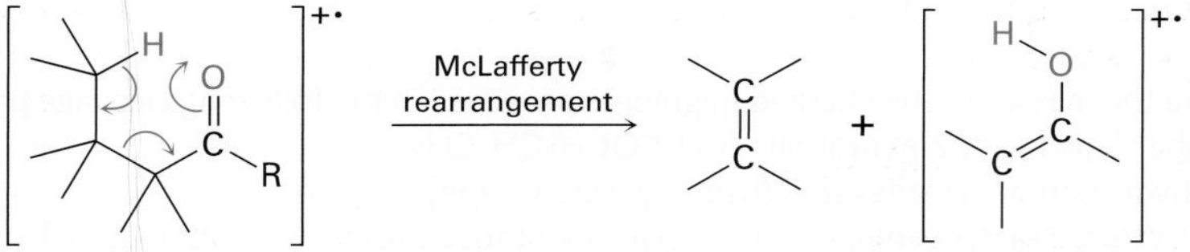

In addition, ketones and aldehydes frequently undergo α cleavage of the bond between the carbonyl group and the neighboring carbon. Alpha cleavage yields a neutral radical and a resonance-stabilized acyl cation.

Worked Example 12.2

Identifying Fragmentation Patterns in a Mass Spectrum

The mass spectrum of 2-methyl-3-pentanol is shown in **Figure 12.8**. What fragments can you identify?

Figure 12.8 Mass spectrum of 2-methyl-3-pentanol, Worked Example 12.2.

Strategy

Calculate the mass of the molecular ion, and identify the functional groups in the molecule. Then write the fragmentation processes you might expect, and compare the masses of the resultant fragments with those peaks present in the spectrum.

Solution

2-Methyl-3-pentanol, an open-chain alcohol, has $M^+ = 102$ and might be expected to fragment by α cleavage and by dehydration. These processes would lead to fragment ions of $m/z = 84$, 73, and 59. Of the three expected fragments, dehydration is not observed (no $m/z = 84$ peak), but both α cleavages take place ($m/z = 73, 59$).

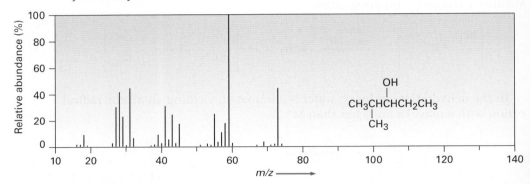

Problem 12.3

What are the masses of the charged fragments produced in the following cleavage pathways?
 (a) Alpha cleavage of 2-pentanone ($CH_3COCH_2CH_2CH_3$)
 (b) Dehydration of cyclohexanol (hydroxycyclohexane)
 (c) McLafferty rearrangement of 4-methyl-2-pentanone [$CH_3COCH_2CH(CH_3)_2$]
 (d) Alpha cleavage of triethylamine [$(CH_3CH_2)_3N$]

Problem 12.4
List the masses of the parent ion and of several fragments you might expect to find in the mass spectrum of the following molecule:

12.4 Mass Spectrometry in Biological Chemistry: Time-of-Flight (TOF) Instruments

Most biochemical analyses by MS use either electrospray ionization (ESI) or matrix-assisted laser desorption ionization (MALDI), typically linked to a time-of-flight (TOF) mass analyzer. Both ESI and MALDI are soft ionization methods that produce charged molecules with little fragmentation, even with biological samples of very high molecular weight.

In an ESI source, the sample is dissolved in a polar solvent and sprayed through a steel capillary tube. As it exits the tube, it is subjected to a high voltage that causes it to become protonated by removing one or more H^+ ions from the solvent. The volatile solvent is then evaporated, giving variably protonated sample molecules ($M+H_n^{n+}$). In a MALDI source, the sample is adsorbed onto a suitable matrix compound, such as 2,5-dihydroxybenzoic acid, which is ionized by a short burst of laser light. The matrix compound then transfers the energy to the sample and protonates it, forming $M+H_n^{n+}$ ions.

Following ion formation, the variably protonated sample molecules are electrically focused into a small packet with a narrow spatial distribution, and the packet is given a sudden kick of energy by an accelerator electrode. Since each molecule in the packet is given the same energy, $E = mv^2/2$, it begins moving with a velocity that depends on the square root of its mass, $v = \sqrt{2E/m}$. Lighter molecules move faster, and heavier molecules move slower. The analyzer itself—the *drift tube*—is simply an electrically grounded metal tube inside which the different charged molecules become separated as they move along at different velocities and take different amounts of time to complete their flight.

The TOF technique is considerably more sensitive than the magnetic sector alternative, and protein samples of up to 100 kilodaltons (100,000 amu) can be separated with a mass accuracy of 3 ppm. **Figure 12.9** shows a MALDI–TOF spectrum of chicken egg-white lysozyme, MW = 14,306.7578 daltons. (Biochemists generally use the unit *dalton*, abbreviated Da, instead of amu.)

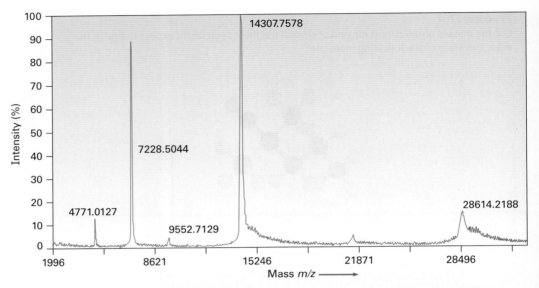

Figure 12.9 MALDI–TOF mass spectrum of chicken egg-white lysozyme. The peak at 14,307.7578 daltons (amu) is due to the monoprotonated protein, $M+H^+$, and the peak at 28,614.2188 daltons is due to an impurity formed by dimerization of the protein. Other peaks are various protonated species, $M+H_n^{n+}$.

12.5 Spectroscopy and the Electromagnetic Spectrum

Infrared, ultraviolet, and nuclear magnetic resonance spectroscopies differ from mass spectrometry in that they are nondestructive and involve the interaction of molecules with electromagnetic energy rather than with an ionizing source. Before beginning a study of these techniques, however, let's briefly review the nature of radiant energy and the electromagnetic spectrum.

Visible light, X rays, microwaves, radio waves, and so forth are all different kinds of *electromagnetic radiation*. Collectively, they make up the **electromagnetic spectrum**, shown in **Figure 12.10**. The electromagnetic spectrum is arbitrarily divided into regions, with the familiar visible region accounting for only a small portion, from 3.8×10^{-7} m to 7.8×10^{-7} m in wavelength. The visible region is flanked by the infrared and ultraviolet regions.

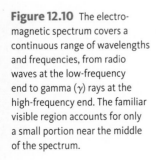

Figure 12.10 The electromagnetic spectrum covers a continuous range of wavelengths and frequencies, from radio waves at the low-frequency end to gamma (γ) rays at the high-frequency end. The familiar visible region accounts for only a small portion near the middle of the spectrum.

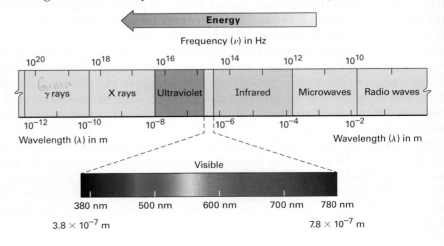

Electromagnetic radiation is often said to have dual behavior. In some respects, it has the properties of a particle, called a *photon*, yet in other respects it behaves as an energy wave. Like all waves, electromagnetic radiation is characterized by a *wavelength*, a *frequency*, and an *amplitude* (**Figure 12.11**). The **wavelength**, λ (Greek lambda), is the distance from one wave maximum to the next. The **frequency**, ν (Greek nu), is the number of waves that pass by a fixed point per unit time, usually given in reciprocal seconds (s^{-1}), or **hertz, Hz** (1 Hz = 1 s^{-1}). The **amplitude** is the height of a wave, measured from midpoint to peak. The intensity of radiant energy, whether a feeble glow or a blinding glare, is proportional to the square of the wave's amplitude.

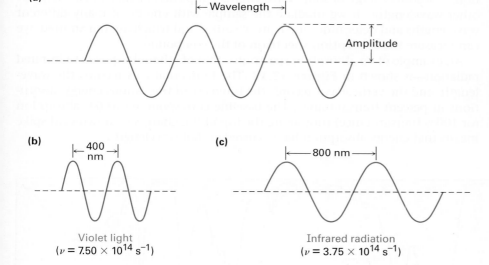

Figure 12.11 Electromagnetic waves are characterized by a wavelength, a frequency, and an amplitude. **(a)** Wavelength (λ) is the distance between two successive wave maxima. Amplitude is the height of the wave measured from the center. **(b)–(c)** What we perceive as different kinds of electromagnetic radiation are simply waves with different wavelengths and frequencies.

Multiplying the wavelength of a wave in meters (m) by its frequency in reciprocal seconds (s^{-1}) gives the speed of the wave in meters per second (m/s). The rate of travel of all electromagnetic radiation in a vacuum is a constant value, commonly called the "speed of light" and abbreviated c. Its numerical value is defined as exactly $2.997\,924\,58 \times 10^8$ m/s, usually rounded off to 3.00×10^8 m/s.

$$\text{Wavelength} \times \text{Frequency} = \text{Speed}$$
$$\lambda \text{ (m)} \times \nu \text{ (s}^{-1}\text{)} = c \text{ (m/s)}$$
$$\lambda = \frac{c}{\nu} \quad \text{or} \quad \nu = \frac{c}{\lambda}$$

Just as matter comes only in discrete units called atoms, electromagnetic energy is transmitted only in discrete amounts called *quanta*. The amount of energy ϵ corresponding to 1 quantum of energy (1 photon) of a given frequency ν is expressed by the Planck equation

$$\varepsilon = h\nu = \frac{hc}{\lambda}$$

where h = Planck's constant (6.62×10^{-34} J · s = 1.58×10^{-34} cal · s).

The Planck equation says that the energy of a given photon varies directly with its frequency ν but inversely with its wavelength λ. High frequencies and short wavelengths correspond to high-energy radiation such as gamma rays; low frequencies and long wavelengths correspond to low-energy radiation such as radio waves. Multiplying ϵ by Avogadro's number N_A gives the same equation in more familiar units, where E represents the energy of Avogadro's number (one "mole") of photons of wavelength λ:

$$E = \frac{N_A hc}{\lambda} = \frac{1.20 \times 10^{-4} \text{ kJ/mol}}{\lambda \text{ (m)}} \quad \text{or} \quad \frac{2.86 \times 10^{-5} \text{ kcal/mol}}{\lambda \text{ (m)}}$$

When an organic compound is exposed to a beam of electromagnetic radiation, it absorbs energy of some wavelengths but passes, or transmits, energy of other wavelengths. If we irradiate the sample with energy of many different wavelengths and determine which are absorbed and which are transmitted, we can measure the **absorption spectrum** of the compound.

An example of an absorption spectrum—that of ethanol exposed to infrared radiation—is shown in **Figure 12.12**. The horizontal axis records the wavelength, and the vertical axis records the intensity of the various energy absorptions in percent transmittance. The baseline corresponding to 0% absorption (or 100% transmittance) runs along the top of the chart, so a downward spike means that energy absorption has occurred at that wavelength.

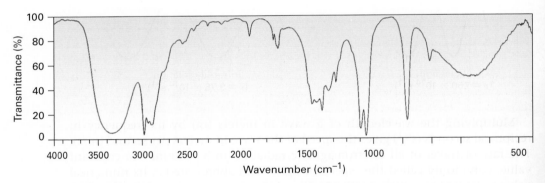

Figure 12.12 An infrared absorption spectrum of ethanol, CH_3CH_2OH. A transmittance of 100% means that all the energy is passing through the sample, whereas a lower transmittance means that some energy is being absorbed. Thus, each downward spike corresponds to an energy absorption.

The energy a molecule gains when it absorbs radiation must be distributed over the molecule in some way. With infrared radiation, the absorbed energy causes bonds to stretch and bend more vigorously. With ultraviolet radiation, the energy causes an electron to jump from a lower-energy orbital to a higher-energy one. Different radiation frequencies affect molecules in different ways, but each provides structural information when the results are interpreted.

There are many kinds of spectroscopies, which differ according to the region of the electromagnetic spectrum used. We'll look at three: infrared spectroscopy, ultraviolet spectroscopy, and nuclear magnetic resonance spectroscopy. Let's begin by seeing what happens when an organic sample absorbs infrared energy.

Worked Example 12.3

Correlating Energy and Frequency of Radiation

Which is higher in energy, FM radio waves with a frequency of 1.015×10^8 Hz (101.5 MHz) or visible green light with a frequency of 5×10^{14} Hz?

Strategy
Remember the equations $\epsilon = h\nu$ and $\epsilon = hc/\lambda$, which say that energy increases as frequency increases and as wavelength decreases.

Solution
Since visible light has a higher frequency than radio waves, it is higher in energy.

Problem 12.5
Which has higher energy, infrared radiation with $\lambda = 1.0 \times 10^{-6}$ m or an X ray with $\lambda = 3.0 \times 10^{-9}$ m? Radiation with $\nu = 4.0 \times 10^9$ Hz or with $\lambda = 9.0 \times 10^{-6}$ m?

Problem 12.6
It's useful to develop a feeling for the amounts of energy that correspond to different parts of the electromagnetic spectrum. Calculate the energies in kJ/mol of each of the following kinds of radiation:

(a) A gamma ray with $\lambda = 5.0 \times 10^{-11}$ m
(b) An X ray with $\lambda = 3.0 \times 10^{-9}$ m
(c) Ultraviolet light with $\nu = 6.0 \times 10^{15}$ Hz
(d) Visible light with $\nu = 7.0 \times 10^{14}$ Hz
(e) Infrared radiation with $\lambda = 2.0 \times 10^{-5}$ m
(f) Microwave radiation with $\nu = 1.0 \times 10^{11}$ Hz

12.6 Infrared Spectroscopy

The **infrared (IR)** region of the electromagnetic spectrum covers the range from just above the visible (7.8×10^{-7} m) to approximately 10^{-4} m, but only the midportion from 2.5×10^{-6} m to 2.5×10^{-5} m is used by organic chemists (**Figure 12.13**). Wavelengths within the IR region are usually given in micrometers (1 μm = 10^{-6} m), and frequencies are given in wavenumbers rather than in hertz. The **wavenumber** ($\tilde{\nu}$) is the reciprocal of the wavelength in centimeters and is therefore expressed in units of cm^{-1}.

$$\text{Wavenumber:} \quad \tilde{\nu} \, (\text{cm}^{-1}) = \frac{1}{\lambda \, (\text{cm})}$$

Thus, the useful IR region is from 4000 to 400 cm^{-1}, corresponding to energies of 48.0 kJ/mol to 4.80 kJ/mol (11.5–1.15 kcal/mol).

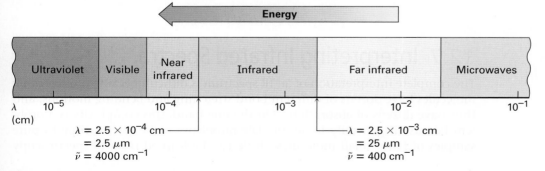

Figure 12.13 The infrared and adjacent regions of the electromagnetic spectrum.

Why does an organic molecule absorb some wavelengths of IR radiation but not others? All molecules have a certain amount of energy and are in constant motion. Their bonds stretch and contract, atoms wag back and forth, and other molecular vibrations occur. Some of the kinds of allowed vibrations are shown below:

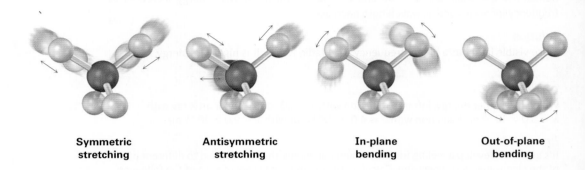

| Symmetric stretching | Antisymmetric stretching | In-plane bending | Out-of-plane bending |

The amount of energy a molecule contains is not continuously variable but is *quantized*. That is, a molecule can stretch or bend only at specific frequencies corresponding to specific energy levels. Take bond stretching, for example. Although we usually speak of bond lengths as if they were fixed, the numbers given are really averages. In fact, a typical C–H bond with an average bond length of 110 pm is actually vibrating at a specific frequency, alternately stretching and contracting as if there were a spring connecting the two atoms.

When a molecule is irradiated with electromagnetic radiation, energy is absorbed if the frequency of the radiation matches the frequency of the vibration. The result of this energy absorption is an increased amplitude for the vibration; in other words, the "spring" connecting the two atoms stretches and compresses a bit further. Since each frequency absorbed by a molecule corresponds to a specific molecular motion, we can find what kinds of motions a molecule has by measuring its IR spectrum. By then interpreting those motions, we can find out what kinds of bonds (functional groups) are present in the molecule.

IR spectrum → What molecular motions? → What functional groups?

12.7 Interpreting Infrared Spectra

The complete interpretation of an IR spectrum is difficult because most organic molecules have dozens of different bond stretching and bending motions, and thus have dozens of absorptions. On the one hand, this complexity is a problem because it generally limits the laboratory use of IR spectroscopy to pure samples of fairly small molecules—little can be learned from IR spectroscopy about large, complex biomolecules. On the other hand, the complexity is useful because an IR spectrum acts as a unique fingerprint of a compound. In fact, the complex region of the IR spectrum from 1500 cm^{-1} to around 400 cm^{-1} is called the *fingerprint region*. If two samples have identical IR spectra, they are almost certainly identical compounds.

Fortunately, we don't need to interpret an IR spectrum fully to get useful structural information. Most functional groups have characteristic IR absorption bands that don't change from one compound to another. The C=O absorption of a ketone is almost always in the range 1680 to 1750 cm^{-1}; the O–H absorption of an alcohol is almost always in the range 3400 to 3650 cm^{-1}; the C=C absorption of an alkene is almost always in the range 1640 to 1680 cm^{-1}; and so forth. By learning where characteristic functional-group absorptions occur, it's possible to get structural information from IR spectra. Table 12.1 lists the characteristic IR bands of some common functional groups.

Table 12.1 Characteristic IR Absorptions of Some Functional Groups

Functional Group	Absorption (cm^{-1})	Intensity	Functional Group	Absorption (cm^{-1})	Intensity
Alkane			Amine		
C–H	2850–2960	Medium	N–H	3300–3500	Medium
Alkene			C–N	1030–1230	Medium
=C–H	3020–3100	Medium	Carbonyl compound		
C=C	1640–1680	Medium	C=O	1670–1780	Strong
Alkyne			Aldehyde	1730	Strong
≡C–H	3300	Strong	Ketone	1715	Strong
C≡C	2100–2260	Medium	Ester	1735	Strong
Alkyl halide			Amide	1690	Strong
C–Cl	600–800	Strong	Carboxylic acid	1710	Strong
C–Br	500–600	Strong	Carboxylic acid		
Alcohol			O–H	2500–3100	Strong, broad
O–H	3400–3650	Strong, broad	Nitrile		
C–O	1050–1150	Strong	C≡N	2210–2260	Medium
Arene			Nitro		
C–H	3030	Weak	NO$_2$	1540	Strong
Aromatic ring	1660–2000	Weak			
	1450–1600	Medium			

Look at the IR spectra of hexane, 1-hexene, and 1-hexyne in **Figure 12.14** to see an example of how IR spectroscopy can be used. Although all three IR spectra contain many peaks, there are characteristic absorptions of the C=C and C≡C functional groups that allow the three compounds to be distinguished. Thus, 1-hexene shows a characteristic C=C absorption at 1660 cm^{-1} and a vinylic =C–H absorption at 3100 cm^{-1}, whereas 1-hexyne has a C≡C absorption at 2100 cm^{-1} and a terminal alkyne ≡C–H absorption at 3300 cm^{-1}.

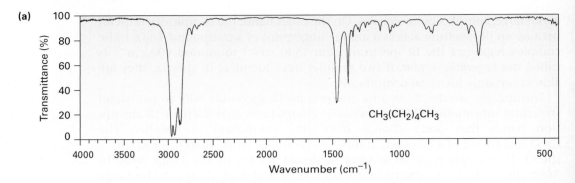

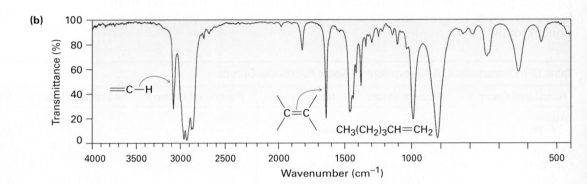

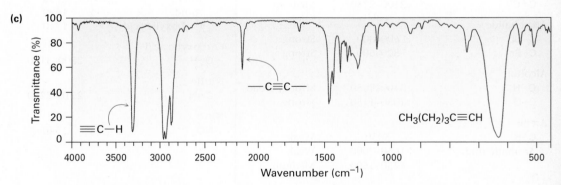

Figure 12.14 IR spectra of **(a)** hexane, **(b)** 1-hexene, and **(c)** 1-hexyne. Spectra like these are easily obtained on submilligram amounts of material in a few minutes using commercially available instruments.

It helps in remembering the position of specific IR absorptions to divide the IR region from 4000 cm^{-1} to 400 cm^{-1} into four parts, as shown in **Figure 12.15**.

- The region from 4000 to 2500 cm^{-1} corresponds to absorptions caused by N–H, C–H, and O–H single-bond stretching motions. N–H and O–H bonds absorb in the 3300 to 3600 cm^{-1} range; C–H bond stretching occurs near 3000 cm^{-1}.

- The region from 2500 to 2000 cm^{-1} is where triple-bond stretching occurs. Both C≡N and C≡C bonds absorb here.
- The region from 2000 to 1500 cm^{-1} is where double bonds (C=O, C=N, and C=C) absorb. Carbonyl groups generally absorb in the range 1680 to 1750 cm^{-1}, and alkene stretching normally occurs in the narrow range 1640 to 1680 cm^{-1}.
- The region below 1500 cm^{-1} is the fingerprint portion of the IR spectrum. A large number of absorptions due to a variety of C–C, C–O, C–N, and C–X single-bond vibrations occur here.

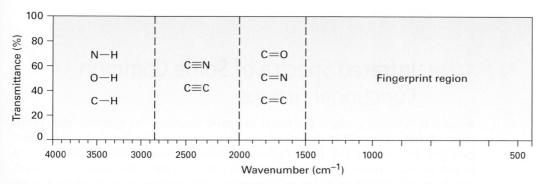

Figure 12.15 The four regions of the infrared spectrum: single bonds to hydrogen, triple bonds, double bonds, and fingerprint.

Why do different functional groups absorb where they do? As noted previously, a good analogy is that of two weights (atoms) connected by a spring (a bond). Short, strong bonds vibrate at a higher energy and higher frequency than do long, weak bonds, just as a short, strong spring vibrates faster than a long, weak spring. Thus, triple bonds absorb at a higher frequency than double bonds, which in turn absorb at a higher frequency than single bonds. In addition, springs connecting small weights vibrate faster than springs connecting large weights. Thus, C–H, O–H, and N–H bonds vibrate at a higher frequency than bonds between heavier C, O, and N atoms.

Distinguishing Isomeric Compounds by IR Spectroscopy

Worked Example 12.4

Acetone (CH$_3$COCH$_3$) and 2-propen-1-ol (H$_2$C=CHCH$_2$OH) are isomers. How could you distinguish them by IR spectroscopy?

Strategy
Identify the functional groups in each molecule, and refer to Table 12.1.

Solution
Acetone has a strong C=O absorption at 1715 cm^{-1}, while 2-propen-1-ol has an –OH absorption at 3500 cm^{-1} and a C=C absorption at 1660 cm^{-1}.

Problem 12.7
What functional groups might the following molecules contain?
(a) A compound with a strong absorption at 1710 cm^{-1}
(b) A compound with a strong absorption at 1540 cm^{-1}
(c) A compound with strong absorptions at 1720 cm^{-1} and at 2500–3100 cm^{-1}

Problem 12.8
How might you use IR spectroscopy to distinguish between the following pairs of isomers?
(a) CH_3CH_2OH and CH_3OCH_3 (b) Cyclohexane and 1-hexene
(c) $CH_3CH_2CO_2H$ and $HOCH_2CH_2CHO$

12.8 Infrared Spectra of Some Common Functional Groups

As each functional group is discussed in future chapters, the spectroscopic properties of that group will be described. For the present, we'll point out some distinguishing features of the hydrocarbon functional groups already studied and briefly preview some other common functional groups. We should also point out, however, that in addition to interpreting absorptions that *are* present in an IR spectrum, it's also possible to get structural information by noticing which absorptions are *not* present. If the spectrum of a compound has no absorptions at 3300 and 2150 cm^{-1}, the compound is not a terminal alkyne; if the spectrum has no absorption near 3400 cm^{-1}, the compound is not an alcohol; and so on.

Alkanes

The IR spectrum of an alkane is fairly uninformative because no functional groups are present and all absorptions are due to C–H and C–C bonds. Alkane C–H bonds show a strong absorption from 2850 to 2960 cm^{-1}, and saturated C–C bonds show a number of bands in the 800 to 1300 cm^{-1} range. Since most organic compounds contain saturated alkane-like portions, most organic compounds have these characteristic IR absorptions. The C–H and C–C bands are clearly visible in the three spectra shown in Figure 12.14.

Alkanes   —C—H   2850–2960 cm^{-1}

—C—C—   800–1300 cm^{-1}

Alkenes

Alkenes show several characteristic stretching absorptions. Vinylic =C–H bonds absorb from 3020 to 3100 cm^{-1}, and alkene C=C bonds usually absorb near 1650 cm^{-1}, although in some cases the peaks can be rather small and difficult to see clearly. Both absorptions are visible in the 1-hexene spectrum in Figure 12.14b.

Monosubstituted and disubstituted alkenes have characteristic =C−H out-of-plane bending absorptions in the 700 to 1000 cm^{-1} range, thereby allowing the substitution pattern on a double bond to be determined. Monosubstituted alkenes such as 1-hexene show strong characteristic bands at 910 and 990 cm^{-1}, and 2,2-disubstituted alkenes ($R_2C=CH_2$) have an intense band at 890 cm^{-1}.

Alkenes =C−H 3020–3100 cm^{-1}

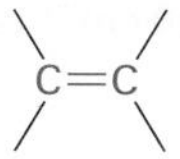

C=C 1640–1680 cm^{-1}

RCH=CH$_2$ 910 and 990 cm^{-1}

R$_2$C=CH$_2$ 890 cm^{-1}

Alkynes

Alkynes show a C≡C stretching absorption at 2100 to 2260 cm^{-1}, an absorption that is much more intense for terminal alkynes than for internal alkynes. In fact, symmetrically substituted triple bonds like that in 3-hexyne show no absorption at all, for reasons we won't go into. Terminal alkynes such as 1-hexyne also have a characteristic ≡C−H stretch at 3300 cm^{-1} (Figure 12.14c). This band is diagnostic for terminal alkynes because it is fairly intense and quite sharp.

Alkynes —C≡C— 2100–2260 cm^{-1}

≡C−H 3300 cm^{-1}

Aromatic Compounds

Aromatic compounds, such as benzene, have a weak C−H stretching absorption at 3030 cm^{-1}, just to the left of a typical saturated C−H band. In addition, they have a series of weak absorptions in the 1660 to 2000 cm^{-1} range and a series of medium-intensity absorptions in the 1450 to 1600 cm^{-1} region. These latter absorptions are due to complex molecular motions of the entire ring. The IR spectrum of phenylacetylene, shown in Figure 12.17 at the end of this section, gives an example.

Aromatic compounds C−H 3030 cm^{-1} (weak)

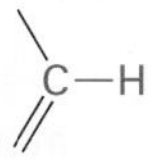

1660–2000 cm^{-1} (weak)
1450–1600 cm^{-1} (medium)

Alcohols

The O−H functional group of alcohols is easy to spot. Alcohols have a characteristic band in the range 3400 to 3650 cm^{-1} that is usually broad and intense. If present, it's hard to miss this band or to confuse it with anything else.

Alcohols —O−H 3400–3650 cm^{-1} (broad, intense)

Amines

The N–H functional group of amines is also easy to spot in the IR, with a characteristic absorption in the 3300 to 3500 cm^{-1} range. Although alcohols absorb in the same range, an N–H absorption is much sharper and less intense than an O–H band.

Amines   —N—H   3300–3500 cm^{-1} (sharp, medium intensity)

Carbonyl Compounds

Carbonyl functional groups are the easiest to identify of all IR absorptions because of their sharp, intense peak in the range 1670 to 1780 cm^{-1}. Most important, the exact position of absorption within the range can often be used to identify the exact kind of carbonyl functional group—aldehyde, ketone, ester, and so forth.

Aldehydes   Saturated aldehydes absorb at 1730 cm^{-1}; aldehydes next to either a double bond or an aromatic ring absorb at 1705 cm^{-1}.

Aldehydes

CH$_3$CH$_2$CHO   CH$_3$CH=CHCHO   PhCHO

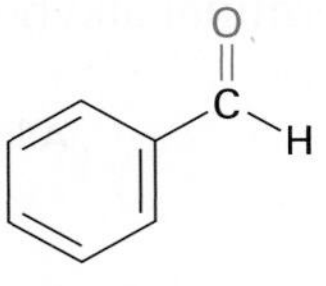

1730 cm^{-1} 1705 cm^{-1} 1705 cm^{-1}

Ketones   Saturated open-chain ketones and six-membered cyclic ketones absorb at 1715 cm^{-1}, five-membered cyclic ketones absorb at 1750 cm^{-1}, and ketones next to a double bond or an aromatic ring absorb at 1690 cm^{-1}.

Ketones

CH$_3$COCH$_3$ cyclopentanone   CH$_3$CH=CHCOCH$_3$ PhCOCH$_3$

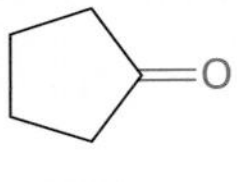

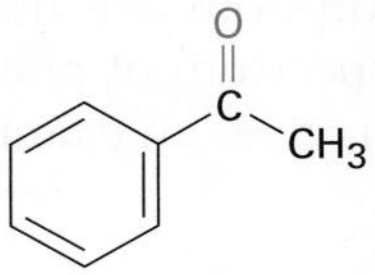

1715 cm^{-1} 1750 cm^{-1} 1690 cm^{-1} 1690 cm^{-1}

Esters   Saturated esters absorb at 1735 cm^{-1}; esters next to either an aromatic ring or a double bond absorb at 1715 cm^{-1}.

Esters

CH$_3$COOCH$_3$ CH$_3$CH=CHCOOCH$_3$ PhCOOCH$_3$

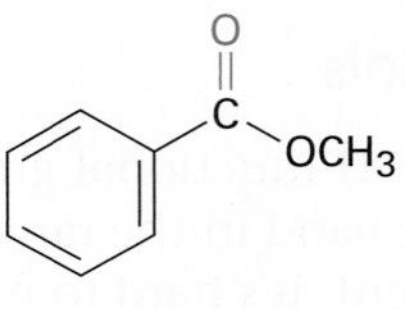

1735 cm^{-1} 1715 cm^{-1} 1715 cm^{-1}

Predicting IR Absorptions of Compounds

Worked Example 12.5

Where might the following compounds have IR absorptions?

(a) [cyclohexene with CH₂OH substituent] (b) $HC\equiv CCH_2CHCH_2COCH_3$ with CH_3 and O substituents

Strategy

Identify the functional groups in each molecule, and then check Table 12.1 to see where those groups absorb.

Solution

(a) *Absorptions:* 3400–3650 cm^{-1} (O–H), 3020–3100 cm^{-1} (=C–H), 1640–1680 cm^{-1} (C=C). This molecule has an alcohol O–H group and an alkene double bond.

(b) *Absorptions:* 3300 cm^{-1} (≡C–H), 2100–2260 cm^{-1} (C≡C), 1735 cm^{-1} (C=O). This molecule has a terminal alkyne triple bond and a saturated ester carbonyl group.

Identifying Functional Groups from an IR Spectrum

Worked Example 12.6

The IR spectrum of an unknown compound is shown in **Figure 12.16**. What functional groups does the compound contain?

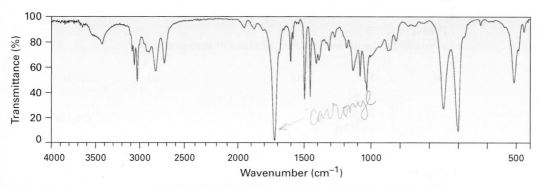

Figure 12.16 IR spectrum for Worked Example 12.6.

Strategy

All IR spectra have many absorptions, but those useful for identifying specific functional groups are usually found in the region from 1500 cm^{-1} to 3300 cm^{-1}. Pay particular attention to the carbonyl region (1670 to 1780 cm^{-1}), the aromatic region (1660 to 2000 cm^{-1}), the triple-bond region (2000 to 2500 cm^{-1}), and the C–H region (2500 to 3500 cm^{-1}).

Solution

The spectrum shows an intense absorption at 1725 cm^{-1} due to a carbonyl group (perhaps an aldehyde, –CHO), a series of weak absorptions from 1800 to 2000 cm^{-1} characteristic of aromatic compounds, and a C–H absorption near 3030 cm^{-1}, also characteristic of aromatic compounds. In fact, the compound is phenylacetaldehyde.

Phenylacetaldehyde

Problem 12.9
The IR spectrum of phenylacetylene is shown in **Figure 12.17**. What absorption bands can you identify?

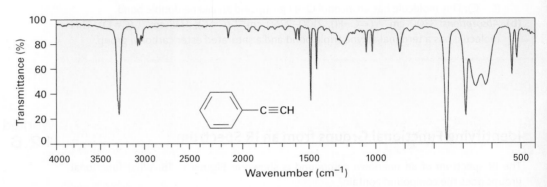

Figure 12.17 The IR spectrum of phenylacetylene, Problem 12.9.

Problem 12.10
Where might the following compounds have IR absorptions?

(a) cyclohexenyl-COCH$_3$ (b) HC≡CCH$_2$CH$_2$CH(=O)   (c) benzene with CO$_2$H and CH$_2$OH

Problem 12.11
Where might the following compound have IR absorptions?

A DEEPER LOOK X-Ray Crystallography

The various spectroscopic techniques described in this and the next two chapters are enormously important in chemistry and have been fine-tuned to such a degree that the structure of almost any molecule can be found. Nevertheless, wouldn't it be nice if you could simply look at a molecule and "see" its structure with your eyes?

Determining the three-dimensional shape of an object around you is easy—you just look at it, let your eyes focus the light rays reflected from the object, and let your brain assemble the data into a recognizable image. If the object is small, you use a microscope and let the microscope lens focus the visible light. Unfortunately, there is a limit to what you can see, even with the best optical microscope. Called the diffraction limit, you can't see anything smaller than the wavelength of light you are using for the observation. Visible light has wavelengths of several hundred nanometers, but atoms in molecules have dimension on the order of 0.1 nm. Thus, to "see" a molecule—whether a small one in the laboratory or a large, complex enzyme with a molecular weight in the tens of thousands—you need wavelengths in the 0.1 nm range, which corresponds to X rays.

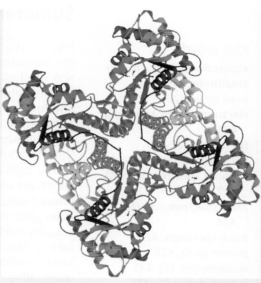

The structure of human muscle fructose-1,6-bisphosphate aldolase, as determined by X-ray crystallography and downloaded from the Protein Data Bank, 1ALD.

Let's say that we want to determine the structure and shape of an enzyme or other biological molecule. The technique used is called *X-ray crystallography*. First, the molecule is crystallized (which often turns out to be the most difficult and time-consuming part of the entire process) and a small crystal with a dimension of 0.4 to 0.5 mm on its longest axis is glued to the end of a glass fiber. The fiber and attached crystal are then mounted in an instrument called an X-ray diffractometer, which consists of a radiation source, a sample positioning and orienting device that can rotate the crystal in any direction, a detector, and a controlling computer.

Once mounted in the diffractometer, the crystal is irradiated with X rays, usually so-called CuK_α radiation with a wavelength of 0.154 nm. When the X rays strike the enzyme crystal, they interact with electrons in the molecule and are scattered into a diffraction pattern which, when detected and visualized, appears as a series of intense spots against a null background.

Manipulation of the diffraction pattern to extract three-dimensional molecular data is a complex process, but the final result is that an electron-density map of the molecule is produced. Because electrons are largely localized around atoms, any two centers of electron density located within bonding distance of each other are assumed to represent bonded atoms, leading to a recognizable chemical structure. So important is this structural information for biochemistry that an online database of more than 66,000 biological substances has been created. Operated by Rutgers University and funded by the U.S. National Science Foundation, the Protein Data Bank (PDB) is a worldwide repository for processing and distributing three-dimensional structural data for biological macromolecules. We'll see how to access the PDB in the Chapter 26 *A Deeper Look*.

Summary

Key words

absorption spectrum, 436
amplitude, 435
base peak, 426
electromagnetic spectrum, 434
frequency, (ν), 435
hertz (Hz), 435
infrared spectroscopy (IR), 437
mass spectrometry (MS), 424
mass spectrum, 426
parent peak, 426
wavelength, (λ), 435
wavenumber ($\tilde{\nu}$), 437

Finding the structure of a new molecule, whether a small one synthesized in the laboratory or a large protein found in living organisms, is central to progress in chemistry and biochemistry. The structure of an organic molecule is usually determined using spectroscopic methods, including mass spectrometry and infrared spectroscopy. **Mass spectrometry (MS)** tells the molecular weight and formula of a molecule; **infrared (IR) spectroscopy** identifies the functional groups present in the molecule.

In small-molecule mass spectrometry, molecules are first ionized by collision with a high-energy electron beam. The ions then fragment into smaller pieces, which are magnetically sorted according to their mass-to-charge ratio (m/z). The ionized sample molecule is called the *molecular ion*, M^+, and measurement of its mass gives the molecular weight of the sample. Structural clues about unknown samples can be obtained by interpreting the fragmentation pattern of the molecular ion. Mass-spectral fragmentations are usually complex, however, and interpretation is often difficult. In biological mass spectrometry, molecules are protonated using either electrospray ionization (ESI) or matrix-assisted laser desorption ionization (MALDI), and the protonated molecules are separated by time-of-flight (TOF).

Infrared spectroscopy involves the interaction of a molecule with **electromagnetic radiation**. When an organic molecule is irradiated with infrared energy, certain **frequencies** are absorbed by the molecule. The frequencies absorbed correspond to the amounts of energy needed to increase the amplitude of specific molecular vibrations such as bond stretchings and bendings. Since every functional group has a characteristic combination of bonds, every functional group has a characteristic set of infrared absorptions. For example, the terminal alkyne ≡C–H bond absorbs IR radiation of 3300 cm^{-1} frequency, and the alkene C=C bond absorbs in the range 1640 to 1680 cm^{-1}. By observing which frequencies of infrared radiation are absorbed by a molecule and which are not, it's possible to determine the functional groups a molecule contains.

Exercises

OWL Interactive versions of these problems are assignable in OWL for Organic Chemistry.

Visualizing Chemistry

(Problems 12.1–12.11 appear within the chapter.)

12.12 Where in the IR spectrum would you expect each of the following molecules to absorb?

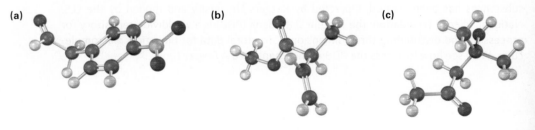

12.13 Show the structures of the likely fragments you would expect in the mass spectra of the following molecules:

(a) (b)

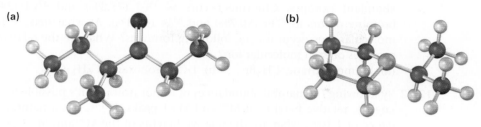

Additional Problems

Mass Spectrometry

12.14 Propose structures for compounds that fit the following mass-spectral data:
(a) A hydrocarbon with $M^+ = 132$
(b) A hydrocarbon with $M^+ = 166$
(c) A hydrocarbon with $M^+ = 84$

12.15 Write molecular formulas for compounds that show the following molecular ions in their high-resolution mass spectra, assuming that C, H, N, and O might be present. The exact atomic masses are: 1.007 83 (^{1}H) 12.000 00 (^{12}C), 14.003 07 (^{14}N), 15.994 91 (^{16}O).
(a) $M^+ = 98.0844$
(b) $M^+ = 123.0320$

12.16 Camphor, a saturated monoketone from the Asian camphor tree, is used among other things as a moth repellent and as a constituent of embalming fluid. If camphor has $M^+ = 152.1201$ by high-resolution mass spectrometry, what is its molecular formula? How many rings does camphor have?

12.17 The *nitrogen rule* of mass spectrometry says that a compound containing an odd number of nitrogens has an odd-numbered molecular ion. Conversely, a compound containing an even number of nitrogens has an even-numbered M^+ peak. Explain.

12.18 In light of the nitrogen rule mentioned in Problem 12.17, what is the molecular formula of pyridine, $M^+ = 79$?

12.19 Nicotine is a diamino compound isolated from dried tobacco leaves. Nicotine has two rings and $M^+ = 162.1157$ by high-resolution mass spectrometry. Give a molecular formula for nicotine, and calculate the number of double bonds.

12.20 The hormone cortisone contains C, H, and O, and shows a molecular ion at $M^+ = 360.1937$ by high-resolution mass spectrometry. What is the molecular formula of cortisone? (The degree of unsaturation of cortisone is 8.)

12.21 Halogenated compounds are particularly easy to identify by their mass spectra because both chlorine and bromine occur naturally as mixtures of two abundant isotopes. Chlorine occurs as ^{35}Cl (75.8%) and ^{37}Cl (24.2%); bromine occurs as ^{79}Br (50.7%) and ^{81}Br (49.3%). At what masses do the molecular ions occur for the following formulas? What are the relative percentages of each molecular ion?
(a) Bromomethane, CH_3Br (b) 1-Chlorohexane, $C_6H_{13}Cl$

12.22 By knowing the natural abundances of minor isotopes, it's possible to calculate the relative heights of M$^+$ and M+1 peaks. If ^{13}C has a natural abundance of 1.10%, what are the relative heights of the M$^+$ and M+1 peaks in the mass spectrum of benzene, C_6H_6?

12.23 Propose structures for compounds that fit the following data:
(a) A ketone with M$^+$ = 86 and fragments at m/z = 71 and m/z = 43
(b) An alcohol with M$^+$ = 88 and fragments at m/z = 73, m/z = 70, and m/z = 59

12.24 2-Methylpentane (C_6H_{14}) has the mass spectrum shown. Which peak represents M$^+$? Which is the base peak? Propose structures for fragment ions of m/z = 71, 57, 43, and 29. Why does the base peak have the mass it does?

12.25 Assume that you are in a laboratory carrying out the catalytic hydrogenation of cyclohexene to cyclohexane. How could you use a mass spectrometer to determine when the reaction is finished?

12.26 What fragments might you expect in the mass spectra of the following compounds?

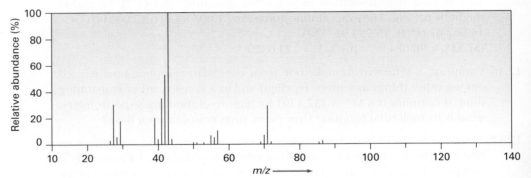

Infrared Spectroscopy

12.27 How might you use IR spectroscopy to distinguish among the three isomers 1-butyne, 1,3-butadiene, and 2-butyne?

12.28 Would you expect two enantiomers such as (R)-2-bromobutane and (S)-2-bromobutane to have identical or different IR spectra? Explain.

12.29 Would you expect two diastereomers such as *meso*-2,3-dibromobutane and (2R,3R)-dibromobutane to have identical or different IR spectra? Explain.

12.30 Propose structures for compounds that meet the following descriptions:
(a) C_5H_8, with IR absorptions at 3300 and 2150 cm^{-1}
(b) C_4H_8O, with a strong IR absorption at 3400 cm^{-1}
(c) C_4H_8O, with a strong IR absorption at 1715 cm^{-1}
(d) C_8H_{10}, with IR absorptions at 1600 and 1500 cm^{-1}

12.31 How could you use infrared spectroscopy to distinguish between the following pairs of isomers?
(a) HC≡CCH$_2$NH$_2$ and CH$_3$CH$_2$C≡N
(b) CH$_3$COCH$_3$ and CH$_3$CH$_2$CHO

12.32 Two infrared spectra are shown. One is the spectrum of cyclohexane, and the other is the spectrum of cyclohexene. Identify them, and explain your answer.

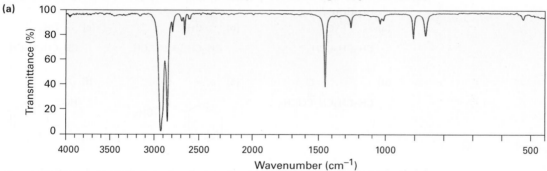

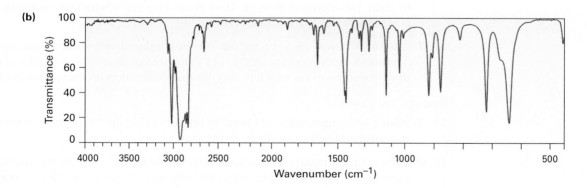

12.33 At what approximate positions might the following compounds show IR absorptions?

(a) 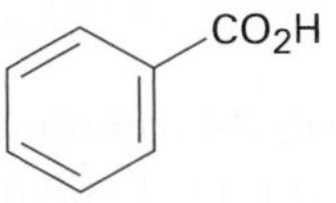 PhCO$_2$H (b) 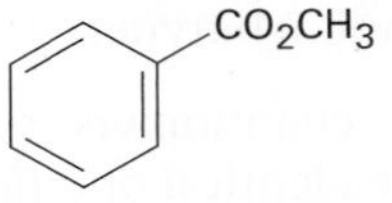 PhCO$_2$CH$_3$ (c) 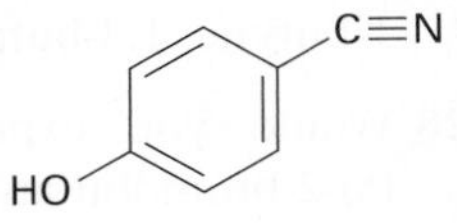 4-HO-C$_6$H$_4$-C≡N

(d) 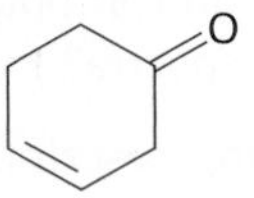 cyclohex-2-enone (e) CH$_3$COCH$_2$CH$_2$COCH$_3$

12.34 How would you use infrared spectroscopy to distinguish between the following pairs of constitutional isomers?

(a) CH$_3$C≡CCH$_3$ and CH$_3$CH$_2$C≡CH

(b) CH$_3$COCH=CHCH$_3$ and CH$_3$COCH$_2$CH=CH$_2$

(c) H$_2$C=CHOCH$_3$ and CH$_3$CH$_2$CHO

12.35 At what approximate positions might the following compounds show IR absorptions?

(a) CH$_3$CH$_2$COCH$_3$

(b) CH$_3$CH(CH$_3$)CH$_2$C≡CH

(c) CH$_3$CH(CH$_3$)CH$_2$CH=CH$_2$

(d) CH$_3$CH$_2$CH$_2$COCH$_3$

(e) 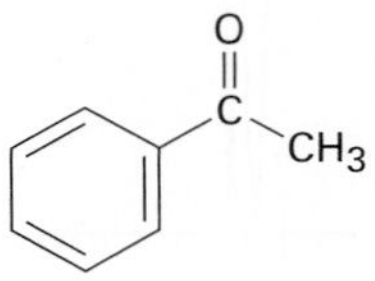 PhCOCH$_3$

(f) 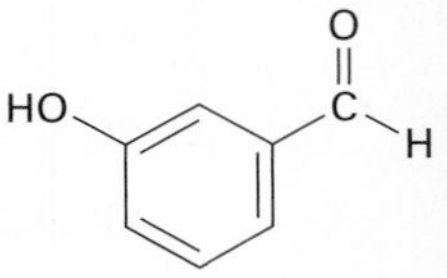 3-HO-C$_6$H$_4$-CHO

12.36 Assume that you are carrying out the dehydration of 1-methylcyclohexanol to yield 1-methylcyclohexene. How could you use infrared spectroscopy to determine when the reaction is complete?

12.37 Assume that you are carrying out the base-induced dehydrobromination of 3-bromo-3-methylpentane (Section 11.7) to yield an alkene. How could you use IR spectroscopy to tell which of two possible elimination products is formed?

General Problems

12.38 Which is stronger, the C=O bond in an ester (1735 cm^{-1}) or the C=O bond in a saturated ketone (1715 cm^{-1})? Explain.

12.39 Carvone is an unsaturated ketone responsible for the odor of spearmint. If carvone has M$^+$ = 150 in its mass spectrum and contains three double bonds and one ring, what is its molecular formula?

12.40 Carvone (Problem 12.39) has an intense infrared absorption at 1690 cm^{-1}. What kind of ketone does carvone contain?

12.41 The mass spectrum (a) and the infrared spectrum (b) of an unknown hydrocarbon are shown. Propose as many structures as you can.

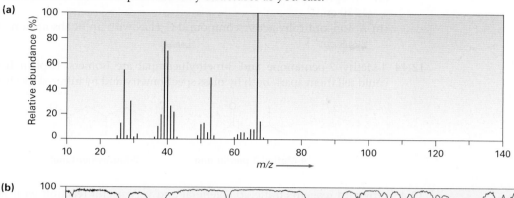

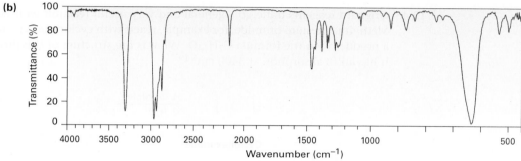

12.42 The mass spectrum (a) and the infrared spectrum (b) of another unknown hydrocarbon are shown. Propose as many structures as you can.

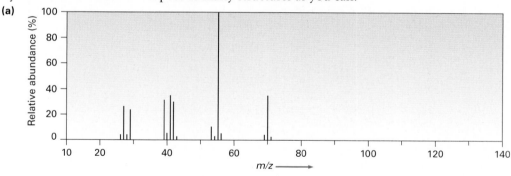

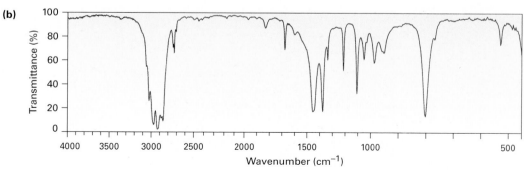

12.43 Propose structures for compounds that meet the following descriptions:
(a) An optically active compound $C_5H_{10}O$ with an IR absorption at 1730 cm^{-1}
(b) A non–optically active compound C_5H_9N with an IR absorption at 2215 cm^{-1}

12.44 4-Methyl-2-pentanone and 3-methylpentanal are isomers. Explain how you could tell them apart, both by mass spectrometry and by infrared spectroscopy.

4-Methyl-2-pentanone   3-Methylpentanal

12.45 Grignard reagents undergo a general and very useful reaction with ketones. Methylmagnesium bromide, for example, reacts with cyclohexanone to yield a product with the formula $C_7H_{14}O$. What is the structure of this product if it has an IR absorption at 3400 cm^{-1}?

Cyclohexanone

12.46 Ketones undergo a reduction when treated with sodium borohydride, $NaBH_4$. What is the structure of the compound produced by reaction of 2-butanone with $NaBH_4$ if it has an IR absorption at 3400 cm^{-1} and $M^+ = 74$ in the mass spectrum?

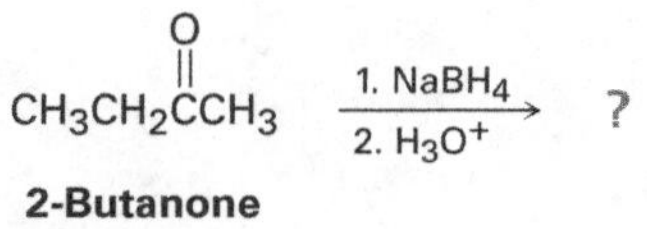

$$CH_3CH_2\overset{\overset{O}{\|}}{C}CH_3 \xrightarrow[\text{2. }H_3O^+]{\text{1. }NaBH_4} \text{ ?}$$

2-Butanone

12.47 Nitriles, R—C≡N, undergo a hydrolysis reaction when heated with aqueous acid. What is the structure of the compound produced by hydrolysis of propanenitrile, $CH_3CH_2C{\equiv}N$, if it has IR absorptions at 2500–3100 cm^{-1} and 1710 cm^{-1} and has $M^+ = 74$?

APPENDIX A

Nomenclature of Polyfunctional Organic Compounds

With more than 40 million organic compounds now known and thousands more being created daily, naming them all is a real problem. Part of the problem is due to the sheer complexity of organic structures, but part is also due to the fact that chemical names have more than one purpose. For Chemical Abstracts Service (CAS), which catalogs and indexes the worldwide chemical literature, each compound must have only one correct name. It would be chaos if half the entries for CH_3Br were indexed under "M" for methyl bromide and half under "B" for bromomethane. Furthermore, a CAS name must be strictly systematic so that it can be assigned and interpreted by computers; common names are not allowed.

People, however, have different requirements than computers. For people—which is to say students and professional chemists in their spoken and written communications—it's best that a chemical name be pronounceable and that it be as easy as possible to assign and interpret. Furthermore, it's convenient if names follow historical precedents, even if that means a particularly well-known compound might have more than one name. People can readily understand that bromomethane and methyl bromide both refer to CH_3Br.

As noted in the text, chemists overwhelmingly use the nomenclature system devised and maintained by the International Union of Pure and Applied Chemistry, or IUPAC. Rules for naming monofunctional compounds were given throughout the text as each new functional group was introduced, and a list of where these rules can be found is given in Table A.1.

Table A.1 Nomenclature Rules for Functional Groups

Functional group	Text section	Functional group	Text section
Acid anhydrides	21.1	Aromatic compounds	15.1
Acid halides	21.1	Carboxylic acids	20.1
Acyl phosphates	21.1	Cycloalkanes	4.1
Alcohols	17.1	Esters	21.1
Aldehydes	19.1	Ethers	18.1
Alkanes	3.4	Ketones	19.1
Alkenes	7.3	Nitriles	20.1
Alkyl halides	10.1	Phenols	17.1
Alkynes	9.1	Sulfides	18.8
Amides	21.1	Thiols	18.8
Amines	24.1	Thioesters	21.1

Naming a monofunctional compound is reasonably straightforward, but even experienced chemists often encounter problems when faced with naming a complex polyfunctional compound. Take the following compound, for instance. It has three functional groups, ester, ketone, and C=C, but how should it be named? As an ester with an *-oate* ending, a ketone with an *-one* ending, or an alkene with an *-ene* ending? It's actually named methyl 3-(2-oxo-6-cyclohexenyl)propanoate.

Methyl 3-(2-oxo-**6-cylohexenyl**)propanoate

The name of a polyfunctional organic molecule has four parts—suffix, parent, prefixes, and locants—which must be identified and expressed in the proper order and format. Let's look at each of the four.

Name Part 1. The Suffix: Functional-Group Precedence

Although a polyfunctional organic molecule might contain several different functional groups, we must choose just one suffix for nomenclature purposes. It's not correct to use two suffixes. Thus, keto ester **1** must be named either as a ketone with an *-one* suffix or as an ester with an *-oate* suffix, but it can't be named as an *-onoate*. Similarly, amino alcohol **2** must be named either as an alcohol (*-ol*) or as an amine (*-amine*), but it can't be named as an *-olamine* or *-aminol*.

1. $CH_3CCH_2CH_2COCH_3$ 2. $CH_3CHCH_2CH_2CH_2NH_2$ (with OH on second carbon)

The only exception to the rule requiring a single suffix is when naming compounds that have double or triple bonds. Thus, the unsaturated acid $H_2C=CHCH_2CO_2H$ is 3-butenoic acid, and the acetylenic alcohol $HC\equiv CCH_2CH_2CH_2OH$ is 5-pentyn-1-ol.

How do we choose which suffix to use? Functional groups are divided into two classes, **principal groups** and **subordinate groups**, as shown in Table A.2. Principal groups can be cited either as prefixes or as suffixes, while subordinate groups are cited only as prefixes. Within the principal groups, an order of priority has been established, with the proper suffix for a given compound determined by choosing the principal group of highest priority. For example, Table A.2 indicates that keto ester **1** should be named as an ester rather than as a ketone because an ester functional group is higher in priority than a ketone. Similarly, amino alcohol **2** should be named as an alcohol rather than as an amine.

APPENDIX A | Nomenclature of Polyfunctional Organic Compounds   A-3

Table A.2 Classification of Functional Groups[a]

Functional group	Name as suffix	Name as prefix
Principal groups		
Carboxylic acids	-oic acid -carboxylic acid	carboxy
Acid anhydrides	-oic anhydride -carboxylic anhydride	—
Esters	-oate -carboxylate	alkoxycarbonyl
Thioesters	-thioate -carbothioate	alkylthiocarbonyl
Acid halides	-oyl halide -carbonyl halide	halocarbonyl
Amides	-amide -carboxamide	carbamoyl
Nitriles	-nitrile -carbonitrile	cyano
Aldehydes	-al -carbaldehyde	oxo
Ketones	-one	oxo
Alcohols	-ol	hydroxy
Phenols	-ol	hydroxy
Thiols	-thiol	mercapto
Amines	-amine	amino
Imines	-imine	imino
Ethers	ether	alkoxy
Sulfides	sulfide	alkylthio
Disulfides	disulfide	—
Alkenes	-ene	—
Alkynes	-yne	—
Alkanes	-ane	—
Subordinate groups		
Azides	—	azido
Halides	—	halo
Nitro compounds	—	nitro

[a]Principal groups are listed in order of decreasing priority; subordinate groups have no priority order.

Thus, the name of **1** is methyl 4-oxopentanoate and the name of **2** is 5-amino-2-pentanol. Further examples are shown:

1. Methyl 4-**oxo**pentanoate
(an ester with a ketone group)

$$CH_3\overset{O}{\underset{\|}{C}}CH_2CH_2\overset{O}{\underset{\|}{C}}OCH_3$$

2. 5-**Amino**-2-pentanol
(an alcohol with an amine group)

$$CH_3\overset{OH}{\underset{|}{C}}HCH_2CH_2CH_2NH_2$$

3. Methyl 5-methyl-6-**oxo**hexanoate
(an ester with an aldehyde group)

$$\overset{CHO}{\underset{|}{CH_3C}}HCH_2CH_2CH_2\overset{O}{\underset{\|}{C}}OCH_3$$

4. 5-Carbamoyl-4-hydroxy**pentanoic acid**
(a carboxylic acid with amide and alcohol groups)

$$H_2N\overset{O}{\underset{\|}{C}}CH_2\overset{OH}{\underset{|}{C}}HCH_2CH_2\overset{O}{\underset{\|}{C}}OH$$

5. 3-**Oxo**cyclohexanecarbaldehyde
(an aldehyde with a ketone group)

Name Part 2. The Parent: Selecting the Main Chain or Ring

The parent, or base, name of a polyfunctional organic compound is usually easy to identify. If the principal group of highest priority is part of an open chain, the parent name is that of the longest chain containing the largest number of principal groups. For example, compounds **6** and **7** are isomeric aldehydo amides, which must be named as amides rather than as aldehydes according to Table A.2. The longest chain in compound **6** has six carbons, and the substance is named 5-methyl-6-oxohexanamide. Compound **7** also has a chain of six carbons, but the longest chain that contains both principal functional groups has only four carbons. Thus, compound **7** is named 4-oxo-3-propylbutanamide.

6. 5-Methyl-6-**oxo**hexanamide

7. 4-**Oxo**-3-propyl**butanamide**

If the highest-priority principal group is attached to a ring, the parent name is that of the ring system. Compounds **8** and **9**, for instance, are isomeric keto nitriles and must both be named as nitriles according to Table A.2. Substance **8** is named as a benzonitrile because the −CN functional group is a substituent on the aromatic ring, but substance **9** is named as an acetonitrile because the −CN functional group is on an open chain. The names are 2-acetyl-(4-bromomethyl)benzonitrile (**8**) and (2-acetyl-4-bromophenyl)acetonitrile (**9**). As further examples, compounds **10** and **11** are both keto acids and must be named as acids, but the parent name in **10** is that of a ring system (cyclohexanecarboxylic acid) and the parent name in **11** is that of an open chain (propanoic acid). The names are *trans*-2-(3-oxopropyl)cyclohexanecarboxylic acid (**10**) and 3-(2-oxocyclohexyl)propanoic acid (**11**).

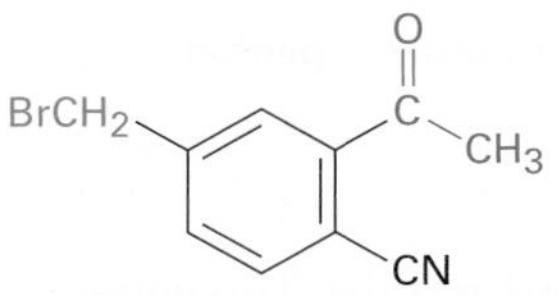

8. 2-Acetyl-(4-bromomethyl)benzonitrile

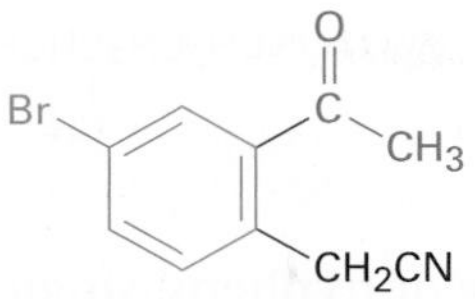

9. (2-Acetyl-4-bromophenyl)acetonitrile

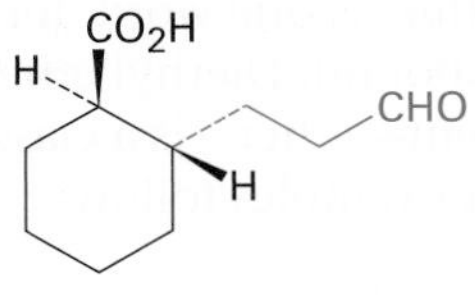

10. *trans*-2-(3-oxopropyl)cyclohexanecarboxylic acid

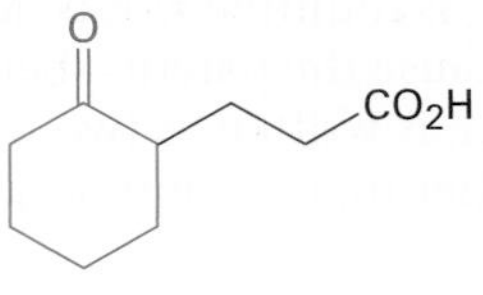

11. 3-(2-Oxocyclohexyl)propanoic acid

Name Parts 3 and 4. The Prefixes and Locants

With the parent name and the suffix established, the next step is to identify and give numbers, or *locants*, to all substituents on the parent chain or ring. The substituents include all alkyl groups and all functional groups other than the one cited in the suffix. For example, compound **12** contains three different functional groups (carboxyl, keto, and double bond). Because the carboxyl group is highest in priority and the longest chain containing the functional groups has seven carbons, compound **12** is a heptenoic acid. In addition, the parent chain has a keto (oxo) substituent and three methyl groups. Numbering from the end nearer the highest-priority functional group gives the name (*E*)-2,5,5-trimethyl-4-oxo-2-heptenoic acid. Look back at some of the other compounds we've named to see other examples of how prefixes and locants are assigned.

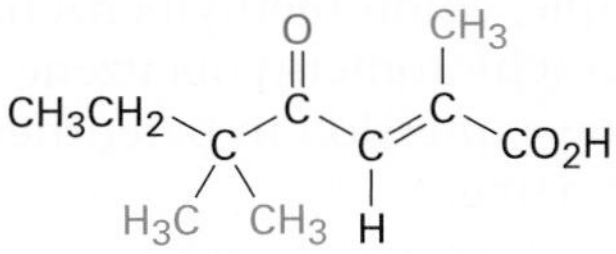

12. (*E*)-2,5,5-Trimethyl-4-oxo-2-heptenoic acid

Writing the Name

With the name parts established, the entire name is then written out. Several additional rules apply:

1. **Order of prefixes.** When the substituents have been identified, the parent chain has been numbered, and the proper multipliers such as *di-* and *tri-* have been assigned, the name is written with the substituents listed in alphabetical, rather than numerical, order. Multipliers such as *di-* and

tri- are not used for alphabetization, but the italicized prefixes *iso-* and *sec-* are used.

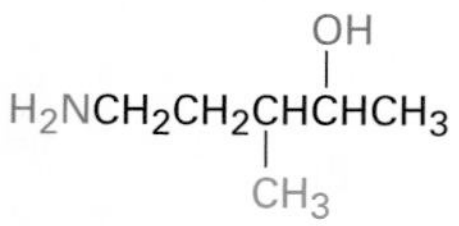

H₂NCH₂CH₂CH(OH)CH(CH₃)CH₃ **13.** 5-Amino-3-methyl-2-**pentanol**

2. **Use of hyphens; single- and multiple-word names.** The general rule is to determine whether the parent is itself an element or compound. If it is, then the name is written as a single word; if it isn't, then the name is written as multiple words. Methylbenzene is written as one word, for instance, because the parent—benzene—is itself a compound. Diethyl ether, however, is written as two words because the parent—ether—is a class name rather than a compound name. Some further examples follow:

H₃C—Mg—CH₃

14. Dimethyl**magnesium**
(one word, because magnesium is an element)

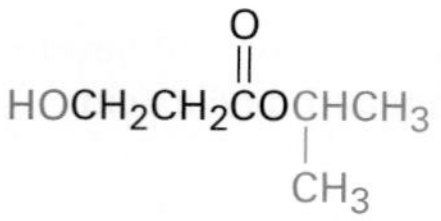

HOCH₂CH₂COCH(CH₃)CH₃

15. Isopropyl 3-hydroxy**propanoate**
(two words, because "propanoate" is not a compound)

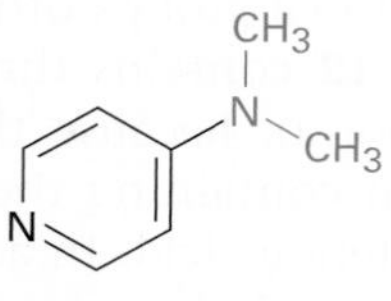

16. 4-(Dimethylamino)**pyridine**
(one word, because pyridine is a compound)

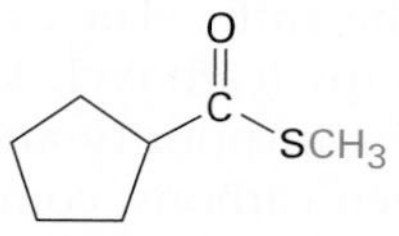

17. Methyl **cyclopentanecarbothioate**
(two words, because "cyclopentanecarbothioate" is not a compound)

3. **Parentheses.** Parentheses are used to denote complex substituents when ambiguity would otherwise arise. For example, chloromethylbenzene has two substituents on a benzene ring, but (chloromethyl)benzene has only one complex substituent. Note that the expression in parentheses is not set off by hyphens from the rest of the name.

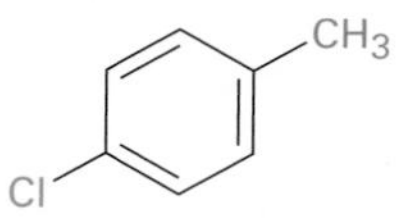

18. *p*-Chloromethyl**benzene**

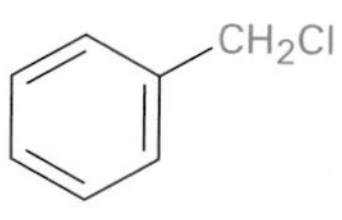

19. (Chloromethyl)**benzene**

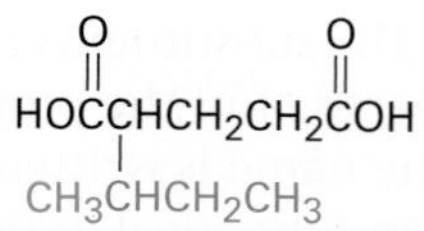

HOCCHCH₂CH₂COH
       |
   CH₃CHCH₂CH₃

20. 2-(1-Methylpropyl)**pentanedioic acid**

Additional Reading

Further explanations of the rules of organic nomenclature can be found online at http://www.acdlabs.com/iupac/nomenclature/ (accessed September 2010) and in the following references:

1. "A Guide to IUPAC Nomenclature of Organic Compounds," CRC Press, Boca Raton, FL, 1993.
2. "Nomenclature of Organic Chemistry, Sections A, B, C, D, E, F, and H," International Union of Pure and Applied Chemistry, Pergamon Press, Oxford, 1979.

APPENDIX B

Acidity Constants for Some Organic Compounds

Compound	pK_a	Compound	pK_a	Compound	pK_a
CH_3SO_3H	−1.8	2-chlorobenzoic acid	3.0	$CH_2BrCH_2CO_2H$	4.0
$CH(NO_2)_3$	0.1	salicylic acid (2-hydroxybenzoic acid)	3.0	2,4-dinitrophenol	4.1
2,4,6-trinitrophenol (picric acid)	0.3	CH_2ICO_2H	3.2	benzoic acid	4.2
CCl_3CO_2H	0.5	$CHOCO_2H$	3.2	$H_2C=CHCO_2H$	4.2
CF_3CO_2H	0.5			$HO_2CCH_2CH_2CO_2H$	4.2; 5.7
CBr_3CO_2H	0.7	4-nitrobenzoic acid	3.4	$HO_2CCH_2CH_2CH_2CO_2H$	4.3; 5.4
$HO_2CC\equiv CCO_2H$	1.2; 2.5				
HO_2CCO_2H	1.2; 3.7	3,4-dinitrobenzoic acid	3.5	pentachlorophenol	4.5
$CHCl_2CO_2H$	1.3				
$CH_2(NO_2)CO_2H$	1.3				
$HC\equiv CCO_2H$	1.9	$HSCH_2CO_2H$	3.5; 10.2	$H_2C=C(CH_3)CO_2H$	4.7
(Z) $HO_2CCH=CHCO_2H$	1.9; 6.3	$CH_2(NO_2)_2$	3.6	CH_3CO_2H	4.8
2-nitrobenzoic acid	2.4	$CH_3OCH_2CO_2H$	3.6	$CH_3CH_2CO_2H$	4.8
		$CH_3COCH_2CO_2H$	3.6	$(CH_3)_3CCO_2H$	5.0
CH_3COCO_2H	2.4	$HOCH_2CO_2H$	3.7	$CH_3COCH_2NO_2$	5.1
$NCCH_2CO_2H$	2.5	HCO_2H	3.7		
$CH_3C\equiv CCO_2H$	2.6				
CH_2FCO_2H	2.7	3-chlorobenzoic acid	3.8	1,3-cyclohexanedione	5.3
CH_2ClCO_2H	2.8				
$HO_2CCH_2CO_2H$	2.8; 5.6	4-chlorobenzoic acid	4.0	$O_2NCH_2CO_2CH_3$	5.8
CH_2BrCO_2H	2.9				

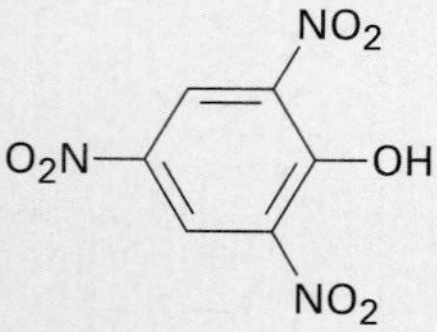

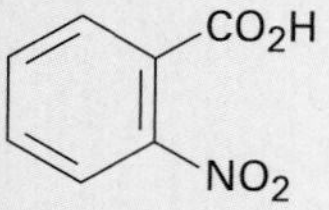

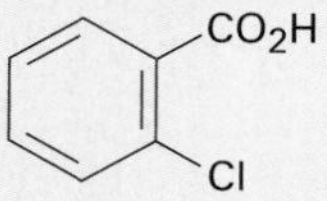

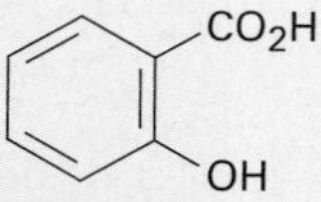

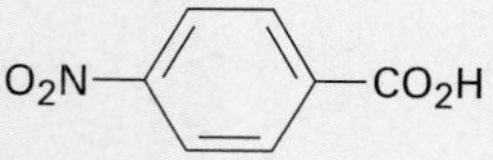

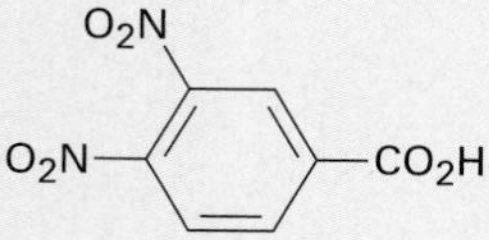

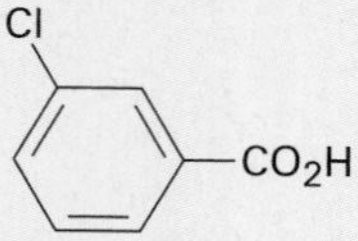

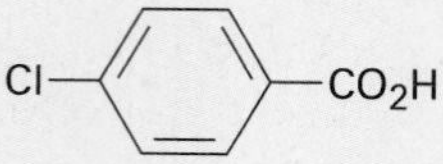

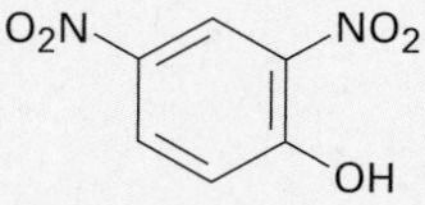

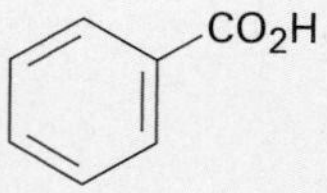

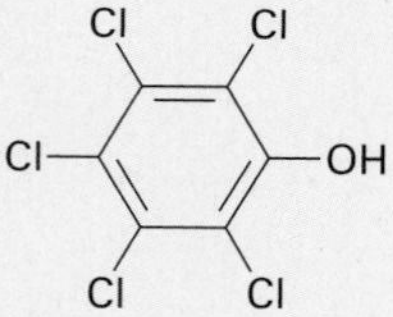

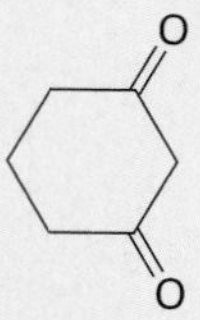

Appendix B | Acidity Constants for Some Organic Compounds

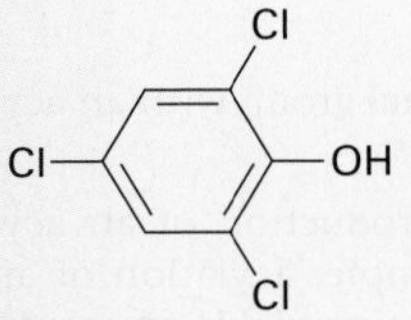

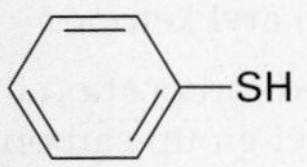

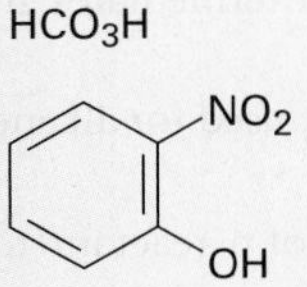

Compound	pK$_a$	Compound	pK$_a$	Compound	pK$_a$
2-formylcyclopentanone	5.8	C$_6$H$_5$CH$_2$SH	9.4	H$_2$C=CHCH$_2$OH	15.5
				CH$_3$CH$_2$OH	16.0
		hydroquinone (1,4-dihydroxybenzene)	9.9; 11.5	CH$_3$CH$_2$CH$_2$OH	16.1
2,4,6-trichlorophenol	6.2			CH$_3$COCH$_2$Br	16.1
		phenol	9.9	cyclohexanone	16.7
C$_6$H$_5$SH	6.6			CH$_3$CHO	17
		CH$_3$COCH$_2$SOCH$_3$	10.0	(CH$_3$)$_2$CHCHO	17
HCO$_3$H	7.1			(CH$_3$)$_2$CHOH	17.1
2-nitrophenol	7.2	2-methylphenol	10.3	(CH$_3$)$_3$COH	18.0
				CH$_3$COCH$_3$	19.3
(CH$_3$)$_2$CHNO$_2$	7.7	CH$_3$NO$_2$	10.3		
		CH$_3$SH	10.3	fluorene	23
2,4-dichlorophenol	7.8	CH$_3$COCH$_2$CO$_2$CH$_3$	10.6	CH$_3$CO$_2$CH$_2$CH$_3$	25
		CH$_3$COCHO	11.0	HC≡CH	25
		CH$_2$(CN)$_2$	11.2	CH$_3$CN	25
		CCl$_3$CH$_2$OH	12.2	CH$_3$SO$_2$CH$_3$	28
CH$_3$CO$_3$H	8.2	Glucose	12.3	(C$_6$H$_5$)$_3$CH	32
2-chlorophenol	8.5	(CH$_3$)$_2$C=NOH	12.4	(C$_6$H$_5$)$_2$CH$_2$	34
		CH$_2$(CO$_2$CH$_3$)$_2$	12.9	CH$_3$SOCH$_3$	35
		CHCl$_2$CH$_2$OH	12.9	NH$_3$	36
CH$_3$CH$_2$NO$_2$	8.5	CH$_2$(OH)$_2$	13.3	CH$_3$CH$_2$NH$_2$	36
4-(trifluoromethyl)phenol	8.7	HOCH$_2$CH(OH)CH$_2$OH	14.1	(CH$_3$CH$_2$)$_2$NH	40
		CH$_2$ClCH$_2$OH	14.3	toluene	41
CH$_3$COCH$_2$COCH$_3$	9.0	cyclopentadiene	15.0		
resorcinol (1,3-dihydroxybenzene)	9.3; 11.1	C$_6$H$_5$CH$_2$OH	15.4	benzene	43
		CH$_3$OH	15.5	H$_2$C=CH$_2$	44
catechol (1,2-dihydroxybenzene)	9.3; 12.6			CH$_4$	~60

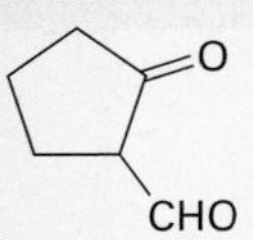

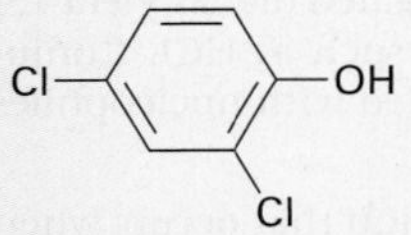

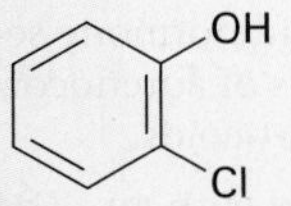

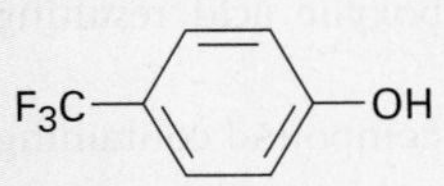

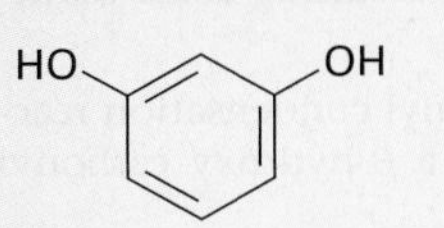

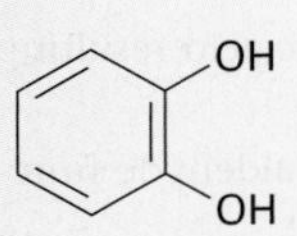

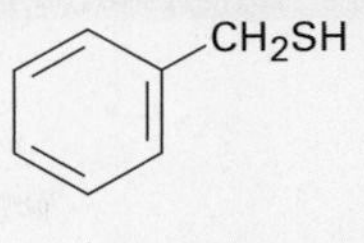

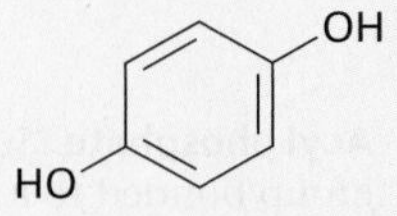

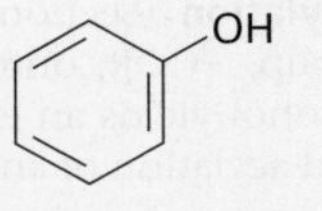

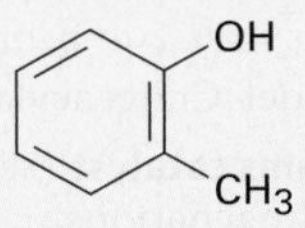

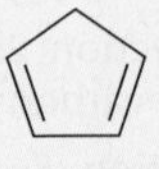

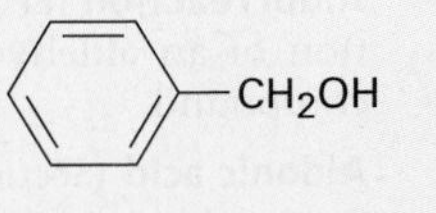

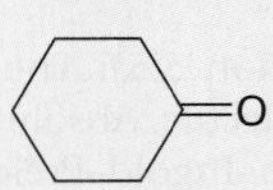

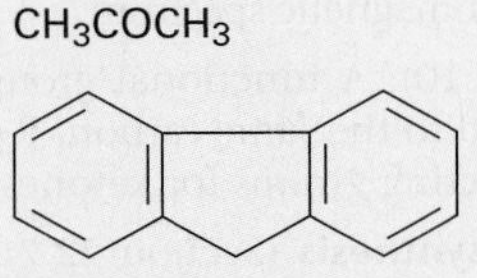

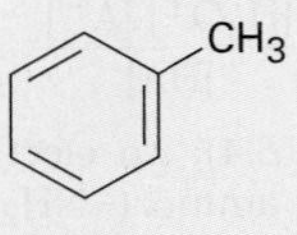

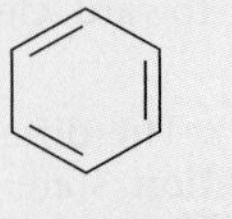

An acidity list covering more than 5000 organic compounds has been published: E.P. Serjeant and B. Dempsey (eds.), "Ionization Constants of Organic Acids in Aqueous Solution," IUPAC Chemical Data Series No. 23, Pergamon Press, Oxford, 1979.

APPENDIX C

Glossary

Absolute configuration (Section 5.5): The exact three-dimensional structure of a chiral molecule. Absolute configurations are specified verbally by the Cahn–Ingold–Prelog R,S convention.

Absorbance (Section 14.7): In optical spectroscopy, the logarithm of the intensity of the incident light divided by the intensity of the light transmitted through a sample; $A = \log I_0/I$.

Absorption spectrum (Section 12.5): A plot of wavelength of incident light versus amount of light absorbed. Organic molecules show absorption spectra in both the infrared and the ultraviolet regions of the electromagnetic spectrum.

Acetal (Section 19.10): A functional group consisting of two —OR groups bonded to the same carbon, $R_2C(OR')_2$. Acetals are often used as protecting groups for ketones and aldehydes.

Acetoacetic ester synthesis (Section 22.7): The synthesis of a methyl ketone by alkylation of an alkyl halide with ethyl acetoacetate, followed by hydrolysis and decarboxylation.

Acetyl group (Section 19.1): The CH_3CO- group.

Acetylide anion (Section 9.7): The anion formed by removal of a proton from a terminal alkyne, $R-C \equiv C:^-$.

Achiral (Section 5.2): Having a lack of handedness. A molecule is achiral if it has a plane of symmetry and is thus superimposable on its mirror image.

Acid anhydride (Section 21.1): A functional group with two acyl groups bonded to a common oxygen atom, RCO_2COR'.

Acid halide (Section 21.1): A functional group with an acyl group bonded to a halogen atom, RCOX.

Acidity constant, K_a (Section 2.8): A measure of acid strength. For any acid HA, the acidity constant is given by the expression

$$K_a = \frac{[H_3O^+][A^-]}{[HA]}.$$

Activating group (Section 16.4): An electron-donating group such as hydroxyl (—OH) or amino (—NH$_2$) that increases the reactivity of an aromatic ring toward electrophilic aromatic substitution.

Activation energy (Section 6.9): The difference in energy between ground state and transition state in a reaction. The amount of activation energy determines the rate at which the reaction proceeds. Most organic reactions have activation energies of 40–100 kJ/mol.

Active site (Sections 6.11, 26.11): The pocket in an enzyme where a substrate is bound and undergoes reaction.

Acyclic diene metathesis (ADMET) (Section 31.5): A method of polymer synthesis that uses the olefin metathesis reaction of an open-chain diene.

Acyl group (Sections 16.3, 19.1): A —COR group.

Acyl phosphate (Section 21.8): A functional group with an acyl group bonded to a phosphate, $RCO_2PO_3^{2-}$.

Acylation (Sections 16.3, 21.4): The introduction of an acyl group, —COR, onto a molecule. For example, acylation of an alcohol yields an ester, acylation of an amine yields an amide, and acylation of an aromatic ring yields an alkyl aryl ketone.

Acylium ion (Section 16.3): A resonance-stabilized carbocation in which the positive charge is located at a carbonyl-group carbon, $R-\overset{+}{C}=O \longleftrightarrow R-C \equiv O^+$. Acylium ions are intermediates in Friedel–Crafts acylation reactions.

Adams catalyst (Section 8.6): The PtO_2 catalyst used for alkene hydrogenations.

1,2 Addition (Sections 14.2, 19.13): Addition of a reactant to the two ends of a double bond.

1,4 Addition (Sections 14.2, 19.13): Addition of a reactant to the ends of a conjugated π system. Conjugated dienes yield 1,4 adducts when treated with electrophiles such as HCl. Conjugated enones yield 1,4 adducts when treated with nucleophiles such as amines.

Addition reaction (Section 6.1): The reaction that occurs when two reactants add together to form a single product with no atoms left over.

Adrenocortical hormone (Section 27.6): A steroid hormone secreted by the adrenal glands. There are two types of adrenocortical hormones: mineralocorticoids and glucocorticoids.

Alcohol (Chapter 17 Introduction): A compound with an —OH group bonded to a saturated, sp^3-hybridized carbon, ROH.

Aldaric acid (Section 25.6): The dicarboxylic acid resulting from oxidation of an aldose.

Aldehyde (Chapter 19 Introduction): A compound containing the —CHO functional group.

Alditol (Section 25.6): The polyalcohol resulting from reduction of the carbonyl group of a sugar.

Aldol reaction (Section 23.1): The carbonyl condensation reaction of an aldehyde or ketone to give a β-hydroxy carbonyl compound.

Aldonic acid (Section 25.6): The monocarboxylic acid resulting from oxidation of the —CHO group of an aldose.

Aldose (Section 25.1): A carbohydrate with an aldehyde functional group.

Alicyclic (Section 4.1): A nonaromatic cyclic hydrocarbon such as a cycloalkane or cycloalkene.

Aliphatic (Section 3.2): A nonaromatic hydrocarbon such as a simple alkane, alkene, or alkyne.

Alkaloid (Chapter 2 *A Deeper Look*): A naturally occurring organic base, such as morphine.

Alkane (Section 3.2): A compound of carbon and hydrogen that contains only single bonds.

Alkene (Chapter 7 Introduction): A hydrocarbon that contains a carbon–carbon double bond, $R_2C=CR_2$.

Alkoxide ion (Section 17.2): The anion RO^- formed by deprotonation of an alcohol.

Alkoxymercuration reaction (Section 18.2): A method for synthesizing ethers by mercuric-ion catalyzed addition of an alcohol to an alkene followed by demercuration on treatment with $NaBH_4$.

Alkyl group (Section 3.3): The partial structure that remains when a hydrogen atom is removed from an alkane.

Alkyl halide (Chapter 10 Introduction): A compound with a halogen atom bonded to a saturated, sp^3-hybridized carbon atom.

Alkylamine (Section 24.1): An amino-substituted alkane, RNH_2, R_2NH, or R_3N.

Alkylation (Sections 9.8, 16.3, 18.2, 22.7): Introduction of an alkyl group onto a molecule. For example, aromatic rings can be alkylated to yield arenes, and enolate anions can be alkylated to yield α-substituted carbonyl compounds.

Alkyne (Chapter 9 Introduction): A hydrocarbon that contains a carbon–carbon triple bond, $RC\equiv CR$.

Allyl group (Section 7.3): A $H_2C=CHCH_2-$ substituent.

Allylic (Section 10.3): The position next to a double bond. For example, $H_2C=CHCH_2Br$ is an allylic bromide.

α-Amino acid (Section 26.1): A difunctional compound with an amino group on the carbon atom next to a carboxyl group, $RCH(NH_2)CO_2H$.

α Anomer (Section 25.5): The cyclic hemiacetal form of a sugar that has the hemiacetal –OH group cis to the –OH at the lowest chirality center in a Fischer projection.

α Helix (Section 26.9): The coiled secondary structure of a protein.

α Position (Chapter 22 Introduction): The position next to a carbonyl group.

α-Substitution reaction (Section 22.2): The substitution of the α hydrogen atom of a carbonyl compound by reaction with an electrophile.

Amide (Chapter 21 Introduction): A compound containing the $-CONR_2$ functional group.

Amidomalonate synthesis (Section 26.3): A method for preparing α-amino acids by alkylation of diethyl amidomalonate with an alkyl halide followed by deprotection and decarboxylation.

Amine (Chapter 24 Introduction): A compound containing one or more organic substituents bonded to a nitrogen atom, RNH_2, R_2NH, or R_3N.

Amino acid (Section 26.1): See α-Amino acid.

Amino sugar (Section 25.7): A sugar with one of its –OH groups replaced by $-NH_2$.

Amphiprotic (Section 26.1): Capable of acting either as an acid or as a base. Amino acids are amphiprotic.

Amplitude (Section 12.5): The height of a wave measured from the midpoint to the maximum. The intensity of radiant energy is proportional to the square of the wave's amplitude.

Amyl group (Section 3.3): An alternative name for a pentyl group.

Anabolic steroid (Section 27.6): A synthetic androgen that mimics the tissue-building effects of natural testosterone.

Anabolism (Section 29.1): The group of metabolic pathways that build up larger molecules from smaller ones.

Androgen (Section 27.6): A male steroid sex hormone.

Angle strain (Section 4.3): The strain introduced into a molecule when a bond angle is deformed from its ideal value. Angle strain is particularly important in small-ring cycloalkanes, where it results from compression of bond angles to less than their ideal tetrahedral values.

Annulation (Section 23.12): The building of a new ring onto an existing molecule.

Anomeric center (Section 25.5): The hemiacetal carbon atom in the cyclic pyranose or furanose form of a sugar.

Anomers (Section 25.5): Cyclic stereoisomers of sugars that differ only in their configuration at the hemiacetal (anomeric) carbon.

Antarafacial (Section 30.5): A pericyclic reaction that takes place on opposite faces of the two ends of a π electron system.

Anti conformation (Section 3.7): The geometric arrangement around a carbon–carbon single bond in which the two largest substituents are 180° apart as viewed in a Newman projection.

Anti periplanar (Section 11.8): Describing the stereochemical relationship in which two bonds on adjacent carbons lie in the same plane at an angle of 180°.

Anti stereochemistry (Section 8.2): The opposite of syn. An anti addition reaction is one in which the two ends of the double bond are attacked from different sides. An anti elimination reaction is one in which the two groups leave from opposite sides of the molecule.

Antiaromatic (Section 15.3): Referring to a planar, conjugated molecule with $4n$ π electrons. Delocalization of the π electrons leads to an increase in energy.

Antibonding MO (Section 1.11): A molecular orbital that is higher in energy than the atomic orbitals from which it is formed.

Anticodon (Section 28.5): A sequence of three bases on tRNA that reads the codons on mRNA and brings the correct amino acids into position for protein synthesis.

Antisense strand (Section 28.4): The template, noncoding strand of double-helical DNA that does not contain the gene.

Arene (Section 15.1): An alkyl-substituted benzene.

Arenediazonium salt (Section 24.8): An aromatic compound $Ar-\overset{+}{N}\equiv N\ X^-$; used in the Sandmeyer reaction.

Aromaticity (Chapter 15 Introduction): The special characteristics of cyclic conjugated molecules, including unusual stability and a tendency to undergo substitution reactions rather than addition reactions on treatment with electrophiles. Aromatic molecules are planar, cyclic, conjugated species with $4n + 2$ π electrons.

Arylamine (Section 24.1): An amino-substituted aromatic compound, $ArNH_2$.

Atactic (Section 31.2): A chain-growth polymer in which the stereochemistry of the substituents is oriented randomly along the backbone.

Atomic mass (Section 1.1): The weighted average mass of an element's naturally occurring isotopes.

Atomic number, Z (Section 1.1): The number of protons in the nucleus of an atom.

ATZ Derivative (Section 26.6): An anilinothiazolinone, formed from an amino acid during Edman degradation of a peptide.

Aufbau principle (Section 1.3): The rules for determining the electron configuration of an atom.

Axial bond (Section 4.6): A bond to chair cyclohexane that lies along the ring axis, perpendicular to the rough plane of the ring.

Azide synthesis (Section 24.6): A method for preparing amines by S_N2 reaction of an alkyl halide with azide ion, followed by reduction.

Azo compound (Section 24.8): A compound with the general structure $R-N=N-R'$.

Backbone (Section 26.4): The continuous chain of atoms running the length of a protein or other polymer.

Base peak (Section 12.1): The most intense peak in a mass spectrum.

Basicity constant, K_b (Section 24.3): A measure of base strength in water. For any base B, the basicity constant is given by the expression

$$B + H_2O \rightleftharpoons BH^+ + OH^-$$

$$K_b = \frac{[BH^+][OH^-]}{[B]}$$

Bent bonds (Section 4.4): The bonds in small rings such as cyclopropane that bend away from the internuclear line and overlap at a slight angle, rather than head-on. Bent bonds are highly strained and highly reactive.

Benzoyl group (Section 19.1): The C_6H_5CO- group.

Benzyl group (Section 15.1): The $C_6H_5CH_2-$ group.

Benzylic (Section 11.5): The position next to an aromatic ring.

Benzyne (Section 16.8): An unstable compound having a triple bond in a benzene ring.

β Anomer (Section 25.5): The cyclic hemiacetal form of a sugar that has the hemiacetal $-OH$ group trans to the $-OH$ at the lowest chirality center in a Fischer projection.

β Diketone (Section 22.5): A 1,3-diketone.

β-Keto ester (Section 22.5): A 3-oxoester.

β Lactam (Chapter 21 *A Deeper Look*): A four-membered lactam, or cyclic amide. Penicillin and cephalosporin antibiotics contain β-lactam rings.

β-Oxidation pathway (Section 29.3): The metabolic pathway for degrading fatty acids.

β-Pleated sheet (Section 26.9): A type of secondary structure of a protein.

Betaine (Section 19.11): A neutral dipolar molecule with nonadjacent positive and negative charges. For example, the adduct of a Wittig reagent with a carbonyl compound is a betaine.

Bicycloalkane (Section 4.9): A cycloalkane that contains two rings.

Bimolecular reaction (Section 11.2): A reaction whose rate-limiting step occurs between two reactants.

Block copolymer (Section 31.3): A polymer in which different blocks of identical monomer units alternate with one another.

Boat cyclohexane (Section 4.5): A conformation of cyclohexane that bears a slight resemblance to a boat. Boat cyclohexane has no angle strain but has a large number of eclipsing interactions that make it less stable than chair cyclohexane.

Boc derivative (Section 26.7): A butyloxycarbonyl N-protected amino acid.

Bond angle (Section 1.6): The angle formed between two adjacent bonds.

Bond dissociation energy, D (Section 6.8): The amount of energy needed to break a bond and produce two radical fragments.

Bond length (Section 1.5): The equilibrium distance between the nuclei of two atoms that are bonded to each other.

Bond strength (Section 1.5): An alternative name for bond dissociation energy.

Bonding MO (Section 1.11): A molecular orbital that is lower in energy than the atomic orbitals from which it is formed.

Branched-chain alkane (Section 3.2): An alkane that contains a branching connection of carbons as opposed to a straight-chain alkane.

Bridgehead atom (Section 4.9): An atom that is shared by more than one ring in a polycyclic molecule.

Bromohydrin (Section 8.3): A 1,2-bromoalcohol; obtained by addition of HOBr to an alkene.

Bromonium ion (Section 8.2): A species with a divalent, positively charged bromine, R_2Br^+.

Brønsted–Lowry acid (Section 2.7): A substance that donates a hydrogen ion (proton; H^+) to a base.

Brønsted–Lowry base (Section 2.7): A substance that accepts H^+ from an acid.

C-terminal amino acid (Section 26.4): The amino acid with a free $-CO_2H$ group at the end of a protein chain.

Cahn–Ingold–Prelog sequence rules (Sections 5.5, 7.5): A series of rules for assigning relative rankings to substituent groups on a chirality center or a double-bond carbon atom.

Cannizzaro reaction (Section 19.12): The disproportionation reaction of an aldehyde on treatment with base to yield an alcohol and a carboxylic acid.

Carbanion (Sections 10.6, 19.7): A carbon anion, or substance that contains a trivalent, negatively charged carbon atom ($R_3C:^-$). Alkyl carbanions are sp^3-hybridized and have eight electrons in the outer shell of the negatively charged carbon.

Carbene (Section 8.9): A neutral substance that contains a divalent carbon atom having only six electrons in its outer shell ($R_2C:$).

Carbinolamine (Section 19.8): A molecule that contains the $R_2C(OH)NH_2$ functional group. Carbinolamines are produced as intermediates during the nucleophilic addition of amines to carbonyl compounds.

Carbocation (Sections 6.5, 7.9): A carbon cation, or substance that contains a trivalent, positively charged carbon atom having six electrons in its outer shell (R_3C^+).

Carbohydrate (Chapter 25 Introduction): A polyhydroxy aldehyde or ketone. Carbohydrates can be either simple sugars, such as glucose, or complex sugars, such as cellulose.

Carbonyl condensation reaction (Section 23.1): A reaction that joins two carbonyl compounds together by a combination of α-substitution and nucleophilic addition reactions.

Carbonyl group (Preview of Carbonyl Chemistry): The C=O functional group.

Carboxyl group (Section 20.1): The $-CO_2H$ functional group.

Carboxylation (Section 20.5): The addition of CO_2 to a molecule.

Carboxylic acid (Chapter 20 Introduction): A compound containing the $-CO_2H$ functional group.

Carboxylic acid derivative (Chapter 21 Introduction): A compound in which an acyl group is bonded to an electronegative atom or substituent that can act as a leaving group in a substitution reaction. Esters, amides, and acid halides are examples.

Catabolism (Section 29.1): The group of metabolic pathways that break down larger molecules into smaller ones.

Catalyst (Section 6.11): A substance that increases the rate of a chemical transformation by providing an alternative mechanism but is not itself changed in the reaction.

Cation radical (Section 12.1): A reactive species, typically formed in a mass spectrometer by loss of an electron from a neutral molecule and having both a positive charge and an odd number of electrons.

Chain-growth polymer (Sections 8.10, 31.1): A polymer whose bonds are produced by chain reaction mechanisms. Polyethylene and other alkene polymers are examples.

Chain reaction (Section 6.3): A reaction that, once initiated, sustains itself in an endlessly repeating cycle of propagation steps. The radical chlorination of alkanes is an example of a chain reaction that is initiated by irradiation with light and then continues in a series of propagation steps.

Chair conformation (Section 4.5): A three-dimensional conformation of cyclohexane that resembles the rough shape of a chair. The chair form of cyclohexane is the lowest-energy conformation of the molecule.

Chemical shift (Section 13.3): The position on the NMR chart where a nucleus absorbs. By convention, the chemical shift of tetramethylsilane (TMS) is set at zero, and all other absorptions usually occur downfield (to the left on the chart). Chemical shifts are expressed in delta units (δ), where 1 δ equals 1 ppm of the spectrometer operating frequency.

Chiral (Section 5.2): Having handedness. Chiral molecules are those that do not have a plane of symmetry and are therefore not superimposable on their mirror image. A chiral molecule thus exists in two forms, one right-handed and one left-handed. The most common cause of chirality in a molecule is the presence of a carbon atom that is bonded to four different substituents.

Chiral environment (Section 5.12): The chiral surroundings or conditions in which a molecule resides.

Chirality center (Section 5.2): An atom (usually carbon) that is bonded to four different groups.

Chlorohydrin (Section 8.3): A 1,2-chloroalcohol; obtained by addition of HOCl to an alkene.

Chromatography (Section 26.5): A technique for separating a mixture of compounds into pure components. Different compounds adsorb to a stationary support phase and are then carried along it at different rates by a mobile phase.

Cis–trans isomers (Sections 4.2, 7.4): Stereoisomers that differ in their stereochemistry about a ring or double bond.

Citric acid cycle (Section 29.7): The metabolic pathway by which acetyl CoA is degraded to CO_2.

Claisen condensation reaction (Section 23.7): The carbonyl condensation reaction of two ester molecules to give a β-keto ester product.

Claisen rearrangement reaction (Sections 18.4, 30.8): The pericyclic conversion of an allyl phenyl ether to an *o*-allylphenol or an allyl vinyl ether to a γ,δ-unsaturated ketone by heating.

Coding strand (Section 28.4): The sense strand of double-helical DNA that contains the gene.

Codon (Section 28.5): A three-base sequence on a messenger RNA chain that encodes the genetic information necessary to cause a specific amino acid to be incorporated into a protein. Codons on mRNA are read by complementary anticodons on tRNA.

Coenzyme (Section 26.10): A small organic molecule that acts as a cofactor in a biological reaction.

Cofactor (Section 26.10): A small nonprotein part of an enzyme that is necessary for biological activity.

Combinatorial chemistry (Chapter 16 *A Deeper Look*): A procedure in which anywhere from a few dozen to several hundred thousand substances are prepared simultaneously.

Complex carbohydrate (Section 25.1): A carbohydrate that is made of two or more simple sugars linked together by glycoside bonds.

Concerted reaction (Section 30.1): A reaction that takes place in a single step without intermediates. For example, the Diels–Alder cycloaddition reaction is a concerted process.

Condensed structure (Section 1.12): A shorthand way of writing structures in which carbon–hydrogen and carbon–carbon bonds are understood rather than shown explicitly. Propane, for example, has the condensed structure $CH_3CH_2CH_3$.

Configuration (Section 5.5): The three-dimensional arrangement of atoms bonded to a chirality center.

Conformation (Section 3.6): The three-dimensional shape of a molecule at any given instant, assuming that rotation around single bonds is frozen.

Conformational analysis (Section 4.8): A means of assessing the energy of a substituted cycloalkane by totaling the steric interactions present in the molecule.

Conformer (Section 3.6): A conformational isomer.

Conjugate acid (Section 2.7): The product that results from protonation of a Brønsted–Lowry base.

Conjugate addition (Section 19.13): Addition of a nucleophile to the β carbon atom of an α,β-unsaturated carbonyl compound.

Conjugate base (Section 2.7): The product that results from deprotonation of a Brønsted–Lowry acid.

Conjugation (Chapter 14 Introduction): A series of overlapping p orbitals, usually in alternating single and multiple bonds. For example, 1,3-butadiene is a conjugated diene, 3-buten-2-one is a conjugated enone, and benzene is a cyclic conjugated triene.

Conrotatory (Section 30.2): A term used to indicate that p orbitals must rotate in the same direction during electrocyclic ring-opening or ring-closure.

Constitutional isomers (Sections 3.2, 5.9): Isomers that have their atoms connected in a different order. For example, butane and 2-methylpropane are constitutional isomers.

Cope rearrangement (Section 30.8): The sigmatropic rearrangement of a 1,5-hexadiene.

Copolymer (Section 31.3): A polymer obtained when two or more different monomers are allowed to polymerize together.

Coupled reactions (Section 29.1): Two reactions that share a common intermediate so that the energy released in the favorable step allows the unfavorable step to occur.

Coupling constant, J (Section 13.11): The magnitude (expressed in hertz) of the interaction between nuclei whose spins are coupled.

Covalent bond (Section 1.5): A bond formed by sharing electrons between atoms.

Cracking (Chapter 3 *A Deeper Look*): A process used in petroleum refining in which large alkanes are thermally cracked into smaller fragments.

Crown ether (Section 18.7): A large-ring polyether; used as a phase-transfer catalyst.

Crystallite (Section 31.6): A highly ordered crystal-like region within a long polymer chain.

Curtius rearrangement (Section 24.6): The conversion of an acid chloride into an amine by reaction with azide ion, followed by heating with water.

Cyanohydrin (Section 19.6): A compound with an −OH group and a −CN group bonded to the same carbon atom; formed by addition of HCN to an aldehyde or ketone.

Cycloaddition reaction (Sections 14.4, 30.5): A pericyclic reaction in which two reactants add together in a single step to yield a cyclic product. The Diels–Alder reaction between a diene and a dienophile to give a cyclohexene is an example.

Cycloalkane (Section 4.1): An alkane that contains a ring of carbons.

D Sugar (Section 25.3): A sugar whose hydroxyl group at the chirality center farthest from the carbonyl group has the same configuration as D-glyceraldehyde and points to the right when drawn in Fischer projection.

d,l form (Section 5.8): The racemic mixture of a chiral compound.

Deactivating group (Section 16.4): An electron-withdrawing substituent that decreases the reactivity of an aromatic ring toward electrophilic aromatic substitution.

Deamination (Section 29.9): The removal of an amino group from a molecule, as occurs with amino acids during metabolic degradation.

Debye, D (Section 2.2): The unit for measuring dipole moments; 1 D = 3.336×10^{-30} coulomb meter (C · m).

Decarboxylation (Section 22.7): The loss of carbon dioxide from a molecule. β-Keto acids decarboxylate readily on heating.

Degenerate orbitals (Section 15.2): Two or more orbitals that have the same energy level.

Degree of unsaturation (Section 7.2): The number of rings and/or multiple bonds in a molecule.

Dehydration (Sections 8.1, 11.10, 17.6): The loss of water from an alcohol to yield an alkene.

Dehydrohalogenation (Sections 8.1, 11.8): The loss of HX from an alkyl halide. Alkyl halides undergo dehydrohalogenation to yield alkenes on treatment with strong base.

Delocalization (Sections 10.4, 15.2): A spreading out of electron density over a conjugated π electron system. For example, allylic cations and allylic anions are delocalized because their charges are spread out over the entire π electron system. Aromatic compounds have $4n + 2$ π electrons delocalized over their ring.

Delta scale (Section 13.3): An arbitrary scale used to calibrate NMR charts. One delta unit (δ) is equal to 1 part per million (ppm) of the spectrometer operating frequency.

Denaturation (Section 26.9): The physical changes that occur in a protein when secondary and tertiary structures are disrupted.

Deoxy sugar (Section 25.7): A sugar with one of its −OH groups replaced by an −H.

Deoxyribonucleic acid (DNA) (Section 28.1): The biopolymer consisting of deoxyribonucleotide units linked together through phosphate–sugar bonds. Found in the nucleus of cells, DNA contains an organism's genetic information.

DEPT-NMR (Section 13.6): An NMR method for distinguishing among signals due to CH_3, CH_2, CH, and quaternary carbons. That is, the number of hydrogens attached to each carbon can be determined.

Deshielding (Section 13.2): An effect observed in NMR that causes a nucleus to absorb toward the left (downfield) side of the chart. Deshielding is caused by a withdrawal of electron density from the nucleus.

Dess–Martin periodinane (Section 17.7): An iodine-based reagent commonly used for the laboratory oxidation of a primary alcohol to an aldehyde or a secondary alcohol to a ketone.

Deuterium isotope effect (Section 11.8): A tool used in mechanistic investigations to establish whether a C−H bond is broken in the rate-limiting step of a reaction.

Dextrorotatory (Section 5.3): A word used to describe an optically active substance that rotates the plane of polarization of plane-polarized light in a right-handed (clockwise) direction.

Diastereomers (Section 5.6): Non–mirror-image stereoisomers; diastereomers have the same configuration at one or more chirality centers but differ at other chirality centers.

Diastereotopic (Section 13.8): Hydrogens in a molecule whose replacement by some other group leads to different diastereomers.

1,3-Diaxial interaction (Section 4.7): The strain energy caused by a steric interaction between axial groups three carbon atoms apart in chair cyclohexane.

Diazonium salt (Section 24.8): A compound with the general structure $RN_2^+ \ X^-$.

Diazotization (Section 24.8): The conversion of a primary amine, RNH_2, into a diazonium ion, RN_2^+, by treatment with nitrous acid.

Dideoxy DNA sequencing (Section 28.6): A biochemical method for sequencing DNA strands.

Dieckmann cyclization reaction (Section 23.9): An intramolecular Claisen condensation reaction of a diester to give a cyclic β-keto ester.

Diels–Alder reaction (Sections 14.4, 30.5): The cycloaddition reaction of a diene with a dienophile to yield a cyclohexene.

Dienophile (Section 14.5): A compound containing a double bond that can take part in the Diels–Alder cycloaddition reaction. The most reactive dienophiles are those that have electron-withdrawing groups on the double bond.

Digestion (Section 29.1): The first stage of catabolism, in which food is broken down by hydrolysis of ester, glycoside (acetal), and peptide (amide) bonds to yield fatty acids, simple sugars, and amino acids.

Dihedral angle (Section 3.6): The angle between two bonds on adjacent carbons as viewed along the C–C bond.

Dipole moment, μ (Section 2.2): A measure of the net polarity of a molecule. A dipole moment arises when the centers of mass of positive and negative charges within a molecule do not coincide.

Dipole–dipole force (Section 2.12): A noncovalent electrostatic interaction between dipolar molecules.

Disaccharide (Section 25.8): A carbohydrate formed by linking two simple sugars through an acetal bond.

Dispersion force (Section 2.12): A noncovalent interaction between molecules that arises because of constantly changing electron distributions within the molecules.

Disrotatory (Section 30.2): A term used to indicate that p orbitals rotate in opposite directions during electrocyclic ring-opening or ring-closing reactions.

Disulfide (Section 18.8): A compound of the general structure RSSR'.

DNA (Section 28.1): Deoxyribonucleic acid.

Double bond (Section 1.8): A covalent bond formed by sharing two electron pairs between atoms.

Double helix (Section 28.2): The structure of DNA in which two polynucleotide strands coil around each other.

Doublet (Section 13.11): A two-line NMR absorption caused by spin–spin splitting when the spin of the nucleus under observation couples with the spin of a neighboring magnetic nucleus.

Downfield (Section 13.3): Referring to the left-hand portion of the NMR chart.

***E* geometry** (Section 7.5): A term used to describe the stereochemistry of a carbon–carbon double bond. The two groups on each carbon are ranked according to the Cahn–Ingold–Prelog sequence rules, and the two carbons are compared. If the higher-ranked groups on each carbon are on opposite sides of the double bond, the bond has *E* geometry.

E1 reaction (Section 11.10): A unimolecular elimination reaction in which the substrate spontaneously dissociates to give a carbocation intermediate, which loses a proton in a separate step.

E1cB reaction (Section 11.10): A unimolecular elimination reaction in which a proton is first removed to give a carbanion intermediate, which then expels the leaving group in a separate step.

E2 reaction (Section 11.8): A bimolecular elimination reaction in which C–H and C–X bond cleavages are simultaneous.

Eclipsed conformation (Section 3.6): The geometric arrangement around a carbon–carbon single bond in which the bonds to substituents on one carbon are parallel to the bonds to substituents on the neighboring carbon as viewed in a Newman projection.

Eclipsing strain (Section 3.6): The strain energy in a molecule caused by electron repulsions between eclipsed bonds. Eclipsing strain is also called torsional strain.

Edman degradation (Section 26.6): A method for N-terminal sequencing of peptide chains by treatment with *N*-phenylisothiocyanate.

Eicosanoid (Section 27.4): A lipid derived biologically from 5,8,11,14-eicosatetraenoic acid, or arachidonic acid. Prostaglandins, thromboxanes, and leukotrienes are examples.

Elastomer (Section 31.6): An amorphous polymer that has the ability to stretch out and spring back to its original shape.

Electrocyclic reaction (Section 30.2): A unimolecular pericyclic reaction in which a ring is formed or broken by a concerted reorganization of electrons through a cyclic transition state. For example, the cyclization of 1,3,5-hexatriene to yield 1,3-cyclohexadiene is an electrocyclic reaction.

Electromagnetic spectrum (Section 12.5): The range of electromagnetic energy, including infrared, ultraviolet, and visible radiation.

Electron configuration (Section 1.3): A list of the orbitals occupied by electrons in an atom.

Electron-dot structure (Section 1.4): A representation of a molecule showing valence electrons as dots.

Electron-transport chain (Section 29.1): The final stage of catabolism in which ATP is produced.

Electronegativity (Section 2.1): The ability of an atom to attract electrons in a covalent bond. Electronegativity increases across the periodic table from left to right and from bottom to top.

Electrophile (Section 6.4): An "electron-lover," or substance that accepts an electron pair from a nucleophile in a polar bond-forming reaction.

Electrophilic addition reaction (Section 7.7): The addition of an electrophile to a carbon–carbon double bond to yield a saturated product.

Electrophilic aromatic substitution reaction (Chapter 16 Introduction): A reaction in which an electrophile (E^+) reacts with an aromatic ring and substitutes for one of the ring hydrogens.

Electrophoresis (Sections 26.2, 28.6): A technique used for separating charged organic molecules, particularly proteins and DNA fragments. The mixture to be separated is placed on a buffered gel or paper, and an electric potential is applied across the ends of the apparatus. Negatively charged molecules migrate toward the positive electrode, and positively charged molecules migrate toward the negative electrode.

Electrostatic potential map (Section 2.1): A molecular representation that uses color to indicate the charge distribution in the molecule as derived from quantum-mechanical calculations.

Elimination reaction (Section 6.1): What occurs when a single reactant splits into two products.

Elution (Section 26.5): The passage of a substance from a chromatography column.

Embden–Meyerhof pathway (Section 29.5): An alternative name for glycolysis.

Enamine (Section 19.8): A compound with the $R_2N-CR=CR_2$ functional group.

Enantiomers (Section 5.1): Stereoisomers of a chiral substance that have a mirror-image relationship. Enantiomers have opposite configurations at all chirality centers.

Enantioselective synthesis (Chapter 19 *A Deeper Look*): A reaction method that yields only a single enantiomer of a chiral product starting from an achiral reactant.

Enantiotopic (Section 13.8): Hydrogens in a molecule whose replacement by some other group leads to different enantiomers.

3′ End (Section 28.1): The end of a nucleic acid chain with a free hydroxyl group at C3′.

5′ End (Section 28.1): The end of a nucleic acid chain with a free hydroxyl group at C5′.

Endergonic (Section 6.7): A reaction that has a positive free-energy change and is therefore nonspontaneous. In an energy diagram, the product of an endergonic reaction has a higher energy level than the reactants.

Endo (Section 14.5): A term indicating the stereochemistry of a substituent in a bridged bicycloalkane. An endo substituent is syn to the larger of the two bridges.

Endothermic (Section 6.7): A reaction that absorbs heat and therefore has a positive enthalpy change.

Energy diagram (Section 6.9): A representation of the course of a reaction, in which free energy is plotted as a function of reaction progress. Reactants, transition states, intermediates, and products are represented, and their appropriate energy levels are indicated.

Enol (Sections 9.4, 22.1): A vinylic alcohol that is in equilibrium with a carbonyl compound, $C=C-OH$.

Enolate ion (Section 22.1): The anion of an enol, $C=C-O^-$.

Enthalpy change, ΔH (Section 6.7): The heat of reaction. The enthalpy change that occurs during a reaction is a measure of the difference in total bond energy between reactants and products.

Entropy change, ΔS (Section 6.7): The change in amount of molecular randomness. The entropy change that occurs during a reaction is a measure of the difference in randomness between reactants and products.

Enzyme (Sections 6.11, 26.10): A biological catalyst. Enzymes are large proteins that catalyze specific biochemical reactions.

Epimers (Section 5.6): Diastereomers that differ in configuration at only one chirality center but are the same at all others.

Epoxide (Section 8.7): A three-membered-ring ether functional group.

Equatorial bond (Section 4.6): A bond to cyclohexane that lies along the rough equator of the ring.

ESI (Section 12.4): Electrospray ionization; a "soft" ionization method used for mass spectrometry of biological samples of very high molecular weight.

Essential amino acid (Section 26.1): One of nine amino acids that are biosynthesized only in plants and microorganisms and must be obtained by humans in the diet.

Essential monosaccharide (Section 25.7): One of eight simple sugars that is best obtained in the diet rather than by biosynthesis.

Essential oil (Chapter 8 *A Deeper Look*): The volatile oil obtained by steam distillation of a plant extract.

Ester (Chapter 21 Introduction): A compound containing the $-CO_2R$ functional group.

Estrogen (Section 27.6): A female steroid sex hormone.

Ether (Chapter 18 Introduction): A compound that has two organic substituents bonded to the same oxygen atom, ROR′.

Exergonic (Section 6.7): A reaction that has a negative free-energy change and is therefore spontaneous. On an energy diagram, the product of an exergonic reaction has a lower energy level than that of the reactants.

Exo (Section 14.5): A term indicating the stereochemistry of a substituent in a bridged bicycloalkane. An exo substituent is anti to the larger of the two bridges.

Exon (Section 28.4): A section of DNA that contains genetic information.

Exothermic (Section 6.7): A reaction that releases heat and therefore has a negative enthalpy change.

Fat (Section 27.1): A solid triacylglycerol derived from an animal source.

Fatty acid (Section 27.1): A long, straight-chain carboxylic acid found in fats and oils.

Fiber (Section 31.6): A thin thread produced by extruding a molten polymer through small holes in a die.

Fibrous protein (Section 26.9): A protein that consists of polypeptide chains arranged side by side in long threads. Such proteins are tough, insoluble in water, and used in nature for structural materials such as hair, hooves, and fingernails.

Fingerprint region (Section 12.7): The complex region of the infrared spectrum from 1500–400 cm^{-1}.

First-order reaction (Section 11.4): A reaction whose rate-limiting step is unimolecular and whose kinetics therefore depend on the concentration of only one reactant.

Fischer esterification reaction (Section 21.3): The acid-catalyzed nucleophilic acyl substitution reaction of a carboxylic acid with an alcohol to yield an ester.

Fischer projection (Section 25.2): A means of depicting the absolute configuration of a chiral molecule on a flat page. A Fischer projection uses a cross to represent the chirality center. The horizontal arms of the cross represent bonds coming out of the plane of the page, and the vertical arms of the cross represent bonds going back into the plane of the page.

Fmoc derivative (Section 26.7): A fluorenylmethyloxycarbonyl N-protected amino acid.

Formal charge (Section 2.3): The difference in the number of electrons owned by an atom in a molecule and by the same atom in its elemental state.

Formyl group (Section 19.1): A —CHO group.

Frequency, ν (Section 12.5): The number of electromagnetic wave cycles that travel past a fixed point in a given unit of time. Frequencies are expressed in units of cycles per second, or hertz.

Friedel–Crafts reaction (Section 16.3): An electrophilic aromatic substitution reaction to alkylate or acylate an aromatic ring.

Frontier orbitals (Section 30.1): The highest occupied (HOMO) and lowest unoccupied (LUMO) molecular orbitals.

FT-NMR (Section 13.4): Fourier-transform NMR; a rapid technique for recording NMR spectra in which all magnetic nuclei absorb at the same time.

Functional group (Section 3.1): An atom or group of atoms that is part of a larger molecule and has a characteristic chemical reactivity.

Functional RNA (Section 28.4): An alternative name for small RNAs.

Furanose (Section 25.5): The five-membered-ring form of a simple sugar.

Gabriel amine synthesis (Section 24.6): A method for preparing an amine by S_N2 reaction of an alkyl halide with potassium phthalimide, followed by hydrolysis.

Gauche conformation (Section 3.7): The conformation of butane in which the two methyl groups lie 60° apart as viewed in a Newman projection. This conformation has 3.8 kJ/mol steric strain.

Geminal (Section 19.5): Referring to two groups attached to the same carbon atom. For example, the hydrate formed by nucleophilic addition of water to an aldehyde or ketone is a geminal diol.

Gibbs free-energy change, ΔG (Section 6.7): The free-energy change that occurs during a reaction, given by the equation $\Delta G = \Delta H - T\Delta S$. A reaction with a negative free-energy change is spontaneous, and a reaction with a positive free-energy change is nonspontaneous.

Gilman reagent (Section 10.7): A diorganocopper reagent, R_2CuLi.

Glass transition temperature, T_g (Section 31.6): The temperature at which a hard, amorphous polymer becomes soft and flexible.

Globular protein (Section 26.9): A protein that is coiled into a compact, nearly spherical shape. Globular proteins, which are generally water-soluble and mobile within the cell, are the structural class to which enzymes belong.

Gluconeogenesis (Section 29.8): The anabolic pathway by which organisms make glucose from simple three-carbon precursors.

Glycal (Section 25.9): An unsaturated sugar with a C1–C2 double bond.

Glycal assembly method (Section 25.9): A method for linking monosaccharides together to synthesize polysaccharides.

Glycerophospholipid (Section 27.3): A lipid that contains a glycerol backbone linked to two fatty acids and a phosphoric acid.

Glycoconjugate (Section 25.6): A molecule in which a carbohydrate is linked through its anomeric center to another biological molecule such as a lipid or protein.

Glycol (Section 8.7): A diol, such as ethylene glycol, $HOCH_2CH_2OH$.

Glycolipid (Section 25.6): A biological molecule in which a carbohydrate is linked through a glycoside bond to a lipid.

Glycolysis (Section 29.5): A series of ten enzyme-catalyzed reactions that break down glucose into 2 equivalents of pyruvate, $CH_3COCO_2^-$.

Glycoprotein (Section 25.6): A biological molecule in which a carbohydrate is linked through a glycoside bond to a protein.

Glycoside (Section 25.6): A cyclic acetal formed by reaction of a sugar with another alcohol.

Graft copolymer (Section 31.3): A copolymer in which homopolymer branches of one monomer unit are "grafted" onto a homopolymer chain of another monomer unit.

Green chemistry (Chapters 11, 24 *A Deeper Look*): The design and implementation of chemical products and processes that reduce waste and minimize or eliminate the generation of hazardous substances.

Grignard reagent (Section 10.6): An organomagnesium halide, RMgX.

Ground state (Section 1.3): The most stable, lowest-energy electron configuration of a molecule or atom.

Haloform reaction (Section 22.6): The reaction of a methyl ketone with halogen and base to yield a haloform (CHX_3) and a carboxylic acid.

Halogenation (Sections 8.2, 16.1): The reaction of halogen with an alkene to yield a 1,2-dihalide addition product or with an aromatic compound to yield a substitution product.

Halohydrin (Section 8.3): A 1,2-haloalcohol, such as that obtained on addition of HOBr to an alkene.

Halonium ion (Section 8.2): A species containing a positively charged, divalent halogen. Three-membered-ring bromonium ions are intermediates in the electrophilic addition of Br_2 to alkenes.

Hammond postulate (Section 7.10): A postulate stating that we can get a picture of what a given transition state looks like by looking at the structure of the nearest stable species. Exergonic reactions have transition states that resemble reactant; endergonic reactions have transition states that resemble product.

Heat of combustion (Section 4.3): The amount of heat released when a compound burns completely in oxygen.

Heat of hydrogenation (Section 7.6): The amount of heat released when a carbon–carbon double bond is hydrogenated.

Heat of reaction (Section 6.7): An alternative name for the enthalpy change in a reaction, ΔH.

Hell–Volhard–Zelinskii (HVZ) reaction (Section 22.4): The reaction of a carboxylic acid with Br_2 and phosphorus to give an α-bromo carboxylic acid.

Hemiacetal (Section 19.10): A functional group having one —OR and one —OH group bonded to the same carbon.

Henderson–Hasselbalch equation (Sections 20.3, 24.5, 26.2): An equation for determining the extent of dissociation of a weak acid at various pH values.

Hertz, Hz (Section 12.5): A unit of measure of electromagnetic frequency, the number of waves that pass by a fixed point per second.

Heterocycle (Sections 15.5, 24.9): A cyclic molecule whose ring contains more than one kind of atom. For example, pyridine is a heterocycle that contains five carbon atoms and one nitrogen atom in its ring.

Heterolytic bond breakage (Section 6.2): The kind of bond-breaking that occurs in polar reactions when one fragment leaves with both of the bonding electrons: $A:B \rightarrow A^+ + B:^-$.

Hofmann elimination reaction (Section 24.7): The elimination reaction of an amine to yield an alkene by reaction with iodomethane followed by heating with Ag_2O.

Hofmann rearrangement (Section 24.6): The conversion of an amide into an amine by reaction with Br_2 and base.

HOMO (Sections 14.7, 30.1): The highest occupied molecular orbital. The symmetries of the HOMO and LUMO are important in pericyclic reactions.

Homolytic bond breakage (Section 6.2): The kind of bond-breaking that occurs in radical reactions when each fragment leaves with one bonding electron: $A:B \rightarrow A\cdot + B\cdot$.

Homopolymer (Section 31.3): A polymer made up of identical repeating units.

Homotopic (Section 13.8): Hydrogens in a molecule that give the identical structure on replacement by X and thus show identical NMR absorptions.

Hormone (Section 27.6): A chemical messenger that is secreted by an endocrine gland and carried through the bloodstream to a target tissue.

HPLC (Section 26.5): High-pressure liquid chromatography; a variant of column chromatography using high pressure to force solvent through very small absorbent particles.

Hückel's rule (Section 15.3): A rule stating that monocyclic conjugated molecules having $4n + 2$ π electrons (n = an integer) are aromatic.

Hund's rule (Section 1.3): If two or more empty orbitals of equal energy are available, one electron occupies each, with their spins parallel, until all are half-full.

Hybrid orbital (Section 1.6): An orbital derived from a combination of atomic orbitals. Hybrid orbitals, such as the sp^3, sp^2, and sp hybrids of carbon, are strongly directed and form stronger bonds than atomic orbitals do.

Hydration (Section 8.4): Addition of water to a molecule, such as occurs when alkenes are treated with aqueous sulfuric acid to give alcohols.

Hydride shift (Section 7.11): The shift of a hydrogen atom and its electron pair to a nearby cationic center.

Hydroboration (Section 8.5): Addition of borane (BH_3) or an alkylborane to an alkene. The resultant trialkylborane products can be oxidized to yield alcohols.

Hydrocarbon (Section 3.2): A compound that contains only carbon and hydrogen.

Hydrogen bond (Sections 2.12, 17.2): A weak attraction between a hydrogen atom bonded to an electronegative atom and an electron lone pair on another electronegative atom.

Hydrogenation (Section 8.6): Addition of hydrogen to a double or triple bond to yield a saturated product.

Hydrogenolysis (Section 26.7): Cleavage of a bond by reaction with hydrogen. Benzylic ethers and esters, for instance, are cleaved by hydrogenolysis.

Hydrophilic (Section 2.12): Water-loving; attracted to water.

Hydrophobic (Section 2.12): Water-fearing; repelled by water.

Hydroquinone (Section 17.10): A 1,4-dihydroxybenzene.

Hydroxylation (Section 8.7): Addition of two —OH groups to a double bond.

Hyperconjugation (Sections 7.6, 7.9): An electronic interaction that results from overlap of a vacant p orbital on one atom with a neighboring C–H σ bond. Hyperconjugation is important in stabilizing carbocations and substituted alkenes.

Imide (Section 24.6): A compound with the —CONHCO— functional group.

Imine (Section 19.8): A compound with the $R_2C=NR$ functional group.

Inductive effect (Sections 2.1, 7.9, 16.5): The electron-attracting or electron-withdrawing effect transmitted through σ bonds. Electronegative elements have an electron-withdrawing inductive effect.

Infrared (IR) spectroscopy (Section 12.6): A kind of optical spectroscopy that uses infrared energy. IR spectroscopy is particularly useful in organic chemistry for determining the kinds of functional groups present in molecules.

Initiator (Sections 6.3, 31.1): A substance that is used to initiate a radical chain reaction or polymerization. For example, radical chlorination of alkanes is initiated when light energy breaks the weak Cl–Cl bond to form Cl· radicals.

Integration (Section 13.10): A technique for measuring the area under an NMR peak to determine the relative number of each kind of proton in a molecule.

Intermediate (Section 6.10): A species that is formed during the course of a multistep reaction but is not the final product. Intermediates are more stable than transition states but may or may not be stable enough to isolate.

Intramolecular, intermolecular (Section 23.6): A reaction that occurs within the same molecule is intramolecular; a reaction that occurs between two molecules is intermolecular.

Intron (Section 28.4): A section of DNA that does not contain genetic information.

Ion pair (Section 11.5): A loose association between two ions in solution. Ion pairs are implicated as intermediates in S_N1 reactions to account for the partial retention of stereochemistry that is often observed.

Ionic bond (Section 1.4): The electrostatic attraction between ions of unlike charge.

Isoelectric point, p*I* (Section 26.2): The pH at which the number of positive charges and the number of negative charges on a protein or an amino acid are equal.

Isomers (Sections 3.2, 5.9): Compounds that have the same molecular formula but different structures.

Isoprene rule (Chapter 8 *A Deeper Look*): An observation to the effect that terpenoids appear to be made up of isoprene (2-methyl-1,3-butadiene) units connected head-to-tail.

Isotactic (Section 31.2): A chain-growth polymer in which the stereochemistry of the substituents is oriented regularly along the backbone.

Isotopes (Section 1.1): Atoms of the same element that have different mass numbers.

IUPAC system of nomenclature (Section 3.4): Rules for naming compounds, devised by the International Union of Pure and Applied Chemistry.

Kekulé structure (Section 1.4): An alternative name for a line-bond structure, which represents a molecule by showing covalent bonds as lines between atoms.

Ketal (Section 19.10): An alternative name for an acetal derived from a ketone rather than an aldehyde and consisting of two —OR groups bonded to the same carbon, $R_2C(OR')_2$. Ketals are often used as protecting groups for ketones.

Keto–enol tautomerism (Sections 9.4, 22.1): The equilibration between a carbonyl form and vinylic alcohol form of a molecule.

Ketone (Chapter 19 Introduction): A compound with two organic substituents bonded to a carbonyl group, $R_2C=O$.

Ketose (Section 25.1): A carbohydrate with a ketone functional group.

Kiliani–Fischer synthesis (Section 25.6): A method for lengthening the chain of an aldose sugar.

Kinetic control (Section 14.3): A reaction that follows the lowest activation energy pathway is said to be kinetically controlled. The product is the most rapidly formed but is not necessarily the most stable.

Kinetics (Section 11.2): Referring to reaction rates. Kinetic measurements are useful for helping to determine reaction mechanisms.

Koenigs–Knorr reaction (Section 25.6): A method for the synthesis of glycosides by reaction of an alcohol with a pyranosyl bromide.

Krebs cycle (Section 29.7): An alternative name for the citric acid cycle, by which acetyl CoA is degraded to CO_2.

L Sugar (Section 25.3): A sugar whose hydroxyl group at the chirality center farthest from the carbonyl group points to the left when drawn in Fischer projection.

Lactam (Section 21.7): A cyclic amide.

Lactone (Section 21.6): A cyclic ester.

Lagging strand (Section 28.3): The complement of the original 3′ → 5′ DNA strand that is synthesized discontinuously in small pieces that are subsequently linked by DNA ligases.

LDA (Section 22.5): Lithium diisopropylamide, $LiN(i\text{-}C_3H_7)_2$, a strong base commonly used to convert carbonyl compounds into their enolate ions.

LD$_{50}$ (Chapter 1 *A Deeper Look*): The amount of a substance per kilogram body weight that is lethal to 50% of test animals.

Leading strand (Section 28.3): The complement of the original 5′ → 3′ DNA strand that is synthesized continuously in a single piece.

Leaving group (Section 11.2): The group that is replaced in a substitution reaction.

Levorotatory (Section 5.3): An optically active substance that rotates the plane of polarization of plane-polarized light in a left-handed (counterclockwise) direction.

Lewis acid (Section 2.11): A substance with a vacant low-energy orbital that can accept an electron pair from a base. All electrophiles are Lewis acids.

Lewis base (Section 2.11): A substance that donates an electron lone pair to an acid. All nucleophiles are Lewis bases.

Lewis structure (Section 1.4): A representation of a molecule showing valence electrons as dots.

Lindlar catalyst (Section 9.5): A hydrogenation catalyst used to convert alkynes to cis alkenes.

Line-bond structure (Section 1.4): An alternative name for a Kekulé structure, which represents a molecule by showing covalent bonds as lines between atoms.

1→4 Link (Section 25.8): A glycoside link between the C1 —OH group of one sugar and the C4 —OH group of another sugar.

Lipid (Chapter 27 Introduction): A naturally occurring substance isolated from cells and tissues by extraction with a nonpolar solvent. Lipids belong to many different structural classes, including fats, terpenoids, prostaglandins, and steroids.

Lipid bilayer (Section 27.3): The ordered lipid structure that forms a cell membrane.

Lipoprotein (Chapter 27 *A Deeper Look*): A complex molecule with both lipid and protein parts that transports lipids through the body.

Locant (Section 3.4): A number in a chemical name that locates the positions of the functional groups and substituents in the molecule.

Lone-pair electrons (Section 1.4): Nonbonding valence-shell electron pairs. Lone-pair electrons are used by nucleophiles in their reactions with electrophiles.

LUMO (Sections 14.7, 30.1): The lowest unoccupied molecular orbital. The symmetries of the LUMO and the HOMO are important in determining the stereochemistry of pericyclic reactions.

Magnetic resonance imaging, MRI (Chapter 13 *A Deeper Look*): A medical diagnostic technique based on nuclear magnetic resonance.

MALDI (Section 12.4): Matrix-assisted laser desorption ionization; a soft ionization method used for mass spectrometry of biological samples of very high molecular weight.

Malonic ester synthesis (Section 22.7): The synthesis of a carboxylic acid by alkylation of an alkyl halide with diethyl malonate, followed by hydrolysis and decarboxylation.

Markovnikov's rule (Section 7.8): A guide for determining the regiochemistry (orientation) of electrophilic addition reactions. In the addition of HX to an alkene, the hydrogen atom bonds to the alkene carbon that has fewer alkyl substituents.

Mass number, A (Section 1.1): The total of protons plus neutrons in an atom.

Mass spectrometry (Section 12.1): A technique for measuring the mass, and therefore the molecular weight (MW), of ions.

McLafferty rearrangement (Section 12.3): A mass-spectral fragmentation pathway for carbonyl compounds.

Mechanism (Section 6.2): A complete description of how a reaction occurs. A mechanism accounts for all starting materials and all products and describes the details of each individual step in the overall reaction process.

Meisenheimer complex (Section 16.7): An intermediate formed by addition of a nucleophile to a halo-substituted aromatic ring.

Melt transition temperature, T_m (Section 31.6): The temperature at which crystalline regions of a polymer melt to give an amorphous material.

Mercapto group (Section 18.8): An alternative name for the thiol group, –SH.

Meso compound (Section 5.7): A compound that contains chirality centers but is nevertheless achiral because it contains a symmetry plane.

Messenger RNA (Section 28.4): A kind of RNA formed by transcription of DNA and used to carry genetic messages from DNA to ribosomes.

Meta, *m*- (Section 15.1): A naming prefix used for 1,3-disubstituted benzenes.

Metabolism (Section 29.1): A collective name for the many reactions that go on in the cells of living organisms.

Metallacycle (Section 31.5): A cyclic compound that contains a metal atom in its ring.

Methylene group (Section 7.3): A –CH$_2$– or =CH$_2$ group.

Micelle (Section 27.2): A spherical cluster of soaplike molecules that aggregate in aqueous solution. The ionic heads of the molecules lie on the outside, where they are solvated by water, and the organic tails bunch together on the inside of the micelle.

Michael reaction (Section 23.10): The conjugate addition reaction of an enolate ion to an unsaturated carbonyl compound.

Molar absorptivity (Section 14.7): A quantitative measure of the amount of UV light absorbed by a sample.

Molecular ion (Section 12.1): The cation produced in a mass spectrometer by loss of an electron from the parent molecule. The mass of the molecular ion corresponds to the molecular weight of the sample.

Molecular mechanics (Chapter 4 *A Deeper Look*): A computer-based method for calculating the minimum-energy conformation of a molecule.

Molecular orbital (MO) theory (Sections 1.11, 14.1): A description of covalent bond formation as resulting from a mathematical combination of atomic orbitals (wave functions) to form molecular orbitals.

Molecule (Section 1.4): A neutral collection of atoms held together by covalent bonds.

Molozonide (Section 8.8): The initial addition product of ozone with an alkene.

Monomer (Sections 8.10, 21.9; Chapter 31 Introduction): The simple starting unit from which a polymer is made.

Monosaccharide (Section 25.1): A simple sugar.

Monoterpenoid (Chapter 8 *A Deeper Look*, Section 27.5): A ten-carbon lipid.

Multiplet (Section 13.11): A pattern of peaks in an NMR spectrum that arises by spin–spin splitting of a single absorption because of coupling between neighboring magnetic nuclei.

Mutarotation (Section 25.5): The change in optical rotation observed when a pure anomer of a sugar is dissolved in water. Mutarotation is caused by the reversible opening and closing of the acetal linkage, which yields an equilibrium mixture of anomers.

n + 1 rule (Section 13.11): A hydrogen with n other hydrogens on neighboring carbons shows $n + 1$ peaks in its ^{1}H NMR spectrum.

N-terminal amino acid (Section 26.4): The amino acid with a free –NH$_2$ group at the end of a protein chain.

Natural gas (Chapter 3 *A Deeper Look*): A naturally occurring hydrocarbon mixture consisting chiefly of methane, along with smaller amounts of ethane, propane, and butane.

Natural product (Chapter 7 *A Deeper Look*): A catchall term generally taken to mean a secondary metabolite found in bacteria, plants, and other living organisms.

Neopentyl group (Section 3.4): The 2,2-dimethylpropyl group, (CH$_3$)$_3$CCH$_2$–.

Neuraminidase (Section 25.11): An enzyme present on the surface of viral particles that cleaves the bond holding the newly formed viral particles to host cells.

New molecular entity, NME (Chapter 6 *A Deeper Look*): A new biologically active chemical substance approved for sale as a drug by the U.S. Food and Drug Administration.

Newman projection (Section 3.6): A means of indicating stereochemical relationships between substituent groups on neighboring carbons. The carbon–carbon bond is viewed end-on, and the carbons are indicated by a circle. Bonds radiating from the center of the circle are attached to the front carbon, and bonds radiating from the edge of the circle are attached to the rear carbon.

Nitration (Section 16.2): The substitution of a nitro group onto an aromatic ring.

Nitrile (Section 20.1): A compound containing the C≡N functional group.

Nitrogen rule (Section 24.10): A compound with an odd number of nitrogen atoms has an odd-numbered molecular weight.

Node (Section 1.2): A surface of zero electron density within an orbital. For example, a *p* orbital has a nodal plane passing through the center of the nucleus, perpendicular to the axis of the orbital.

Nonbonding electrons (Section 1.4): Valence electrons that are not used in forming covalent bonds.

Noncoding strand (Section 28.4): An alternative name for the antisense strand of DNA.

Noncovalent interaction (Section 2.12): One of a variety of nonbonding interactions between molecules, such as dipole–dipole forces, dispersion forces, and hydrogen bonds.

Nonessential amino acid (Section 26.1): One of the eleven amino acids that are biosynthesized by humans.

Normal alkane (Section 3.2): A straight-chain alkane, as opposed to a branched alkane. Normal alkanes are denoted by the suffix *n*, as in *n*-C_4H_{10} (*n*-butane).

NSAID (Chapter 15 *A Deeper Look*): A nonsteroidal anti-inflammatory drug, such as aspirin or ibuprofen.

Nuclear magnetic resonance, NMR (Chapter 13 Introduction): A spectroscopic technique that provides information about the carbon–hydrogen framework of a molecule. NMR works by detecting the energy absorptions accompanying the transitions between nuclear spin states that occur when a molecule is placed in a strong magnetic field and irradiated with radio-frequency waves.

Nucleic acid (Section 28.1): Deoxyribonucleic acid (DNA) and ribonucleic acid (RNA); biological polymers made of nucleotides joined together to form long chains.

Nucleophile (Section 6.4): An electron-rich species that donates an electron pair to an electrophile in a polar bond-forming reaction. Nucleophiles are also Lewis bases.

Nucleophilic acyl substitution reaction (Section 21.2): A reaction in which a nucleophile attacks a carbonyl compound and substitutes for a leaving group bonded to the carbonyl carbon.

Nucleophilic addition reaction (Section 19.4): A reaction in which a nucleophile adds to the electrophilic carbonyl group of a ketone or aldehyde to give an alcohol.

Nucleophilic aromatic substitution reaction (Section 16.7): The substitution reaction of an aryl halide by a nucleophile.

Nucleophilic substitution reaction (Section 11.1): A reaction in which one nucleophile replaces another attached to a saturated carbon atom.

Nucleophilicity (Section 11.3): The ability of a substance to act as a nucleophile in an S_N2 reaction.

Nucleoside (Section 28.1): A nucleic acid constituent, consisting of a sugar residue bonded to a heterocyclic purine or pyrimidine base.

Nucleotide (Section 28.1): A nucleic acid constituent, consisting of a sugar residue bonded both to a heterocyclic purine or pyrimidine base and to a phosphoric acid. Nucleotides are the monomer units from which DNA and RNA are constructed.

Nylon (Section 21.9): A synthetic polyamide step-growth polymer.

Okazaki fragment (Section 28.3): A short segment of a DNA lagging strand that is biosynthesized discontinuously and then linked by DNA ligases.

Olefin (Chapter 7 Introduction): An alternative name for an alkene.

Olefin metathesis polymerization (Section 31.5): A method of polymer synthesis based on using an olefin metathesis reaction.

Olefin metathesis reaction (Section 31.5): A reaction in which two olefins (alkenes) exchange substituents on their double bonds.

Oligonucleotide (Section 28.7): A short segment of DNA.

Optical activity (Section 5.3): The rotation of the plane of polarization of plane-polarized light by a chiral substance in solution.

Optical isomers (Section 5.4): An alternative name for enantiomers. Optical isomers are isomers that have a mirror-image relationship.

Orbital (Section 1.2): A wave function, which describes the volume of space around a nucleus in which an electron is most likely to be found.

Organic chemistry (Chapter 1 Introduction): The study of carbon compounds.

Organohalide (Chapter 10 Introduction): A compound that contains one or more halogen atoms bonded to carbon.

Organometallic compound (Section 10.6): A compound that contains a carbon–metal bond. Grignard reagents, RMgX, are examples.

Organophosphate (Section 1.10): A compound that contains a phosphorus atom bonded to four oxygens, with one of the oxygens also bonded to carbon.

Ortho, *o*- (Section 15.1): A naming prefix used for 1,2-disubstituted benzenes.

Oxidation (Section 10.8): A reaction that causes a decrease in electron ownership by carbon, either by bond formation between carbon and a more electronegative atom (usually oxygen, nitrogen, or a halogen) or by bond-breaking between carbon and a less electronegative atom (usually hydrogen).

Oxime (Section 19.8): A compound with the $R_2C=NOH$ functional group.

Oxirane (Section 8.7): An alternative name for an epoxide.

Oxymercuration (Section 8.4): A method for double-bond hydration by reaction of an alkene with aqueous mercuric acetate followed by treatment with $NaBH_4$.

Ozonide (Section 8.9): The product initially formed by addition of ozone to a carbon–carbon double bond. Ozonides are usually treated with a reducing agent, such as zinc in acetic acid, to produce carbonyl compounds.

Para, *p*- (Section 15.1): A naming prefix used for 1,4-disubstituted benzenes.

Paraffin (Section 3.5): A common name for alkanes.

Parent peak (Section 12.1): The peak in a mass spectrum corresponding to the molecular ion. The mass of the parent peak therefore represents the molecular weight of the compound.

Pauli exclusion principle (Section 1.3): No more than two electrons can occupy the same orbital, and those two must have spins of opposite sign.

Peptide (Chapter 26 Introduction): A short amino acid polymer in which the individual amino acid residues are linked by amide bonds.

Peptide bond (Section 26.4): An amide bond in a peptide chain.

Pericyclic reaction (Chapter 30 Introduction): A reaction that occurs in a single step by a reorganization of bonding electrons in a cyclic transition state.

Periplanar (Section 11.8): A conformation in which bonds to neighboring atoms have a parallel arrangement. In an eclipsed conformation, the neighboring bonds are syn periplanar; in a staggered conformation, the bonds are anti periplanar.

Peroxide (Section 18.1): A molecule containing an oxygen–oxygen bond functional group, ROOR' or ROOH.

Peroxyacid (Section 8.7): A compound with the $-CO_3H$ functional group. Peroxyacids react with alkenes to give epoxides.

Phenol (Chapter 17 Introduction): A compound with an $-OH$ group directly bonded to an aromatic ring, ArOH.

Phenoxide ion (Section 17.2): The anion of a phenol, ArO^-.

Phenyl group (Section 15.1): The name for the $-C_6H_5$ unit when the benzene ring is considered as a substituent. A phenyl group is abbreviated as $-Ph$.

Phosphine (Section 5.10): A trivalent phosphorus compound, R_3P.

Phosphite (Section 28.7): A compound with the structure $P(OR)_3$.

Phospholipid (Section 27.3): A lipid that contains a phosphate residue. For example, glycerophospholipids contain a glycerol backbone linked to two fatty acids and a phosphoric acid.

Phosphoramidite (Section 28.7): A compound with the structure $R_2NP(OR)_2$.

Phosphoric acid anhydride (Section 29.1): A substance that contains PO_2PO link, analogous to the CO_2CO link in carboxylic acid anhydrides.

Physiological pH (Section 20.3): The pH of 7.3 that exists inside cells.

Photochemical reaction (Section 30.2): A reaction carried out by irradiating the reactants with light.

Pi (π) bond (Section 1.8): The covalent bond formed by sideways overlap of atomic orbitals. For example, carbon–carbon double bonds contain a π bond formed by sideways overlap of two p orbitals.

PITC (Section 26.6): Phenylisothiocyanate; used in the Edman degradation.

pK_a (Section 2.8): The negative common logarithm of the K_a; used to express acid strength.

Plane of symmetry (Section 5.2): A plane that bisects a molecule such that one half of the molecule is the mirror image of the other half. Molecules containing a plane of symmetry are achiral.

Plane-polarized light (Section 5.3): Light that has its electromagnetic waves oscillating in a single plane rather than in random planes. The plane of polarization is rotated when the light is passed through a solution of a chiral substance.

Plasticizer (Section 31.6): A small organic molecule added to polymers to act as a lubricant between polymer chains.

Polar aprotic solvent (Section 11.3): A polar solvent that can't function as a hydrogen ion donor. Polar aprotic solvents such as dimethyl sulfoxide (DMSO) and dimethylformamide (DMF) are particularly useful in S_N2 reactions because of their ability to solvate cations.

Polar covalent bond (Section 2.1): A covalent bond in which the electron distribution between atoms is unsymmetrical.

Polar reaction (Section 6.4): A reaction in which bonds are made when a nucleophile donates two electrons to an electrophile and in which bonds are broken when one fragment leaves with both electrons from the bond.

Polarity (Section 2.1): The unsymmetrical distribution of electrons in a molecule that results when one atom attracts electrons more strongly than another.

Polarizability (Section 6.4): The measure of the change in a molecule's electron distribution in response to changing electrostatic interactions with solvents or ionic reagents.

Polycarbonate (Section 31.4): A polyester in which the carbonyl groups are linked to two $-OR$ groups, $[O=C(OR)_2]$.

Polycyclic aromatic compound (Section 15.6): A compound with two or more benzene-like aromatic rings fused together.

Polycyclic compound (Section 4.9): A compound that contains more than one ring.

Polymer (Sections 8.10, 21.9; Chapter 31 Introduction): A large molecule made up of repeating smaller units. For example, polyethylene is a synthetic polymer made from repeating ethylene units and DNA is a biopolymer made of repeating deoxyribonucleotide units.

Polymerase chain reaction, PCR (Section 28.8): A method for amplifying small amounts of DNA to produce larger amounts.

Polysaccharide (Section 25.9): A carbohydrate that is made of many simple sugars linked together by glycoside (acetal) bonds.

Polyunsaturated fatty acid (Section 27.1): A fatty acid that contains more than one double bond.

Polyurethane (Section 31.4): A step-growth polymer prepared by reaction between a diol and a diisocyanate.

Posttranslational modification (Section 28.6): A chemical modification of a protein that occurs after translation from DNA.

Primary, secondary, tertiary, quaternary (Section 3.3): Terms used to describe the substitution pattern at a specific site. A primary site has one organic substituent attached to it, a secondary site has two organic substituents, a tertiary site has three, and a quaternary site has four.

	Carbon	Carbocation	Hydrogen	Alcohol	Amine
Primary	RCH_3	RCH_2^+	RCH_3	RCH_2OH	RNH_2
Secondary	R_2CH_2	R_2CH^+	R_2CH_2	R_2CHOH	R_2NH
Tertiary	R_3CH	R_3C^+	R_3CH	R_3COH	R_3N
Quaternary	R_4C				

Primary structure (Section 26.9): The amino acid sequence in a protein.

pro-R (Section 5.11): One of two identical atoms or groups of atoms in a compound whose replacement leads to an *R* chirality center.

pro-S (Section 5.11): One of two identical atoms or groups of atoms in a compound whose replacement leads to an *S* chirality center.

Prochiral (Section 5.11): A molecule that can be converted from achiral to chiral in a single chemical step.

Prochirality center (Section 5.11): An atom in a compound that can be converted into a chirality center by changing one of its attached substituents.

Promotor sequence (Section 28.4): A short sequence on DNA located upstream of the transcription start site and recognized by RNA polymerase.

Propagation step (Section 6.3): A step in a radical chain reaction that carries on the chain. The propagation steps must yield both product and a reactive intermediate.

Prostaglandin (Section 27.4): A lipid derived from arachidonic acid. Prostaglandins are present in nearly all body tissues and fluids, where they serve many important hormonal functions.

Protecting group (Sections 17.8, 19.10, 26.7): A group that is introduced to protect a sensitive functional group toward reaction elsewhere in the molecule. After serving its protective function, the group is removed.

Protein (Chapter 26 Introduction): A large peptide containing 50 or more amino acid residues. Proteins serve both as structural materials and as enzymes that control an organism's chemistry.

Protein Data Bank (Chapter 19 *A Deeper Look*): A worldwide online repository of X-ray and NMR structural data for biological macromolecules. To access the Protein Data Bank, go to http://www.rcsb.org/pdb/.

Protic solvent (Section 11.3): A solvent such as water or alcohol that can act as a proton donor.

Pyramidal inversion (Section 24.2): The rapid stereochemical inversion of a trivalent nitrogen compound.

Pyranose (Section 25.5): The six-membered, cyclic hemiacetal form of a simple sugar.

Quartet (Section 13.11): A set of four peaks in an NMR spectrum, caused by spin–spin splitting of a signal by three adjacent nuclear spins.

Quaternary: *See* **Primary.**

Quaternary ammonium salt (Section 24.1): An ionic compound containing a positively charged nitrogen atom with four attached groups, $R_4N^+ \, X^-$.

Quaternary structure (Section 26.9): The highest level of protein structure, involving an ordered aggregation of individual proteins into a larger cluster.

Quinone (Section 17.10): A 2,5-cyclohexadiene-1,4-dione.

R configuration (Section 5.5): The configuration at a chirality center as specified using the Cahn–Ingold–Prelog sequence rules.

R group (Section 3.3): A generalized abbreviation for an organic partial structure.

Racemate (Section 5.8): A mixture consisting of equal parts (+) and (−) enantiomers of a chiral substance; also called a racemic mixture.

Radical (Section 6.2): A species that has an odd number of electrons, such as the chlorine radical, Cl·.

Radical reaction (Section 6.3): A reaction in which bonds are made by donation of one electron from each of two reactants and in which bonds are broken when each fragment leaves with one electron.

Rate constant (Section 11.2): The constant *k* in a rate equation.

Rate equation (Section 11.2): An equation that expresses the dependence of a reaction's rate on the concentration of reactants.

Rate-limiting step (Section 11.4): The slowest step in a multistep reaction sequence; also called the rate-determining step. The rate-limiting step acts as a kind of bottleneck in multistep reactions.

Re face (Section 5.11): One of two faces of a planar, sp^2-hybridized atom.

Rearrangement reaction (Section 6.1): What occurs when a single reactant undergoes a reorganization of bonds and atoms to yield an isomeric product.

Reducing sugar (Section 25.6): A sugar that reduces silver ion in the Tollens test or cupric ion in the Fehling or Benedict tests.

Reduction (Section 10.8): A reaction that causes an increase of electron ownership by carbon, either by bond-breaking between carbon and a more electronegative atom or by bond formation between carbon and a less electronegative atom.

Reductive amination (Sections 24.6, 26.3): A method for preparing an amine by reaction of an aldehyde or ketone with ammonia and a reducing agent.

Refining (Chapter 3 *A Deeper Look*): The process by which petroleum is converted into gasoline and other useful products.

Regiochemistry (Section 7.8): A term describing the orientation of a reaction that occurs on an unsymmetrical substrate.

Regiospecific (Section 7.8): A term describing a reaction that occurs with a specific regiochemistry to give a single product rather than a mixture of products.

Replication (Section 28.3): The process by which doublestranded DNA uncoils and is replicated to produce two new copies.

Replication fork (Section 28.3): The point of unraveling in a DNA chain where replication occurs.

Residue (Section 26.4): An amino acid in a protein chain.

Resolution (Section 5.8): The process by which a racemate is separated into its two pure enantiomers.

Resonance effect (Section 16.4): The donation or withdrawal of electrons through orbital overlap with neighboring π bonds. For example, an oxygen or nitrogen substituent donates electrons to an aromatic ring by overlap of the O or N orbital with the aromatic ring *p* orbitals.

Resonance form (Section 2.4): An individual structural form of a resonance hybrid.

Resonance hybrid (Section 2.4): A molecule, such as benzene, that can't be represented adequately by a single Kekulé structure but must instead be considered as an average of two or more resonance forms. The resonance forms themselves differ only in the positions of their electrons, not their nuclei.

Restriction endonuclease (Section 28.6): An enzyme that is able to cleave a DNA molecule at points in the chain where a specific base sequence occurs.

Retrosynthetic (Sections 9.9, 16.11): Planning an organic synthesis by working backward from the final product to the starting material.

Ribonucleic acid (RNA) (Section 28.1): The biopolymer found in cells that serves to transcribe the genetic information found in DNA and uses that information to direct the synthesis of proteins.

Ribosomal RNA (Section 28.4): A kind of RNA used in the physical makeup of ribosomes.

Ring current (Section 15.7): The circulation of π electrons induced in aromatic rings by an external magnetic field. This effect accounts for the downfield shift of aromatic ring protons in the ^{1}H NMR spectrum.

Ring-flip (Section 4.6): A molecular motion that interconverts two chair conformations of cyclohexane. The effect of a ring-flip is to convert an axial substituent into an equatorial substituent.

Ring-opening metathesis polymerization (ROMP): A method of polymer synthesis that uses an olefin metathesis reaction of a cycloalkene.

RNA (Section 28.1): Ribonucleic acid.

Robinson annulation reaction (Section 23.12): A method for synthesis of cyclohexenones by sequential Michael reaction and intramolecular aldol reaction.

S configuration (Section 5.5): The configuration at a chirality center as specified using the Cahn–Ingold–Prelog sequence rules.

s-Cis conformation (Section 14.5): The conformation of a conjugated diene that is cis-like around the single bond.

Saccharide (Section 25.1): A sugar.

Salt bridge (Section 26.9): An ionic attraction between two oppositely charged groups in a protein chain.

Sandmeyer reaction (Section 24.8): The nucleophilic substitution reaction of an arenediazonium salt with a cuprous halide to yield an aryl halide.

Sanger dideoxy method (Section 28.6): A commonly used method of DNA sequencing.

Saponification (Section 21.6): An old term for the base-induced hydrolysis of an ester to yield a carboxylic acid salt.

Saturated (Section 3.2): A molecule that has only single bonds and thus can't undergo addition reactions. Alkanes are saturated, but alkenes are unsaturated.

Sawhorse structure (Section 3.6): A manner of representing stereochemistry that uses a stick drawing and gives a perspective view of the conformation around a single bond.

Schiff base (Sections 19.8, 29.5): An alternative name for an imine, $R_2C=NR'$, used primarily in biochemistry.

Second-order reaction (Section 11.2): A reaction whose rate-limiting step is bimolecular and whose kinetics are therefore dependent on the concentration of two reactants.

Secondary: *See* **Primary.**

Secondary metabolite (Chapter 7 *A Deeper Look*): A small naturally occurring molecule that is not essential to the growth and development of the producing organism and is not classified by structure.

Secondary structure (Section 26.9): The level of protein substructure that involves organization of chain sections into ordered arrangements such as β-pleated sheets or α helices.

Semiconservative replication (Section 28.3): The process by which DNA molecules are made containing one strand of old DNA and one strand of new DNA.

Sense strand (Section 28.4): The coding strand of double-helical DNA that contains the gene.

Sequence rules (Sections 5.5, 7.5): A series of rules for assigning relative rankings to substituent groups on a double-bond carbon atom or on a chirality center.

Sesquiterpenoid (Section 27.5): A 15-carbon lipid.

Sharpless epoxidation (Chapter 19 *A Deeper Look*): A method for enantioselective synthesis of a chiral epoxide by treatment of an allylic alcohol with *tert*-butyl hydroperoxide, $(CH_3)_3C$—OOH, in the presence of titanium tetraisopropoxide and diethyl tartrate.

Shell (electron) (Section 1.2): A group of an atom's electrons with the same principal quantum number.

Shielding (Section 13.2): An effect observed in NMR that causes a nucleus to absorb toward the right (upfield) side of the chart. Shielding is caused by donation of electron density to the nucleus.

***Si* face** (Section 5.11): One of two faces of a planar, sp^2-hybridized atom.

Sialic acid (Section 25.7): One of a group of more than 300 carbohydrates based on acetylneuramic acid.

Side chain (Section 26.1): The substituent attached to the α carbon of an amino acid.

Sigma (σ) bond (Section 1.5): A covalent bond formed by head-on overlap of atomic orbitals.

Sigmatropic reaction (Section 30.8): A pericyclic reaction that involves the migration of a group from one end of a π electron system to the other.

Silyl ether (Section 17.8): A substance with the structure R_3Si—O—R. The silyl ether acts as a protecting group for alcohols.

Simmons–Smith reaction (Section 8.9): The reaction of an alkene with CH_2I_2 and Zn–Cu to yield a cyclopropane.

Simple sugar (Section 25.1): A carbohydrate that cannot be broken down into smaller sugars by hydrolysis.

Single bond (Section 1.8): A covalent bond formed by sharing one electron pair between atoms.

Skeletal structure (Section 1.12): A shorthand way of writing structures in which carbon atoms are assumed to be at each intersection of two lines (bonds) and at the end of each line.

Small RNAs (Section 28.4): A type of RNA that has a variety of functions within the cell, including silencing transcription and catalyzing chemical modifications of other RNA molecules.

S_N1 reaction (Section 11.4): A unimolecular nucleophilic substitution reaction.

S_N2 reaction (Section 11.2): A bimolecular nucleophilic substitution reaction.

Solid-phase synthesis (Section 26.8): A technique of synthesis whereby the starting material is covalently bound to a solid polymer bead and reactions are carried out on the bound substrate. After the desired transformations have been effected, the product is cleaved from the polymer.

Solvation (Section 11.3): The clustering of solvent molecules around a solute particle to stabilize it.

sp Hybrid orbital (Section 1.9): A hybrid orbital derived from the combination of an *s* and a *p* atomic orbital. The two *sp* orbitals that result from hybridization are oriented at an angle of 180° to each other.

sp^2 Hybrid orbital (Section 1.8): A hybrid orbital derived by combination of an *s* atomic orbital with two *p* atomic orbitals. The three sp^2 hybrid orbitals that result lie in a plane at angles of 120° to each other.

sp^3 Hybrid orbital (Section 1.6): A hybrid orbital derived by combination of an *s* atomic orbital with three *p* atomic orbitals. The four sp^3 hybrid orbitals that result are directed toward the corners of a regular tetrahedron at angles of 109° to each other.

Specific rotation, $[\alpha]_D$ (Section 5.3): The optical rotation of a chiral compound under standard conditions.

Sphingomyelin (Section 27.3): A phospholipid that has sphingosine as its backbone rather than glycerol.

Spin–spin splitting (Section 13.11): The splitting of an NMR signal into a multiplet because of an interaction between nearby magnetic nuclei whose spins are coupled. The magnitude of spin–spin splitting is given by the coupling constant, *J*.

Staggered conformation (Section 3.6): The three-dimensional arrangement of atoms around a carbon–carbon single bond in which the bonds on one carbon bisect the bond angles on the second carbon as viewed end-on.

Statin (Chapter 29 *A Deeper Look*): A drug that controls cholesterol biosynthesis in the body by blocking the HMG-CoA reductase enzyme.

Step-growth polymer (Sections 21.9, 31.4): A polymer in which each bond is formed independently of the others. Polyesters and polyamides (nylons) are examples.

Stereocenter (Section 5.2): An alternative name for a chirality center.

Stereochemistry (Chapters 3, 4, 5): The branch of chemistry concerned with the three-dimensional arrangement of atoms in molecules.

Stereogenic center (Section 5.2): An alternative name for a chirality center.

Stereoisomers (Section 4.2): Isomers that have their atoms connected in the same order but have different three-dimensional arrangements. The term *stereoisomer* includes both enantiomers and diastereomers.

Stereospecific (Section 8.9): A term indicating that only a single stereoisomer is produced in a given reaction rather than a mixture.

Steric strain (Sections 3.7, 4.7): The strain imposed on a molecule when two groups are too close together and try to occupy the same space. Steric strain is responsible both for the greater stability of trans versus cis alkenes and for the greater stability of equatorially substituted versus axially substituted cyclohexanes.

Steroid (Section 27.6): A lipid whose structure is based on a tetracyclic carbon skeleton with three 6-membered and one 5-membered ring. Steroids occur in both plants and animals and have a variety of important hormonal functions.

Stork enamine reaction (Section 23.11): The conjugate addition of an enamine to an α,β-unsaturated carbonyl compound, followed by hydrolysis to yield a 1,5-dicarbonyl product.

STR loci (Chapter 28 *A Deeper Look*): Short tandem repeat sequences of noncoding DNA that are unique to every individual and allow DNA fingerprinting.

Straight-chain alkane (Section 3.2): An alkane whose carbon atoms are connected without branching.

Substitution reaction (Section 6.1): What occurs when two reactants exchange parts to give two new products. S_N1 and S_N2 reactions are examples.

Sulfide (Section 18.8): A compound that has two organic substituents bonded to the same sulfur atom, RSR'.

Sulfonation (Section 16.2): The substitution of a sulfonic acid group ($-SO_3H$) onto an aromatic ring.

Sulfone (Section 18.8): A compound of the general structure RSO_2R'.

Sulfonium ion (Section 18.8): A species containing a positively charged, trivalent sulfur atom, R_3S^+.

Sulfoxide (Section 18.8): A compound of the general structure RSOR'.

Suprafacial (Section 30.5): A word used to describe the geometry of pericyclic reactions. Suprafacial reactions take place on the same side of the two ends of a π electron system.

Suzuki–Miyaura reaction (Section 10.7): The palladium-catalyzed coupling reaction of an aromatic or vinylic halide with an aromatic or vinylic boronic acid.

Symmetry-allowed, symmetry-disallowed (Section 30.2): A symmetry-allowed reaction is a pericyclic process that has a favorable orbital symmetry for reaction through a concerted pathway. A symmetry-disallowed reaction is one that does not have favorable orbital symmetry for reaction through a concerted pathway.

Symmetry plane (Section 5.2): A plane that bisects a molecule such that one half of the molecule is the mirror image of the other half. Molecules containing a plane of symmetry are achiral.

Syn periplanar (Section 11.8): Describing a stereochemical relationship in which two bonds on adjacent carbons lie in the same plane and are eclipsed.

Syn stereochemistry (Section 8.5): The opposite of anti. A syn addition reaction is one in which the two ends of the double bond react from the same side. A syn elimination is one in which the two groups leave from the same side of the molecule.

Syndiotactic (Section 31.2): A chain-growth polymer in which the stereochemistry of the substituents alternates regularly on opposite sides of the backbone.

Tautomers (Sections 9.4, 22.1): Isomers that interconvert spontaneously, usually with the change in position of a hydrogen.

Template strand (Section 28.4): The strand of double-helical DNA that does not contain the gene.

Terpenoid (Chapter 8 *A Deeper Look*, Section 27.5): A lipid that is formally derived by head-to-tail polymerization of isoprene units.

Tertiary: *See* **Primary.**

Tertiary structure (Section 26.9): The level of protein structure that involves the manner in which the entire protein chain is folded into a specific three-dimensional arrangement.

Thermodynamic control (Section 14.3): An equilibrium reaction that yields the lowest-energy, most stable product is said to be thermodynamically controlled.

Thermoplastic (Section 31.6): A polymer that has a high T_g and is hard at room temperature but becomes soft and viscous when heated.

Thermosetting resin (Section 31.6): A polymer that becomes highly cross-linked and solidifies into a hard, insoluble mass when heated.

Thioester (Section 21.8): A compound with the RCOSR′ functional group.

Thiol (Section 18.8): A compound containing the —SH functional group.

Thiolate ion (Section 18.8): The anion of a thiol, RS⁻.

TMS (Section 13.3): Tetramethylsilane; used as an NMR calibration standard.

TOF (Section 12.4): Time-of-flight mass spectrometry; a sensitive method of mass detection accurate to about 3 ppm.

Tollens' reagent (Section 25.6): A solution of Ag_2O in aqueous ammonia; used to oxidize aldehydes to carboxylic acids.

Torsional strain (Section 3.6): The strain in a molecule caused by electron repulsion between eclipsed bonds. Torsional strain is also called eclipsing strain.

Tosylate (Section 11.1): A *p*-toluenesulfonate ester; useful as a leaving group in nucleophilic substitution reactions.

Transamination (Section 29.9): The exchange of an amino group and a keto group between reactants.

Transcription (Section 28.4): The process by which the genetic information encoded in DNA is read and used to synthesize RNA in the nucleus of the cell. A small portion of double-stranded DNA uncoils, and complementary ribonucleotides line up in the correct sequence for RNA synthesis.

Transfer RNA (Section 28.4): A kind of RNA that transports amino acids to the ribosomes, where they are joined together to make proteins.

Transimination (Section 29.9): The exchange of an amino group and an imine group between reactants.

Transition state (Section 6.9): An activated complex between reactants, representing the highest energy point on a reaction curve. Transition states are unstable complexes that can't be isolated.

Translation (Section 28.5): The process by which the genetic information transcribed from DNA onto mRNA is read by tRNA and used to direct protein synthesis.

Tree diagram (Section 13.12): A diagram used in NMR to sort out the complicated splitting patterns that can arise from multiple couplings.

Triacylglycerol (Section 27.1): A lipid, such as those found in animal fat and vegetable oil, that is a triester of glycerol with long-chain fatty acids.

Tricarboxylic acid cycle (Section 29.7): An alternative name for the citric acid cycle by which acetyl CoA is degraded to CO_2.

Triple bond (Section 1.9): A covalent bond formed by sharing three electron pairs between atoms.

Triplet (Section 13.11): A symmetrical three-line splitting pattern observed in the ^1H NMR spectrum when a proton has two equivalent neighbor protons.

Turnover number (Section 26.10): The number of substrate molecules acted on by an enzyme molecule per unit time.

Twist-boat conformation (Section 4.5): A conformation of cyclohexane that is somewhat more stable than a pure boat conformation.

Ultraviolet (UV) spectroscopy (Section 14.7): An optical spectroscopy employing ultraviolet irradiation. UV spectroscopy provides structural information about the extent of π electron conjugation in organic molecules.

Unimolecular reaction (Section 11.4): A reaction that occurs by spontaneous transformation of the starting material without the intervention of other reactants. For example, the dissociation of a tertiary alkyl halide in the S_N1 reaction is a unimolecular process.

Unsaturated (Section 7.2): A molecule that has one or more multiple bonds.

Upfield (Section 13.3): The right-hand portion of the NMR chart.

Urethane (Section 31.4): A functional group in which a carbonyl group is bonded to both an —OR and an —NR₂.

Uronic acid (Section 25.6): A monocarboxylic acid formed by oxidizing the —CH₂OH end of an aldose without affecting the —CHO end.

Valence bond theory (Section 1.5): A bonding theory that describes a covalent bond as resulting from the overlap of two atomic orbitals.

Valence shell (Section 1.4): The outermost electron shell of an atom.

van der Waals forces (Section 2.12): Intermolecular forces that are responsible for holding molecules together in the liquid and solid states.

Vegetable oil (Section 27.1): A liquid triacylglycerol derived from a plant source.

Vicinal (Section 9.2): A term used to refer to a 1,2-disubstitution pattern. For example, 1,2-dibromoethane is a vicinal dibromide.

Vinyl group (Section 7.3): A $H_2C=CH-$ substituent.

Vinyl monomer (Sections 8.10, 31.1): A substituted alkene monomer used to make a chain-growth polymer.

Vinylic (Section 9.3): A term that refers to a substituent at a double-bond carbon atom. For example, chloroethylene is a vinylic chloride, and enols are vinylic alcohols.

Virion (Section 25.11): A viral particle.

Vitamin (Section 26.10): A small organic molecule that must be obtained in the diet and is required in trace amounts for proper growth and function.

Vulcanization (Section 14.6): A technique for cross-linking and hardening a diene polymer by heating with a few percent by weight of sulfur.

Walden inversion (Section 11.1): The inversion of configuration at a chirality center that accompanies an S_N2 reaction.

Wave equation (Section 1.2): A mathematical expression that defines the behavior of an electron in an atom.

Wave function (Section 1.2): A solution to the wave equation for defining the behavior of an electron in an atom. The square of the wave function defines the shape of an orbital.

Wavelength, λ (Section 12.5): The length of a wave from peak to peak. The wavelength of electromagnetic radiation is inversely proportional to frequency and inversely proportional to energy.

Wavenumber, $\tilde{\nu}$ (Section 12.6): The reciprocal of the wavelength in centimeters.

Wax (Section 27.1): A mixture of esters of long-chain carboxylic acids with long-chain alcohols.

Williamson ether synthesis (Section 18.2): A method for synthesizing ethers by S_N2 reaction of an alkyl halide with an alkoxide ion.

Wittig reaction (Section 19.11): The reaction of a phosphorus ylide with a ketone or aldehyde to yield an alkene.

Wohl degradation (Section 25.6): A method for shortening the chain of an aldose sugar by one carbon.

Wolff–Kishner reaction (Section 19.9): The conversion of an aldehyde or ketone into an alkane by reaction with hydrazine and base.

X-ray crystallography (Chapter 12 *A Deeper Look*): A technique that uses X rays to determine the structure of molecules.

Ylide (Section 19.11): A neutral species with adjacent + and − charges, such as the phosphoranes used in Wittig reactions.

Z geometry (Section 7.5): A term used to describe the stereochemistry of a carbon–carbon double bond. The two groups on each carbon are ranked according to the Cahn–Ingold–Prelog sequence rules, and the two carbons are compared. If the higher ranked groups on each carbon are on the same side of the double bond, the bond has *Z* geometry.

Zaitsev's rule (Section 11.7): A rule stating that E2 elimination reactions normally yield the more highly substituted alkene as major product.

Ziegler–Natta catalyst (Section 31.2): A catalyst of an alkylaluminum and a titanium compound used for preparing alkene polymers.

Zwitterion (Section 26.1): A neutral dipolar molecule in which the positive and negative charges are not adjacent. For example, amino acids exist as zwitterions, $H_3\overset{+}{N}-CHR-CO_2^-$.

APPENDIX D

Answers to In-Text Problems

The following answers are meant only as a quick check while you study. Full answers for all problems are provided in the accompanying *Study Guide and Solutions Manual*.

Chapter 1

1.1 (a) $1s^2\ 2s^2\ 2p^4$ (b) $1s^2\ 2s^2\ 2p^3$
(c) $1s^2\ 2s^2\ 2p^6\ 3s^2\ 3p^4$

1.2 (a) 2 (b) 2 (c) 6

1.3 [structure: CHCl₃]

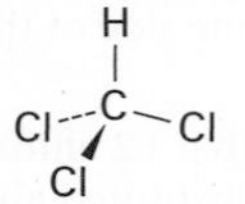

1.4 [structure: C₂H₆]

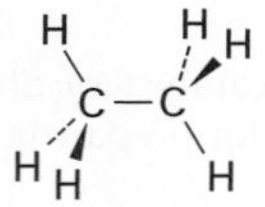

1.5 (a) CCl_4 (b) AlH_3 (c) CH_2Cl_2
(d) SiF_4 (e) CH_3NH_2

1.6 (a) [Lewis structures of CH₂Cl₂ and CHCl₃]
(b) [Lewis structures of H₂S and SH₂]
(c) [Lewis structures of CH₃NH₂]
(d) [Lewis structures of CH₃Li]

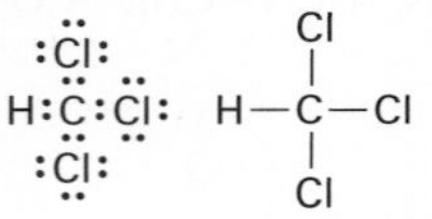

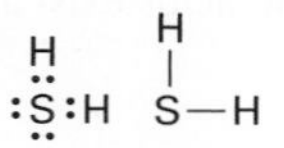

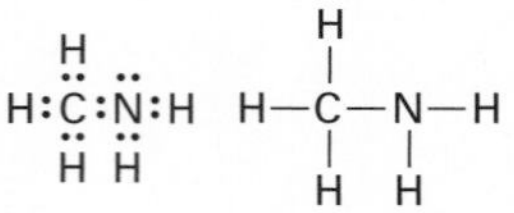

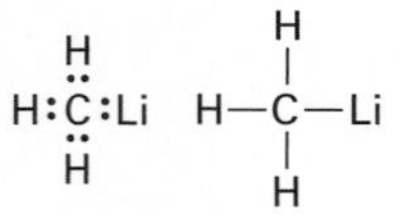

1.7 C_2H_7 has too many hydrogens for a compound with two carbons.

1.8 [structures of propane and cyclopropane]
All bond angles are near 109°.

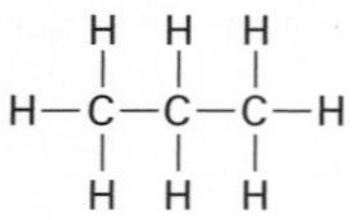

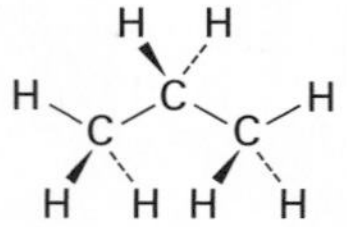

1.9 [structure of hexane]

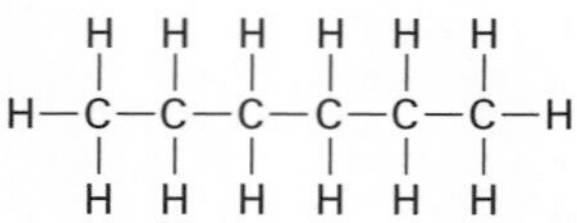

1.10 The CH₃ carbon is sp^3; the double-bond carbons are sp^2; the C=C–C and C=C–H bond angles are approximately 120°; other bond angles are near 109°.

[structure of propene]

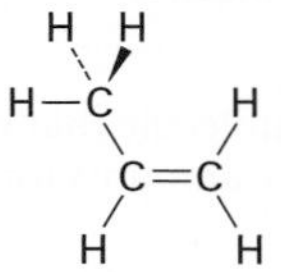

1.11 All carbons are sp^2, and all bond angles are near 120°.

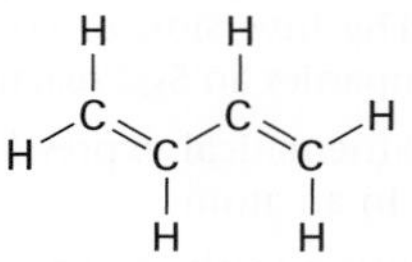

[structure of 1,3-butadiene]

1.12 All carbons except CH₃ are sp^2.

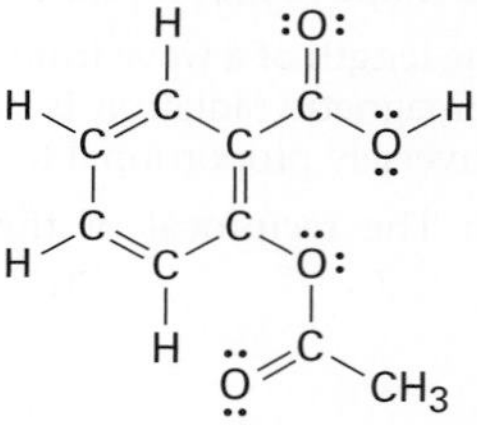

[structure of aromatic compound]

1.13 The CH₃ carbon is sp^3; the triple-bond carbons are sp; the C≡C–C and H–C≡C bond angles are approximately 180°.

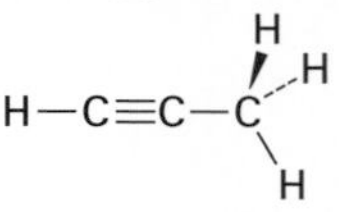

[structure of propyne]

1.14 (a) O has 2 lone pairs and is sp^3-hybridized.
(b) N has 1 lone pair and is sp^3-hybridized.
(c) P has 1 lone pair and is sp^3-hybridized.
(d) S has 2 lone pairs and is sp^3-hybridized.

A-28

APPENDIX D | **Answers to In-Text Problems**   A-29

1.15 (a) Adrenaline—$C_9H_{13}NO_3$

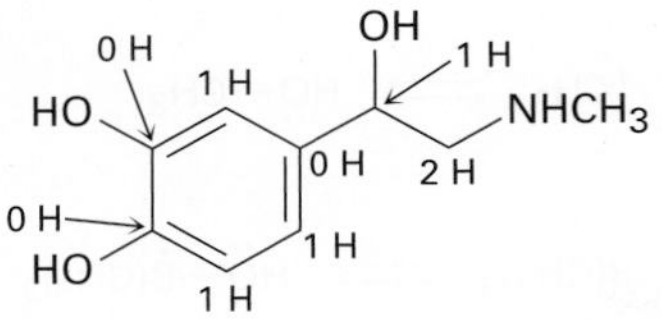

(b) Estrone—$C_{18}H_{22}O_2$

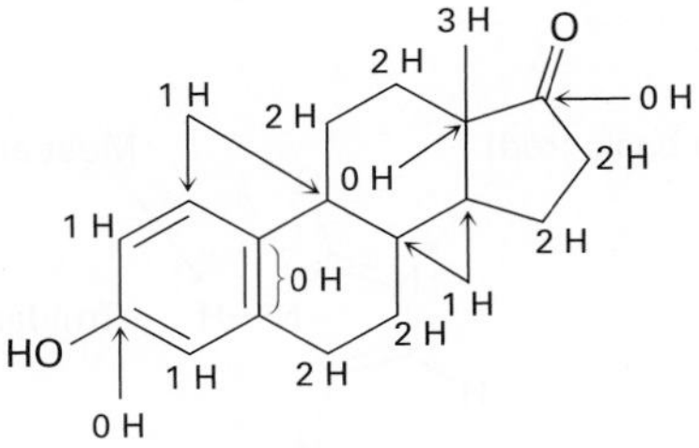

1.16 There are numerous possibilities, such as:

(a) C_5H_{12} $CH_3CH_2CH_2CH_2CH_3$ $CH_3CH_2CHCH_3$ CH_3CCH_3
                                              | |
                                              CH_3 CH_3
                                                               |
                                                               CH_3

(b) C_2H_7N $CH_3CH_2NH_2$ CH_3NHCH_3

(c) C_3H_6O $CH_3CH_2\overset{O}{\overset{\|}{C}}H$ $H_2C=CHCH_2OH$ $H_2C=CHOCH_3$

(d) C_4H_9Cl $CH_3CH_2CH_2CH_2Cl$ $CH_3CH_2CHCH_3$ CH_3CHCH_2Cl
                                             | |
                                             Cl CH_3

1.17

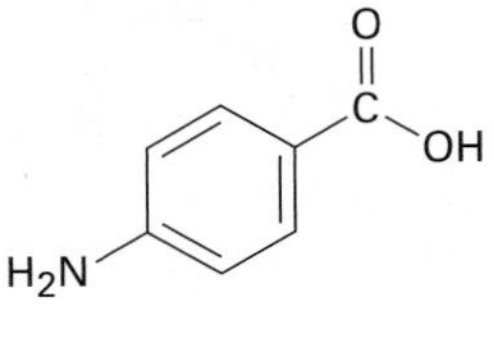

Chapter 2

2.1 (a) H (b) Br (c) Cl (d) C

2.2

(a) $\overset{\delta+}{H_3C}-\overset{\delta-}{Cl}$ (b) $\overset{\delta+}{H_3C}-\overset{\delta-}{NH_2}$ (c) $\overset{\delta-}{H_2N}-\overset{\delta+}{H}$

(d) H_3C-SH (e) $\overset{\delta-}{H_3C}-\overset{\delta+}{MgBr}$ (f) $\overset{\delta+}{H_3C}-\overset{\delta-}{F}$

Carbon and sulfur have identical electronegativities.

2.3 $H_3C-OH < H_3C-MgBr < H_3C-Li = H_3C-F < H_3C-K$

2.4 The chlorine is electron-rich, and the carbon is electron-poor.

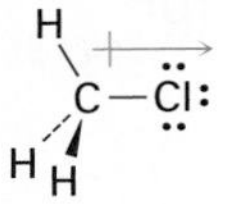

2.5 The two C–O dipoles cancel because of the symmetry of the molecule:

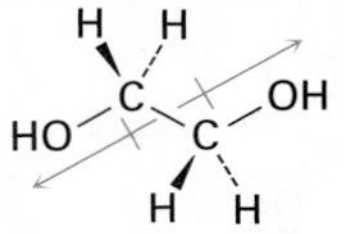

2.6 (a) $H\overset{}{\underset{}{C}}=C\overset{H}{\underset{H}{}}$ **(b)**

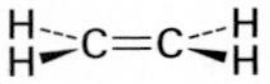

No dipole moment

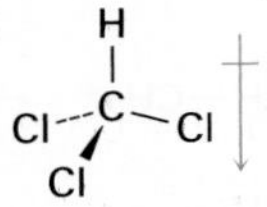

(c)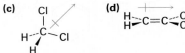

2.7 (a) For carbon: FC = 4 − 8/2 − 0 = 0
For the middle nitrogen: FC = 5 − 8/2 − 0 = +1
For the end nitrogen: FC = 5 − 4/2 − 4 = −1
(b) For nitrogen: FC = 5 − 8/2 − 0 = +1
For oxygen: FC = 6 − 2/2 − 6 = −1
(c) For nitrogen: FC = 5 − 8/2 − 0 = +1
For the triply bonded carbon: FC = 4 − 6/2 − 2 = −1

2.8

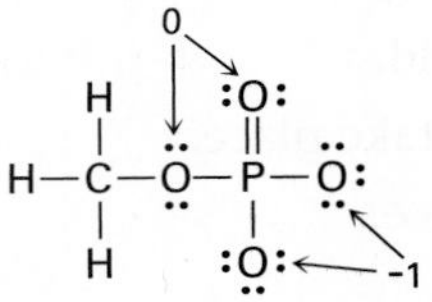

2.9 The structures in **(a)** are resonance forms.

APPENDIX D | Answers to In-Text Problems

2.10

(a) [resonance structures of methyl phosphate]

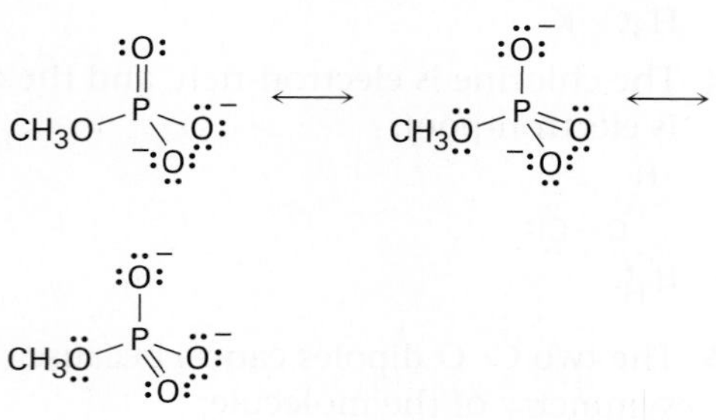

(b) [resonance structures of nitrate]

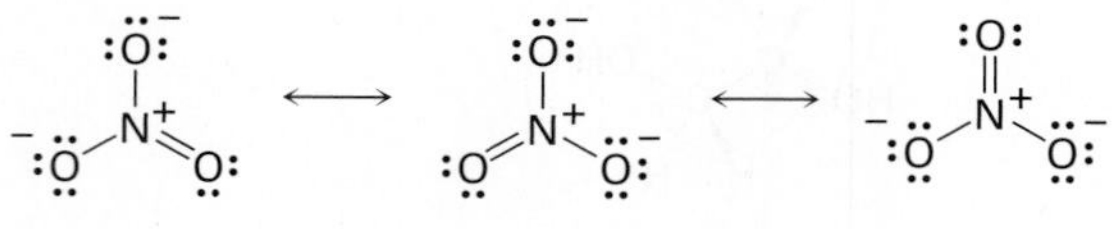

(c) $H_2C=CH-CH_2^+ \longleftrightarrow H_2\overset{+}{C}-CH=CH_2$

(d) [resonance structures of benzoate]

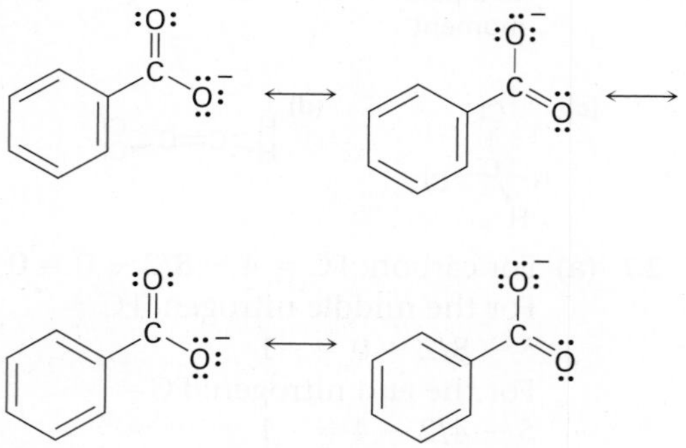

2.11 $HNO_3 + NH_3 \longrightarrow NH_4^+ + NO_3^-$
  Acid   Base   Conjugate   Conjugate
                      acid   base

2.12 Phenylalanine is stronger.

2.13 Water is a stronger acid.

2.14 Neither reaction will take place.

2.15 Reaction will take place.

2.16 $K_a = 4.9 \times 10^{-10}$

2.17

(a)
$CH_3CH_2\overset{..}{\underset{..}{O}}H + H-Cl \rightleftharpoons CH_3CH_2\overset{+}{\underset{|}{O}}\underset{H}{H} + Cl^-$

$H\overset{..}{N}(CH_3)_2 + H-Cl \rightleftharpoons H\overset{+}{N}(CH_3)_2\underset{|}{\overset{H}{}} + Cl^-$

$\overset{..}{P}(CH_3)_3 + H-Cl \rightleftharpoons H-\overset{+}{P}(CH_3)_3 + Cl^-$

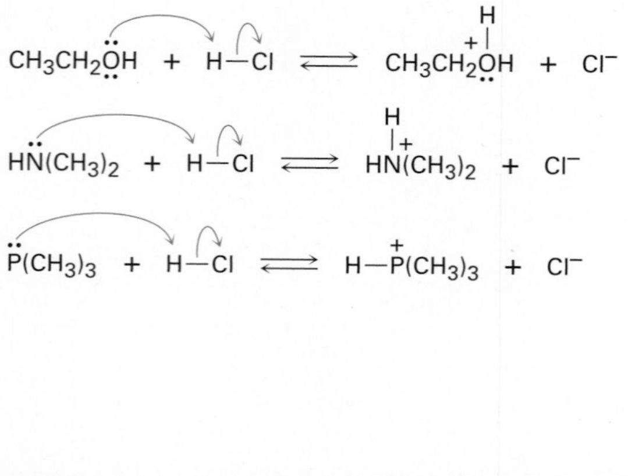

(b)
$H\overset{..}{\underset{..}{O}}{:}^- + {}^+CH_3 \rightleftharpoons H\overset{..}{\underset{..}{O}}-CH_3$

$H\overset{..}{\underset{..}{O}}{:}^- + B(CH_3)_3 \rightleftharpoons H\overset{..}{\underset{..}{O}}-\bar{B}(CH_3)_3$

$H\overset{..}{\underset{..}{O}}{:}^- + MgBr_2 \rightleftharpoons H\overset{..}{\underset{..}{O}}-\bar{M}gBr_2$

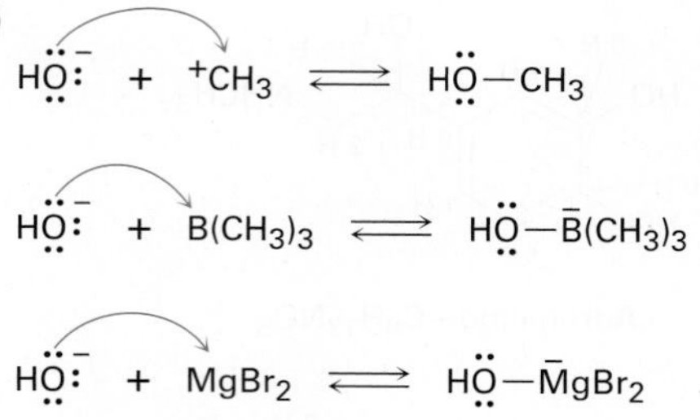

2.18

(a) More basic (red) ↓   Most acidic (blue) ↓
[imidazole structure]   Imidazole

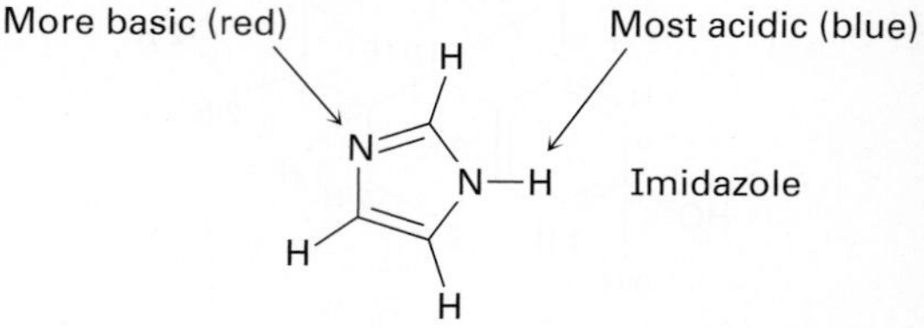

(b) [imidazole protonation resonance structures]

[imidazole deprotonation resonance structures]

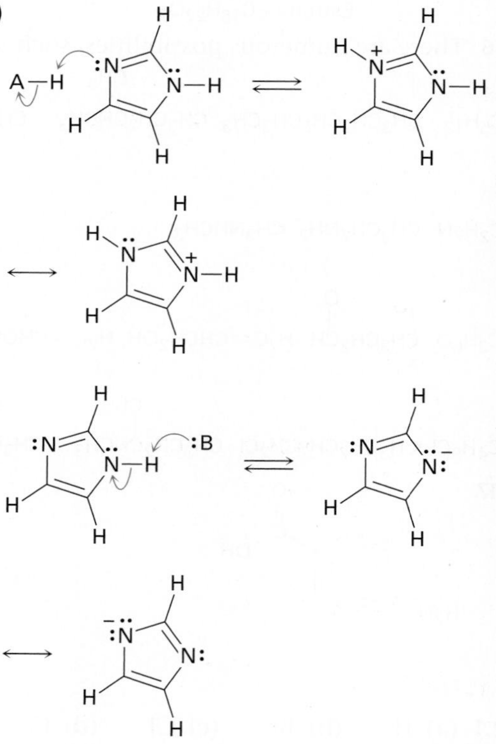

2.19 Vitamin C is water-soluble (hydrophilic); vitamin A is fat-soluble (hydrophilic).

APPENDIX D | Answers to In-Text Problems   A-31

Chapter 3

3.1 (a) Sulfide, carboxylic acid, amine
(b) Aromatic ring, carboxylic acid
(c) Ether, alcohol, aromatic ring, amide, C=C bond

3.2 (a) CH$_3$OH (b) C$_6$H$_5$CH$_3$ (c) CH$_3$COOH
(d) CH$_3$NH$_2$ (e) CH$_3$COCH$_2$NH$_2$ (f) CH$_2$=CHCH=CH$_2$

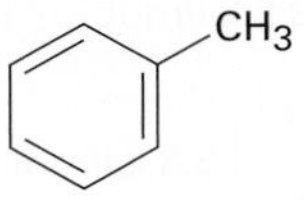

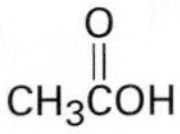

3.3 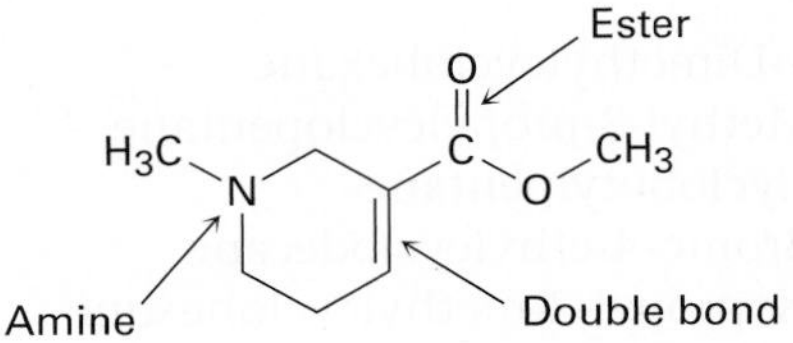 Ester, Amine, Double bond; C$_8$H$_{13}$NO$_2$

3.4
CH$_3$CH$_2$CH$_2$CH$_2$CH$_2$CH$_3$
CH$_3$CHCH$_2$CH$_2$CH$_3$ with CH$_3$
CH$_3$CH$_2$CHCH$_2$CH$_3$ with CH$_3$
CH$_3$CCH$_2$CH$_3$ with two CH$_3$
CH$_3$CHCHCH$_3$ with two CH$_3$

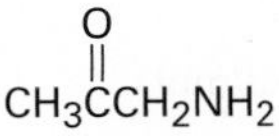

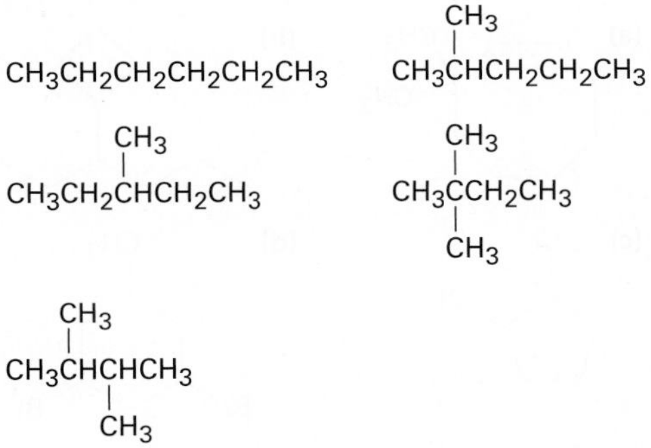

3.5 Part (a) has nine possible answers.
(a) CH$_3$CH$_2$CH$_2$COCH$_3$ CH$_3$CH$_2$COCH$_2$CH$_3$ CH$_3$COCHCH$_3$ with CH$_3$
(b) CH$_3$CHC≡N with CH$_3$;   CH$_3$CH$_2$CH$_2$C≡N
(c) CH$_3$CH$_2$SSCH$_2$CH$_3$ CH$_3$SSCH$_2$CH$_2$CH$_3$ CH$_3$SSCHCH$_3$ with CH$_3$

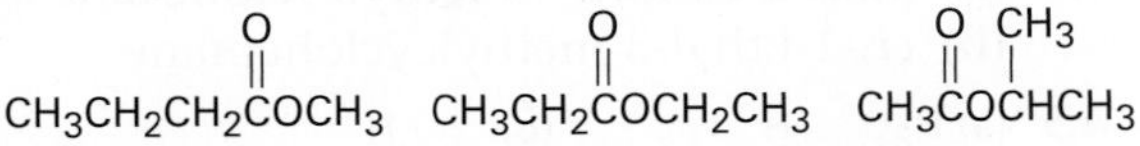

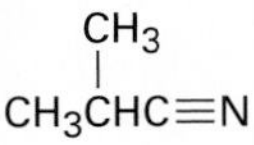

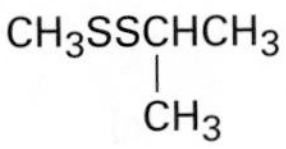

3.6 (a) Two (b) Four (c) Four

3.7
CH$_3$CH$_2$CH$_2$CH$_2$CH$_2$—   CH$_3$CH$_2$CH$_2$CH— with CH$_3$
CH$_3$CH$_2$CH— with CH$_2$CH$_3$ CH$_3$CH$_2$CHCH$_2$— with CH$_3$
CH$_3$CHCH$_2$CH$_2$— with CH$_3$ CH$_3$CH$_2$C— with two CH$_3$
CH$_3$CHCH— with two CH$_3$ CH$_3$CCH$_2$— with two CH$_3$

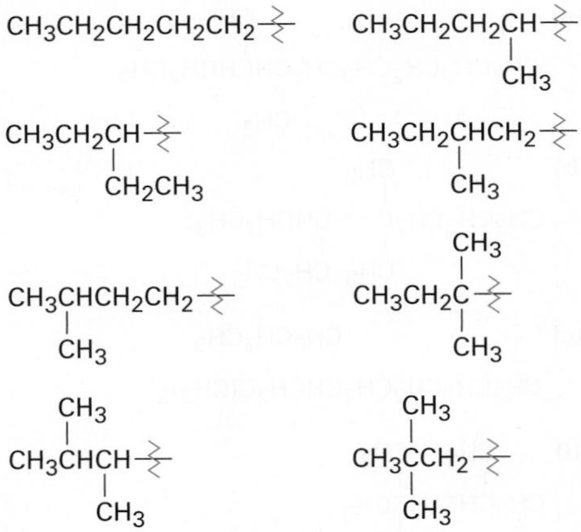

3.8 (a) CH$_3$CHCH$_2$CH$_2$CH$_3$ with CH$_3$; p, t, s, s, p
(b) CH$_3$CHCH$_3$ / CH$_3$CH$_2$CHCH$_2$CH$_3$; p, t, p / p, s, t, s, p
(c) CH$_3$CHCH$_2$—C—CH$_3$ with CH$_3$, CH$_3$, CH$_3$; p, t, s, q, p, p

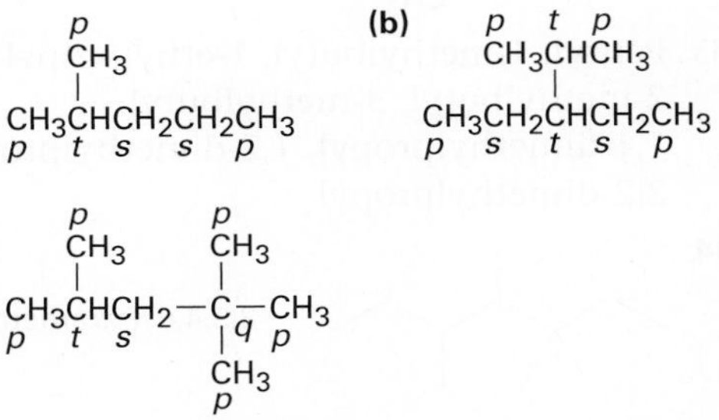

3.9 Primary carbons have primary hydrogens, secondary carbons have secondary hydrogens, and tertiary carbons have tertiary hydrogens.

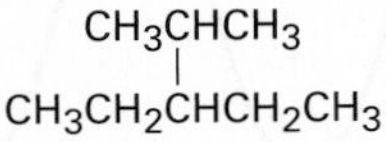

3.10 (a) CH$_3$CHCHCH$_3$ with two CH$_3$
(b) CH$_3$CHCH$_3$ / CH$_3$CH$_2$CHCH$_2$CH$_3$
(c) CH$_3$CCH$_2$CH$_3$ with two CH$_3$

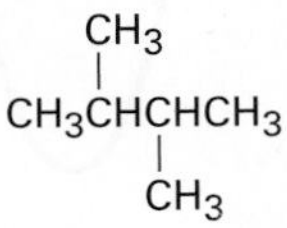

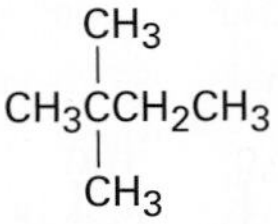

3.11 (a) Pentane, 2-methylbutane, 2,2-dimethylpropane
(b) 2,3-Dimethylpentane
(c) 2,4-Dimethylpentane
(d) 2,2,5-Trimethylhexane

3.12 (a) CH₃CH₂CH₂CH₂CH₂CH(CH₃)CH₂CH₃ with CH₃ branch

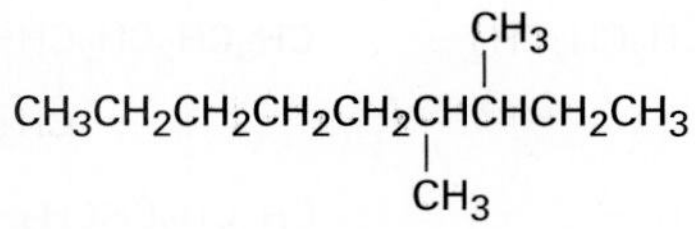

(b) CH₃CH₂CH₂C(CH₃)₂—CH(CH₂CH₃)CH₃

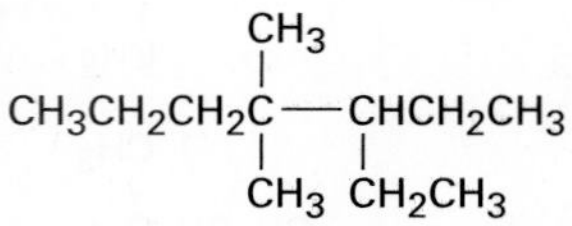

(c) CH₃CH₂CH₂CH₂CH(CH₂CH₂CH₃)CH₂C(CH₃)₃

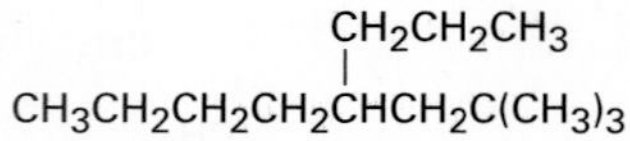

(d) CH₃CH(CH₃)CH₂C(CH₃)₂CH₃

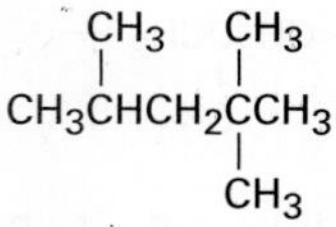

3.13 Pentyl, 1-methylbutyl, 1-ethylpropyl, 2-methylbutyl, 3-methylbutyl, 1,1-dimethylpropyl, 1,2-dimethylpropyl, 2,2-dimethylpropyl

3.14 3,3,4,5-Tetramethylheptane

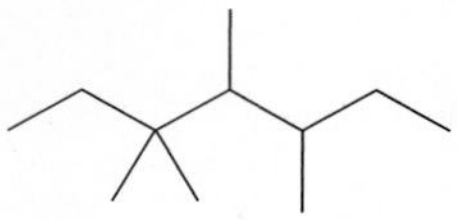

3.15 14 kJ/mol

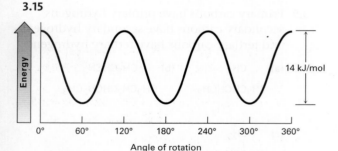

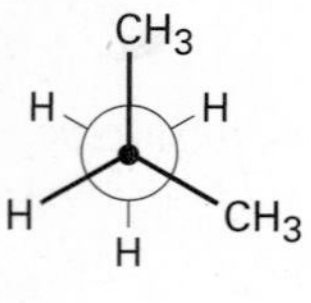

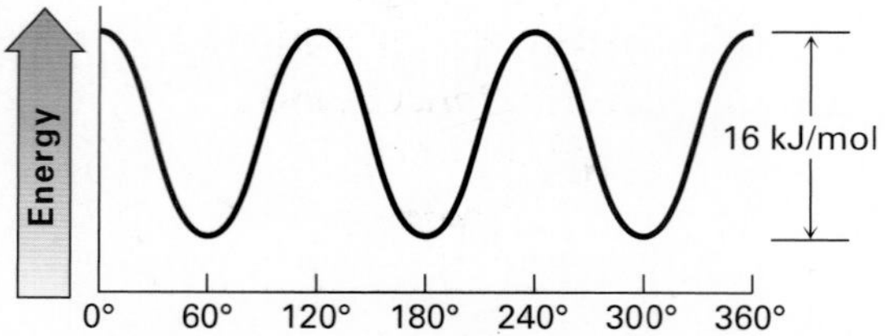

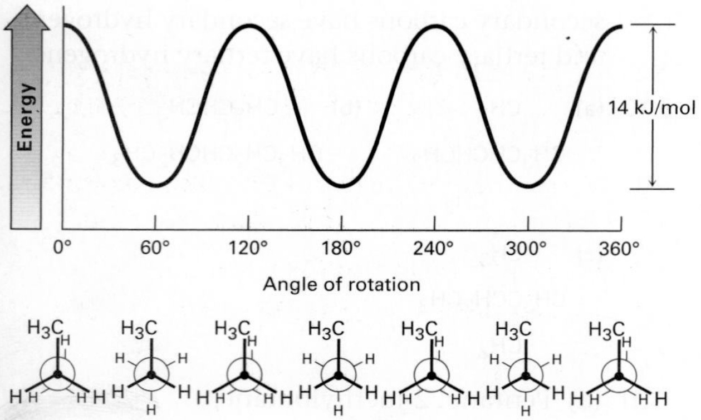

3.16 (a) Newman projection of ethane, staggered
(b) 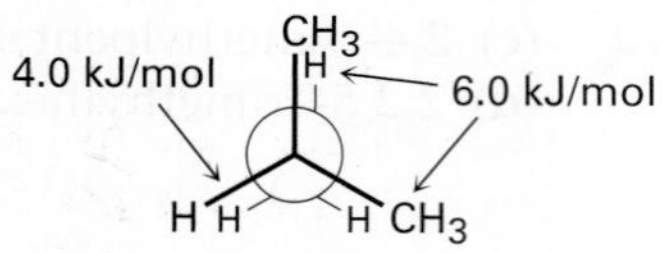 4.0 kJ/mol, 6.0 kJ/mol labels on eclipsed conformation
(c), (d) Energy diagram, 16 kJ/mol

3.17 Newman projection: H₃C, H, H, CH₃, CH₃, H

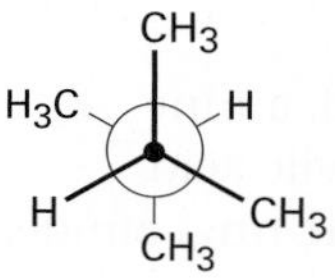

3.18 3.8 kJ/mol, 3.8 kJ/mol, 3.8 kJ/mol; Total: 11.4 kJ/mol

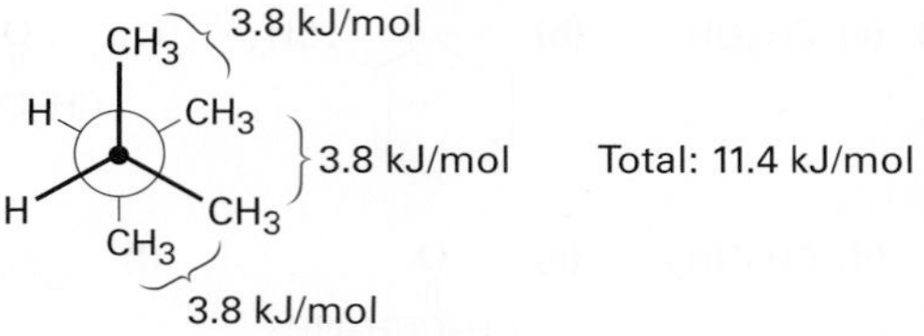

Chapter 4

4.1 (a) 1,4-Dimethylcyclohexane
(b) 1-Methyl-3-propylcyclopentane
(c) 3-Cyclobutylpentane
(d) 1-Bromo-4-ethylcyclodecane
(e) 1-Isopropyl-2-methylcyclohexane
(f) 4-Bromo-1-*tert*-butyl-2-methylcycloheptane

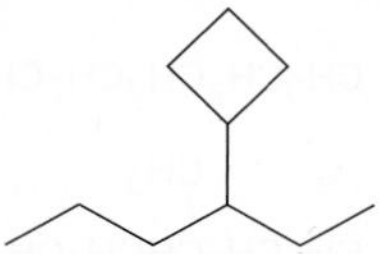

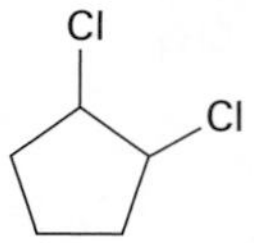

4.2 (a) cyclooctane with two CH₃ groups on same carbon
(b) cyclobutyl group on pentane
(c) 1,2-dichlorocyclopentane
(d) 1-methyl-3,5-dibromocyclohexane

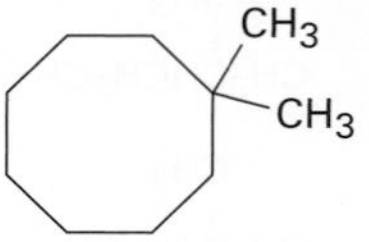

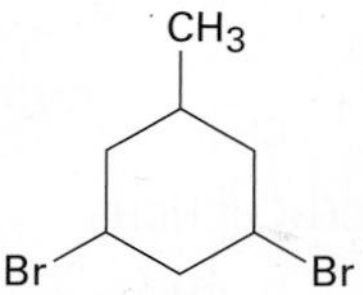

4.3 3-Ethyl-1,1-dimethylcyclopentane

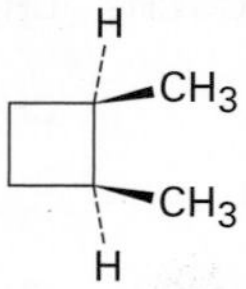

4.4 (a) *trans*-1-Chloro-4-methylcyclohexane
(b) *cis*-1-Ethyl-3-methylcycloheptane

4.5 (a) cyclohexane with H₃C, H up and Br, H down
(b) cyclobutane with CH₃, CH₃ and H, H
(c) cyclohexane with CH₂CH₃, H and C(CH₃)₃, H

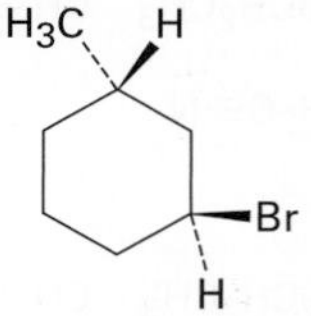

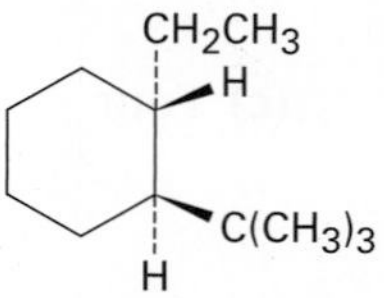

4.6 The two hydroxyl groups are cis. The two side chains are trans.

4.7 (a) *cis*-1,2-Dimethylcyclopentane
(b) *cis*-1-Bromo-3-methylcyclobutane

4.8 Six interactions; 21% of strain

4.9 The cis isomer is less stable because the methyl groups nearly eclipse each other.

4.10 Ten eclipsing interactions; 40 kJ/mol; 35% is relieved.

4.11 Conformation (a) is more stable because the methyl groups are farther apart.

4.12 [chair conformations with OH axial (a) and OH equatorial (e)]

4.13 [chair conformations showing CH3 groups]

4.14 Before ring-flip, red and blue are equatorial and green is axial. After ring-flip, red and blue are axial and green is equatorial.

4.15 4.2 kJ/mol

4.16 Cyano group points straight up.

4.17 Equatorial = 70%; axial = 30%

4.18 (a) 2.0 kJ/mol (axial Cl)
(b) 11.4 kJ/mol (axial CH3)
(c) 2.0 kJ/mol (axial Br)
(d) 8.0 kJ/mol (axial CH2CH3)

4.19 [chair conformation] 1-Chloro-2,4-dimethyl-cyclohexane (less stable chair form)

4.20 *trans*-Decalin is more stable because it has no 1,3-diaxial interactions.

4.21 Both ring-fusions are trans.

Chapter 5

5.1 Chiral: screw, shoe

5.2 (a) [piperidine with CH2CH2CH3]
(b) [cyclohexane with HO, CH3, and isopropyl substituents]
(c) [CH3O-substituted morphinan-like structure with N—CH3]

5.3 [Fischer-like structures of two enantiomers with CO2H, CH3, NH2, H]

5.4 (a) [structure with HO, H, OH groups and C=O]
(b) [structure with F, H, Cl, O]

5.5 Levorotatory

5.6 +16.1°

5.7 (a) —Br (b) —Br
(c) —CH2CH3 (d) —OH
(e) —CH2OH (f) —CH=O

5.8 (a) —OH, —CH2CH2OH, —CH2CH3, —H
(b) —OH, —CO2CH3, —CO2H, —CH2OH
(c) —NH2, —CN, —CH2NHCH3, —CH2NH2
(d) —SSCH3, —SH, —CH2SCH3, —CH3

5.9 (a) S (b) R (c) S

5.10 (a) S (b) S (c) R

5.11 [structure with HO, H, CH2CH2CH3, H3C around central C]

5.12 S

5.13 Compound (a) is D-erythrose 4-phosphate, (d) is its enantiomer, and (b) and (c) are diastereomers.

5.14 Five chirality centers and $2^5 = 32$ stereoisomers

5.15 *S,S*

5.16 Compounds (a) and (d) are meso.

5.17 Compounds (a) and (c) have meso forms.

5.18 [cyclopentane with H3C, CH3, and OH] Meso

5.19 The product retains its *S* stereochemistry because the chirality center is not affected.

5.20 Two diastereomeric salts: (*R*)-lactic acid plus (*S*)-1-phenylethylamine and (*S*)-lactic acid plus (*S*)-1-phenylethylamine

5.21 (a) Constitutional isomers
(b) Diastereomers

5.22 (a) 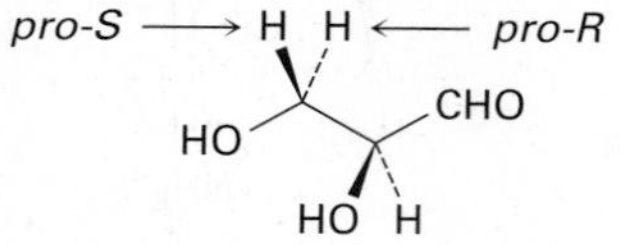 pro-S ⟶ H H ⟵ pro-R

HO—C(H)(CHO) with HO and H

(b) 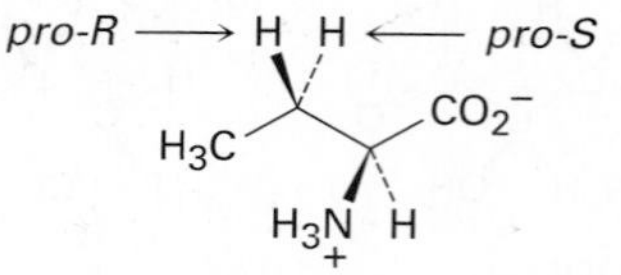 pro-R ⟶ H H ⟵ pro-S

H₃C—C—CO₂⁻ with H₃N⁺ and H

5.23 (a) 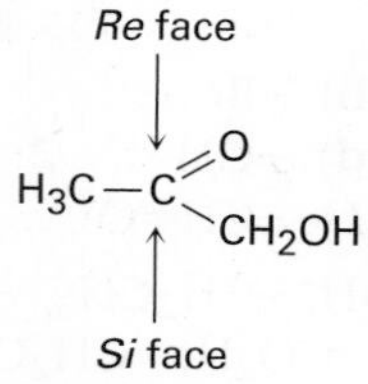 Re face / Si face: H₃C—C(=O)—CH₂OH

(b) 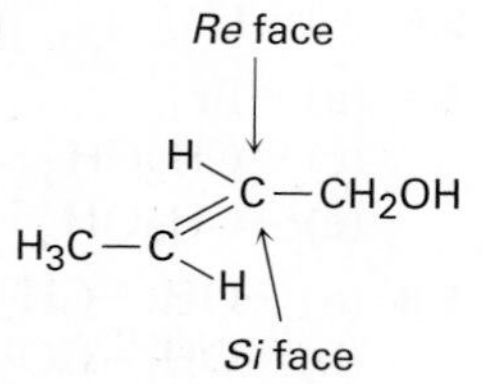 Re face / Si face: H₃C—C(H)=C(H)—CH₂OH

5.24 (*S*)-Lactate

5.25 The —OH adds to the *Re* face of C2, and —H adds to the *Re* face of C3. The overall addition has anti stereochemistry.

Chapter 6

6.1 (a) Substitution (b) Elimination (c) Addition

6.2 1-Chloro-2-methylpentane, 2-chloro-2-methylpentane, 3-chloro-2-methylpentane, 2-chloro-4-methylpentane, 1-chloro-4-methylpentane

6.3 [structure showing peroxide cyclization mechanism with CO₂H-containing chain]

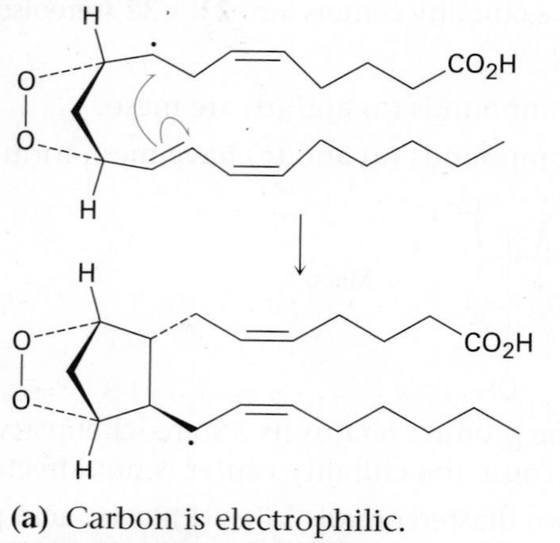

6.4 (a) Carbon is electrophilic.
(b) Sulfur is nucleophilic.
(c) Nitrogens are nucleophilic.
(d) Oxygen is nucleophilic; carbon is electrophilic.

6.5 BF₃ — Electrophilic; vacant *p* orbital

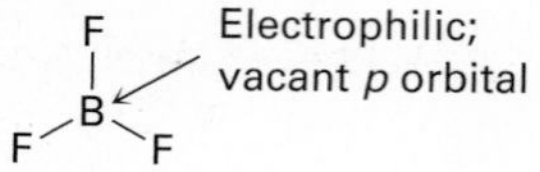

6.6 Bromocyclohexane; chlorocyclohexane

6.7 $(CH_3)_3C^+$ (tert-butyl cation)

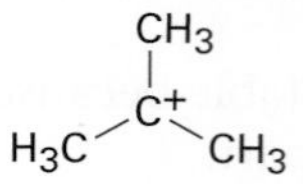

6.8
(a) Cl—Cl + :NH₃ ⇌ ClNH₃⁺ + Cl⁻
(b) CH₃Ö:⁻ + H₃C—Br ⟶ CH₃ÖCH₃ + Br⁻

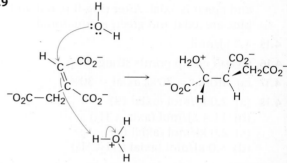

(c) tetrahedral intermediate from addition of methoxide to acid chloride, collapsing to methyl ester + Cl⁻

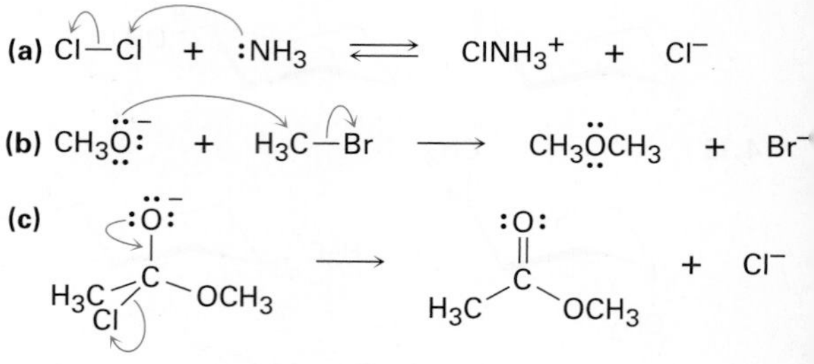

6.9

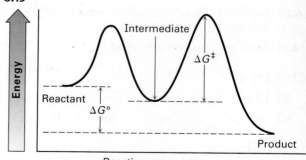

6.10 Negative $\Delta G°$ is more favored.
6.11 Larger K_{eq} is more exergonic.
6.12 Lower $\Delta G^\ddagger$ is faster.
6.13 [Energy diagram: Reactant → Intermediate → Product, showing $\Delta G°$ and $\Delta G^\ddagger$]

Chapter 7

7.1 (a) 1 (b) 2 (c) 2
7.2 (a) 5 (b) 5 (c) 3 (d) 1 (e) 6 (f) 5
7.3 $C_{16}H_{13}ClN_2O$
7.4 (a) 3,4,4-Trimethyl-1-pentene
(b) 3-Methyl-3-hexene
(c) 4,7-Dimethyl-2,5-octadiene
(d) 6-Ethyl-7-methyl-4-nonene

7.5 (a) 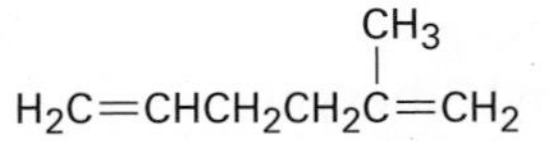 H₂C=CHCH₂CH₂C(CH₃)=CH₂
(b) 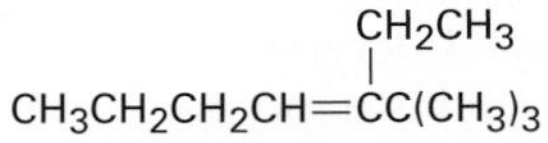 CH₃CH₂CH₂CH=CC(CH₃)₃ with CH₂CH₃ branch
(c) 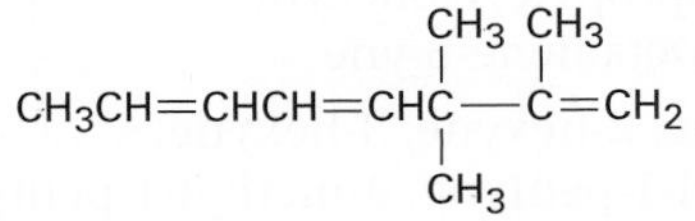 CH₃CH=CHCH=CHC(CH₃)(CH₃)C(CH₃)=CH₂
(d) 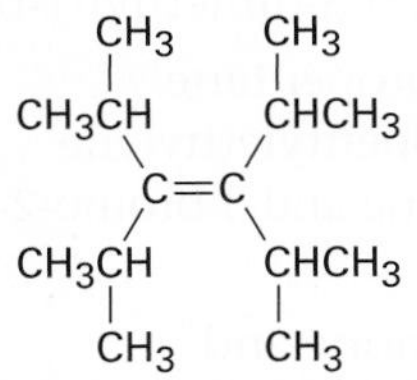 (CH₃CH(CH₃))(CH₃CH(CH₃))C=C(CH(CH₃)CH₃)(CH(CH₃)CH₃)

7.6 (a) 1,2-Dimethylcyclohexene
(b) 4,4-Dimethylcycloheptene
(c) 3-Isopropylcyclopentene

7.7 (a) 2,5,5-Trimethylhex-2-ene
(b) 2,3-Dimethylcyclohexa-1,3-diene

7.8

7.9 Compounds (c), (e), and (f) have cis–trans isomers.

7.10 (a) cis-4,5-Dimethyl-2-hexene
(b) trans-6-Methyl-3-heptene

7.11 (a) –CH₃ (b) –Cl (c) –CH=CH₂
(d) –OCH₃ (e) –CH=O (f) –CH=O

7.12 (a) –Cl, –OH, –CH₃, –H
(b) –CH₂OH, –CH=CH₂, –CH₂CH₃, –CH₃
(c) –CO₂H, –CH₂OH, –C≡N, –CH₂NH₂
(d) –CH₂OCH₃, –C≡N, –C≡CH, –CH₂CH₃

7.13 (a) Z (b) E (c) Z (d) E

7.14 Z isomer with =C(CO₂CH₃)(CH₂OH) and =C(CH(CH₃)₂)(H)

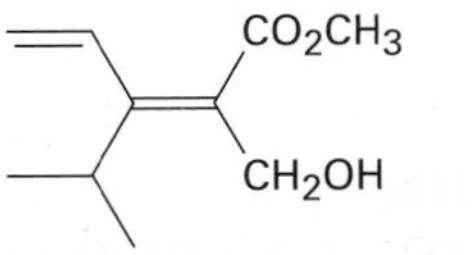

7.15 (a) 2-Methylpropene is more stable than 1-butene.
(b) trans-2-Hexene is more stable than cis-2-hexene.
(c) 1-Methylcyclohexene is more stable than 3-methylcyclohexene.

7.16 (a) Chlorocyclohexane
(b) 2-Bromo-2-methylpentane
(c) 4-Methyl-2-pentanol
(d) 1-Bromo-1-methylcyclohexane

7.17 (a) Cyclopentene
(b) 1-Ethylcyclohexene or ethylidene-cyclohexane
(c) 3-Hexene
(d) Vinylcyclohexane (cyclohexylethylene)

7.18 (a) 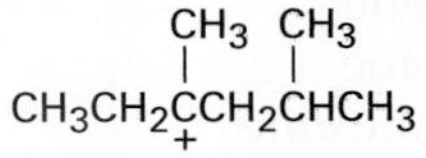 CH₃CH₂C⁺(CH₃)CH₂CH(CH₃)CH₃
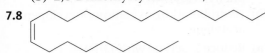

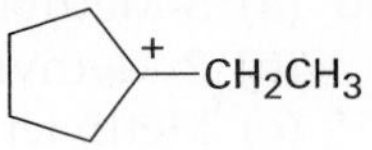

7.19 In the conformation shown, only the methyl-group C–H that is parallel to the carbocation *p* orbital can show hyperconjugation.

7.20 The second step is exergonic; the transition state resembles the carbocation.

7.21

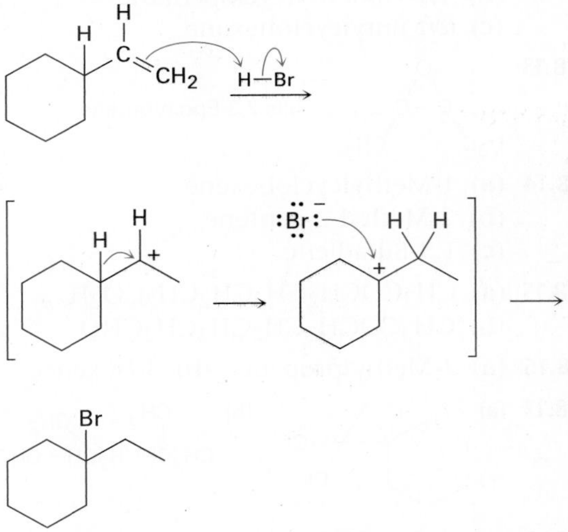

Chapter 8

8.1 2-Methyl-2-butene and 2-methyl-1-butene
8.2 Five
8.3 trans-1,2-Dichloro-1,2-dimethylcyclohexane
8.4 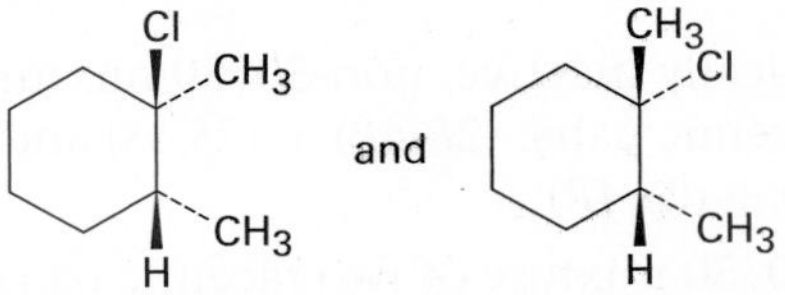
8.5 trans-2-Bromocyclopentanol
8.6 Markovnikov
8.7 (a) 2-Pentanol (b) 2-Methyl-2-pentanol

8.8 (a) Oxymercuration of 2-methyl-1-hexene or 2-methyl-2-hexene
(b) Oxymercuration of cyclohexylethylene or hydroboration of ethylidenecyclohexane

8.9 (a) 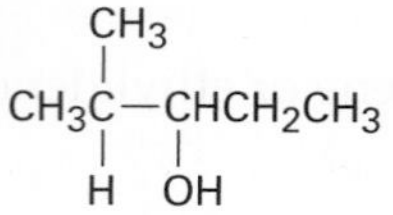 $CH_3\underset{\underset{H}{|}}{\overset{\overset{CH_3}{|}}{C}}-\underset{\underset{OH}{|}}{CH}CH_2CH_3$ (b) 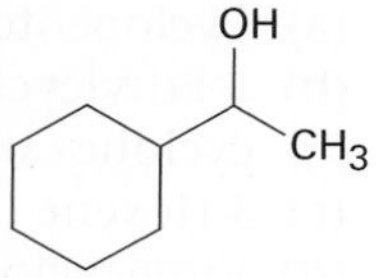 cyclohexyl–CH(OH)–CH$_3$

8.10 (a) 3-Methyl-1-butene
(b) 2-Methyl-2-butene
(c) Methylenecyclohexane

8.11 [two cyclopentane stereoisomer structures with CH$_3$ and OH groups, shown "and"]

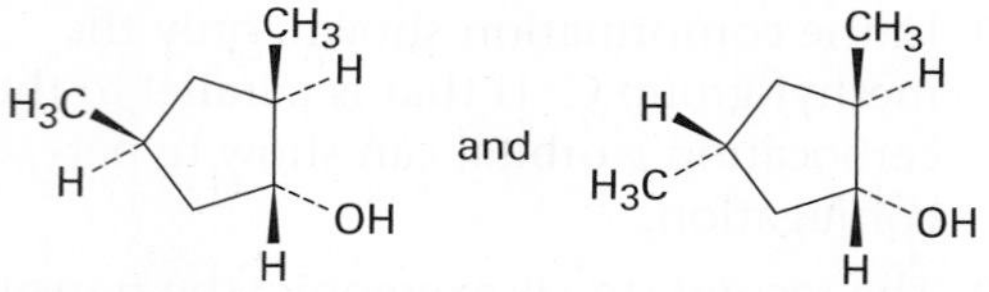

8.12 (a) 2-Methylpentane
(b) 1,1-Dimethylcyclopentane
(c) *tert*-Butylcyclohexane

8.13 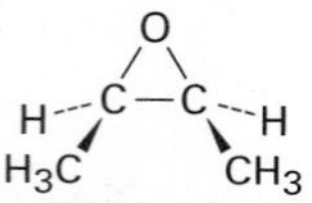 *cis*-2,3-Epoxybutane [structure shown]

8.14 (a) 1-Methylcyclohexene
(b) 2-Methyl-2-pentene
(c) 1,3-Butadiene

8.15 (a) $CH_3COCH_2CH_2CH_2CH_2CO_2H$
(b) $CH_3COCH_2CH_2CH_2CH_2CHO$

8.16 (a) 2-Methylpropene (b) 3-Hexene

8.17 (a) 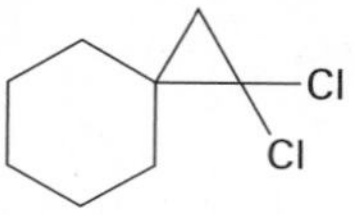 spiro cyclohexane-cyclopropane with two Cl (b) 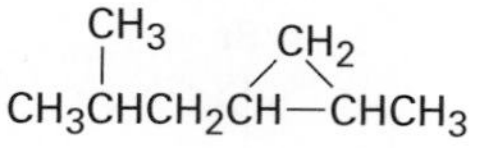 $CH_3CHCH_2CH-CHCH_3$ with CH$_3$ and CH$_2$ substituents

8.18 (a) $H_2C=CHOCH_3$ (b) $ClCH=CHCl$

8.19 $\cdot CH_2CH_2 + \cdot CH-CH_2(H) \longrightarrow \cdot CH_2CH_3 + \cdot CH=CH_2$

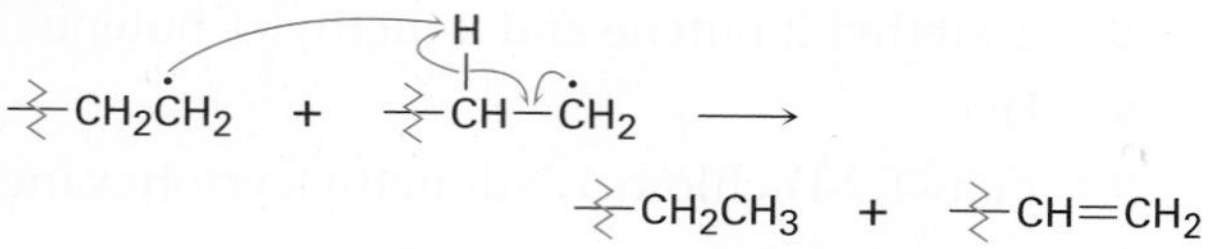

8.20 An optically inactive, non-50:50 mixture of two racemic pairs: (2R,4R) + (2S,4S) and (2R,4S) + (2S,4R)

8.21 Non-50:50 mixture of two racemic pairs: (1S,3R) + (1R,3S) and (1S,3S) + (1R,3R)

Chapter 9

9.1 (a) 2,5-Dimethyl-3-hexyne
(b) 3,3-Dimethyl-1-butyne
(c) 3,3-Dimethyl-4-octyne
(d) 2,5,5-Trimethyl-3-heptyne
(e) 6-Isopropylcyclodecyne
(f) 2,4-Octadiene-6-yne

9.2 1-Hexyne, 2-hexyne, 3-hexyne, 3-methyl-1-pentyne, 4-methyl-1-pentyne, 4-methyl-2-pentyne, 3,3-dimethyl-1-butyne

9.3 (a) 1,1,2,2-Tetrachloropentane
(b) 1-Bromo-1-cyclopentylethylene
(c) 2-Bromo-2-heptene and 3-bromo-2-heptene

9.4 (a) 4-Octanone
(b) 2-Methyl-4-octanone and 7-methyl-4-octanone

9.5 (a) 1-Pentyne (b) 2-Pentyne

9.6 (a) $C_6H_5C\equiv CH$ (b) 2,5-Dimethyl-3-hexyne

9.7 (a) Mercuric sulfate–catalyzed hydration of phenylacetylene
(b) Hydroboration/oxidation of cyclopentylacetylene

9.8 (a) Reduce 2-octyne with Li/NH$_3$
(b) Reduce 3-heptyne with H$_2$/Lindlar catalyst
(c) Reduce 3-methyl-1-pentyne

9.9 No: (a), (c), (d); yes: (b)

9.10 (a) 1-Pentyne + CH$_3$I, or propyne + CH$_3$CH$_2$CH$_2$I
(b) 3-Methyl-1-butyne + CH$_3$CH$_2$I
(c) Cyclohexylacetylene + CH$_3$I

9.11 $CH_3C\equiv CH \xrightarrow[\text{2. CH}_3\text{I}]{\text{1. NaNH}_2} CH_3C\equiv CCH_3$

$\xrightarrow[\text{Lindlar cat.}]{H_2}$ *cis*-$CH_3CH=CHCH_3$

9.12 (a) KMnO$_4$, H$_3$O$^+$
(b) H$_2$/Lindlar
(c) 1. H$_2$/Lindlar; 2. HBr
(d) 1. H$_2$/Lindlar; 2. BH$_3$; 3. NaOH, H$_2$O$_2$
(e) 1. H$_2$/Lindlar; 2. Cl$_2$
(f) O$_3$

9.13 (a) 1. HC≡CH + NaNH$_2$;
2. CH$_3$(CH$_2$)$_6$CH$_2$Br; 3. 2 H$_2$/Pd
(b) 1. HC≡CH + NaNH$_2$;
2. (CH$_3$)$_3$CCH$_2$CH$_2$I; 3. 2 H$_2$/Pd
(c) 1. HC≡CH + NaNH$_2$;
2. CH$_3$CH$_2$CH$_2$CH$_2$I; 3. BH$_3$; 4. H$_2$O$_2$
(d) 1. HC≡CH + NaNH$_2$;
2. CH$_3$CH$_2$CH$_2$CH$_2$CH$_2$I; 3. HgSO$_4$, H$_3$O$^+$

Chapter 10

10.1 (a) 1-Iodobutane
(b) 1-Chloro-3-methylbutane
(c) 1,5-Dibromo-2,2-dimethylpentane
(d) 1,3-Dichloro-3-methylbutane
(e) 1-Chloro-3-ethyl-4-iodopentane
(f) 2-Bromo-5-chlorohexane

10.2 (a) CH$_3$CH$_2$CH$_2$C(CH$_3$)$_2$CH(Cl)CH$_3$
(b) CH$_3$CH$_2$CH$_2$C(Cl)$_2$CH(CH$_3$)$_2$
(c) CH$_3$CH$_2$C(Br)(CH$_2$CH$_3$)$_2$
(d) [structure: cyclohexane with two Br on same carbon, isopropyl substituent]
(e) CH$_3$CH$_2$CH$_2$CH$_2$CH$_2$CHCH$_2$CHCH$_3$ with CH$_3$CHCH$_2$CH$_3$ branch and Cl
(f) [structure: cyclohexane with two Br on same carbon, tert-butyl substituent]

10.3 Chiral: 1-chloro-2-methylpentane, 3-chloro-2-methylpentane, 2-chloro-4-methylpentane
Achiral: 2-chloro-2-methylpentane, 1-chloro-4-methylpentane

10.4 1-Chloro-2-methylbutane (29%), 1-chloro-3-methylbutane (14%), 2-chloro-2-methylbutane (24%), 2-chloro-3-methylbutane (33%)

10.5 [three cyclohexadienyl radical resonance structures]

10.6 The intermediate allylic radical reacts at the more accessible site and gives the more highly substituted double bond.

10.7 (a) 3-Bromo-5-methylcycloheptene and 3-bromo-6-methylcycloheptene
(b) Four products

10.8 (a) 2-Methyl-2-propanol + HCl
(b) 4-Methyl-2-pentanol + PBr$_3$
(c) 5-Methyl-1-pentanol + PBr$_3$
(d) 3,3-Dimethyl-cyclopentanol + HF, pyridine

10.9 Both reactions occur.

10.10 React Grignard reagent with D$_2$O.

10.11 (a) 1. NBS; 2. (CH$_3$)$_2$CuLi
(b) 1. Li; 2. CuI; 3. CH$_3$CH$_2$CH$_2$CH$_2$Br
(c) 1. BH$_3$; 2. H$_2$O$_2$, NaOH; 3. PBr$_3$; 4. Li, then CuI; 5. CH$_3$(CH$_2$)$_4$Br

10.12
(a) [cyclohexane] < [cyclohexanone] = [cyclohexenone with =O] < [chlorocyclohexene] < [chlorobenzene]
(b) CH$_3$CH$_2$NH$_2$ < H$_2$NCH$_2$CH$_2$NH$_2$ < CH$_3$C≡N

10.13 (a) Reduction (b) Neither

Chapter 11

11.1 (R)-1-Methylpentyl acetate, CH$_3$CO$_2$CH(CH$_3$)CH$_2$CH$_2$CH$_2$CH$_3$

11.2 (S)-2-Butanol

11.3 (S)-2-Bromo-4-methylpentane ⟶ (R) CH$_3$CHCH$_2$CHCH$_3$ with CH$_3$ and SH groups

11.4 (a) 1-Iodobutane (b) 1-Butanol (c) 1-Hexyne (d) Butylammonium bromide

11.5 (a) (CH$_3$)$_2$N$^-$ (b) (CH$_3$)$_3$N (c) H$_2$S

11.6 CH$_3$OTos > CH$_3$Br > (CH$_3$)$_2$CHCl > (CH$_3$)$_3$CCl

11.7 Similar to protic solvents

11.8 Racemic 1-ethyl-1-methylhexyl acetate

11.9 90.1% racemization, 9.9% inversion

11.10
(S)-Bromide ⟶ [structure: phenyl group bonded to C with H$_3$C, OH, CH$_2$CH$_3$]
Racemic

11.11 H$_2$C=CHCH(Br)CH$_3$ > CH$_3$CH(Br)CH$_3$ > CH$_3$CH$_2$Br > H$_2$C=CHBr

11.12 The same allylic carbocation intermediate is formed.

11.13 (a) S$_N$1 (b) S$_N$2

11.14

Linalyl diphosphate → [intermediate] → Limonene

11.15 (a) Major: 2-methyl-2-pentene; minor: 4-methyl-2-pentene
(b) Major: 2,3,5-trimethyl-2-hexene; minor: 2,3,5-trimethyl-3-hexene and 2-isopropyl-4-methyl-1-pentene
(c) Major: ethylidenecyclohexane; minor: cyclohexylethylene

11.16 (a) 1-Bromo-3,6-dimethylheptane
(b) 4-Bromo-1,2-dimethylcyclopentane

11.17 (Z)-1-Bromo-1,2-diphenylethylene

11.18 (Z)-3-Methyl-2-pentene

11.19 Cis isomer reacts faster because the bromine is axial.

11.20 (a) S$_N$2 (b) E2 (c) S$_N$1 (d) E1cB

Chapter 12

12.1 C$_{19}$H$_{28}$O$_2$

12.2 (a) 2-Methyl-2-pentene (b) 2-Hexene

12.3 (a) 43, 71 (b) 82 (c) 58 (d) 86

12.4 102 (M$^+$), 84 (dehydration), 87 (alpha cleavage), 59 (alpha cleavage)

12.5 X-ray energy is higher; $\lambda = 9.0 \times 10^{-6}$ m is higher in energy.

12.6 (a) 2.4×10^6 kJ/mol (b) 4.0×10^4 kJ/mol
(c) 2.4×10^3 kJ/mol (d) 2.8×10^2 kJ/mol
(e) 6.0 kJ/mol (f) 4.0×10^{-2} kJ/mol

12.7 (a) Ketone or aldehyde (b) Nitro compound
(c) Carboxylic acid

12.8 (a) CH$_3$CH$_2$OH has an −OH absorption.
(b) 1-Hexene has a double-bond absorption.
(c) CH$_3$CH$_2$CO$_2$H has a very broad −OH absorption.

12.9 1450–1600 cm^{-1}: aromatic ring; 2100 cm^{-1}: C≡C; 3300 cm^{-1}: C≡C−H

12.10 (a) 1715 cm^{-1} (b) 1730, 2100, 3300 cm^{-1}
(c) 1720, 2500–3100, 3400–3650 cm^{-1}

12.11 1690, 1650, 2230 cm^{-1}

Chapter 13

13.1 7.5×10^{-5} kJ/mol for ^{19}F; 8.0×10^{-5} kJ/mol for ^{1}H

13.2 1.2×10^{-4} kJ/mol

13.3 The vinylic C−H protons are nonequivalent.

13.4 (a) 7.27 δ (b) 3.05 δ (c) 3.46 δ (d) 5.30 δ

13.5 (a) 420 Hz (b) 2.1 δ (c) 1050 Hz

13.6 (a) 4 (b) 7 (c) 4 (d) 5 (e) 5 (f) 7

13.7 (a) 1,3-Dimethylcyclopentene
(b) 2-Methylpentane
(c) 1-Chloro-2-methylpropane

13.8 −CH$_3$, 9.3 δ; −CH$_2$−, 27.6 δ; C=O, 174.6 δ; −OCH$_3$, 51.4 δ

13.9 23, 26 δ; 132 δ; 124 δ; 39 δ; 24 δ; 68 δ; 18 δ; OH

13.10 DEPT-135 (+), DEPT-135 (−), DEPT-135 (+), DEPT-135 (+) → H$_3$C, H ← DEPT-90, DEPT-135 (+)

13.11 [structure: phenyl-CH$_2$-C(CH$_3$)$_2$-CH$_3$ with CH$_3$ groups]

13.12 A DEPT-90 spectrum would show two absorptions for the non-Markovnikov product (RCH=CHBr) but no absorptions for the Markovnikov product (RBrC=CH$_2$).

13.13 (a) Enantiotopic (b) Diastereotopic
  (c) Diastereotopic (d) Diastereotopic
  (e) Diastereotopic (f) Homotopic

13.14 (a) 2 (b) 4 (c) 3 (d) 4 (e) 5 (f) 3

13.15 4

13.16 (a) 1.43 δ (b) 2.17 δ (c) 7.37 δ
  (d) 5.30 δ (e) 9.70 δ (f) 2.12 δ

13.17 Seven kinds of protons

13.18 Two peaks; 3:2 ratio

13.19 (a) –CHBr$_2$, quartet; –CH$_3$, doublet
  (b) CH$_3$O–, singlet; –OCH$_2$–, triplet;
      –CH$_2$Br, triplet
  (c) ClCH$_2$–, triplet; –CH$_2$–, quintet
  (d) CH$_3$–, triplet; –CH$_2$–, quartet;
      –CH–, septet; (CH$_3$)$_2$, doublet
  (e) CH$_3$–, triplet; –CH$_2$–, quartet;
      –CH–, septet; (CH$_3$)$_2$, doublet
  (f) =CH, triplet, –CH$_2$–, doublet,
      aromatic C–H, two multiplets

13.20 (a) CH$_3$OCH$_3$ (b) CH$_3$CH(Cl)CH$_3$
  (c) ClCH$_2$CH$_2$OCH$_2$CH$_2$Cl
  (d) CH$_3$CH$_2$CO$_2$CH$_3$ or CH$_3$CO$_2$CH$_2$CH$_3$

13.21 CH$_3$CH$_2$OCH$_2$CH$_3$

13.22 J_{1-2} = 16 Hz; J_{2-3} = 8 Hz

13.23 1-Chloro-1-methylcyclohexane has a singlet methyl absorption.

Chapter 14

14.1 Expected $\Delta H°_{hydrog}$ for allene is –252 kJ/mol. Allene is less stable than a nonconjugated diene, which is less stable than a conjugated diene.

14.2 1-Chloro-2-pentene, 3-chloro-1-pentene, 4-chloro-2-pentene

14.3 4-Chloro-2-pentene predominates in both.

14.4 1,2 Addition: 6-bromo-1,6-dimethylcyclohexene
  1,4 Addition: 3-bromo-1,2-dimethylcyclohexene

14.5 Interconversion occurs by S$_N$1 dissociation to a common intermediate cation.

14.6 The double bond is more highly substituted.

14.7

14.8 Good dienophiles: (a), (d)

14.9 Compound (a) is s-cis. Compound (c) can rotate to s-cis.

14.10

14.11

14.12

14.13 300–600 kJ/mol; UV energy is greater than IR or NMR energy.

14.14 1.46 × 10^{-5} M

14.15 All except (a) have UV absorptions.

Chapter 15

15.1 (a) Meta (b) Para (c) Ortho

15.2 (a) m-Bromochlorobenzene
  (b) (3-Methylbutyl)benzene
  (c) p-Bromoaniline
  (d) 2,5-Dichlorotoluene
  (e) 1-Ethyl-2,4-dinitrobenzene
  (f) 1,2,3,5-Tetramethylbenzene

15.3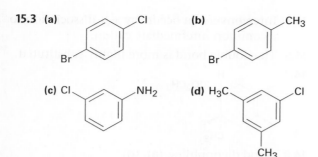

15.4 Pyridine has an aromatic sextet of electrons.

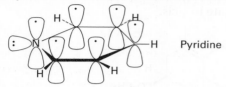

15.5 Cyclodecapentaene is not flat because of steric interactions.

15.6 All C—C bonds are equivalent; one resonance line in both ^{1}H and ^{13}C NMR spectra.

15.7 The cyclooctatetraenyl dianion is aromatic (ten π electrons) and flat.

15.8

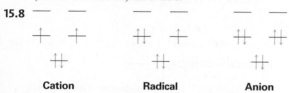

15.9

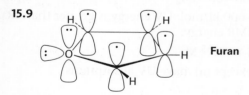

15.10 The thiazolium ring has six π electrons.

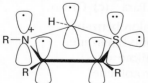

15.11

15.12 The three nitrogens in double bonds each contribute one; the remaining nitrogen contributes two.

Chapter 16

16.1 o-, m-, and p-Bromotoluene

16.2

16.3 o-Xylene: 2; p-xylene: 1; m-xylene: 3

16.4 D$^+$ does electrophilic substitutions on the ring.

16.5 No rearrangement: (a), (b), (e)

16.6 tert-Butylbenzene

16.7 (a) (CH$_3$)$_2$CHCOCl (b) PhCOCl

16.8 (a) Phenol > Toluene > Benzene > Nitrobenzene
(b) Phenol > Benzene > Chlorobenzene > Benzoic acid
(c) Aniline > Benzene > Bromobenzene > Benzaldehyde

16.9 (a) o- and p-Bromonitrobenzene
(b) m-Bromonitrobenzene
(c) o- and p-Chlorophenol
(d) o- and p-Bromoaniline

16.10 Alkylbenzenes are more reactive than benzene itself, but acylbenzenes are less reactive.

16.11 Toluene is more reactive; the trifluoromethyl group is electron-withdrawing.

16.12 The nitrogen electrons are donated to the nearby carbonyl group by resonance and are less available to the ring.

16.13 The meta intermediate is most favored.

16.14 (a) Ortho and para to —OCH$_3$
(b) Ortho and para to —NH$_2$
(c) Ortho and para to —Cl

16.15 (a) Reaction occurs ortho and para to the —CH$_3$ group.
(b) Reaction occurs ortho and para to the —OCH$_3$ group.

16.16 The phenol is deprotonated by KOH to give an anion that carries out a nucleophilic acyl substitution reaction on the fluoronitrobenzene.

16.17 Only one benzyne intermediate can form from *p*-bromotoluene; two different benzyne intermediates can form from *m*-bromotoluene.

16.18 (a) *m*-Nitrobenzoic acid
(b) *p*-tert-Butylbenzoic acid

16.19 A benzyl radical is more stable than a primary alkyl radical by 52 kJ/mol and is similar in stability to an allyl radical.

16.20 1. CH_3CH_2Cl, $AlCl_3$; 2. NBS; 3. KOH, ethanol

16.21 1. PhCOCl, $AlCl_3$; 2. H_2/Pd

16.22 (a) 1. HNO_3, H_2SO_4; 2. Cl_2, $FeCl_3$
(b) 1. CH_3COCl, $AlCl_3$; 2. Cl_2, $FeCl_3$; 3. H_2/Pd
(c) 1. CH_3CH_2COCl, $AlCl_3$; 2. Cl_2, $FeCl_3$; 3. H_2/Pd; 4. HNO_3, H_2SO_4
(d) 1. CH_3Cl, $AlCl_3$; 2. Br_2, $FeBr_3$; 3. SO_3, H_2SO_4

16.23 (a) Friedel–Crafts acylation does not occur on a deactivated ring.
(b) Rearrangement occurs during Friedel–Crafts alkylation with primary halides; chlorination occurs ortho to the alkyl group.

Chapter 17

17.1 (a) 5-Methyl-2,4-hexanediol
(b) 2-Methyl-4-phenyl-2-butanol
(c) 4,4-Dimethylcyclohexanol
(d) *trans*-2-Bromocyclopentanol
(e) 4-Bromo-3-methylphenol
(f) 2-Cyclopenten-1-ol

17.2 (a) [structure: H_3C, CH_2OH / C=C / H, CH_2CH_3]
(b) [structure: cyclohexenol with OH]
(c) [structure: cycloheptane with OH (wedge) and Cl (dash)]
(d) [structure: $CH_3CHCH_2CH_2CH_2OH$ with OH on second carbon]
(e) [structure: 2,6-dimethylphenol, H_3C, CH_3, OH]
(f) [structure: 2-(2-hydroxyphenyl)ethanol, OH, CH_2CH_2OH]

17.3 Hydrogen-bonding is more difficult in hindered alcohols.

17.4 (a) HC≡CH < $(CH_3)_2$CHOH < CH_3OH < $(CF_3)_2$CHOH
(b) *p*-Methylphenol < Phenol < *p*-(Trifluoromethyl)phenol
(c) Benzyl alcohol < Phenol < *p*-Hydroxybenzoic acid

17.5 The electron-withdrawing nitro group stabilizes an alkoxide ion, but the electron-donating methoxyl group destabilizes the anion.

17.6 (a) 2-Methyl-3-pentanol
(b) 2-Methyl-4-phenyl-2-butanol
(c) *meso*-5,6-Decanediol

17.7 (a) $NaBH_4$ (b) $LiAlH_4$ (c) $LiAlH_4$

17.8 (a) Benzaldehyde or benzoic acid (or ester)
(b) Acetophenone (c) Cyclohexanone
(d) 2-Methylpropanal or 2-methylpropanoic acid (or ester)

17.9 (a) 1-Methylcyclopentanol
(b) 1,1-Diphenylethanol
(c) 3-Methyl-3-hexanol

17.10 (a) Acetone + CH_3MgBr, or ethyl acetate + 2 CH_3MgBr
(b) Cyclohexanone + CH_3MgBr
(c) 3-Pentanone + CH_3MgBr, or 2-butanone + CH_3CH_2MgBr, or ethyl acetate + 2 CH_3CH_2MgBr
(d) 2-Butanone + PhMgBr, or ethyl phenyl ketone + CH_3MgBr, or acetophenone + CH_3CH_2MgBr
(e) Formaldehyde + PhMgBr
(f) Formaldehyde + $(CH_3)_2CHCH_2MgBr$

17.11 Cyclohexanone + CH_3CH_2MgBr

17.12 1. *p*-TosCl, pyridine; 2. NaCN

17.13 (a) 2-Methyl-2-pentene
(b) 3-Methylcyclohexene
(c) 1-Methylcyclohexene
(d) 2,3-Dimethyl-2-pentene
(e) 2-Methyl-2-pentene

17.14 (a) 1-Phenylethanol
(b) 2-Methyl-1-propanol
(c) Cyclopentanol

17.15 (a) Hexanoic acid, hexanal
(b) 2-Hexanone
(c) Hexanoic acid, no reaction

17.16 S_N2 reaction of F^- on silicon with displacement of alkoxide ion.

17.17 Protonation of 2-methylpropene gives the *tert*-butyl cation, which carries out an electrophilic aromatic substitution reaction.

17.18 Disappearance of —OH absorption; appearance of C=O

17.19 (a) Singlet (b) Doublet (c) Triplet (d) Doublet (e) Doublet (f) Singlet

Chapter 18

18.1 (a) Diisopropyl ether
(b) Cyclopentyl propyl ether
(c) *p*-Bromoanisole or 4-bromo-1-methoxybenzene
(d) 1-Methoxycyclohexene
(e) Ethyl isobutyl ether
(f) Allyl vinyl ether

18.2 A mixture of diethyl ether, dipropyl ether, and ethyl propyl ether is formed in a 1:1:2 ratio.

18.3 (a) $CH_3CH_2CH_2O^- + CH_3Br$
(b) $PhO^- + CH_3Br$
(c) $(CH_3)_2CHO^- + PhCH_2Br$
(d) $(CH_3)_3CCH_2O^- + CH_3CH_2Br$

18.4

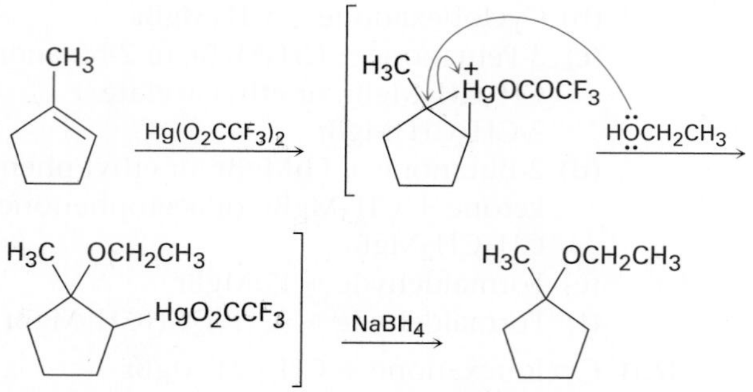

18.5 (a) Either method (b) Williamson (c) Alkoxymercuration (d) Williamson

18.6 (a) Bromoethane > 2-Bromopropane > Bromobenzene
(b) Bromoethane > Chloroethane > 1-Iodopropene

18.7 (a)

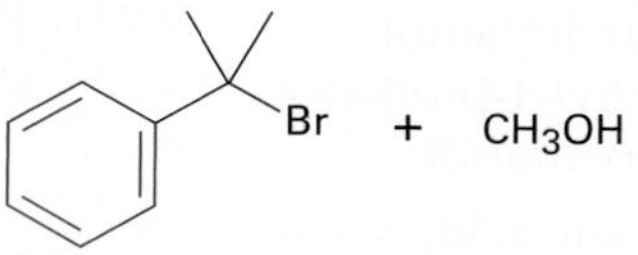

(b) CH_3CH_2CHOH with CH_3 + $CH_3CH_2CH_2Br$

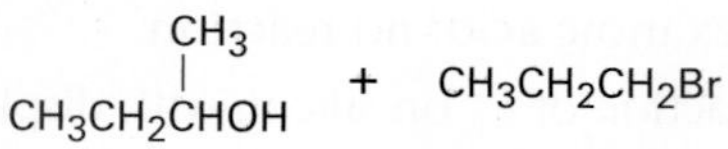

18.8 Protonation of the oxygen atom, followed by E1 reaction

18.9 Br^- and I^- are better nucleophiles than Cl^-.

18.10 *o*-(1-Methylallyl)phenol

18.11 Epoxidation of *cis*-2-butene yields *cis*-2,3-epoxybutane, while epoxidation of *trans*-2-butene yields *trans*-2,3-epoxybutane.

18.12 (a) (b)

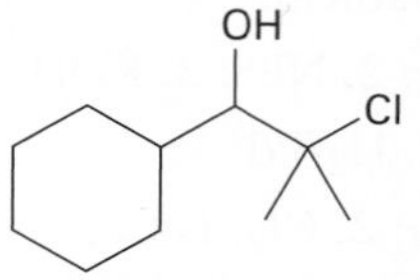

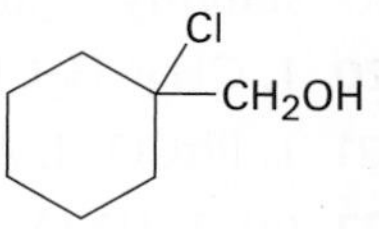

18.13 (a) 1-Methylcyclohexene + OsO_4; then $NaHSO_3$
(b) 1-Methylcyclohexene + *m*-chloroperoxybenzoic acid, then H_3O^+

18.14 (a) (b)

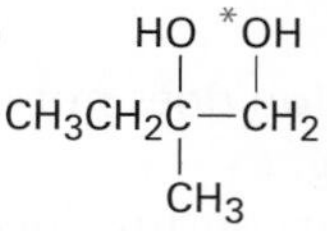

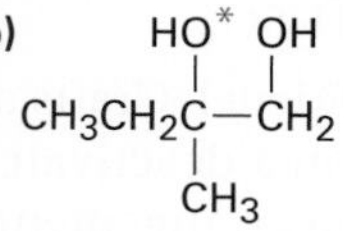

(c)

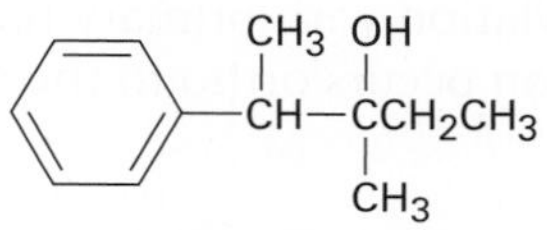

18.16 (a) 2-Butanethiol
(b) 2,2,6-Trimethyl-4-heptanethiol
(c) 2-Cyclopentene-1-thiol
(d) Ethyl isopropyl sulfide
(e) *o*-Di(methylthio)benzene
(f) 3-(Ethylthio)cyclohexanone

18.17 (a) 1. $LiAlH_4$; 2. PBr_3; 3. $(H_2N)_2C=S$; 4. H_2O, NaOH
(b) 1. HBr; 2. $(H_2N)_2C=S$; 3. H_2O, NaOH

18.18 1,2-Epoxybutane

Preview of Carbonyl Chemistry

1. Acetyl chloride is more electrophilic than acetone.

2.

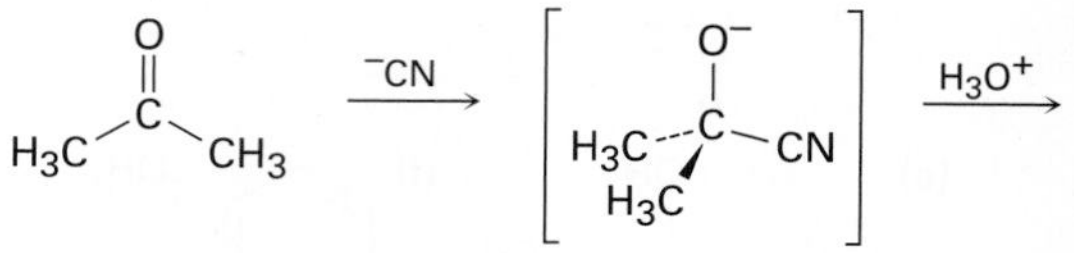

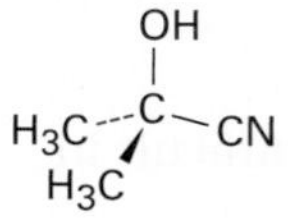

3. (a) Nucleophilic acyl substitution
(b) Nucleophilic addition
(c) Carbonyl condensation

Chapter 19

19.1 (a) 2-Methyl-3-pentanone
(b) 3-Phenylpropanal
(c) 2,6-Octanedione
(d) trans-2-Methylcyclohexanecarbaldehyde
(e) 4-Hexenal
(f) cis-2,5-Dimethylcyclohexanone

19.2
(a) CH_3CHCH_2CHO with CH_3 substituent
(b) $CH_3CHCH_2CCH_3$ with Cl and =O
(c) Ph-CH_2CHO
(d) cyclohexane with $(CH_3)_3C$ and CHO (both H shown)
(e) $H_2C=CCH_2CHO$ with CH_3
(f) $CH_3CH_2CHCH_2CH_2CHCHO$ with CH_3 and CH_3CHCl

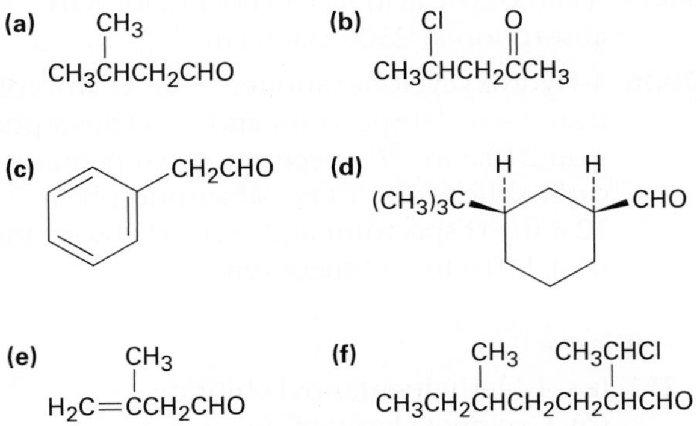

19.3 (a) Dess–Martin periodinane (b) 1. O_3; 2. Zn
(c) DIBAH
(d) 1. BH_3, then H_2O_2, NaOH;
2. Dess–Martin periodinane

19.4 (a) $HgSO_4$, H_3O^+
(b) 1. CH_3COCl, $AlCl_3$; 2. Br_2, $FeBr_3$
(c) 1. Mg; 2. CH_3CHO; 3. H_3O^+; 4. CrO_3
(d) 1. BH_3; 2. H_2O_2, NaOH; 3. CrO_3

19.5 cyclohexane with CN and OH

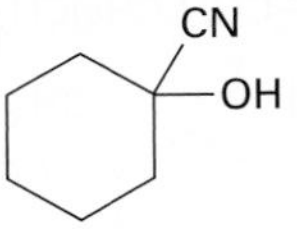

19.6 The electron-withdrawing nitro group in p-nitrobenzaldehyde polarizes the carbonyl group.

19.7 $CCl_3CH(OH)_2$

19.8 Labeled water adds reversibly to the carbonyl group.

19.9 The equilibrium is unfavorable for sterically hindered ketones.

19.10 cyclohexane=NCH_2CH_3 and cyclohexene-$N(CH_2CH_3)_2$

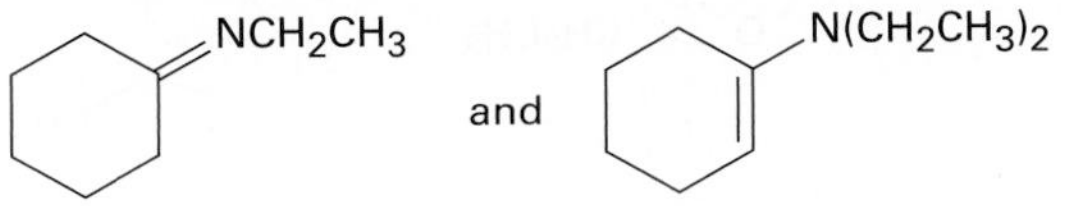

19.11 The steps are the exact reverse of the forward reaction, shown in Figure 19.6.

19.12 cyclopentanone + $(CH_3CH_2)_2NH$ → cyclopentene-$N(CH_2CH_3)_2$

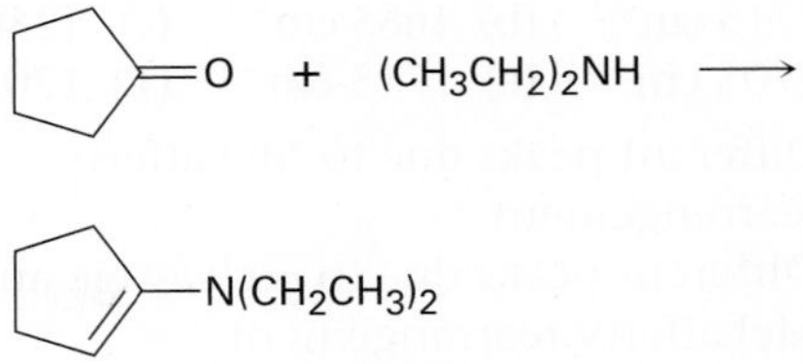

19.13 (a) H_2/Pd (b) N_2H_4, KOH
(c) 1. H_2/Pd; 2. N_2H_4, KOH

19.14 The mechanism is identical to that between a ketone and 2 equivalents of a monoalcohol, shown in Figure 19.10.

19.15 benzene with CH_3O_2C and $CH(CH_3)CHO$ + CH_3OH

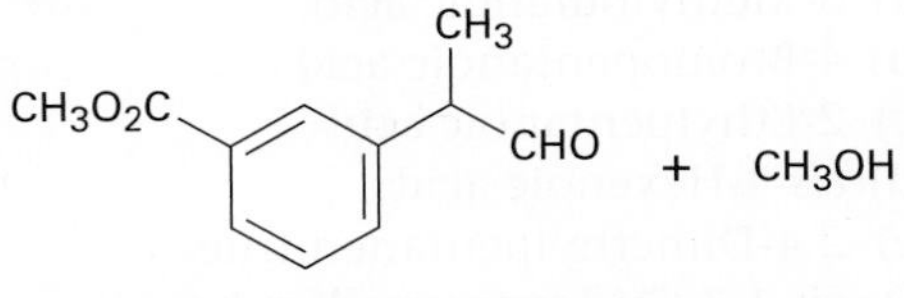

19.16 (a) Cyclohexanone + $(Ph)_3P=CHCH_3$
(b) Cyclohexanecarbaldehyde + $(Ph)_3P=CH_2$
(c) Acetone + $(Ph)_3P=CHCH_2CH_2CH_3$
(d) Acetone + $(Ph)_3P=CHPh$
(e) $PhCOCH_3$ + $(Ph)_3P=CHPh$
(f) 2-Cyclohexenone + $(Ph)_3P=CH_2$

19.17 β-Carotene structure

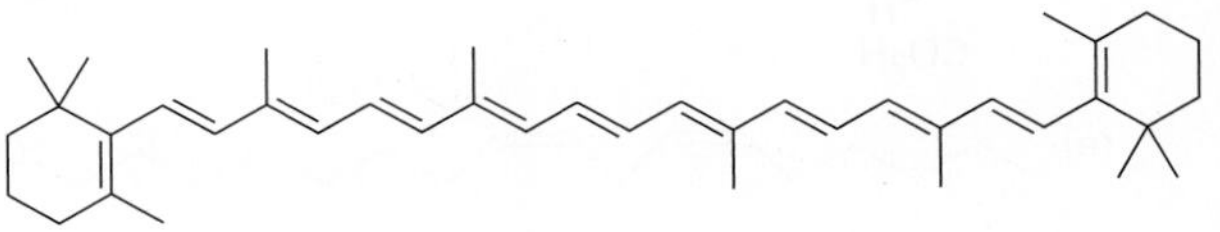

19.18 Intramolecular Cannizzaro reaction

19.19 Addition of the pro-R hydrogen of NADH takes place on the Re face of pyruvate.

19.20 The –OH group adds to the Re face at C2, and –H adds to the Re face at C3, to yield (2R,3S)-isocitrate.

19.21 cyclohexane with O and CN

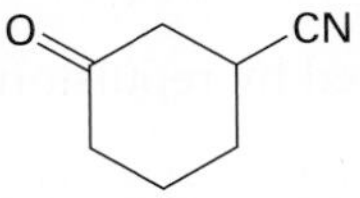

19.22 (a) 3-Buten-2-one + $(CH_3CH_2CH_2)_2CuLi$
(b) 3-Methyl-2-cyclohexenone + $(CH_3)_2CuLi$
(c) 4-tert-Butyl-2-cyclohexenone + $(CH_3CH_2)_2CuLi$
(d) Unsaturated ketone + $(H_2C=CH)_2CuLi$

19.23 Look for appearance of either an alcohol or a saturated ketone in the product.

19.24 (a) 1715 cm⁻¹ (b) 1685 cm⁻¹ (c) 1750 cm⁻¹
(d) 1705 cm⁻¹ (e) 1715 cm⁻¹ (f) 1705 cm⁻¹

19.25 (a) Different peaks due to McLafferty rearrangement
(b) Different peaks due to α cleavage and McLafferty rearrangement
(c) Different peaks due to McLafferty rearrangement

19.26 IR: 1750 cm⁻¹; MS: 140, 84

Chapter 20

20.1 (a) 3-Methylbutanoic acid
(b) 4-Bromopentanoic acid
(c) 2-Ethylpentanoic acid
(d) *cis*-4-Hexenoic acid
(e) 2,4-Dimethylpentanenitrile
(f) *cis*-1,3-Cyclopentanedicarboxylic acid

20.2
(a) CH₃CH₂CH₂CH(CH(CH₃)₂)CO₂H — H₃C CH₃ / CH₃CH₂CH₂CHCHCO₂H
(b) CH₃CHCH₂CH₂CO₂H with CH₃
(c) cyclobutane with H and CO₂H (cis), H and CO₂H
(d) benzene with CO₂H and OH (ortho, salicylic acid)
(e) long chain diene with CO₂H
(f) CH₃CH₂CH=CHCN

20.3 Dissolve the mixture in ether, extract with aqueous NaOH, separate and acidify the aqueous layer, and extract with ether.

20.4 43%

20.5 (a) 82% dissociation (b) 73% dissociation

20.6 Lactic acid is stronger because of the inductive effect of the −OH group.

20.7 The dianion is destabilized by repulsion between charges.

20.8 More reactive

20.9 (a) *p*-Methylbenzoic acid < Benzoic acid < *p*-Chlorobenzoic acid
(b) Acetic acid < Benzoic acid < *p*-Nitrobenzoic acid

20.10 (a) 1. Mg; 2. CO₂; 3. H₃O⁺
(b) 1. Mg; 2. CO₂; 3. H₃O⁺ or 1. NaCN; 2. H₃O⁺

20.11 1. NaCN; 2. H₃O⁺; 3. LiAlH₄

20.12 1. PBr₃; 2. NaCN; 3. H₃O⁺; 4. LiAlH₄

20.13 (a) Propanenitrile + CH₃CH₂MgBr, then H₃O⁺
(b) *p*-Nitrobenzonitrile + CH₃MgBr, then H₃O⁺

20.14 1. NaCN; 2. CH₃CH₂MgBr, then H₃O⁺

20.15 A carboxylic acid has a very broad −OH absorption at 2500–3300 cm⁻¹.

20.16 4-Hydroxycyclohexanone: H−C−O absorption near 4 δ in ¹H spectrum and C=O absorption near 210 δ in ¹³C spectrum. Cyclopentanecarboxylic acid: −CO₂H absorption near 12 δ in ¹H spectrum and −CO₂H absorption near 170 δ in ¹³C spectrum.

Chapter 21

21.1 (a) 4-Methylpentanoyl chloride
(b) Cyclohexylacetamide
(c) Isopropyl 2-methylpropanoate
(d) Benzoic anhydride
(e) Isopropyl cyclopentanecarboxylate
(f) Cyclopentyl 2-methylpropanoate
(g) *N*-Methyl-4-pentenamide
(h) (*R*)-2-Hydroxypropanoyl phosphate
(i) Ethyl 2,3-Dimethyl-2-butenethioate

21.2
(a) C₆H₅CO₂C₆H₅ (b) CH₃CH₂CH₂CON(CH₃)CH₂CH₃
(c) (CH₃)₂CHCH₂CH(CH₃)COCl
(d) cyclohexane with CH₃ and CO₂CH₃
(e) CH₃CH₂C(=O)CH₂C(=O)OCH₂CH₃
(f) 4-bromobenzoyl thiomethyl ester (Br-C₆H₄-C(=O)SCH₃)
(g) H−C(=O)−O−C(=O)−CH₂CH₃
(h) cyclopentane with H/COBr and H/CH₃

21.3

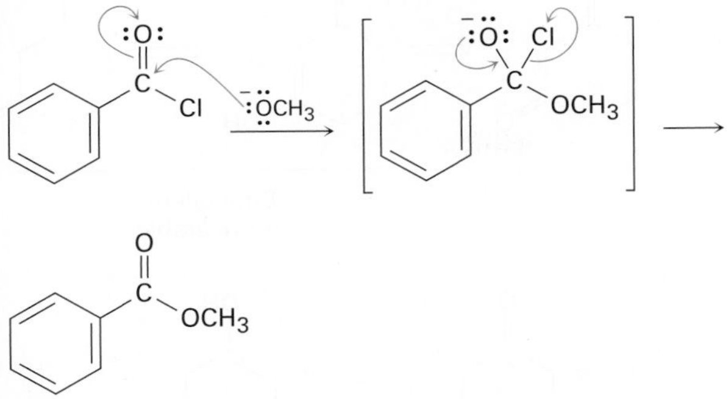

[Reaction mechanism: benzoyl chloride + methanol (via tetrahedral intermediate) → methyl benzoate]

21.4 (a) Acetyl chloride > Methyl acetate > Acetamide
(b) Hexafluoroisopropyl acetate > 2,2,2-Trichloroethyl acetate > Ethyl acetate

21.5 (a) $CH_3CO_2^- Na^+$ (b) CH_3CONH_2
(c) $CH_3CO_2CH_3 + CH_3CO_2^- Na^+$
(d) $CH_3CONHCH_3$

21.6

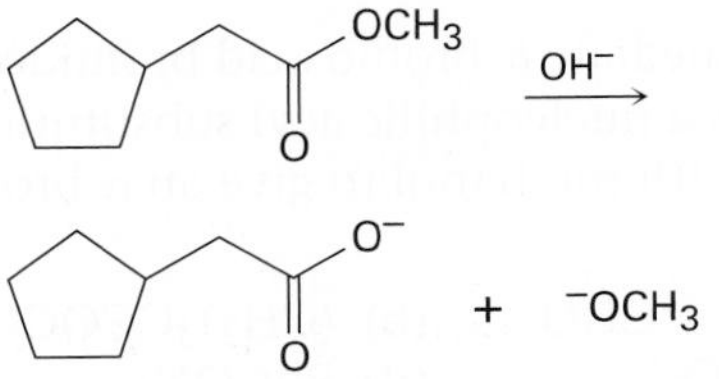

[Cyclopentylacetate methyl ester + OH⁻ → cyclopentylacetate anion + ⁻OCH₃]

21.7 (a) Acetic acid + 1-butanol
(b) Butanoic acid + methanol
(c) Cyclopentanecarboxylic acid + isopropyl alcohol

21.8

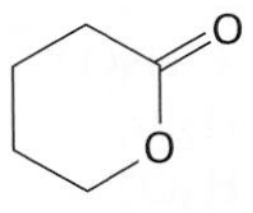

[δ-valerolactone structure]

21.9 (a) Propanoyl chloride + methanol
(b) Acetyl chloride + ethanol
(c) Benzoyl chloride + ethanol

21.10 Benzoyl chloride + cyclohexanol

21.11 This is a typical nucleophilic acyl substitution reaction, with morpholine as the nucleophile and chloride as the leaving group.

21.12 (a) Propanoyl chloride + methylamine
(b) Benzoyl chloride + diethylamine
(c) Propanoyl chloride + ammonia

21.13 (a) Benzoyl chloride + [(CH₃)₂CH]₂CuLi, or 2-methylpropanoyl chloride + Ph₂CuLi
(b) 2-Propenoyl chloride + (CH₃CH₂CH₂)₂CuLi, or butanoyl chloride + (H₂C=CH)₂CuLi

21.14 This is a typical nucleophilic acyl substitution reaction, with p-hydroxyaniline as the nucleophile and acetate ion as the leaving group.

21.15 Monomethyl ester of benzene-1,2-dicarboxylic acid

21.16 Reaction of a carboxylic acid with an alkoxide ion gives the carboxylate ion.

21.17 $LiAlH_4$ gives $HOCH_2CH_2CH_2CH_2OH$; DIBAH gives $HOCH_2CH_2CH_2CHO$

21.18 (a) $CH_3CH_2CH_2CH(CH_3)CH_2OH + CH_3OH$
(b) $PhOH + PhCH_2OH$

21.19 (a) Ethyl benzoate + 2 CH_3MgBr
(b) Ethyl acetate + 2 $PhMgBr$
(c) Ethyl pentanoate + 2 CH_3CH_2MgBr

21.20 (a) H_2O, NaOH (b) Benzoic acid + $LiAlH_4$
(c) $LiAlH_4$

21.21 1. Mg; 2. CO_2, then H_3O^+; 3. $SOCl_2$; 4. $(CH_3)_2NH$; 5. $LiAlH_4$

21.22

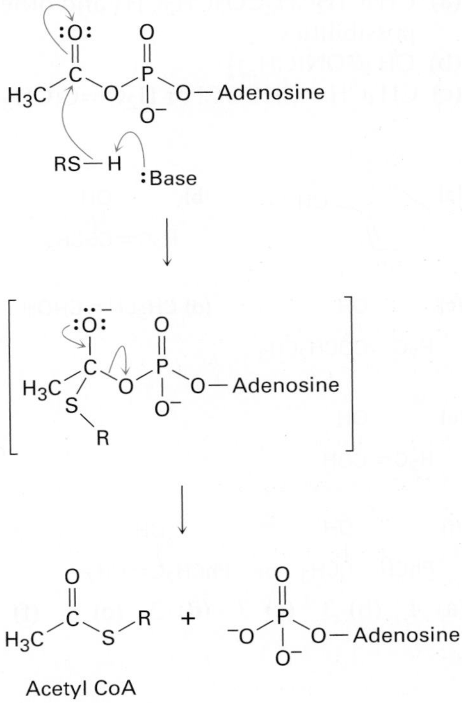

[Mechanism of acetyl CoA formation: acetyl phosphate-adenosine + RS-H, base → tetrahedral intermediate → Acetyl CoA (H₃C-C(=O)-S-R) + phosphate-adenosine]

Acetyl CoA

21.23
(a) $-(\text{OCH}_2\text{CH}_2\text{CH}_2\text{OCH}_2\text{CH}_2\text{CH}_2)_n-$

(b) $-(\text{OCH}_2\text{CH}_2\text{OC}(=\text{O})(\text{CH}_2)_6\text{C}(=\text{O}))_n-$

(c) $-(\text{NH}(\text{CH}_2)_6\text{NHC}(=\text{O})(\text{CH}_2)_4\text{C}(=\text{O}))_n-$

21.24
$-(\text{NH}-\text{C}_6\text{H}_4-\text{NH}-\text{C}(=\text{O})-\text{C}_6\text{H}_4-\text{C}(=\text{O}))_n-$

21.25
(a) Ester (b) Acid chloride
(c) Carboxylic acid
(d) Aliphatic ketone or cyclohexanone

21.26
(a) $\text{CH}_3\text{CH}_2\text{CH}_2\text{CO}_2\text{CH}_2\text{CH}_3$ and other possibilities
(b) $\text{CH}_3\text{CON}(\text{CH}_3)_2$
(c) $\text{CH}_3\text{CH}=\text{CHCOCl}$ or $\text{H}_2\text{C}=\text{C}(\text{CH}_3)\text{COCl}$

Chapter 22

22.1
(a) cyclopentenol with OH

(b) $\text{H}_2\text{C}=\text{C}(\text{OH})\text{SCH}_3$

(c) $\text{H}_2\text{C}=\text{C}(\text{OH})\text{CH}_2\text{CH}_3$

(d) $\text{CH}_3\text{CH}=\text{CHOH}$

(e) $\text{H}_2\text{C}=\text{C}(\text{OH})\text{H}$

(f) $\text{PhCH}=\text{C}(\text{OH})\text{CH}_3$ or $\text{PhCH}_2\text{C}(\text{OH})=\text{CH}_2$

22.2
(a) 4 (b) 3 (c) 3 (d) 2 (e) 4 (f) 5

22.3
Equivalent; more stable

Equivalent; less stable

22.4
Acid-catalyzed formation of an enol is followed by deuteronation of the enol double bond and dedeuteronation of oxygen.

22.5
1. Br_2; 2. Pyridine, heat

22.6
The intermediate α-bromo acid bromide undergoes a nucleophilic acyl substitution reaction with methanol to give an α-bromo ester.

22.7
(a) $\text{CH}_3\text{CH}_2\text{CHO}$ (b) $(\text{CH}_3)_3\text{CCOCH}_3$
(c) $\text{CH}_3\text{CO}_2\text{H}$ (d) PhCONH_2
(e) $\text{CH}_3\text{CH}_2\text{CH}_2\text{CN}$ (f) $\text{CH}_3\text{CON}(\text{CH}_3)_2$

22.8
$^-\!:\!\text{CH}_2\text{C}\!\equiv\!\text{N}: \longleftrightarrow \text{H}_2\text{C}=\text{C}=\ddot{\text{N}}:^-$

22.9
Acid is regenerated, but base is used stoichiometrically.

22.10
(a) 1. $\text{Na}^+\ ^-\text{OEt}$; 2. PhCH_2Br; 3. H_3O^+
(b) 1. $\text{Na}^+\ ^-\text{OEt}$; 2. $\text{CH}_3\text{CH}_2\text{CH}_2\text{Br}$; 3. $\text{Na}^+\ ^-\text{OEt}$; 4. CH_3Br; 5. H_3O^+
(c) 1. $\text{Na}^+\ ^-\text{OEt}$; 2. $(\text{CH}_3)_2\text{CHCH}_2\text{Br}$; 3. H_3O^+

22.11
Malonic ester has only two acidic hydrogens to be replaced.

22.12
1. $\text{Na}^+\ ^-\text{OEt}$; 2. $(\text{CH}_3)_2\text{CHCH}_2\text{Br}$; 3. $\text{Na}^+\ ^-\text{OEt}$; 4. CH_3Br; 5. H_3O^+

22.13
(a) $(\text{CH}_3)_2\text{CHCH}_2\text{Br}$ (b) $\text{PhCH}_2\text{CH}_2\text{Br}$

22.14
None can be prepared.

22.15
1. 2 $\text{Na}^+\ ^-\text{OEt}$; 2. $\text{BrCH}_2\text{CH}_2\text{CH}_2\text{CH}_2\text{Br}$; 3. H_3O^+

22.16 (a) Alkylate phenylacetone with CH$_3$I
(b) Alkylate pentanenitrile with CH$_3$CH$_2$I
(c) Alkylate cyclohexanone with H$_2$C=CHCH$_2$Br
(d) Alkylate cyclohexanone with excess CH$_3$I
(e) Alkylate C$_6$H$_5$COCH$_2$CH$_3$ with CH$_3$I
(f) Alkylate methyl 3-methylbutanoate with CH$_3$CH$_2$I

Chapter 23

23.1 (a) CH$_3$CH$_2$CH$_2$CH(OH)CH(CH$_2$CH$_3$)CHO

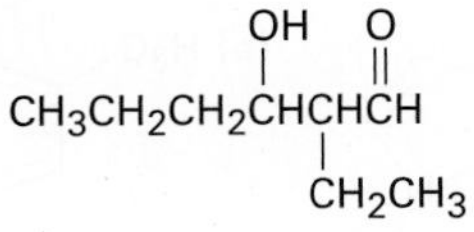

(b) PhCOCH$_2$C(OH)(CH$_3$)Ph

(c) 1-(1-hydroxycyclopentyl)-2-cyclopentanone

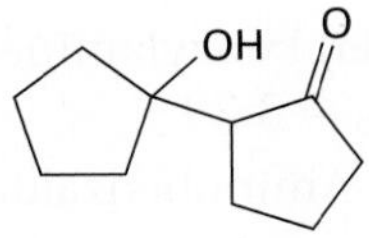

23.2 The reverse reaction is the exact opposite of the forward reaction, shown in Figure 23.1.

23.3
(a) cyclopentylidenecyclopentanone
(b) PhC(CH$_3$)=CHC(O)Ph
(c) (CH$_3$)$_2$CHCH$_2$CH=C(CH(CH$_3$)$_2$)CHO

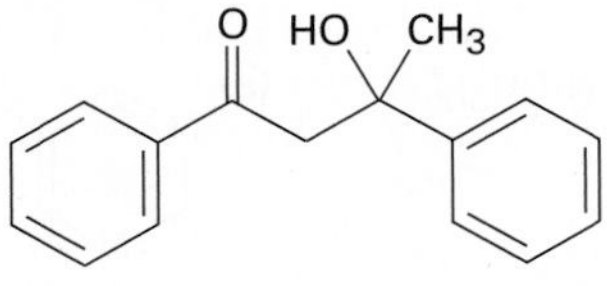

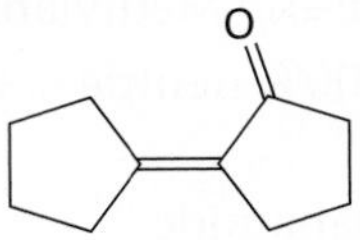

23.4 two isomeric cyclohexylidene cyclohexanone products with methyl substituents

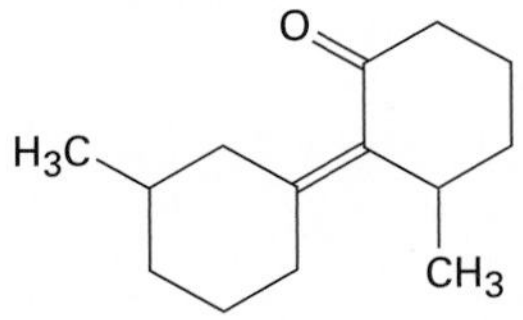

23.5 (a) Not an aldol product (b) 3-Pentanone

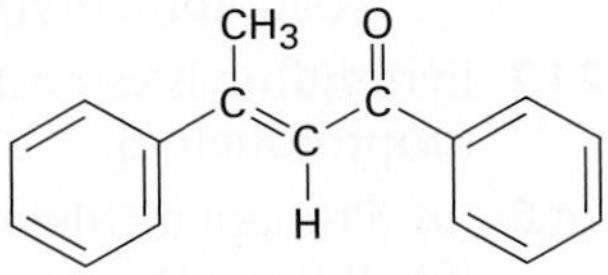

23.6 1. NaOH; 2. LiAlH$_4$; 3. H$_2$/Pd

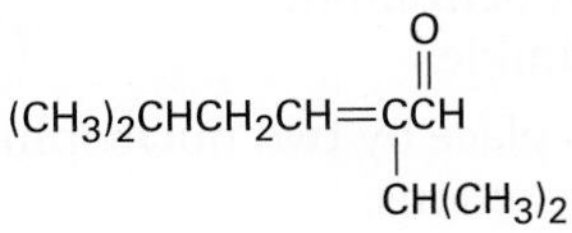

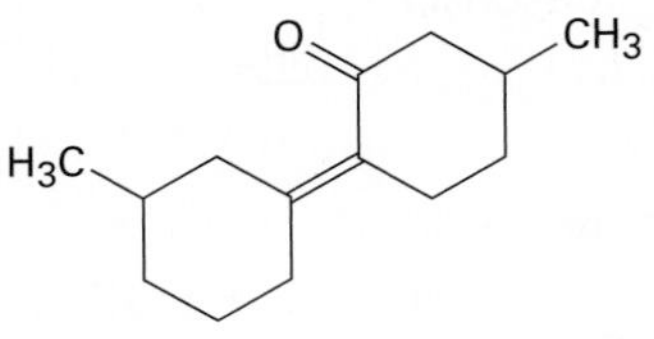

23.7 spirocyclopropane aldol condensation product with NaOH

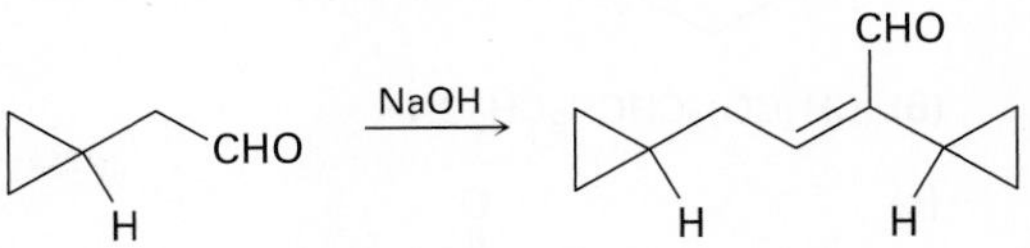

23.8 (a) C$_6$H$_5$CHO + CH$_3$COCH$_3$
(b), (c) Not easily prepared

23.9 The CH$_2$ position between the two carbonyl groups is so acidic that it is completely deprotonated to give a stable enolate ion.

23.10 bicyclic enone

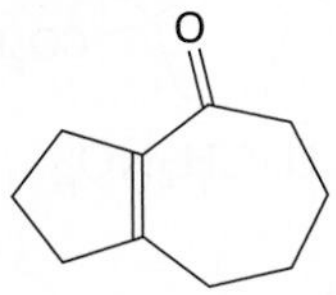

23.11 (a) CH$_3$CH(CH$_3$)CH$_2$C(O)CH(CH(CH$_3$)$_2$)CO$_2$Et
(b) PhCH$_2$C(O)CH(Ph)CO$_2$Et
(c) C$_6$H$_{11}$CH$_2$C(O)CH(C$_6$H$_{11}$)CO$_2$Et

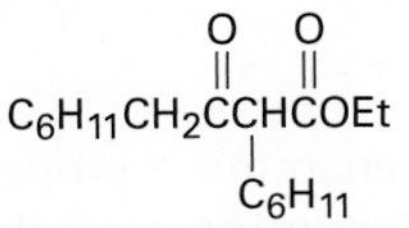

23.12 The cleavage reaction is the exact reverse of the forward reaction.

23.13 2,2-dimethyl-6-(2-oxopropanoyl)cyclohexanone

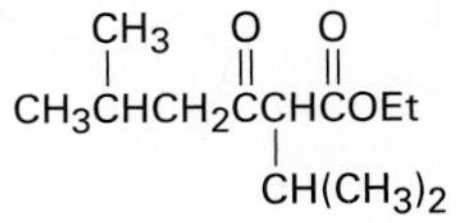

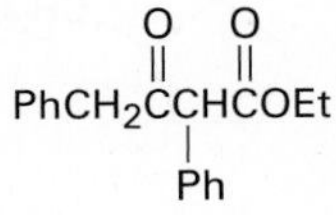

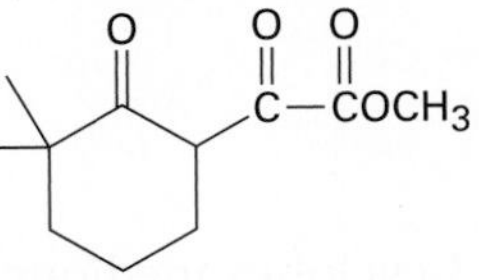

23.14 4-methyl-2-(ethoxycarbonyl)cyclohexanone

23.15 5-methyl-2-(ethoxycarbonyl)cyclohexanone + 3-methyl-2-(ethoxycarbonyl)cyclohexanone

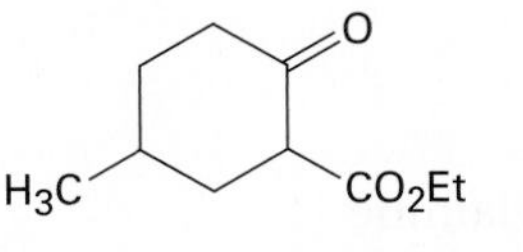

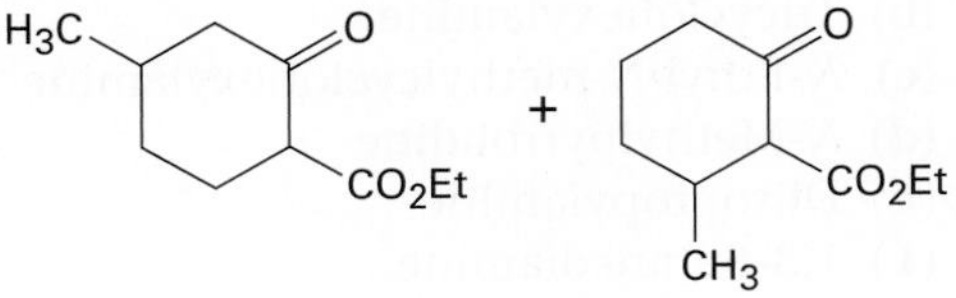

23.16 (a) 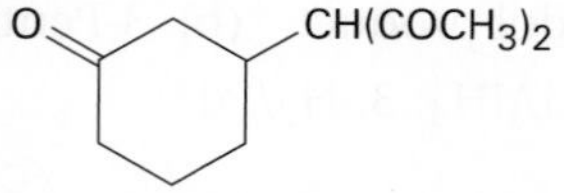 O=C(cyclohexyl)-CH(COCH₃)₂

(b) (CH₃CO)₂CHCH₂CH₂CN

(c) (CH₃CO)₂CHCHCH₂COEt with CH₃ branch

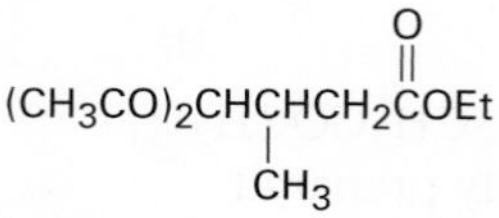

23.17
(a) 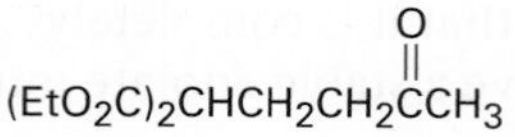 (EtO₂C)₂CHCH₂CH₂CCH₃ (with C=O)

(b) 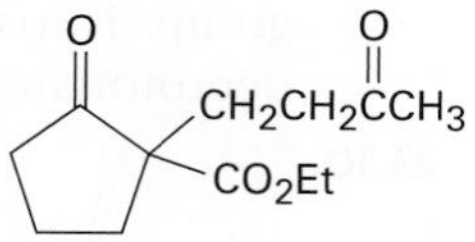 cyclopentanone with CH₂CH₂CCH₃ and CO₂Et substituents

23.18 CH₃CH₂COCH=CH₂ + CH₃CH₂NO₂

23.19
(a) 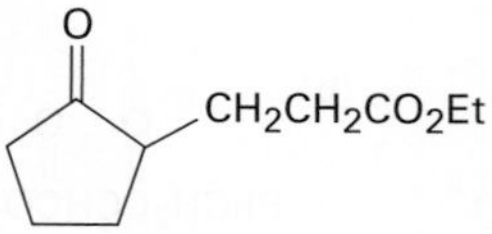 cyclopentanone with CH₂CH₂CO₂Et

(b) 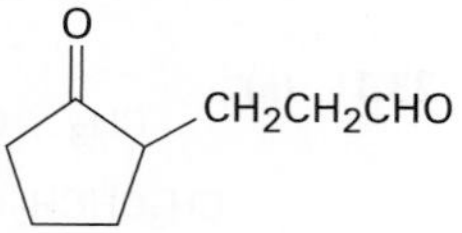 cyclopentanone with CH₂CH₂CHO

(c) 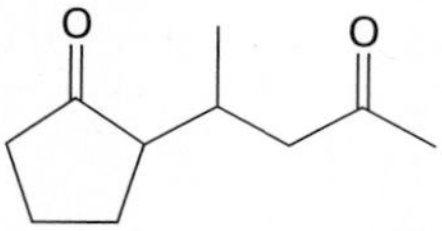 cyclopentanone with CH(CH₃)CH₂COCH₃ group

23.20 (a) Cyclopentanone enamine + propenenitrile
(b) Cyclohexanone enamine + methyl propenoate

23.21 bicyclic diketone structure

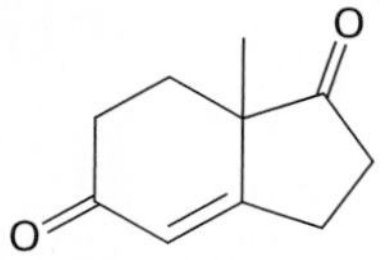

23.22 2,5,5-Trimethyl-1,3-cyclohexanedione + 1-penten-3-one

Chapter 24

24.1 (a) *N*-Methylethylamine
(b) Tricyclohexylamine
(c) *N*-Ethyl-*N*-methylcyclohexylamine
(d) *N*-Methylpyrrolidine
(e) Diisopropylamine
(f) 1,3-Butanediamine

24.2 (a) [(CH₃)₂CH]₃N   (b) (H₂C=CHCH₂)₃N

(c) 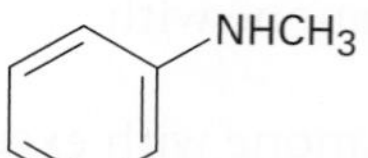 C₆H₅–NHCH₃

(d) 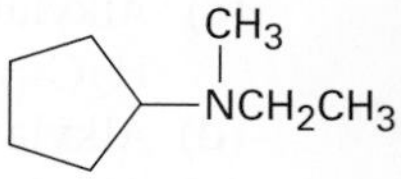 cyclopentyl–N(CH₃)(CH₂CH₃)

(e) 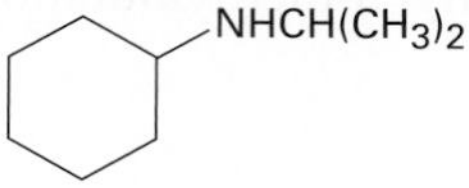 cyclohexyl–NHCH(CH₃)₂

(f) 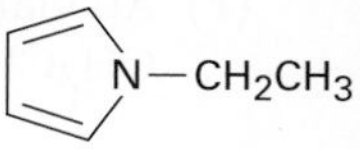 pyrrole N–CH₂CH₃

24.3 (a) 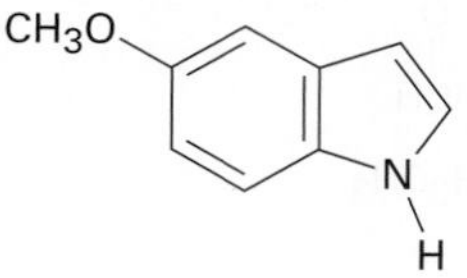 5-methoxyindole (CH₃O-indole)

(b) 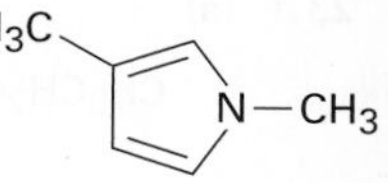 3-methyl-1-methylpyrrole (H₃C on ring, N–CH₃)

(c) 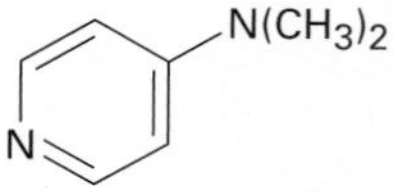 4-(dimethylamino)pyridine, N(CH₃)₂

(d) 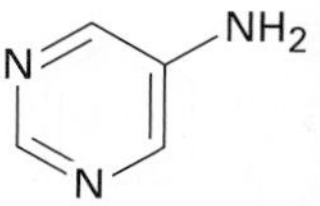 5-aminopyrimidine, NH₂

24.4 (a) CH₃CH₂NH₂   (b) NaOH   (c) CH₃NHCH₃

24.5 Propylamine is stronger; benzylamine pK_b = 4.67; propylamine pK_b = 3.29

24.6 (a) *p*-Nitroaniline < *p*-Aminobenzaldehyde < *p*-Bromoaniline
(b) *p*-Aminoacetophenone < *p*-Chloroaniline < *p*-Methylaniline
(c) *p*-(Trifluoromethyl)aniline < *p*-(Fluoromethyl)aniline < *p*-Methylaniline

24.7 Pyrimidine is essentially 100% neutral (unprotonated).

24.8 (a) Propanenitrile or propanamide
(b) *N*-Propylpropanamide
(c) Benzonitrile or benzamide
(d) *N*-Phenylacetamide

24.9 The reaction takes place by two nucleophilic acyl substitution reactions.

24.10 3,4-dihydroxyphenyl-CH₂CH₂Br $\xrightarrow{\text{NH}_3}$

or

3,4-dihydroxyphenyl-CH₂Br $\xrightarrow[\text{2. LiAlH}_4]{\text{1. NaCN}}$

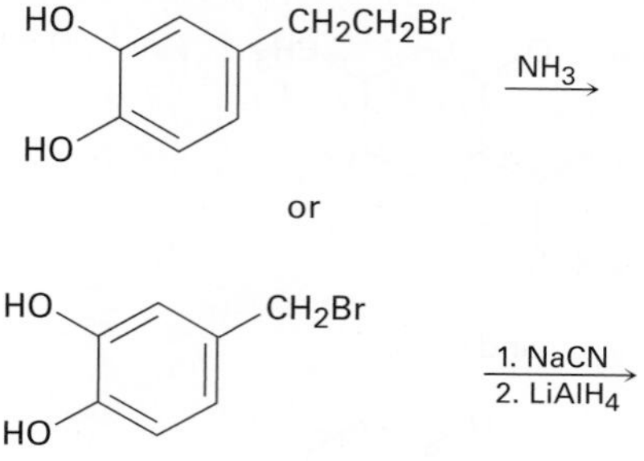

24.11 (a) Ethylamine + acetone, or isopropylamine + acetaldehyde
(b) Aniline + acetaldehyde
(c) Cyclopentylamine + formaldehyde, or methylamine + cyclopentanone

24.12 3-methylbenzaldehyde + (CH$_3$)$_2$NH →[NaBH$_4$]

24.13 (a) 4,4-Dimethylpentanamide or 4,4-dimethylpentanoyl azide
(b) *p*-Methylbenzamide or *p*-methylbenzoyl azide

24.14 (a) 3-Octene and 4-octene
(b) Cyclohexene   (c) 3-Heptene
(d) Ethylene and cyclohexene

24.15 H$_2$C=CHCH$_2$CH$_2$CH$_2$N(CH$_3$)$_2$

24.16 1. HNO$_3$, H$_2$SO$_4$; 2. H$_2$/PtO$_2$; 3. (CH$_3$CO)$_2$O; 4. HOSO$_2$Cl; 5. aminothiazole; 6. H$_2$O, NaOH

24.17 (a) 1. HNO$_3$, H$_2$SO$_4$; 2. H$_2$/PtO$_2$; 3. 2 CH$_3$Br
(b) 1. HNO$_3$, H$_2$SO$_4$; 2. H$_2$/PtO$_2$; 3. (CH$_3$CO)$_2$O; 4. Cl$_2$; 5. H$_2$O, NaOH
(c) 1. HNO$_3$, H$_2$SO$_4$; 2. Cl$_2$, FeCl$_3$; 3. SnCl$_2$
(d) 1. HNO$_3$, H$_2$SO$_4$; 2. H$_2$/PtO$_2$; 3. (CH$_3$CO)$_2$O; 4. 2 CH$_3$Cl, AlCl$_3$; 5. H$_2$O, NaOH

24.18 (a) 1. CH$_3$Cl, AlCl$_3$; 2. HNO$_3$, H$_2$SO$_4$; 3. SnCl$_2$; 4. NaNO$_2$, H$_2$SO$_4$; 5. CuBr; 6. KMnO$_4$, H$_2$O
(b) 1. HNO$_3$, H$_2$SO$_4$; 2. Br$_2$, FeBr$_3$; 3. SnCl$_2$, H$_3$O$^+$; 4. NaNO$_2$, H$_2$SO$_4$; 5. CuCN; 6. H$_3$O$^+$
(c) 1. HNO$_3$, H$_2$SO$_4$; 2. Cl$_2$, FeCl$_3$; 3. SnCl$_2$; 4. NaNO$_2$, H$_2$SO$_4$; 5. CuBr
(d) 1. CH$_3$Cl, AlCl$_3$; 2. HNO$_3$, H$_2$SO$_4$; 3. SnCl$_2$; 4. NaNO$_2$, H$_2$SO$_4$; 5. CuCN; 6. H$_3$O$^+$
(e) 1. HNO$_3$, H$_2$SO$_4$; 2. H$_2$/PtO$_2$; 3. (CH$_3$CO)$_2$O; 4. 2 Br$_2$; 5. H$_2$O, NaOH; 6. NaNO$_2$, H$_2$SO$_4$; 7. CuBr

24.19 1. HNO$_3$, H$_2$SO$_4$; 2. SnCl$_2$; 3a. 2 equiv. CH$_3$I; 3b. NaNO$_2$, H$_2$SO$_4$; 4. product of 3a + product of 3b

24.20
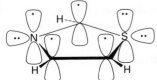

24.21 4.1% protonated

24.22
Attack at C2:

[pyridine with E$^+$ attacking C2]

[three resonance structures] Unfavorable

Attack at C3:

[pyridine with E$^+$ attacking C3]

[three resonance structures]

Attack at C4:

[pyridine with E$^+$ attacking C4]

[three resonance structures] Unfavorable

24.23 The side-chain nitrogen is more basic than the ring nitrogen.

APPENDIX D | Answers to In-Text Problems

24.24 Reaction at C2 is disfavored because the aromaticity of the benzene ring is lost.

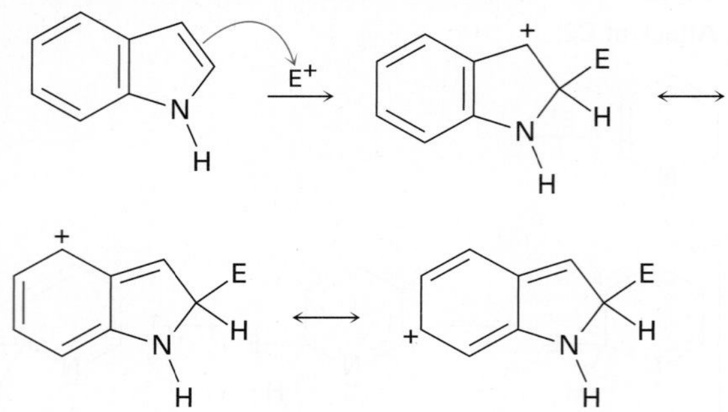

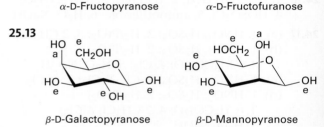

24.25 (CH₃)₃CCOCH₃ → (CH₃)₃CCH(NH₂)CH₃

Chapter 25

25.1 (a) Aldotetrose (b) Ketopentose (c) Ketohexose (d) Aldopentose

25.2 (a) S (b) R (c) S

25.3 A, B, and C are the same.

25.4

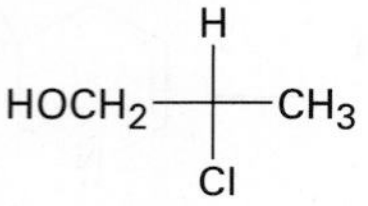

HOCH₂—C(H)(CH₃)(Cl) R

25.5

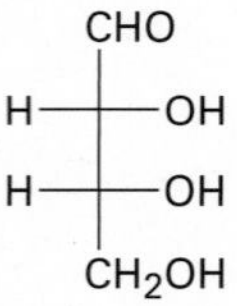

CHO
H—OH   R
H—OH   R
CH₂OH

25.6 (a) L-Erythrose; 2S,3S (b) D-Xylose; 2R,3S,4R (c) D-Xylulose; 3S,4R

25.7

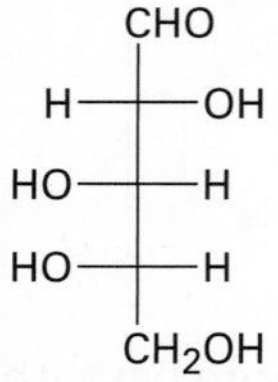

CHO
H—OH
HO—H   L-(+)-Arabinose
HO—H
CH₂OH

25.8

(a)
CHO
HO—H
H—OH
HO—H
CH₂OH

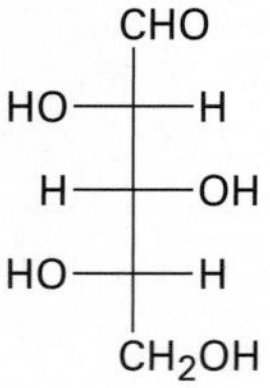

(b)
CHO
HO—H
H—OH
H—OH
HO—H
CH₂OH

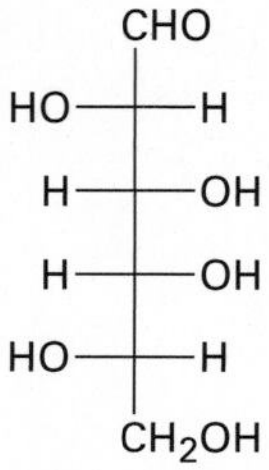

(c)
CHO
HO—H
H—OH
HO—H
HO—H
CH₂OH

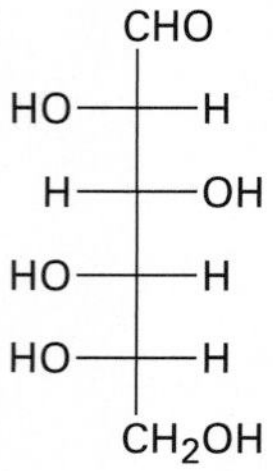

25.9 16 D and 16 L aldoheptoses

25.10

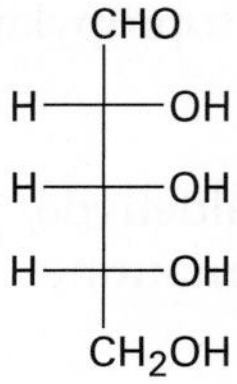

CHO
H—OH
H—OH   D-Ribose
H—OH
CH₂OH

25.11

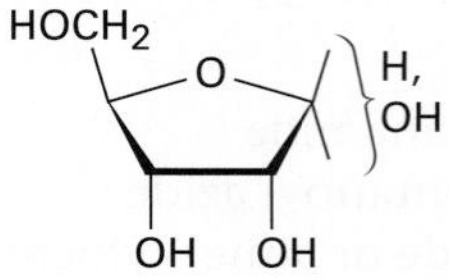

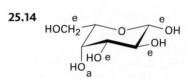

25.12

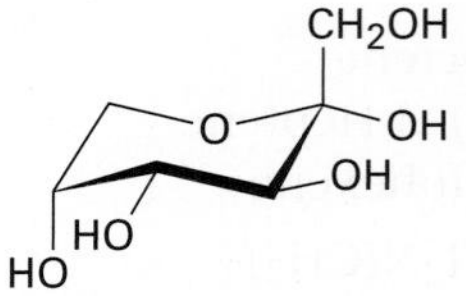

α-D-Fructopyranose 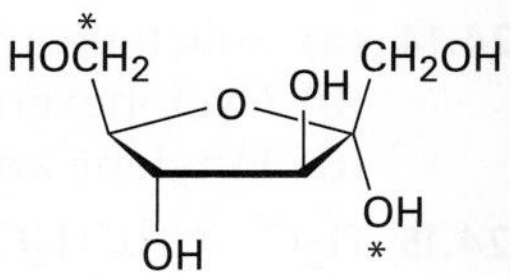   α-D-Fructofuranose

25.13

β-D-Galactopyranose   β-D-Mannopyranose

25.14

25.15 α-D-Allopyranose

25.16 CH₃OCH₂ ... OCH₃ / OCH₃ OCH₃   AcOCH₂ ... OAc / OAc OAc

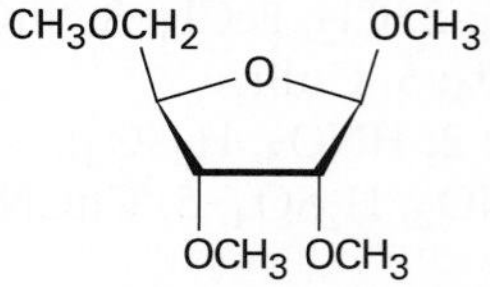

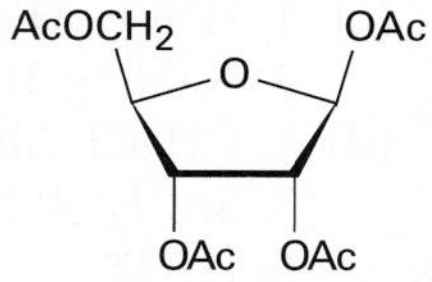

25.17 D-Galactitol has a plane of symmetry and is a meso compound, whereas D-glucitol is chiral.

25.18 The —CHO end of L-gulose corresponds to the —CH₂OH end of D-glucose after reduction.

25.19 D-Allaric acid has a symmetry plane and is a meso compound, but D-glucaric acid is chiral.

25.20 D-Allose and D-galactose yield meso aldaric acids; the other six D-hexoses yield optically active aldaric acids.

25.21 D-Allose + D-altrose

25.22 L-Xylose

25.23 D-Xylose and D-lyxose

25.24

[Fischer projection showing conversion: CO₂⁻, C=O, H₂C—H with :Base, arrows showing enolization via H—C=O: with CH₃CONH—H, HO—H, H—OH, H—OH, CH₂OH → CO₂⁻, C=O, CH₂, H—OH, CH₃CONH—H, HO—H, H—OH, H—OH, CH₂OH]

25.25 (a) The hemiacetal ring is reduced.
(b) The hemiacetal ring is oxidized.
(c) All hydroxyl groups are acetylated.

Chapter 26

26.1 Aromatic: Phe, Tyr, Trp, His; sulfur-containing: Cys, Met; alcohols: Ser, Thr; hydrocarbon side chains: Ala, Ile, Leu, Val, Phe

26.2 The sulfur atom in the —CH₂SH group of cysteine makes the side chain higher in ranking than the —CO₂H group.

26.3

[Three Fischer projections: L-Threonine (H₃N⁺—S—H, H—R—OH, CH₃ at bottom, CO₂⁻ at top); Diastereomers of L-threonine (H₃N⁺—S—H, HO—S—H, CH₃) and (H—R—NH₃⁺, H—R—OH, CH₃)]

26.4 Net positive at pH = 5.3; net negative at pH = 7.3

26.5 (a) Start with 3-phenylpropanoic acid:
1. Br₂, PBr₃; 2. NH₃
(b) Start with 3-methylbutanoic acid:
1. Br₂, PBr₃; 2. NH₃

26.6 (a) (CH₃)₂CHCH₂Br (b) [imidazole-CH₂Br structure]
(c) [indole-3-CH₂Br structure] (d) CH₃SCH₂CH₂Br

26.7

[Structure: (CH₃)₂CH and CO₂H on C=C with H and NHCOCH₃]

1. H₂, [Rh(DiPAMP)(COD)]⁺ BF₄⁻
2. NaOH, H₂O

[Product: isobutyl group, CO₂⁻, H₃N⁺, H]

26.8 Val-Tyr-Gly (VYG), Tyr-Gly-Val (YGV), Gly-Val-Tyr (GVY), Val-Gly-Tyr (VGY), Tyr-Val-Gly (YVG), Gly-Tyr-Val (GYV)

26.9

H₃N⁺CHC—N—CHC—NHCHC—NHCH₂CO⁻
    | | |
CH₃SCH₂CH₂ (pyrrolidine ring) CH(CH₃)₂

26.10

HOCCH₂—SCH₂CHCO⁻
               |
              ⁺NH₃

26.11

[Ninhydrin-derived purple product structure with N linking two indanedione units] + (CH₃)₂CHCHO + CO₂

26.12 Trypsin: Asp-Arg + Val-Tyr-Ile-His-Pro-Phe
Chymotrypsin: Asp-Arg-Val-Tyr + Ile-His-Pro-Phe

26.13 Methionine

26.14

[PTH derivative: C₆H₅—N—C=O ring with S=C, N—H, and C—CH₂CO₂H with H]

26.15 (a) Arg-Pro-Leu-Gly-Ile-Val
(b) Val-Met-Trp-Asp-Val-Leu (VMWNVL)

26.16 This is a typical nucleophilic acyl substitution reaction, with the amine of the amino acid as the nucleophile and *tert*-butyl carbonate as the leaving group. The *tert*-butyl carbonate then loses CO₂ and gives *tert*-butoxide, which is protonated.

26.17 (1) Protect the amino group of leucine.
(2) Protect the carboxylic acid group of alanine.
(3) Couple the protected amino acids with DCC.
(4) Remove the leucine protecting group.
(5) Remove the alanine protecting group.

26.18 (a) Lyase (b) Hydrolase (c) Oxidoreductase

Chapter 27

27.1 $CH_3(CH_2)_{18}CO_2CH_2(CH_2)_{30}CH_3$

27.2 Glyceryl tripalmitate is higher melting.

27.3 $[CH_3(CH_2)_7CH=CH(CH_2)_7CO_2^-]_2 \, Mg^{2+}$

27.4 Glyceryl dioleate monopalmitate → glycerol + 2 sodium oleate + sodium palmitate

27.5

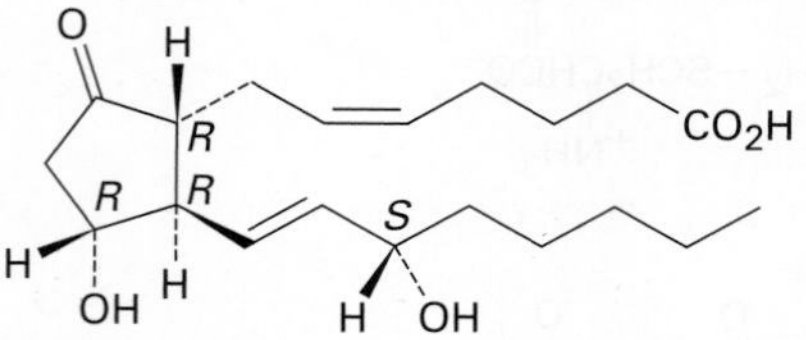

27.6 The *pro-S* hydrogen is cis to the $-CH_3$ group; the *pro-R* hydrogen is trans.

27.7
(a)

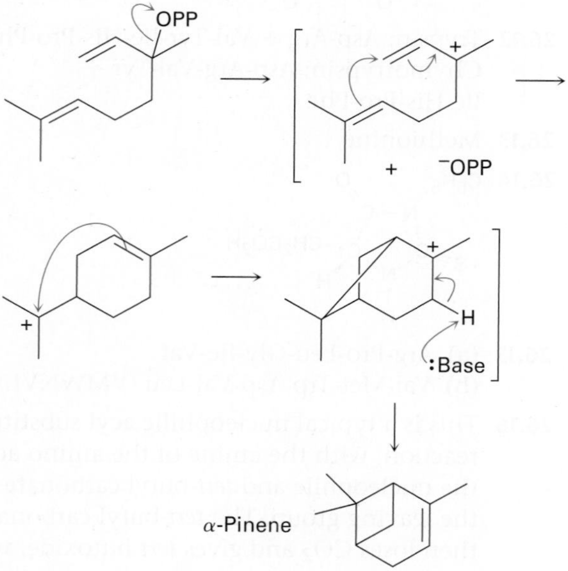

α-Pinene

(b)

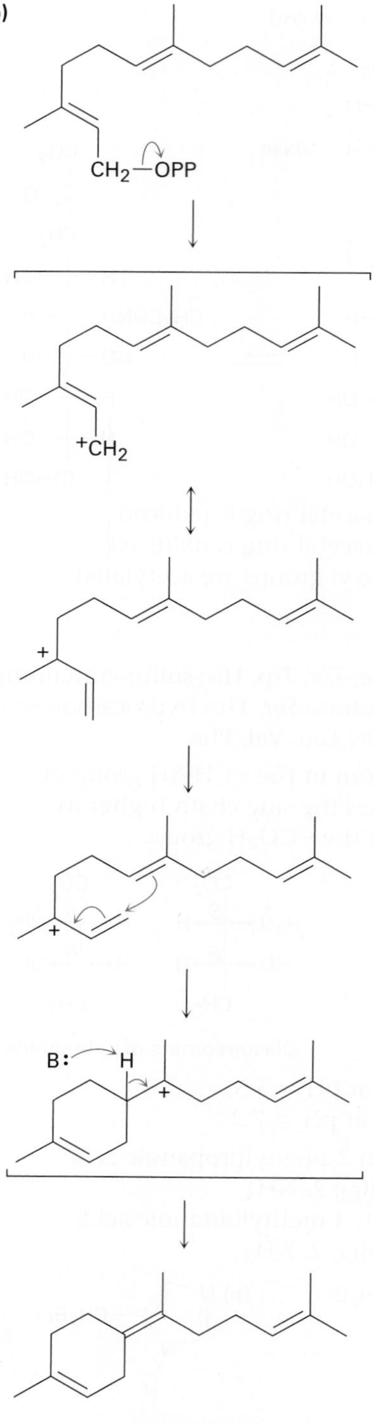

γ-Bisabolene

27.8

(a) [structure: decalin with axial H's and equatorial CH3 labeled e]

(b) [structure: decalin with axial H's and axial CH3 labeled a]

27.9

[steroid structure with CH3 groups, side chain ending in CO2H, and equatorial OH labeled e]

27.10 Three methyl groups are removed, the side-chain double bond is reduced, and the double bond in the B ring is migrated.

Chapter 28

28.3 (5') ACGGATTAGCC (3')

28.4 [base pair structure showing hydrogen bonding between uracil and adenine]

28.5 (3') CUAAUGGCAU (5')

28.6 (5') ACTCTGCGAA (3')

28.7
(a) GCU, GCC, GCA, GCG
(b) UUU, UUC
(c) UUA, UUG, CUU, CUC, CUA, CUG
(d) UAU, UAC

28.8
(a) AGC, GGC, UGC, CGC
(b) AAA, GAA
(c) UAA, CAA, GAA, GAG, UAG, CAG
(d) AUA, GUA

28.9 Leu-Met-Ala-Trp-Pro-Stop

28.10 (5') TTA-GGG-CCA-AGC-CAT-AAG (3')

28.11 The cleavage is an S_N1 reaction that occurs by protonation of the oxygen atom followed by loss of the stable triarylmethyl carbocation.

28.12 [structure showing RO-P(=O)(OR')-O-CH2-CHC≡N with NH3 attacking; E2 reaction]

Chapter 29

29.1 $HOCH_2CH(OH)CH_2OH + ATP \rightarrow HOCH_2CH(OH)CH_2OPO_3^{2-} + ADP$

29.2 Caprylyl CoA → Hexanoyl CoA → Butyryl CoA → 2 Acetyl CoA

29.3 (a) 8 acetyl CoA; 7 passages
(b) 10 acetyl CoA; 9 passages

29.4 The dehydration is an E1cB reaction.

29.5 At C2, C4, C6, C8, and so forth

29.6 The *Si* face

29.7 Steps 7 and 10

29.8 Steps 1, 3: Phosphate transfers; steps 2, 5, 8: isomerizations; step 4: retro-aldol reaction; step 5: oxidation and nucleophilic acyl substitution; steps 7, 10: phosphate transfers; step 9: E1cB dehydration

29.9 C1 and C6 of glucose become $-CH_3$ groups; C3 and C4 become CO_2.

29.10 Citrate and isocitrate

29.11 E1cB elimination of water, followed by conjugate addition

29.12 *pro-R*; anti geometry

29.13 The reaction occurs by two sequential nucleophilic acyl substitutions, the first by a cysteine residue in the enzyme, with phosphate as leaving group, and the second by hydride donation from NADH, with the cysteine residue as leaving group.

29.14 Initial imine formation between PMP and α-ketoglutarate is followed by double-bond rearrangement to an isomeric imine and hydrolysis.

29.15 $(CH_3)_2CHCH_2COCO_2^-$

29.16 Asparagine

Chapter 30

30.1 Ethylene: ψ_1 is the HOMO and ψ_2^* is the LUMO in the ground state; ψ_2^* is the HOMO and there is no LUMO in the excited state. 1,3-Butadiene: ψ_2 is the HOMO and ψ_3^* is the LUMO in the ground state; ψ_3^* is the HOMO and ψ_4^* is the LUMO in the excited state.

30.2 Disrotatory: *cis*-5,6-dimethyl-1,3-cyclohexadiene; conrotatory: *trans*-5,6-dimethyl-1,3-cyclohexadiene. Disrotatory closure occurs.

30.3 The more stable of two allowed products is formed.

30.4 *trans*-5,6-Dimethyl-1,3-cyclohexadiene; *cis*-5,6-dimethyl-1,3-cyclohexadiene

30.5 *cis*-3,6-Dimethylcyclohexene; *trans*-3,6-dimethylcyclohexene

30.6 A [6 + 4] suprafacial cycloaddition

30.7 An antarafacial [1,7] sigmatropic rearrangement

30.8 A series of [1,5] hydrogen shifts occur.

30.9 Claisen rearrangement is followed by a Cope rearrangement.

30.10 (a) Conrotatory (b) Disrotatory
(c) Suprafacial (d) Antarafacial
(e) Suprafacial

Chapter 31

31.1 $H_2C=CHCO_2CH_3 < H_2C=CHCl < H_2C=CHCH_3 < H_2C=CH-C_6H_5$

31.2 $H_2C=CHCH_3 < H_2C=CHC_6H_5 < H_2C=CHC\equiv N$

31.3 The intermediate is a resonance-stabilized benzylic carbanion, $Ph-\ddot{C}HR$.

31.4 The polymer has no chirality centers.

31.5 The polymers are racemic and have no optical rotation.

31.6 [structure]

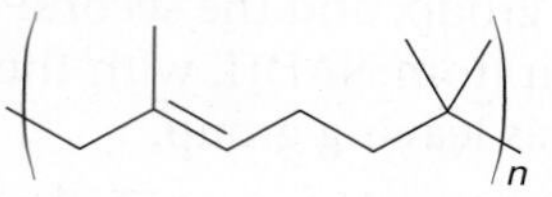

31.7 [structure] Polybutadiene chain / Polystyrene chain

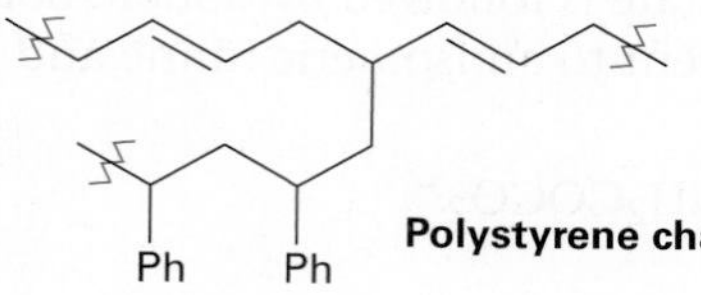

31.8 [structure]

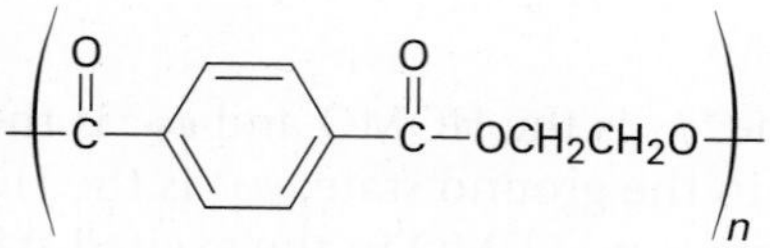

31.9 [mechanism structures]

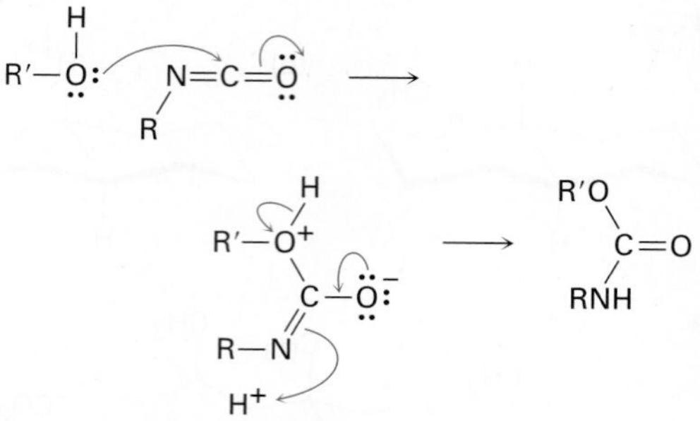

31.10 Vestenamer: ADMET polymerization of 1,9-decadiene or ROMP of cyclooctene; Norsorex: ROMP of norbornene.

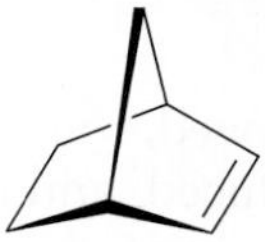

Norbornene

31.11 Atactic

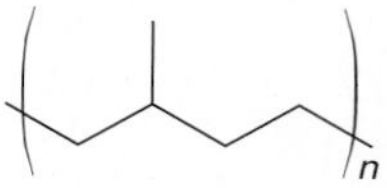

31.12 [mechanism structures]

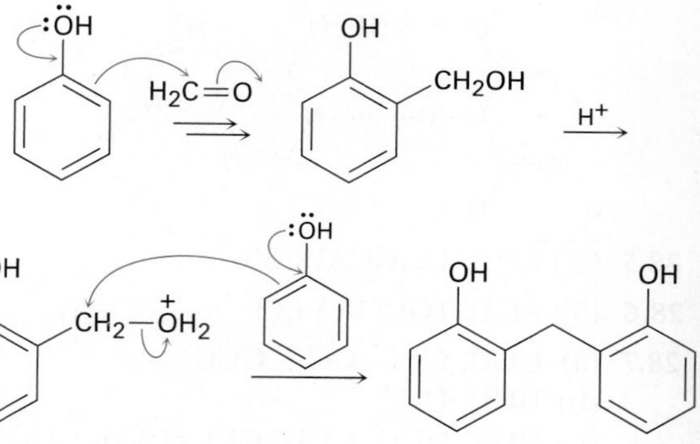

Index

Boldface references refer to pages where terms are defined.

α, *see* Alpha
ABS polymer, structure and uses of, 1247
Absolute configuration, **154**
Absorbance (UV), **519**
Absorption spectrum, **436**
Acesulfame-K, structure of, 1034
  sweetness of, 1033
Acetal(s), **742**
  from aldehydes, 742–744
  from ketones, 742–744
  hydrolysis of, 743–744
  mechanism of formation of, 743–744
Acetaldehyde, aldol reaction of, 905–906
  bond angles in, 714
  bond lengths in, 714
  electrostatic potential map of, 714, 820, 951
  ^{13}C NMR absorptions of, 758
  ^{1}H NMR spectrum of, 758
Acetaminophen, molecular model of, 27
  synthesis of, 836
Acetanilide, electrophilic aromatic substitution of, 967
Acetate ion, bond lengths in, 42
  electrostatic potential map of, 42, 52, 55, 784
  resonance in, 42–43
Acetic acid, bond angles in, 782
  bond lengths in, 782
  dimer of, 782
  dipole moment of, 38
  electrostatic potential map of, 52, 54
  industrial synthesis of, 779
  pK_a of, 51, 783
  properties of, 782
  protonation of, 59
  uses of, 779
Acetic acid dimer, electrostatic potential map of, 782
Acetic anhydride, electrostatic potential map of, 820
  reaction with amines, 835–836
  reaction with monosaccharides, 1015–1016
  synthesis of, 824

Acetoacetic ester, alkylation of, 885–886
  ketones from, 885–886
  mixed aldol reactions of, 913
Acetoacetic ester synthesis, **885–886**
Acetoacetyl CoA, biosynthesis of, 1101
Acetone, electrostatic potential map of, 54, 56, 79
  enol content of, 871
  hydrate of, 731
  industrial synthesis of, 722–723
  pK_a of, 53, 877
  uses of, 722
Acetone anion, electrostatic potential map of, 55
  resonance in, 45
Acetonitrile, electrostatic potential map of, 794
Acetophenone, ^{13}C NMR absorptions of, 758
  structure of, 724
Acetyl ACP, structure of, 1169
Acetyl azide, electrostatic potential map of, 859
Acetyl chloride, electrostatic potential map of, 820
  reaction with alcohols, 831
  reaction with amines, 832
Acetyl CoA, *see* Acetyl coenzyme A
Acetyl coenzyme A, carbonyl condensation reactions of, 930
  carboxylation of, 1170
  catabolism of, 1185–1190
  citric acid cycle and, 1185–1190
  fat catabolism and, 1162–1166
  fatty acids from, 1167–1173
  from pyruvate, 1181–1185
  function of, 846
  reaction with glucosamine, 846
  structure of, 1156
  thioester in, 846
Acetyl group, **724**
Acetylene, bond angles in, 17
  bond lengths in, 17, 317
  bond strengths in, 17, 317
  electrostatic potential map of, 317
  molecular model of, 16
  pK_a of, 53, 326
  sp hybrid orbitals in, 16
  structure of, 16–17, 316–317
  uses of, 314

N-Acetylgalactosamine, structure of, 1024
N-Acetylglucosamine, biosynthesis of, 846
  structure of, 1024
Acetylide anion, **325**
  alkylation of, 327–328
  electrostatic potential map of, 326
  formation of, 325
  stability of, 326
N-Acetylneuraminic acid, structure of, 1024
Achiral, **144**
Acid, Brønsted–Lowry, **49**
  Lewis, **56–58**
  organic, 54–55
  strengths of, 50–51
Acid anhydride(s), **814**
  amides from, 835–836
  electrostatic potential map of, 820
  esters from, 835
  from acid chlorides, 831
  from carboxylic acids, 824
  IR spectroscopy of, 851
  naming, 815
  NMR spectroscopy of, 852
  nucleophilic acyl substitution reactions of, 835–836
  reaction with alcohols, 835
  reaction with amines, 835–836
Acid bromide, enol of, 877
  from carboxylic acid, 830
Acid chloride(s), acid anhydrides from, 831
  alcohols from, 833
  alcoholysis of, 831
  amides from, 832
  amines from, 960, 962
  aminolysis of, 832
  carboxylic acids from, 830–831
  electrostatic potential map of, 820
  esters from, 831
  from carboxylic acids, 823–824
  Grignard reaction of, 833
  hydrolysis of, 830–831
  IR spectroscopy of, 851
  ketones from, 833–834
  naming, 815
  NMR spectroscopy of, 852
  nucleophilic acyl substitution reactions of, 830–834
  pK_a of, 879

Acid chloride(s)—cont'd
  reaction with alcohols, 831
  reaction with amines, 832
  reaction with ammonia, 832
  reaction with diorganocopper reagents, 833–834
  reaction with Gilman reagents, 833–834
  reaction with Grignard reagents, 833
  reaction with LiAlH$_4$, 833
  reaction with water, 830–831
  reduction of, 833
Acid halide(s), **814**
  naming, 815
  nucleophilic acyl substitution reactions of, 830–834
  see also Acid chloride
Acidity, alcohols and, 624–626
  amines and, 951–952
  carbonyl compounds and, 877–880
  carboxylic acids and, 782–784
  phenols and, 624–626
Acidity constant (K_a), **50**
Acid–base reactions, prediction of, 52–53
Acifluorfen, synthesis of, 710
Acrolein, structure of, 724
Acrylic acid, pK_a of, 783
  structure of, 780
Activating group (aromatic substitution), **581**
  acidity and, 787
Activation energy, **206**
  reaction rate and, 206–207
  typical values for, 207
Active site (enzyme), **210**–211
  citrate synthase, 1074
  hexokinase, 211
  HMG-CoA reductase, 1204
Acyclic diene metathesis polymerization (ADMET), 1252
  mechanism of, 1251
Acyl adenosyl phosphate, from carboxylic acids, 828–830
  mechanism of formation of, 828–830
Acyl adenylate, from carboxylic acids, 828–830
  mechanism of formation of, 828–830
Acyl azide, amines from, 960, 962
Acyl carrier protein, function of, 1169
Acyl cation, electrostatic potential map of, 578
  Friedel–Crafts acylation reaction and, 577–578
  resonance in, 577–578

Acyl group, **577, 712**
  names of, 780
Acyl phosphate, **814**
  naming, 817
Acylation (aromatic), *see* Friedel–Crafts reaction
Adams, Roger, 277
Adams catalyst, 277
Addition reaction, **184**–185
1,2-Addition reaction (carbonyl), **751**–755
1,2-Addition reaction (conjugated diene), **505**
1,4-Addition reaction (carbonyl), **751**–755
1,4-Addition reaction (conjugated diene), **505**
  kinetic control of, 508–509
  thermodynamic control of, 508–509
Adenine, electrostatic potential map of, 1132
  molecular model of, 66
  protection of, 1142–1143
  structure of, 1129
Adenosine diphosphate (ADP), structure and function of, 205, 1156–1157
Adenosine triphosphate (ATP), bond dissociation energy and, 204
  coupled reactions and, 1157–1158
  function of, 204–205
  reaction with glucose, 1158
  structure and function of, 205, 1072, 1156–1157
S-Adenosylmethionine, from methionine, 694
  function of, 396
  structure of, 1073
(S)-S-Adenosylmethionine, stereochemistry of, 166
Adipic acid, structure of, 780
ADMET, *see* acyclic diene metathesis polymerization, 1252
ADP, *see* Adenosine diphosphate
Adrenaline, biosynthesis of, 396
  molecular model of, 175
  structure of, 23
Adrenocortical steroid, 1111
-al, aldehyde name suffix, 723
Alanine, configuration of, 153–154
  electrostatic potential map of, 1045
  molecular model of, 26, 1044
  structure and properties of, 1046
  titration curve for, 1051
  zwitterion form of, 56
Alanylserine, molecular model of, 1056

Alcohol(s), **620**
  acetals from, 742–744
  acidity of, 624–626
  aldehydes from, 645–646
  alkenes from, 263–264, 641–643
  alkoxide ions from, 624
  alkyl halides from, 354–355, 391–392, 639
  α cleavage of, 431, 657
  biological dehydration of, 643
  boiling points of, 623
  carbonyl compounds from, 645–646
  carboxylic acids from, 645–646
  common names of, 622
  dehydration of, 263–264, 641–643
  electrostatic potential map of, 78
  esters from, 644
  ethers from, 678–680
  from acid chlorides, 833
  from aldehydes, 630–631, 734–735
  from alkenes, 269–274
  from carbonyl compounds, 630–636
  from carboxylic acids, 632–633, 827–828
  from esters, 632–633, 840–842
  from ethers, 681–682
  from ketones, 630–631, 734–735
  hybrid orbitals in, 18
  hydrogen bonds in, 623
  IR spectroscopy of, 443, 654
  ketones from, 645–646
  mass spectrometry of, 431, 657
  mechanism of dehydration of, 641–642
  mechanism of oxidation of, 646
  naming, 621–622
  NMR spectroscopy of, 655–656
  oxidation of, 645–647
  primary, 621
  properties of, 623–627
  protecting group for, 648–650
  reaction with acid, 641–642
  reaction with acid anhydrides, 835
  reaction with acid chlorides, 831
  reaction with aldehydes, 742–744
  reaction with alkenes, 680
  reaction with alkyl halides, 678–679
  reaction with ATP, 1156–1157
  reaction with carboxylic acids, 644, 824–826
  reaction with chlorotrimethylsilane, 648–649
  reaction with CrO$_3$, 645–646
  reaction with Dess–Martin periodinane, 645–646
  reaction with Grignard reagents, 626

reaction with HX, 354, 391–392, 639
reaction with ketones, 742–744
reaction with KMnO$_4$, 645
reaction with Na$_2$Cr$_2$O$_7$, 645–646
reaction with NaH, 626
reaction with NaNH$_2$, 626
reaction with PBr$_3$, 355, 639
reaction with POCl$_3$, 641–643
reaction with potassium, 626
reaction with SOCl$_2$, 355, 639
reaction with p-toluenesulfonyl chloride, 639–640
secondary, 621
synthesis of, 629–636
tertiary, 621
tosylates from, 639–640
trimethylsilyl ethers of, 648–650
Alcoholysis, 821
Aldaric acid, **1021**
from aldoses, 1021
Aldehyde, **722**
acetals from, 742–744
alcohols from, 630–631, 734–735
aldol reaction of, 905–906
alkanes from, 741–742
alkenes from, 746–748
α cleavage of, 432, 759
amines from, 958–959
biological reduction of, 631–632, 750–751
bromination of, 874–876
Cannizzaro reaction of, 750
carbonyl condensation reactions of, 905–906
carboxylic acids from, 727
common names of, 724
cyanohydrins from, 733
2,4-dinitrophenylhydrazones from, 739
enamines from, 736–739
enols of, 871–872
enones from, 908–909
from acetals, 743–744
from alcohols, 645–646
from alkenes, 284–286
from alkynes, 321–322
from esters, 725–726, 841
hydrate of, 727, 731–732
imines from, 736–739
IR spectroscopy of, 444, 756–757
mass spectrometry of, 431–432, 758–759
McLafferty rearrangement of, 431, 758
mechanism of hydration of, 731–732
mechanism of reduction of, 734
naming, 723–724

NMR spectroscopy of, 757–758
oxidation of, 727
oximes from, 737–738
pK_a of, 879
protecting groups for, 745
reaction with alcohols, 742–744
reaction with amines, 736–739
reaction with Br$_2$, 874–876
reaction with CrO$_3$, 727
reaction with 2,4-dinitrophenylhydrazine, 739
reaction with Grignard reagents, 635, 735
reaction with H$_2$O, 731–732
reaction with HCN, 733
reaction with HX, 732–733
reaction with hydrazine, 741–742
reaction with LiAlH$_4$, 630, 734
reaction with NaBH$_4$, 630, 734
reaction with NH$_2$OH, 737–738
reactivity of versus ketones, 729–730
reduction of, 630–631, 734
reductive amination of, 958–959
Wittig reaction of, 746–748
Wolff–Kishner reaction of, 741–742
Aldehyde group, directing effect of, 588–589
Alditol(s), **1020**
from aldoses, 1020
Aldol reaction, **905–906**
biological example of, 928–929
cyclohexenones from, 913–915
cyclopentenones from, 913–915
dehydration in, 908–909
enones from, 908–909
equilibrium in, 906
intramolecular, 913–915
mechanism of, 905–906
mixed, 912–913
reversibility of, 905–906
steric hindrance to, 906
uses of, 910–911
Aldolase, mechanism of, 928–929, 1177–1178
type I, 928–929
type II, 928–929
Aldonic acid(s), **1020**
from aldoses, 1020–1021
Aldose(s), **1002**
aldaric acids from, 1021
alditols from, 1020
aldonic acids from, 1020
Benedict's test on, 1020
chain-lengthening of, 1022
chain-shortening of, 1023
configurations of, 1008–1010
Fehling's test on, 1020

Kiliani–Fischer synthesis on, 1022
names of, 1010
natural occurrence of, 1008
oxidation of, 1020–1021
reaction with Br$_2$, 1020
reaction with HCN, 1022
reaction with HNO$_3$, 1021
reaction with NaBH$_4$, 1020
reduction of, 1020
see also Carbohydrate, Monosaccharide
Tollens' test on, 1020
uronic acids from, 1021
Wohl degradation of, 1023
Aldosterone, structure and function of, 1111
Algae, chloromethane from, 344
Alicyclic, **109**
Aliphatic, **81**
Alitame, structure of, 1034
sweetness of, 1033
Alkaloid(s), **63–64**
history of, 63
number of, 63
Alkane(s), **80**
boiling points of, 93
branched-chain, **82**
combustion of, 93
conformations of, 98–99
dispersion forces in, 61, 93
from aldehydes, 741–742
from alkyl halides, 356
from Grignard reagents, 356
from ketones, 741–742
general formula of, 81
IR spectroscopy of, 442
isomers of, 81–82
mass spectrometry of, 428–429
melting points of, 93
naming, 87–91
Newman projections of, 94
normal (*n*), **82**
parent names of, 83
pK_a of, 326
properties of, 92–94
reaction with Br$_2$, 347
reaction with Cl$_2$, 93, 347–349
sawhorse representations of, 94
straight-chain, **82**
Alkene(s), **222**
alcohols from, 269–274
aldehydes from, 284–286
alkoxymercuration of, 680
allylic bromination of, 350–351
biological addition reactions of, 294–295
bond rotation in, 229
bromohydrins from, 267–269
bromonium ion from, 265–266

Alkene(s)—cont'd
  cis–trans isomerism in, 229–230
  cleavage of, 284–286
  common names of, 228
  cyclopropanes from, 287–289
  1,2-dihalides from, 264–266
  diols from, 282–284
  electron distribution in, 195
  electrophilic addition reactions of, 237–238
  electrostatic potential map of, 78, 194
  epoxides from, 281–282
  ethers from, 680
  E,Z configuration of, 231–232
  from alcohols, 263–264, 641–643
  from aldehydes, 746–748
  from alkyl halides, 263
  from alkynes, 322–325
  from amines, 964–965
  from ketones, 746–748
  general formula of, 224
  halogenation of, 264–266
  halohydrins from, 267–269
  hydration of, 269–274
  hydroboration of, 272–274
  hydrogenation of, 276–280
  hydroxylation of, 282–284
  hyperconjugation in, 235–236
  industrial preparation of, 223
  IR spectroscopy of, 442–443
  ketones from, 284–286
  Markovnikov's rule and, 240–241
  mechanism of hydration of, 270
  naming, 226–227
  new naming system for, 227
  nucleophilicity of, 195
  old naming system for, 226–227
  organoboranes from, 272–274
  oxidation of, 281–286
  oxymercuration of, 271–272
  ozonolysis of, 284–285
  pK_a of, 326
  polymerization of, 290–292
  reaction with alcohols, 680
  reaction with borane, 272–274
  reaction with
    N-bromosuccinimide, 350–351
  reaction with Br_2, 264–266
  reaction with carbenes, 287–289
  reaction with Cl_2, 264–266
  reaction with halogen, 264–267
  reaction with HBr, 238
  reaction with HCl, 238
  reaction with HI, 238
  reaction with hydrogen, 276–280
  reaction with $KMnO_4$, 285
  reaction with mercuric ion, 271–272
  reaction with OsO_4, 283–284
  reaction with ozone, 284–285
  reaction with peroxyacids, 281–282
  reaction with radicals, 291–292
  reduction of, 276–280
  Sharpless epoxidation of, 761
  Simmons–Smith reaction of, 288–289
  stability of, 234–236
  steric strain in, 234–235
  synthesis of, 263–264
  uses of, 223
Alkoxide ion, **624**
  solvation of, 625
Alkoxymercuration, **680**
  mechanism of, 680
Alkyl group(s), **84**
  directing effect of, 585–586
  inductive effect of, 583
  naming, 84–85, 89–90
  orienting effect of, 581
  table of, 85
Alkyl halide(s), **344**
  alkenes from, 263
  amines from, 956–957
  amino acids from, 1054
  carboxylic acids from, 790
  coupling reactions of, 357–358
  dehydrohalogenation of, 263
  electrostatic potential map of, 78
  ethers from, 678–679
  from alcohols, 354–355, 391–392, 639
  from ethers, 681–682
  Grignard reagents from, 355–356
  malonic ester synthesis with, 883–884
  naming, 345–346
  phosphonium salts from, 747
  polarity of, 346
  polarizability of, 191
  reaction with alcohols, 678–679
  reaction with amines, 956
  reaction with azide ion, 956–957
  reaction with carboxylate ions, 824
  reaction with Gilman reagents, 357–358
  reaction with HS$^-$, 692
  reaction with phthalimide ion, 957
  reaction with sulfides, 694
  reaction with thiols, 693
  reaction with thiourea, 692
  reaction with tributyltin hydride, 370
  reaction with triphenylphosphine, 747
  see also Organohalide
  structure of, 346
  thiols from, 692
Alkyl shift, carbocations and, 250
Alkylamine(s), **944**
  basicity of, 950
Alkylation (aromatic), **575**–577
  see also Friedel–Crafts reaction
Alkylation (carbonyl), 882–889
  acetoacetic ester, 885–886
  acetylide anions and, 327–328
  biological example of, 889–890
  ester, 888
  ketone, 887–890
  lactone, 888
  malonic ester, 883–884
  nitrile, 888
Alkylbenzene, biological oxidation of, 597
  from aryl alkyl ketones, 599–600
  reaction with $KMnO_4$, 596–597
  reaction with NBS, 597–598
  side-chain bromination of, 597–598
  side-chain oxidation of, 596–597
Alkylthio group, 693
Alkyne(s), **314**
  acetylide anions from, 325–326
  acidity of, 325–326
  aldehydes from, 321–322
  alkenes from, 322–325
  alkylation of, 327–328
  cleavage of, 325
  electrostatic potential map of, 78
  from dihalides, 316
  hydration of, 319–321
  hydroboration of, 321–322
  hydrogenation of, 322–323
  IR spectroscopy of, 443
  ketones from, 319–321
  naming, 314–315
  oxidation of, 325
  pK_a of, 326
  reaction with BH_3, 321–322
  reaction with Br_2, 317–318
  reaction with Cl_2, 317–318
  reaction with HBr, 317–318
  reaction with HCl, 317–318
  reaction with $KMnO_4$, 325
  reaction with lithium, 323–325
  reaction with $NaNH_2$, 325
  reaction with O_3, 325
  reduction of, 322–325
  structure of, 316–317
  synthesis of, 316
  vinylic carbocation from, 318
  vinylic halides from, 317–318
Alkynyl group, 315
Allene, heat of hydrogenation of, 259

Allinger, Norman Louis, 132
Allose, configuration of, 1009
Allyl aryl ether, Claisen
  rearrangement of, 683–684
Allyl carbocation, electrostatic
  potential map of, 390
Allyl group, **228**
Allylic, **350**
Allylic bromination, 350–351
  mechanism of, 350–351
Allylic carbocation, electrostatic
  potential map of, 506
  resonance in, 506
  S_N1 reaction and, 389–390
  stability of, 506
Allylic halide, S_N1 reaction and,
    389–390
  S_N2 reaction and, 391
Allylic protons, ^{1}H NMR spectroscopy
  and, 474–475
Allylic radical, molecular orbital of,
    351
  resonance in, 351–352
  spin density surface of, 352
  stability of, 351–352
Alpha amino acid, **1045**
  see Amino acid
Alpha anomer, **1012**
Alpha cleavage, alcohol mass
    spectrometry and, 431, 657
  aldehyde mass spectrometry and,
    432, 759
  amine mass spectrometry and,
    431, 981–982
  ketone mass spectrometry and,
    432, 759
Alpha farnesene, structure of, 256
Alpha helix (protein), **1066–1067**
Alpha-keratin, molecular model of,
    1067
  secondary structure of, 1066–1067
Alpha-keto acid, amino acids from,
    1054
  reductive amination of, 1054
Alpha pinene, structure of, 222
Alpha substitution reaction, **718, 870**
  carbonyl condensation reactions
    and, 907–908
  evidence for mechanism of, 876
  mechanism of, 874
Altrose, configuration of, 1009
Aluminum chloride, Friedel–Crafts
    reaction and, 575
Amantadine, structure of, 139
Amide(s), **814**
  amines from, 844–845, 960–962
  basicity of, 951
  carboxylic acids from, 843–844
  electrostatic potential map of, 820

from acid anhydrides, 835–836
from acid chlorides, 832
from carboxylic acids, 826–827
from esters, 840
from nitriles, 795–796
hydrolysis of, 843–844
IR spectroscopy of, 851
mechanism of hydrolysis of,
    843–844
mechanism of reduction of, 845
naming, 816
nitriles from, 793–794
NMR spectroscopy of, 852
nucleophilic acyl substitution
    reactions of, 843–845
occurrence of, 842
pK_a of, 879
reaction with Br_2, 960, 961
reaction with $LiAlH_4$, 844–845
reaction with $SOCl_2$, 793–794
reduction of, 844–845
restricted rotation in, 1057
Amidomalonate synthesis, **1054**
-*amine*, name suffix, 945
Amine(s), **944**
  acidity of, 951–952
  alkenes from, 964–965
  α cleavage of, 431, 981–982
  basicity of, 948–950
  chirality of, 165–166, 947
  conjugate carbonyl addition
    reaction of, 753
  electronic structure of, 947
  electrostatic potential map of, 79
  from acid chlorides, 960, 962
  from acyl azides, 960, 962
  from aldehydes, 958–959
  from alkyl azides, 956–957
  from alkyl halides, 956–957
  from amides, 844–845, 960–962
  from ketones, 958–959
  from lactams, 845
  from nitriles, 796
  Henderson–Hasselbalch equation
    and, 954
  heterocyclic, 946
  Hofmann elimination of, 964–965
  hybrid orbitals in, 17–18
  hydrogen-bonding in, 948
  IR spectroscopy of, 444, 979
  mass spectrometry of, 431,
    981–982
  naming, 944–946
  nitrogen rule and, 981–982
  occurrence of, 944
  odor of, 948
  primary, **944**
  properties of, 948
  purification of, 951

pyramidal inversion in, 947
reaction with acid anhydrides,
    835–836
reaction with acid chlorides, 832
reaction with aldehydes, 736–739
reaction with alkyl halides, 956
reaction with carboxylic acids,
    826–827
reaction with enones, 753
reaction with epoxides, 689–690
reaction with esters, 840
reaction with ketones, 736–739
secondary, **944**
synthesis of, 955–962
tertiary, **944**
uses of, 948
Amino acid(s), **1044**
  abbreviations for, 1046–1047
  acidic, 1049
  amidomalonate synthesis of, 1054
  amphiprotic behavior of, 1045
  basic, 1049
  biosynthesis of, 1054
  Boc derivatives of, 1062–1063
  catabolism of, 1197–1201
  configuration of, 1048–1049
  electrophoresis of, 1053
  enantioselective synthesis of, 1055
  essential, 1049
  esters of, 1062
  Fmoc derivatives of, 1062–1063
  from alkyl halides, 1054
  from α-keto acids, 1054
  from carboxylic acids, 1053
  Henderson–Hasselbalch equation
    and, 1050–1051
  isoelectric points of, 1046–1047
  molecular weights of, 1046–1047
  neutral, 1049
  nonprotein, 1048
  pK_a's of, 1046–1047
  protecting groups for, 1062–1063
  reaction with di-*tert*-butyl
    dicarbonate, 1062–1063
  reaction with ninhydrin, 1058
  resolution of, 1054
  synthesis of, 1053–1055
  table of, 1046–1047
  C-terminal, **1056**
  N-terminal, **1056**
  transamination of, 1198–1201
  zwitterion form of, 1045
Amino acid analyzer, 1058–1059
  Ion-exchange chromatography
    and, 1058–1059
Amino group, **945**
  directing effect of, 586–587
  orienting effect of, 581
Amino sugar, **1024**

p-Aminobenzoic acid, molecular model of, 23
Aminolysis, **821**
Ammonia, dipole moment of, 38
  electrostatic potential map of, 192
  pK_a of, 53, 879
  reaction with acid chlorides, 832
  reaction with carboxylic acids, 826–827
Ammonium cyanate, urea from, 1
Ammonium ion, acidity of, 949–950
Amobarbital, synthesis of, 891
Amphetamine, structure of, 217
  synthesis of, 958
Amplitude, **435**
Amylopectin, 1→6-α-links in, 1029
  structure of, 1029
Amylose, 1→4-α-links in, 1028
  structure of, 1028
Anabolism, **1154**
  fatty acids, 1167–1173
  glucose, 1191–1197
Analgesic, **554**
Androgen, **1110**
  function of, 1110
Androstenedione, structure and function of, 1110
Androsterone, structure and function of, 1110
-*ane*, alkane name suffix, 83
Anesthetics, dental, 63–64
Angle strain, **114**
Angstrom, **3**
Anhydride, *see* Acid anhydride
Aniline, basicity of, 950
  electrostatic potential map of, 953
  from nitrobenzene, 572
  synthesis of, 572
Anilinium ion, electrostatic potential map of, 953
Anilinothiazolinone, Edman degradation and, 1059–1061
Anionic polymerization, 1243
Anisole, electrostatic potential map of, 804
  molecular model of, 676
  ^{13}C NMR spectrum of, 696
Annulation reaction, 927–928
[18]Annulene, electrostatic potential map of, 553
  ring current in, 552–553
Anomer, **1012**
Anomeric center, **1012**
Ant, sex attractant of, 834
Antarafacial geometry, **1223**
Anti conformation, **96**

Anti periplanar geometry, **401**
  E2 reaction and, 401–402
  molecular model of, 401
Anti stereochemistry, **265**
Antiaromaticity, **541**
Antibiotic, β-lactam, 853–854
Antibonding molecular orbital, **20**
Anticodon (tRNA), **1138**
Antisense strand (DNA), **1136**
Arabinose, configuration of, 1009
  Kiliani–Fischer synthesis on, 1022
Arachidic acid, structure of, 1090
Arachidonic acid, eicosanoids from, 1097–1098
  prostaglandins from, 188–189, 294–295, 1097–1098
  structure of, 1090
Arecoline, molecular model of, 80
Arene(s), **536**
  electrostatic potential map of, 78
  from arenediazonium salts, 970
  from aryl alkyl ketones, 599–600
  see also Aromatic compound
Arenediazonium salt(s), **968**
  arenes from, 970
  aryl bromides from, 969
  aryl chlorides from, 969
  aryl iodides from, 969
  coupling reactions of, 972
  from arylamines, 968–969
  nitriles from, 969
  phenols from, 970
  reaction with arylamines, 972
  reaction with CuBr, 969
  reaction with CuCl, 969
  reaction with CuCN, 970
  reaction with Cu_2O, 970
  reaction with H_3PO_2, 970
  reaction with NaI, 969
  reaction with phenols, 972
  reduction of, 970
  substitution reactions of, 969–970
Arginine, structure and properties of, 1047
epi-Aristolochene, biosynthesis of, 261
Aromatic compound(s), **534**
  acylation of, 577–578
  alkylation of, 575–577
  biological hydroxylation of, 573–574
  bromination of, 567–569
  characteristics of, 541
  chlorination of, 570
  coal tar and, 535
  common names for, 535–536
  fluorination of, 570
  Friedel–Crafts acylation of, 577–578

  Friedel–Crafts alkylation of, 575–577
  halogenation of, 567–571
  hydrogenation of, 599
  iodination of, 570–571
  IR spectroscopy of, 443, 551
  naming, 535–537
  nitration of, 571–572
  NMR ring current and, 552–553
  NMR spectroscopy of, 552–554
  nucleophilic aromatic substitution reaction of, 592–593
  oxidation of, 596–598
  reduction of, 599
  see also Aromaticity
  sources of, 535
  sulfonation of, 572–573
  trisubstituted, 600–604
  UV spectroscopy of, 552
Aromatic protons, ^1H NMR spectroscopy and, 474–475
Aromaticity, cycloheptatrienyl cation and, 544–545
  cyclopentadienyl anion and, 544–545
  Hückel $4n + 2$ rule and, 541–543
  imidazole and, 547
  indole and, 550
  ions and, 544–545
  isoquinoline and, 550
  naphthalene and, 550
  polycyclic aromatic compounds and, 549–550
  purine and, 550
  pyridine and, 546–547
  pyrimidine and, 546–547
  pyrrole and, 547
  quinoline and, 550
  requirements for, 541
Arrow, electron movement and, 44–45, 57–58, 197–199
  fishhook, 186, 291
  see also Curved arrow
Arsenic trioxide, LD_{50} of, 24
Aryl alkyl ketone, reduction of, 599–600
Aryl boronic acid, Suzuki–Miyaura reaction of, 359
Aryl halide, S_N2 reaction and, 379–380
  Suzuki–Miyaura reaction of, 359
Arylamine(s), **944**
  basicity of, 950, 952–953
  diazotization of, 968–969
  electrophilic aromatic substitution of, 966–968
  from nitroarenes, 955
  reaction with arenediazonium salts, 972

reaction with HNO_2, 968–969
resonance in, 952
table of basicity in, 953
Ascorbic acid, see Vitamin C
-ase, enzyme name suffix, 1070
Asparagine, structure and properties of, 1046
Aspartame, molecular model of, 27
structure of, 1034
sweetness of, 1033
Aspartic acid, structure and properties of, 1047
Asphalt, composition of, 100
Aspirin, history of, 554
LD_{50} of, 24
molecular model of, 15
synthesis of, 835
toxicity of, 555
Asymmetric center, **145**
Atactic polymer, **1245**
-ate, ester name suffix, 816
Atom, atomic mass of, 3
atomic number of, 3
electron configurations of, 5
electron shells in, 4
isotopes of, 3
orbitals in, 3–5
quantum mechanical model of, 3–5
size of, 2
structure of, 2–5
Atomic mass, **3**
Atomic number (Z), **3**
Atomic weight, **3**
Atorvastatin, structure of, 1, 534
statin drugs and, 1203–1204
ATP, see Adenosine triphosphate
Atrazine, LD_{50} of, 24
ATZ, see Anilinothiazolinone, 1059–1061
Aufbau principle, **5**
Avian flu, 1032
Axial bonds (cyclohexane), **120**
drawing, 121
Azide, amines from, 956–957
reduction of, 956–957
Azo compound, **971**
synthesis of, 972
uses of, 971
Azulene, dipole moment of, 558
electrostatic potential map of, 558
structure of, 551

β, see Beta
Backbone (protein), **1056**
Backside displacement, S_N2 reaction and, 376–377
von Baeyer, Adolf, 114
Baeyer strain theory, 114
Bakelite, structure of, 1256

Banana, esters in, 836
Barbiturates, 890–891
history of, 890
synthesis of, 891
Base, Brønsted–Lowry, **49**
Lewis, **56–59**
organic, 56
strengths of, 50–51
Base pair (DNA), 1131–1132
electrostatic potential maps of, 1132
hydrogen-bonding in, 1131–1132
Base peak (mass spectrum), **426**
Basicity, alkylamines, 950
amides, 951
amines, 948–950
arylamines, 950, 952–953
heterocyclic amines, 950
nucleophilicity and, 381
Basicity constant (K_b), **949**
Beeswax, components of, 1088–1089
Benedict's test, 1020
Bent bond, cyclopropane, 116
Benzaldehyde, electrostatic potential map of, 583, 730
IR spectrum of, 756
mixed aldol reactions of, 912
^{13}C NMR absorptions of, 758
Benzene, acylation of, 577–578
alkylation of, 575–577
bond lengths in, 539
bromination of, 567–569
chlorination of, 570
discovery of, 536
electrostatic potential map of, 43, 539, 583
fluorination of, 570
Friedel–Crafts reactions of, 575–579
heat of hydrogenation of, 539
Hückel $4n + 2$ rule and, 542
iodination of, 570–571
molecular orbitals of, 540–541
nitration of, 571–572
^{13}C NMR absorption of, 554
reaction with Br_2, 567–569
reaction with Cl_2, 570
reaction with F-TEDA-BF_4, 570
reaction with HNO_3, 571–572
reaction with H_2SO_4/HNO_3, 572
reaction with I_2, 570–571
resonance in, 43, 539–540
stability of, 538–539
structure of, 538–541
sulfonation of, 572
toxicity of, 534
UV absorption of, 520
Benzenediazonium ion, electrostatic potential map of, 972

Benzenesulfonic acid, synthesis of, 572
Benzodiazepine, combinatorial library of, 605
Benzoic acid, ^{13}C NMR absorptions in, 798
pK_a of, 783
substituent effects on acidity of, 787
Benzophenone, structure of, 724
Benzoquinone, electrostatic potential map of, 653
Benzoyl group, **724**
Benzoyl peroxide, ethylene polymerization and, 291
Benzo[a]pyrene, carcinogenicity of, 549
structure of, 549
Benzyl ester, hydrogenolysis of, 1062
Benzyl group, **536**
Benzylic, **390**
Benzylic acid rearrangement, 867
Benzylic carbocation, electrostatic potential map of, 390
resonance in, 390
S_N1 reaction and, 389–390
Benzylic halide, S_N1 reaction and, 389–390
S_N2 reaction and, 391
Benzylic radical, resonance in, 598
spin-density surface of, 598
Benzylpenicillin, discovery of, 853
structure of, 1
Benzyne, **595**
Diels–Alder reaction of, 595
electrostatic potential map of, 595
evidence for, 595
structure of, 595
Bergström, Sune K., 1095
Beta anomer, **1012**
Beta-carotene, industrial synthesis of, 748
structure of, 222
UV spectrum of, 521
Beta-diketone, Michael reactions and, 922–923
Beta-keto ester, alkylation of, 885–886
cyclic, 919–920
decarboxylation of, 886
Michael reactions and, 922–923
pK_a of, 879
synthesis of, 919–920
Beta-lactam antibiotics, 853–854
Beta-oxidation pathway, **1162–1167**
mechanisms in, 1162–1167
steps in, 1162
Beta-pleated sheet (protein), **1066–1067**
molecular model of, 1067
secondary protein structure and, 1066–1067

Betaine, **747**
Bextra, structure of, 562
BHA, synthesis of, 652
BHT, synthesis of, 652
Bicycloalkane, **130**
Bimolecular, **376**
Biodegradable polymers, 850, 1256–1257
Biological acids, Henderson–Hasselbalch equation and, 785–786
Biological carboxylic acid derivative, 845–847
Biological mass spectrometry, 433–434
Biological reaction, alcohol dehydration, 264
  alcohol oxidation, 647
  aldehyde reduction, 631–632, 750–751
  aldol reaction, 928–929
  alkene halogenation, 267
  alkene hydration, 270–271
  alkene hydrogenation, 280
  α-substitution reaction, 889–890
  aromatic hydroxylation, 573–574
  aromatic iodination, 571
  benzylic oxidation, 597
  bromohydrin formation, 269
  carbonyl condensations, 928–930
  carboxylation, 790
  characteristics of, 210–212
  Claisen condensation, 930
  Claisen rearrangement, 684, 1229–1230
  comparison with laboratory reaction, 210–212
  conclusions about, 1202
  conventions for writing, 211, 239
  decarboxylation, 1181–1185
  dehydration, 643
  electrophilic aromatic substitution, 571
  elimination reactions, 407
  energy diagram of, 209
  fat hydrolysis, 839–840
  Friedel–Crafts alkylation, 578–579
  ketone alkylation, 889–890
  ketone reduction, 631–632, 750–751
  nucleophilic acyl substitution, 828–830
  nucleophilic substitutions, 395–396
  oxidation, 647
  protein hydrolysis, 844
  radical additions, 294–295
  reduction, 631–632, 750–751
  reductive amination, 959
  S_N1 reaction, 395–396

S_N2 reaction, 396
  thioester reduction, 847
Biological substitution reactions, diphosphate leaving group in, 395–396
Biomass, carbohydrates and, 1000
Bioprospecting, 251–252
Biosynthesis, fatty acids, 1167–1173
Biot, Jean Baptiste, 147
Biotin, fatty acid biosynthesis and, 1170
  stereochemistry of, 178
  structure of, 1073
Bird flu, 1032
Bisphenol A, epoxy resins from, 697–698
  polycarbonates from, 849–850
Block copolymer, **1248**
  synthesis of, 1248
Boat conformation (cyclohexane), steric strain in, 119–120
Boc (*tert*-butoxycarbonyl amide), 1062–1063
  amino acid derivatives of, 1062–1063
Bond, covalent, 9–10
  molecular orbital theory of, 19–21
  pi, **14**
  sigma, **10**
  valence bond description of, 9–10
Bond angle, **12**
Bond dissociation energy (D), **203**
  table of, 204
Bond length, **10**
Bond rotation, alkanes, 94–95
  alkenes, 229
  butane, 96–98
  ethane, 94–95
  propane, 96
Bond strength, **10**
Bonding molecular orbital, **20**
Borane, electrophilicity of, 273
  electrostatic potential map of, 273
  reaction with alkenes, 272–274
  reaction with alkynes, 321–322
  reaction with carboxylic acids, 828
Boron trifluoride, electrostatic potential map of, 57, 194
Branched-chain alkane, **82**
Breathalyzer test, 658
Bridgehead atom (polycyclic compound), **129**
Broadband-decoupled NMR, 467
Bromine, reaction with aldehydes, 874–876
  reaction with alkanes, 347
  reaction with alkenes, 264–266
  reaction with alkynes, 317–318
  reaction with aromatic compounds, 567–569

reaction with carboxylic acids, 876–877
  reaction with enolate ions, 881–882
  reaction with ketones, 874–876
  reactions with aldoses, 1020
Bromo group, directing effect of, 587–588
p-Bromoacetophenone, molecular model of, 466
  ^{13}C NMR spectrum of, 465
  symmetry plane in, 466
p-Bromobenzoic acid, pK_a of, 787
Bromocyclohexane, molecular model of, 122
  ring-flip in, 122
Bromoethane, electrostatic potential maps of, 194
  ^1H NMR spectrum of, 477
  spin–spin splitting in, 477–478
Bromohydrin(s), **267**
  from alkenes, 267–269
  mechanism of formation of, 268
Bromomethane, bond length of, 346
  bond strength of, 346
  dipole moment of, 346
  electrostatic potential map of, 192
Bromonium ion, **265**
  electrostatic potential map of, 266
  from alkenes, 265–266
  stability of, 266
2-Bromopropane, ^1H NMR spectrum of, 478
  spin–spin splitting in, 478
N-Bromosuccinimide, bromohydrin formation with, 268
  reaction with alkenes, 268, 350–351
  reaction with alkylbenzenes, 597–598
p-Bromotoluene, ^1H NMR spectrum of, 553
Brønsted–Lowry acid, **49**
  conjugate base of, **49**
  strengths of, 50–51
Brønsted–Lowry base, **49**
  conjugate acid of, **49**
  strengths of, 50–51
Brown, Herbert Charles, 272
Bupivacaine, structure of, 64
Butacetin, structure of, 865
1,3-Butadiene, 1,2-addition reactions of, 505–506
  1,4-addition reactions of, 505–506
  bond lengths in, 502
  electrophilic addition reactions of, 505–506
  electrostatic potential map of, 504
  heat of hydrogenation of, 503

molecular orbitals in, 503–504, 1215
polymerization of, 516
reaction with Br$_2$, 506
reaction with HBr, 505–506
stability of, 502–504
UV spectrum of, 519
Butanal, 2-ethyl-1-hexanol from, 910–911
Butane, anti conformation of, 96
bond rotation in, 96–98
conformations of, 96–98
gauche conformation of, 97
molecular model of, 81
Butanoic acid, IR spectrum of, 798
1-Butanol, mass spectrum of, 657
2-Butanone, ^{13}C NMR spectrum of, 465
3-Buten-2-one, electrostatic potential map of, 752
UV absorption of, 520
1-Butene, heat of hydrogenation of, 236
cis-2-Butene, heat of hydrogenation of, 235
molecular model of, 230, 234
steric strain in, 234–235
trans-2-Butene, heat of hydrogenation of, 235
molecular model of, 230, 234
Butoxycarbonyl (Boc) protecting group, 1062–1063
Butter, composition of, 1090
tert-Butyl alcohol, pK_a of, 625
tert-Butyl carbocation, electrostatic potential map of, 245
molecular model of, 244
Butyl group, 85
Butyl rubber polymer, structure and uses of, 1247
Butyllithium, electrostatic potential map of, 357

c (Speed of light), 435
C-terminal amino acid, **1056**
Cadaverine, odor of, 948
Caffeine, structure of, 32
Cahn–Ingold–Prelog sequence rules, 150–152
enantiomers and, 150–154
E,*Z* alkene isomers and, 231–232
Caine anesthetics, 63–64
Calicene, dipole moment of, 561
Camphor, molecular model of, 131
specific rotation of, 149
structure of, 1099
Cannizzaro reaction, **750**
mechanism of, 750
Caprolactam, nylon 6 from, 1249

Capsaicin, structure of, 80
-carbaldehyde, aldehyde name suffix, 723
Carbamic acid, 1250
Carbanion, electrostatic potential map of, 326
stability of, 326
Carbene(s), **287**
electronic structure of, 288
reaction with alkenes, 287–289
Carbenoid, **289**
Carbinolamine, **736**
Carbocaine, structure of, 64
Carbocation(s), **196**
alkyl shift in, 250
E1 reaction and, 405
electronic structure of, 243–244
electrophilic addition reactions and, 196, 237–238
electrophilic aromatic substitution and, 568–569
electrostatic potential map of, 245, 288
Friedel–Crafts reaction and, 576–577
Hammond postulate and, 248
hydride shift in, 249–250
hyperconjugation in, 245
inductive effects on, 245
Markovnikov's rule and, 241
rearrangements of, 249–250, 576–577
S$_N$1 reactions and, 389–390
solvation of, 393
stability of, 243–245, 390
vinylic, 318
Carbocation rearrangement, lanosterol biosynthesis and, 1113–1117
Carbohydrate(s), **1000**
amount of in biomass, 1000
anomers of, 1011–1013
catabolism of, 1173–1181
classification of, 1001–1002
complex, **1001**
essential, 1023–1025
Fischer projections and, 1004–1005
glycosides and, 1016–1018
1→4-links in, 1025–1026
name origin of, 1000
photosynthesis of, 1000–1001
see also Aldose, Monosaccharide
Carbon atom, 3-dimensionality of, 6
ground-state electron configuration of, 5
tetrahedral geometry of, 6
Carbonate ion, resonance in, 47
Carbonic anhydrase, turnover number of, 1069

-carbonitrile, nitrile name suffix, 781
Carbonyl compound(s), acidity of, 877–880
alcohols from, 630–636
alkylation of, 882–889
electrostatic potential map of, 79, 192
from alcohols, 645–646
general reactions of, 714–719
IR spectroscopy of, 444
kinds of, 79, 712–713
mass spectrometry of, 431–432
Carbonyl condensation reaction, **719, 904**–906
α-substitution reactions and, 907–908
biological examples of, 928–930
mechanism of, 904–905
Carbonyl group, **712**
bond angles in, 714
bond length of, 714
bond strength of, 714
directing effect of, 588–589
inductive effect of, 583
orienting effect of, 581
resonance effect of, 584
structure of, 714
-carbonyl halide, acid halide name suffix, 815
-carbothioate, thioester name suffix, 816
-carboxamide, amide name suffix, 816
Carboxybiotin, fatty acid biosynthesis and, 1170
Carboxyl group, **779**
-carboxylate, ester name suffix, 816
Carboxylate ion, reaction with alkyl halides, 824
resonance in, 784
Carboxylation, **790**
biological example of, 790
-carboxylic acid, name suffix, 779
Carboxylic acid(s), **55, 778**
acid anhydrides from, 824
acid bromide from, 830
acid chlorides from, 823–824
acidity of, 782–784
alcohols from, 632–633, 827–828
amides from, 826–827
amino acids from, 1053
biological, 785–786
bromination of, 876–877
common names of, 779–780
derivatives of, 814
dimers of, 782
dissociation of, 782–783
esters from, 824–826
from acid chlorides, 830–831
from alcohols, 645–646

Carboxylic acid(s)—cont'd
  from aldehydes, 727
  from alkyl halides, 790, 883–884
  from amides, 843–844
  from esters, 837–840
  from Grignard reagents, 790
  from malonic ester, 883–884
  from nitriles, 789–790, 795–796
  Hell–Volhard–Zelinskii reaction of, 876–877
  hydrogen-bonding in, 782
  inductive effects in, 786–787
  IR spectroscopy of, 797–798
  naming, 779–780
  NMR spectroscopy of, 798–799
  nucleophilic acyl substitution reactions of, 823–830
  occurrence of, 778
  pK_a table of, 783
  properties of, 782–784
  reaction with alcohols, 644, 824–826
  reaction with amines, 826–827
  reaction with ammonia, 826–827
  reaction with borane, 828
  reaction with Br_2, 876–877
  reaction with diazomethane, 866
  reaction with $LiAlH_4$, 632–633
  reaction with PBr_3, 830
  reaction with $SOCl_2$, 823–824
  reduction of, 632–633, 827–828
  synthesis of, 789–791
Carboxylic acid derivative(s), **814**
  biological, 845–847
  electrostatic potential maps of, 820
  interconversions of, 820–821
  IR spectroscopy of, 851
  kinds of, 814
  naming, 815–817
  NMR spectroscopy of, 852
  nucleophilic acyl substitution reactions of, 820–821
  relative reactivity of, 819–821
  table of names for, 817
Cardiolipin, structure of, 1121
Caruthers, Wallace Hume, 849
Carvone, chirality of, 146
  structure of, 23
Caryophyllene, structure of, 1123
Catabolism, **1154**
  acetyl CoA, 1185–1190
  amino acids, 1197–1201
  ATP and, 1156–1157
  carbohydrates, 1173–1181
  fats, 1158–1167
  fatty acids, 1162–1167
  glucose, 1173–1181
  glycerol, 1158–1161

  overview of, 1155
  protein, 1197–1201
  pyruvate, 1181–1185
  triacylglycerols, 1158–1167
Catalytic cracking, 101
Catalytic hydrogenation, *see* Hydrogenation
Cation radical, mass spectrometry and, 425–426
Cationic polymerization, 1243
Celebrex, 555
Celecoxib, NSAIDs and, 555
Cell membrane, lipid bilayer in, 1095
Cellobiose, 1→4-β-link in, 1026
  molecular model of, 1026
  mutarotation of, 1026
  structure of, 1026
Cellulose, 1→4-β-links in, 1028
  function of, 1028
  structure of, 1028
  uses of, 1028
Cellulose nitrate, 1028
Cephalexin, stereochemistry of, 181
  structure of, 854
Cephalosporin, structure of, 854
Chain, Ernst, 853
Chain-growth polymer, **291–292**, 847, 1242–1244
Chain reaction (radical), **188**
Chair conformation (cyclohexane), **118**
  drawing, 119
  molecular model of, 119
  see also Cyclohexane
Chemical Abstracts, 74
Chemical shift (NMR), **462**
  ^{13}C NMR spectroscopy and, 464–465
  ^1H NMR spectroscopy and, 474–475
Chemical structure, drawing, 21–22
Chevreul, Michel-Eugène, 1
Chiral, **144**
Chiral drugs, 172–173
Chiral environment, **171**–172
  prochirality and, 171–172
Chiral methyl group, 422
Chirality, amines and, 947
  cause of, 145
  electrophilic addition reactions and, 296–298
  naturally occurring molecules and, 170–172
  tetrahedral carbon and, 144–146
Chirality center, **145**
  detection of, 145–146
  Fischer projections and, 1002–1005
  inversion of configuration of, 373–374
  R,S configuration of, 150–154

Chitin, structure of, 1031
Chloramphenicol, stereochemistry of, 181
Chlorine, reaction with alkanes, 93, 347–349
  reaction with alkenes, 264–266
  reaction with alkynes, 317–318
  reaction with aromatic compounds, 570
Chloro group, directing effect of, 587–588
Chloroalkanes, dissociation enthalpy of, 244
Chlorobenzene, electrostatic potential map of, 583
  ^{13}C NMR absorptions of, 554
  phenol from, 594
p-Chlorobenzoic acid, pK_a of, 787
2-Chlorobutanoic acid, pK_a of, 787
3-Chlorobutanoic acid, pK_a of, 787
4-Chlorobutanoic acid, pK_a of, 787
Chloroethane, dissociation enthalpy of, 244
Chloroform, dichlorocarbene from, 287
  LD_{50} of, 24
Chloromethane, bond length of, 346
  bond strength of, 346
  dipole moment of, 346
  dissociation enthalpy of, 244
  electrostatic potential map of, 37, 190, 346
  natural sources of, 344
2-Chloro-2-methylbutane, dissociation enthalpy of, 244
Chloronium ion, **265**
p-Chlorophenol, pK_a of, 625
Chlorophyll, biosynthesis of, 994
Chloroprene, polymerization of, 516
2-Chloropropane, dissociation enthalpy of, 244
Chlorosulfite, 823–824
Chlorotrimethylsilane, bonds lengths in, 649
  reaction with alcohols, 648–649
Cholecalciferol, structure of, 1232
Cholestanol, structure of, 158
Cholesterol, amount of in body, 1203
  biosynthesis of, 1112–1117
  carbocation rearrangements and, 250
  heart disease and, 1118
  molecular model of, 1109
  specific rotation of, 149
  statin drugs and, 1203–1204
  stereochemistry of, 1109
Cholic acid, molecular model of, 778
Chorismate, Claisen rearrangement of, 1229–1230

Chromium trioxide, reaction with aldehydes, 727
Chrysanthemic acid, structure of, 108
Chymotrypsin, peptide cleavage with, 1061
trans-Cinnamaldehyde, ¹H NMR spectrum of, 482
   tree diagram for, 483
cis–trans Isomers, **113**
   alkenes and, 229–230
   cycloalkanes and, 111–113
   requirements for, 230
Citanest, structure of, 64
Citrate, prochirality of, 1187
Citrate synthase, active site of, 1074
   function of, 1071
   mechanism of action of, 1071, 1074–1075
   molecular model of, 1074
Citric acid, molecular model of, 27
Citric acid cycle, **1185**–1190
   mechanisms in, 1187–1190
   requirements for, 1187
   result of, 1190
   steps in, 1186
Claisen condensation reaction, **915**–917
   biological example of, 930
   intramolecular, 919–920
   mechanism of, 915–916
   mixed, 917–918
Claisen rearrangement, 683–684, **1229**–1230
   biological example of, 684, 1229–1230
   mechanism of, 683–684
   suprafacial geometry of, 1229–1230
   transition state of, 683–684
Clomiphene, structure of, 257
Clopidogrel, structure of, 32
Clostridium perfringens, DNA bases in, 1131
Coal, structure of, 535
Coal tar, compounds from, 535
Cocaine, specific rotation of, 149
   structure of, 63, 944
   structure proof of, 901
   synthesis of, 943
Coconut oil, composition of, 1090
Coding strand (DNA), 1136
CODIS, DNA fingerprint registry, 1146
Codon (mRNA), **1137**–1138
   table of, 1137
Coenzyme, **212**, **1071**
   table of, 1072–1073
Coenzyme A, structure of, 846, 1072
Coenzyme Q, 653–654
Cofactor (enzyme), **1071**

Color, perception of, 522
   UV spectroscopy and, 521–522
Combinatorial chemistry, **605**–606
   kinds of, 605
Combinatorial library, **605**–606
Complex carbohydrate, **1001**
Computer chip, manufacture of, 523–524
Concanavalin A, secondary structure of, 1066–1067
Concerted reaction, **1214**
Condensation reaction, **908**
Condensed structure, **21**
Cone cells, vision and, 522
Configuration, **150**
   assignment of, 150–154
   chirality centers and, 150–154
   Fischer projections and, 1004
   inversion of, 373–374
   R, 152
   S, 152
Conformation, **94**
   anti, **96**
   calculating energy of, 132
   eclipsed, **95**
   gauche, **97**
   staggered, **95**
Conformational analysis (cyclohexane), **127**–128
Conformer, **94**
Coniine, chirality of, 147
   molecular model of, 26
Conjugate acid, **49**
Conjugate base, **49**
Conjugate carbonyl addition reaction, **751**–755
   amines and, 753
   enamines and, 925–926
   Gilman reagents and, 754–755
   mechanism of, 752
   Michael reactions and, 921–923
   water and, 753
Conjugated compound, **500**
Conjugated diene, 1,2-addition reactions of, 505–506
   1,4-addition reactions of, 505–506
   allylic carbocations from, 506
   bond lengths in, 502
   electrocyclic reactions of, 1217
   electrophilic addition reactions of, 505–506
   electrostatic potential map of, 504
   heats of hydrogenation of, 503
   molecular orbitals in, 503–504
   polymers of, 516–517
   reaction with Br_2, 506
   reaction with HBr, 505–506
   stability of, 502–504
   synthesis of, 501

Conjugated polyene, electrocyclic reactions of, 1217–1222
   molecular orbitals of, 1215–1216
Conjugated triene, electrocyclic reactions of, 1217
Conjugation, ultraviolet spectroscopy and, 520
Conrotatory motion, **1219**
Consensus sequence (DNA), 1135
Constitutional isomers, **82**
   kinds of, 82
Contraceptive, steroid, 1111
Cope rearrangement, **1229**–1230
   suprafacial geometry of, 1229–1230
Copolymer, **1246**–1248
   block, **1248**
   graft, **1248**
   table of, 1247
Copper(II) chloride, aromatic iodination and, 570–571
Coprostanol, structure of, 158
Corn oil, composition of, 1090
Coronary heart disease, cholesterol and, 1203–1204
   statin drugs and, 1203–1204
Coronene, structure of, 549
Cortisone, structure of, 108
Couper, Archibald Scott, 6
Coupled reactions, **1157**–1158
   ATP and, 1157–1158
Coupling (NMR), **477**
   see also Spin–spin splitting
Coupling constant, **478**
   size of, 478
   use of, 478–479
Covalent bond(s), **7**
   bond angle in, 12
   bond length in, 10
   bond strength in, 10
   molecular orbital theory of, 19–21
   polar, **34**–35
   rotation around, 94, 112
   sigma, 10
   valence bond theory of, 9–10
COX-2 inhibitors, 555, 1097
Cracking, steam, 223–224
Crick, Francis H. C., 1131
Crotonaldehyde, structure of, 724
Crotonic acid, ¹³C NMR absorptions in, 798
Crown ether, **690**–691
   electrostatic potential map of, 691
   S_N2 reactions and, 691
   solvation of cations by, 691
Crystallite, **1253**
Crystallization, fractional, 161
Cumene, phenol from, 650
Cumulene, structure of, 342

Curtius rearrangement, 960, 962
  mechanism of, 962
Curved arrow, electron movement
    and, 44–45, 57–58
  guidelines for using, 197–198
  polar reactions and, 192, 197–199
  radical reactions and, 186, 291
p-Cyanobenzoic acid, pK_a of, 787
Cyanocycline A, structure of, 793
Cyanogenic glycoside, **793**
Cyanohydrin(s), **733**
  from aldehydes, 733
  from ketones, 733
  mechanism of formation of, 733
  uses of, 733–734
Cycloaddition reaction, **510, 1222–1225**
  antarafacial geometry of, 1223–1225
  cyclobutane synthesis and, 1225
  photochemical, 1225
  see also Diels–Alder reaction
  stereochemical rules for, 1225
  stereochemistry of, 1224–1225
  suprafacial geometry of, 1223–1225
  thermal, 1224
Cycloalkane(s), **109**
  angle strain in, 114–115
  Baeyer strain theory and, 114
  cis–trans isomerism in, 111–113
  heats of combustion of, 115
  naming, 109–111
  skeletal structures of, 109
  strain energies of, 115
Cycloalkene, naming, 227–228
Cyclobutadiene, antiaromaticity of, 542
  electrostatic potential map of, 542
  Hückel $4n + 2$ rule and, 542
  reactivity of, 542
Cyclobutane, angle strain in, 117
  conformation of, 117
  molecular model of, 117
  photochemical synthesis of, 1225
  strain energy of, 115
  torsional strain in, 117
Cyclodecane, strain energy of, 115
1,3,5,7,9-Cyclodecapentaene, molecular model of, 543, 557
Cycloheptane, strain energy of, 115
Cycloheptatriene, reaction with Br$_2$, 544–545
Cycloheptatrienyl cation, aromaticity of, 544–545
  electrostatic potential map of, 545
  Hückel $4n + 2$ rule and, 544–545
Cycloheptatrienylium bromide, synthesis of, 544–545

1,3-Cyclohexadiene, heat of hydrogenation of, 539
  UV absorption of, 520
Cyclohexane, axial bonds in, 120–122
  barrier to ring flip in, 122
  bond angles in, 118
  chair conformation of, 118–119
  conformational analysis of, 126–128
  1,3-diaxial interactions in, 124–125
  drawing chair form of, 119
  equatorial bonds in, 120–122
  IR spectrum of, 451
  rate of ring-flip in, 460–461
  ring-flip in, **122**
  strain energy of, 115
  twist-boat conformation of, 120
Cyclohexane conformation, E2 reactions and, 403–404
Cyclohexanol, IR spectrum of, 654
  ^{13}C NMR spectrum of, 655
Cyclohexanone, aldol reaction of, 906
  enol content of, 871
  enolate ion of, 878
  IR spectrum of, 756
  ^{13}C NMR absorptions of, 758
Cyclohexene, heat of hydrogenation of, 539
  IR spectrum of, 451
Cyclohexenones, from 1,5-diketones, 913–915
Cyclohexylamine, IR spectrum of, 979
Cyclohexylmethanol, ^{1}H NMR spectrum of, 485
Cyclononane, strain energy of, 115
Cyclooctane, strain energy of, 115
Cyclooctatetraene, bond lengths in, 543
  dianion of, 546
  electrostatic potential map of, 542–543
  Hückel $4n + 2$ rule and, 542–543
  ^{1}H NMR absorption of, 553
  reactivity of, 542
1,3-Cyclopentadiene, Diels–Alder reactions of, 514
  electrostatic potential map of, 974
  pK_a of, 544
Cyclopentadienyl anion, aromaticity of, 544–545
  electrostatic potential map of, 545
  Hückel $4n + 2$ rule and, 544–545
Cyclopentane, angle strain in, 117
  conformation of, 117
  molecular model of, 117

strain energy of, 115
  torsional strain in, 117
Cyclopentanone, IR spectroscopy of, 757
Cyclopentenones, from 1,4-diketones, 913–915
Cyclopropane, angle strain in, 116
  bent bonds in, 116
  from alkenes, 287–289
  molecular model of, 112, 116
  strain energy of, 115
  torsional strain in, 116
Cystathionine, cysteine from, 1213
Cysteine, biosynthesis of, 1213
  disulfide bridges from, 1057
  structure and properties of, 1046
Cytosine, electrostatic potential map of, 1132
  molecular model of, 66
  protection of, 1142–1143
  structure of, 1129

D (Bond dissociation energy), **203**
D (Debye), 37
D Sugar, **1007**
  Fischer projections of, 1007
Dacron, structure of, 849
Darzens reaction, 942
DCC (dicyclohexylcarbodiimide), 826
  amide bond formation with, 826–827
  mechanism of amide formation with, 826–827
  peptide synthesis with, 1062–1063
Deactivating group (aromatic substitution), **581**
  acidity and, 787
Debye (D), **37**
cis-Decalin, conformation of, 130
  molecular model of, 130, 1108
trans-Decalin, conformation of, 130
  molecular model of, 130, 1108
Decarboxylation, **883**
  β-keto esters and, 886
  biological example of, 1181–1185
  malonic esters and, 883–884
  pyruvate and, 1181–1185
DEET, structure of, 865
Degenerate orbitals, **541**
Degree of unsaturation, **224**
  calculation of, 224–226
Dehydration, **263**
  alcohol mass spectrum and, 657
  alcohols, 263–264, 641–643
  aldol reaction and, 908–909
  biological example of, 264, 643
7-Dehydrocholesterol, vitamin D from, 1232
Dehydrohalogenation, **263**

Delocalized, 352
Delta scale (NMR), 462
Denature (protein), **1068**
Dental anesthetics, 63–64
Deoxy sugar, **1024**
Deoxyribonucleic acid (DNA), **1128**
  antisense strand of, **1136**
  base-pairing in, 1131–1132
  bases in, 1129
  cleavage of, 1140
  consensus sequence in, 1135
  double helix in, 1131–1132
  3′ end of, 1131
  5′ end of, 1131
  exons in, 1136
  fingerprinting with, 1146–1147
  heredity and, 1133
  hydrogen-bonding in, 62, 1131–1132
  introns in, 1136
  lagging strand in, 1135
  leading strand in, 1135
  major groove in, 1132
  minor groove in, 1132
  molecular model of, 62, 1132
  Okazaki fragments in, 1135
  polymerase chain reaction and, 1145–1146
  promotor sequence in, 1135
  replication fork in, 1134
  replication of, 1133–1135
  sense strand of, **1136**
  sequencing of, 1140–1141
  size of, 1129
  structure of, 1130–1131
  synthesis of, 1142–1144
  transcription of, 1135–1136
  Watson–Crick model of, 1131–1132
Deoxyribonucleotide, structures of, 1130
2′-Deoxyribose, structure of, 1129
  equilibrium forms of, 1031
1-Deoxyxylulose 5-phosphate pathway, terpenoid biosynthesis and, 1099
DEPT-NMR, 467–469
  uses of, 467–468
DEPT-NMR spectrum, 6-methyl-5-hepten-2-ol, 468
Dermabond, structure of, 1244
Dess–Martin periodinane, alcohol oxidations with, 645–646
  reaction with alcohols, 645
  structure of, 645
Detergent, structure of, 1093
Deuterium isotope effect, **400**
  E1 reaction and, 406
  E2 reaction and, 400

Dewar benzene, 1238
Dextromethorphan, chirality of, 147
Dextrorotatory, **148**
Dextrose, see Glucose
Dialkylamine, pK_a of, 879
Diastereomers, **157**
  kinds of, 164–165
Diastereotopic protons (NMR), **473**
1,3-Diaxial interactions, 124–125
  table of, 125
Diazepam, degree of unsaturation in, 224
Diazomethane, reaction with carboxylic acids, 866
Diazonio group, **969**
Diazonium coupling reaction, 972
Diazoquinone–novolac resist, 523–524
Diazotization reaction, **968–969**
DIBAH, see Diisobutylaluminum hydride
Dibutyl phthalate, use as plasticizer, 837
Dichlorocarbene, electronic structure of, 288
  electrostatic potential map of, 288
  from chloroform, 287
  mechanism of formation of, 287
1,2-Dichloroethane, synthesis of, 264–265
cis-1,2-Dichloroethylene, electrostatic potential map of, 66
trans-1,2-Dichloroethylene, electrostatic potential map of, 66
2,4-Dichlorophenoxyacetic acid, synthesis of, 652
Dideoxy DNA sequencing, **1140–1141**
2′,3′-Dideoxyribonucleotide, 1140–1141
Dieckmann cyclization, **919–920**
  mechanism of, 919–920
Diels–Alder reaction, **510**
  characteristics of, 511–512
  dienes in, 513–514
  dienophiles in, 511
  electrostatic potential map of, 510
  endo stereochemistry of, 512
  HOMO in, 1224
  LUMO in, 1224
  mechanism of, 510
  s-cis diene conformation in, 513–514
  stereochemistry of, 512, 1224
  suprafacial geometry of, 1224
Diene polymers, 516–517
  vulcanization of, 517
Dienophile, **511**
  requirements for, 511

Diethyl ether, IR spectrum of, 695
  molecular model of, 676
  synthesis of, 678
Diethyl malonate, alkylation of, 883–884
  carboxylic acids from, 883–884
  Michael reactions and, 922–923
  pK_a of, 879
  see also Malonic ester
Diethyl propanedioate, see Diethyl malonate
Digitoxigenin, structure of, 1125
Digitoxin, structure of, 1017
Dihedral angle, **95**
Diiodomethane, Simmons–Smith reaction with, 288–289
Diisobutylaluminum hydride, reaction with esters, 841
  structure of, 726
Diisopropylamine, pK_a of, 878, 952
1,3-Diketone, pK_a of, 879
Dimethyl disulfide, bond angles in, 19
Dimethyl ether, electrostatic potential map of, 57, 677
Dimethyl sulfide, molecular model of, 19
Dimethyl sulfoxide, electrostatic potential map of, 40
  formal charges in, 40–41
  S_N2 reaction and, 384
Dimethylallyl diphosphate, geraniol biosynthesis and, 395–396
  biosynthesis of, 1103–1105
cis-1,2-Dimethylcyclohexane, conformational analysis of, 126–127
  molecular model of, 126, 112
trans-1,2-Dimethylcyclohexane, conformational analysis of, 127
  molecular model of, 127, 112
Dimethylformamide, S_N2 reaction and, 384
2,2-Dimethylpropane, mass spectrum of, 427
  molecular model of, 82
N,N-Dimethyltryptamine, electrostatic potential map of, 979
2,4-Dinitrophenylhydrazone, **738**
  from aldehydes, 739
  from ketones, 739
Diol, **282**
1,2-Diol, cleavage of, 285–286
  from alkenes, 282–284
  from epoxides, 282–283, 686–687
  reaction with HIO$_4$, 285–286

Diorganocopper reagent, conjugate carbonyl addition reactions of, 754–755
　reaction with acid chlorides, 833–834
　see also Gilman reagent
Diovan, synthesis of, 359
Dioxane, use of, 685
DiPAMP ligand, amino acid synthesis and, 1055
Diphosphate, as leaving group, 395–396
Dipole moment (μ), 37
　halomethanes, 346
　polar covalent bonds and, 37–38
　table of, 38
Dipole–dipole forces, 60–61
Dipropyl ether, ^{1}H NMR spectrum of, 696
Disaccharide, **1025**–1027
　1→4-link in, 1025–1026
　synthesis of, 1029–1030
Dispersion forces, **61**
　alkanes and, 93
Disrotatory motion, **1218**
Distortionless enhancement by polarization transfer, see DEPT-NMR
Disulfide(s), **692**
　electrostatic potential map of, 79
　from thiols, 692
　hybridization of, 19
　reduction of, 692
　thiols from, 692
Disulfide bridge, peptides and, 1057
Diterpene, **300**
Diterpenoid, **1098**
DMAPP, see Dimethylallyl diphosphate
DMF, see Dimethylformamide
DMSO, see Dimethyl sulfoxide
DMT (dimethoxytrityl ether), DNA synthesis and, 1142
DNA, see Deoxyribonucleic acid
DNA fingerprinting, 1146–1147
　reliability of, 1147
　STR loci and, 1146
Dopamine, molecular model of, 957
Double bond, **13**
　electronic structure of, 14
　length of, 15
　molecular orbitals in, 21
　see also Alkene
　strength of, 14
Double helix (DNA), **1131**–1132
Doublet (NMR), 478
Downfield (NMR), **461**

Drugs, approval procedure for, 213–214
　chiral, 172–173
　origin of, 213

E configuration, 231–**232**
　assignment of, 231–232
E1 reaction, 398, **405**–406
　carbocations and, 405
　deuterium isotope effect and, 406
　kinetics of, 406
　mechanism of, 405
　rate-limiting step in, 406
　stereochemistry of, 406
　Zaitsev's rule and, 406
E1cB reaction, 398, **406**–407
　carbanion intermediate in, 406–407
　mechanism of, 406–407
E2 reaction, 398, **399**–402
　alcohol oxidation and, 646
　anti periplanar geometry of, 401–402
　cyclohexane conformation and, 403–404
　deuterium isotope effect and, 400
　kinetics of, 400
　mechanism of, 400
　menthyl chloride and, 404
　neomenthyl chloride and, 404
　rate law for, 400
　stereochemistry of, 401–402
　Zaitsev's rule and, 403
Easter Island, rapamycin from, 251–252
Eclipsed conformation, ethane and, **95**
　molecular model of, 95
Edman degradation, **1059**–1061
　mechanism of, 1059–1061
Eicosanoid(s), **1096**–1098
　biosynthesis of, 1097–1098
　naming, 1096–1097
Elaidic acid, from vegetable oil, 1091
Elastomer, **1255**
　characteristics of, 1255
　cross-links in, 1255
　T_g of, 1255
Electrocyclic reaction, **1217**–1222
　conrotatory motion in, 1219
　disrotatory motion in, 1218
　examples of, 1217–1218
　HOMO and, 1219–1221
　photochemical, 1221–1222
　stereochemical rules for, 1222
　stereochemistry of, 1219–1222
　thermal, 1219–1220

Electromagnetic radiation, **434**–436
　amplitude of, 435
　characteristics of, 435
　energy of, 435–436
　frequency of, 435
　kinds of, 434
　wavelength of, 435
Electromagnetic spectrum, **434**
　regions in, 434
Electron, lone-pair, **8**
　nonbonding, **8**
Electron configuration, ground state, **5**
　rules for assigning, 5
　table of, 5
Electron movement, curved arrows and, 44–45, 57–58, 197–198
　fishhook arrows and, 186, 291
Electron shell, **4**
Electron-dot structure, **7**
Electronegativity, **35**
　inductive effects and, 36
　polar covalent bonds and, 35–36
　table of, 35
Electrophile, **192**
　characteristics of, 197–198
　curved arrows and, 197–198
　electrostatic potential maps of, 192
　examples of, 192
Electrophilic addition reaction, **194**–196, **237**–238
　carbocation rearrangements in, 249–250
　chirality and, 296–298
　energy diagram of, 206, 208, 238
　Hammond postulate and, 248
　intermediate in, 208
　Markovnikov's rule and, 240–241
　mechanism of, 195–196, 237–238
　regiospecificity of, 240–241
　transition state in, 248
Electrophilic aromatic substitution reaction, 566
　arylamines and, 966–968
　biological example of, 571
　inductive effects in, 583
　kinds of, 566–567
　mechanism of, 568–569
　orientation in, 580–581
　pyridine and, 976
　pyrrole and, 974–975
　resonance effects in, 584
　substituent effects in, 580–581
Electrophoresis, 1052–**1053**
　DNA sequencing and, 1141
Electrospray ionization (ESI) mass spectrometry, 433

Index

Electrostatic potential map, 36
  acetaldehyde, 714
  acetamide, 820, 951
  acetate ion, 42, 52, 55, 784
  acetic acid, 52, 54
  acetic acid dimer, 782
  acetic anhydride, 820
  acetone, 54, 56, 79
  acetone anion, 55
  acetonitrile, 794
  acetyl azide, 859
  acetyl chloride, 820
  acetylene, 317
  acetylide anion, 326
  acid anhydride, 820
  acid chloride, 820
  acyl cation, 578
  adenine, 1132
  alanine, 1045
  alcohol, 78
  alkene, 78, 194
  alkyl halide, 78
  alkyne, 78
  allylic carbocation, 390, 506
  amide, 820
  amine, 79
  amine hydrogen-bonding, 948
  ammonia, 192
  aniline, 953
  anilinium ion, 953
  anisole, 804
  [18]annulene, 553
  arene, 78
  azulene, 558
  benzaldehyde, 583, 730
  benzene, 43, 539, 583
  benzenediazonium ion, 972
  benzoquinone, 653
  benzyl carbocation, 390
  benzyne, 595
  borane, 273
  boron trifluoride, 57, 194
  bromoethane, 194
  bromomethane, 192
  bromonium ion, 266
  1,3-butadiene, 504
  3-buten-2-one, 752
  tert-butyl carbocation, 245
  butyllithium, 357
  carbanion, 326
  carbocation, 245, 288
  carbonyl compound, 79, 192
  carboxylic acid derivatives, 820
  chlorobenzene, 583
  chloromethane, 37, 190, 346
  conjugated diene, 504
  crown ether, 691
  cyclobutadiene, 542
  cycloheptatrienyl cation, 545
  cyclooctatetraene, 542–543
  1,3-cyclopentadiene, 974
  cyclopentadienyl anion, 545
  cytosine, 1132
  dichlorocarbene, 288
  cis-1,2-dichloroethylene, 66
  trans-1,2-dichloroethylene, 66
  Diels–Alder reaction, 510
  dimethyl ether, 57, 677
  dimethyl sulfoxide, 40
  N,N-dimethyltryptamine, 979
  disulfide, 79
  DNA base pairs, 1132
  electrophiles, 192
  enamine, 925
  enol, 873
  enolate ion, 878, 881
  ester, 820
  ether, 78
  ethoxide ion, 784
  ethyl carbocation, 245
  ethylene, 75, 194
  ethylene oxide, 685
  fatty acid carboxylate, 1093
  formaldehyde, 216, 730
  formate ion, 784
  Grignard reagent, 356
  guanine, 1132
  histidine, 1049
  HSO_3^+ ion, 572
  hydrogen bond, 61, 623
  hydronium ion, 192
  hydroxide ion, 52, 192
  imidazole, 60, 547
  isopropyl carbocation, 245
  menthene, 75
  methanethiol, 216
  methanol, 36, 54, 56, 190, 623
  methoxide ion, 55, 627
  methyl acetate, 820
  methyl anion, 326
  methyl carbocation, 245
  methyl thioacetate, 820
  9-methyladenine, 1149
  methylamine, 56, 951
  N-methylguanine, 1149
  methyllithium, 36, 190
  methylmagnesium iodide, 356
  naphthalene, 550
  nitrile, 794
  nitronium ion, 572
  nucleophiles, 192
  1,3-pentadiene, 504
  phenol, 583
  phenoxide ion, 627
  phosphate, 78
  polar covalent bonds and, 36
  propenal, 511
  propenenitrile, 511
  protonated methanol, 190
  purine, 978
  pyridine, 546
  pyrimidine, 546
  pyrrole, 547, 974
  pyrrolidine, 974
  S_N2 reaction, 377
  sulfide, 79
  thioanisole, 804
  thioester, 820
  thiol, 79
  thymine, 1132
  toluene, 585
  trifluoromethylbenzene, 585
  trimethylamine, 949
  2,4,6-trinitrochlorobenzene, 592
  vinylic anion, 326
  vinylic carbocation, 318
  water, 52
  zwitterion, 1045
Elimination reaction, **185**, **397**
  biological examples of, 407
  summary of kinds of, 407
Embden–Meyerhof pathway, 1173–1181
  see also Glycolysis
Enamido acid, amino acids from, 1055
Enamine(s), **736**
  conjugate addition reactions of, 925–926
  electrostatic potential map of, 925
  from aldehydes, 736–739
  from ketones, 736–739
  mechanism of formation of, 738–739
  nucleophilicity of, 924–925
  pH dependence of formation, 739
  reaction with enones, 925–926
  Stork reaction of, 925–926
Enantiomeric excess, **761**
Enantiomers, **143**
  discovery of, 150
  resolution of, 161–163
Enantioselective synthesis, 173, **760–761**
Enantiotopic protons (NMR), **472**
Endergonic, **201**
Endergonic reaction, Hammond postulate and, 247
Endo stereochemistry, Diels–Alder reaction and, **512**
Endothermic, **202**
-*ene*, alkene name suffix, 226
Energy diagram, **206–207**
  activation energy in, 206
  biological reactions and, 209
  electrophilic addition reactions and, 206, 208

Energy diagram—cont'd
  endergonic reactions and, 206–207
  exergonic reactions and, 206–207
  intermediates and, 208
  reaction coordinate in, 206
  transition state in, 206
Energy difference, equilibrium
    position and, 123–124
Enflurane, molecular model of, 147
Enol, 319, **871**
  electrostatic potential map of, 873
  from acid bromides, 877
  from aldehydes, 871–872
  from ketones, 871–872
  mechanism of formation of, 871–872
  reactivity of, 873–874
Enolate ion, **872**
  alkylation of, 882–889
  electrostatic potential map of, 878, 881
  halogenation of, 881–882
  reaction with Br_2, 881–882
  reactivity of, 881–882
  resonance in, 878
  stability of, 878
Enone, conjugate carbonyl addition reactions of, 751–755
  from aldehydes, 908–909
  from aldol reaction, 908–909
  from ketones, 908–909
  IR spectroscopy of, 757
  Michael reactions of, 922–923
  molecular orbitals of, 909
  reaction with amines, 753
  reaction with enamines, 925–926
  reaction with Gilman reagents, 754–755
  reaction with water, 753
  synthesis of, 876
Enthalpy change (ΔH), 202
  explanation of, 202
Entropy change (ΔS), 202
  explanation of, 202
Enzyme, 210, **1068**–1070
  active site in, 210–211
  classification of, 1070
  naming, 1070
  Protein Data Bank and, 1076–1077
  rate acceleration of, 1068–1069
  specificity of, 1069
  substrate of, 1069
  turnover number of, 1069
  X-ray crystal structures of, 447
Enzyme–substrate complex, 1069
Ephedrine, structure of, 63
Epichlorohydrin, epoxy resins from, 697–698
Epimer, **157**–158

Epoxidation, enantioselective method of, 761
Epoxide(s), **281**
  acid-catalyzed cleavage of, 282–283, 686–688
  base-catalyzed cleavage of, 689–690
  1,2-diols from, 282–283, 686–687
  from alkenes, 281–282
  from halohydrins, 282
  mechanism of acid-catalyzed cleavage of, 282–283, 686–688
  NMR spectroscopy of, 696
  reaction with acids, 282–283, 686–688
  reaction with amines, 689–690
  reaction with base, 689–690
  reaction with Grignard reagents, 690
  reaction with HX, 687–688
  reaction with $LiAlH_4$, 705
  reduction of, 705
  S_N2 reactions of, 383
  synthesis of, 281–282
Epoxy resin, preparation of, 697–698
  prepolymer for, 697–698
1,2-Epoxypropane, ^1H NMR spectrum of, 696
Equatorial bonds (cyclohexane), **120**
  drawing, 121
Equilibrium constant (K_{eq}), **200**
  free-energy change and, 201
Equilibrium position, energy difference and, 123–124
Ergocalciferol, structure of, 1232
Ergosterol, UV absorption of, 532
  vitamin D from, 1232
Erythronolide B, structure of, 176
Erythrose, configuration of, 1009
Eschenmoser, Albert, 333
Essential amino acid, **1049**
Essential carbohydrate, 1023–1025
  function of, 1024
Essential oil, **299**
Ester(s), **814**
  acid-catalyzed hydrolysis of, 839
  alcohols from, 632–633, 840–842
  aldehydes from, 725–726, 841
  alkylation of, 888
  amides from, 840
  aminolysis of, 840
  base-catalyzed hydrolysis of, 838
  β-keto esters from, 919–920
  carbonyl condensation reactions of, 915–917
  carboxylic acids from, 837–840
  electrostatic potential map of, 820
  from acid anhydrides, 835
  from acid chlorides, 831

  from alcohols, 644
  from carboxylates, 824
  from carboxylic acids, 824–826
  hydrolysis of, 837–840
  IR spectroscopy of, 444, 851
  mechanism of hydrolysis of, 838–839
  mechanism of reduction of, 840–841
  naming, 816
  NMR spectroscopy of, 852
  nucleophilic acyl substitution reactions of, 837–842
  occurrence of, 836
  partial reduction of, 841
  pK_a of, 879
  reaction with amines, 840
  reaction with DIBAH, 841
  reaction with Grignard reagents, 635, 842
  reaction with LDA, 888
  reaction with $LiAlH_4$, 632–633, 840–841
  reduction of, 632–633, 840–841
  saponification of, 838
  uses of, 837
Ester group, directing effect of, 588–589
Estradiol, structure and function of, 1110
Estrogen, **1110**
  function of, 1110
Estrone, conformation of, 131
  structure and function of, 1110
  synthesis of, 927–928, 1240
Ethane, bond angles in, 12
  bond lengths in, 12
  bond rotation in, 94–95
  bond strengths in, 12
  conformations of, 94–95
  eclipsed conformation of, 95
  molecular model of, 13, 81
  rotational barrier in, 95
  sp^3 hybrid orbitals in, 12–13
  staggered conformation of, 95
  structure of, 12–13
  torsional strain in, 95
Ethanol, history of, 658
  industrial synthesis of, 270, 620–621
  IR spectrum of, 436
  LD_{50} of, 24
  metabolism of, 658
  physiological effects of, 658
  pK_a of, 51, 625
  toxicity of, 658
Ethene, see Ethylene
Ether(s), **676**
  alcohols from, 681–682
  alkyl halides from, 681–682

boiling points of, 677
bond angles in, 677
Claisen rearrangement of, 683–684
cleavage of, 681–682
electrostatic potential map of, 78
from alcohols, 678–680
from alkenes, 680
from alkyl halides, 678–679
IR spectroscopy of, 695
naming, 677
NMR spectroscopy of, 696
peroxides from, 678
properties of, 677–678
reaction with HBr, 681–682
synthesis of, 678–680
uses of, 676
Ethoxide ion, electrostatic potential map of, 784
Ethyl acetate, ethyl acetoacetate from, 916
^{1}H NMR spectrum of, 852
Ethyl acetoacetate, mixed aldol reactions of, 913
see Acetoacetic ester
Ethyl acrylate, ^{13}C NMR absorptions in, 466
Ethyl alcohol, see Ethanol
Ethyl benzoate, mixed Claisen condensation reaction of, 917–918
^{13}C NMR spectrum of, 492
Ethyl carbocation, electrostatic potential map of, 245
Ethyl formate, mixed Claisen condensation reaction of, 917–918
Ethyl group, **84**
Ethylcyclopentane, mass spectrum of, 429
Ethylene, bond angles in, 14
bond lengths in, 14
bond strengths in, 14–15
Ethylene, electrostatic potential maps of, 194
electrostatic potential map of, 75
ethanol from, 270
heat of hydrogenation of, 236
hormonal activity of, 222
industrial preparation of, 223
molecular model of, 14
molecular orbitals of, 21, 1215
pK_a of, 326
polymerization of, 290–292
reaction with HBr, 194–196
sp^2 hybrid orbitals in, 13–14
structure of, 13–15
uses of, 223
Ethylene dichloride, synthesis of, 264–265

Ethylene glycol, acetals from, 745
manufacture of, 282
uses of, 282
Ethylene oxide, electrostatic potential map of, 685
industrial synthesis of, 685
uses of, 685
2-Ethyl-1-hexanol, synthesis of, 910–911
N-Ethylpropylamine, mass spectrum of, 982
Ethynylestradiol, structure and function of, 1111
von Euler, Ulf Svante, 1095
Exergonic, **201**
Exergonic reaction, Hammond postulate and, 247
Exo stereochemistry, Diels–Alder reaction and, **512**
Exon (DNA), **1136**
Exothermic, **202**

FAD, see Flavin adenine dinucleotide, 1163
FADH$_2$, Flavin adenine dinucleotide (reduced), 1163
Faraday, Michael, 536
Farnesyl diphosphate, biosynthesis of, 1105
Fat(s), **1089**–1090
catabolism of, 1158–1167
composition of, 1090
hydrolysis of, 839–840, 1158–1161
saponification of, 1092
Fatty acid(s), **1089**
acetyl CoA from, 1162–1167
anabolism of, 1167–1173
biosynthesis of, 1167–1173
catabolism of, 1162–1167
melting point trends in, 1091
number of, 1089
polyunsaturated, **1089**
table of, 1090
Favorskii reaction, 901
Fehling's test, 1020
Fen–Phen, structure of, 962
Fiber, **1255**
crystallites in, 1255
manufacture of, 1255
Fieser, Louis F., 1010
Fingerprint region (IR), **439**
First-order reaction, **386**
Fischer, Emil, 1002, 1022
Fischer esterification reaction, **824**–826
mechanism of, 824–825
Fischer projection, **1002**–1005
carbohydrates and, 1004–1005
conventions for, 1003

D sugars, 1007
L, sugars, 1007
rotation of, 1003–1004
R,S configuration of, 1004
Fishhook arrow, radical reactions and, 186, 291
Flavin adenine dinucleotide, biological hydroxylation and, 573–574
mechanism of, 1163–1164
structure and function of, 1072, 1163–1164
Flavin adenine dinucleotide (reduced), structure of, 1163
Fleming, Alexander, 853
Flexibilene, structure of, 1123
Florey, Howard, 853
Fluorenylmethyloxycarbonyl (Fmoc) protecting group, 1062–1063
Fluorination (aromatic), 570
Fluoromethane, bond length of, 346
bond strength of, 346
dipole moment of, 346
Fluoxetine, molecular model of, 170
stereochemistry of, 170
synthesis of, 709
Fmoc (fluorenylmethyloxycarbonyl amide), 1062–1063
amino acid derivatives of, 1062–1063
Food, catabolism of, 1155
Food and Drug Administration (FDA), 213
Formal charge, 39–**41**
calculation of, 40–41
summary table of, 41
Formaldehyde, dipole moment of, 38
electrostatic potential map of, 216, 730
hydrate of, 731
industrial synthesis of, 722–723
mixed aldol reactions of, 912
reaction with Grignard reagents, 635
uses of, 722
Formate ion, bond lengths in, 784
electrostatic potential map of, 784
Formic acid, bond lengths in, 784
pK_a of, 783
Formyl group, **724**
p-Formylbenzoic acid, pK_a of, 787
Fourier-transform NMR spectroscopy (FT-NMR), 463–464
Fractional crystallization, resolution and, 161
Fragmentation (mass spectrum), 426–429
Free radical, **187**
see also Radical

Free-energy change (ΔG), 201
    standard, **201**
Fremy's salt, 653
Frequency (ν), 435
Friedel–Crafts acylation reaction, 577–578
    acyl cations in, 577–578
    arylamines and, 967
    mechanism of, 577–578
Friedel–Crafts alkylation reaction, **575**–577
    arylamines and, 967
    biological example of, 578–579
    limitations of, 575–576
    mechanism of, 575
    polyalkylation in, 576
    rearrangements in, 576–577
Frontier orbitals, **1216**
Fructose, anomers of, 1012–1013
    furanose form of, 1012–1013
    sweetness of, 1033
Fructose-1,6-bisphosphate aldolase, X-ray crystal structure of, 447
Fucose, biosynthesis of, 1043
    structure of, 1024
Fukui, Kenichi, 1216
Fumaric acid, structure of, 780
Functional group, **74**–79
    carbonyl compounds and, 79
    electronegative atoms in, 78–79
    importance of, 74–75
    IR spectroscopy of, 439–444
    multiple bonds in, 75, 78
    polarity patterns of, 191
    table of, 76–77
Functional RNAs, **1135**
Furan, industrial synthesis of, 973
Furanose, **1012**–1013
    fructose and, 1012–1013

γ, see Gamma
Gabriel amine synthesis, **957**
Galactose, biosynthesis of, 1040
    configuration of, 1009
    Wohl degradation of, 1023
Gamma-aminobutyric acid, structure of, 1048
Gamma rays, electromagnetic spectrum and, 434
Ganciclovir, structure and function of, 1153
Gasoline, manufacture of, 100–101
    octane number of, 100–101
Gatterman–Koch reaction, 617
Gauche conformation, **97**
    butane and, 97
    steric strain in, 97
Gel electrophoresis, DNA sequencing and, 1141

Geminal (gem), **731**
Genome, size of in humans, 1135
Gentamicin, structure of, 1031
Geraniol, biosynthesis of, 395–396
Geranyl diphosphate, biosynthesis of, 1105
    monoterpenes from, 1106
Gibbs free-energy change (ΔG), 201
    standard, **201**
    equilibrium constant and, 201
Gilman reagent, **357**
    conjugate carbonyl addition reactions of, 754–755
    organometallic coupling reactions of, 357–358
    reaction with acid chlorides, 833–834
    reaction with alkyl halides, 357–358
    reaction with enones, 754–755
Glass transition temperature (polymers), **1254**
Glucocorticoid, **1111**
Gluconeogenesis, **1191**–1197
    mechanisms in, 1192–1197
    overall result of, 1197
    steps in, 1192–1193
Glucosamine, biosynthesis of, 1040
    structure of, 1031
Glucose, α anomer of, 1012
    anabolism of, 1191–1197
    anomers of, 1012
    β anomer of, 1012
    biosynthesis of, 1191–1197
    catabolism of, 1173–1181
    chair conformation of, 120, 128
    configuration of, 1009
    Fischer projection of, 1005
    from pyruvate, 1191–1197
    glycosides of, 1016–1018
    keto-enol tautomerization of, 1175–1176
    Koenigs–Knorr reaction of, 1017–1018
    molecular model of, 120, 128, 1012
    mutarotation of, 1013
    pentaacetyl ester of, 1015–1016
    pentamethyl ether of, 1016
    pyranose form of, 1011–1012
    pyruvate from, 1173–1181
    reaction with acetic anhydride, 1015–1016
    reaction with ATP, 1158
    reaction with iodomethane, 1016
    sweetness of, 1033
    Williamson ether synthesis with, 1016
Glutamic acid, structure and properties of, 1047

Glutamine, structure and properties of, 1046
Glutaric acid, structure of, 780
Glutathione, function of, 692
    prostaglandin biosynthesis and, 1097–1098
    structure of, 693
Glycal, **1030**
Glycal assembly method, **1030**
(+)-Glyceraldehyde, absolute configuration of, 1006–1007
(−)-Glyceraldehyde, configuration of, 153–154
(R)-Glyceraldehyde, Fischer projection of, 1003
    molecular model of, 1003
Glyceric acid, structure of, 780
Glycerol, catabolism of, 1158–1161
sn-Glycerol 3-phosphate, naming of, 1161
Glycerophospholipid, **1094**
Glycine, structure and properties of, 1046
Glycoconjugate, **1019**
    influenza virus and, 1032
Glycogen, structure and function of, 1029
Glycol, **282**, **686**
Glycolic acid, structure of, 780
Glycolipid, **1019**
Glycolysis, **1173**–1181
    mechanisms in, 1174–1181
    overall result of, 1181
    steps in, 1174–1175
Glycoprotein, **1019**
    biosynthesis of, 1019
Glycoside, **1016**
    occurrence of, 1017
    synthesis of, 1017–1018
Glyptal, structure of, 1260
GPP, see Geranyl diphosphate
Graft copolymer, **1248**
    synthesis of, 1248
Grain alcohol, 620
Green chemistry, **409**–410, **982**–984
    ibuprofen synthesis by, 410
    ionic liquids and, 982–984
    principles of, 409–410
Grignard, François Auguste Victor, 355
Grignard reaction, aldehydes and, 635
    carboxylic acids and, 636
    esters and, 635
    formaldehyde and, 635
    ketones and, 635
    limitations of, 636
    mechanism of, 735
    strategy for, 637

Grignard reagent, **355**
  alkanes from, 356
  carboxylation of, 790
  carboxylic acids from, 790
  electrostatic potential map of, 356
  from alkyl halides, 355–356
  reaction with acids, 356
  reaction with aldehydes, 635, 735
  reaction with carboxylic acids, 636
  reaction with CO_2, 790
  reaction with epoxides, 690
  reaction with esters, 635, 842
  reaction with formaldehyde, 635
  reaction with ketones, 635, 735
  reaction with nitriles, 796
  reaction with oxetanes, 708
Grubbs catalyst, olefin metathesis polymerization and, 1251
Guanine, electrostatic potential map of, 1132
  protection of, 1142–1143
  structure of, 1129
Gulose, configuration of, 1009
Guncotton, 1028
Gutta-percha, structure of, 516

$\Delta H°_{hydrog}$ (heat of hydrogenation), 235
Hagemann's ester, synthesis of, 940
Halo group, directing effect of, 587–588
  inductive effect of, 583
  orienting effect of, 581
  resonance effect of, 584
Haloalkane, see Alkyl halide
Haloform reaction, **882**
Halogen, inductive effect of, 583
  resonance effect of, 584
Halogenation, aldehydes and, 874–876
  alkenes and, 264–266
  alkynes and, 317–318
  aromatic compounds and, 567–571
  biological example of, 267
  carboxylic acids and, 876–877
  ketones and, 874–876
Halohydrin, **267**
  epoxides from, 282
  reaction with base, 282
Halomon, anticancer activity of, 362–363
  biosynthesis of, 267
Haloperoxidase, bromohydrin formation and, 269
Hammond, George Simms, 247
Hammond postulate, **247**–248
  carbocation stability and, 248
  endergonic reactions and, 247

exergonic reactions and, 247
  Markovnikov's rule and, 248
  S_N1 reaction and, 389
Handedness, molecular, 143–146
HDL, heart disease and, 1118
Heart disease, cholesterol and, 1118
  statin drugs and, 1203–1204
Heat of combustion, **115**
Heat of hydrogenation, **235**
  table of, 236
Heat of reaction, **202**
Helicase, DNA replication and, 1133
Hell–Volhard–Zelinskii reaction, 876–877
  amino acid synthesis and, 1053
  mechanism of, 877
Heme, biosynthesis of, 994
  structure of, 973
Hemiacetal, **743**
Hemiketal, 743
Hemithioacetal, **1179**
Henderson–Hasselbalch equation, amines and, 954
  amino acids and, 1050–1051
  biological acids and, **785**–786
Hertz (Hz), **435**
Heterocycle, **546**, **972**
  aromatic, 546–548
  polycyclic, 977–978
Heterocyclic amine, **946**
  basicity of, 950
  names for, 946
Heterolytic bond-breaking, **186**
Hevea brasieliensis, rubber from, 516
Hexachlorophene, synthesis of, 616, 652
Hexamethylphosphoramide, S_N2 reaction and, 384
Hexane, IR spectrum of, 440
  mass spectrum of, 428
1,3,5-Hexatriene, molecular orbitals of, 1216
  UV absorption of, 520
1-Hexene, IR spectrum of, 440
2-Hexene, mass spectrum of, 430
Hexokinase, active site in, 211
  molecular model of, 211
1-Hexyne, IR spectrum of, 440
High-density polyethylene, synthesis of, 1246
High-molecular-weight polyethylene, uses of, 1246
High-pressure liquid chromatography, amino acid analyzer and, 1058–1059
Highest occupied molecular orbital (HOMO), **518**, **1216**
  cycloaddition reactions and, 1224–1225

electrocyclic reactions and, 1219–1221
  UV spectroscopy and, 518
Histamine, structure of, 989
Histidine, electrostatic potential map of, 1049
  structure and properties of, 1047
HMG-CoA reductase, active site in, 1204
  statin drugs and, 1203–1204
HMPA, *see*, Hexamethylphosphoramide
Hoffmann, Roald, 1215
Hoffmann-LaRoche Co., vitamin C synthesis and, 800–801
Hofmann elimination reaction, **964**–965
  biological example of, 965
  mechanism of, 964–965
  molecular model of, 965
  regiochemistry of, 964–965
  Zaitsev's rule and, 964–965
Hofmann rearrangement, **960**–962
  mechanism of, 960, 962
HOMO, see Highest occupied molecular orbital
Homocysteine, structure of, 1048
Homolytic bond-breaking, **186**
Homopolymer, **1246**
Homotopic protons (NMR), **472**
Honey, sugars in, 1027
Hormone, **1110**
  sex, 1110
Hückel, Erich, 541
Hückel $4n + 2$ rule, 541
  cyclobutadiene and, 542
  cycloheptatrienyl cation and, 544–545
  cyclooctatetraene and, 542–543
  cyclopentadienyl anion and, 544–545
  explanation of, 543
  imidazole and, 547
  molecular orbitals and, 543
  pyridine and, 546–547
  pyrimidine and, 546–547
  pyrrole and, 547
Hughes, Edward Davies, 376
Human fat, composition of, 1090
Human genome, size of, 1135, 1142
Humulene, structure of, 299
Hund's rule, **5**
sp Hybrid orbitals, **16**–17
sp^2 Hybrid orbitals, **14**
sp^3 Hybrid orbitals, **11**–13
Hydrate, from aldehydes, 731–732
  from ketones, 731–732
Hydration, alkene, **269**–274
  alkyne, 319–321

Hydrazine, reaction with aldehydes, 741–742
  reaction with ketones, 741–742
Hydride shift, **249**–250
Hydroboration, alkene, **272**–274
  alkyne, 321–322
  mechanism of, 274
  regiochemistry of, 273–274, 484–485
  stereochemistry of, 273–274
Hydrocarbon, **81**
  acidity of, 326
  saturated, 81
  unsaturated, 224
Hydrochloric acid, pK_a of, 51
Hydrocortisone, conformation of, 136
  structure and function of, 1111
Hydrogen bond, **61**–62
  alcohols and, 623
  amines and, 948
  biological consequences of, 62
  carboxylic acids and, 782
  DNA base pairs and, 1131–1132
  electrostatic potential map of, 61, 623
Hydrogen molecule, bond length in, 10
  bond strength in, 10
  molecular orbitals in, 20
Hydrogen peroxide, reaction with organoboranes, 272–273
[1,5] Hydrogen shift, 1228
Hydrogenation, **276**
  alkenes, 276–280
  alkynes, 322–323
  aromatic compounds, 599
  biological example of, 280
  catalysts for, 277
  mechanism of, 277–278
  stereochemistry of, 277–278
  steric hindrance and, 278–279
  trans fatty acids from, 279–280
  vegetable oil, 1091
Hydrogenolysis, benzyl esters and, 1062
Hydrolase, 1070
Hydrolysis, **821**
  amides, 843–844
  biological, 839–840, 844
  esters, 837–840
  fats, 839–840
  nitriles, 795–796
  proteins, 844
Hydronium ion, electrostatic potential map of, 192
Hydrophilic, **62**
Hydrophobic, **62**
Hydroquinone, **653**
  from quinones, 653

Hydroxide ion, electrostatic potential map of, 52, 192
Hydroxyacetic acid, pK_a of, 783
p-Hydroxybenzaldehyde, pK_a of, 628
p-Hydroxybenzoic acid, pK_a of, 787
Hydroxyl group, directing effect of, 586–587
  inductive effect of, 583
  orienting effect of, 581
  resonance effect of, 584
Hydroxylation (alkene), **282**–284
Hydroxylation (aromatic), 573–574
Hyperconjugation, **235**–236
  alkenes and, 235–236
  carbocation stability and, 245

Ibuprofen, chirality and, 173
  green synthesis of, 410
  molecular model of, 66, 173
  NSAIDs and, 555
  stereochemistry of, 172–173
  synthesis of, 790
Idose, configuration of, 1009
Imidazole, aromaticity of, 547
  basicity of, 950, 975
  electrostatic potential map of, 60, 547
  Hückel $4n + 2$ rule and, 547
Imide(s), **957**
  hydrolysis of, 957
Imine(s), **736**
  from aldehydes, 736–739
  from ketones, 736–739
  mechanism of formation of, 736–737
  pH dependence of formation, 739
  see also Schiff base
IND, *see* Investigational new drug, 213
Indole, aromaticity of, 550
  electrophilic substitution reaction of, 978
  structure of, 946
Indolmycin, biosynthesis of, 889–890
Inductive effect, **36**, **583**
  alcohol acidity and, 625
  carbocation stability and, 245
  carboxylic acid strength and, 786–787
  electronegativity and, 36
  electrophilic aromatic substitution and, 583
  polar covalent bonds and, 36
Influenza virus, classification of, 1032
  glycoconjugates and, 1032
Infrared radiation, electromagnetic spectrum and, 434
  energy of, 437
  frequencies of, 437
  wavelengths of, 437

Infrared spectroscopy, **437**–444
  acid anhydrides, 851
  acid chlorides, 851
  alcohols, 443, 654
  aldehydes, 444, 756–757
  alkanes, 442
  alkenes, 442–443
  alkynes, 443
  amides, 851
  amines, 444, 979
  aromatic compounds, 443, 551
  bond stretching in, 438
  carbonyl compounds, 444
  carboxylic acid derivatives, 851
  carboxylic acids, 797–798
  esters, 444, 851
  ethers, 695
  explanation of, 438
  fingerprint region in, 439–441
  ketones, 444, 756–757
  lactones, 851
  molecular motions in, 438
  nitriles, 798
  phenols, 655
  regions in, 440–441
  table of absorptions in, 439–441
  vibrations in, 438
Infrared spectrum, benzaldehyde, 756
  butanoic acid, 798
  cyclohexane, 451
  cyclohexanol, 654
  cyclohexanone, 756
  cyclohexene, 451
  cyclohexylamine, 979
  diethyl ether, 695
  ethanol, 436
  hexane, 440
  1-hexene, 440
  1-hexyne, 440
  phenol, 655
  phenylacetaldehyde, 445
  phenylacetylene, 446
  toluene, 551
Ingold, Christopher Kelk, 376
Initiation step (radical reaction), **188**
Insulin, disulfide bridges in, 1057
  structure of, 1057
Integration (NMR), **476**
Intermediate, *See* Reaction intermediate
Intoxilyzer test, 658
Intramolecular aldol reaction, 913–915
  mechanism of, 914–915
Intramolecular Claisen condensation, *see* Dieckmann cyclization
Intron (DNA), **1136**
Invert sugar, **1027**
Investigational new drug (IND), 213

Iodination (aromatic), 570–571
Iodoform reaction, 882
Iodomethane, bond length of, 346
　　bond strength of, 346
　　dipole moment of, 346
Ion pair, 388
　　S_N1 reaction and, 388
Ion-exchange chromatography,
　　　amino acid analyzer and,
　　　1058–1059
Ionic bond, 7
Ionic liquids, green chemistry and,
　　982–984
　　properties of, 983
　　structures of, 982–983
IPP, see Isopentenyl diphosphate
IR, see Infrared
Iron, reaction with nitroarenes, 955
Iron(III) bromide, aromatic
　　　bromination and, 568
Iron sulfate, LD_{50} of, 24
Isoamyl group, 90
Isobutane, molecular model of, 81
Isobutyl group, 85
Isobutylene, polymerization of, 1243
Isoelectric point (pI), 1051–1052
　　calculation of, 1052
　　table of, 1046–1047
Isoleucine, metabolism of, 940
　　molecular model of, 159
　　structure and properties of, 1046
Isomerase, 1070
Isomers, 82
　　alkanes, 81–82
　　cis–trans cycloalkanes, 113
　　cis–trans alkenes, 229–230
　　conformational, 94
　　constitutional, 82
　　diastereomers and, 157
　　enantiomers and, 143–144
　　epimers and, 157–158
　　kinds of, 164–165
　　review of, 164–165
　　stereoisomers, 113
Isopentenyl diphosphate,
　　　biosynthesis of, 1099–1103
　　geraniol biosynthesis and,
　　　395–396
　　isomerization of, 1103–1105
　　terpenoids from, 1103–1106
Isoprene, heat of hydrogenation of,
　　503
　　industrial synthesis of, 501
　　structure of, 228
　　UV absorption of, 520
Isoprene rule, terpenes and, 299–300
Isopropyl carbocation, electrostatic
　　　potential map of, 245
Isopropyl group, 85

Isoquinoline, aromaticity of, 550
　　electrophilic substitution reaction
　　　of, 978
Isotactic polymer, 1245
Isotope, 3
IUPAC nomenclature, 87
　　new system, 227–228
　　old system, 227

J, see Coupling constant, 478

K_a (acidity constant), 50
K_b (basicity constant), 949
K_{eq} (equilibrium constant), 200
Kekulé, Friedrich August, 6
Kekulé structure, 7
Keratin, α helix in, 1066–1067
Kerosene, composition of, 100
Ketal, 742
　　see also Acetal
Keto–enol tautomerism, 319,
　　871–872
Ketone(s), 722
　　acetals from, 742–744
　　acidity of, 877–880
　　alcohols from, 630–631, 734–735
　　aldol reaction of, 906
　　alkanes from, 741–742
　　alkenes from, 746–748
　　alkylation of, 887–890
　　α cleavage of, 432, 759
　　amines from, 958–959
　　biological reduction of, 631–632,
　　　750–751
　　bromination of, 874–876
　　carbonyl condensation reactions
　　　of, 906
　　common names of, 724
　　cyanohydrins from, 733
　　2,4-dinitrophenylhydrazones
　　　from, 739
　　enamines from, 736–739
　　enols of, 871–872
　　enones from, 908–909
　　from acetals, 743–744
　　from acetoacetic ester, 885–886
　　from acid chlorides, 833–834
　　from alcohols, 645–646
　　from alkenes, 284–286
　　from alkynes, 319–321
　　from nitriles, 796
　　hydrates of, 731–732
　　imines from, 736–739
　　IR spectroscopy of, 444,
　　　756–757
　　mass spectrometry of, 431–432,
　　　758–759
　　McLafferty rearrangement of, 431,
　　　758

　　mechanism of hydration of,
　　　731–732
　　mechanism of reduction of, 734
　　naming, 724
　　NMR spectroscopy of, 757–758
　　oxidation of, 728
　　oximes from, 737–738
　　pK_a of, 879
　　protecting groups for, 745
　　reaction with alcohols, 742–744
　　reaction with amines, 736–739
　　reaction with Br_2, 874–876
　　reaction with
　　　2,4-dinitrophenylhydrazine,
　　　739
　　reaction with Grignard reagents,
　　　635, 735
　　reaction with HCN, 733
　　reaction with H_2O, 731–732
　　reaction with HX, 732–733
　　reaction with hydrazine, 741–742
　　reaction with $KMnO_4$, 728
　　reaction with LDA, 888
　　reaction with $LiAlH_4$, 630, 734
　　reaction with lithium
　　　diisopropylamide, 878
　　reaction with $NaBH_4$, 630, 734
　　reaction with NH_2OH, 737–738
　　reactivity versus aldehydes,
　　　729–730
　　reduction of, 630–631, 734
　　reductive amination of, 958–959
　　Wittig reaction of, 746–748
　　Wolff–Kishner reaction of, 741–742
Ketone bodies, origin of, 1209
Ketose, 1002
Kiliani, Heinrich, 1022
Kiliani–Fischer synthesis, 1022
Kimball, George, 265
Kinetic control, 508–509
　　1,4-addition reactions and,
　　　508–509
Kinetics, 375
　　E1 reaction and, 406
　　E2 reaction and, 400
　　S_N1 reaction and, 386–387
　　S_N2 reaction and, 375–376
Knoevenagel reaction, 941
Knowles, William S., 760, 1055
Kodel, structure of, 1259
Koenigs–Knorr reaction, 1017–1018
　　mechanism of, 1018
　　neighboring-group effect in, 1018
Krebs, Hans Adolf, 1185
Krebs cycle, see Citric acid cycle

L Sugar, 1007
　　Fischer projections of, 1007
Labetalol, synthesis of, 948

Laboratory reaction, comparison with biological reaction, 210–212
Lactam(s), **845**
  amines from, 845
  reaction with LiAlH$_4$, 845
Lactic acid, configuration of, 153
  enantiomers of, 143
  molecular model of, 145
  resolution of, 162–163
  structure of, 780
Lactone(s), **837**
  alkylation of, 888
  IR spectroscopy of, 851
  reaction with LDA, 888
Lactose, molecular model of, 1027
  occurrence of, 1027
  structure of, 1027
  sweetness of, 1033
Lagging strand, DNA replication and, **1135**
Lanosterol, biosynthesis of, 1112–1117
  carbocation rearrangements and, 1115, 1117
  structure of, 300
Lapworth, Arthur, 733
Lard, composition of, 1090
Latex, rubber from, 516
Laurene, synthesis of, 902
Lauric acid, structure of, 1090
LD$_{50}$, **24**
  table of, 24
LDA, *see* Lithium diisopropylamide
LDL, heart disease and, 1118
Le Bel, Joseph Achille, 6
Leading strand, DNA replication and, **1135**
Leaving group, **382**
  biological reactions and, 395–396
  S$_N$1 reaction and, 391–392
  S$_N$2 reactions and, 382–383
Leucine, biosynthesis of, 941, 1213
  metabolism of, 940
  structure and properties of, 1046
Leukotriene E$_4$, structure of, 1096
Leuprolide, structure of, 1083
Levorotatory, **148**
Lewis, Gilbert Newton, 7
Lewis acid, **56–58**
  examples of, 58
  reactions of, 57–58
Lewis base, **56–59**
  examples of, 59
  reactions of, 58–59
Lewis structure, **7**
  resonance and, 42–43
Lewis Y hexasaccharide, synthesis of, 1030

Lexan, structure and uses of, 849–850, 1249
Lidocaine, molecular model of, 102
  structure of, 64
Ligase, 1070
Light, plane-polarized, 147–148
  speed of, 435
Limonene, biosynthesis of, 261, 1106
  biosynthesis of, 1106
  enantiomers of, 170
  molecular model of, 170
  odor of enantiomers of, 170
Lindlar catalyst, **323**
Line-bond structure, **7**
1→4-Link, **1025**
Linoleic acid, structure of, 1090
Linolenic acid, molecular model of, 1091
  structure of, 1090
Lipase, mechanism of, 1158–1161
Lipid, **1088**
  classification of, 1088
Lipid bilayer, **1095**
Lipitor, structure of, 1, 534
  statin drugs and, 1203–1204
Lipoamide, structure and function of, 1184
Lipoic acid, structure and function of, 1073, 1184
Lipoprotein, heart disease and, 1118
Lithium, reaction with alkynes, 323–325
Lithium aluminum hydride, reaction with carboxylic acids, 632–633
  reaction with esters, 632–633
  reaction with ketones and aldehydes, 630
Lithium diisopropylamide (LDA), formation of, 878
  properties of, 878
  reaction with cyclohexanone, 878
  reaction with esters, 888
  reaction with ketones, 878, 888
  reaction with lactones, 888
  reaction with nitriles, 888
Lithium diorganocopper reagent, *see* Gilman reagent
Lithocholic acid, structure of, 1109
Locant, IUPAC naming and, 87–**88**
Lone-pair electrons, 8
Loratadine, structure of, 255, 570
Lotaustralin, structure of, 793
Lovastatin, biosynthesis of, 515
  statin drugs and, 1203–1204
  structure of, 515
Low-density polyethylene, synthesis of, 1246
Lowest unoccupied molecular orbital (LUMO), **518**, **1216**

cycloaddition reactions and, 1224–1225
LUMO, *see* Lowest unoccupied molecular orbital
Lyase, 1070
Lycopene, structure of, 501
Lysergic acid diethylamide, structure of, 989
Lysine, structure and properties of, 1047
Lysozyme, MALDI-TOF mass spectrum of, 433–434
  p*I* of, 1052
Lyxose, configuration of, 1009

Magnetic field, NMR spectroscopy and, 457–458
Magnetic resonance imaging, **486**
  uses of, 486
Major groove (DNA), 1132
MALDI mass spectrometry, 433
Maleic acid, structure of, 780
Malic acid, structure of, 780
  Walden inversion of, 372–373
Malonic ester, carboxylic acids from, 883–884
  decarboxylation of, 883–884
  pK_a of, 879
Malonic ester synthesis, **883–884**
  intramolecular, 884
Maltose, 1→4-α-link in, 1026
  molecular model of, 1026
  mutarotation of, 1026
  structure of, 1026
Manicone, synthesis of, 834
Mannich reaction, 943
Mannose, biosynthesis of, 1040
  chair conformation of, 128
  configuration of, 1009
  molecular model of, 128
Marcaine, structure of, 64
Margarine, manufacture of, 279–280, 1091
Markovnikov, Vladimir Vassilyevich, 240
Markovnikov's rule, **240–241**
  alkene additions and, 240–241
  alkyne additions and, 317
  carbocation stability and, 241, 243–245
  Hammond postulate and, 248
  hydroboration and, 273–274
  oxymercuration and, 271
Mass number (*A*), **3**
Mass spectrometer, double-focusing, 427
  exact mass measurement in, 427
  kinds of, 425
  operation of, 425–426

Mass spectrometry (MS), **424**
  alcohols, 431, 657
  aldehydes, 431–432, 758–759
  alkanes, 428–429
  α cleavage of alcohols in, 431
  α cleavage of amines in, 431
  amines, 431, 981–982
  base peak in, 426
  biological, 433–434
  carbonyl compounds, 431–432
  cation radicals in, 425–426
  dehydration of alcohols in, 431
  electron-impact ionization in, 425–426
  electrospray ionization in, 433
  fragmentation in, 426–429
  ketones and, 431–432, 758–759
  MALDI ionization in, 433
  McLafferty rearrangement in, 431, 758
  molecular ion in, 426
  nitrogen rule and, 981–982
  parent peak in, 426
  soft ionization in, 427
  time-of-flight, 433
Mass spectrum, **426**
  1-butanol, 657
  computer matching of, 428
  2,2-dimethylpropane, 427
  ethylcyclopentane, 429
  N-ethylpropylamine, 982
  hexane, 428
  2-hexene, 430
  interpretation of, 426–429
  lysozyme, 434
  methylcyclohexane, 429
  5-methyl-2-hexanone, 759
  2-methylpentane, 450
  2-methyl-2-pentanol, 432
  2-methyl-2-pentene, 430
  propane, 426
Maxam–Gilbert DNA sequencing, 1140
McLafferty rearrangement, **431, 758**
Mechanism (reaction), **186**
  acetal formation, 743–744
  acetylide alkylation, 327
  acid chloride formation with $SOCl_2$, 823–824
  acid-catalyzed epoxide cleavage, 282–283, 686–688
  acid-catalyzed ester hydrolysis, 839
  alcohol dehydration with acid, 641–642
  alcohol dehydration with $POCl_3$, 642–643
  alcohol oxidation, 646
  aldehyde hydration, 731–732

aldehyde oxidation, 727
aldehyde reduction, 734
aldol reaction, 905–906
aldolase catalyzed reactions, 928–929, 1177–1178
alkane chlorination, 347–348
alkene epoxidation, 281–282
alkene halogenation, 265–266
alkene hydration, 270
alkene polymerization, 291–292
alkoxymercuration, 680
alkylbenzene bromination, 597–598
alkyne addition reactions, 317–318
alkyne hydration, 319–320
alkyne reduction with Li/NH_3, 324–325
α-bromination of ketones, 874–876
α-substitution reaction, 874
allylic bromination, 350–351
amide formation with DCC, 826–827
amide hydrolysis, 843–844
amide reduction, 845
amino acid transamination, 1198–1201
aromatic bromination, 568–569
aromatic chlorination, 570
aromatic fluorination, 570
aromatic iodination, 570–571
aromatic nitration, 571–572
aromatic sulfonation, 572
base-catalyzed epoxide cleavage, 689
base-catalyzed ester hydrolysis, 838
β-oxidation pathway, 1162–1166
biological hydroxylation, 573–574
biotin-mediated carboxylation, 1170
bromohydrin formation, 268
bromonium ion formation, 265
Cannizzaro reaction, 750
carbonyl condensation reaction, 904–905
citrate synthase, 1074–1075
Claisen condensation reaction, 915–916
Claisen rearrangement, 683–684
conjugate carbonyl addition reaction, 752
Curtius rearrangement, 962
cyanohydrin formation, 733
dichlorocarbene formation, 287
Dieckmann cyclization reaction, 919–920
Diels–Alder reaction, 510

E1 reaction, 405
E1cB reaction, 406–407
E2 reaction, 400
Edman degradation, 1059–1061
electrophilic addition reaction, 195–196, 237–238
electrophilic aromatic substitution, 568–569
enamine formation, 738–739
enol formation, 871–872
ester hydrolysis, 838–839
ester reduction, 840–841
FAD reactions, 1163–1164
fat catabolism, 1162–1166
fat hydrolysis, 1158–1161
Fischer esterification reaction, 824–825
Friedel–Crafts acylation reaction, 577–578
Friedel–Crafts alkylation reaction, 575
glycolysis, 1173–1181
Grignard carboxylation, 790
Grignard reaction, 735
Hell–Volhard–Zelinskii reaction, 877
Hofmann elimination reaction, 964–965
Hofmann rearrangement, 960, 962
hydroboration, 274
hydrogenation, 277–278
imine formation, 736–737
intramolecular aldol reaction, 914–915
isopentenyl diphosphate biosynthesis, 1099–1103
ketone hydration, 731–732
ketone reduction, 734
Koenigs–Knorr reaction, 1018
Michael reaction, 921–922
mutarotation, 1013
nitrile hydrolysis, 795–796
nucleophilic acyl substitution reaction, 819
nucleophilic addition reaction, 728
nucleophilic aromatic substitution reaction, 592–593
olefin metathesis polymerization, 1251
diorganocopper conjugate addition, 755
organometallic coupling reaction, 358–359
oxidative decarboxylation, 1181–1185
oxymercuration, 271–272
phenol from cumene, 650–651

I-24 Index

Mechanism (reaction)—cont'd
  polar reactions, 190–193
  prostaglandin biosynthesis, 294–295
  radical reactions, 187–189
  reductive amination, 958
  Robinson annulation reaction, 927
  Sandmeyer reaction, 970
  saponification, 838
  S_N1 reaction, 386–387
  S_N2 reaction, 376–377
  Stork enamine reaction, 925
  Suzuki–Miyaura reaction, 359
  Williamson ether synthesis, 678–679
  Wittig reaction, 746–747
  Wolff–Kishner reaction, 741–742
Meerwein–Ponndorf–Verley reaction, 772
Meerwein's reagent, 707
Melmac, structure of, 1260
Melt transition temperature (polymers), **1254**
Menthene, electrostatic potential map of, 75
  functional groups in, 75
Menthol, chirality of, 147
  molecular model of, 118
  structure of, 118
Menthyl chloride, E1 reaction of, 406
  E2 reaction of, 404
Mepivacaine, structure of, 64
Mercapto group, **691**
Mercuric trifluoroacetate, alkoxymercuration with, 680
Mercurinium ion, **271**
Merrifield, Robert Bruce, 1064
Merrifield solid-phase peptide synthesis, 1064–1066
Meso compound, **159**
  plane of symmetry in, 159–160
Messenger RNA, **1135**
  codons in, 1137–1138
  translation of, 1137–1139
Mestranol, structure of, 342
Meta (*m*) prefix, **537**
Meta-directing group, **581**
Metabolism, **1154**
Methacrylic acid, structure of, 780
Methamphetamine, synthesis of, 994
Methandrostenolone, structure and function of, 1111
Methane, bond angles in, 12
  bond lengths in, 12
  bond strengths in, 12
  chlorination of, 347–348
  molecular model of, 12, 81
  pK_a of, 326

  reaction with Cl_2, 187–188
  sp^3 hybrid orbitals in, 11–12
  structure of, 12
Methanethiol, bond angles in, 18, 19
  dipole moment of, 38
  electrostatic potential map of, 36, 54, 56, 190, 216, 623
  industrial synthesis of, 620
  molecular model of, 18, 19
  pK_a of, 625
  polar covalent bond in, 36
  sp^3 hybrid orbitals in, 18
  toxicity of, 620
  uses of, 620
1,6-Methanonaphthalene, molecular model of, 557
Methionine, *S*-adenosylmethionine from, 694
  biosynthesis of, 770
  molecular model of, 156
  structure and properties of, 1046
Methoxide ion, electrostatic potential map of, 55, 627
p-Methoxybenzoic acid, pK_a of, 787
p-Methoxypropiophenone, ^{1}H NMR spectrum of, 480
Methyl acetate, electrostatic potential map of, 820
  ^{13}C NMR spectrum of, 458
  ^{1}H NMR spectrum of, 458
Methyl α-cyanoacrylate, polymerization of, 1244
Methyl anion, electrostatic potential map of, 326
  stability of, 326
Methyl carbocation, electrostatic potential map of, 245
Methyl 2,2-dimethylpropanoate, ^{1}H NMR spectrum of, 476
Methyl group, **84**
  chiral, 422
  directing effect of, 585–586
  inductive effect of, 583
  orienting effect of, 581
Methyl phosphate, bond angles in, 18
  molecular model of, 18
  structure of, 18
Methyl propanoate, ^{13}C NMR spectrum of, 467
Methyl propyl ether, ^{13}C NMR spectrum of, 696
Methyl salicylate, as flavoring agent, 621
Methyl shift, carbocations and, 250

Methyl thioacetate, electrostatic potential map of, 820
9-Methyladenine, electrostatic potential map of, 1149
Methylamine, bond angles in, 17
  dipole moment of, 38
  electrostatic potential map of, 56, 951
  molecular model of, 18
  sp^3 hybrid orbitals in, 17–18
Methylarbutin, synthesis of, 1017–1018
p-Methylbenzoic acid, pK_a of, 787
2-Methylbutane, molecular model of, 82
2-Methyl-2-butanol, ^{1}H NMR spectrum of, 481
Methylcyclohexane, 1,3-diaxial interactions in, 124–125
  conformations of, 124–125
  mass spectrum of, 429
  molecular model of, 124, 146
1-Methylcyclohexanol, ^{1}H NMR spectrum of, 485
2-Methylcyclohexanone, chirality of, 146
  molecular model of, 146
1-Methylcyclohexene, ^{13}C NMR spectrum of, 471
N-Methylcyclohexylamine, ^{13}C NMR spectrum of, 980
  ^{1}H NMR spectrum of, 980
Methylene group, **228**
Methylerythritol phosphate pathway, terpenoid biosynthesis and, 1099
N-Methylguanine, electrostatic potential map of, 1149
6-Methyl-5-hepten-2-ol, DEPT-NMR spectra of, 468
5-Methyl-2-hexanone, mass spectrum of, 759
Methyllithium, electrostatic potential map of, 36, 190
  polar covalent bond in, 36
Methylmagnesium iodide, electrostatic potential map of, 356
N-Methylmorpholine *N*-oxide, reaction with osmates, 283–284
2-Methylpentane, mass spectrum of, 450
2-Methyl-3-pentanol, mass spectrum of, 432
2-Methyl-2-pentene, mass spectrum of, 430
p-Methylphenol, pK_a of, 625
2-Methylpropane, molecular model of, 81

2-Methyl-1-propanol, ^{13}C NMR spectrum of, 469
2-Methylpropene, heat of hydrogenation of, 236
Metoprolol, synthesis of, 690
Mevacor, structure of, 515
Mevalonate, decarboxylation of, 1102
Mevalonate pathway, terpenoid biosynthesis and, 1099–1103
Micelle, **1092–1093**
Michael reaction, **921**–923
    acceptors in, 922–923
    donors in, 922–923
    mechanism of, 921–922
    partners in, 922–923
    Robinson annulation reactions and, 927
Microwaves, electromagnetic spectrum and, 434
Mineralocorticoid, **1111**
Minor groove (DNA), 1132
Mitomycin C, structure of, 998
Mixed aldol reaction, 912–913
Mixed Claisen condensation reaction, 917–918
Molar absorptivity (UV), **519**
Molecular ion (M$^+$), **426**
Molecular mechanics, 132
Molecular model, acetaminophen, 27
    acetylene, 16
    adenine, 66
    adrenaline, 175
    alanine, 26, 1044
    alanylserine, 1056
    α helix, 1067
    p-aminobenzoic acid, 23
    anisole, 676
    anti periplanar geometry, 401
    arecoline, 80
    aspartame, 27
    aspirin, 15
    β-pleated sheet, 1067
    p-bromoacetophenone, 466
    bromocyclohexane, 122
    butane, 81
    cis-2-butene, 230, 234
    trans-2-butene, 230e, 234
    tert-butyl carbocation, 244
    camphor, 131
    cellobiose, 1026
    chair cyclohexane, 119
    cholesterol, 1109
    cholic acid, 778
    citrate synthase, 1074
    citric acid, 27
    coniine, 26
    cyclobutane, 117
    1,3,5,7,9-cyclodecapentaene, 543, 557

cyclohexane ring flip, 122
cyclopentane, 117
cyclopropane, 112, 116
cytosine, 66
cis-decalin, 130, 1108
trans-decalin, 130, 1108
diethyl ether, 676
dimethyl disulfide, 19
cis-1,2-dimethylcyclohexane, 126
trans-1,2-dimethylcyclohexane, 127
cis-1,2-dimethylcyclopropane, 112
trans-1,2-dimethylcyclopropane, 112
dimethylpropane, 82
DNA, 62, 1132
dopamine, 957
eclipsed ethane conformation, 95
enflurane, 147
ethane, 13, 81
ethylene, 14
fluoxetine, 170
glucose, 120, 128
(R)-glyceraldehyde, 1003
hexokinase, 211
Hofmann elimination, 965
ibuprofen, 66, 173
isobutane, 81
isoleucine, 159
lactic acid, 145
lactose, 1027
lidocaine, 102
(−)-limonene, 170
(+)-limonene, 170
linolenic acid, 1091
maltose, 1026
mannose, 128
menthol, 118
meso-tartaric acid, 160
methane, 12, 81
methanethiol, 19
methanol, 18
1,6-methanonaphthalene, 557
methionine, 156
methyl phosphate, 18
methylamine, 18
2-methylbutane, 82
methylcyclohexane, 124, 146
2-methylcyclohexanone, 146
2-methylpropane, 81
naphthalene, 65
Newman projections, 94
norbornane, 131
omega-3 fatty acid, 1091
oseltamivir phosphate, 132
pentane, 82
phenylalanine, 102
piperidine, 966
propane, 81

propane conformations, 96
pseudoephedrine, 175
serylalanine, 1056
staggered ethane conformation, 95
stearic acid, 1090
steroid, 1107
sucrose, 1027
syn periplanar geometry, 401
Tamiflu, 132
testosterone, 130
tetrahydrofuran, 676
threose, 147
trimethylamine, 947
tRNA, 1138
twist boat cyclohexane, 120
vitamin C, 800
Molecular orbital, **19**
    allylic radical, 351
    antibonding, **20**
    benzene, 540–541
    bonding, **20**
    1,3-butadiene, 503–504, 1215
    conjugated diene, 503–504
    conjugated enone, 909
    degenerate, 541
    ethylene, 1215
    1,3,5-hexatriene, 1216
Molecular orbital (MO) theory, **19–21**
    Hückel $4n + 2$ rule and, 543
Molecular weight, mass spectral determination of, 426–427
Molecule(s), **7**
    condensed structures of, 21
    electron-dot structures of, 7
    Kekulé structures of, 7
    line-bond structures of, 7
    skeletal structures of, 21–22
Molozonide, **284**
Monomer, **289**
Monosaccharide(s), **1001**
    anomers of, 1011–1013
    configurations of, 1008–1010
    cyclic forms of, 1011–1013
    essential, 1023–1025
    esters of, 1015–1016
    ethers of, 1016
    Fischer projections and, 1004–1005
    glycosides of, 1016–1018
    hemiacetals of, 1011–1013
    osazones from, 1042
    oxidation of, 1020–1021
    phosphorylation of, 1019
    reaction with acetic anhydride, 1015–1016
    reaction with iodomethane, 1016
    reduction of, 1020
    see also Aldose
Monoterpene, **300**

Monoterpenoid, **1098**
Moore, Stanford, 1058
Morphine, biosynthesis of, 997
   specific rotation of, 149
   structure of, 63
MRI, see Magnetic resonance
     imaging, 486
mRNA, see Messenger RNA
MS, see Mass spectrometry
Mullis, Kary Banks, 1145
Multiplet (NMR), **476**
   table of, 479
Muscalure, synthesis of, 358
Mutarotation, **1013**
   glucose and, 1013
   mechanism of, 1013
Mycomycin, stereochemistry of, 181
Mylar, structure of, 849
myo-Inositol, structure of, 139
Myrcene, structure of, 299
Myristic acid, catabolism of, 1166
   structure of, 1090

n (normal), 82
$n + 1$ rule, 478
N-terminal amino acid, **1056**
Naming, acid anhydrides, 815
   acid chlorides, 815
   acid halides, 815
   acyl phosphate, 817
   alcohols, 621–622
   aldehydes, 723–724
   aldoses, 1010
   alkanes, 87–91
   alkenes, 226–227
   alkyl groups, 84–85, 89–90
   alkyl halides, 345–346
   alkynes, 314–315
   alphabetizing and, 91
   amides, 816
   amines, 944–946
   aromatic compounds, 535–537
   carboxylic acid derivatives,
     815–817
   carboxylic acids, 779–780
   cycloalkanes, 109–111
   cycloalkenes, 227–228
   eicosanoids, 1096–1097
   enzymes, 1070
   esters, 816
   ethers, 677
   heterocyclic amines, 946
   ketones, 724
   new IUPAC system for, 227–228
   nitriles, 781
   old IUPAC system for, 227
   phenols, 622
   prostaglandins, 1096–1097
   sulfides, 693

   thioesters, 816
   thiols, 691
Naphthalene, aromaticity of, 550
   electrostatic potential map of, 550
   Hückel $4n + 2$ rule and, 550
   molecular model of, 65
   ^{13}C NMR absorptions of, 554
   orbitals picture of, 550
   reaction with Br_2, 549
   resonance in, 549
Naproxen, NSAIDs and, 555
   structure of, 33
Natural gas, composition of, 100
Natural product, **251**
   drugs from, 213
   number of, 251
NBS, see *N*-Bromosuccinimide
NDA, see New drug application, 214
Neighboring-group effect, **1018**
Neomenthyl chloride, E2 reaction of,
     404
Neopentyl group, **90**
   S_N2 reaction and, 379
Neoprene, synthesis and uses of, 516
New drug application (NDA), 214
New molecular entity (NME), number
     of, 213
Newman, Melvin S., 94
Newman projection, **94**
   molecular model of, 94
Nicotinamide adenine dinucleotide,
     biological oxidations with, 647
   biological reductions with,
     631–632
   reactions of, 751
   structure of, 751, 1072
Nicotinamide adenine dinucleotide
     phosphate, biological
     reductions and, 280
Nicotine, structure of, 28, 944
Ninhydrin, reaction with amino
     acids, 1058
Nitration (aromatic), 571–572
Nitric acid, pK_a of, 51
Nitrile(s), **781**
   alkylation of, 888
   amides from, 795–796
   amines from, 796
   carboxylic acids from, 789–790,
     795–796
   electrostatic potential map of, 794
   from amides, 793–794
   from arenediazonium salts, 969
   hydrolysis of, 789–790, 795–796
   IR spectroscopy of, 798
   ketones from, 796
   naming, 781
   naturally occurrence of, 793
   NMR spectroscopy of, 798

   pK_a of, 879
   reaction with Grignard reagents,
     796
   reaction with LDA, 888
   reaction with $LiAlH_4$, 796
   reduction of, 796
   synthesis of, 793–794
Nitrile group, directing effect of,
     588–589
   inductive effect of, 583
   orienting effect of, 581
   resonance effect of, 584
Nitrile rubber polymer, structure and
     uses of, 1247
Nitro compound, Michael reactions
     and, 922–923
Nitro group, directing effect of,
     588–589
   inductive effect of, 583
   orienting effect of, 581
   resonance effect of, 584
Nitroarene, arylamines from, 955
   reaction with iron, 955
   reaction with $SnCl_2$, 955
   reduction of, 955
Nitrobenzene, aniline from, 572
   reduction of, 572
   synthesis of, 572
p-Nitrobenzoic acid, pK_a of, 787
Nitrogen, hybridization of, 17–18
Nitrogen rule of mass spectrometry,
     981–982
Nitronium ion, 571–572
   electrostatic potential map of, 572
p-Nitrophenol, pK_a of, 625
Nitrous acid, reaction with amines,
     968–969
NME, see New molecular entity, 213
NMO, see *N*-Methylmorpholine
     N-oxide
NMR, see Nuclear magnetic resonance
Node, **4**
Nomenclature, see Naming
Nomex, structure of, 1259
Nonbonding electrons, **8**
Noncovalent interaction, 60–62
Nonequivalent protons, spin–spin
     splitting and, 482–483
   tree diagram of, 483
Nootkatone, chirality of, 146
Norbornane, molecular model of, 131
Norepinephrine, adrenaline from,
     396
   biosynthesis of, 597
Norethindrone, structure and
     function of, 1111
Normal (*n*) alkane, **82**
Norsorex, synthesis of, 1253
Novocaine, structure of, 64

Novolac resin, 523–524
Noyori, Ryoji, 760
NSAID, **554**
Nuclear magnetic resonance
 spectrometer, field strength of, 458
 operation of, 460
Nuclear magnetic resonance
 spectroscopy (NMR), **456**
 acid anhydrides, 852
 acid chlorides, 852
 alcohols, 655–656
 aldehydes, 757–758
 allylic protons and, 474–475
 amides, 852
 amines, 979–980
 aromatic compounds, 552–554
 aromatic protons and, 474–475
 calibration peak for, 462
 carboxylic acid derivatives, 852
 carboxylic acids, 798–799
 chart for, 461
 ^{13}C chemical shifts in, 464–465
 ^1H chemical shifts in, 474–475
 coupling constants in, 478
 delta scale for, 462
 DEPT-NMR and, 467–469
 diastereotopic protons and, 473
 enantiotopic protons and, 472
 energy levels in, 458
 epoxides, 696
 esters, 852
 ethers, 696
 field strength and, 457–458
 FT-NMR and, 463–464
 homotopic protons and, 472
 integration of, 476
 ketones, 757–758
 multiplets in, 477–479
 $n + 1$ rule and, 478
 nitriles, 798
 overlapping signals in, 482
 ^{13}C peak assignments in, 467–469
 ^1H peak size in, 483–484
 phenols, 656
 principle of, 456–458
 proton equivalence and, 471–473
 pulsed, 463–464
 radiofrequency energy and, 457–458
 ring current and, 552–553
 shielding in, 458
 signal averaging in, 463–464
 spin-flips in, 457
 spin–spin splitting in, 477–480
 time scale of, 460–461
 uses of ^{13}C, 470–471
 uses of ^1H, 484–485
 vinylic protons and, 457–458

^{13}C Nuclear magnetic resonance
 spectrum, acetaldehyde, 758
 acetophenone, 758
 anisole, 696
 benzaldehyde, 758
 benzoic acid, 798
 p-bromoacetophenone, 465
 2-butanone, 465, 758
 crotonic acid, 798
 cyclohexanol, 655
 cyclohexanone, 758
 ethyl benzoate, 492
 methyl acetate, 458
 methyl propanoate, 467
 methyl propyl ether, 696
 1-methylcyclohexene, 471
 N-methylcyclohexylamine, 980
 2-methyl-1-propanol, 469
 1-pentanol, 464
 propanenitrile, 798
 propanoic acid, 798
 propionic acid, 798
^1H Nuclear magnetic resonance
 spectrum, acetaldehyde, 758
 bromoethane, 477
 2-bromopropane, 478
 p-bromotoluene, 553
 trans-cinnamaldehyde, 482
 cyclohexylmethanol, 485
 dipropyl ether, 696
 1,2-epoxypropane, 696
 ethyl acetate, 852
 methyl acetate, 458
 methyl 2,2-dimethylpropanoate, 476
 p-methoxypropiophenone, 480
 2-methyl-2-butanol, 481
 1-methylcyclohexanol, 485
 N-methylcyclohexylamine, 980
 phenylacetic acid, 799
 1-propanol, 656
 toluene, 482
Nuclear spin, common nuclei and, 458
 NMR and, 456–457
Nucleic acid, **1128**
 see Deoxyribonucleic acid, Ribonucleic acid
Nucleophile(s), **192**
 characteristics of, 197–198
 curved arrows and, 197–198
 electrostatic potential maps of, 192
 examples of, 192
 S_N1 reaction and, 392
 S_N2 reaction and, 380–381
Nucleophilic acyl substitution
 reaction, **717**, **818**–819
 abbreviated mechanism for, 1169

acid anhydrides, 835–836
acid chlorides, 830–834
acid halides, 830–834
amides, 843–845
carboxylic acids and, 823–830
esters, 837–842
kinds of, 821
mechanism of, 819
reactivity in, 819–821
Nucleophilic addition reaction, **715**, **728**–730
 acid catalysis of, 731–732
 base catalysis of, 731–732
 mechanism of, 728
 steric hindrance in, 729
 trajectory of, 729
 variations of, 729
Nucleophilic aromatic substitution
 reaction, 592–594
 mechanism of, 592–593
Nucleophilic substitution reaction, **373**
 biological examples of, 395–396
 see S_N1 reaction, S_N2 reaction
 summary of, 407–408
Nucleophilicity, **380**
 basicity and, 381
 table of, 381
 trends in, 381
Nucleoside, **1128**
Nucleotide, **1128**
 3' end of, 1131
 5' end of, 1131
Nucleus, size of, 2
Nylon, **847**–849
 manufacture of, 849
 naming, 849
 uses of, 849
Nylon 6, structure of, 848
 synthesis of, 1249
Nylon 6,6, structure of, 849
 synthesis of, 1249
Nylon 10,10, uses of, 1259

Ocimene, structure of, 256
Octane number (fuel), **100**–101
Octet rule, 6
-oic acid, carboxylic acid name suffix, 779
Okazaki fragments, DNA replication and, 1135
-ol, alcohol name suffix, 622
Olah, George Andrew, 266
Olefin, **222**
Olefin metathesis polymerization, **1251**–1253
 Grubbs catalyst for, 1251
 kinds of, 1252
 mechanism of, 1251
Oleic acid, structure of, 1090

Oligonucleotide, **1142**
   synthesis of, 1142–1144
Olive oil, composition of, 1090
Omega-3 fatty acid, **1089**
   molecular model of, 1091
-one, ketone name suffix, 724
-onitrile, nitrile name suffix, 781
Optical activity, 147–149
   measurement of, 148
Optical isomers, **150**
Optically active, **148**
Orbital, **3**
   energies of, 4
   hybridization of, 11–18
   shapes of, 3–4
p Orbital, nodes in, 4
   shape of, 3–4
d Orbital, shape of, 4
s Orbital, shape of, 3
Organic chemicals, number of, 74
   toxicity of, 24–25
Organic chemistry, **1**
   foundations of, 1–2
Organic compound(s), elements found in, 2
   number of, 1
   oxidation level of, 361
   polar covalent bonds in, 190–191
   size of, 1
Organic foods, 24–25
Organic reactions, chirality and, 296–298
   conventions for writing, 239
   kinds of, 184–185
Organic synthesis, enantioselective, 760–761
   strategy of, 329
Organoborane, from alkenes, 272–274
   reaction with H_2O_2, 272–273
Organocopper reagent, *see* Diorganocopper reagent, Gilman reagent
Organodiphosphate, biological substitution reactions and, 395–396
Organohalide(s), **344**
   biological uses of, 362–363
   naturally occurring, 362–363
   number of, 362
   reaction with Gilman reagents, 357–358
   see also Alkyl halide
   uses of, 344
Organomagnesium halide, *see* Grignard reagent
Organomercury compounds, reaction with $NaBH_4$, 271–272
Organometallic compound, **356**

Organometallic coupling reaction, 357–359
Organopalladium compound, Suzuki–Miyaura reaction of, 359
Organophosphate, bond angles in, 18
   hybrid orbitals in, 18
Orlon, structure and uses of, 293
Ortho (o) prefix, **537**
Ortho- and para-directing group, **581**
Osazone, 1042
-ose, carbohydrate name suffix, 1002
Oseltamivir phosphate, mechanism of, 1032–1033
   molecular model of, 132
   structure of, 32
Osmate, **283**
Osmium tetroxide, reaction with alkenes, 283–284
   toxicity of, 283
Oxalic acid, structure of, 780
Oxaloacetic acid, structure of, 780
Oxaphosphatane, **746**
Oxetane, reaction with Grignard reagents, 708
Oxidation, **281**
   alcohols, 645–647
   aldehydes, 727
   aldoses, 1020–1021
   alkenes, 281–286
   biological, 647
   organic, 360
   phenols, 653
   sulfides, 694
   thiols, 692
Oxidation level, table of, 361
Oxidative decarboxylation, pyruvate catabolism and, 1181–1185
   steps in, 1182
Oxidoreductase, 1070
Oxime, **738**
   from aldehydes and ketones, 737–738
Oxirane, **281**
Oxo group, **725**
Oxycodone, structure of, 1
OxyContin, structure of, 1
Oxyfluorfen, synthesis of, 594
Oxygen, hybridization of, 18
Oxymercuration, **271–272**
   mechanism of, 271–272
   regiochemistry of, 271
Ozone, preparation of, 284
   reaction with alkenes, 284–285
   reaction with alkynes, 325
Ozonide, **284**
   danger of, 285

Paclitaxel, structure of, 333
Palmitic acid, structure of, 1090
Palmitoleic acid, structure of, 1090

PAM resin, solid-phase peptide synthesis and, 1065
Para (p) prefix, **537**
Paraffin, 92
Parallel synthesis, **605**
Parent peak (mass spectrum), **426**
Partial charge, 35
Pasteur, Louis, enantiomers and, 150
   resolution of enantiomers and, 161
Patchouli alcohol, structure of, 1099
Paternity, DNA test for, 1146–1147
Pauli exclusion principle, **5**
Pauling, Linus Carl, 11
PCR, *see* Polymerase chain reaction, 1145–1146
PDB, *see* Protein Data Bank, 1076–1077
Peanut oil, composition of, 1090
Pedersen, Charles John, 690
Penicillin, discovery of, 853
Penicillin V, specific rotation of, 149
   stereochemistry of, 172
Penicillium notatum, penicillin from, 853
Pentachlorophenol, synthesis of, 652
1,4-Pentadiene, electrostatic potential map of, 504
Pentadienyl radical, resonance in, 47
Pentalene, 561
Pentane, molecular model of, 82
2,4-Pentanedione, pK_a of, 880
2,4-Pentanedione anion, resonance in, 46
1-Pentanol, ^{13}C NMR spectrum of, 464
Pentobarbital, synthesis of, 891
Pentose phosphate pathway, 1208, 1210–1211
Pepsin, pI of, 1052
Peptide(s), **1044**
   amino acid sequencing of, 1059–1061
   backbone of, 1056
   covalent bonding in, 1056–1057
   disulfide bonds in, 1057
   Edman degradation of, 1059–1061
   reaction with phenylisothiocyanate, 1059–1060
   solid-phase synthesis of, 1064–1066
   synthesis of, 1062–1066
Peptide bond, **1056**–1057
   DCC formation of, 826–827, 1062–1063
   restricted rotation in, 1057
Pericyclic reaction, **1214**
   frontier orbitals and, 1216
   kinds of, 1214

stereochemical rules for, 1231
Woodward–Hoffmann rules for, 1215–1216
Periodic acid, reaction with 1,2-diols, 285–286
Periplanar, **401**
Perlon, structure of, 848
Peroxide, **678**
Peroxyacid, **281**
   reaction with alkenes, 281–282
PET, see Polyethylene terephthalate, 1254
Petit, Rowland, 542
Petroleum, catalytic cracking of, 101
   composition of, 100
   gasoline from, 100–101
   history of, 100
   refining of, 100–101
   reforming of, 101
Pfu DNA polymerase, PCR and, 1145
Pharmaceuticals, approval procedure for, 213–214
   origin of, 213
Phenol(s), **620**
   acidity of, 624–626
   Bakelite from, 1256
   Dow process for, 650
   electrophilic aromatic substitution reactions of, 652
   electrostatic potential map of, 583
   from arenediazonium salts, 970
   from chlorobenzene, 594
   from cumene, 650
   hydrogen bonds in, 623
   IR spectroscopy of, 655
   IR spectrum of, 655
   mechanism of synthesis of, 650–651
   naming, 622
   NMR spectroscopy of, 656
   oxidation of, 653
   phenoxide ions from, 624
   pK_a of, 625
   properties of, 623–627
   quinones from, 653
   reaction with arenediazonium salts, 972
   uses of, 621, 650, 652
Phenolic resin, 1256
Phenoxide ion, **624**
   electrostatic potential map of, 627
   resonance in, 627
Phentermine, synthesis of, 962
Phenyl group, **536**
Phenylacetaldehyde, aldol reaction of, 906
   IR spectrum of, 445
Phenylacetic acid, ^{1}H NMR spectrum of, 799

Phenylacetylene, IR spectrum of, 446
Phenylalanine, biosynthesis of, 684, 1229–1230
   molecular model of, 102
   pK_a of, 52
   structure and properties of, 1046
Phenylisothiocyanate, Edman degradation and, 1059–1060
Phenylthiohydantoin, Edman degradation and, 1059–1061
Phosphate, electrostatic potential map of, 78
Phosphatidic acid, glycerophospholipids from, 1094
Phosphatidylcholine, structure of, 1094
Phosphatidylethanolamine, structure of, 1094
Phosphatidylserine, structure of, 1094
Phosphine(s), chirality of, 166
Phosphite, DNA synthesis and, **1143**
   oxidation of, 1144
Phospholipid, **1094–1095**
   classification of, 1094
Phosphopantetheine, coenzyme A from, 846, 1156
Phosphoramidite, DNA synthesis and, **1143**
Phosphorane, **746**
Phosphoric acid, pK_a of, 51
Phosphoric acid anhydride, 1156
Phosphorus, hybridization of, 18
Phosphorus oxychloride, alcohol dehydration with, 641–643
Phosphorus tribromide, reaction with alcohols, 355, 639
Photochemical reaction, **1217**
Photolithography, **523–524**
   resists for, 523–524
Photon, **435**
   energy of, 435–436
Photosynthesis, 1000–1001
Phthalates, use as plasticizers, 837
Phthalic acid, structure of, 780
Phthalimide, Gabriel amine synthesis and, 957
Phylloquinone, biosynthesis of, 578–579
Pi (π) bond, **14**
   acetylene and, 16
   ethylene and, 14
   molecular orbitals in, 21
Picometer, **3**
Picric acid, synthesis of, 650
Pinacol rearrangement, 672
Pineapple, esters in, 836
Piperidine, molecular model of, 966
   structure of, 946

PITC, see Phenylisothiocyanate, 1059–1060
pK_a, **50**
   table of, 51
Planck equation, 435–436
Plane of symmetry, **144**
   meso compounds and, 159–160
Plane-polarized light, 147–148
Plasmalogen, structure of, 1121
Plastic, recyclable, 1256–1257
   see also Polymer
Plasticizer, **837, 1254**
   structure and function of, 1254
   toxicity of, 1254
Plavix, structure of, 32
Plexiglas, structure of, 293
Poison ivy, urushiols in, 621
Polar aprotic solvent, **383**
   S_N1 reaction and, 393
   S_N2 reaction and, 383–384
Polar covalent bond, **34–35**
   dipole moments and, 37–38
   electronegativity and, 35–36
   electrostatic potential maps and, 36
   inductive effects and, 36
Polar reaction, **187, 190–193**
   characteristics of, 190–193
   curved arrows in, 192, 197–199
   electrophiles in, 192
   example of, 194–196
   nucleophiles in, 192
Polarimeter, 148
Polarizability, **191**
Poly(ethylene terephthalate), structure of, 1254
Poly(glycolic acid), biodegradability of, 1257
   uses of, 850
Poly(hydroxybutyrate), biodegradability of, 1257
   uses of, 850
Poly(lactic acid), biodegradability of, 1257
   uses of, 850
Poly(methyl methacrylate), uses of, 293
Poly(vinyl acetate), uses of, 293
Poly(vinyl butyral), uses of, 1260
Poly(vinyl chloride), plasticizers in, 1254
   uses of, 293
Polyacrylonitrile, uses of, 293
Polyalkylation, Friedel–Crafts reaction and, 576
Polyamide, **847**
Polybutadiene, synthesis of, 516
   vulcanization of, 517
Polycarbonate, 849–850, **1249**

Polycyclic aromatic compound, **549**
  aromaticity of, 549–550
Polycyclic compound, **129**
  bridgehead atoms in, 129
  conformations of, 129–130
Polycyclic heterocycle, 977–978
Polyester, **847**
  manufacture of, 849
  uses of, 849
Polyethylene, crystallites in, 1253
  high-density, 1246
  high-molecular-weight, 1246
  kinds of, 1246
  low-density, 1246
  synthesis of, 291–292
  ultrahigh-molecular-weight, 1246
  uses of, 293
  Ziegler–Natta catalysts and, 1246
Polyimide, structure of, 862
Polymer(s), **289**
  atactic, **1245**
  biodegradable, 850, 1256–1257
  biological, 289–290
  chain-growth, **291–292, 1242–1244**
  classification of, 1242
  crystallites in, 1253
  elastomer, 1255
  fiber, 1255
  glass transition temperature of, 1254
  isotactic, **1245**
  kinds of, 1254
  melt transition temperature of, 1254
  plasticizers in, 1254
  recycling codes for, 1257
  representation of, 1242
  step-growth, **847–850, 1248–1250**
  syndiotactic, **1245**
  table of, 293
  thermoplastic, 1254
  thermosetting resin, 1256
  van der Waals forces in, 1253
Polymerase chain reaction (PCR), **1145–1146**
  amplification factor in, 1145
  Pfu DNA polymerase in, 1145
  Taq DNA polymerase in, 1145
Polymerization, anionic, 1243
  cationic, 1243
  mechanism of, 291–292
  Ziegler–Natta catalysts for, 1245–1246
Polypropylene, polymerization of, 1245
  stereochemical forms of, 1245
  uses of, 293

Polysaccharide(s), **1028–1029**
  synthesis of, 1029–1030
Polystyrene, uses of, 293
Polytetrafluoroethylene, uses of, 293
Polyunsaturated fatty acid, **1089**
Polyurethane, **1250**
  foam, 1250
  kinds of, 1250
  stretchable, 1250
Polyynes, occurrence of, 314
Posttranslational modification, protein, 1142
Potassium nitrosodisulfonate, reaction with phenols, 653
Potassium permanganate, reaction with alcohols, 645
  reaction with alkenes, 285
  reaction with alkylbenzenes, 596–597
  reaction with ketones, 728
Pravachol, structure of, 107
Pravadoline, green synthesis of, 983
Pravastatin, structure of, 107
  statin drugs and, 1203–1204
Prepolymer, epoxy resins and, 697–698
Prilocaine, structure of, 64
Primary alcohol, 621
Primary amine, **944**
Primary carbon, **86**
Primary hydrogen, **86**
Primary structure (protein), 1066
pro-R prochirality center, 167–168
pro-S prochirality center, 167–168
Problems, how to work, 26
Procaine, structure of, 30, 64
Prochirality, **167–169**
  assignment of, 167–168
  chiral environments and, 171–172
  naturally occurring molecules and, 168–169
  re descriptor for, 167
  si descriptor for, 167
Prochirality center, **167–168**
  pro-R, 167–168
  pro-S, 167–168
Progesterone, structure and function of, 1110
  structure of, 501
Progestin, **1110**
  function of, 1110
Proline, biosynthesis of, 959
  structure and properties of, 1046
Promotor sequence (DNA), 1135
Propagation step (radical), **188**
Propane, bond rotation in, 96
  conformations of, 96
  mass spectrum of, 426
  molecular model of, 81, 96

Propanenitrile, ^{13}C NMR absorptions in, 798
Propanoic acid, ^{13}C NMR absorptions in, 798
1-Propanol, ^{1}H NMR spectrum of, 656
Propenal, electrostatic potential map of, 511
Propene, *see* Propylene
Propenenitrile, electrostatic potential map of, 511
Propionic acid, *see* Propanoic acid
Propyl group, **85**
Propylene, heat of hydrogenation of, 236
  industrial preparation of, 223
  uses of, 223
Prostaglandin(s), **1095–1098**
  biosynthesis of, 294–295, 188–189, 1097–1098
  functions of, 188, 1095–1096
  naming, 1096–1097
  occurrence of, 1095
  see also Eicosanoid
Prostaglandin E_1, structure of, 108, 1096
Prostaglandin E_2, biosynthesis of, 1097–1098
Prostaglandin F_{2a}, structure of, 114
Prostaglandin H_2, biosynthesis of, 188–189, 1097–1098
Prostaglandin I_2, structure of, 1096
Protecting group, **648**
  alcohols, 648–650
  aldehydes, 745
  ketones, 745
  nucleic acid synthesis and, 1142–1143
  peptide synthesis and, 1062–1063
Protein(s), **1044**
  α helix in, 1066–1067
  backbone of, 1056
  biosynthesis of, 1137–1139
  denaturation of, 1068
  isoelectric point of, 1052
  mechanism of hydrolysis of, 844
  number of in humans, 1137
  primary structure of, 1066
  quaternary structure of, 1066
  secondary structure of, 1066–1067
  see also Peptide
  C-terminal amino acid in, 1056
  N-terminal amino acid in, 1056
  tertiary structure of, 1066, 1068
Protein Data Bank (PDB), 1076–1077
  downloading structures from, 1076
  number of structures in, 1076

Protic solvent, 383
  S_N1 reaction and, 393
  S_N2 reaction and, 383–384
Proton equivalence, ^1H NMR
    spectroscopy and, 471–473
Protonated methanol, electrostatic
    potential map of, 190
Protosteryl cation, lanosterol
    biosynthesis and, 1114, 1117
Prozac, structure of, 170
Pseudoephedrine, molecular model
    of, 175
PTH, see Phenylthiohydantoin,
    1059–1061
Purine, aromaticity of, 550
  electrostatic potential map of, 978
  nucleotides from, 1129
  structure of, 978
Pyramidal inversion, amines and, 947
  energy barrier to, 947
Pyranose, **1011**–1013
  glucose and, 1011–1012
Pyridine, aromaticity of, 546–547, 976
  basicity of, 950, 976
  dipole moment of, 976–977
  electrophilic substitution reactions
    of, 976
  electrostatic potential map of, 546
  Hückel $4n + 2$ rule and, 546–547
Pyridoxal phosphate, amino acid
    catabolism and, 1198
  imines from, 736
  structure of, 30, 1073
Pyridoxamine phosphate,
    transamination and, 1198
Pyrimidine, aromaticity of, 546–547
  basicity of, 950, 977
  electrostatic potential map of, 546
  Hückel $4n + 2$ rule and, 546–547
  nucleotides from, 1129
Pyrrole, aromaticity of, 547, 973–974
  basicity of, 950, 973
  electrophilic substitution reactions
    of, 974–975
  electrostatic potential map of, 547,
    974
  Hückel $4n + 2$ rule and, 547
  industrial synthesis of, 973
Pyrrolidine, electrostatic potential
    map of, 974
  structure of, 946
Pyrrolysine, structure of, 1048
Pyruvate, acetyl CoA from,
    1181–1185
  catabolism of, 1181–1185
  from glucose, 1173–1181
  glucose from, 1191–1197
  oxidative decarboxylation of,
    1181–1185

  reaction with thiamin
    diphosphate, 1181–1183
Pyruvate dehydrogenase complex,
    1181
Pyruvic acid, structure of, 780

Qiana, structure of, 861
Quantum mechanical model, 3–5
Quartet (NMR), 477
Quaternary ammonium salt, **945**
  Hofmann elimination and,
    964–965
Quaternary carbon, **86**
Quaternary structure (protein), **1066**
Quetiapine, structure of, 31
Quinine, structure of, 550, 977
Quinoline, aromaticity of, 550
  electrophilic substitution reaction
    of, 978
  Lindlar catalyst and, 323
Quinone(s), **653**
  from phenols, 653
  hydroquinones from, 653
  reduction of, 653

R configuration, **152**
  assignment of, 152
R group, **86**
Racemate, **161**
Racemic mixture, **161**
Radical, **187**
  reactivity of, 187
  stability of, 349, 351
Radical chain reaction, **188**
  initiation steps in, 188
  propagation steps in, 188
  termination steps in, 188
Radical reaction(s), **187**–189
  addition, 187
  biological example of, 294–295
  characteristics of, 188
  fishhook arrows and, 186
  prostaglandin biosynthesis and,
    188–189, 1097–1098
  substitution, 187
Radio waves, electromagnetic
    spectrum and, 434
Radiofrequency energy, NMR
    spectroscopy and, 457–458
Rapamycin, discovery of, 252
  structure of, 251
Rate equation, 376
Rate-determining step, **386**
Rate-limiting step, **386**
Rayon, 1028
Re prochirality, **167**
Reaction (polar), **187**, 190–193
Reaction (radical), **187**–189
Reaction coordinate, **206**

Reaction intermediate, **208**
Reaction mechanism, **186**
Reaction rate, activation energy and,
    206–207
Rearrangement reaction, **185**
Reducing sugar, **1020**
Reduction, **277**
  acid chlorides, 833
  aldehydes, 630–631, 734
  aldoses, 1020
  alkene, 276–280
  alkyne, 322–325
  amides, 844–845
  arenediazonium salt, 970
  aromatic compounds, 599–600
  carboxylic acids, 632–633,
    827–828
  disulfides, 692
  esters, 632–633, 840–841
  ketones, 630–631, 734
  lactams, 845
  nitriles, 796
  organic, **360**
  quinones, 653
Reductive amination, **958**–959
  amino acid synthesis and, 1054
  biological example of, 959
  mechanism of, 958
Refining (petroleum), 100–101
Regiospecific, **240**
Registry of Mass Spectral Data, 428
Relenza, mechanism of, 1032–1033
Replication (DNA), **1133**–1135
  direction of, 1134–1135
  error rate during, 1135
  lagging strand in, 1135
  leading strand in, 1135
  Okazaki fragments in, 1135
  replication fork in, 1134
Replication fork (DNA), **1134**
Residue (protein), **1056**
Resist, photolithography and,
    523–524
Resolution (enantiomers), 161–163
Resonance, **42**–46
  acetate ion and, 42–43
  acetone anion and, 45
  acyl cations and, 577–578
  allylic carbocations and, 506
  allylic radical and, 351–352
  arylamines and, 952
  benzene and, 43, 539–540
  benzylic carbocation and, 390
  benzylic radical and, 598
  carbonate ion and, 47
  carboxylate ions and, 784
  enolate ions and, 878
  Lewis structures and, 42–43
  naphthalene and, 549

Resonance—cont'd
  pentadienyl radical and, 47
  2,4-pentanedione anion and, 46
  phenoxide ions and, 627
Resonance effect, electrophilic
    aromatic substitution and, **584**
Resonance form, **43**
  drawing, 45–46
  electron movement and, 43–45
  rules for, 43–45
  stability and, 45
  three-atom groupings in, 46
Resonance hybrid, **43**
Restriction endonuclease, **1140**
  number of, 1140
  palindrome sequences in, 1140
Retin A, structure of, 259
Retinal, vision and, 522
Retrosynthetic analysis, **329**
Rhodium, aromatic hydrogenation
    catalyst, 599
Rhodopsin, isomerization of, 522
  vision and, 522
Ribavirin, structure of, 561
Ribonucleic acid (RNA), **1128**
  bases in, 1129
  biosynthesis of, 1135–1136
  3' end of, 1131
  5' end of, 1131
  kinds of, 1135
  messenger, **1135**
  ribosomal, **1135**
  size of, 1130
  small, **1135**
  structure of, 1130–1131
  transfer, **1135**
  translation of, 1137–1139
Ribonucleotide(s), structures of,
    1130
Ribose, configuration of, 1009
Ribosomal RNA, **1135**
  function of, 1135
Ring current (NMR), **552**
  [18]annulene and, 552–553
Ring-expansion reaction, 901
Ring-flip (cyclohexane), **122**
  energy barrier to, 122
  molecular model of, 122
Ring-opening metathesis
    polymerization (ROMP), 1252
Risk, chemicals and, 24–25
RNA, *see* Ribonucleic acid
Roberts, Irving, 265
Robinson annulation reaction,
    927–928
  mechanism of, 927
Rod cells, vision and, 522
Rofecoxib, NSAIDs and, 555
  structure of, 1

ROMP, *see* ring-opening metathesis
    polymerization, 1252
rRNA, *see* Ribosomal RNA
Rubber, production of, 516
  structure of, 516
  vulcanization of, 517

S configuration, **152**
  assignment of, 152
s-cis conformation, **513**
  Diels–Alder reaction and, 513–514
Saccharin, structure of, 1034
  sweetness of, 1033
Safrole, structure of, 707
Samuelsson, Bengt, 1095
Sandmeyer reaction, **969**–970
  mechanism of, 970
Sanger, Frederick, 1063
Sanger dideoxy DNA sequencing,
    1140–1141
Sanger's reagent, 592
β-Santalene, structure of, 299
Saponification, **838**, **1092**
  mechanism of, 838
Saran, structure and uses of,
    1246–1247
Sativene, synthesis of, 902
Saturated, 81
Saturated hydrocarbon, **81**
Sawhorse representation, **94**
SBR polymer, structure and uses of,
    1247
Schiff base, **736**, **1177**
  see also Imine
Scurvy, vitamin C and, 800
sec-, name prefix, **85**
sec-Butyl group, **85**
Secobarbital, synthesis of, 891
Second-order reaction, **376**
Secondary alcohol, 621
Secondary amine, **944**
Secondary carbon, **86**
Secondary hydrogen, **86**
Secondary metabolite, **251**
  number of, 251
Secondary structure (protein),
    1066–1067
Sedoheptulose, structure of, 1002
Selenocysteine, structure of, 1048
Semiconservative replication (DNA),
    1134
Sense strand (DNA), **1136**
Sequence rules, 150–152
  E,*Z* alkene isomers and, 231–232
  enantiomers and, 150–154
Serine, biosynthesis of, 1212
  structure and properties of, 1047
Seroquel, structure of, 31
Serum lipoprotein, table of, 1118

Serylalanine, molecular model of,
    1056
Sesquiterpene, **300**
Sesquiterpenoid, **1098**
Sex hormone, 1110
Sharpless, K. Barry, 760
Sharpless epoxidation, 761
Shell (electron), **4**
  capacity of, 4
Shielding (NMR), **458**
Si prochirality, **167**
Sialic acid, 1024
Side chain (amino acid), **1048**
Sigma (σ) bond, **10**
  symmetry of, 10
Sigmatropic rearrangement, **1226**–1230
  antarafacial geometry of, 1227
  examples of, 1228–1230
  [1,5] hydrogen shift and, 1228
  notation for, 1227
  stereochemical rules for, 1227
  suprafacial geometry of, 1227
  vitamin D and, 1232
Signal averaging, FT-NMR
    spectroscopy and, 463–464
Sildenafil, structure of, 973
Silver oxide, Hofmann elimination
    reaction and, 964–965
Simmons–Smith reaction, **288**–289
Simple sugar, **1001**
Simvastatin, structure of, 107
Single bond, **13**
  electronic structure of, 12–13
  length of, 12
  see also Alkane
  strength of, 12
Sirolimus, structure of, 251
Skeletal structure, **21**
  rules for drawing, 21–22
Skunk scent, cause of, 692
Small RNAs, **1135**
S_N1 reaction, **385**–388
  biological examples of, 395–396
  carbocation stability and, 389–390
  characteristics of, 389–394
  energy diagram for, 387
  epoxide cleavage and, 688
  ion pairs in, 388
  kinetics of, 386
  leaving groups in, 391–392
  mechanism of, 386–387
  nucleophiles and, 392
  racemization in, 387–388
  rate law for, 386
  rate-limiting step in, 386–387
  solvent effects on, 392–393
  stereochemistry of, 387–388
  substrate structure and, 389–390
  summary of, 394

S_N2 reaction, **376–377**
   allylic halides in, 391
   amines and, 956
   benzylic halides in, 391
   biological example of, 396
   characteristics of, 378–385
   crown ethers and, 691
   electrostatic potential maps of, 377
   energy diagrams for, 385
   epoxide cleavage and, 383, 688, 689
   inversion of configuration in, 376–377
   kinetics of, 375–376
   leaving groups and, 382–383
   mechanism of, 376–377
   nucleophiles in, 380–381
   rate law for, 376
   solvent effects and, 383–384
   stereochemistry of, 376–377
   steric hindrance in, 378–379
   substrate structure and, 378–380
   summary of, 384–385
   table of, 381
   tosylates and, 382
   Williamson ether synthesis and, 678–679
Soap, 1092–1093
   history of, 1092
   manufacture of, 1092
   mechanism of action of, 1093
   micelles of, 1092–1093
Sodium amide, reaction with alcohols, 626
Sodium bisulfite, osmate reduction with, 283
Sodium borohydride, reaction with ketones and aldehydes, 630
   reaction with organomercury compounds, 271–272
Sodium chloride, dipole moment of, 38
Sodium cyclamate, LD_{50} of, 24
Sodium hydride, reaction with alcohols, 626
Solid-phase peptide synthesis, 1064–1066
   PAM resin in, 1065
   Wang resin in, 1065
Solvation, **383**
   carbocations and, 393
   S_N2 reaction and, 383
Solvent, polar aprotic, **383**
   protic, **383**
   S_N1 reaction and, 392–393
   S_N2 reaction and, 383–384
Sorbitol, structure of, 1020
Spandex, synthesis of, 1250

Specific rotation, **148–149**
   table of, 149
Sphingomyelin, **1094–1095**
Sphingosine, structure of, 1095
Spin density surface, allylic radical, 352
   benzylic radical, 598
Spin-flip, NMR spectroscopy and, 457
Spin–spin splitting, **477**
   alcohols and, 656
   bromoethane and, 477–478
   2-bromopropane and, 478
   $n + 1$ rule and, 478
   ^{13}C NMR spectroscopy and, 480
   ^1H NMR spectroscopy and, 477–480
   nonequivalent protons and, 482–483
   origin of, 477–478
   rules for, 479
   tree diagrams and, 483
Split synthesis, **605–606**
Squalene, epoxidation of, 1112–1113
   from farnesyl diphosphate, 1112
   steroid biosynthesis and, 1112–1113
Squalene oxide, cyclization of, 1114, 1116–1117
Staggered conformation, ethane and, **95**
   molecular model of, 95
Stannous chloride, reaction with nitroarenes, 955
Starch, 1→4-α-links in, 1028
   structure of, 1028
Statin drugs, heart disease and, 1203–1204
   mechanism of action of, 1203–1204
   sales of, 106
   structure of, 107
Steam cracking, 223–224
Steam distillation, **299**
Stearic acid, molecular model of, 1090
   structure of, 1090
Stein, William, 1058
Step-growth polymer, **847–850, 1248–1250**
   table of, 848
Stereocenter, **145**
Stereochemistry, **94, 112**
   absolute configuration and, 154
   Diels–Alder reaction and, 512
   E1 reaction and, 406
   E2 reaction and, 401–402
   electrophilic addition reactions and, 296–298
   R,S configuration and, 150–154

   S_N1 reaction and, 387–388
   S_N2 reactions and, 376–377
Stereogenic center, **145**
Stereoisomers, **112**
   kinds of, 164–165
   number of, 156
   properties of, 160
Stereospecific, **288, 512**
Stereospecific numbering, sn-glycerol 3-phosphate and, 1161
Steric hindrance, S_N2 reaction and, 378–379
Steric strain, **97**
   cis alkenes and, 234–235
   substituted cyclohexanes and, 124–125
Steroid(s), **1107–1111**
   adrenocortical, 1111
   anabolic, 1111
   androgens, 1110
   biosynthesis of, 1112–1117
   cis A–B ring fusion in, 1108
   conformation of, 1107
   contraceptive, 1111
   estrogens, 1110
   glucocorticoid, 1111
   mineralocorticoid, 1111
   molecular model of, 1107
   numbering of, 1107
   progestins, 1110
   stereochemistry of, 1108–1109
   synthetic, 1111
   trans A–B ring fusion in, 1108
Stork enamine reaction, **925–926**
   advantages of, 926
   mechanism of, 925
STR loci, DNA fingerprinting and, 1146
Straight-chain alkane, **82**
Strecker synthesis, 999
Structure, condensed, **21**
   electron-dot, 7
   Kekulé, 7
   Lewis, 7
   line-bond, 7
   skeletal, 21
Strychnine, LD_{50} of, 24
Styrene, anionic polymerization of, 1244
Substituent, **87**
Substituent effect, additivity of, 590
   electrophilic aromatic substitution and, 580–581
   explanation of, 582–589
   summary of, 589
Substitution reaction, **185**
Substrate (enzyme), **1069**
Succinic acid, structure of, 780

Sucralose, structure of, 1034
   sweetness of, 1033
Sucrose, molecular model of, 1027
   specific rotation of, 149
   structure of, 1027
   sweetness of, 1033
Sugar, complex, **1001**
   D, **1007**
   L, **1007**
   see also Aldose, Carbohydrate
   simple, **1001**
Sulfa drug, **968**
   synthesis of, 573
Sulfanilamide, structure of, 573
   synthesis of, 968
Sulfathiazole, structure of, 968
Sulfide(s), **676**, 693–695
   electrostatic potential map of, 79
   from thiols, 693
   naming, 693
   oxidation of, 694
   reaction with alkyl halides, 694
   sulfoxides from, 694
Sulfonation (aromatic), 572–573
Sulfone, **694**
   from sulfoxides, 694
Sulfonium ion(s), **694**
   chirality of, 166
Sulfoxide(s), **694**
   from sulfides, 694
   oxidation of, 694
Sunshine vitamin, 1232
Super glue, structure of, 1244
Suprafacial geometry, **1223**
Suture, polymers in, 850
Suzuki–Miyaura reaction, **359**
   mechanism of, 359
Sweeteners, synthetic, 1033–1034
Swine flu, 1032
Symmetry plane, **144**
Symmetry-allowed reaction, **1215**
Symmetry-disallowed reaction, **1215**
Syn periplanar geometry, **401**
   molecular model of, 401
Syn stereochemistry, **273**
Syndiotactic polymer, **1245**
Synthase, **1167**
Synthesis, strategy of, 329

Table sugar, see Sucrose
Tagatose, structure of, 1002
Talose, configuration of, 1009
Tamiflu, mechanism of, 1032–1033
   molecular model of, 132
   structure of, 32
Tamoxifen, structure of, 257
   synthesis of, 771
Taq DNA polymerase, PCR and, 1145
Tartaric acid, stereoisomers of, 159

Tautomer, **319**, **871**
Taxol, structure of, 333
Tazobactam, 867
Teflon, structure and uses of, 293
Template strand (DNA), 1136
Terephthalic acid, synthesis of, 596
Termination step (radical), **188**
Terpene, 299–300
Terpenoid, 299–300, **1098**–1106
   biosynthesis of, 299–300,
     1098–1106
   classification of, 300, 1098
   isoprene rule and, 299–300
   mevalonate biosynthetic pathway
     for, 1099–1103
tert-, name prefix, **85**
tert-Amyl group, **90**
tert-Butyl group, **85**
Tertiary alcohol, 621
Tertiary amine, **944**
Tertiary carbon, **86**
Tertiary hydrogen, **86**
Tertiary structure (protein), **1066**,
   1068
Testosterone, conformation of, 130
   molecular model of, 130
   structure and function of, 1110
Tetracaine, structure of, 995
Tetrahedral geometry, conventions
   for drawing, 6
Tetrahydrofolate, structure of, 1073
Tetrahydrofuran, as reaction solvent,
   263
   molecular model of, 676
Tetramethylsilane, NMR spectroscopy
   and, 462
Thermodynamic control, 508–**509**
   1,4-addition reactions and,
     508–509
Thermoplastic polymer, **1254**
   characteristics of, 1254
   examples of, 1254
   T_g of, 1254
   uses of, 1254
Thermosetting resin, **1256**
   cross-linking in, 1256
   uses of, 1256
Thiamin diphosphate, pK_a of, 1183
   reaction with pyruvate, 1181–1183
   structure of, 1183
   ylide from, 1183
Thiamin, structure of, 549, 975, 1073
   thiazolium ring in, 549
Thiazole, basicity of, 975
thio-, thioester name suffix, 816
Thioacetal, synthesis of, 770
Thioanisole, electrostatic potential
   map of, 804
-*thioate*, thioester name suffix, 816

Thioester(s), **814**
   biological reduction of, 847
   electrostatic potential map of, 820
   naming, 816
   pK_a of, 879
Thiol(s), **676**, 691–693
   disulfides from, 692
   electrostatic potential map of, 79
   from alkyl halides, 692
   hybridization of, 19
   naming, 691
   odor of, 692
   oxidation of, 692
   pK_a of, 625
   polarizability of, 191
   reaction with alkyl halides, 693
   reaction with Br_2, 692
   reaction with NaH, 693
   sulfides from, 693
   thiolate ions from, 693
Thiolate ion, **693**
Thionyl chloride, reaction with
   alcohols, 355, 639
   reaction with amides, 793–794
   reaction with carboxylic acids,
     823–824
Thiophene, aromaticity of, 548
Thiourea, reaction with alkyl halides,
   692
Threonine, stereoisomers of, 156
   structure and properties of, 1047
Threose, configuration of, 1009
   molecular model of, 147
Thromboxane B_2, structure of, 1096
Thymine, electrostatic potential map
   of, 1132
   structure of, 1129
Thyroxine, biosynthesis of, 571
   structure of, 1048
Time-of-flight (TOF) mass
   spectrometry, 433
Titration curve, alanine, 1051
TMS, see Tetramethylsilane,
   Trimethylsilyl ether
Tollens' test, 1020
Toluene, electrostatic potential map
   of, 585
   IR spectrum of, 551
   ^{13}C NMR absorptions of, 554
   ^1H NMR spectrum of, 482
Toluene-2,4-diisocyanate,
   polyurethanes from, 1250
p-Toluenesulfonyl chloride, reaction
   with alcohols, 639–640
Torsional strain, **95**
Tosylate, 373–374
   from alcohols, 639–640
   S_N2 reactions and, 382, 640
   uses of, 640

Index

Toxicity, chemicals and, 24–25
Trans fatty acid, from vegetable oil, 1091
   from hydrogenation of fats, 279–280
Transamination, **1198–1201**
   mechanisms in, 1198–1201
   steps in, 1198–1201
Transcription (DNA), **1135–1136**
   antisense strand and, 1136
   consensus sequence and, 1135
   promoter sequence and, 1135
   sense strand and, 1136
Transfer RNA, **1135**
   anticodons in, 1138–1139
   function of, 1138–1139
   molecular model of, 1138
   shape of, 1138
Transferase, 1070
Transition state, **206**
   Hammond postulate and, 247
Translation (RNA), **1137–1139**
Tranylcypromine, synthesis of, 962
Tree diagram (NMR), 483
Triacylglycerol, **1089**
   catabolism of, 1158–1167
Trialkylsulfonium ion(s), alkylations with, 694
   chirality of, 166
Tributyltin hydride, reaction with alkyl halides, 370
Tricarboxylic acid cycle, *see* Citric acid cycle
Trifluoroacetic acid, pK_a of, 783
Trifluoromethylbenzene, electrostatic potential map of, 585
Triglyceride, *see* Triacylglycerol, 1089
Trimethylamine, bond angles in, 947
   bond lengths in, 947
   electrostatic potential map of, 949
   molecular model of, 947
Trimethylsilyl ether, cleavage of, 649
   from alcohols, 648–649
   synthesis of, 648–649
Trimetozine, synthesis of, 832
2,4,6-Trinitrochlorobenzene, electrostatic potential map of, 592
Triphenylphosphine, reaction with alkyl halides, 747
Triple bond, **16**
   electronic structure of, 16
   length of, 17
   see also Alkyne
   strength of, 17
Triplet (NMR), 477
Trisubstituted aromatic compound, synthesis of, 600–604
Triterpenoid, **1098**

tRNA, *see* Transfer RNA
Trypsin, peptide cleavage with, 1061
Tryptophan, pK_a of, 52
   structure and properties of, 1047
Turnover number (enzyme), **1069**
Twist-boat conformation (cyclohexane), **119–120**
   steric strain in, 119–120
   molecular model of, 120
Tyrosine, biosynthesis of, 643
   catabolism of, 1211
   iodination of, 571
   structure and properties of, 1047

Ubiquinones, function of, 654
   structure of, 653
Ultrahigh-molecular-weight polyethylene, uses of, 1246
Ultraviolet light, electromagnetic spectrum and, 434
   wavelength of, 517–518
Ultraviolet spectroscopy, **517–520**
   absorbance and, 519
   aromatic compounds, 552
   conjugation and, 520
   HOMO–LUMO transition in, 518
   molar absorptivity and, 519
Ultraviolet spectrum, benzene, 520
   β-carotene, 521
   1,3-butadiene, 519
   3-buten-2-one, 520
   1,3-cyclohexadiene, 520
   ergosterol, 532
   1,3,5-hexatriene, 520
   isoprene, 520
Unimolecular, **386**
Unsaturated, **224**
Unsaturated aldehyde, conjugate addition reactions of, 751–755
Unsaturated ketone, conjugate addition reactions of, 751–755
Unsaturation, degree of, **224**
Upfield, (NMR), **461**
Uracil, structure of, 1129
Urea, from ammonium cyanate, 1
Urethane, **1250**
Uric acid, pK_a of, 805
Uronic acid, **1021**
   from aldoses, 1021
Urushiols, structure of, 621
UV, *see* Ultraviolet

Valence bond theory, 9–10
Valence shell, **6**
Valganciclovir, structure and function of, 1153
Valine, structure and properties of, 1047

Valsartan, synthesis of, 359
Van der Waals force, polymers and, 1253
van't Hoff, Jacobus Hendricus, 6
Vegetable oil, **1089–1090**
   composition of, 1090
   hydrogenation of, 279–280, 1091
Veronal, synthesis of, 890
Vestenamer, synthesis of, 1252–1253
Vicinal, 316, **686**
Vinyl group, **228**
Vinyl monomer, **292**
Vinylcyclopropane, rearrangement of, 1237
Vinylic anion, electrostatic potential map of, 326
   stability of, 326
Vinylic carbocation, electronic structure of, 318
   electrostatic potential map of, 318
   from alkynes, 318
   stability of, 318
Vinylic halide, alkynes from, 316
   S_N2 reaction and, 379–380
Vinylic protons, ^{1}H NMR spectroscopy and, 474–475
Vinylic radical, alkyne reduction and, 324–325
Vioxx, 1, 555
Visible light, electromagnetic spectrum and, 434
Vision, chemistry of, 522
   retinal and, 522
Vitalistic theory, 1
Vitamin, **799**
Vitamin A, industrial synthesis of, 323
   structure of, 62
   synthesis of, 748
Vitamin B_1, structure of, 975
Vitamin B_{12}, structure of, 333
   synthesis of, 333
Vitamin C, industrial synthesis of, 800–801
   molecular model of, 800
   scurvy and, 800
   stereochemistry of, 178
   structure of, 62
   uses of, 800
Vitamin D, sigmatropic rearrangements and, 1232
Vitamin K_1, biosynthesis of, 578–579
Viton polymer, structure and uses of, 1247
VLDL, heart disease and, 1118
Volcano, chloromethane from, 344
Vulcanization, **517**

Walden, Paul, 372
Walden inversion, 372–375
Wang resin, solid-phase peptide synthesis and, 1065
Water, acid–base behavior of, 50
  dipole moment of, 38
  electrostatic potential map of, 52
  nucleophilic addition reaction of, 731–732
  pK_a of, 51–52
  reaction with aldehydes, 731–732
  reaction with ketones, 731–732
Watson, James Dewey, 1131
Watson–Crick DNA model, 1131–1132
Wave equation, 3
Wave function, 3
  molecular orbitals and, 19–20
Wavelength (λ), 435
Wavenumber, **437**
Wax, **1088**–1089
Whale blubber, composition of, 1090
Whitmore, Frank C., 249
Wieland–Miescher ketone, synthesis of, 938

Williamson ether synthesis, **678**–679
  carbohydrates and, 1016
  mechanism of, 678–679
Willstätter, Richard, 542
Winstein, Saul, 388
Wittig reaction, **746**–748
  mechanism of, 746–747
  uses of, 747–748
  vitamin A synthesis using, 748
Wohl degradation, **1023**
Wöhler, Friedrich, 1
Wolff–Kishner reaction, **741**–742
  mechanism of, 741–742
Wood alcohol, 620
Woodward, Robert Burns, 333, 1215
Woodward–Hoffmann rules, 1215–1216

X rays, electromagnetic spectrum and, 434
X-Ray crystallography, **447**
X-Ray diffractometer, 447
o-Xylene, ozonolysis of, 559
Xylocaine, structure of, 64
Xylose, configuration of, 1009

-*yl*, alkyl group name suffix, 84
-*yl phosphate,* acyl phosphate name suffix, 817
Ylide, **746**
-*yne,* alkyne name suffix, 314

Z configuration, **231**–**232**
  assignment of, 231–232
Zaitsev, Alexander M., 397
Zaitsev's rule, **397**
  alcohol dehydration and, 641
  E1 reaction and, 406
  E2 reaction and, 403
  Hofmann elimination and, 964–965
  proof for, 470–471
Zanamivir, mechanism of, 1032–1033
Zeisel method, 708
Ziegler–Natta catalyst, **1245**
Zinc–copper, Simmons–Smith reaction and, 288–289
Zocor, structure of, 107
Zwitterion, **1045**
  electrostatic potential map of, 1045

Structures of Some Common Functional Groups

Name	Structure*	Name ending	Example
Alkene (double bond)	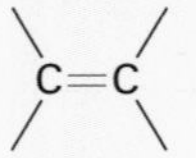 C=C	-ene	$H_2C=CH_2$ Ethene
Alkyne (triple bond)	—C≡C—	-yne	HC≡CH Ethyne
Arene (aromatic ring)	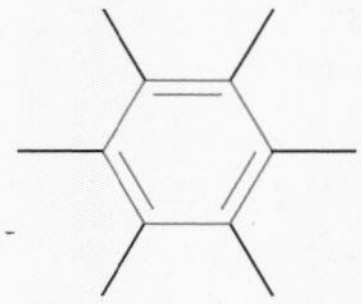 (benzene ring)	None	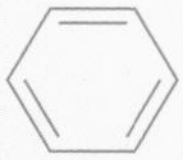 Benzene
Halide	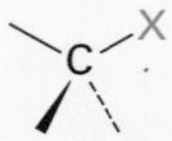 C—X (X = F, Cl, Br, I)	None	CH_3Cl Chloromethane
Alcohol	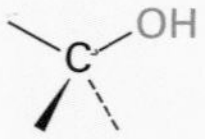 C—OH	-ol	CH_3OH Methanol
Ether	 C—O—C	ether	CH_3OCH_3 Dimethyl ether
Monophosphate	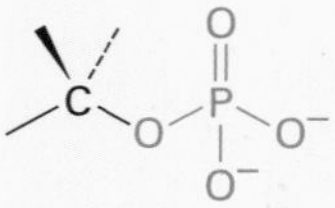 C—O—P(=O)(O⁻)(O⁻)	phosphate	$CH_3OPO_3^{2-}$ Methyl phosphate
Diphosphate	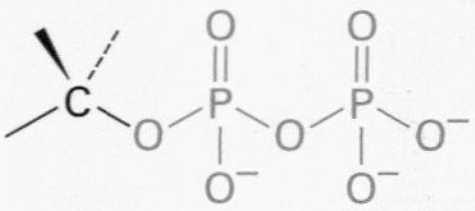 C—O—P(=O)(O⁻)—O—P(=O)(O⁻)(O⁻)	diphosphate	$CH_3OP_2O_6^{3-}$ Methyl diphosphate
Amine	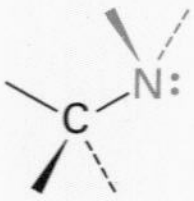 C—N:	-amine	CH_3NH_2 Methylamine
Imine (Schiff base)	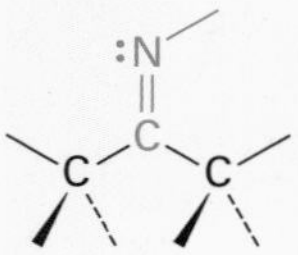 C=N:	None	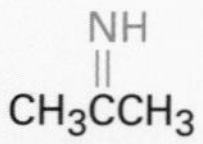 NH ‖ CH_3CCH_3 Acetone imine
Nitrile	—C≡N	-nitrile	$CH_3C≡N$ Ethanenitrile
Thiol	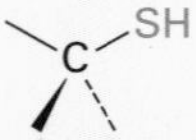 C—SH	-thiol	CH_3SH Methanethiol

*The bonds whose connections aren't specified are assumed to be attached to carbon or hydrogen atoms in the rest of the molecule.